Proceedings

Today, a steadily growing store of information is called for in order to understand the increasingly complex technologies used in modern automobiles. Functions, modes of operation, components and systems are rapidly evolving, while at the same time the latest expertise is disseminated directly from conferences, congresses and symposia to the professional world in ever-faster cycles. This series of proceedings offers rapid access to this information, gathering the specific knowledge needed to keep up with cutting-edge advances in automotive technologies, employing the same systematic approach used at conferences and congresses and presenting it in print (available at Springer.com) and electronic (at SpringerLink and Springer Professional) formats.

The series addresses the needs of automotive engineers, motor design engineers and students looking for the latest expertise in connection with key questions in their field, while professors and instructors working in the areas of automotive and motor design engineering will also find summaries of industry events they weren't able to attend. The proceedings also offer valuable answers to the topical questions that concern assessors, researchers and developmental engineers in the automotive and supplier industry, as well as service providers.

Peter E. Pfeffer
Editor

9th International Munich Chassis Symposium 2018

chassis.tech plus

Springer Vieweg

Editor

Prof. Dr. Peter E. Pfeffer
Munich University of Applied Sciences
Munich, Germany

ISSN 2198-7432 ISSN 2198-7440 (electronic)
Proceedings
ISBN 978-3-658-22049-5 ISBN 978-3-658-22050-1 (eBook)
https://doi.org/10.1007/978-3-658-22050-1
Springer Heidelberg Dordrecht London New York

Springer Vieweg

Printed on acid-free paper

This Springer Vieweg imprint is published by the registered company Springer Fachmedien Wiesbaden GmbH part of Springer Nature.
The registered company address is: Abraham-Lincoln-Str. 46, 65189 Wiesbaden, Germany

WELCOME

By forming the link between the road surface and the vehicle, the chassis plays a key role in enhancing vehicle dynamics and ride comfort. With its control systems, it provides the basis for the further development of driver assistance systems which support the driver in the task of driving the vehicle. This applies to an even greater extent to autonomous vehicles. Electromechanical steering and steer-by-wire systems are one solution available. At the same time, the brake system as a safety component needs to be developed in such a way that it fulfills the requirements of powertrain hybridization and electrification.

As a platform for the exchange of experience and constructive discussions, the 9th International Munich Chassis symposium chassis.tech plus will bring together numerous experts in chassis systems, steering, brakes, and tires/wheels from all over the world on 12 and 13 June 2018.

High-profile keynote speakers will provide a comprehensive overview of new solutions to these challenges. Speakers from industry and research will discuss the very latest developments in four parallel strands: chassis systems, steering, brakes, and tires/wheels. The program will be rounded off by two plenary sessions at the beginning and end of the conference.

We look forward to welcoming you at the Hotel Bayerischer Hof in the center of Munich and hope that you enjoy a stimulating conference.

Prof. Dr. Peter E. Pfeffer
Munich University of Applied Sciences
Scientific Director of the Symposium

INDEX

CHASSIS.TECH PLUS SECTION

KEYNOTE LECTURES I

KEYNOTE LECTURES II

KEYNOTE LECTURES III

PARALLEL STRAND I

NEW CHASSIS SYSTEMS

AUTONOMOUS VEHICLES

PARALLEL STRAND II

RIDE COMFORT

CHASSIS CONTROL SYSTEMS

CHASSIS.TECH SECTION

DEVELOPMENT METHODS

ELASTOMERIC BUSHING AND WHEEL SUSPENSION

INNOVATIVE MATERIALS AND METHODS

STEERING.TECH SECTION

STEERING FEEL AND HUMAN-MACHINE INTERFACE (HMI)

NEW STEERING CONCEPTS

DEVELOPMENT METHODS

BRAKE.TECH SECTION

FUTURE BRAKE SYSTEMS

BRAKE DUST

BRAKE CONTROL SYSTEMS AND VIBRATION

TIRE.WHEEL.TECH SECTION

WHEEL TECHNOLOGIES AND TRENDS

TIRE TESTING AND SIMULATION

TIRE PERFORMANCE AND TIRE SLIP ANALYSES

SPEAKERS

Masaya Akita
Lexus International
(Toyota Motor Corporation),
Japan

Manuel Armbrüster
MAN Truck & Bus AG

Seda Aydogdu
MdynamiX AG

Klaus Baltruschat
TÜV SÜD
Product Service GmbH

Eric Begenau
Dr. Ing. h.c. F. Porsche AG

Johannes Betz
Institute of Automotive
Technology (FTM), TU Munich

Albert Biermann
Hyundai Motor Group, South Korea

Mattia Bruschetta
Department of Information
Engineering, University of Padova,
Italy

Stefan Büchner
Institute of Automotive
Technology (FTM), TU Munich

Gerd Burk
Dr. Ing. h.c. F. Porsche AG

Francesco Calabrese
Fraunhofer Institute for
Industrial Mathematics (ITWM)

Xabier Carrera Akutain PhD
Toyota Motor Europe, Belgium

Ronnie Dessort
TESIS GmbH

Miguel D'Haens
Tenneco Automotive
Europe bvba, Belgium

Dario Düsterloh
Dr. Ing. h.c. F. Porsche AG

Ralf Duning
Maxion Wheels EAAP
Holding GmbH

Luca Dusini
Maserati S.p.A., Italy

Riccardo Ficca
Jaguar Land Rover Limited, UK

Martin Förster
Automotive Engineering,
Munich University
of Applied Sciences

Dr. Uwe Folchert
Continental Teves
AG & Co. oHG

Dr. Sebastian Gramstat
AUDI AG

Roland Greul
Robert Bosch
Automotive Steering GmbH

Urs Gunzert
MAN Truck & Bus AG

Dr. Falk Hecker
Knorr-Bremse Systeme
für Nutzfahrzeuge GmbH

David Hesse
Department of
Automotive Engineering,
TU Ilmenau

Dr. Thorsten Hestermeyer
Ford-Werke GmbH

Dr. Andree Hohm
Continental Teves
AG & Co. oHG

Yuriy Kandinov
ZF Group, USA

Etsuo Katsuyama
Toyota Motor Corporation,
Japan

Dr. Holger Klein
ZF Friedrichshafen AG

Joe Klesing
Nexteer Automotive Corp.,
USA

Tobias Köb,
thyssenkrupp Presta AG,
Liechtenstein

Dr. Lars König
Bosch Engineering GmbH

Jan-Dominik Korus
BMW Group

Dr. Stefan Kurtenbach
Automotive Center
South Westfalia

Dr. Alisa Lindner
Robert Bosch
Automotive Steering GmbH

Per Magnusson
NIRA Dynamics AB, Sweden

Thomas Maur
ZF Group

Mathias Metzler
Centre for
Automotive Engineering,
University of Surrey, UK

Dr. Gerd Müller
Department of
Vehicle Engineering,
TU Berlin

Thomas Müller
AUDI AG

Dr. Eckehard Münch
ZF Friedrichshafen AG

Stefano Murgia
FCA Italy S.p.A.

Lars Netsch
TÜV SÜD
Auto Service GmbH

Donatus Neudeck
Dr. Ing. h.c. F. Porsche AG

Prof. Dr. Andreas Nevoigt
Laboratory for
Suspension Systems,
South Westfalia
University of Applied Sciences

Maciej Obszarny
ZF Group, Poland

Lutz Pander
Vehicle System Design,
TU Chemnitz

Stefan Randak
Atreus GmbH

Maximilian Georg Reisner
Chair of
Automotive Engineering,
TU Dresden

Frieder Riedel
Daimler AG

Oswin Röder
AUDI AG

Steve Sattler
MAN Truck & Bus AG

Prof. Bernhard Schick
Research Center
Allgäu (FZA),
Kempten University
of Applied Sciences

Michael Schmidt
Institute of Automotive
Technology (FTM), TU Munich

Beniamin Szewczyk
Brembo S.p.A, Italy

Dr. Martin Treimer
BMW Group

Dr. Vasilis Tsinias
Renault Sport Racing
Limited, UK

Daisuke Umetsu
Mazda Motor Corporation,
Japan

Christian Vey
Continental Teves
AG & Co. oHG

Daniel Walser
Michelin Recherche
et Technique SA,
Switzerland

Dr. Michael Wein
AUDI AG

Alexander Zech
BMW Group

Thomas Zehelein
Institute of Automotive
Technology (FTM), TU Munich

CHASSIS.TECH PLUS SECTION

KEYNOTE LECTURES I

Chassis 2025 – strategic key competence or commodity?

Thomas Müller, M. C. Paefgen, AUDI AG

This manuscript is not available according to publishing restriction. Thank you for your understanding.

P. Pfeffer (Ed.), *9th International Munich Chassis Symposium 2018*, Proceedings,
https://doi.org/10.1007/978-3-658-22050-1_1

Networked, electric, autonomous – the intelligent chassis of the future

Dr. Holger Klein, ZF Friedrichshafen AG

P. Pfeffer (Ed.), *9th International Munich Chassis Symposium 2018*, Proceedings,
https://doi.org/10.1007/978-3-658-22050-1_2

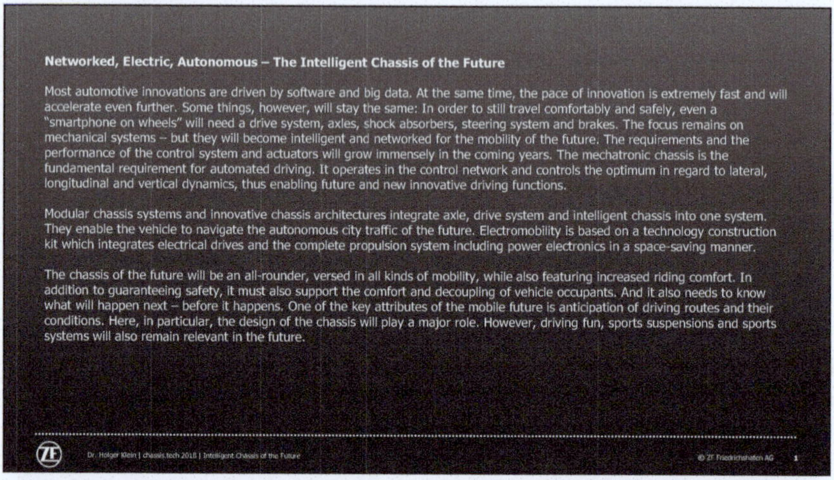

Networked, Electric, Autonomous – The Intelligent Chassis of the Future

Most automotive innovations are driven by software and big data. At the same time, the pace of innovation is extremely fast and will accelerate even further. Some things, however, will stay the same: In order to still travel comfortably and safely, even a "smartphone on wheels" will need a drive system, axles, shock absorbers, steering system and brakes. The focus remains on mechanical systems – but they will become intelligent and networked for the mobility of the future. The requirements and the performance of the control system and actuators will grow immensely in the coming years. The mechatronic chassis is the fundamental requirement for automated driving. It operates in the control network and controls the optimum in regard to lateral, longitudinal and vertical dynamics, thus enabling future and new innovative driving functions.

Modular chassis systems and innovative chassis architectures integrate axle, drive system and intelligent chassis into one system. They enable the vehicle to navigate the autonomous city traffic of the future. Electromobility is based on a technology construction kit which integrates electrical drives and the complete propulsion system including power electronics in a space-saving manner.

The chassis of the future will be an all-rounder, versed in all kinds of mobility, while also featuring increased riding comfort. In addition to guaranteeing safety, it must also support the comfort and decoupling of vehicle occupants. And it also needs to know what will happen next – before it happens. One of the key attributes of the mobile future is anticipation of driving routes and their conditions. Here, in particular, the design of the chassis will play a major role. However, driving fun, sports suspensions and sports systems will also remain relevant in the future.

Dr. Holger Klein | chassis.tech 2018 | Intelligent Chassis of the Future
© ZF Friedrichshafen AG 1

Dr. Holger Klein | chassis.tech 2018 | Intelligent Chassis of the Future
© ZF Friedrichshafen AG 2

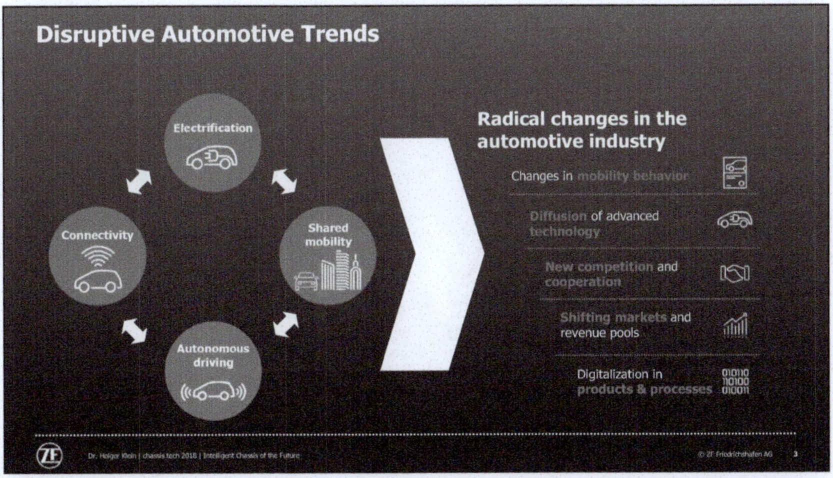
Disruptive Automotive Trends
Electrification
Connectivity
Shared mobility
Autonomous driving
Radical changes in the automotive industry
Changes in mobility behavior
Diffusion of advanced technology
New competition and cooperation
Shifting markets and revenue pools
Digitalization in products & processes
Dr. Holger Klein | chassis.tech 2018 | Intelligent Chassis of the Future
© ZF Friedrichshafen AG
3

On the Road to Vision Zero
enabled by
Intelligent Mechanical Systems
Dr. Holger Klein | chassis.tech 2018 | Intelligent Chassis of the Future
© ZF Friedrichshafen AG
4

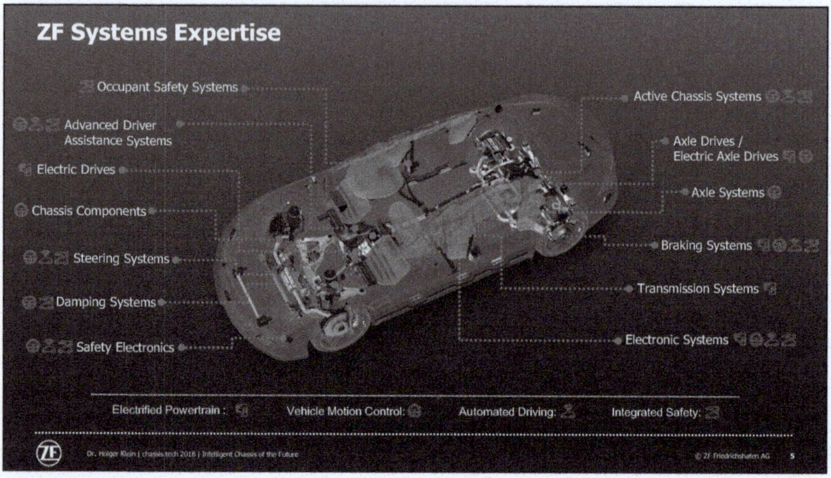
ZF Systems Expertise
Occupant Safety Systems
Advanced Driver Assistance Systems
Electric Drives
Chassis Components
Steering Systems
Damping Systems
Safety Electronics
Active Chassis Systems
Axle Drives / Electric Axle Drives
Axle Systems
Braking Systems
Transmission Systems
Electronic Systems
Electrified Powertrain :
Vehicle Motion Control:
Automated Driving:
Integrated Safety:
Dr. Holger Klein | chassis.tech 2018 | Intelligent Chassis of the Future
© ZF Friedrichshafen AG
5

Dr. Holger Klein | chassis.tech 2018 | Intelligent Chassis of the Future
© ZF Friedrichshafen AG
6

Disruptive changes – a challenge for product and corporate strategies

Stefan Randak, Atreus GmbH

P. Pfeffer (Ed.), *9th International Munich Chassis Symposium 2018*, Proceedings,
https://doi.org/10.1007/978-3-658-22050-1_3

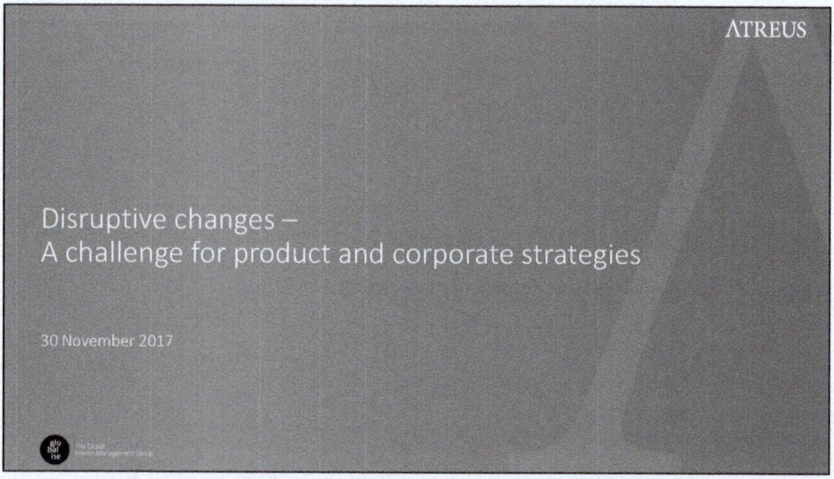
ATREUS
Disruptive changes –
A challenge for product and corporate strategies
30 November 2017

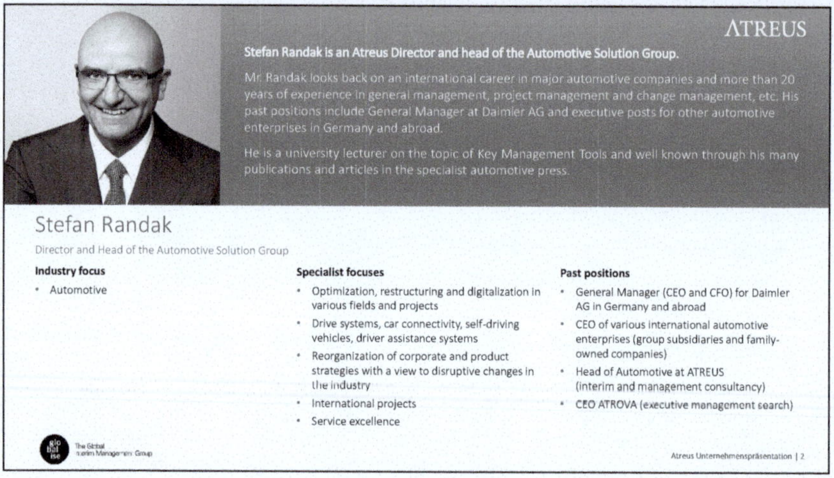
ATREUS
Stefan Randak is an Atreus Director and head of the Automotive Solution Group.
Mr. Randak looks back on an international career in major automotive companies and more than 20 years of experience in general management, project management and change management, etc. His past positions include General Manager at Daimler AG and executive posts for other automotive enterprises in Germany and abroad.
He is a university lecturer on the topic of Key Management Tools and well known through his many publications and articles in the specialist automotive press.
Stefan Randak
Director and Head of the Automotive Solution Group
Industry focus
• Automotive
Specialist focuses
• Optimization, restructuring and digitalization in various fields and projects
• Drive systems, car connectivity, self-driving vehicles, driver assistance systems
• Reorganization of corporate and product strategies with a view to disruptive changes in the industry
• International projects
• Service excellence
Past positions
• General Manager (CEO and CFO) for Daimler AG in Germany and abroad
• CEO of various international automotive enterprises (group subsidiaries and family-owned companies)
• Head of Automotive at ATREUS (interim and management consultancy)
• CEO ATROVA (executive management search)
Atreus Unternehmenspräsentation | 2

ATREUS

Press publications

manager magazin

Zuletzt aktualisiert: 25. Juli 2016, 15:12 Uhr

Der Traum vom autonomen Fahren

Diese Hürden müssen selbstfahrende Autos noch nehmen

1. Wir brauchen leistungsfähige Mobilfunknetze

„Starres Hierarchiedenken ist kontraproduktiv"

12. Juli 2016

Die Digitale Transformation ermöglicht und erfordert neue Geschäftsmodelle ebenso wie neues Know-How im Unternehmen und einen neuen Führungsstil. Im Interview mit automotiveIT spricht Stefan Randak, Leiter der Praxisgruppe Automotive beim Interim-Management-Provider Atreus, über die Schaffung

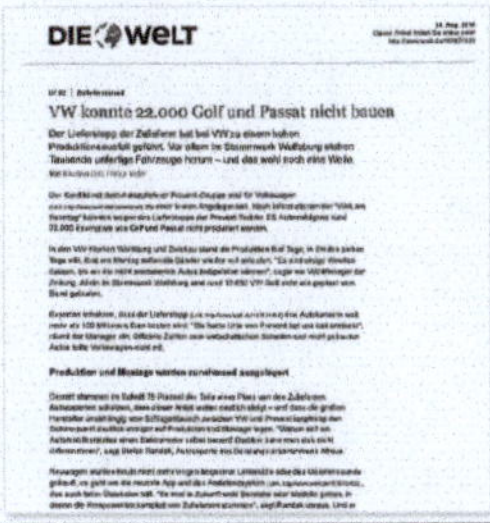

DIE WELT

VW konnte 22.000 Golf und Passat nicht bauen

The Global Interim Management Group

ATREUS

Statement 1

This is how we see the future of our industry ...
The companies that benefit will be the ones that occupy themselves early on with the new mobility trends and adjust their business strategies accordingly – like Webasto.

Source: Karriere – Mehr wissen. Mehr erreichen

The Global Interim Management Group

ATREUS

Statement 2

The Chinese government's huge measures package and new offers will continue to fuel the market for e-vehicles and further underscore **China's pioneering role as the world's most important market for automobiles**.

Source: Rheinmetall Automotive / Heart-Beat

Atreus Unternehmenspräsentation | 5

ATREUS

Statement 3

The union IG Metall fears that of the 880,000 jobs at the German automakers and their suppliers, the 250,000 in the powertrain segment will be the ones most affected.

Source: Süddeutsche Zeitung

Atreus Unternehmenspräsentation | 6

ATREUS

Statement 4

The European automotive suppliers industry is under pressure from the growth of electromobility and fears a large-scale exodus of jobs to China. Europe lags behind in the production of sensors, microchips and batteries. It is risky to be dependent on Chinese deliveries ...

Source: Deutsche Wirtschaft Nachrichten

globalise The Global Interim Management Group

Atreus Unternehmenspräsentation | 7

ATREUS

Global automotive supplier study 2016

The industry is faced with a radical transformation

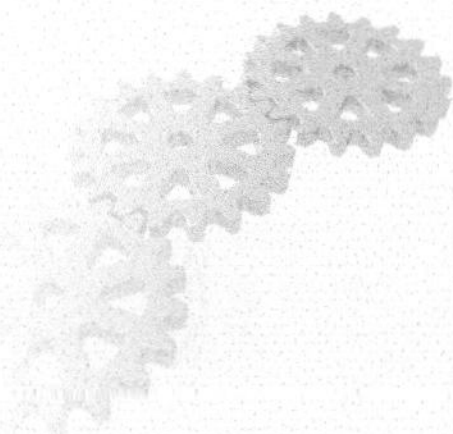

The current developments very clearly show that the **global automotive industry is faced with the biggest upheaval in its history**.

Disruptive changes mean **big opportunities** for the suppliers in the next 10 years – but there is also **enormous uncertainty** about when and where these opportunities will arise.

According to the study, the **market volume for vehicle components** will **increase** from around € 700 billion (2015) to over € 850 billion (2025). But this will go hand in hand with **major profit shifts between the segments and to some degree to new market players**.

Source: Roland Berger

globalise The Global Interim Management Group

Atreus Unternehmenspräsentation | 8

Disruptive changes – a challenge for product and corporate strategies

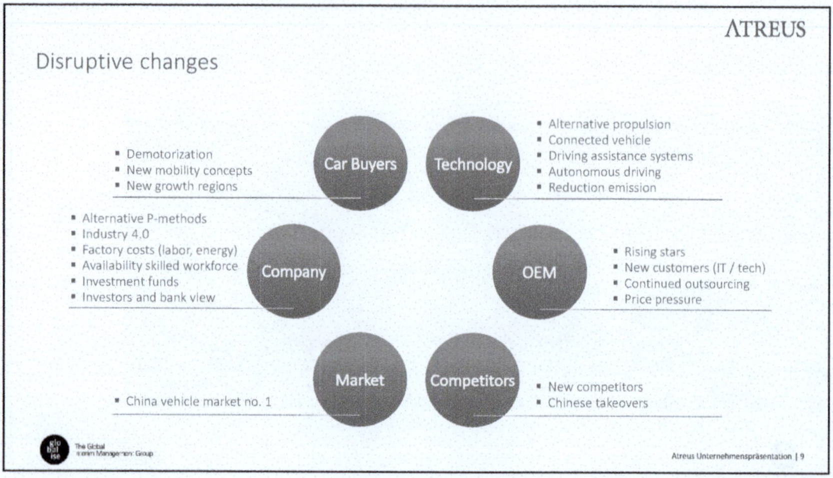

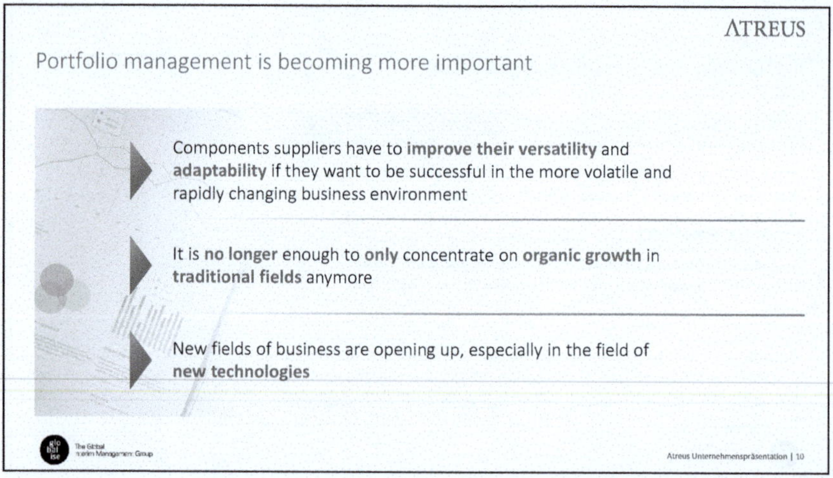

ATREUS

Electric powertrain technology is on the advance

Compliance strategies of OEMs will lead to increased production of alternative powertrains – Electrification is expected to play a key role

Implications on powertrains (1/2): Alternative powertrains

Global powertrain production [m units] | **Global xEV production** [m units]

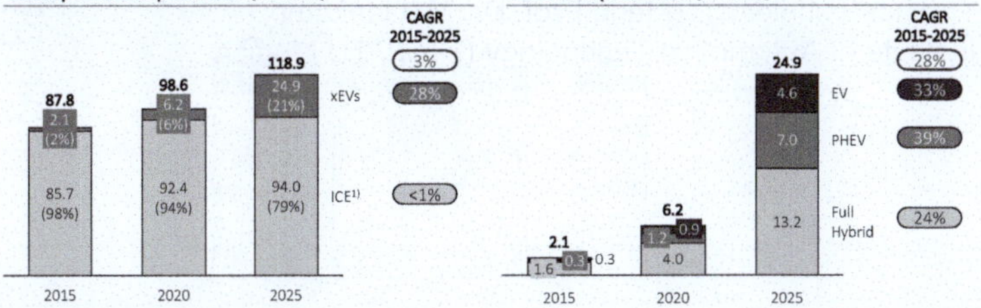

1) Including mild hybrid vehicles (up to 20 kW) and ICE start-stop

Source: Roland Berger

The Global Interim Management Group

Atreus Unternehmenspräsentation | 11

ATREUS

Diesel motors are on the decline

Diesel powertrain still expected to hold dominant position in upper passenger car segments, but to decline across all segments until 2030

New car diesel forecast by segment in EU-28 until 2030[1]

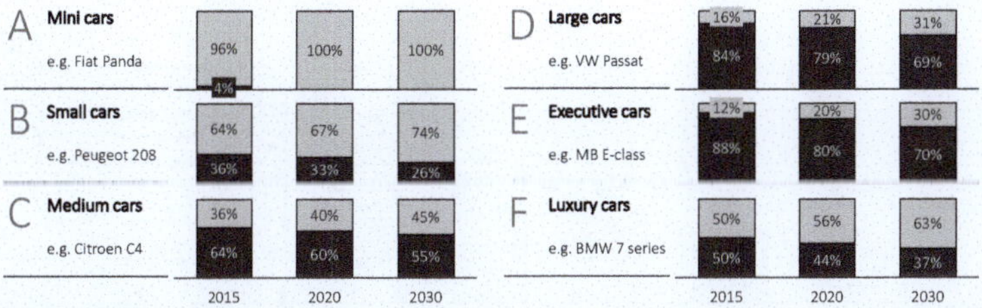

Other powertrains | Diesel (incl. MH diesel)

1) In % of new car sales

Source: Roland Berger

The Global Interim Management Group

Atreus Unternehmenspräsentation | 12

Disruptive changes – a challenge for product and corporate strategies

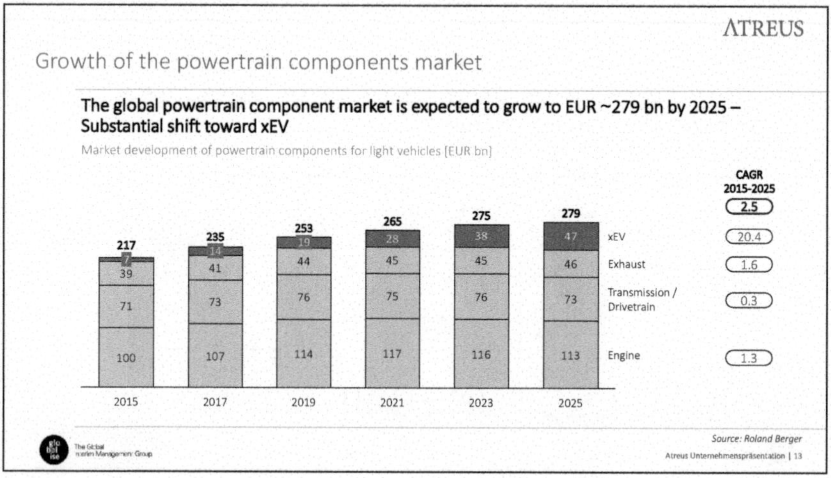

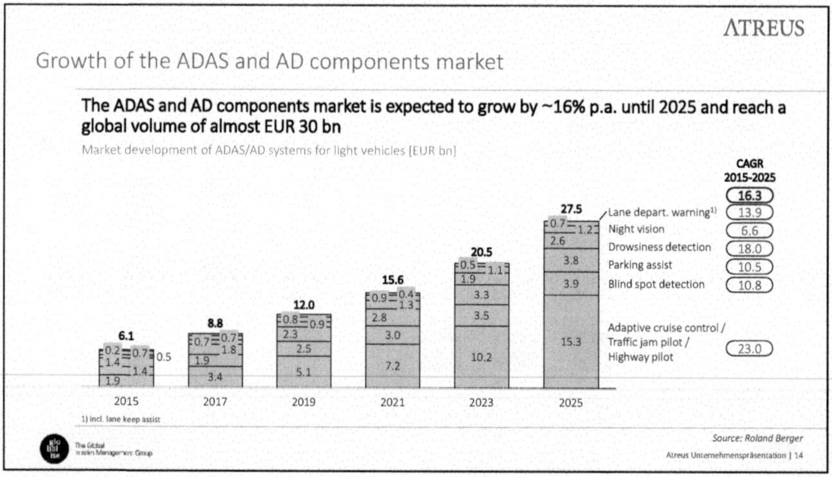

ATREUS

Winners and losers in the components market

The market for auto suppliers is expected to yield substantial further growth opportunities – Innovation will be key to capturing them

Global automotive component market development 2015 vs. 2025

Component market value[1] [EUR bn]

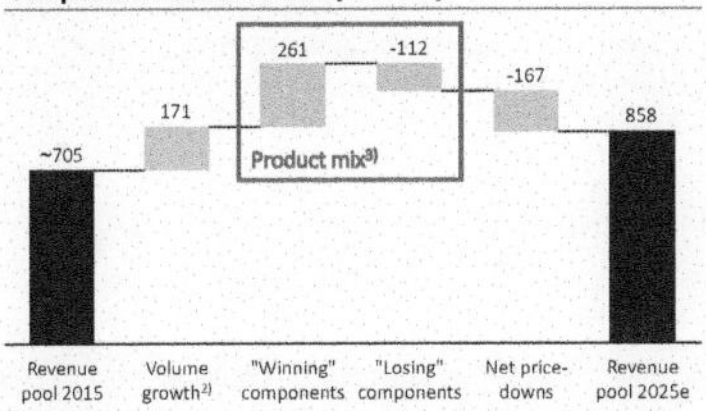

1) Light vehicle OE market, excluding commercial vehicle and aftermarket portion
2) Change driven by vehicle production volume (volume per car at 2015 level)
3) Additional growth/decline caused by change in product/technology content per car

Winning and losing components

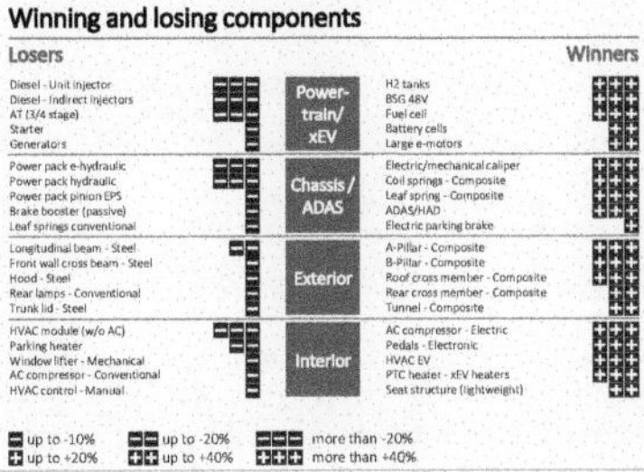

Source: Roland Berger

The Global ... Management Group

Atreus Unternehmenspräsentation | 15

ATREUS

Consequences for the auto supplier industry (1/2)

The increasing regulation of emissions is increasing the complexity and costs of exhaust systems, especially diesel	Short-term potential for some suppliers
Diesel is becoming less and less important in Europe	Growing challenge for suppliers that have specialized in this kind of engine
The market for e-mobility will grow rapidly over the next decade	Increasing competition for traditional suppliers due to new market players on the market and OEMs that want to build up their own expertise
Sudden regulatory pressure, possibly politically motivated, is conceivable	Disruptive breakthrough of e-mobility combined with drastic declines in parts of the traditional automotive supplier world

The Global ... Management Group

Atreus Unternehmenspräsentation | 16

ATREUS

Consequences for the auto supplier industry (2/2)

Drastic increase in the number of ADAS/AD features over the next decade	Rapid increase in sales for suppliers of these features
Conventional features are being pushed out of the market (switches, levers, mirrors)	Order and sales collapse among the suppliers of conventional features
Many (new) automotive features are software driven	Enormous potential for software-focused suppliers
ADAS/AD features are becoming part of the new vehicle concept and hence also of the integrated mobility solutions	New entry of global players (Google) and other new service providers

The Global Interim Management Group

Atreus Unternehmenspräsentation | 17

ATREUS

Key actions for automotive suppliers (1/2)

Make a **scenario plan** to prepare for a disruptive breakthrough of e-mobility and autonomous driving in the medium term

Review the medium to long-term chances of your own **product portfolios**

Focus investment on key product innovations (xEV, composites, ADAS/AD software, HMI, etc.) without endangering the development of successful core products

More **speed in the development process**; install innovation teams outside the normal process, in order to work with new ideas and products

The Global Interim Management Group

Atreus Unternehmenspräsentation | 18

ATREUS

Key actions for automotive suppliers (2/2)

Estimate investment resources; form co-operations; include financial backers and investors (PE); sell shares

Structure the company holistically under the heading **"Digitalization"**
(What do my customers expect of me?)

Do a **core expertise analysis**:
What are my strengths, what direction can I develop in, do I have the staff, funding and partners I require to do so?

Structure the entire **product and corporate strategy** based on these new insights

glo bal ise The Global Interim Management Group

Atreus Unternehmenspräsentation | 19

ATREUS

We design success

glo bal ise The Global Interim Management Group

KEYNOTE LECTURES II

Steer-by-wire – next-generation steering system for future mobility

Roland Greul, B. Hauser, Robert Bosch Automotive Steering GmbH;
A. Gaedke, Robert Bosch GmbH

This manuscript is not available according to publishing restriction.
Thank you for your understanding.

P. Pfeffer (Ed.), *9th International Munich Chassis Symposium 2018*, Proceedings,
https://doi.org/10.1007/978-3-658-22050-1_4

Driving experience vs. mental stress with automated lateral guidance from the customer's point of view

The relation between moderate customer reviews of the lane keeping assistant, reduced confidence and high mental stress

Authors:

Prof. Bernhard Schick[1], M.Sc. Psych. Corinna Seidler[1], Dipl. -Wirtsch. -Ing. (FH) Seda Aydogdu[2], B.Sc. Yu-Jeng Kuo[1]

1 University of Applied Sciences/Research Center Allgaeu
2 MdynamiX AG Munich

P. Pfeffer (Ed.), *9th International Munich Chassis Symposium 2018*, Proceedings,
https://doi.org/10.1007/978-3-658-22050-1_5

Abstract

Driver Assistance Systems (ADAS) and Automated Driving (AD) are a megatrend in the automotive industry. The desire to use the travel time, to increase comfort and safety creates interest among customers. According to a recent study [1] (Bitkom, VdTÜV), three-quarters of surveyed, at least in a certain situation, want automatic driving functions and self-driven-vehicles.

ADAS for longitudinal control, such as ACC, plus additional functions, have today already a high degree of maturity and customer acceptance [2]. The lateral guidance, such as the lane keeping assistance system (LKAS), is comparatively at the beginning here, but is a key to the successful introduction of (highly) automated driving. Ultimately the breakthrough of automated driving will decide on customer acceptance.

In the classic objective driving comfort evaluation only the impact on the human body or the physical effort are assessed. In contrast, what does the customer feel by using the LKAS and what mental and physical stress do the different driving situations trigger? What expectations does he have of a driving experience and what inspires him? Which behaviors are rated as good and which as disappointed? What is felt to be comfortable and uncomfortable? What is accepted and which behavior and attributes are "no-go"?

University of Applied Sciences Kempten and MdynamiX AG, analyzed the LKAS in broad-based studies on these and other issues. With a total of N> 120 subjects, surveys, stress research and vehicle benchmark tests were conducted on public roads. Based on the QFD – Quality Function Deployment and Kano methods as well as TAM – Technology Acceptance Model the customer requirements were translated and weak points in the customer acceptance were analyzed.

As a result, it must be noted that the work load and the stress with LKAS, both subjectively and objectively, are significantly higher, quite in contrast to the claim. The stress at 120 kph with LKAS is significantly higher than at 160 kph without LKAS [3]. Even the expectations of the subjects the current maturity level could also not be fulfilled by far. Subjects estimate maturity level between 41.5–58.9 % and poor evaluation ratings in the premium benchmark were disappointed. The clear and always transparent communication between "man and machine", a positive subjective driving experience (for driver and passengers), reliable availability and predictability form the basis of a good customer assessment. Considerable potential for improvement was worked out here.

To reduce the feeling of comfort with ADAS/AD only to the physical stress is by far not sufficient according to [7]. In addition to the physical stress, the consideration of mental stress is of immense importance for a holistic comfort assessment in this case.

Challenges for Vehicle Development

Important impulses for the further automobile development will originate, in medium and long term, from legislative authorities as well as from the end customers in the different markets. The mandatory reduction of CO_2 fleet and local emissions are significantly stricter than in the past. At the same time, the European Union is committed to improve road safety and to reduce the number of road fatalities every decade by half and the digital revolution is running. In addition, the automotive industry strives to improve comfort and emotional values in order to make mobility even more attractive. Further decisive purchasing factors are purely subjective criteria like social trends, living in a net-worked world and, most important a "positive driving experience". This means that next to the technical targets performance, the subjective customer experience becomes very important.

The customer percepts the characteristics and values of a vehicle, e.g. styling, ergonomics, usability, infotainment and assistance systems, sense of security, driving comfort, agility and drivability, in the overall context of the full vehicle behavior. The technical answers are mechatronic systems, software as an added value with increasing importance, the electrification of the powertrain, advanced driver assistance systems and automated driving functions as well as the networked vehicle in a cyber-physical system. The trend towards advanced driver assistance systems (ADAS) and automated driving (AD) is a megatrend in the vehicle industry from these basic conditions. Nearly all manufacturers have announced highly-automated driving according to Level 3 (Figure 1) from 2020 onwards.

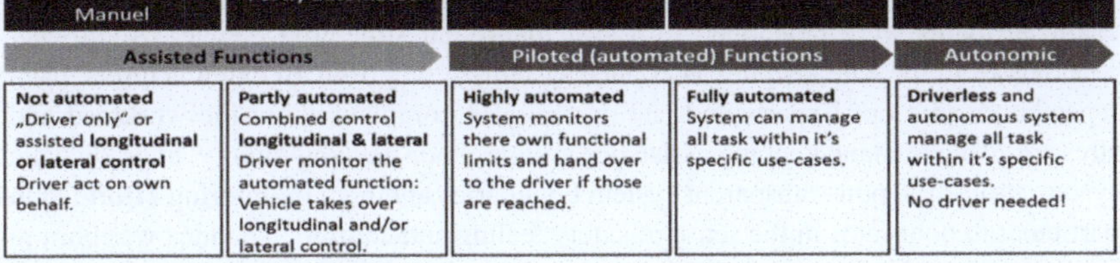

Figure 1: Definition of autonomous driving according SAE J3016 [4]

Due to the enormous complexity of automated systems, the manufacturers have huge effort to provide safe functions with millions of test kilometer. In this case, the automotive industry should not forget the customer who ultimately has to buy the product.

Customer acceptance of the automatic driving functions in level 1 and 2 (ADAS) is of immense importance and the key to a successful introduction of (highly) automated driving to Level 3+. Finally, the customer acceptance will be won with benefits, the ease of use and the positive driving experience.

Motivation

So far, the development of ADAS/AD in the fireworks of technologies such as environment sensors, environment model, artificial intelligence and driving functions are almost exclusively function-based. This means that the function is in focus and which application can be generated with it. At the beginning of a development, hardly any vehicle manufacturers define clear driving attribute targets and the driving experience "in front of the customer" and deduce the requirements for the vehicle systems and components, as anchored, for example, in modern attribute based vehicle dynamics development.

The University of Applied Sciences Kempten and MdynamiX have set themselves the goal of researching the driving characteristics and driving experience of ADAS/AD and describing them with a clear evaluation and target metric. Both of them also want to answer the question: How to transfer todays fun-to-drive into future fun-to-be-driven?

The subjective & objective evaluation and target metric required for this purpose should be consistently assessable and traceable in all test instances – whether validation test in MIL/SIL/HIL simulation, on the steering-in-the-loop system test bench [5], in the driving simulator, on proving ground and on public road. This allows the desired effectiveness and efficiency of a customer-oriented ADAS/AD development – for vehicles that customers love.

In numerous expert workshops, benchmark tests and measurement campaigns, the relevant attributes were developed in a systematic and structured manner. Significant weaknesses in the lane keeping assistance system, in fact from all manufacturers, have repeatedly came up. Experts and test drivers also complain about the workload and stress of using the lane keeping assistance system. Poor tracking quality, unpredictable system drop-offs, non-transparent system boundaries and high monitoring effort led in particular to poor rates in the sense of safety feeling and comfort. The idea was born to interview normal customers and take a closer look at workload and stress in relation of the lane keeping assistance system usage.

State of the Art LKAS Functions

The lane keeping assistance system is a lateral control assistance system. This can be divided into the Lane Departure Warning System (LDWS) as well as the Lane Keeping Assistance System (LKAS) [2].

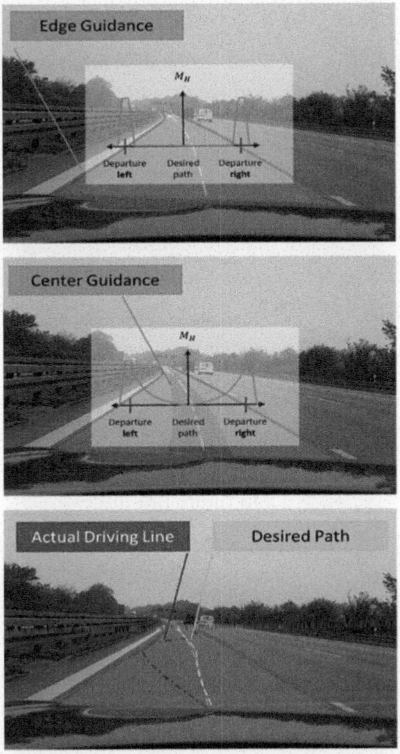

Figure 2: Above LKAS type 1, middle type 2, below drift oscillation issues

There are two types of LKAS. Type 1 does not support steering torque in the lane center, whereas the warning and the intervention on the edge of the lane are important – known as the edge guidance. Therefore, there is a wide control-free-corridor and the vehicle (Figure 2 above) is returned only when approaching the lane boundary by an abrupt steering torque intervention. In contrast, the LKAS Type 2 is also supported in the center area by low steering assistance while tracking – known as the center guidance. Here, the control corridor is kept narrow (Figure 2 middle). The steering torque intervention is similar to a half pipe or a V-profile. The LKAS tries to guide the vehicle in the center of the lane by means of permanent steering torque interventions. The test vehicles used in the studies were all equipped with LKAS Type 2.

During the expert workshops and driving events, the poor tracking performance and the drift oscillation at almost all benchmark vehicles were especially noticeable (Figure 2 below). It may happen that the vehicle builds up so much momentum that the LKAS gets to its system limits, such as the torque limit.

The result was unforeseen breakthroughs in the lane boundary. Furthermore, surprising system drops-offs were noticed. As a consequence, increased stress and reduced customer acceptance were suspected.

Study Methods and Procedure

In numerous expert workshops, benchmark tests and measuring campaigns, the relevant attributes for the lane LKAS with center guidance were systematically and structurally developed. The subjective and objective characteristics were transferred and linked to a so-called level model, consisting of subjective customer evaluation, subjective expert evaluation and objective characteristic values (KPIs – key performance indicators).

At the top customer level are the main criteria such as lane tracking quality, edge guidance, driver-vehicle interaction, availability, de-stress, sense of safety and HMI, which are further detailed at the expert level. Figure 3 shows the overall level model.

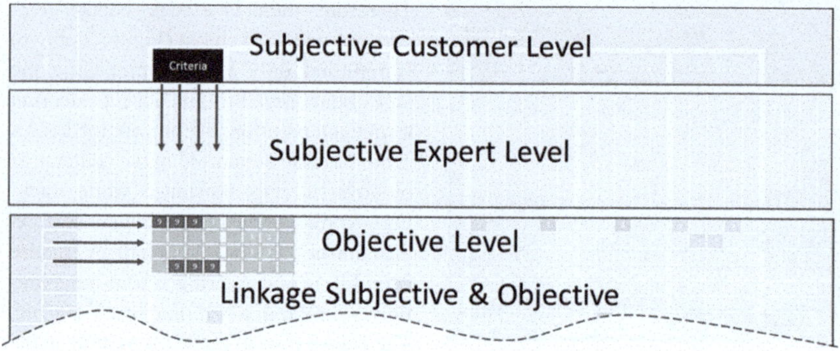

Figure 3: Evaluation Level Model

In order of objectification, new measurement and test methods were developed to create representative KPI's for the individual subjective expert criteria. In particular, the measurement method focused on the exact location of the vehicle on the track based on digital "ground truth" maps and the measurement of the steering feel. In order to capture as many relevant driving situations as possible, a comprehensive route and maneuver catalog have been developed to determine the desired characteristics there. More details to this will be published soon.

On the basis of these findings, two subject studies (Figure 4) were planned and carried out on public roads:

1. Stress study with survey and objective measurement of physiological parameters on the basis of one vehicle.
2. Benchmark study with survey, vehicle evaluation on the basis of three premium vehicles

Figure 4: Two different subject studies with customer test rides

In order to work on the topic as holistically as possible, a team formation was made with specialists from psychology, vehicle dynamics evaluation, ADAS measurement technology, product management and data analysis. To this end, the procedure, methodologies, survey catalog, measurements concept, public routes and processes were worked out together. With a pure engineering approach, the customer experience, the mental stress and above all the "unspoken" customer wishes – the truth would not have come out.

But what is human stress? This is a psychological and physical reaction caused by specific external stimuli, which enable the human to perform difficult tasks. According to [6], it is a "stimulus-dependent emotional response syndrome that refers to affective experience, expressive behavior, activation processes and instrumental action". In general, it is a protective reaction that brings us to higher performance, but which is perceived as uncomfortable and can even make us sick. In contrast, reduced stress in the context of the driving task is the reduction of the difficulty of the driving task, reducing the driver workload, increasing the ride comfort, increasing the maximum driver performance and thus increasing driving safety. This would be desired by the LKAS.

Stress Study

For the preparation of the study, working hypotheses were first developed on the basis of the expert's experience, which had to be analyzed and checked. The rating of the LKAS is influenced by:

1. the subjectively experienced stress
2. the drivers workload
3. the experienced driving pleasure
4. the experienced driving comfort

Additional working hypotheses: The subjective stress and driver's workload are associated with the use of LKAS. These are significantly higher with the use of the LKAS as without use of the LKAS, in contrast to the objectives of the LKAS. The physiological values and objective stress indicators are related to the use of LKAS [3].

The sample consisting of 50 subjects (M_{age} = 37.86, SD_{age} = 14.42, 72% male, range 18-65 years) was won through a call for tenders. They consisted of one pupil, 12 students and the rest full-time / part-time working people, including officials. The absolute majority (52%) stated that the highest level of education was completed, with 14% of the test drivers receiving a doctorate. Almost half of the subjects had already gained experience with LKAS [3].

The used test vehicle was a current premium luxury class vehicle, which was equipped with a high-end IMU – Inertia Measurement Unit with RTK D-GPS to record the precise vehicle motion. Additionally CAN/FlexRay bus signals were recorded, such as steering angle, steering torque and the camera's object information beside others.

Likewise, a BIOPAC system was used, with which the physiological parameters of the test persons were recorded (Figure 5), such as heart rate variability (ECG), skin conductibility/wetness (EDA), pulse rate (PPG) and relative depths of breathing (RSP).

For the purposes of the intended use, the A7 / A980 motorways and the well-developed B19 highway were selected as routes. The test ride took place with and without LKAS and in defined sections with 120 and 160 kph respectively. In order to correlate the human data with the vehicle and road data, according to the principle "driver-vehicle environment", trigger points were set on the specific road sections and driving events [3].

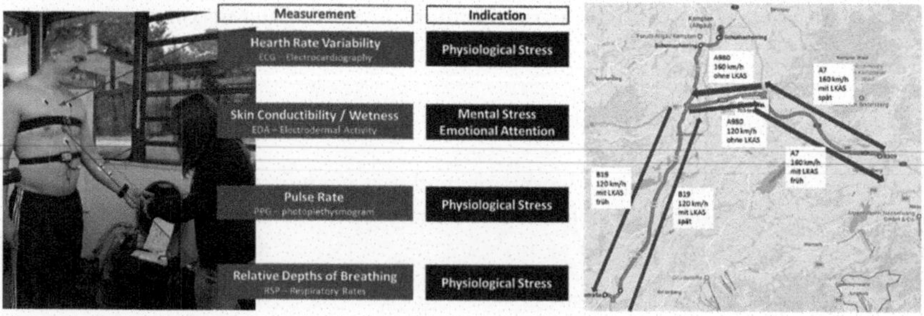

Figure 5: Measurement of physiological parameters and test routes

To combine the physiological parameters with the behavior and experience of the driver, a questionnaire for the test persons were created. In addition to existing questionnaires, such as NASA-TLX and the questionnaire on driving pleasure and comfort of Anna Engelbrecht [7], additional own research questions were included in the survey, in order to get a full survey on, for example, driving skills and conditions for the current driving behavior. In addition to the subjective stress experience, driving style and technical affinity as well as other descriptive information on the driver were recorded [3].

All subjects passed through a uniform and defined process with greeting, explanation, presentation of the LKAS, questionnaire before the test ride, instrumentation of the physiological measuring device, recovery phase, familiarization to the vehicle, driving with and without LKAS, one subjective stress rating while driving after defined road sections and an evaluation sheet as well as questionnaire after the test ride [3]. Questions during the driving task had not taken place, as this could falsified the physiological measurements.

Benchmark Study

As can be seen in the results chapter, the working hypotheses from stress studies have been largely confirmed. In addition, the expert evaluation of all benchmark vehicles was consistently rated as poor and with a lot of room for improvement. For this reason, a benchmark study with subjects should answer the following questions: What does the customer want? What are the differences between the systems in the tested vehicles? Which characteristics are rated as good and therefore accepted? Which ones disappoint and which ones are no-go? What is the interaction of the customers with the systems? At what maturity level is the customer ready to buy the system? Is it possible to predict the customer rating with a customer-oriented rating system for experts (level model)?

Three current premium vehicles from different manufacturers were used. The test program was set up with a similar process as in the stress study. However, physiological measurements and psychological questions have been omitted. For this purpose, a questioning process based on the QFD and KANO methods [8] were developed. An impact chain analysis were carried out in advance in order to be able to work out the customer requirements and customer wishes of the LKAS with open and closed questions and inquiries. In order to be able to evaluate the individual criteria at the customer level, the subjects were instructed during the test ride on relevant situations and maneuvers. After each experience and evaluation, the requirements and wishes were recorded directly in an open questionnaire dialog.

The 50 selected subjects were largely (about 50%) from the stress study to be already familiar with the LKAS and the reference vehicle. To avoid the inaccuracy caused by tiredness, the test subjects were randomly divided into two groups, each comparing two of the three vehicles. In total, 100 test drives over 4000 km were conducted. From the stress study we had the experience that 2 hours as a whole process is well feasible but already comes to the limit. For this reason, the reference vehicle (1) had been tested and evaluated one half of the subjects against vehicle (2) and the other half with respect to vehicle (3). Thus, a comparison of the vehicles (2) and (3) is possible. Here, the differences, strengths and weaknesses and the resulting optimization potential for current and future automatic lateral driving functions should be derived. The question about the importance of the criteria was carried out before and after the test ride in order to detect a possible sensitization of the subjects on certain criteria.

In order to place the human at the center of the development, methods such as Quality Function Deployment (QFD) and KANO [8], as well as the Technology Acceptance Model (TAM) were applied. With QFD, the customer's wishes could be identified, differentiated with KANO and classified in the TAM and finally translated with QFD into technical features and properties (Figure 6).

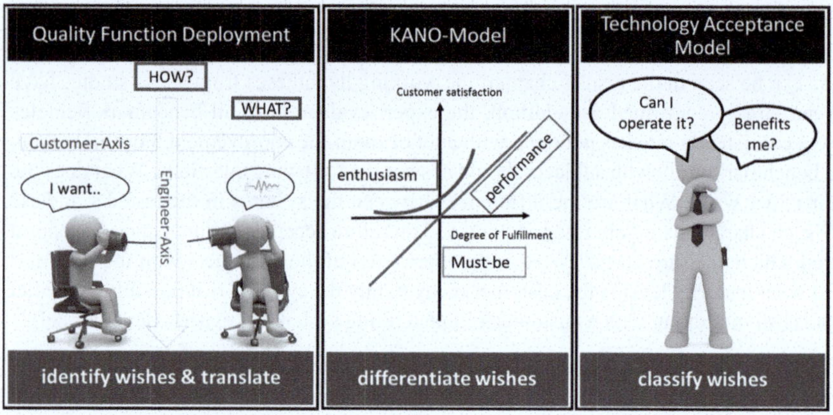

Figure 6: Applied methods for the benchmark study

In addition, an online survey was conducted with > 250 participants. The majority of the survey participants are between 20-40 years (approx. 80%) and therefore representative for potential customers of the next years. The level of information, the previous knowledge and the expectations of the subjects regarding the topic of the LKAS and ADAS in general were determined. Beside others this would ask the key question: "How important is it for you to have a lane keeping assistant in your next car?" each time before and after a LKAS explanatory video.

At the same time, 11 car dealers, in particular the car sellers, were interviewed, to obtain as much information as possible through an open discussion. In the selection process, manufacturers of almost all categories from small cars, lower mid-size cars and upper mid-size cars up to the premium class were selected.

Results

Stress Study

The following section presents the results of the stress study. Only a small outline of the results is presented in this paper. More detailed analyzes and results from the stress study will be published shortly. In the following, in addition to the report of the descriptive results, the measurement results of the NASA-TLX are described in more detail. Likewise, the physiological parameters are described and related to the evaluation of the LKAS. Furthermore, the assessment of the LKAS is made and its relationships with the other variables are investigated [3].

The correlation of the load, measured with the NASA-TLX, and the use of the LKAS was calculated to support the previous hypothesis. The analysis of variance with repeated measurements with Greenhouse-Geisser correction shows that the stress experiences differs significantly depending on the test run. A Bonferroni corrected post-hoc test showed a significant difference in stress between driving with LKAS and without LKAS, both at 120 and 160 kph. The other post-hoc tests also showed significant differences. The following Figure 7 shows an example of the stress experienced by type of test track. The increase of the experienced stress with the LKAS becomes clear. The statistical analysis of all objective data on the stress indicator EDA peaks also shows a significant increase and hotspot on certain sections of the route. The subjects feel consistently more stressed with LKAS [3].

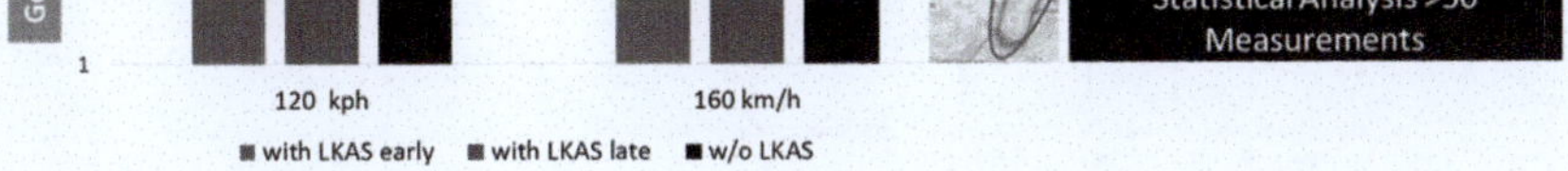

Figure 7: Subjectively perceived (left) and objective stress (EDA peaks) with vs. w/o LKAS

Significant correlations was also found in the items and overall comfort factor with the LKAS rating. As evaluation, the customer level of the level model was evaluated by the test persons. The items "calming" (r = -.409, p = .003), "relieving" (r = -.473, p = .001), "relaxing" (r = -.449, p = .001), "uncomplicated" (r = -.491, p = .000) and "supportive" (r = -.617, p = .000) correlate significantly with the score of the LKAS. The total comfort factor, which is composed of ten items, also correlates significantly with the score (r = -.461, p = .001). If the LKAS is rated better, the comfort experience increases. The negative sign results from different rating system of the questionnaire sheets (small or bigger number are good).

The feedback from the subjects and the evaluation of the questionnaires make it very clear that the perceived stress is due to a lack of trust in LKAS. These are the results of sudden system drop-offs without warning in seemingly normal driving situations, unforeseen system limitations, malfunctions (true-negatives), inconsistent feedback, necessary ad-hoc takeovers and the lack of transparency and high monitoring effort.

Figure 8 shows an example of a typical case that has occurred repeatedly and several times per trip with all vehicles. The system drop-off for some reason or stops working without being displayed. The driver realized this only when he left the lane. He intervenes and leads the vehicle by strong steering back into the lane. The EDA value as a stress indicator increases very quickly.

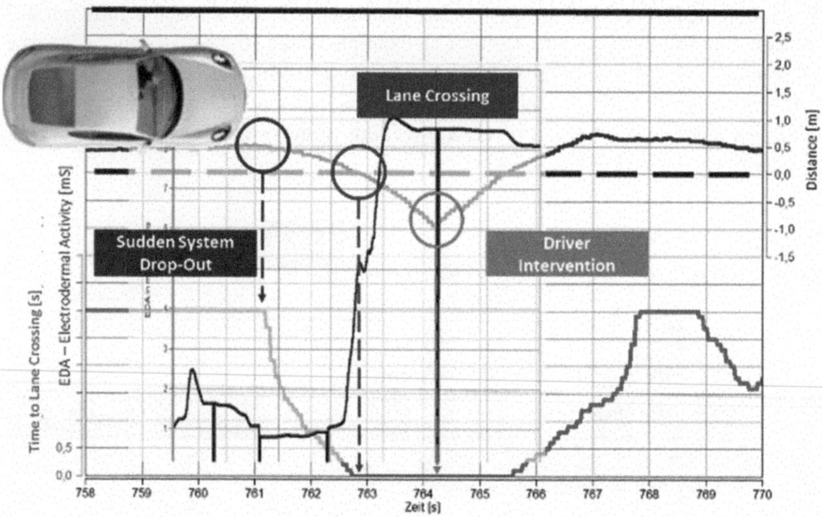

Figure 8: Unexpected situation leads to strong increase in skin conductivity

Benchmark Study

As mentioned above, an online survey and a survey of various car dealers were carried out in advance. 80% of respondents from the online survey said they knew the term LKAS. In the equipment question ("which ADAS would you equip your next car?") however, the subjects favor other assistance systems. For example, participants reported Brake Assist (73%), ACC (43%) and Blind Spot Assist (47%) as more important features than LKAS (34%).

Before the LKAS instructional video, nearly a third of respondents (31%) said that they had a very poor or poor knowledge about LKAS. The importance of the LKAS was also rated as low. After the instructional video, however, the assessment changed significantly. After Rodgers' theory "Diffusion of Innovation", the customer first became aware of the importance of the product. Now 75% spoke in favor of a LKAS in their next vehicle.

The fact that the subjects classify the LKAS as an enthusiasm factor according to Kano represents another important finding for car manufacturers. Enthusiasm factors according to Kano have the greatest influence on customer satisfaction and purchasing decisions. It therefore offers car manufacturers the potential to bind customers to their own brand or product. This is also confirmed by results from a survey of the 11 car dealers in Kempten and surrounding areas. During the discussions it became clear that the topic of LKAS is not well known by customers. There is a need for explanation. For car dealers this is a chance to positive point to the LKAS during the test drive.

In the benchmark study, the average age of the subjects was from 19 to 65 years. 30% of the participants were women. The participating test persons cover an average of almost 17,000 kilometers per year. As in the online survey, the subjects estimate the LKAS first as not very important.

The majority of participants would like to see increased road safety (65%) in the context of the LKAS. Thus, the customers expect that the LKAS is able to keep the lane reliable at all times of the purpose of use. As mentioned, some test persons had had negative experiences in the previous LKAS study. Nevertheless, according to Kano, this was identified as a characteristic of enthusiasm.

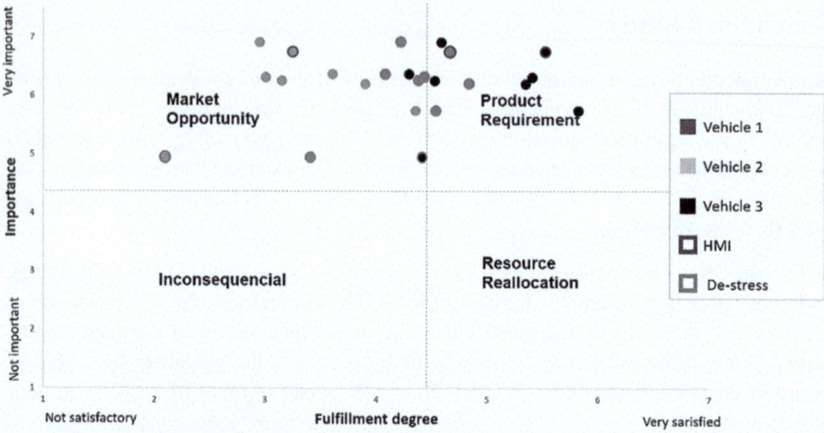

Figure 9: Market Opportunity Map of rated criteria

On the given Likert scale (1=not important to 7=very important) six of the eight criteria were evaluated with an importance of at least six. It was noticeable that all criteria, with the exception of tracking quality, became increasingly important after the test drive was completed. The criteria with the highest importance were the feeling of safety with 6.9 out of 7.0 points, the HMI with 6.8 points and the edge guidance with 6.5 points. As in the stress study the importance of the edge guidance became clear to most test persons only after the self-experience with the test drive. Original statement: "If the vehicle cannot drive exactly in the driving line, then it should at least prevent the vehicle from leaving the lane".

This result is in line with the experience gained during the expert test rides. Also subjects complain the workload and stress by using LKAS. Strong tracking offset to the outside of the curves were perceived as extremely unpleasant. In contrast, sudden unpredictable system drop-offs were rated as absolute no-go. Non-transparent system boundaries and high monitoring effort were rates poor in the sense of safety feeling and comfort. In addition to the importance, the respective degree of fulfilment of the individual criteria of the customer level of the evaluation level model was queried by the subjects. This shows considerable deficits and improvement potential to all criteria's, such as lane tracking quality, edge guidance, driver-vehicle interaction, availability, de-stress, sense of safety and HMI.

Figure 9 clearly shows the results in a MOM – Market Opportunity Map. The MOM is divided into 4 areas. The decisive factor is to shift the different criteria in the upper left to the upper right area. The graph shows that each vehicle has deficits.

Only one vehicle was deliberately equipped with a head-up display (vehicle 3). The HMI evaluation for this vehicle was rated the best as well as its degree of de-stress. This suggests that there is a correlation here. Overall, it is assumed that a good head-up display has a positive interaction with other criteria because of the necessary function transparency and simple monitoring. According to subject statements, drivers can relax more and focus on traffic environment and events.

The subjective assessment of the test persons' about the degree of maturity (Figure 10) showed poor values of 41.5–58.9% for all three vehicles. Analyses have shown that a positive purchase motivation and technology acceptance could only be recognized from a subjective maturity level of well over 75%.

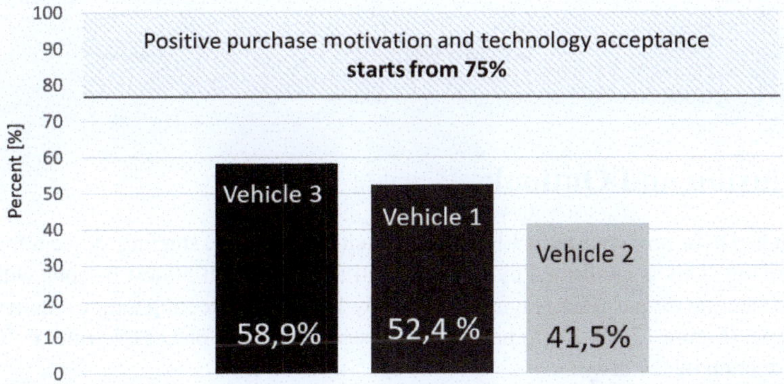

Figure 10: Maturity level of all 3 vehicles rated by the subjects

The strong relationship between customer and expert evaluation can finally be proven in a correlation diagram (Figure 11). It should be noted that the group of experts evaluated the vehicles with their expert criteria. These were transferred to the customer criterion as an average value. The test persons had rated on the customer level as usual. The degree of fulfilment values of the criteria of the Likert scale were transferred to a typical expert scale of 10.

This makes it clear that the evaluation scheme with the level model is valid. On the one hand, the evaluation criteria have proven themselves, as the test persons are able to handle the terminology well and evaluate the vehicles in a comprehensible way. On the other hand, it makes it clear that the experts can predict the customer rating with their evaluations. If the vehicles in the sub-criteria are good, the probability that they will be rated as good by the customer is extremely high. This correlation is of inestimable value for the LKAS development. In further steps, a weighting of the individual criteria could be introduced here.

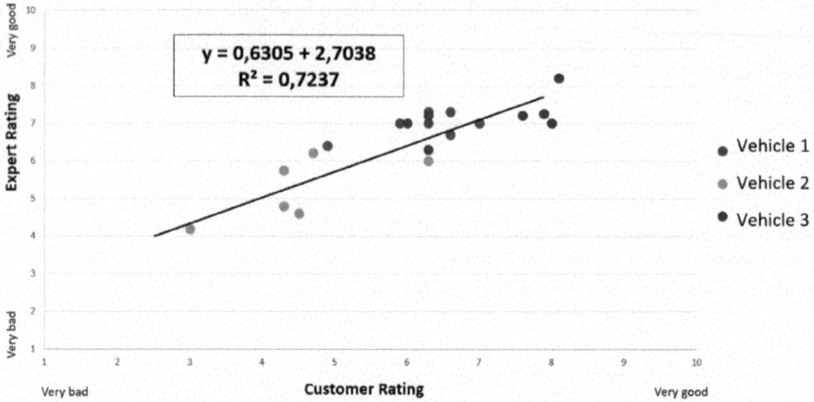

Figure 11: Correlation between customer and expert ratings

Conclusion and Outlook

In fact, it had to be proven that the LKAS significantly reduces the steering torque effort while driving. This is confirmed by the statistical analysis of all 50 test persons with regard to steering torque effort (Figure 12) in study 1. The high steering torque required with LKAS (Figure 12) is due to unexpected steering interventions or overrule of the LKAS function by the driver.

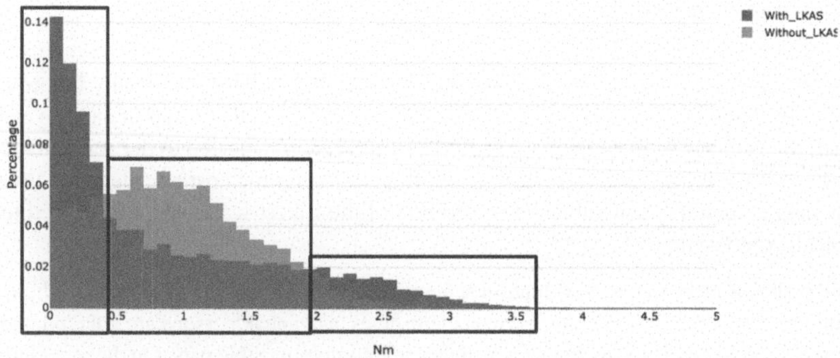

Figure 12: Probability density of steering torque with and without LKAS

In the classic objective driving comfort evaluation only the impact on the human body or the physical effort are assessed. In contrast, to reduce the feeling of comfort with ADAS only to the physical stress is by far not sufficient according to [7]. In addition to the physical stress, the consideration of mental stress is of immense importance for a holistic comfort assessment in this case.

In order to investigate the acclimatization effect and the influence of the test sequence in more detail as well to confirm the results of study 1, a study 3 was conducted. Only test 20 subjects from studies 1 and 2 were admitted and drove exactly the same routes with and without LKAS. One group started with and continued without LKAS and vice versa. Figure 13 shows the level of mental stress based on EDA peaks. This clearly shows an increase with LKAS, no matter in which order and confirming the results of study 1. More about study 3 will be published in the near future.

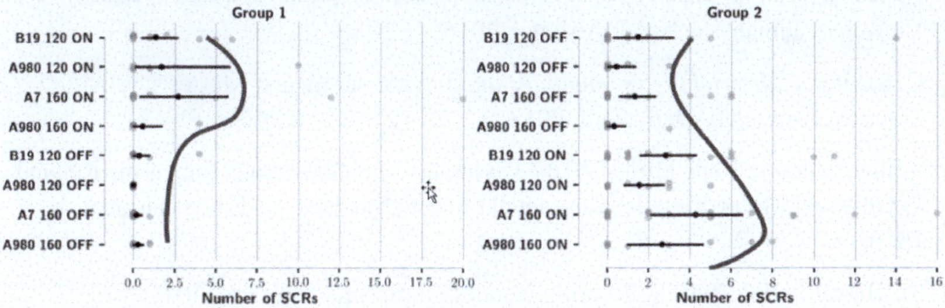

Figure 13: More EDA peaks appear on LKAS ON than LKAS OFF

An acclimatization effect can only be observed very little, if at all. Even experts and subjects complain the mental stress with used LKAS. Unpredictable system drop-offs, non-transparent system boundaries, high monitoring effort and poor tracking quality, led in particular to poor rates in the sense of safety feeling and comfort. It is very difficult to get familiar to such things, according to the subjects, so they are classified as basic requirements according to Kano.

Since humans hand over control to the vehicle, trust and the associated acceptance play a central role. Ultimately the breakthrough of automated driving will decide on customer acceptance. "If only the engineer realizes the differences and understands the system, the customer has no benefit". The findings of the two studies show that in ADAS/AD development, the human being, or rather the customer, should be placed much more at the center of development. Furthermore, it is necessary to focus on driving attributes and the driving experience in the sense of an attribute-based development.

Therefore, measures must be derived that increase the customer's confidence and driving characteristics. The clear and always transparent communication between "man and machine", a positive subjective driving experience (for driver and passengers), reliable availability and predictability form the basis of a good customer assessment. A good HMI based on the principle – trust is good, control is better – can be very effective here, too. In future studies, effective measures to increase trust and acceptance are to be developed by taking a holistic view.

Literatur

[1] R. Bönsch, „Autonom verändert den Markt," VDI-Nachrichten, Nr. 17, 27.04.2018.

[2] H. Winner, S. Hakuli, F. Lotz und S. C. , Handbuch Fahrerassistenzsysteme, Vieweg+Teubner Verlag, 3. Auflage; 2015.

[3] C. Seidler, *Fahrerlebnis vs. mentaler und physischer Stress,* Master Thesis Tchnical University Darmstadt, 2018.

[4] S. International, SAE J3016-20169 Taxonomy and Definitions for Terms Related to On-Road Motor Vehicle Automated Driving Systems, SAE International, 2016.

[5] M. Harrer und P. Pfeffer, Steering Handbook, Springer Verlag, 2016.

[6] A. Martin und K. Schieber, Psychologische Grundkonzepte der Verhaltensmedizin, U. Ehlert (Ed.), Springer-Lehrbuch. Verhaltensmedizin: Springer, 2016.

[7] A. Engelbrecht, Fahrkomfort und Fahrspaß bei Einsatz von Fahrerassistenzsystemen, Hamburg: Disserta Verlag, 2013.

[8] B. Schick, S. Resch, M. Yamamoto, I. Kushiro und N. Hagiwara, *Optimization of steering behavior through systematic implementation of customer requirements in technical targets on the basis of quality function deployment,* Yokohama/Japan: FISITA, 2006.

KEYNOTE LECTURES III

The chassis of the all-new LS/LC – Lexus luxury prestige Sedan/Coupé

Masaya Akita,
Department General Manager, Chassis Engineering,
Lexus International, (Toyota Motor Corporation), Japan

P. Pfeffer (Ed.), *9th International Munich Chassis Symposium 2018*, Proceedings,
https://doi.org/10.1007/978-3-658-22050-1_6

1 Introduction

Lexus is moving to the next new phase by revolutionizing all elements of R&D, Engineering and Design. The revolution started from developing an all-new FR platform that enhances the driving dynamics, and realizes a beautiful design.

"What does FR-like characteristic mean?" The answer was in creating a platform that significantly increases the potential of the driving dynamics. "Back to Basics" was the key phrase. We went back to the fundamentals and focused our efforts on improving the inertia mass specifications and structural rigidity.

The new LC utilizes the newly developed GA-L, the new Lexus Global Architecture – Luxury platform (Fig. 1). GA-L ensured a low centre of gravity and optimum weight distribution, contributing to the vehicle's essential stability and handling agility, which in turn delivers increased driver rewards. Adoption of the GA-L platform realizes "Even Sharper", significantly more agile, and "More Refined" driving experience.

The new LS is built on the new GA-L platform, which realizes "Exhilarating Performance" (Fig. 2). Moreover, LS500 F SPORT model features the Vehicle Dynamics Integrated Management (VDIM) Step 6 to deliver an ultimate level of stability, controllability, and ride comfort.

Lexus designers took full advantage of the opportunity provided by the low centre of gravity and optimum weight distribution from the new GA-L platform to produce a lower profile and a long wheelbase, giving the vehicle a stretched, ground-hugging appearance.

Fig. 1: New LC500

Fig. 2: New LS500

To realize this driving dynamics and design, fully renewed the suspension system and achieved a high level of both handling and ride comfort. This paper shows outlines of the chassis system development for the all-new LS/LC.

2 New Platform

The new LS/LC adopted the newly developed GA-L platform – the basis platform for future Lexus FR line-up.

Fastidious attention was paid to factors impacting the vehicle's inertia, together with the body rigidity and weight reduction, which significantly contributes to the achievement of the driving appeal and design of the FR. For the new platform, chassis system, body, and powertrain were renewed. As for the chassis system, the whole system was newly developed, the suspension, steering, and brake.

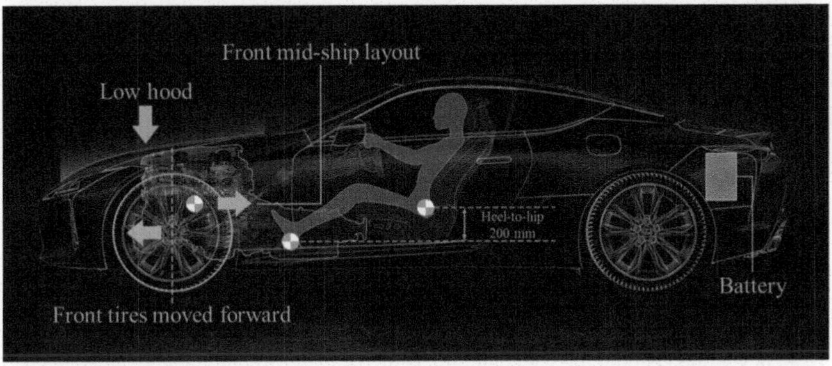

Fig. 3: New platform package

To optimize the inertia specifications the driving position was redesigned, the centre of the engine position was lowered and moved further rearwards by using the new generation powertrain, front tires were placed further forwards. Furthermore, run-flat tires were adopted to remove the spare tire and to place the battery to the luggage room. These efforts realized an optimized front to rear weight balance (Fig. 3).

At the development of GA-L, the suspension system was fully renewed to enhance the driving dynamics and design. High level of both handling and ride comfort was achieved.

3 Chassis System outline

"Even Sharper, More Refined" driving dynamics pursued by the LC. The development of an all-new chassis system started in order to realize this goal. "Even Sharper" means an excellent response and linear steering characteristics. "More Refined" means a high vehicle stability at any situation and flat ride.

The keyword for LS was "Exhilarating Performance", focusing on the delightful driving experience with steering accuracy and body control. The renewed suspension system enabled both stability and controllability at any situation.

Figure 4 shows the key targeting performances to achieve these goals.

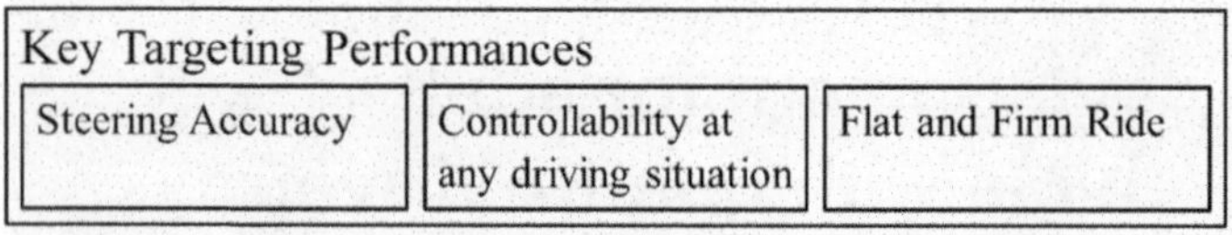

Fig. 4: Key targeting performances for LC/LS

3.1 Steering Accuracy

To achieve precise steering, the first point is to focus on the increase of the yaw eigen frequency while keeping the yaw damping (Fig. 5).

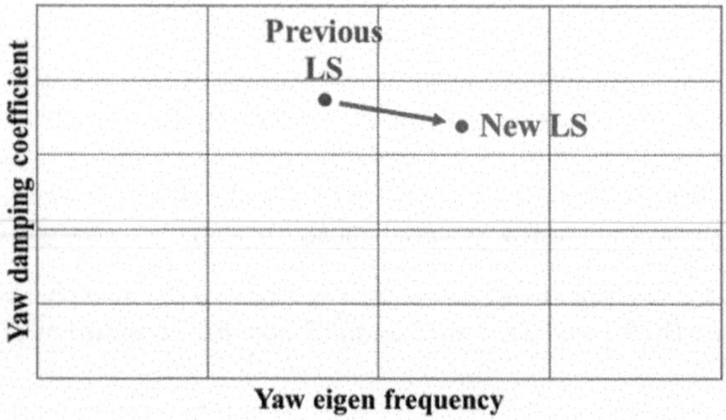

Fig. 5: Yaw eigen frequency and yaw damping

To increase yaw eigen frequency:

- Increased the normalized cornering stiffness (Cp/W), optimizing the tire to a larger size than the previous model (Fig. 6).

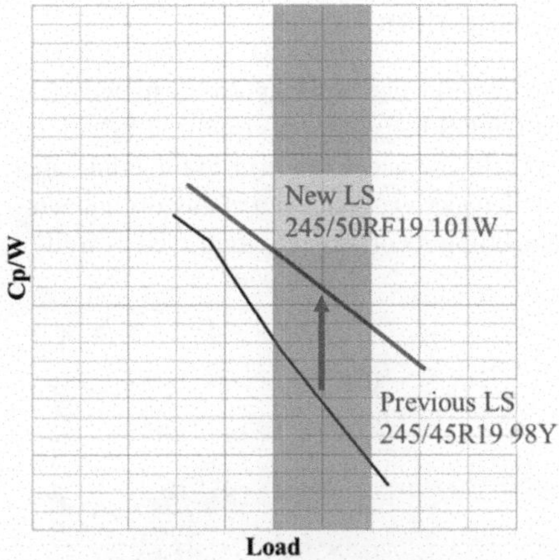

Fig. 6: Normalized cornering stiffness of tire

- Increased the suspension stiffness – front camber and front/rear torsional stiffness – to suppress the increase in time constant of cornering force related to the increase in Cp/W, ensuring the yaw damping.

The second point is to define target values for the on-centre and off-centre steering feel by focusing on the steering torque. Conforming to the target value was achieved by the friction control of the steering gear and tuning of EPS (Fig. 7 and Fig. 8).

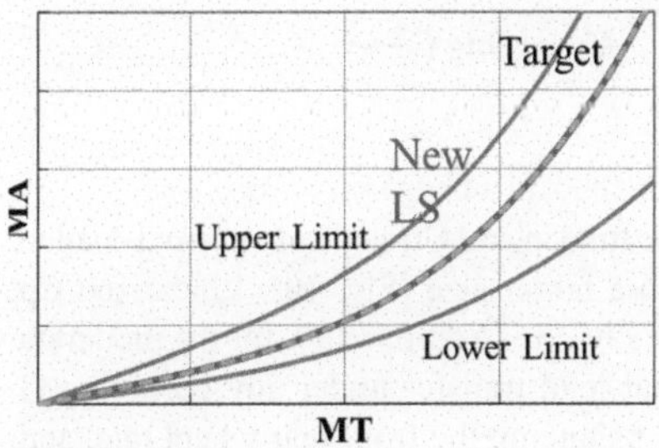

Fig. 7: On-centre steering feel

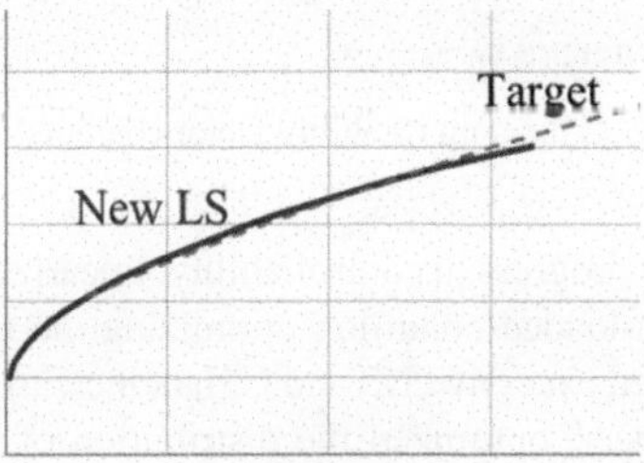

Fig. 8: Off-centre steering feel

As the third point, the rack bush stiffness and steering column support rigidity are increased while reducing the inertial moment of the steering wheel, improving the road surface feedback. Adverse effect of causing steering vibrations under braking was suppressed by the filter in the EPS.

3.2 Controllability at any driving situation

To ensure controllability at any situation, the dynamic performance was designed using the G-G diagram as shown in figure 9. The US/OS behaviour is visualized by colour. The G-G diagram uses two axes: cornering G (lateral acceleration) and longitudinal G (acceleration and deceleration). Moderate understeer was achieved at any driving situation. Compared to the previous model the maximum cornering G improved as well.

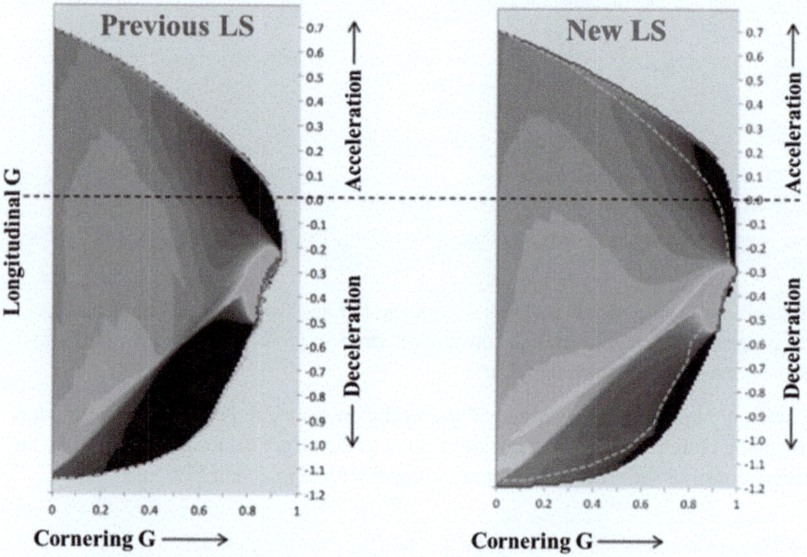

Fig. 9: G-G diagram (Stability factor contours)

For the progress in controllability, linearized the tire Cp and Cfmax from normal-loaded to high-loaded condition optimizing the tire to a larger size (Fig. 10). Linearized Cp removed excessive US and Cfmax increased steady grip. Additionally, for the moderate understeer, optimized the distribution of lateral load transfer under any cornering G condition by increasing the front roll stiffness, balancing the front/rear wheel rate, and smoothing the front/rear wheel rate gradient.

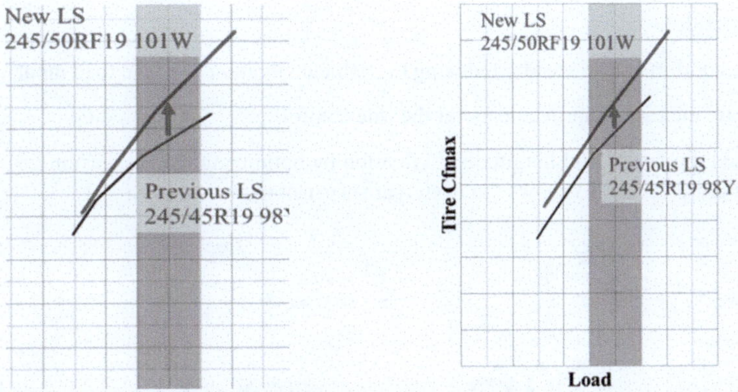

Fig. 10: Linearized Tire Cp and Cfmax

3.3 Flat and Firm Ride

- Primary ride

Flat body movement is achieved on undulation road input. Figure 11 shows the floor acceleration of LS for body control and isolation compared to key competitors.

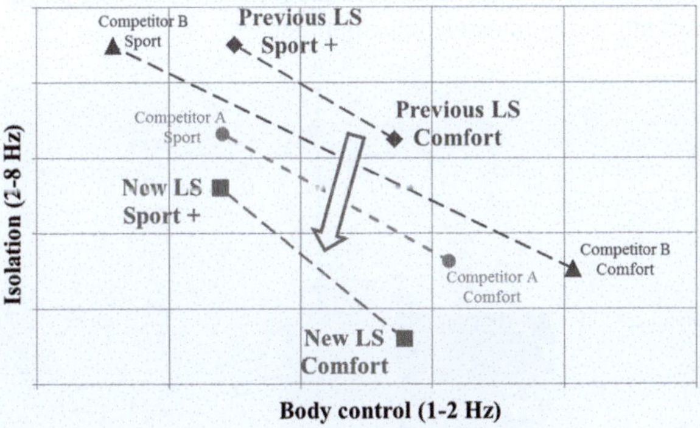

Fig. 11: Floor acceleration for body control and isolation

Improvements to achieve the flat body movement are:

- Reduced wheel rate during in-phase motion while keeping the roll stiffness by increasing the stabilizer bar stiffness.
- Newly developed absorber controlling system: the intelligent Body Control (iBC).

- Secondary ride

Refined ride comfort is achieved by reducing the vehicle vibration from the road input.

Followings are the main improvements for the ride comfort:

- Suppressed the lateral and longitudinal vibration by optimizing the suspension geometry. Figure 12 and 13 shows the front/rear suspension layout.

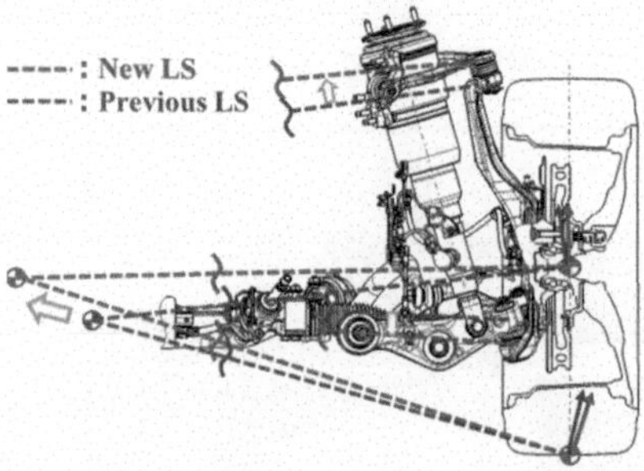

Fig. 12: Front suspension layout (rear view)

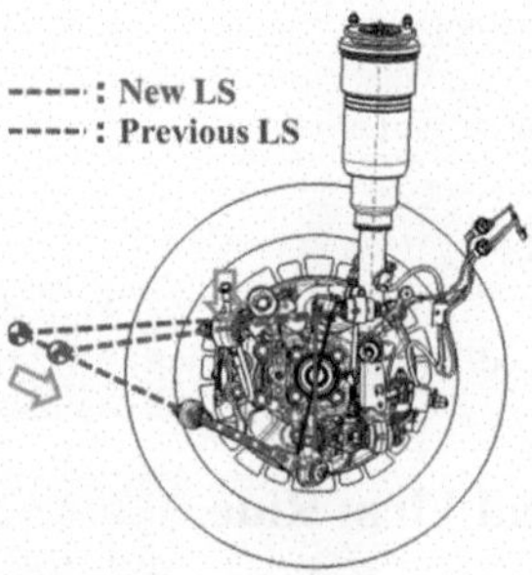

Fig. 13: Rear suspension layout (side view)

- Supressed the shock from small protrusions on the road by reducing the spring rate at small amplitude, reducing friction of shock absorber and replacing rear ball joints to bushes. Figure 14 shows the spring rate at small amplitude.

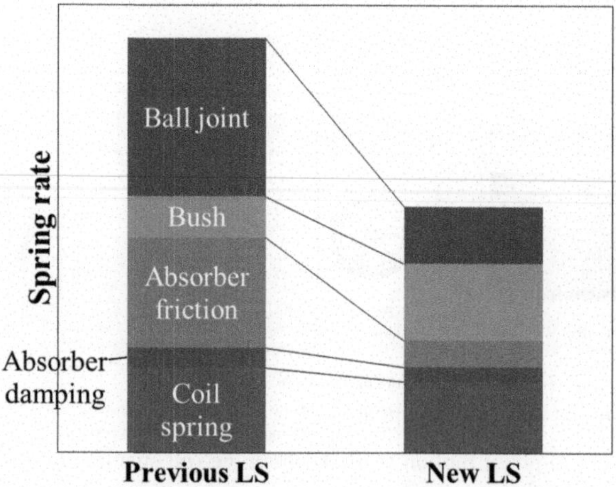

Fig. 14: Spring rate at small amplitude

4 Vehicle Dynamics Integrated Management Step 6

The latest generation of a chassis control technology was developed to provide a higher level of both handling and ride comfort for the new LS500 F SPORT: the VDIM Step 6.

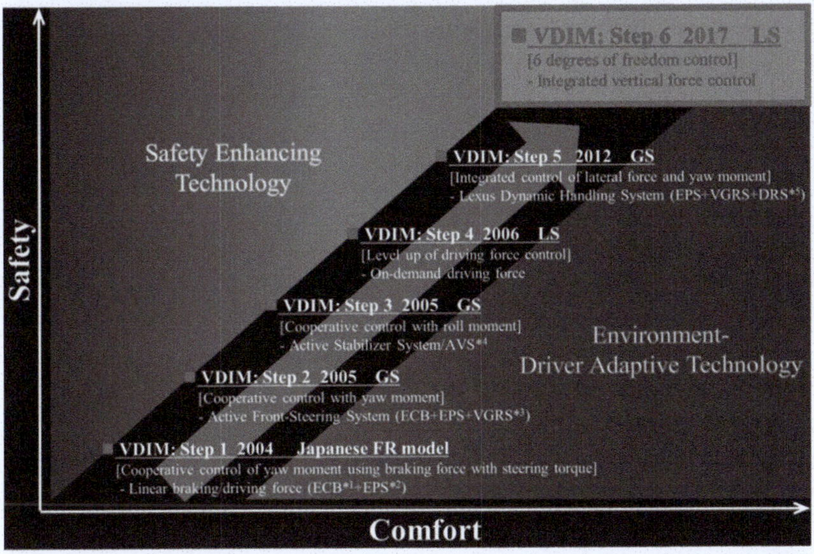

Fig. 15: Evolution of VDIM

"Six degrees of freedom control". the freedom control of longitudinal, lateral, vertical, yaw, roll and pitch motion. To achieve this goal, Lexus improved the vehicle control systems and integrated them step by step since the first VDIM. (Fig. 15).

In 2004, VDIM was revealed as the world's first integrated control system which combined the previously independent ABS, TRC, VSC, EPS, and other functions into a single system to enhance the vehicle dynamic performance: "driving, turning, and stopping".

---

*1: Electronically Controlled Brake System
*2: Electric Power Steering
*3: Variable Gear Ratio Steering
*4: Adaptive Variable Suspension System
*5: Dynamic Rear Steering

In 2012, the current GS350/450h adopted the four-wheel active steering integrated control system – the Lexus Dynamic Handling (LDH) system – for enhanced safety and driving dynamics that responds to the driver's intention. The LDH enabled great steering response from initial steering to cornering at normal speed, and confident with non-delay response while high speed cornering.

Today, the new LS500 F SPORT achieved the six degrees of freedom control by integrating the LDH and active stabilizer system*6, the unique vehicle controlling technology of Lexus. The VDIM has reached one of the goals of vehicle dynamic control, making the LS an emotional driver's car.

However, adding the active stabilizer system to the LDH wasn't so simple. Optimized six degrees of freedom vehicle motion control won't be achieved by independently operating the two systems, because controlling one motion may adversely affect others. In particular, the roll motion couple with the yaw/lateral motion and influence each other. Therefore, the optimized motion control requires the development of an integrated control system which considers interaction between each motion control.

The new LS realized the optimization control of yaw, lateral, and roll motion for normal driving situation by the VGRS ECU which operates integrated control algorithm.

On the other hand, the brake ECU is responsible for the vehicle control at the stability limit region. LS achieved seamless control of the six degrees of freedom from normal driving to the stability limit, by cooperatively controlling the VGRS/brake ECU.

VDIM Step 6 achieved a high level of both steering response and flat vehicle posture without sacrificing natural handling while any cornering G and speed (Fig 16 and 17).

Fig. 16: With VDIM

Fig. 17: Without VDIM

*6: Active stabilizer system: installed in GS (2005) as shown in figure 15.

5 Summary

The development process for the new LS/LC yielded the entirely new GA-L platform which will be the basis platform for future Lexus FR models. The platform made its debut in the LC flagship coupe and now underpins the new LS flagship sedan with a longer wheelbase. The developing keyword for LC/LS was "Even Sharper, More Refined" and "Exhilarating Performance". To achieve these goals, we newly developed the whole chassis system with three key targeting performances: "Steering accuracy", "Controllability at any driving situation", and "Flat and firm ride". As a result, the new LS/LC opened a future chapter in Lexus history with emotional driving and advanced technologies. Ultimate level of stability, controllability, and ride comfort were achieved.

Fig. 18: New LC500

Fig. 19: New LS500h

Driverless robocabs – challenges and solutions regarding chassis technology

Dr. Andree Hohm, N. Balbierer, S. Pla, R. Syrnik,
Continental Teves AG & Co. oHG

This manuscript is not available according to publishing restriction. Thank you for your understanding.

P. Pfeffer (Ed.), *9th International Munich Chassis Symposium 2018*, Proceedings,
https://doi.org/10.1007/978-3-658-22050-1_7

i30N – the first high-performance vehicle development from Hyundai

A. Biermann, J.H. Park, K. Koester

P. Pfeffer (Ed.), *9th International Munich Chassis Symposium 2018*, Proceedings,
https://doi.org/10.1007/978-3-658-22050-1_8

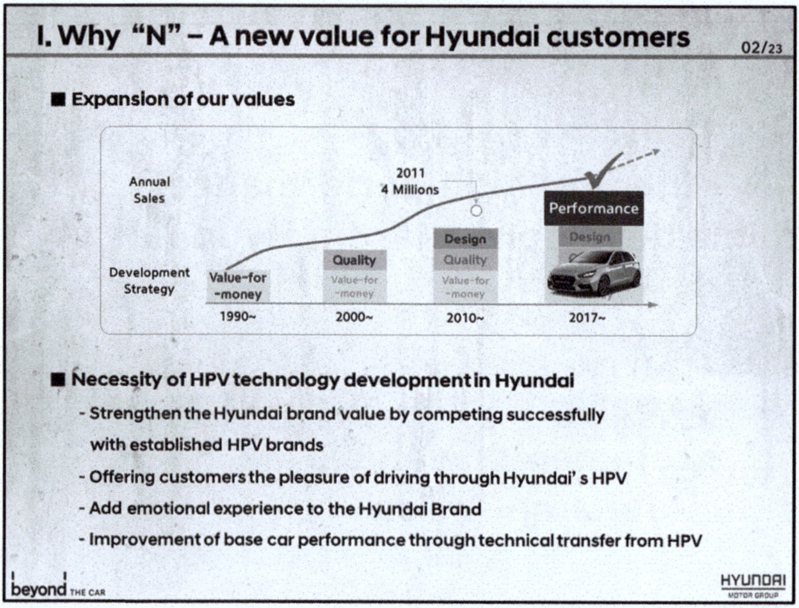
I. Why "N" – A new value for Hyundai customers
02/23
Expansion of our values
Annual Sales
2011
4 Millions
Performance
Design
Design
Quality
Quality
Development Strategy
Value-for -money
Value-for -money
Value-for -money
1990~
2000~
2010~
2017~
Necessity of HPV technology development in Hyundai
- Strengthen the Hyundai brand value by competing successfully with established HPV brands
- Offering customers the pleasure of driving through Hyundai' s HPV
- Add emotional experience to the Hyundai Brand
- Improvement of base car performance through technical transfer from HPV
beyond THE CAR
HYUNDAI MOTOR GROUP

I. History of Hyundai "N" roots in motorsport
03/23
Genesis coupe (2008)
Rear wheel driven sporty car
i20 WRC (2014~)
Back to WRC
2014 : Germany rally win
RM14 (2014)
Busan Motor show
i20 WRC (2015)
RM15 (2015)
Seoul Motor show
NBR 24h (2017)
4th and 9th place with i30N
NBR 24h (2016)
Completion with i30/ Veloster model
RM16 (2016)
Busan Motor show
i20 WRC (2016)
Five rally wins
"N sub-brand Launch" (2015)
Frankfurt motor show
i30 N Launch
(2017.9 Europe, '18.3 Australia)
i30N TCR (2017)
TCR international debut and win at China Zhejiang
NBR 24h + WTCR (2018)
Veloster N
(2018. 5 Korea/North America)
beyond THE CAR
HYUNDAI MOTOR GROUP

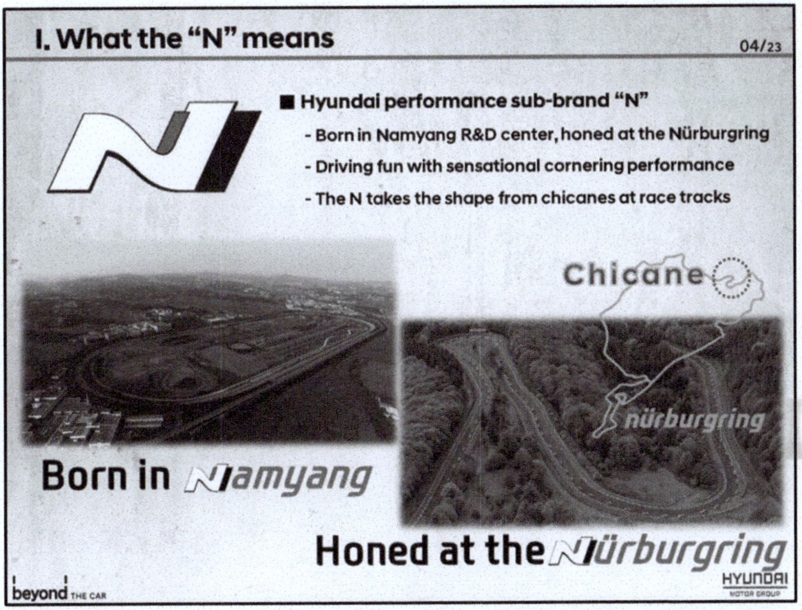
I. What the "N" means
04/23
Hyundai performance sub-brand "N"
- Born in Namyang R&D center, honed at the Nürburgring
- Driving fun with sensational cornering performance
- The N takes the shape from chicanes at race tracks
Chicane
nürburgring
Born in Namyang
Honed at the Nürburgring
beyond THE CAR
HYUNDAI
MOTOR GROUP

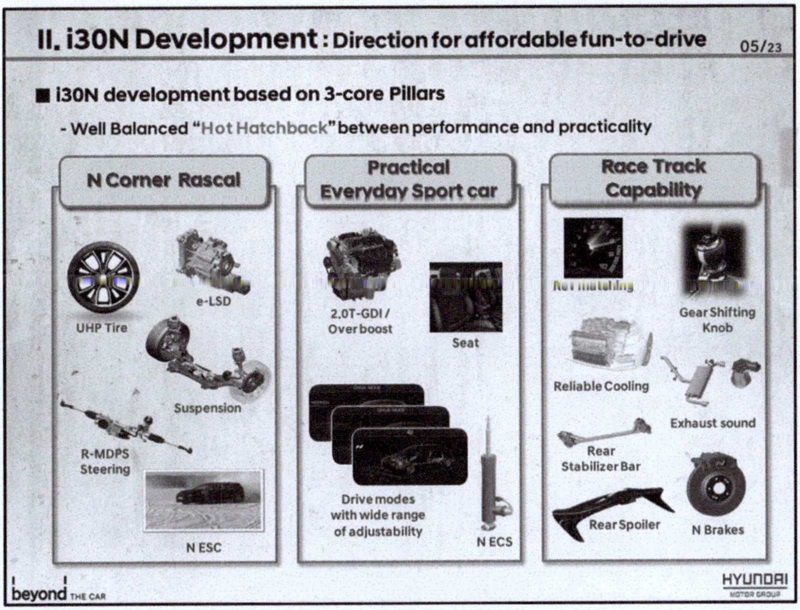
II. i30N Development : Direction for affordable fun-to-drive
05/23
i30N development based on 3-core Pillars
- Well Balanced "Hot Hatchback" between performance and practicality
N Corner Rascal
UHP Tire
e-LSD
Suspension
R-MDPS Steering
N ESC
Practical Everyday Sport car
2.0T-GDI / Over boost
Seat
Drive modes with wide range of adjustability
N ECS
Race Track Capability
Gear Shifting Knob
Reliable Cooling
Exhaust sound
Rear Stabilizer Bar
Rear Spoiler
N Brakes
beyond THE CAR
HYUNDAI
MOTOR GROUP

II. i30N Development : Powertrain

06/23

➢ **Powerful turbo gasoline direct injection engine**

- Key focus : High dynamic response for precise limit handling

System Information	
Engine	Θ-II 2.0 T-GDi
Displacement	1,998 cc
Bore x Stroke	86.0 x 86.0 mm
Compression ratio	9.5
Maximum Power	250 PS [standard package], 275 PS [performance package]
Maximum Torque	353 Nm (Overboost 378 Nm)

Twin scroll turbo charger

0 to 100 kph
in 6.1 seconds
250kph max. speed

Launch control

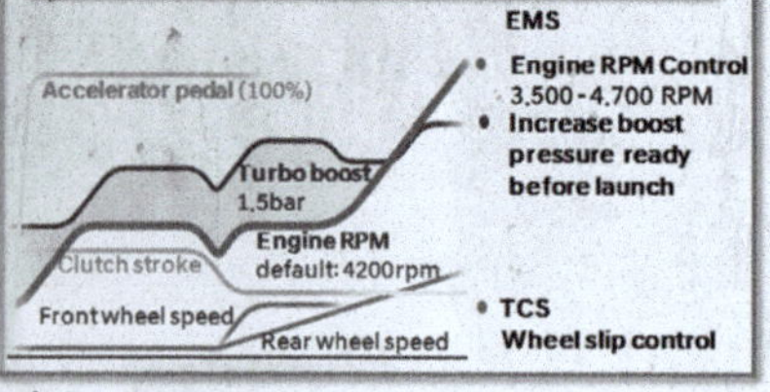

Overboost control

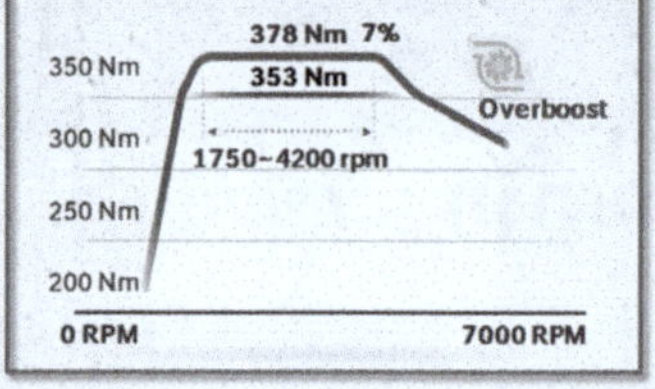

beyond THE CAR   HYUNDAI MOTOR GROUP

II. i30N Development : Powertrain

07/23

➢ **N Transmission**

- Key focus : Accurate & precise shift feel, direct & fast clutch response

System Information	
T/M type	6 speed manual
Clutch type	DMF
Maximum torque	385 Nm
Final gear ratio (1/2)	4.154/3.176 [Standard] 4.333/3.250 [Performance]
Step gear ratio ($1^{st}/2^{nd}/3^{rd}/4^{th}/5^{th}/6^{th}$)	3.08/1.93/1.70/1.28/1.03 /0.85 (rev 3.59)

Gear material reinforcement for robust durability

Increased burst strength

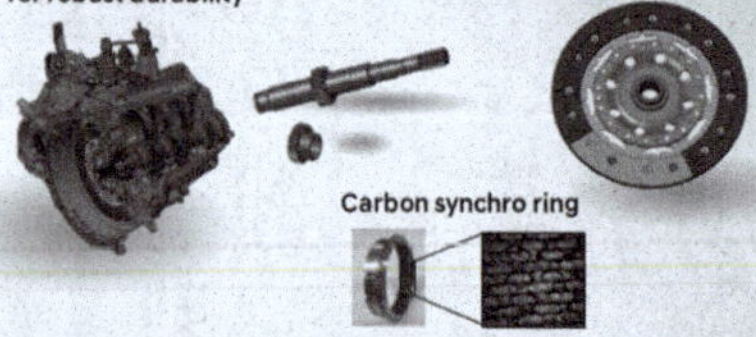

Double layer facing clutch disk

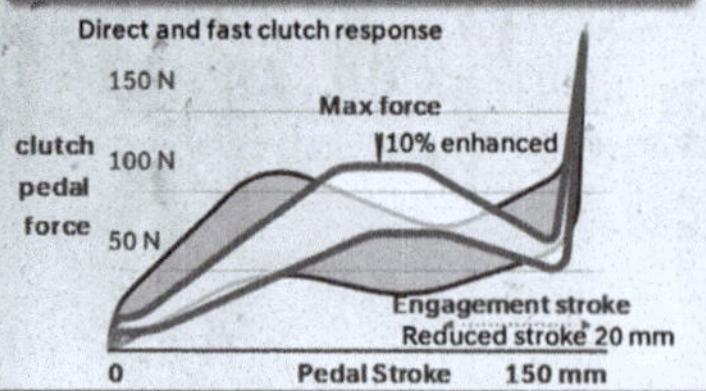

Precise shifting

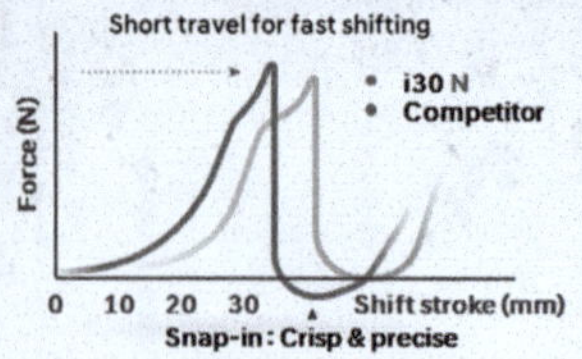

beyond THE CAR   HYUNDAI MOTOR GROUP

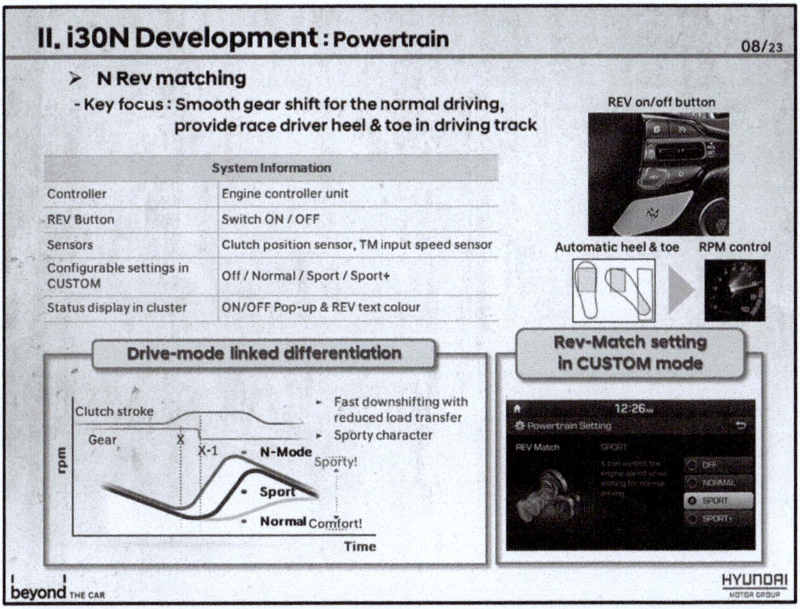
II. i30N Development : Powertrain
08/23
➤ N Rev matching
- Key focus : Smooth gear shift for the normal driving, provide race driver heel & toe in driving track
System Information
Controller | Engine controller unit
REV Button | Switch ON / OFF
Sensors | Clutch position sensor, TM input speed sensor
Configurable settings in CUSTOM | Off / Normal / Sport / Sport+
Status display in cluster | ON/OFF Pop-up & REV text colour
REV on/off button
Automatic heel & toe
RPM control
Drive-mode linked differentiation
Clutch stroke
Gear
X
X-1
rpm
N-Mode
Sport
Normal
Sporty!
Comfort!
Time
▸ Fast downshifting with reduced load transfer
▸ Sporty character
Rev-Match setting in CUSTOM mode
12:26
Powertrain Setting
REV Match
SPORT
OFF
NORMAL
SPORT
SPORT+
beyond THE CAR
HYUNDAI MOTOR GROUP

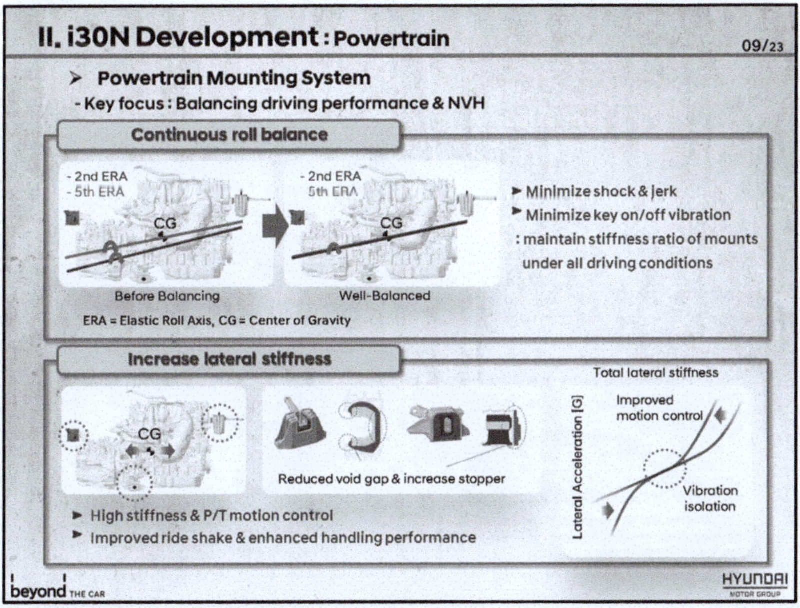
II. i30N Development : Powertrain
09/23
➤ Powertrain Mounting System
- Key focus : Balancing driving performance & NVH
Continuous roll balance
- 2nd ERA
- 5th ERA
CG
Before Balancing
Well-Balanced
ERA = Elastic Roll Axis, CG = Center of Gravity
▸ Minimize shock & jerk
▸ Minimize key on/off vibration
: maintain stiffness ratio of mounts under all driving conditions
Increase lateral stiffness
CG
Reduced void gap & increase stopper
▸ High stiffness & P/T motion control
▸ Improved ride shake & enhanced handling performance
Total lateral stiffness
Lateral Acceleration [G]
Improved motion control
Vibration isolation
beyond THE CAR
HYUNDAI MOTOR GROUP

II. i30N Development : Brake

10/23

➤ **N Brakes**

- Key focus : Prevent brake fade, consistency in brake feel

System Information	
Supplier	Hyundai Mobis (Korea)
Front caliper	Floating, 2-piston, cast iron
Front disk	Φ345, 30T, Vented
Rear caliper	Floating, 1-piston, cast iron
Rear disk	Φ314, 22T, Vented
Parking brake	Mechanical

Brake Cooling Enhancement

Brake cooling air duct

Enhancement of brake cooling by air curtain

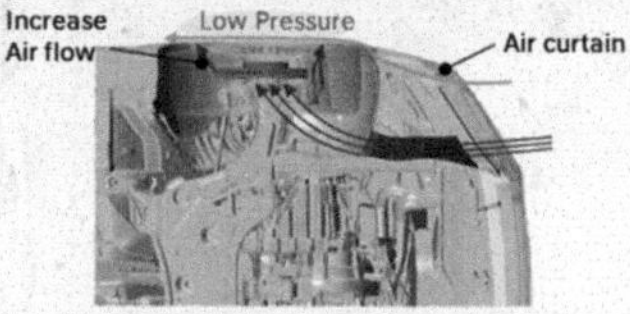

beyond THE CAR

HYUNDAI MOTOR GROUP

II. i30N Development : Aero/Cooling

11/23

➤ **N Aerodynamics and Cooling**

- Key focus : Aerodynamic balance focused on high speed driving stability
  Cooling performance for robust race track driving

System Information	
Drag (C_D)	0.32
Front Lift (C_{LF})	-0.02
Rear Lift (C_{LR})	-0.03

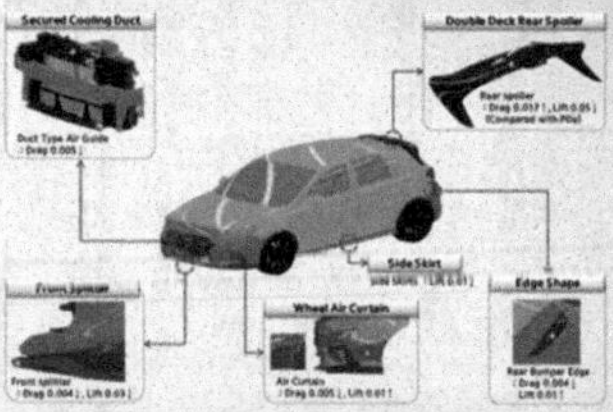

Down force balance

For securing stability in straight and cornering, developed downforce both front and rear and more downforce on rear

Reliable cooling

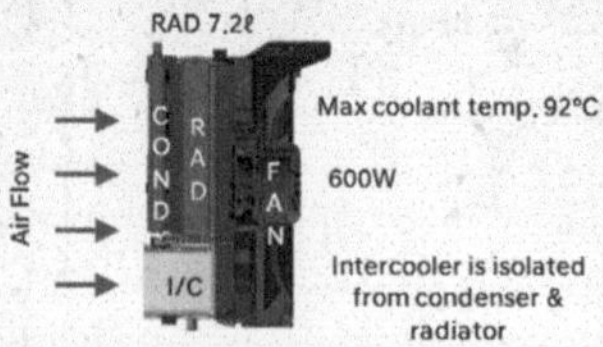

beyond THE CAR

HYUNDAI MOTOR GROUP

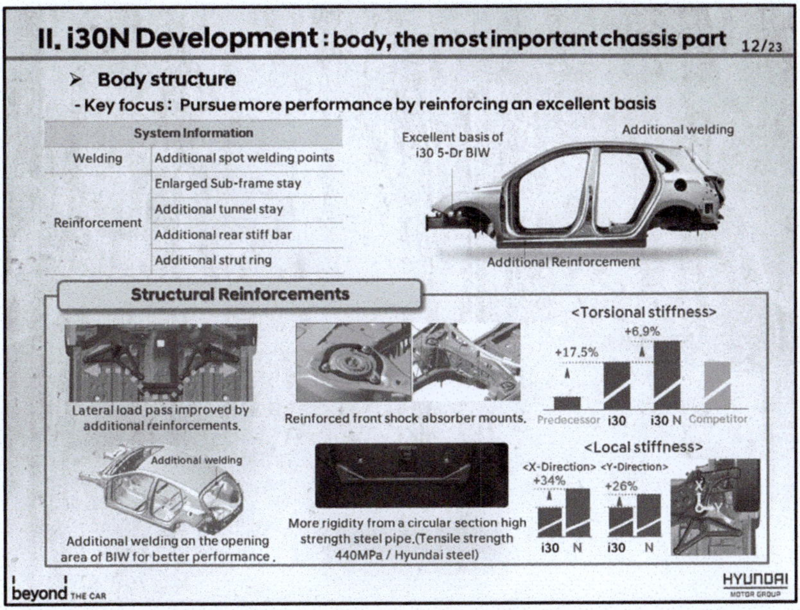
II. i30N Development : body, the most important chassis part
12/23
➢ Body structure
- Key focus : Pursue more performance by reinforcing an excellent basis
System Information
Welding
Additional spot welding points
Reinforcement
Enlarged Sub-frame stay
Additional tunnel stay
Additional rear stiff bar
Additional strut ring
Excellent basis of i30 5-Dr BIW
Additional welding
Additional Reinforcement
Structural Reinforcements
Lateral load pass improved by additional reinforcements.
Reinforced front shock absorber mounts.
<Torsional stiffness>
+17.5%
+6.9%
Predecessor
i30
i30 N
Competitor
Additional welding
Additional welding on the opening area of BIW for better performance.
More rigidity from a circular section high strength steel pipe.(Tensile strength 440MPa / Hyundai steel)
<Local stiffness>
<X-Direction>
<Y-Direction>
+34%
+26%
i30
N
i30
N
beyond THE CAR
HYUNDAI
MOTOR GROUP

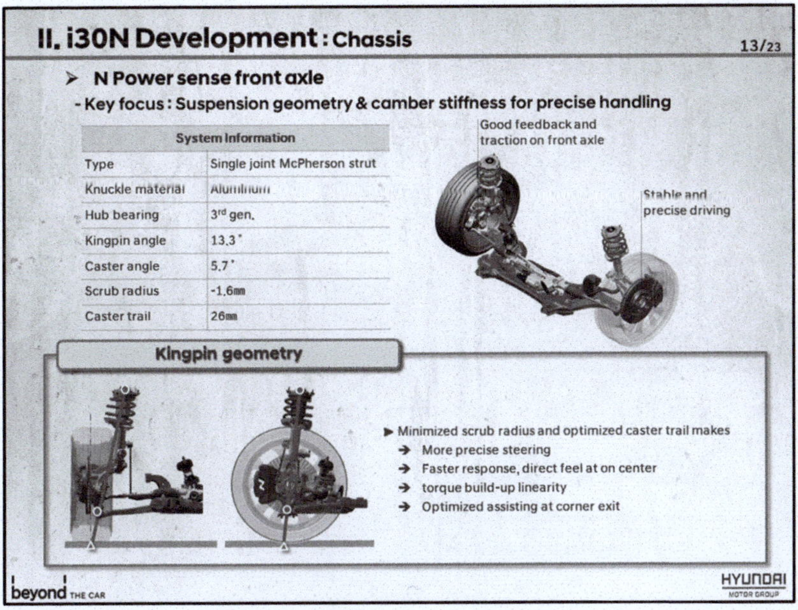
II. i30N Development : Chassis
13/23
➢ N Power sense front axle
- Key focus : Suspension geometry & camber stiffness for precise handling
System Information
Type
Single joint McPherson strut
Knuckle material
Aluminium
Hub bearing
3rd gen.
Kingpin angle
13.3˚
Caster angle
5.7˚
Scrub radius
-1.6㎜
Caster trail
26㎜
Good feedback and traction on front axle
Stable and precise driving
Kingpin geometry
▶ Minimized scrub radius and optimized caster trail makes
➔ More precise steering
➔ Faster response, direct feel at on center
➔ torque build-up linearity
➔ Optimized assisting at corner exit
beyond THE CAR
HYUNDAI
MOTOR GROUP

II. i30N Development : Chassis
14/23

➢ **N Power sense front axle**

Roll center height increased to reduce the roll angle and improve traction

▶ Camber stiffness increased by knuckle stiffness optimization

Optimized knuckle

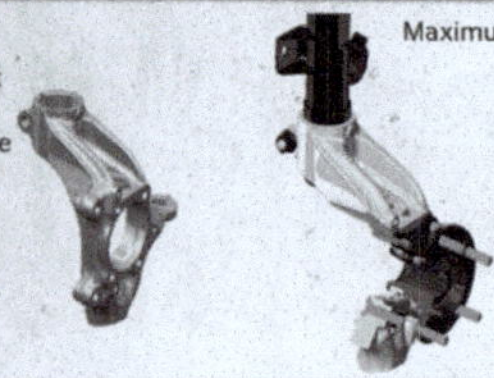

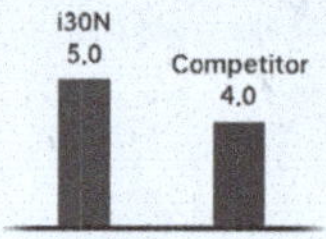

beyond THE CAR — HYUNDAI MOTOR GROUP

II. i30N Development : Chassis
15/23

➢ **N Rear axle**

- Key focus : Optimized Kinematic and & Compliance (K&C) and increased stiffness for quick response and precision

System Information	
Type	Multi-link
Knuckle material	Steel
Hub bearing	3rd gen.
Roll center height	148㎜ (@CVW)
Lateral steer	0.040 deg/kN
Lateral stiffness	749 N/㎜

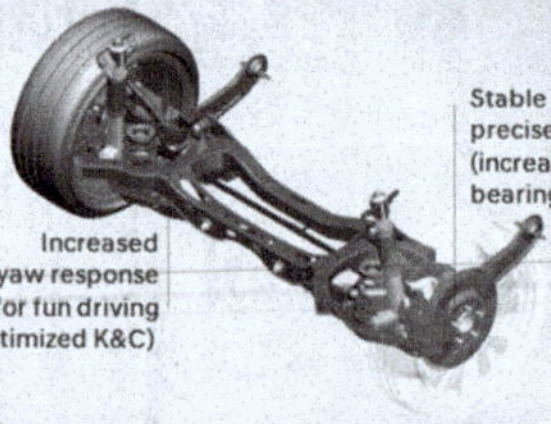

Optimized K&C characteristics

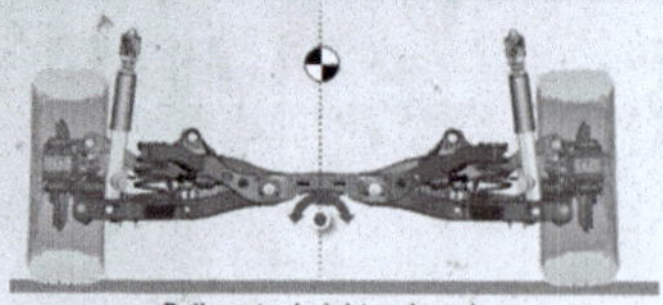

Roll center height reduced for roll linearity and front/rear balance

Reduced steer under lateral force to increase yaw response

beyond THE CAR — HYUNDAI MOTOR GROUP

II. i30N Development : Chassis

16/23

➢ **R-MDPS (Rack mounted Motor Driven Power Steering)**

- Key focus : Fast and precise steering, intense steering feedback

System Information	
Supplier	Mando (Korea)
Assist type	Dual pinion rack
G/Ratio type	Constant ratio
TGR	12.2(@360°)
Motor type	Brushless AC
Motor capacity	720W
Specific logics	CAC, TSC, etc

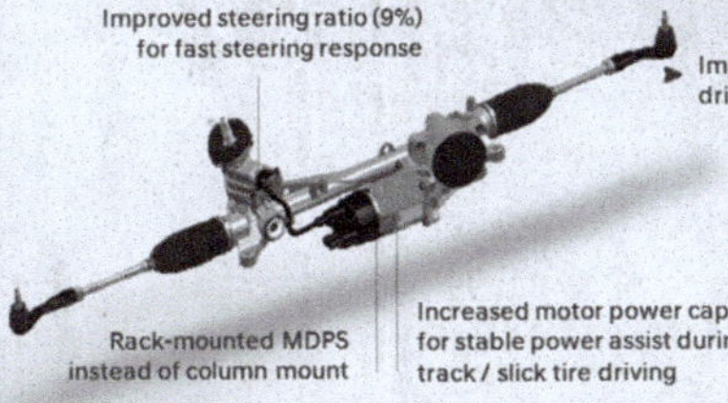

CAC (Cornering Acceleration Control)

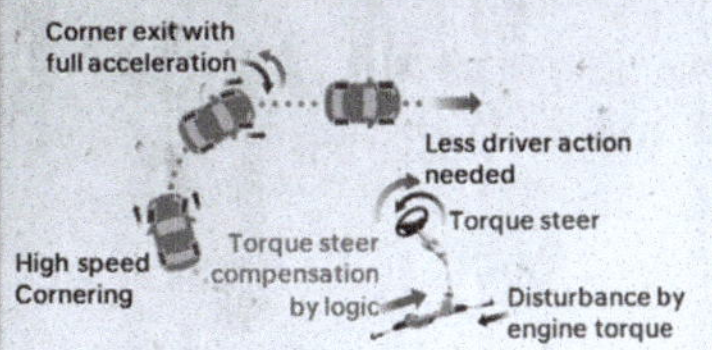

TSC (Torque Steer Control)

beyond THE CAR — HYUNDAI MOTOR GROUP

II. i30N Development : Chassis

17/23

➢ **N Tire**

- Specially developed for high performance

System Information	
	Performance Pack
Supplier	Pirelli
Patten	P Zero
Size	235/35R19

Pirelli
P Zero

OE Marking – HN code

Developed to match the chassis dynamics of "Hyundai N" cars.

beyond THE CAR — HYUNDAI MOTOR GROUP

II. i30N Development : Chassis control

18/23

➢ **ESC : Electronic Stability Control**

- Key focus : Maximized agility and stability without interference during cornering

System Information	
Supplier	Hyundai Mobis (Korea)
Model	MEB4-ESC_{AEB}
Motor capacity	315W
ESC Mode	Normal / Sport / OFF
Value added function	TVBB, Brake pre-fill Karussell detection

Karussell detection

ESC detect banked "Karussell" at Nurburgring and control adjust

Carraciola Karussell

TVBB (Torque vectoring by Brake)

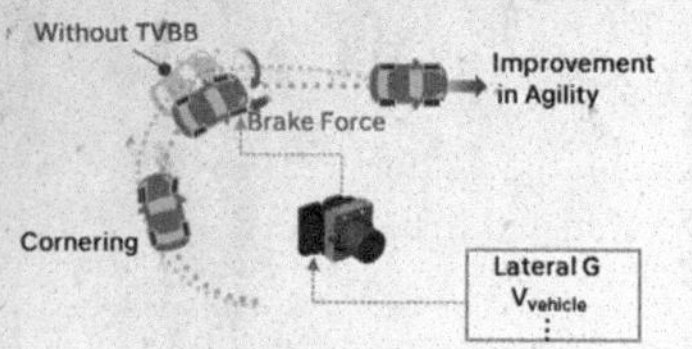

Brake Pre-fill

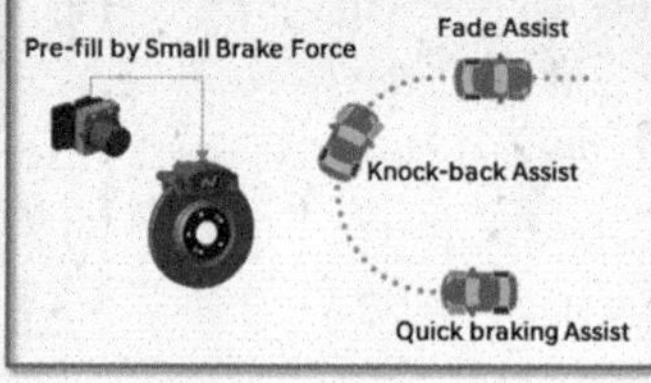

beyond THE CAR

HYUNDAI MOTOR GROUP

II. i30N Development : Chassis control

19/23

➢ **N Corner Carving Differential : electronically controlled LSD (e-LSD)**

- Key focus : Fast & controllable, consistency & carving feel

System Information	
Supplier	Hyundai-WIA (Korea)
Max. Capacity	1,400Nm
Type	FF type
Actuator	Hydraulic type
Weight	9.8Kgf

Electronic controlled LSD System

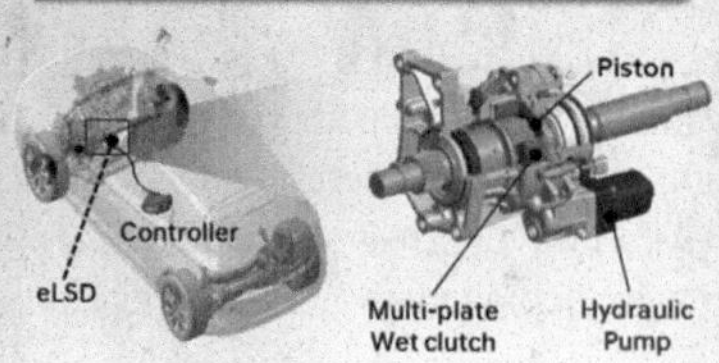

Increase of carving feel

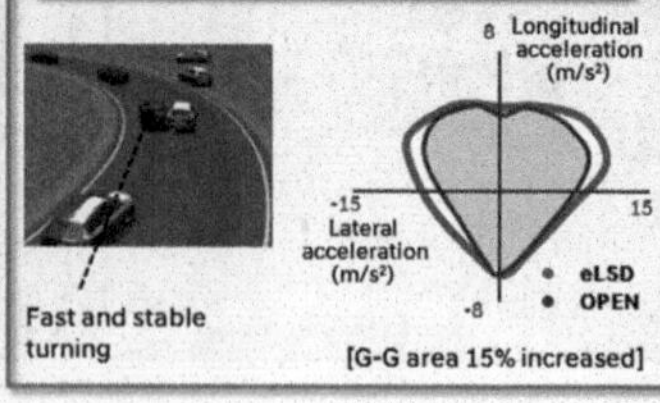

beyond THE CAR

HYUNDAI MOTOR GROUP

II. i30N Development : Chassis control

➤ **ECS : Electronic Controlled Suspension**

- Key focus : Independent control for good ride comfort and load transfer

System Information	
Supplier	Mando (Korea)
Model	SDC 50
Control Type	4 Independent Continuously variable valves
Sensors	2 Wheel G-Sensors 3 Body G-Sensors
Features	· Enhancement of ride comfort · Fast and natural system response · Diversified tuning parameters for valves

Continuous damping control

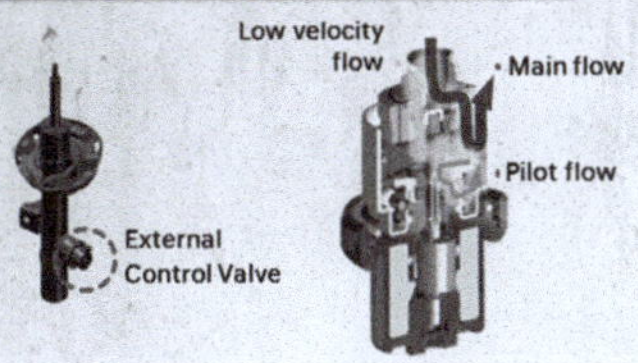

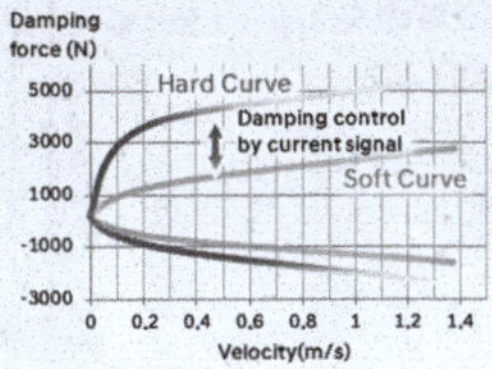

II. i30N Development : Sound

➤ **Dynamic N Sound control**

- Key focus : Feel the emotional N Performance

System Information	
Exhaust Valve	Faurecia (France)
Actuator	Küster (Germany)
Control	by ECU

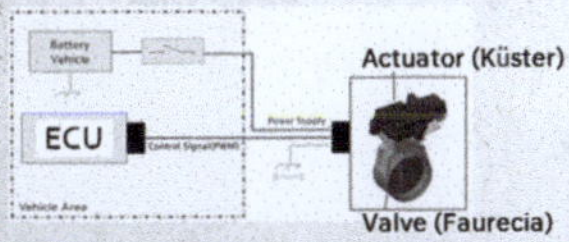

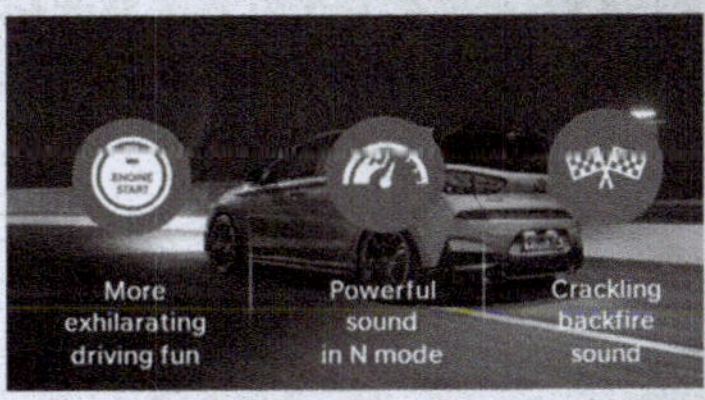

Variable exhaust valve

► Different sounds depending on drive mode

► Powerful sound in N mode

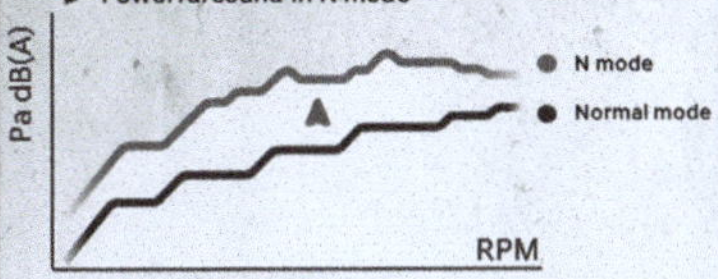

Afterburn sound (N mode)

► Create explosive sound as in race cars by retarding ignition timing

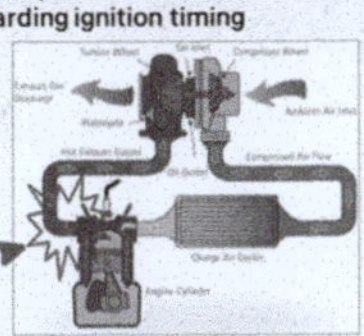

III. The Road to High Performance: Nürburgring 24H

22/23

■ **2016 NBR 24h (Engine verification with previous i30)**

: High performance engine has been quenched in Nürburgring 24h 2016 to enable it to perform strongly under extreme driving conditions and in various situation

participated in the 44th ADAC Zurich 24h Race with the existing i30 with 2.0 turbo engine being developed for the first model of high performance N.

Namyang R&D Center tested the Engine for HPV delivered outstanding Performance

■ **2017 NBR24h (Vehicle verification with i30 N)**

: the successful verification of "Total Performance"

beyond THE CAR

HYUNDAI MOTOR GROUP

23/23

Thank you for your attentioN.

beyond THE CAR

HYUNDAI MOTOR GROUP

PARALLEL STRAND I

NEW CHASSIS SYSTEMS

Chassis development for a fully electric vehicle with quattro drivetrain

Oswin Röder, Dr. Michael Wein, AUDI AG

P. Pfeffer (Ed.), *9th International Munich Chassis Symposium 2018*, Proceedings,
https://doi.org/10.1007/978-3-658-22050-1_9

Chassis development for a fully electric vehicle with quattro drivetrain

chassis.tech plus 12.06.2018

O. Röder, Dr. M. Wein

Index

- Product mission Audi e-tron prototype
- Definition of technical target values
- Vehicle concept
- Systems
  - Driving performance
  - Efficiency technologies
- Functions
  - Traction Control System (TCS)
- quattro
- Target accomplishment
  - Measurement results
- Summary

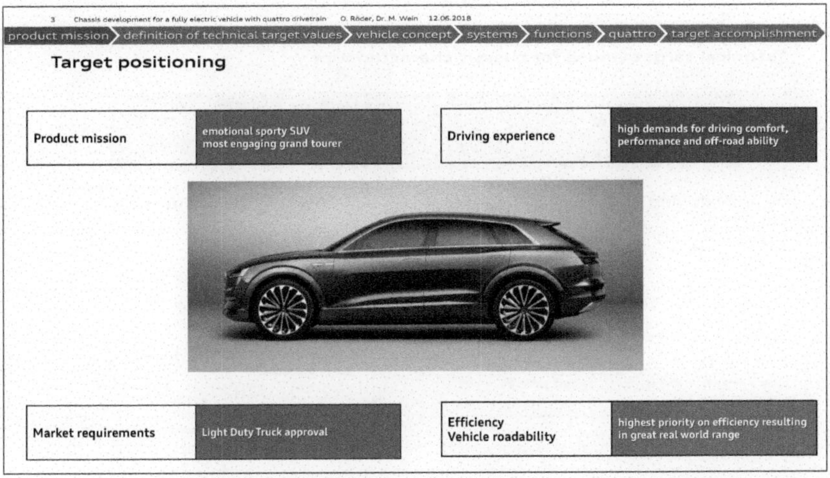
3 Chassis development for a fully electric vehicle with quattro drivetrain O. Röder, Dr. M. Wein 12.06.2018
product mission
definition of technical target values
vehicle concept
systems
functions
quattro
target accomplishment
Target positioning
Product mission
emotional sporty SUV
most engaging grand tourer
Driving experience
high demands for driving comfort, performance and off-road ability
Market requirements
Light Duty Truck approval
Efficiency
Vehicle roadability
highest priority on efficiency resulting in great real world range

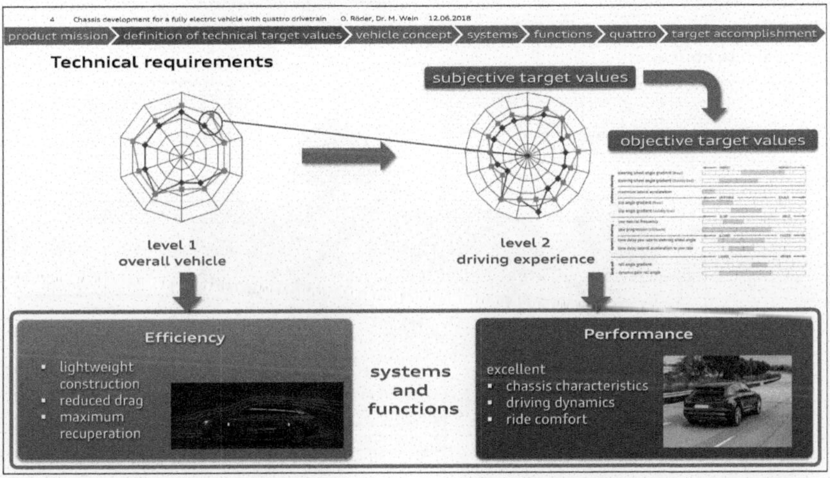
4 Chassis development for a fully electric vehicle with quattro drivetrain O. Röder, Dr. M. Wein 12.06.2018
product mission
definition of technical target values
vehicle concept
systems
functions
quattro
target accomplishment
Technical requirements
subjective target values
objective target values
level 1
overall vehicle
level 2
driving experience
Efficiency
lightweight construction
reduced drag
maximum recuperation
systems and functions
Performance
excellent
chassis characteristics
driving dynamics
ride comfort

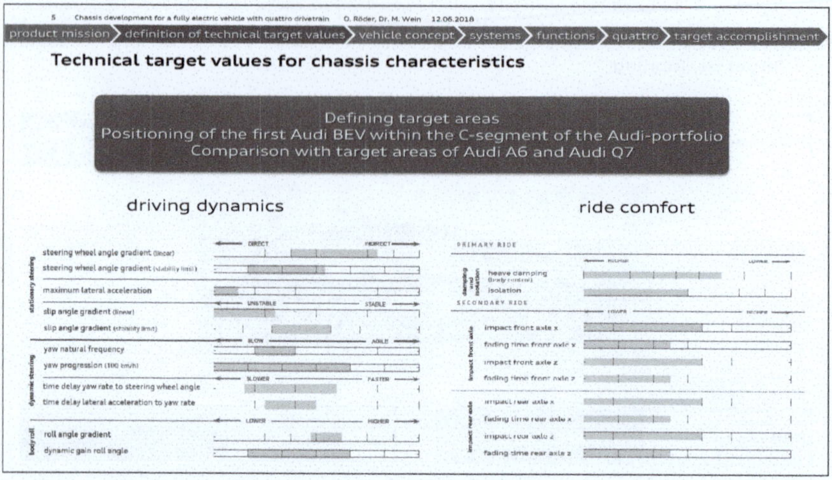
5 Chassis development for a fully electric vehicle with quattro drivetrain O. Röder, Dr. M. Wein 12.06.2018
product mission
definition of technical target values
vehicle concept
systems
functions
quattro
target accomplishment
Technical target values for chassis characteristics
Defining target areas
Positioning of the first Audi BEV within the C-segment of the Audi-portfolio
Comparison with target areas of Audi A6 and Audi Q7
driving dynamics
stationary steering
steering wheel angle gradient (linear)
steering wheel angle gradient
maximum lateral acceleration
slip angle gradient (linear)
slip angle gradient
dynamic steering
yaw natural frequency
yaw progression
time delay yaw rate to steering wheel angle
time delay lateral acceleration to yaw rate
body roll
roll angle gradient
dynamic gain roll angle
DIRECT
UNSTABLE
STABLE
SLOWER
FASTER
LOWER
HIGHER
ride comfort
PRIMARY RIDE
damping and isolation
heave damping
isolation
SECONDARY RIDE
impact front axle
impact front axle x
fading time front axle x
impact front axle z
fading time front axle z
impact rear axle
fading time rear axle x
fading time rear axle z

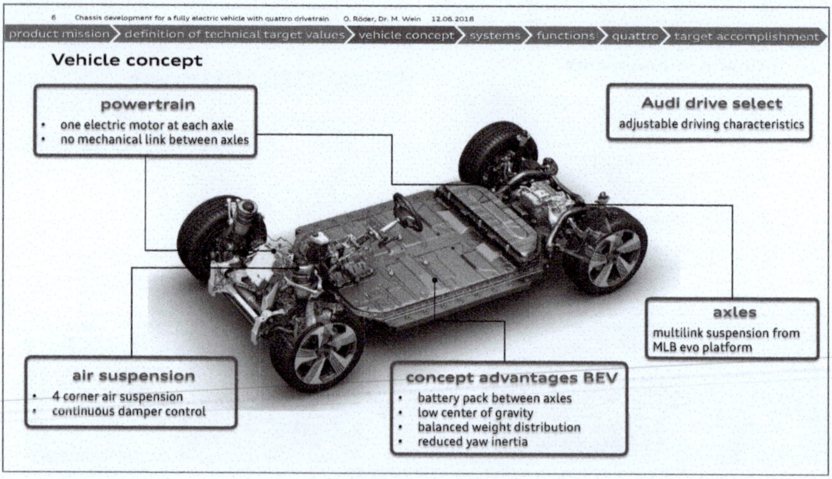
6 Chassis development for a fully electric vehicle with quattro drivetrain O. Röder, Dr. M. Wein 12.06.2018
product mission
definition of technical target values
vehicle concept
systems
functions
quattro
target accomplishment
Vehicle concept
powertrain
• one electric motor at each axle
• no mechanical link between axles
Audi drive select
adjustable driving characteristics
axles
multilink suspension from MLB evo platform
air suspension
• 4 corner air suspension
• continuous damper control
concept advantages BEV
• battery pack between axles
• low center of gravity
• balanced weight distribution
• reduced yaw inertia

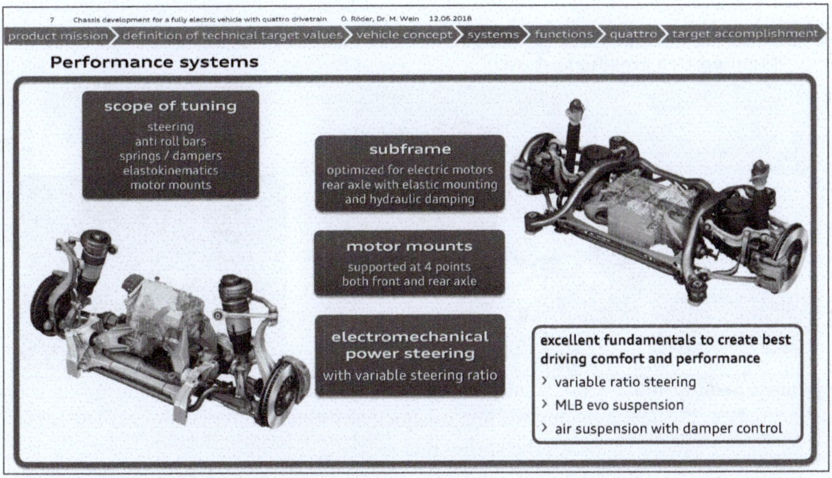
7 Chassis development for a fully electric vehicle with quattro drivetrain O. Röder, Dr. M. Wein 12.06.2018
product mission
definition of technical target values
vehicle concept
systems
functions
quattro
target accomplishment
Performance systems
scope of tuning
steering
anti roll bars
springs / dampers
elastokinematics
motor mounts
subframe
optimized for electric motors
rear axle with elastic mounting
and hydraulic damping
motor mounts
supported at 4 points
both front and rear axle
electromechanical
power steering
with variable steering ratio
excellent fundamentals to create best
driving comfort and performance
variable ratio steering
MLB evo suspension
air suspension with damper control

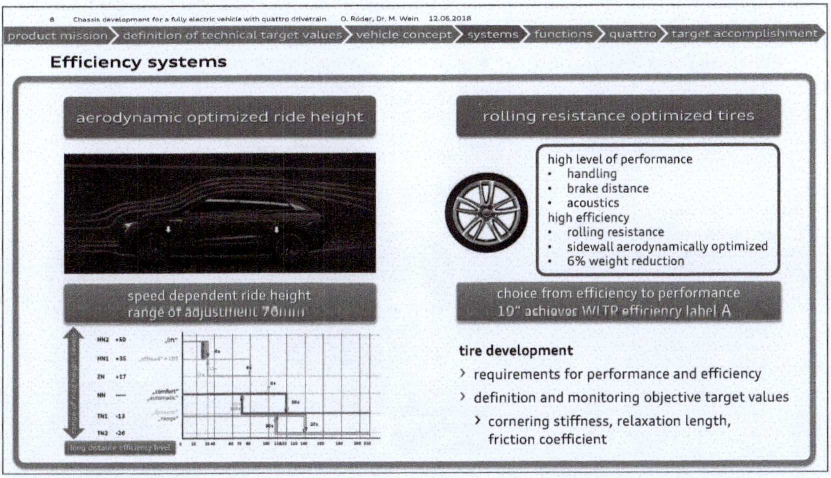
8 Chassis development for a fully electric vehicle with quattro drivetrain O. Röder, Dr. M. Wein 12.06.2018
product mission
definition of technical target values
vehicle concept
systems
functions
quattro
target accomplishment
Efficiency systems
aerodynamic optimized ride height
rolling resistance optimized tires
high level of performance
handling
brake distance
acoustics
high efficiency
rolling resistance
sidewall aerodynamically optimized
6% weight reduction
speed dependent ride height
range of adjustment 76mm
choice from efficiency to performance
19" achieves WLTP efficiency label A
HN2 +50
HN1 +35
ZN +17
NN
TN1 -13
TN2 -26
„lift"
„comfort"
„automatic"
„range"
long distance efficiency level
tire development
requirements for performance and efficiency
definition and monitoring objective target values
cornering stiffness, relaxation length,
friction coefficient

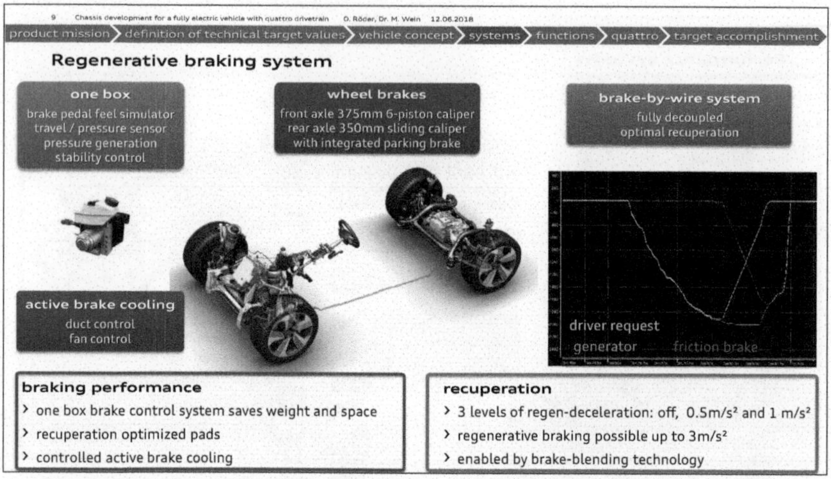
9 Chassis development for a fully electric vehicle with quattro drivetrain O. Röder, Dr. M. Wein 12.06.2018
product mission
definition of technical target values
vehicle concept
systems
functions
quattro
target accomplishment
Regenerative braking system
one box
brake pedal feel simulator
travel / pressure sensor
pressure generation
stability control
wheel brakes
front axle 375mm 6-piston caliper
rear axle 350mm sliding caliper
with integrated parking brake
brake-by-wire system
fully decoupled
optimal recuperation
active brake cooling
duct control
fan control
driver request
generator
friction brake
braking performance
› one box brake control system saves weight and space
› recuperation optimized pads
› controlled active brake cooling
recuperation
› 3 levels of regen-deceleration: off, 0.5m/s² and 1 m/s²
› regenerative braking possible up to 3m/s²
› enabled by brake-blending technology

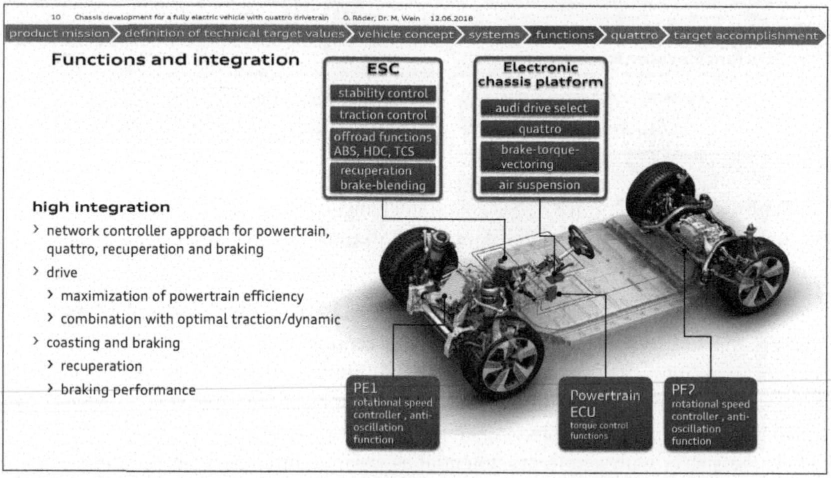
10 Chassis development for a fully electric vehicle with quattro drivetrain O. Röder, Dr. M. Wein 12.06.2018
product mission
definition of technical target values
vehicle concept
systems
functions
quattro
target accomplishment
Functions and integration
ESC
stability control
traction control
offroad functions ABS, HDC, TCS
recuperation brake-blending
Electronic chassis platform
audi drive select
quattro
brake-torque-vectoring
air suspension
high integration
› network controller approach for powertrain, quattro, recuperation and braking
› drive
› maximization of powertrain efficiency
› combination with optimal traction/dynamic
› coasting and braking
› recuperation
› braking performance
PE1
rotational speed controller , anti-oscillation function
Powertrain ECU
torque control functions
PE2
rotational speed controller , anti-oscillation function

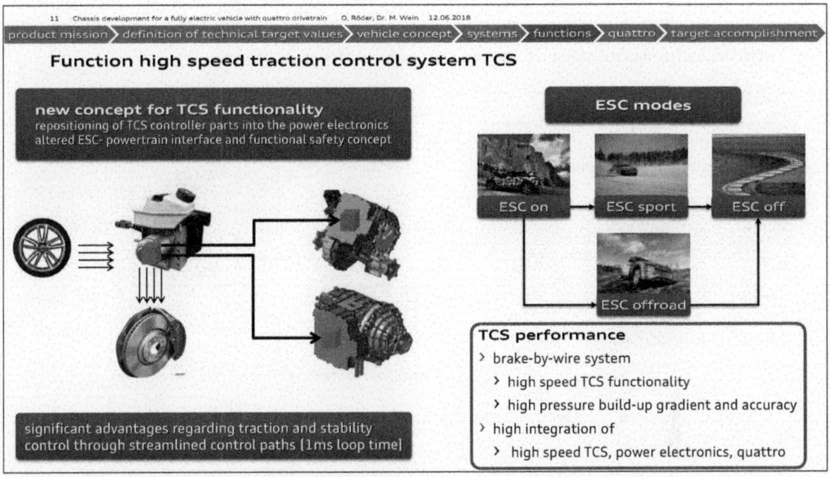
11 Chassis development for a fully electric vehicle with quattro drivetrain O. Röder, Dr. M. Wein 12.06.2018
product mission
definition of technical target values
vehicle concept
systems
functions
quattro
target accomplishment
Function high speed traction control system TCS
new concept for TCS functionality
repositioning of TCS controller parts into the power electronics
altered ESC- powertrain interface and functional safety concept
ESC modes
ESC on
ESC sport
ESC off
ESC offroad
TCS performance
› brake-by-wire system
› high speed TCS functionality
› high pressure build-up gradient and accuracy
› high integration of
› high speed TCS, power electronics, quattro
significant advantages regarding traction and stability control through streamlined control paths [1ms loop time]

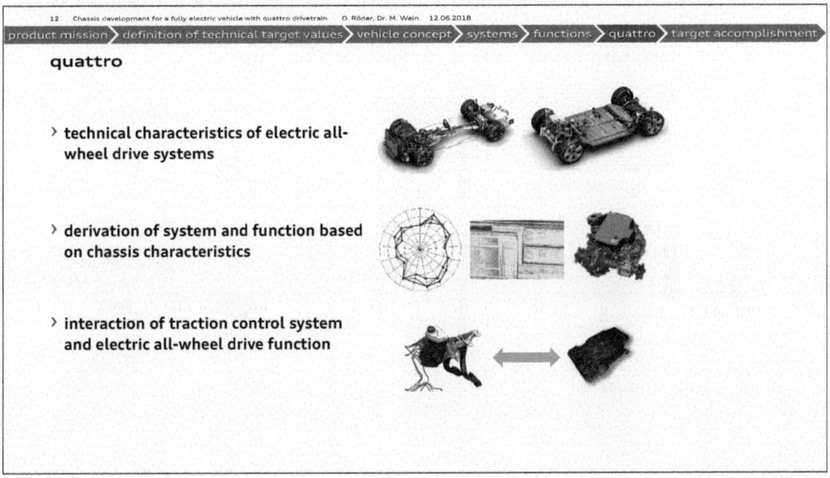
12 Chassis development for a fully electric vehicle with quattro drivetrain O. Röder, Dr. M. Wein 12.06.2018
product mission
definition of technical target values
vehicle concept
systems
functions
quattro
target accomplishment
quattro
› technical characteristics of electric all-wheel drive systems
› derivation of system and function based on chassis characteristics
› interaction of traction control system and electric all-wheel drive function

Technical characteristics of electric all-wheel drive systems

- **distribution area**
  - is a criteria for the performance of an all-wheel drive system
  - depends on the installed power at each axle
  - varies with the driver's power demand

summary electric all-wheel drive
- torque can be shifted independent of rotation speed
- the installed power limits the variability

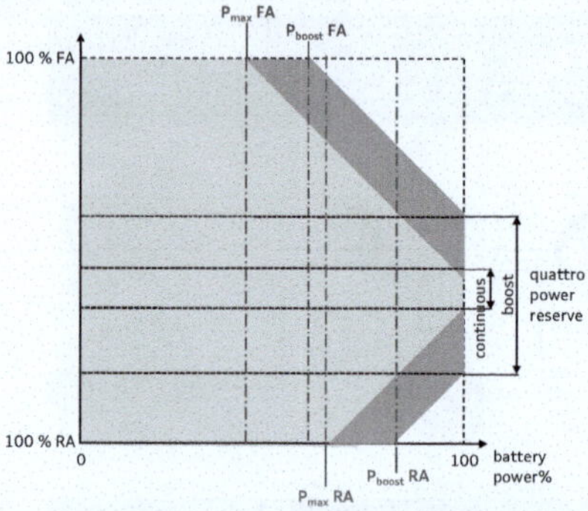

Technical characteristics of electric all-wheel drive systems

- **speed**
  - depends on the time constant of the electric machines
  - is limited by the communication architecture and signal propagation delay
  - requires a constant rotational speed control

summary electric all-wheel drive
- can shift torque very fast
- feed forward by torque
- feed forward rotational speed
- control rotational speed

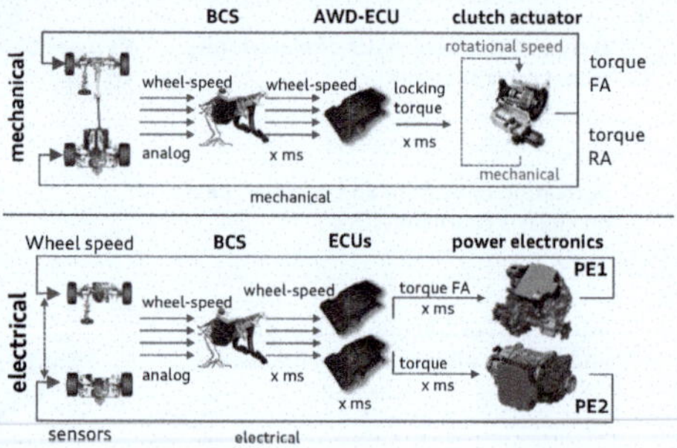

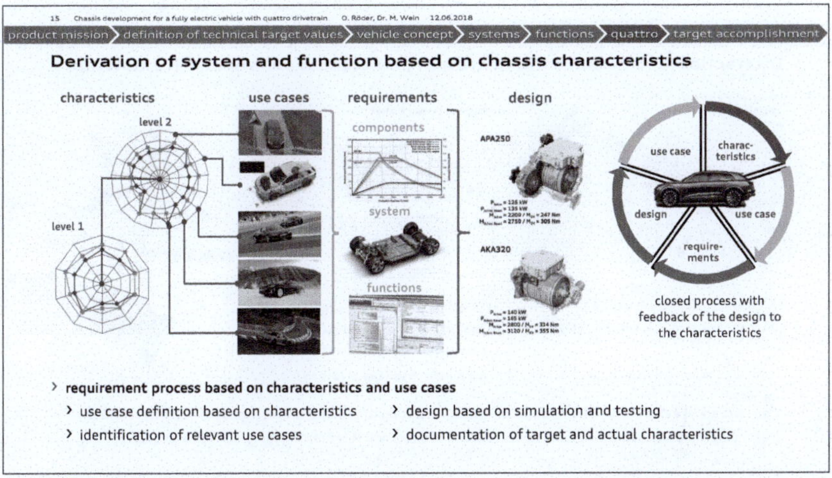
15 Chassis development for a fully electric vehicle with quattro drivetrain O. Röder, Dr. M. Wein 12.06.2018
product mission
definition of technical target values
vehicle concept
systems
functions
quattro
target accomplishment
Derivation of system and function based on chassis characteristics
characteristics
use cases
requirements
design
level 2
level 1
components
system
functions
APA250
AKA320
use case
charac-teristics
design
use case
require-ments
closed process with feedback of the design to the characteristics
requirement process based on characteristics and use cases
use case definition based on characteristics
identification of relevant use cases
design based on simulation and testing
documentation of target and actual characteristics

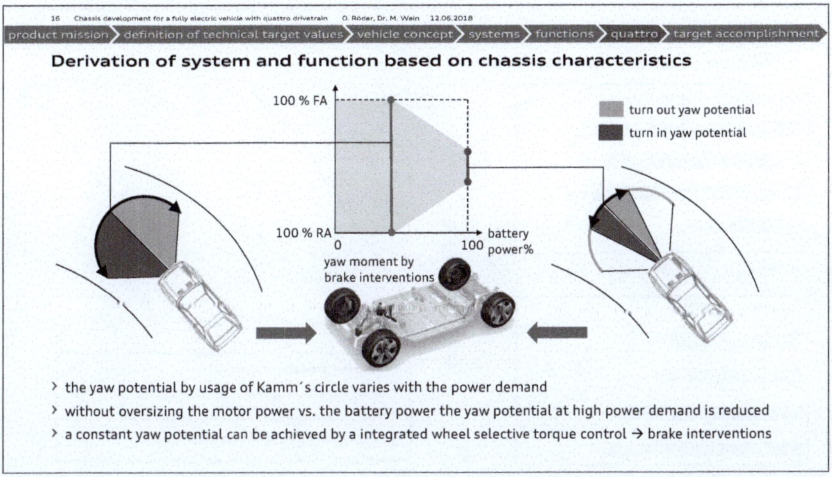
16 Chassis development for a fully electric vehicle with quattro drivetrain O. Röder, Dr. M. Wein 12.06.2018
product mission
definition of technical target values
vehicle concept
systems
functions
quattro
target accomplishment
Derivation of system and function based on chassis characteristics
100 % FA
100 % RA
0
100
battery power%
turn out yaw potential
turn in yaw potential
yaw moment by brake interventions
the yaw potential by usage of Kamm´s circle varies with the power demand
without oversizing the motor power vs. the battery power the yaw potential at high power demand is reduced
a constant yaw potential can be achieved by a integrated wheel selective torque control → brake interventions

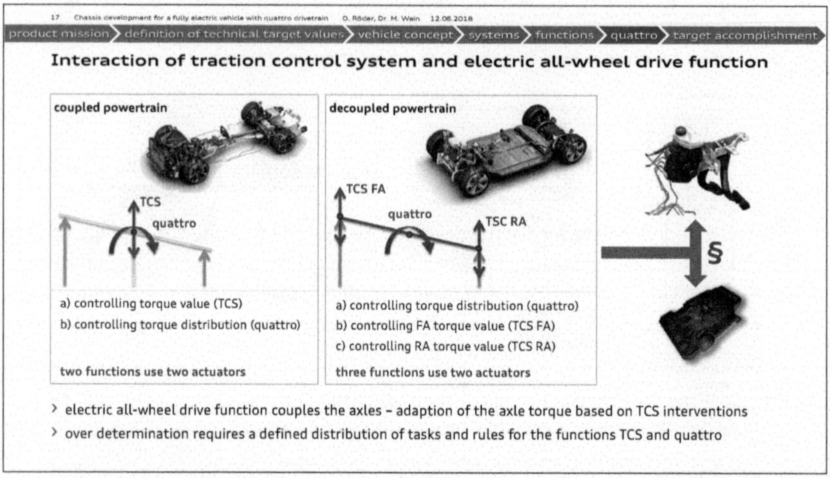
17 Chassis development for a fully electric vehicle with quattro drivetrain O. Röder, Dr. M. Wein 12.06.2018
product mission
definition of technical target values
vehicle concept
systems
functions
quattro
target accomplishment
Interaction of traction control system and electric all-wheel drive function
coupled powertrain
TCS
quattro
a) controlling torque value (TCS)
b) controlling torque distribution (quattro)
two functions use two actuators
decoupled powertrain
TCS FA
quattro
TSC RA
a) controlling torque distribution (quattro)
b) controlling FA torque value (TCS FA)
c) controlling RA torque value (TCS RA)
three functions use two actuators
§
› electric all-wheel drive function couples the axles - adaption of the axle torque based on TCS interventions
› over determination requires a defined distribution of tasks and rules for the functions TCS and quattro

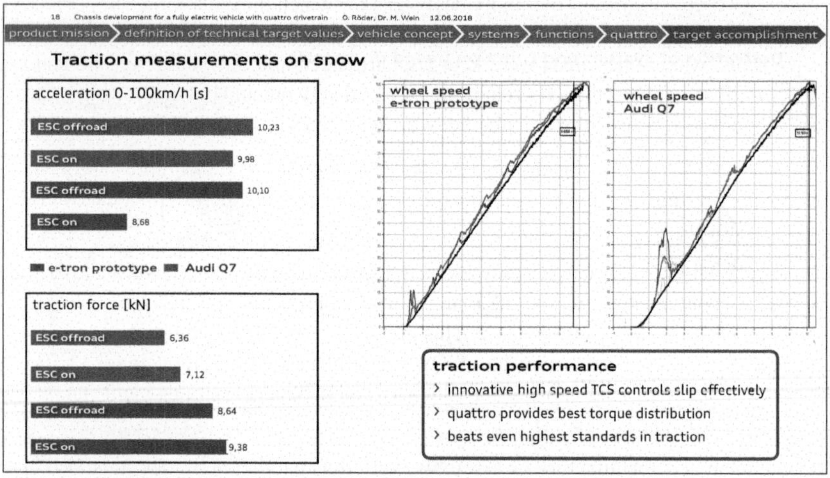
18 Chassis development for a fully electric vehicle with quattro drivetrain O. Röder, Dr. M. Wein 12.06.2018
product mission
definition of technical target values
vehicle concept
systems
functions
quattro
target accomplishment
Traction measurements on snow
acceleration 0-100km/h [s]
ESC offroad 10,23
ESC on 9,98
ESC offroad 10,10
ESC on 8,68
e-tron prototype
Audi Q7
traction force [kN]
ESC offroad 6,36
ESC on 7,12
ESC offroad 8,64
ESC on 9,38
wheel speed
e-tron prototype
wheel speed
Audi Q7
traction performance
› innovative high speed TCS controls slip effectively
› quattro provides best torque distribution
› beats even highest standards in traction

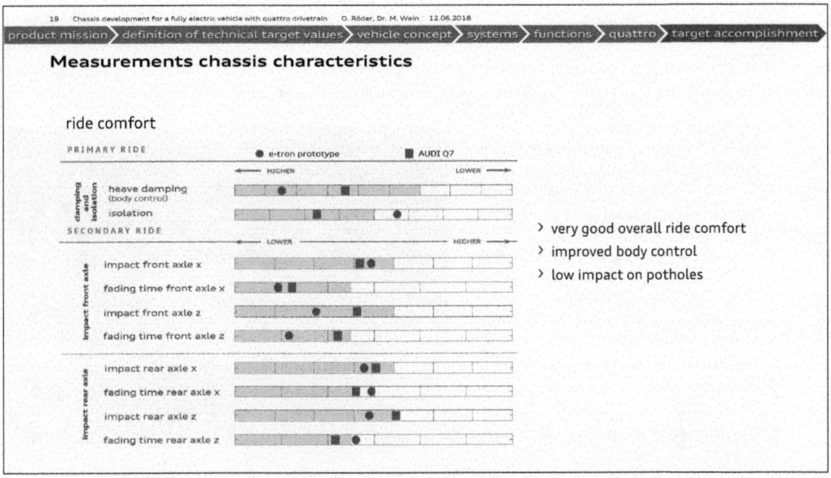
19 Chassis development for a fully electric vehicle with quattro drivetrain O. Röder, Dr. M. Wein 12.06.2018
product mission
definition of technical target values
vehicle concept
systems
functions
quattro
target accomplishment
Measurements chassis characteristics
ride comfort
PRIMARY RIDE
e-tron prototype
AUDI Q7
HIGHER
LOWER
damping and isolation
heave damping (body control)
isolation
SECONDARY RIDE
LOWER
HIGHER
impact front axle
impact front axle x
fading time front axle x
impact front axle z
fading time front axle z
impact rear axle
impact rear axle x
fading time rear axle x
impact rear axle z
fading time rear axle z
very good overall ride comfort
improved body control
low impact on potholes

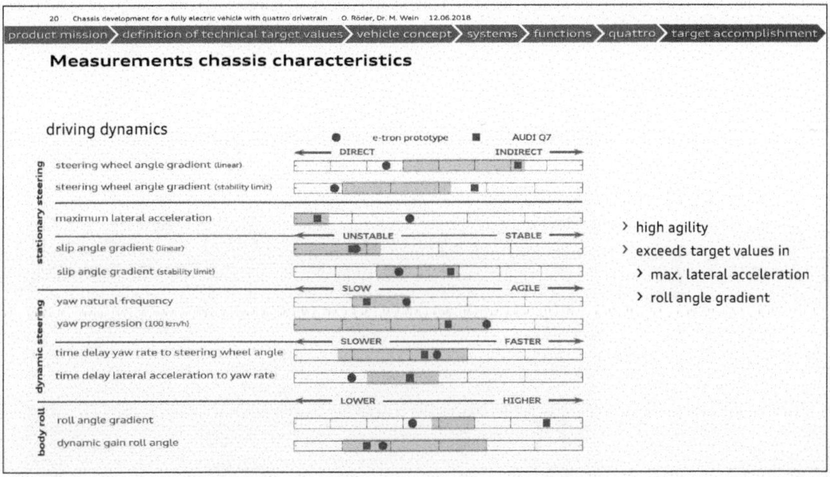
20 Chassis development for a fully electric vehicle with quattro drivetrain O. Röder, Dr. M. Wein 12.06.2018
product mission
definition of technical target values
vehicle concept
systems
functions
quattro
target accomplishment
Measurements chassis characteristics
driving dynamics
e-tron prototype
AUDI Q7
DIRECT
INDIRECT
stationary steering
steering wheel angle gradient (linear)
steering wheel angle gradient (stability limit)
maximum lateral acceleration
UNSTABLE
STABLE
slip angle gradient (linear)
slip angle gradient (stability limit)
SLOW
AGILE
dynamic steering
yaw natural frequency
yaw progression (100 km/h)
SLOWER
FASTER
time delay yaw rate to steering wheel angle
time delay lateral acceleration to yaw rate
LOWER
HIGHER
body roll
roll angle gradient
dynamic gain roll angle
high agility
exceeds target values in
max. lateral acceleration
roll angle gradient

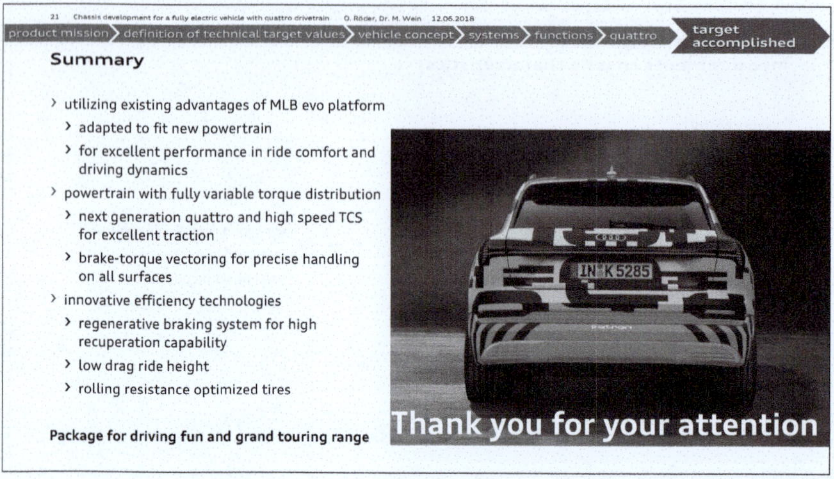
21 Chassis development for a fully electric vehicle with quattro drivetrain O. Röder, Dr. M. Wein 12.06.2018
product mission
definition of technical target values
vehicle concept
systems
functions
quattro
target accomplished
Summary
› utilizing existing advantages of MLB evo platform
› adapted to fit new powertrain
› for excellent performance in ride comfort and driving dynamics
› powertrain with fully variable torque distribution
› next generation quattro and high speed TCS for excellent traction
› brake-torque vectoring for precise handling on all surfaces
› innovative efficiency technologies
› regenerative braking system for high recuperation capability
› low drag ride height
› rolling resistance optimized tires
Package for driving fun and grand touring range
IN K 5285
Thank you for your attention

Chassis design of the aCar – a light commercial vehicle for Sub-Saharan Africa

Michael Schmidt, M. Sc.
Thomas Zehelein, M. Sc.
Prof. Dr.-Ing. Markus Lienkamp

Chair of Automotive Technology, Technical University of Munich

P. Pfeffer (Ed.), *9th International Munich Chassis Symposium 2018*, Proceedings,
https://doi.org/10.1007/978-3-658-22050-1_10

Abstract

The "aCar mobility" project is a publicly funded research project with the aim to provide the rural population of Sub-Saharan Africa with an attractive mobility concept, which serves the basic need for mobility and facilitates the connection of rural areas to urban infrastructure. Therefore, multiple usage scenarios need to be covered by the vehicle with a focus on transportation of people and goods. Some of the mobility concept requirements are design for local production, for easy maintenance and repairing, as well as for durability and maneuverability in near-off-road conditions. All design decisions were made with a special focus on manufacturing costs to provide an affordable vehicle to the customer. Combining all these requirements leads to an electric all-wheel-drive vehicle with a maximum speed of 60 km/h. Two 8 kW motors, one per axle, ensure proper propulsion of the 800 kg unladen weight also with a payload of 1000 kg.

This paper shows the process and challenges of chassis system development for a vehicle that is strictly developed for its functional range. We define the term chassis system as the suspension linkage, springs and dampers, braking- and steering systems, as well as wheels and tyres. In addition to the basic requirements of safety and functionality, the aspects of spare part availability in the targeted markets, minimal development and production invest, as well as strict package requirements were strong drivers for the chassis design. As a result, we based the chassis system on a donor vehicle. With the help of multi-body simulation, we analyzed the effects of the changed vehicle setup. We quantified the chassis system loads as well as modifications to the kinematics. We tested the first prototype on a two-step basis. First, we ensured driving safety and basic functionality by standardized test maneuvers. In the second step, we tested the prototype in its intended range of use in Ghana.

We incorporated findings from testing the first prototype into the design of the second prototype. To maximize the load capability, the rear axle system is a rigid axle, suspended by a leaf spring for simplicity reasons. The front axle is a MacPherson axle with a rack-and-pinion steering system. Design alterations to mechanical parts were made only where necessary and to a minimum extent, mainly to increase the vehicles' off-road capabilities and its safe driving behavior on unpaved ground. Dampers and springs were customized and the steering system was redesigned. The paper closes with an illustration of the final chassis system.

1 The aCar mobility Project & Vehicle Concept

The aCar mobility project was an interdisciplinary research project of the Technical University of Munich (TUM), funded by the Bavarian Research Foundations (BFS). This project dealt with the mobility needs in rural areas of Sub-Saharan Africa. Its goal was to provide the rural population with an attractive mobility concept, which helps to avoid the rural exodus and strengthens the independence of the rural regions. The Institute of Automotive Technology cooperated with other institutes of TUM, as well as with other universities in Germany (Hochschule Rosenheim, Universität Bayreuth). Partnerships with local universities in Africa, especially Kwame Nkrumah University of Science and Technology (KNUST) in Kumasi, Ghana, provided valuable insight to the local community and everyday life. The expertise and resources of various industry partners also supported the project.

Figure 1: Illustration of the aCar and two application types (2nd prototype)

The aCar concept serves multiple usage scenarios with a focus on transportation of people and goods. Its load area is designed to carry different interchangeable modules, such as medical supplies for mobile health services, a water-cleaning device or a cargo-/passenger platform. General requirements of the concept include a simple construction for easy maintenance, using locally available materials to allow local production as far as possible, and minimum production costs to achieve a maximum price of 10.000 €. Extensive studies in the target markets lead to a 48 V battery electric vehicle concept that fulfils the requirements of the EU's L7e-C U vehicle class. Its overall dimensions are 3700 x 1500 x 2100 mm (length x width x height). The unladen weight including the battery is 800 kg; the payload capacity is 1000 kg. The maximum velocity is 60 km/h.

The drivetrain is an all-wheel drive concept with two 8 kW electric motors, one per axle, for reasons of environmental and technical sustainability, as well as production- and maintenance costs. The battery capacity of 20 kWh provides a range of 80 km. Due to the aCar's usage in mostly rural areas with accordingly tough road conditions the suspension must be robust and provide proper off-road capabilities.

2 Development Process

Developing the aCar's chassis system was a special challenge due to three conditions. First, project resources were limited. Second, hardly any products on the market or their manufacturers could serve as best practice regarding mobility catering specifically to rural Sub-Saharan Africa. Third, requirements to the chassis system only could be derived from the overall concept's requirements due to a severe lack of detailed information about the exact driving-, road- and usage conditions in the targeted markets. The classic V-model development process therefore was not applicable, and we chose to follow a rapid-prototyping-inspired approach. The goal was to build a fully functional first prototype within one year after project start. This first prototype (P1) should verify the vehicle- and its system concepts, as well as serve as the main test vehicle. Testing was planned in two phases: functional testing in Germany and total vehicle testing under real usage conditions in Sub-Sahara Africa. Testing results should serve as input for developing the second prototype (P2) and building it roughly one and a half years after P1, following the V-model.

The limitations in project resources were a major boundary condition. For the development of P1, we therefore chose to use the chassis system of a donor vehicle with as little modifications as possible. Modifications were considered necessary if either packaging requirements or safety-related driving behavior requirements could not be met. To anticipate the driving behavior characteristics and to investigate which modifications, e.g. to the suspension kinematics, had which effect, we built a multi-body-simulation (MBS) model. Another output from the MBS-model were load estimations for frame design. The necessary and feasible modifications to the kinematics were realized for P1, which was subsequently fitted with measuring equipment and then tested in suitable standard maneuvers. The following chapters describe the process steps in detail.

2.1 Deriving Chassis Requirements from Concept Parameters and Project Goals

We initially derived chassis requirements from the aCar's concept specifications. Basic requirements were robustness, maintainability and safety in all operating conditions. In the robustness category, we regarded the payload capabilities and appropriate off-road capabilities. Maintainability was covered by simplicity of the suspension concept as well as the donor vehicle's spare part availability in the targeted markets. To ensure safety, we paid special attention to understeering driving behavior in all payload conditions, and to braking performance in high- and low friction scenarios. In terms of packaging and compatibility to the total vehicle, we set the required maximum width of the suspension including wheels and tyres to 1500 mm as defined by the L7e-C U. Rims and tyres must carry the maximum payload and provide traction in muddy terrain. While the general requirements regarding robustness, maintainability and packaging compatibility were sufficient input for the donor vehicle selection, the safety- and driving behavior aspects needed to be detailed later in the development process, once quantified information was attainable from simulation and testing.

2.2 Identifying a Suitable Donor Vehicle

Based on the initial chassis requirements, we selected a suitable donor vehicle. The premise was to use as many parts of the chassis system and to make as little modifications to them as possible. Reasons being cost effectiveness and homologation concerns, but also the ability to rapidly build a first prototype. As for suspension types, we defined three principle chassis concepts: (1) "simple and robust" with rigid axles front and rear combined with a ball and nut steering; (2) "ideal package and off-road-capabilities" with double wishbones all-around and rack-and-pinion steering; and (3) "cost-efficient trade-off" with a rigid axle in the rear and a MacPherson-type suspension in the front with rack-and-pinion steering. We rated the three concepts based on [1, p. 461] and [2, p. 742] with regard to our requirements. Concept (3) turned out to be the most favorable as it allows high payloads in the rear without compromising kinematics as well as packaging advantages in the front end, overall at a low cost level.

After collecting information about donor vehicle candidates, such as vehicle dimensions, curb- and gross weight, axle and steering concepts, drivetrain layout options, wheel dimensions, clearance, maximum velocity, and price, we narrowed it down to the Piaggio Porter. The Porter is a licensed version of the Daihatsu Hijet and is mainly used as municipal utility vehicle in Europe. Vehicle dimensions, suspension concept, curb-weight-payload-ratios, and spare part availability in African countries due to its Daihatsu relative were the best match to the aCar concept. Consequently, the braking system could be transferred to the aCar as well. The following table provides a comparison of the Porter and the aCar concept.

Table 1: Specification comparison of the aCar concept and the Piaggio Porter

	aCar Concept	Piaggio Porter Topdeck Version
Length x width x height	3700 x 1500 x 2100 mm	3775 x 1460 x 1730 mm
Wheel base	2400 mm	1830 mm
Track width front / rear	1266 / 1320 mm	1210 / 1220 mm
Curb weight / gross weight	800 / 1800 kg	850 / 1700 kg
Max. velocity	60 km/h	130 km/h
Ground clearance	220 mm	180 mm
Wheels	175/80 R15	155/80 R13
Price	Max. 10.000 €	14.745 € (2018)

2.3 Identification of Modifications for the 1st Prototype

After the identification of a suitable donor vehicle for the chassis system, it was necessary to analyze the vehicle dynamics and driving behavior of the aCar. To get quick results, we used simulation methods before the first prototype existed. Therefore, a simulation tool was required that allowed:

- Accurate modelling of the donor chassis system
- Evaluation of the vehicle's driving behaviour
- Analysis of impacts on vehicle dynamics from modifications at the component level
- Analysis of further vehicle development interfaces (e.g. chassis system loads, collision detection etc.)

A bicycle model as well as a two-track model do not allow modifications on a component basis. Furthermore, an analysis of development interfaces is not possible. Therefore, we applied multi-body simulation for the further vehicle dynamics analysis.

As CAD-data of the donor vehicle was not available at that point, we created an optical 3-D-scan of the donor vehicle's chassis system to identify the kinematic hard points and component connection points. As the overall goal was to carry over as many components from the donor vehicle as possible, the initial chassis system was modeled with as few modifications as possible. Therefore, the hard points of each axle were not changed relative to each other. Only the rear axle was moved horizontally towards the rear to adapt the wheelbase of the donor chassis system to the specification of the aCar.

Besides robustness and simplicity, a main task of the chassis system is guaranteeing a safe vehicle behavior during driving. Therefore, classic driving maneuvers such as constant radius cornering, step steer, sinus steering and a double lane change according to ISO3888-2 were simulated to gain insights into the vehicle's driving behavior. The results of those maneuvers were used to iteratively adapt the chassis system to the modifications mentioned in the following paragraphs.

While the front and rear axle could be carried over nearly without any modifications, this was different for other sub-systems that are relevant for the overall vehicle appearance. Those sub-systems are mainly the steering system as well as the tyres.

Tyres have a large influence on the vehicle's off-road capability as they directly influence ground clearance, ramp angles as well as the angle of approach. The donor vehicle is equipped with a tyre size of 155/80 R13. For normal road vehicles, the tyre forces during extreme driving maneuvers define the required tyre width. This does not apply to the aCar. As the aCar's expected usage scenarios feature mainly dirt roads [3], tyre width and diameter are defined by a trade-off between off-road capability and package requirements. A greater tyre width increases the traction on sandy- and gravel roads; a greater tyre diameter increases ground clearance, ramp angles and the angle of approach with an increased driving comfort while decreasing the rolling resistance. The selected tyre size is 175/80 R15. The reduction of braking performance and the increased loads on the suspension components as a consequence of increasing the tyre diameter by 14 % need to be considered during testing. Consulting component manufacturer experts showed that the increased load and torque on bearings and the suspension linkage was no problem for prototype testing. However, the effects on fatigue life and durability need to be evaluated further for the series production version of the vehicle.

Increasing the tyre size also required changing the rim size. Hereby, it is also possible to adapt the track width by changing the wheel offset. While there are no major negative effects at the rear axle, the increased scrub radius at the front axle had to be considered.

The steering system connects the driver's seating position to the actual location of the front axle. The donor vehicle is designed with the driver sitting on top of the front axle with the front tyres directly below the seat. This is realized with a steering system that consists of a one-sided steering gearbox, which actuates a central steering plate, which acts as a link to both tie rods. The aCar, however, is designed to have its front axle in front of the driver. Therefore, the steering system needed to be adapted. To minimize problems with component interfaces, we decided to take over as many parts of the donor vehicle's steering system as possible, even though the uncommon system design. The steering wheel position as well as the steering column were predetermined according to the ergonomic requirements of the driver. The position of the steering plate was adapted to change the driving behavior to increased understeer. This was necessary because the donor vehicle tends to oversteering on its own, and because the aCar's payload mainly rests on the rear axle, which further increases oversteering tendencies in the laden condition. The steering gearbox was positioned to compensate the cardan error of the intermediate shaft cardan joints. The new position of the steering plate and the steering gearbox required redesigning the steering plate. The intermediate shaft's length was adapted to the required distance between steering column and steering gearbox.

Overall, it was possible to adapt the donor vehicle's chassis system to P1 of the aCar with only a few changes.

2.4 Remaining Steps within the Development Process

While the development process of the chassis system itself was sped up by taking over a donor vehicle's chassis system, it is still necessary to comply with the overall vehicle development process. This means the interfaces between development tasks – especially across different vehicle systems (e.g. frame and chassis system) – need to be coordinated explicitly.

Chassis system loads need to be quantified for the development of the frame. This is especially challenging for the aCar project, as there are no quantified requirements regarding the range of application of the vehicle. In addition, there are no representative load spectra of road excitations. The chassis system loads were identified by applying the 3-2-1g-rule, which is know from motorsport applications. The resulting load cases of the 3-2-1-g-rule are similar to the load cases mentioned by [1, p. 268]. This method applies different combinations of ±3g in vertical, ±2g in longitudinal and ±1g in lateral direction to the vehicle's center of gravity and determines the corresponding tyre forces. Those tyre forces were applied to the tyres in the next step and the resulting chassis system loads could be calculated.

Another aspect that needed to be covered for the overall vehicle development process is generating envelopes of the chassis components' range of motion to enable a collision detection in CAD. An additional safety margin was considered as the envelopes are only based on kinematic motion and elastokinematic motion is not accounted.

3 Chassis System Testing

The testing procedure in the aCar project can be divided in two different categories:

- vehicle dynamics and handling
- off-road capabilities

Off-road capabilities were tested in Germany as well as in Ghana. This paper, however, deals with the vehicle dynamics and handling.

3.1 Overall Testing Procedure

The testing goals for the chassis system were derived from the overall project goals and from designing the first prototype. As the aCar is a lightweight commercial vehicle and the first prototype was mainly used as a functional prototype for quantifying more precise requirements of the vehicle's range of application, the main task of the chassis system was to provide driving safety.

Therefore, the first tests were chosen to ensure that the vehicle can be operated safely and there are no component breakdowns. The conducted maneuvers were:

- steering stress test
- braking performance test

Afterwards, classic vehicle dynamics maneuvers such as

- constant radius cornering
- step steering
- sinus steering
- braking/accelerating in a turn
- lane change

were selected to be tested. However, the focus of the vehicle behavior is not lateral or longitudinal dynamic performance but providing the driver with a safe driving behavior. We performed the classic vehicle dynamics tests at a testing facility in Germany as well as on an unpaved area in Ghana. Another aspect of testing was to ensure that there are no component failures due to using them in a different vehicle than the donor vehicle.

Measurement setup

For conducting the testing maneuvers, the vehicle was equipped with a measurement steering wheel that measures steering wheel angle as well as steering wheel torque. Furthermore, four-wheel speed sensors and suspension travel sensors as well as the vehicle dynamics measurement device OxTS RT2500 were deployed as measuring equipment.

Steering stress test

The steering stress test checks the capability of the steering system to withstand steering forces. During the test, the applied forces should be greater than the forces generated during normal driving or misuse. This test was necessary as parts of the steering system were redesigned and were manufactured on a prototype basis. In the test, the front wheels were locked by applying the brakes while turning the steering wheel. The maximum applied steering wheel torque was 65 Nm, which is five times above the average steering torque during normal driving.

Braking performance test

The braking performance test was conducted according to the ECE-13H regulation. At first, the cold brake performance was evaluated by decelerating the vehicle from 80 % of its maximum speed to a complete stop. 15 consecutive decelerations at medium pedal force heated up the brakes before a last full stop was performed to measure the hot braking performance. Finally, the recovery performance was evaluated after another series of four stops with 1.5 km of driving between each stop to cool down the brakes. Table 1 shows the results of the brake performance tests for the unladen vehicle. The actual values of deceleration and stopping distance were always within the prescribed values except for the cold performance deceleration. However, the applied brake pedal force during this test didn't reach the maximal prescribed pedal force. Therefore, the compliance with this requirement is likely.

Table 2: Results of brake performance tests (*test conditions in italic*)

	Cold Performance		Hot Performance		Recovery Performance	
	Prescribed	Actual	Prescribed	Actual	Prescribed	Actual
Speed	*32 km/h*	*29 km/h*	*32 km/h*	*32 km/h*	*32 km/h*	*37 km/h*
Pedal Force	*65 – 500 N*	*290 N*	*65 – 500 N*	*292 N*	*65 – 500 N*	*378 N*
Deceleration	≥ 5,76 m/s²	5,0 m/s²	≥ 4,32 m/s²	5,78 m/s²	≥ 4,03 m/s²	6,21 m/s²
Stopping distance	8 m	6 m	9,3 m	7 m	12 m	9 m

Constant radius cornering

Constant radius cornering was performed because it gives many indications of the vehicle's driving behavior. It was conducted with different vehicle setups and on different pavements. Special consideration was given to the results regarding the understeering gradient and regarding the steering wheel torque as shown in Figure 3 and Figure 2. As the maneuvers were conducted with different radii, the shown measured steering wheel angle is corrected by the required Ackermann steering wheel angle. As expected, the vehicle behavior tends to less understeer in the laden condition. Driving with rear wheel drive (RWD) leads to less understeering tendencies as well. The test drive in Ghana on unpaved ground resulted in more understeering compared to the vehicle performance on paved ground. The steering wheel torque is an acceptable range even though there is no electric power steering available in the aCar. Due to a defective steering wheel torque sensor, there is no data for the unpaved ground in Ghana.

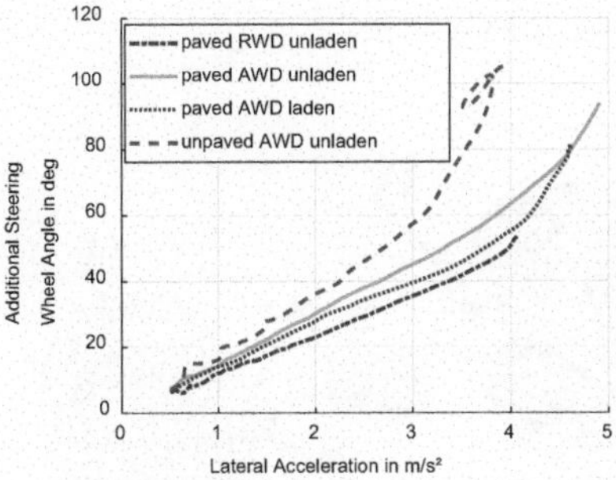

Figure 2: Understeering gradient for different vehicle setups

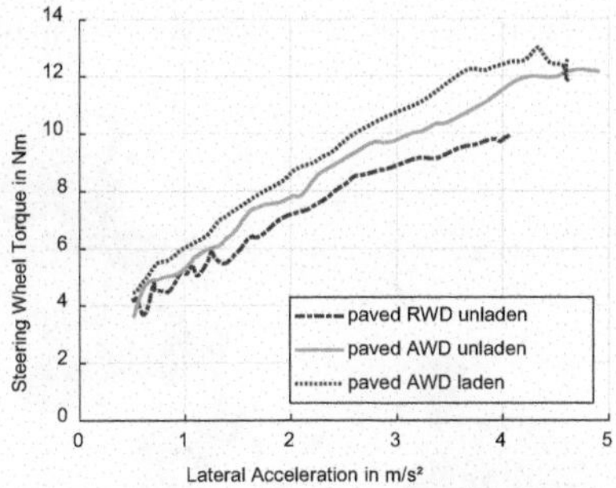

Figure 3: Steering wheel torque for different vehicle setups

3.2 Identified Modifications for the 2nd Prototype

From building and testing the P1, we could draw conclusions on which components and characteristics we should prioritize for an advanced second prototype. We identified the steering system as a major concern regarding packaging in the front end, part reliability, cost and weight, as well as drawbacks to vehicle dynamics due to the high elasticity of the steering mechanism. Every single component of the steering system was reconsidered to reduce part count, package space requirements, and to improve steering- and front axle kinematics.

The suspension needed optimization towards greater wheel travel and roll stiffness at the front axle, as well as improved damping performance for the aCar's typical usage scenarios. Furthermore, to achieve a more understeering driving behavior overall, we analyzed the front- and rear axle kinematics for optimization potential via our MBS model. With minimal modifications to the physical parts as a boundary condition, we could identify new hard point positions for the existing linkage components.

The braking system proved to perform well in the aCar concept, and we could keep modifications to slight changes in component positioning.

4 Chassis System Components of the 2nd Prototype

In this chapter, we give a summary of the chassis components as realized in P2. Figure 1 shows an overview of the chassis system of the second prototype.

Figure 4: Overview of the chassis system of the second prototype of the aCar (CAD rendering)

4.1 Suspension Linkage & Kinematics

The linkages of the P2's front and rear axle are very similar to the original Piaggio parts. At the front axle, the lower control arm of the MacPherson suspension is a carry-over-part, while the strut is a modified part. Interfaces to the braking system are similar while the damper has a slightly larger outer diameter and increased travel. Consequently, the top mount is positioned roughly 60 mm higher while its physical parts are carry-over-parts from the Porter. The front suspension's kinematic hard points were moved relative to the frame, so that the increased wheel travel in combination with the new steering geometry results in an optimized wheel trajectory.

The rear axle features the same axle beam and leaf springs as the standard Porter. The leaf springs' back link length was increased and the position of the front mount was lowered relative to the frame. This leads to a different inclination of the leaf springs, thus a different wheel trajectory. Ultimately, we could modify the rear axle kinematics so that its roll steering behavior shows an understeering tendency, turning the vehicle out of the corner.

4.2 Springs & Dampers

We modified the front axles springs and dampers to increase wheel travel by 60 %. We further optimized the damper characteristics to the aCar's specifications and driving scenarios. Custom dampers were designed featuring the new damping characteristics and wheel travel while maintaining the original interfaces to the Porter's upright and braking system. The damping performance was increased by roughly 45 % (rebound) and 200 % (compression) to withstand the aCar's tougher usage scenarios. The springs were custom designed to match the new wheel travel and damper characteristics, while being compatible to the Porter's top mount and the installation space overall.

The rear axle springs remained original, as mentioned in the previous section. The dampers were custom built to match the optimal damping characteristics based on [4, p. 340] and on our additional MBS-analysis. This resulted in a monotube damper with a linear damping curve, designed for the much tougher usage scenarios compared to the Porter. The rear axle damping performance was increased by roughly 160 % (rebound) and 130 % (compression).

Figure 5 and Figure 6 show the front and rear damper characteristics of P2 in comparison to the Porter/P1.

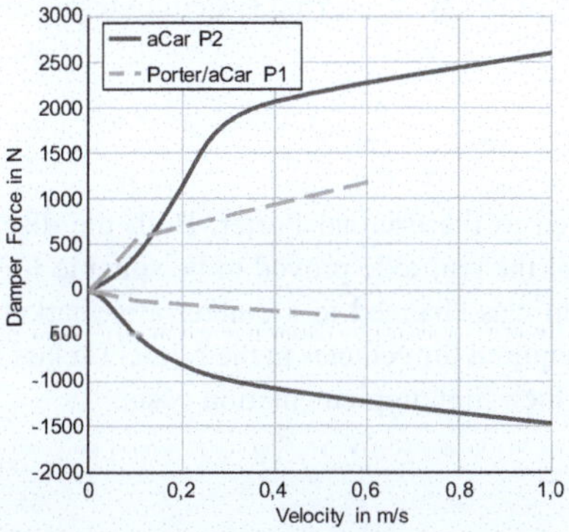

Figure 5: Front axle damper characteristics

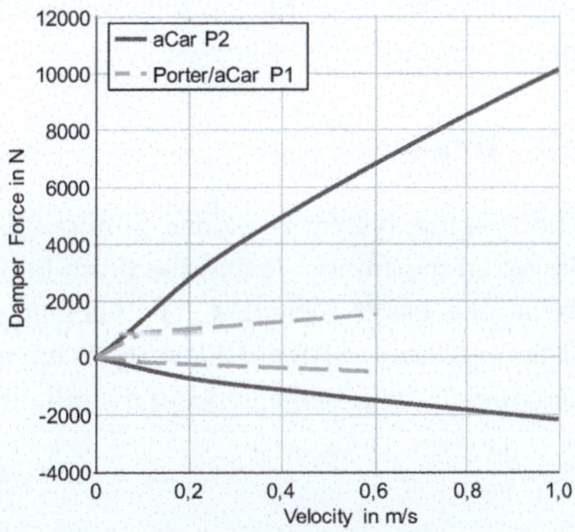

Figure 6: Rear axle damper characteristics

4.3 Steering System

The steering concept of the donor vehicle consists of more parts than a classic steering system. This leads to a higher weight and therefore higher costs of the system. More packaging effort is required as well. Therefore, the steering system of the second prototype was redesigned to a rack-and-pinion steering concept. To comply with the overall vehicle concept goals regarding cost efficiency and parts availability, the new steering system is based on existing parts.

Four different steering gearboxes were identified to be suitable in principal and were further analyzed regarding weight, transmission ratio, kinematic effects and package. Finally, the steering gearbox of a Smart ForTwo 450 was selected based on its technical details. Even though, an electric power steering (EPS) is not necessary for the aCar, the Smart steering gearbox would also be available as an EPS version, which leads to more flexibility for further development activities.

A major criteria for selecting the steering column besides package and weight is flexibility regarding the choice of the steering wheel and steering wheel levers as those components are a major human-machine-interface and therefore relevant in terms of design and ergonomics. The steering column of a Volkswagen Polo 6R showed to be best suited for those requirements.

With a defined position of the steering gearbox and the steering column, the cardan error was compensated by rotating the steering gearbox and the intermediate shaft. As those requirements are unique for every vehicle, there is no solution for an intermediate shaft from stock. Therefore, the Polo's intermediate shaft was adapted together with the original manufacturer to meet the requirements of the aCar steering system in terms of length and cardan joint angles.

4.4 Brakes

The braking system remained almost unaltered to the original Porter. Both the disc brakes of the front axle and the drum brakes of the rear axle proved to be suitable for the aCar's usage scenarios. The vacuum pump was changed to a quieter aftermarket design. Because ABS or ESP systems are not required for vehicles in the L7e-C U class, we omitted them in the concept overall due to their high implementation costs.

5 Conclusion and Outlook

This paper showed the chassis system development of the the aCar – a light commercial vehicle for Sub-Saharan Africa. The main difference compared to the normal development process was the unavailability of quantified vehicle- and chassis system requirements. Therefore, a first functional prototype was developed to be able to detail the requirements. The chassis system was based on a donor vehicle and as few modifications as possible were conducted to keep development- and product costs as low as possible. Additional modifications for a second prototype were identified by analyzing data from testing in Germany and Ghana. Finally, the second prototype was developed, manufactured and assembled. It was presented at the International Motor Show (IAA) in Frankfurt/Main in September 2017.

At this point, the reasonability of using a donor vehicle's chassis system as a basis for a new vehicle concept may arise. The major advantage of this approach is the ability to build and test a prototype rapidly. It ensured the development and assembly of two prototypes within the project's timespan. However, the effort for identifying required chassis system modification should not be underestimated. Especially the steering system is barely transferable without modifications due to different packaging and ergonomics of each vehicle.

Further steps that need to follow the research project are the detailed analysis and quantification of chassis system loads based on actual driving data, and the final validation of existing component design, especially regarding their long-term durability. For that matter, extensive further testing needs to be conducted especially with regard to the vehicle's actual usage scenarios in Africa.

Contributions

Michael Schmidt and Thomas Zehelein conducted research and development work as heads of driving dynamics in the aCar mobility project, including simulation, testing, and collaborating with the project's industry partners. Mr. Schmidt and Mr. Zehelein also initiated and wrote this paper. Markus Lienkamp contributed to the conception of the research project and revised the paper critically for important intellectual content.

Acknowledgements

This paper draws from research and project work funded by the Bavarian Research Foundation covering a timespan of two and a half years, including many students' bachelor's and master's theses as well as term projects. Figures and data are partly taken from the works of Oliver Cameron, Eduardo Capriotti, Heiko Desor, Maximilian Eiba, Ioannis Ganotis, Tobias Hierlmeier, Laura Müller, Felix Schwencke, Benjamin Vorwerg, and Martin Weber – we sincerely thank you all for your great work.

References

[1] M. Ersoy and S. Gies, *Fahrwerkhandbuch: Grundlagen – Fahrdynamik – Fahrverhalten- Komponenten – Elektronische Systeme – Fahrerassistenz – Autonomes Fahren- Perspektiven*. Wiesbaden: Springer Fachmedien Wiesbaden, 2017.

[2] H.-H. Braess and U. Seiffert, Eds., *Vieweg Handbuch Kraftfahrzeugtechnik,* 7th ed. Wiesbaden: Springer Vieweg, 2013.

[3] K. Gwilliam, V. Foster, R. Archondo-Callao, C. Briceño-Garmendia, A. Nogales, K. Sethi, "The Burden of Maintenance: Roads in Sub-Saharan Africa: Background Paper 14 (Phase I)," Washington D.C., 2009.

[4] M. Trzesniowski, Ed., *Handbuch Rennwagentechnik*. Wiesbaden: Springer Vieweg, 2017.

Development of a disruptive semitrailer tractor in lightweight

Jürgen Hintereder Dipl.-Ing. (FH)
MAN Truck & Bus AG
Research Department EZRC – vehicle concepts
Complete vehicle concepts, project manager
research-vehicle light-weight tractor

Steve Sattler Dipl.-Ing. (FH)
MAN Truck & Bus AG
Research Department EZRC – vehicle concepts
Chassis concepts, design, CAE-methods

Urs Gunzert Dipl.-Ing. Univ.
MAN Truck & Bus AG
Research Senior Department Manager EZRC – vehicle concepts

P. Pfeffer (Ed.), *9th International Munich Chassis Symposium 2018*, Proceedings,
https://doi.org/10.1007/978-3-658-22050-1_11

1 Motivation

The most widely used supporting structure in truck design today is the ladder-type frame. This principle offers many advantages and meets requirements outstandingly, particularly at the beginning of truck development. Aside from the big payload, the concept is simple and robust while enabling a high degree of variability, usually in combination with a modular system. The relatively low torsional rigidity around the vehicle's longitudinal axis (by comparison with a body) is an advantage in what was previously a frequent type of deployment, namely operation on uneven, unpaved surfaces.

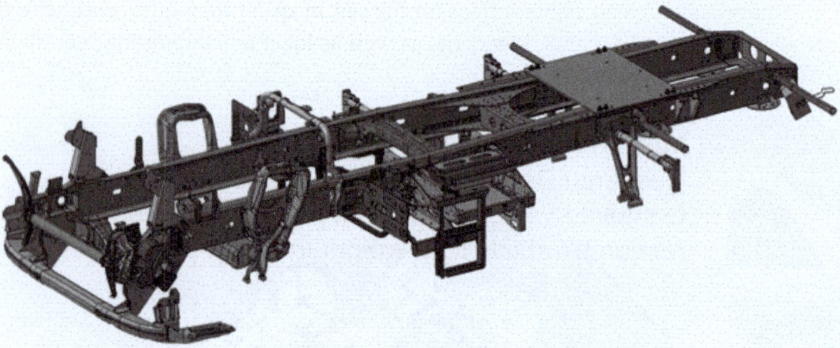

Figure 1: Reference vehicle: 18-tonne semitrailer tractor ladder frame

Over the course of well over a century of development, however, design engineers have had to rise to the challenge of a great many changes in their market. To name just the most important changes, these include more drive power, more stringent emission standards with simultaneous prevention of CO_2, more complex and highly variable vehicle systems for a variety of transport tasks, more volume and comfort in the driver's cab and specialisation in long-haul transport on predominantly smooth roads.

In consequence, different boundary conditions for the further development of the vehicle have defined themselves. Just for example, continually increased mileage and enhanced equipment with simultaneously reduced energy demands and improved environmental balance. The potential for optimisation here is mainly in the areas of energy management, driving strategy, driving resistances and lightweight construction. The lightweight construction area is under extreme cost pressure, especially in the commercial vehicles sector where the relatively low volumes are problematic. As a result, developments in this direction require significant resources because cost-effective lightweight construction is difficult to achieve by implementing isolated measures.

A further reason is directly related to ladder frames:

an additional consequence of the changes in the market is that the installation space available for the ladder frame has become severely restricted in all spatial directions. Given the inflexible ladder-frame concept, it was not possible to adapt the vehicle architecture to any significant extent. A consequence of the current frame concept for modern trucks is that developers are forced to make considerable compromises in defining installation space and in the subsequently achievable mechanical properties of the ladder frames.

These compromises have an impact on the potential of existing components to be constructed in lightweight design and on the consequent costs.

Similar constraints led to passenger cars being produced with self-supporting bodies from around 1930 and buses and coaches from around 1954. However, significantly higher requirements with respect to product variability and to the associated logistics and production have until now hindered any similar development from taking place in trucks. However, thanks to the growing CO2 debate, alternative drive concepts for trucks are again being more sharply focused on. This development has brought to the fore an advantage of the body concept that until now has not been given much attention in the sector: the gain of previously unusable installation space. The assignment of the research vehicle presented here is to demonstrate the magnitude of existing potential in lightweight construction and the gain in installation space on a semitrailer tractor with a short wheelbase.

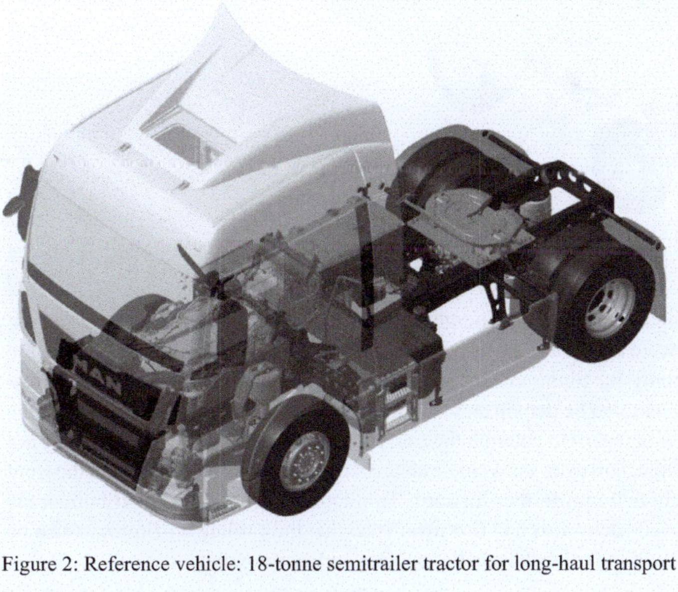

Figure 2: Reference vehicle: 18-tonne semitrailer tractor for long-haul transport

2 Conceptualisation

Preliminary studies regarding the potential of lightweight construction of ladder frames indicated to MAN Research that we would have to abandon the current installation space of the frame in order to achieve the necessary improvement in dead weight. Seen together with the general changes in market conditions outlined in the introduction, the time had come for the structuring of a vehicle with a new vehicle architecture. The Research Department decided on a semitrailer tractor with equipment that is typical in long-haul transport as a reference vehicle. Additional air suspension on the front axle makes for better comparison of later measurements with the new vehicle architecture.

Figure 3: Complete vehicle concept with highlighted running-gear components

The cab, in its original location with mount, and the diesel driveline with engine and automatic gearbox were adopted from series production. Although the supporting structure is predestined mainly for future vehicle concepts with alternative drive systems, fundamental feasibility can also be demonstrated with the classic configuration. Because of the current situation with regard to entering the cab, the location of the front axle was also not changed. In principle, however, our vehicle concept makes it possible to relocate the front axle approximately 600 mm further forward. To improve the axle-load distribution, the driveline was moved backwards and downwards so that the cooling air-flow can also be optimised, which will in turn make smaller cooling surfaces possible.

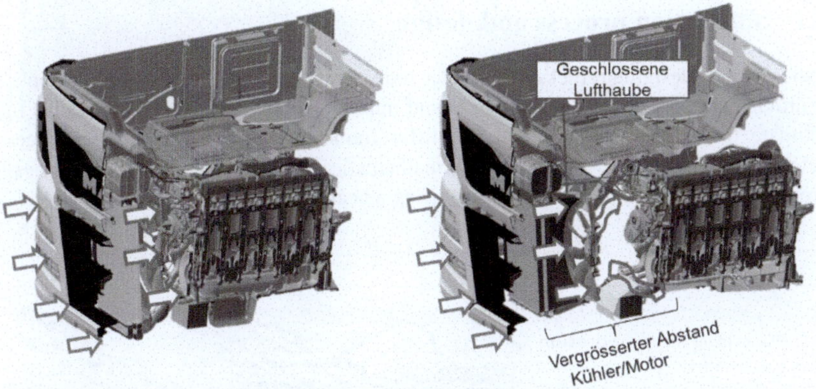

Abbildung 4: Increased distance cooler to engine compared to reference

Although the vehicle's dimensions remain unchanged with a wheelbase of 3,600 mm, the supporting structure ends approximately 250 mm behind the two-bellows air suspension on the rear axle that enables this construction.

The aim of building a research vehicle that has a considerably more rigid supporting structure was to study and evaluate additional potential in handling and comfort as well as conceptual lightweight construction with a simultaneous gain in installation space. The first installation-space concepts already made it clear that independent wheel suspension and rack-and-pinion steering on the front axle exhibited advantages in terms of installation space, with additional potential as positive influences on handling. Prototypes are fitted with rack-and-pinion steering and independent wheel suspension.

The freed-up installation space is to be used to demonstrate possible improvements to the front of the vehicle in the area of active and passive safety.

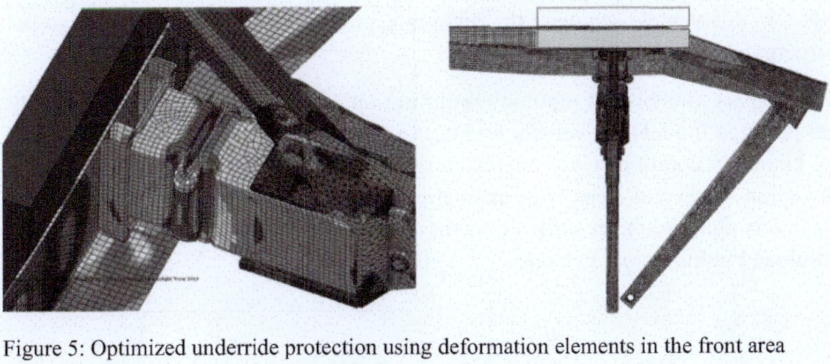

Figure 5: Optimized underride protection using deformation elements in the front area

2.1 Simulation process and design

Experience in designing trucks with ladder frames cannot be applied to body concepts. With the help of some preliminary tests and multibody simulations, it was possible to estimate the effect of a substantially stiffer frame system on running-gear forces beforehand, both experimentally and mathematically. For this purpose a series-production frame was globally stiffened using a bearer system.

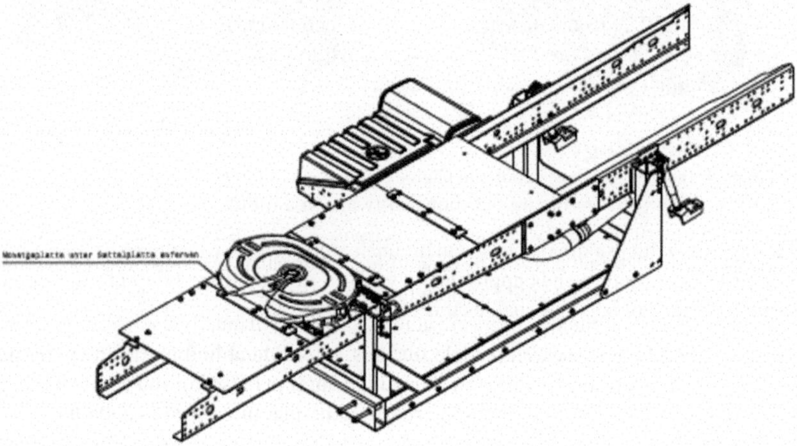

Figure 6: Bearer system for global stiffening of series-production frame

The loads for deployment in long-haul and distribution transport increase only minimally, which allows the use of corresponding near-series running-gear components in the initial approach. For the FE analyses of the concepts, a simplified method of static and dynamic stresses was used, the same as in the early concept phase of coach design. Moreover, in this vehicle segment, the running gear and thc rigidities exhibited in the floor structure are comparable.

Installation-space models for topology optimisation are derived from the adequately detailed package models for vehicle architecture. The method is then applied – with varying boundary conditions – to manufacturing restrictions. The geometry is defined by calculation in several steps with manual control taking place between steps. Processing in this phase requires skills in the disciplines of design and simulation, preferably combined in individual persons.

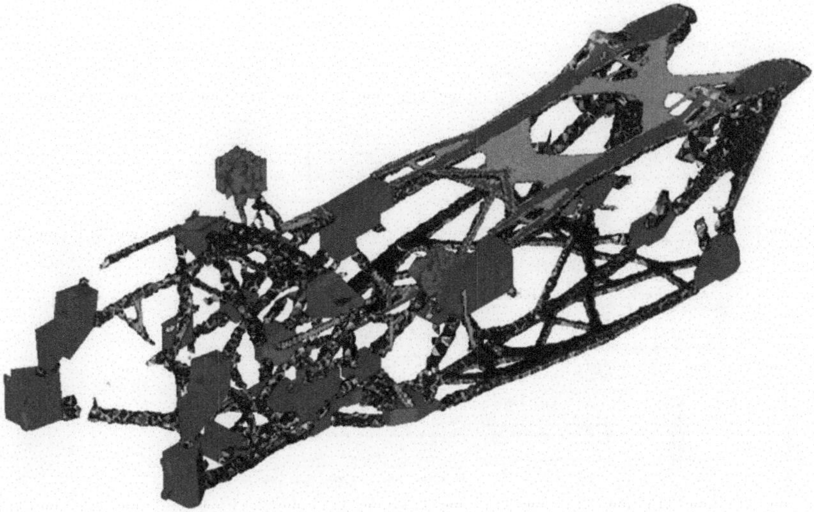

Figure 7: Interim results of topology optimisation of the complete supporting structure

Of central importance in conceptualisation is the earliest possible use of simulation methods: in our case, they were already applied to the structures from topology optimisation. The mechanical properties of the smoothed geometry are checked using FEA directly and without complex reduction into CAD. Besides static rigidities and eigenvalues, only nominal stresses are checked. In this way, fewer resources are invested in the detailing of designs that hold little promise of success.

The concepts or designs for supporting structures generated here are given corresponding version numbers depending on the package status within the different virtual concept vehicles. In our case these represent different drive concepts.

This means that it is possible to find many concepts that are equally worth following and with which the target values being aimed at are achievable. This modus operandi gives rise to a database with many and various solution approaches to manufacturing and production – an ideal pool of pre-development projects.

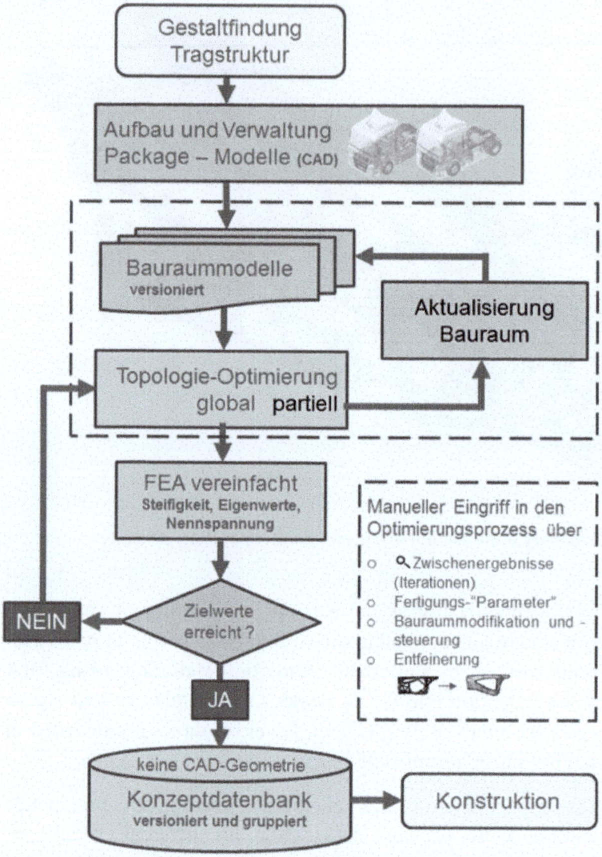

Figure 8: Flow chart of simulation process

2.2 Concept selection, degree of detail and verification by calculation

For the prioritised tasks of building a research vehicle and clarifying feasibility, the solution approaches selected were primarily those that are linked with sound manufacturing know-how. The approach finally chosen was one that combined locally reinforced tubular segments and sandwich structures with insert components. In this context, extensive development parameters are available to us, especially with regard to our own prototype construction.

Further processing and detailing of the design were carried out using CAD as the master system. This also makes it easier to apply a standardised process in verification by calculation. Besides the local verification of structural strength, a simplified examination of fatigue strength also takes place here. Target values for rigidity and eigenvalues continue to be checked here, as does the weight balance.

The project was completely processed in an agile manner by a small team possessing all the requisite skills. We were able to use this to our advantage to detail the entire supporting structure from individual cells and then complete it seamlessly.

In concrete terms this means that individual areas for which the environmental geometry had already been set down were initially detailed and parallel to this were also constructed as segments, while other areas were kept less structured and thus variable. The consequence of this method is that one achieves one's goal very quickly but that the big picture only crystallises little by little. Put another way: it emerges only during processing.

With respect to achieving the target values and the ongoing FEA, this represents a challenge only to a limited degree, in that the influence on global properties by local changes is not significant.

Valuable development time can thus be saved, especially in the manufacturing area. An additional positive side-effect is the rising learning curve and with it, engineering designs that become increasingly viable. This processing phase saw continued and extensive use of local topology optimisation. Over long phases, the FE models for verification consisted of hybrid structures, for which a good networking strategy and standards are a prerequisite.

Development and manufacture or assembly were thus parallel activities; this also applies to the complete vehicle body and the cable routing.

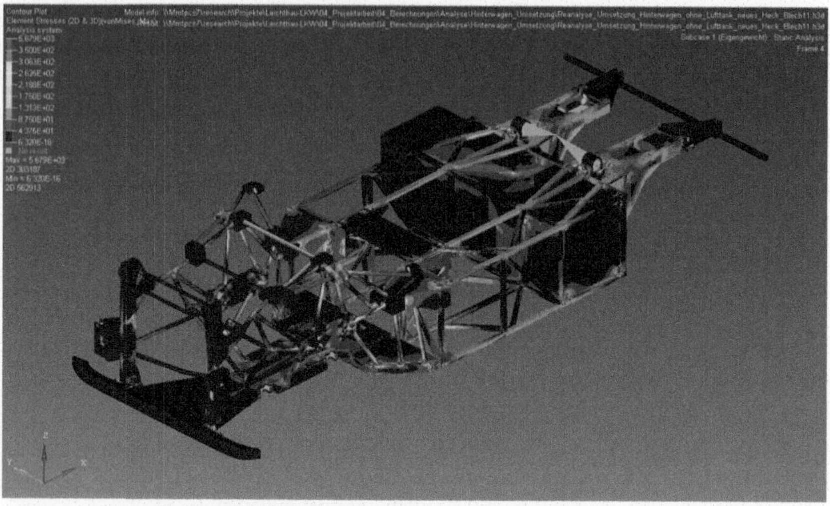

Figure 9: Calculation result with hybride structures

3 Prototype construction

As already mentioned, in the case under discussion, a chronological separation of development, prototype construction and assembly of the complete vehicle was not possible.

Nevertheless, this is an attempt to provide a cohesive description of the contents of the phase dealing purely with construction and assembly. This must be understood in the context of the development method described.

As elements of the supporting structure, open and closed semi-finished materials, flat or simply formed lasered sheets and solid components machined from full material were used. The components were connected to one other using manual MAG welding. Construction was performed on assembly bases in combination with existing jigs and angle brackets. With few exceptions, manufacturing utilised comparatively simple manufacturing machinery.

The evolutionary method resulted in the emergence of a main front segment and a main rear segment, into each of which as many functions and adapters as possible could be integrated. For process-related reasons, certain segments were also formed, constructed separately and joined to the main structure using bolted connections. The interface between the two main segments was additionally influenced by production-related parameters, in this specific case by the size of the dipping plant available for cathodic dip coating (KTL). Arising from the concept, this also serves to indicate a possible modularisation of several vehicle derivatives within a vehicle group. In the event of commercialisation, the current interface would be relocated elsewhere.

For optimal coating of the segments the dipping plant, simulations of the dipping process and wetting were run continually.

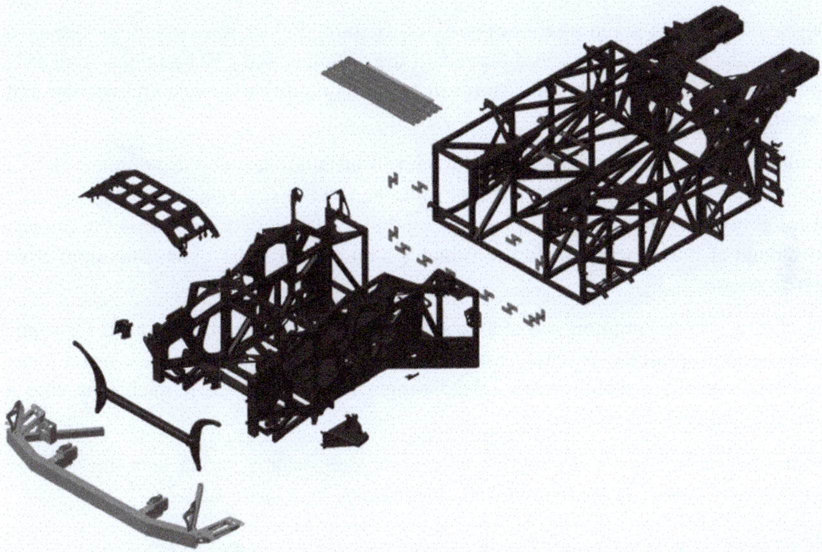

Figure 10: Exploded view of the supporting frame segments

4 Complete vehicle and results

The initial measurement result showed a weight saving of 400 kg for the supporting structure with top coating. As expected, deviations from the CAD model were slight – under 2 kg, despite manual welding and painting.

During development, an attempt was also made to simplify the arrangement and connection of the units (control units, valves, compressed-air tanks and so on) by means of optimised cabling and piping. In the braking system, for example, it was possible to eliminate several metres of cable and air piping.

The remainder of the complete vehicle construction may be described as unproblematic, which more than justifies the great amount of effort put into the package models (largely complete and with a high degree of detail).

The complete vehicle in a roadworthy state has a dead weight < 6,000 kg, which translates to a weight saving of up to 1,300 kg or at least 1,100 kg, depending on the running gear and tyre variants.

The weight saving forecast was a minimum of 1,000 kg. It was exceeded because of the many small parts, which were difficult to calculate in advance, and because the masses of some of the new prototype units were not available beforehand.

Besides the 400 kg saved on the supporting structure, 250 kg were saved by measures taken on the rear axle, 200 kg on the rest of the driveline and 150 kg on the front axle and steering. The remaining 300 kg result from the modified vehicle architecture and overall concept.

A further, highly important result is the gain in installation space of approximately a cubic metre. Even on the prototype this could be used to realise a tank with a volume of roughly 1,300 litres. In our opinion, the additional installation space is the biggest advantage of this concept for future vehicles with electrical drivelines and alternative drive systems.

As of now, our assumption is that a dead weight of < 5.3 t can be achieved for semitrailer tractor applications where weight is an important factor. With the usual long-haul equipment this would mean a dead weight of around 5.8 t, in each case with a conventional diesel driveline.

With a factor of 5 to 20, the rigidity values are significantly higher than those of the comparison vehicle, as are the relevant eigenmodes, by a factor of 1.7 to 2.5.

Figure 11: First-order torsion around the vehicle's longitudinal axis

5 Start-up and testing

The research vehicle was registered as a test vehicle. Complete homologation was required for the new rack-and-pinion steering.

This also applied to the brake system due to the wide-ranging modifications in the installation. However, homologation was unproblematic in both cases. Agreements were defined for each of the components relevant to safety, optionally together with the respective supplier. The agreements include separate inspection and replacement intervals.

No separate insulation measures were necessary for the acoustic test despite the relatively open framework construction in the area of the engine and gearbox. This shows that excellent values for sound emission could be achieved with comparatively little effort.

Appropriate adjustments to the running gear air suspension were needed. Good tuning was very quickly realised with minor measures. There is still more potential, which will be tapped in the course of follow-up projects.

Numerous measuring runs were carried out on various test tracks with different variants of running gear, above all with changes to the roll rate. Somewhat more extensive tuning was needed because the greatest stiffening by contrast with the ladder frame was reached here.

Moreover, different tyre variants were assessed, including single tyres on the rear axle. These are necessary on vehicles that react sensitively to weight and on vehicles equipped with electrical drivelines, for example e-axles.

Figure 12: Testing around Munich

6 Summary and outlook

With the research vehicle under discussion it proved possible to generate a product idea outstandingly capable of taking into consideration both the current boundary conditions and the future developments necessary to the truck market.

We were also able to verify the feasibility of this vehicle idea by experiment. A paradigm change – state of the art for passenger cars and buses/coaches since the beginning and middle of the last century respectively – is thus also sensible and possible for trucks.

Because construction is comparable, advanced projects will be able to draw on a state of the art with already proven and highly developed know-how from the field of body construction. However, the specific boundary conditions and target values must be taken into account here. Relative to the prototypical implementation, there is thus further potential for structural optimisation and lightweight construction.

From our point of view the biggest challenge lies in replacing the product structure of trucks with ladder frames, which has been cultivated for over one hundred years and has seen disproportionately high growth. We believe that it is both sensible and necessary to take a completely new type of approach in order to realise modularisation for the product variance that is essential.

To this end, MAN has already been deliberating ways in which the conceptual potential of the product idea can be fittingly unlocked, also in the areas of logistics and production.

The construction principle can be applied to vehicles for distribution and long-haul transport. Consequently, municipal and construction-site vehicles that operate to only a small extent on poor surfaces, at least in highly developed countries, are also conceivable.

As far as off-road vehicles, WorldWide construction-site vehicles and high-mobility vehicles are concerned, our experience indicates that the ladder frame remains the better supporting structure for trucks.

Figure 13: Testing at MAN Test-area

AUTONOMOUS VEHICLES

What can we learn from autonomous level-5 motorsport?

* Johannes Betz, M. Sc.

** Alexander Wischnewski, M. Sc.

* Alexander Heilmeier, M. Sc.

* Felix Nobis, M. Sc.

* Tim Stahl, M. Sc.

* Leonhard Hermansdorfer, M. Sc.

** Prof. Dr.-Ing. Boris Lohmann

* Prof. Dr.-Ing. Markus Lienkamp

---

* Chair of Automotive Technology, Technical University of Munich

** Chair of Automatic Control, Technical University of Munich

P. Pfeffer (Ed.), *9th International Munich Chassis Symposium 2018*, Proceedings,
https://doi.org/10.1007/978-3-658-22050-1_12

1 Introduction

Whether BMW, VW or Google: Almost all leading automobile and technology companies are researching and developing the multi-stage autonomy of vehicles, which enables a completely self-driven vehicle without a driver in autonomy level 5. Based on his assessment and experience, the driver had previously carried out environmental detection, localization and vehicle control. The elimination of the driver creates numerous challenges in the development of level 5 vehicles. However, in order to achieve an efficient and safe driving style, the driving strategy of the vehicle must be adapted to the current environmental conditions.

With the beginning of the third Formula E series an additional support series called *Roborace* will take place on the tracks currently used by the Formula E [2]. The goal of Roborace is to provide the first racing series for autonomous vehicles. The teams that take part at this competition will develop only the software for the provided autonomous cars (Robocars) [1]. The following paper is showing an overview of the Roborace project and the different software parts that have to be developed for a car that has to drive autonomously. Afterwards an evaluation of about what we can learn from autonomous level 5 motorsport is done.

2 State of the Art

Urbanization and globalization are two reasons for the increase in mobility demand and eventually traffic in the next decades [3]. Vehicles operating in these congested traffic systems must be more efficient, more environmentally friendly and safer than current generation vehicles [3]. In order to meet this challenge, the largest role for solving this problem is attributed to the autonomous vehicle in addition to component optimization [4]. The evolutionary introduction of driver assistance systems (ADAS) enables more and more automated functions in the vehicle, which increases the degree of automation in the vehicle. After the definition of SAE [5], the levels of autonomous driving are defined from level 0 (the driver drives himself) to level 5 (no driver required in the vehicle anymore).

The state of the art in today's passenger cars shows almost only level 1 functions. For example, the manufacturers Volkswagen, Mercedes-Benz and BMW offer the assistance functions Lane Departure Warning (LDW), Adaptive Cruise Control (ACC) or Parking Control in high-priced vehicles [6, 7]. As one of the few vehicles available today, the *Tesla Model S* has an autopilot that enables automatic parking as well as a longitudinal guidance and lane keeping function on the motorway [8]. As the world's first production car with level 3 automation, the *Audi A8*, presented in June 2017, enables autonomous longitudinal and transverse control in slow-flowing traffic at up to 60 km/h (congestion pilot) [9, 10].

The driver does not have to monitor the vehicle constantly and the car is setting the indicators, is accelerating, braking and steering automatically.

In addition to these already production-ready vehicles, autonomous driving is the subject of numerous research projects. Industry in particular has shown that the development of complete vehicle concepts is being driven forward. Mercedes Benz presented the *S 500 Intelligent Drive*, an autonomous research vehicle that implements level 3 functions for city and long-distance journeys [11]. With the *RS7 Piloted Driving Concept* Audi showed an autonomous high-speed driving at 240 km/h on a race track [12, 13]. In addition to classic car manufacturers, numerous technology companies are also researching in the field of autonomous driving. Emerging from research vehicles of the DARPA Challenge [14], *Alphabet Inc.* (formerly Google) showed the first self-driven vehicle (level 5) intended for public transport 2011, which is being researched in its own subsidiary *Waymo* [15]. *Nvidia* provides the technology for the development of autonomous driving functions with a software and hardware portfolio [16]. The latest state of development of Nvidia's autonomous driving functions is the publication of an "End-to-End Deep Learning Approach" for the complete control of the vehicle through an artificial neuronal network (ANN) [17]. As a start-up spin-off of Stanford University, *drive.ai* [18] is researching software development for autonomous vehicles based on deep learning approaches. The focus here is also on the perception of the environment and localization.

The majority of universities are focusing mainly on the development and research of isolated driving functions for autonomous vehicles. The Center for Automotive Research (CARS) at Stanford University [19] is focusing on functions for image processing (automatic camera calibration, vehicle and object detection [20, 21] and methods for localization [22, 23]. Using a driving simulator, the University of Leeds [24] is investigating both human factors and functions to increase safety. UC Berkeley [25] is researching in the development of algorithms based on artificial intelligence for the detection of vehicles from videos, a vehicle control system based on real driver input as well as security and privacy in the deep learning area. In Germany, the joint project *Pegasus* [26] deserves special mention, further research institutions for autonomous driving are located at the FU Berlin [27], at the University of Toronto [28] and at the Massachusetts Institute of Technology (MIT) [29].

This overview of the current activities of industry and universities shows the enormous relevance of autonomous driving. The Roborace racing series, which provides an electrically powered, automated level 5 racing vehicle, serves as a platform for practical testing and real validation of developed autonomous driving functions. The Technical University of Munich (TUM) decided to participate in this racing series with its own team based on the knowledge of different institutes. The Roborace platform offers an autonomous level 5 vehicle that has to use high-performance, robust and secure algorithms at high clock rates in real time due to its high speed. The absence of a driver additionally eliminates a limiting factor in the vehicle and thus enables the technical potential of the algorithms to be fully tested on safe and reproducible terrain (racetrack) in various scenarios. The racing car thus represents an extreme case. All the knowledge gained from the driving tests in this project can thus be transferred to the development of future conventional passenger cars.

The ultimate motivation for the intended project comes from the knowledge that the driving strategy of the autonomous vehicle must be adapted to the current environmental conditions for a safe and efficient driving style. This means that in addition to a de-detection of the environment (objects, road course, etc.), a detection of the road conditions and a classification of the friction coefficients based on this must also be carried out.. The prediction of the coefficient of friction and optimum energy management in an autonomous vehicle are therefore functions of essential importance. The potential of these processes in the vehicle functions mentioned above is thus largely unknown.

2 Robocar Vehicle Concept

2.1 Hardware

The Roborace company provides for each team two fully autonomous level 5 racing vehicles which are equipped with identical hardware (figure 1).

Figure 1: The Robocar with sensors and actors [1]

The Robocar is a battery electric vehicle with a Lithium Polymer battery capacity of 62 kWh. The car is powered by four electric motors that provide a power of 300 kW and motor torque of 300 Nm for each electrical motor. Beside the vehicle control systems, the car consists of different sensors for perceiving the environment and localizing the vehicle. The Robocar has two front and four surround cameras, five lidar systems (four around the front wheels and one in the back of the car), two radars (one in the front and one in the back) as well as 17 ultrasonic sensors all around the Robocar.

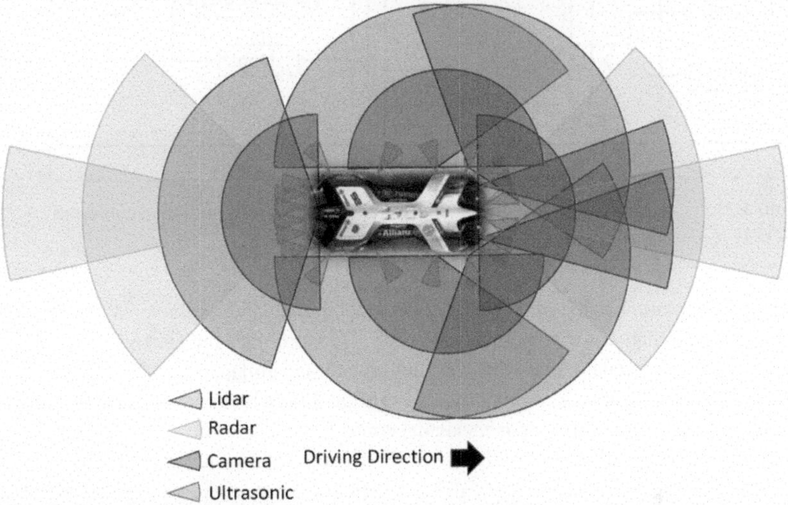

Figure 2: Sensors for perception of the Robocar

2.2 Software

When talking about autonomous driving, there are three different tasks that have to be accomplished for driving the car autonomously: Perception, planning and control [30]. A detailed overview is displayed in figure 3 [31].

Firstly, the data from the different sensors in the vehicle have to be unified in the perception functions. A lane detection algorithm ([32–34]), a function for detection and tracking objects ([35]) and a free space detection leads to holistic information for the current environment where the car is currently driving. With this information a detailed modeling of the environment based on objects or occupancy grids [36] can be done. Secondly, in the field of perception an additional localization of the vehicle (e. g. GPS or lidar-only localization) has to be done.

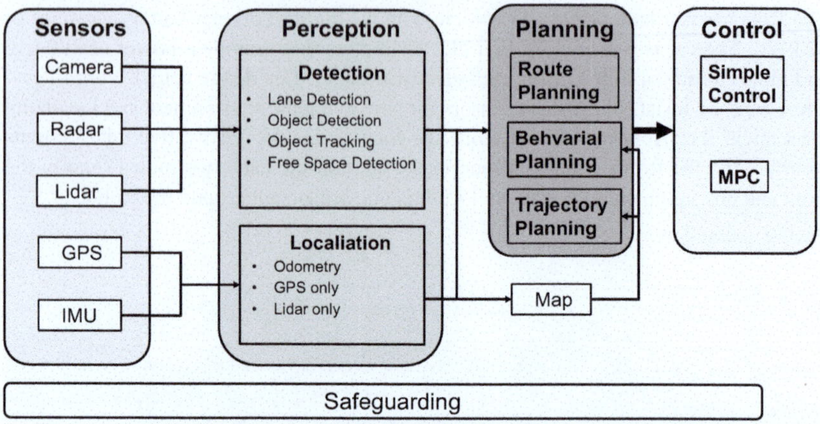

Figure 3: Holistic overview of the software structure for autonomous driving in conformity with [31]

In the field of planning first of all we have to plan the route of the vehicle. When the route is planned a prediction for velocities and acceleration provides the general behavioral of the car. In an additional behavioral planning, which is similar to a state machine, time critical decisions the car e.g. overtaking, lane change, safety braking have to be made. The last field of autonomous driving is the actual control of the car. The control includes both longitudinal and lateral control of the car for both acceleration and curvature.

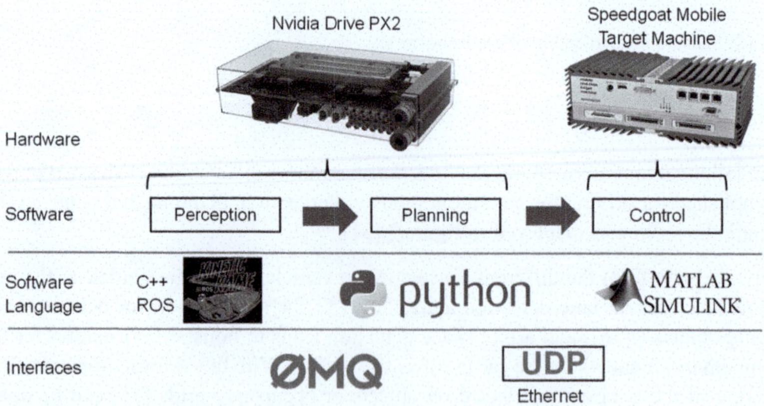

Figure 4: Software structure and interfaces for the ECUs in the Robocar

For all functions, there is the need for an additional overarching safeguarding which is surveying limit values of the data. This software structure has then to be integrated on both electrical control systems (ECUs), that are available in the Roborace (figure 4)

All the information from the sensors is united in one ECU, the *Nvidia Drive PX2* [37]. With this ECU, a platform for deep learning, sensor fusion and surround vision is provided for the software developers of the Robocar. Afterwards, the functions developed for the Nvidia Drive PX2 provide the information for the real time motion controller (*Speedgoat mobile target machine* [38]) that calculates the needed control output values to steer and accelerate the vehicle.

The functions for the ECUs has to be programmed in different languages because of the different hardware architectures. The Speedgoat Mobile Target machine is programmed with MATLAB/Simulink. The Ubuntu based Nvidia Drive PX2 is capable of running C, C++, Cuda and python code. In addition, the *ROS* (Robot Operating System [39]) packages, which are implemented in C++, can be used for perception and planning development. The whole strategy of the racecar can be implemented in *python* code. Possible solutions for the communication between the two ECUs and the different software parts are *ZeroMQ* sockets, *UDP* (Ethernet) connection or ROS messages

3 What can we learn from autonomous motorsport?

3.1 What we learned so far from motorsport

Motorsport itself developed in the late 19th century from races of the first owners of motor vehicles. Due to the poor road conditions and insufficiently developed vehicle technology, the reliability and resistance of the motor vehicles in particular had to be confirmed. Races against riders, cyclists and railways have achieved higher top speeds, but endurance speeds have not yet been satisfactory in view of frequent problems [40].

In the next 100 years, the motorsport developed ever further and new racing series were added. Today, the most famous motor sports are Formula 1, Rallye, Le Mans Prototype and MotoGP. Many vehicle manufacturers such as Ferrari, Mercedes and BMW or also vehicle suppliers such as Mclaren have been established participants in this motorsport series for years. In addition to obvious reasons, such as prestige and marketing, progress in technical development plays a major role. For many manufacturers, the following points are the reasons for participating in motorsport races:

1. Progress in vehicle technology (e. g. new engine technologies)
2. Progress in production technology (e. g. new casting technologies)
3. Progress in automotive processes (e. g. CAD)

Looking to the future of motorsport and the introduction of autonomous racing series such as Roborace, the question now arises: What can we learn from autonomous level 5 motorsport?

The following pages will provide a detailed answer to this question by shedding more light on the individual areas in which knowledge can be built up.

3.2 Artificial Intelligence

When talking about Artificial Intelligence (AI) in the field of autonomous driving, we are always talking about algorithms. Machine learning deals with the extraction of information from data, which is subsequently converted into "artificial" knowledge or experience [17, 41, 42]. The algorithms developed based on these machine learning methods are then able to execute intelligent behavior. Intelligent behavior refers to the optimal solution of problems for which the algorithm was developed and trained [43, 44]. Due to the increase in computing capability (multi-core processors, GPU programming) and the availability of large data sets, these methods can easily be used from the state of the art of hardware and software. Current areas of application include speech and face recognition, image classification and data prediction (e.g. financial sector) [45]. Figure 5 shows an overview of the machine learning methods, divided according to the final use of the method for classification, regression or clustering of data.

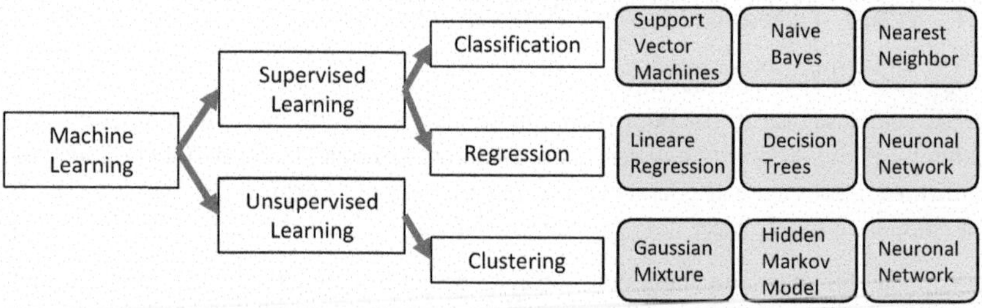

Figure 5: Overview machine learning algorithms and methods in conformity with [44]

Due to the complexity of autonomous road travel (markings, changing light and weather conditions, construction sites, static and dynamic objects, etc.), an enormous amount of data is generated in the vehicle, which must be processed quickly and efficiently. Machine learning methods show advantages in the accuracy and speed of these processes, for example in the area of perceiving the environment compared to classical computer vision methods.

Looking at the autonomous motorsport, there are three big tasks for AI integration:

1. **AI as a function:** Firstly, AI Algorithms can be used as a normal function in the autonomous vehicle that overtakes the tasks of the driver. Right now, mainly in the field of perception ANNs are used as computer vision functions [46–49]. Especially for the detection and tracking of traffic signs [50, 51] or other vehicles [52, 53]. The current research is focusing on image segmentation [54] where each pixel of an image can be classified as an object. The state of the art for ANNs is broad in the perception area, right now the tasks of planning and control are still lacking of machine learning approaches and so providing a large area for research.
2. **AI for support:** Secondly, AI can be used as a conventional ADAS function, which supports the driver while driving manually. This means, we do not have to focus on autonomous driving only. Conventional vehicles will still be existing but need better ADAS functions. This can be either for safety functions like safety stop assistant, energy management for saving energy or a comfort system for planning better routes [55, 56].
3. **AI for prediction:** The last task for AI functions is for predicting both the behavioral of the own vehicle and other vehicles and objects (e. g. humans). This means that in addition to a detection of the environment (objects, road course, etc.), a detection of the road conditions and a classification of the friction coefficients must also be carried out. This also means that the self-driven vehicle must travel the planned route in an energy-optimal manner and thus make a prediction of the energy consumption and a final optimal distribution of the energy in the vehicle. For both functions the field of AI provides promisingly methods.

In addition to these tasks there is always the discussion about the AI development pipeline. When we talk about the development of artificial neural networks, it foremost means the *Training* of the ANNs. The state of the art and science has already provided networks such as LeNEt, GoogleNet [57] etc. which have too many parameters (weights, bias) to be trained during a training epoch. For this reason, training on graphical processor units (GPUs) is indispensable. The pipeline for the development, training, testing and final inference of the AI algorithms based on ANN is shown in figure 6 and is mainly based on the hardware NVIDIA is offering today.

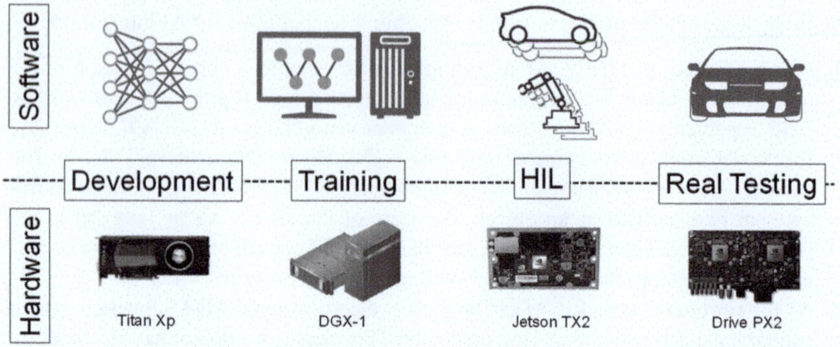

Figure 6: AI Development workflow with Nvidia Hardware

The mainly base development can be done on classical consumer GPUs like an *Nvidia Titan Xp*. For further development, a training of hundreds and thousands of epochs has to be done on a multiple GPU server like the *Nvidia DGX-1*. For testing the trained ANNs in an inference in an HIL or robotic system we can use the *Nvidia Jetson*, for real tests in the vehicle with all the bus communications (e. g. CAN) the Nvidia Drive PX2 can be used. We highly recommend this pipeline because all the known frameworks for training and developing ANNs (e. g. tensorflow) are optimized for those GPU based hardware.

3.3 Perception

The autonomous vehicle can use different sensors like cameras, lidar or radar for perceiving information from the environment. Based on the data gathered from the different sensors the car has to make a representation of the environment in a virtual map which is called mapping algorithm [58, 59].

The current state of the art is mainly based on 2D mapping algorithms which are gathering laser scan data from a lidar [60, 61]. With using ROS, different packages are provided in the ROS environment for mapping the environment in an occupancy grid like *hectorSLAM* or *gmapping*. Although the algorithms are well-know and there are publications for how to tune the algorithm [62], the different algorithms are still lacking in speed and accuracy [63].

3.4 Planning

Given the assumption that the situation has been understood to a sufficient level of detail, it remains to plan reasonable actions for the car. In an autonomous racing setting, this includes generation of a global raceline, as well as local replanning for overtaking and evasion maneuver (figure 7).

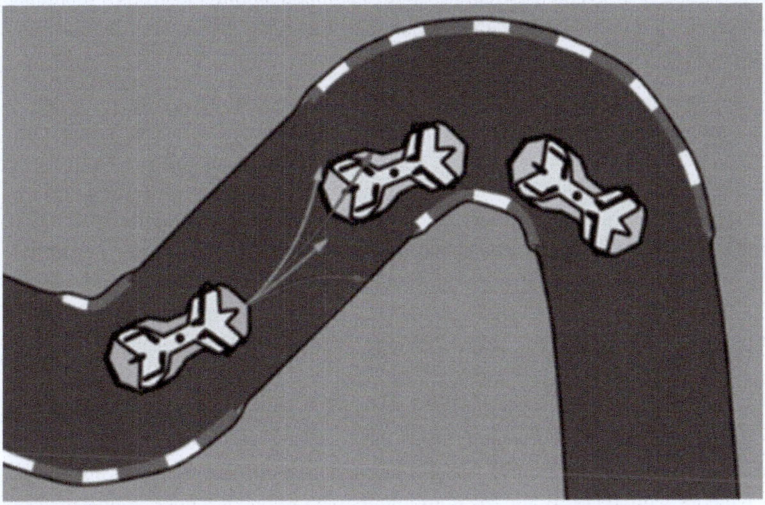

Figure 7: Trajectory Planning – Different options for choosing the raceline

In general, it is not necessary to have explicit high-fidelity knowledge about vehicle physics while a general understanding of the physical capabilities is for sure beneficial at this level of the driving stack. Two main concepts are used throughout literature to represent the maneuvers designed at the planning [64, 65]: Either the combination of a geometric path and a velocity profile or the specification of a trajectory in Cartesian coordinates. Since the latter can lead to unpredictable results at the limits of handling due to control constraints, we focus on the velocity profile version.

The motion planning was tackled numerous times for slow and medium lateral acceleration driving situations by geometrical approaches, graph search or tree search. However, it remains an open question how to plan trajectories which are capable of meeting all constraints of the vehicle dynamics at real-time at the speeds necessary for autonomous racing (200 kph and more). This can be considered highly relevant to emergency driving situations in standard road scenarios. Several authors [66, 67] propose to use optimal control for this purpose, however it lacks robustness due to its high dependencies on model quality and poses difficulties in the numerical efficient and reliable computation of the solutions. A promising approach is the combination of offline optimization for preplanned maneuvers [68] which can then be composed online. However, this method also highly depends on the model precision and is therefore difficult to apply in series production vehicles, since it needs lots of tuning to the actual operating conditions. From a software architecture point of view, it is therefore of interest to provide high fidelity model knowledge to the vehicle dynamics control component only, instead of including it into the planning process. This enables high reusability of the algorithms along a wide variety of vehicles.

We therefore need to derive "easy to check"-conditions for geometric path and velocity profile planning. They should ensure that the trajectory is drivable and is "nicely controllable" in the sense, that an underlying control algorithm is capable of stabilizing it under a wide set of disturbances. Further, they must avoid crashes with static and dynamic objects. It is desirable, that these conditions can be derived based on past measurement data of the vehicle, since this enables the possibility to incorporate perception and control information during the race to optimize the planning process. Similar to the need in everyday driving scenarios, this allows to deal with sudden changes of environment variables (most important weather) or tire degradation. A promising approach is presented in [69] which applies optimal control to check whether a model conforms with the real vehicle behavior, however no online adaption strategy is proposed. In addition, the deterministic nature of the approach might lead to overly conservative solutions.

Further, the competitive environment poses the challenge to improve the global racing line in each consecutive lap. While the minimum curvature line is a good approximation of the best solution, it is clearly not the global optimum of the minimum time problem which has to be solved during racing [70]. Recent approaches, which are capable to solve this problem in a suitable time [67, 71], introduce assumptions which contradicts basic intuition of a racecar driver, e.g. a constant friction coefficient at the complete racetrack and non-combined tire models. It is an open research question, how this additional information can be included effectively with moderate additional computational load, possibly using an online learning algorithm instead of an offline optimization problem.

Finally, a method to judge the upcoming possibilities during the planning phase based on their corresponding risks and advantages has to be defined. On one side, this is a very difficult task, while on the other side having high relevancy to series production vehicles. Costumers will not accept vehicles which do not take the chance to do an evasion maneuver, but at the same time they want them to react reasonable and not overly aggressive.

3.5 Control

We refer to the control part of the autonomous driving task, as everything which is related to tracking the path and velocity profile specified by the planning part. It is a standard approach to use a two-degrees of freedom structure to ensure sufficient control performance. One can therefore separate the problem in two steps: Feedforward control generation and feedback control.

For both there exists a variety of working concepts in the automotive sector, which have proven valid in a wide range of research and industrial applications [65, 72, 73]. However, there are still several open questions arising from the need to approach high velocities while maintaining a sufficiently good control performance in the nonlinear range of vehicle dynamics.

The latter requires a sufficient amount of feedback to suppress model uncertainty and random disturbances from influencing the control performance. In an ideal scenario, state of the art sensors could deliver the necessary information quality to apply this. However, autonomous vehicles usually rely on a variety of different sensor devices and decentralized networked postprocessing algorithms, which leads to uncertain time delays and a high chance of a single failure in the system. Since standard sensor fusion approaches pose strict assumptions (most important mean-free errors) the control performance can degrade significantly if they are not met. To illustrate the effects, figure 8 shows the influence of unknown time delay on the lateral tracking error of an autonomous vehicle in a simulation environment.

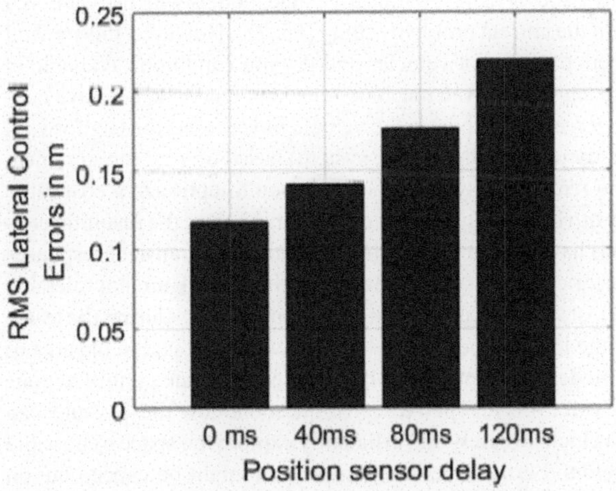

Figure 8: Lateral Control Error and sensors delay

While partial solutions to the problem are available [74–76] less effort is spend on self-configuration and self-calibration in terms of quality and delay. This makes it difficult to reach the robustness and availability necessary to be considered production level ready.

Most vehicle dynamics control systems for autonomous driving focus on the linear range of the vehicle dynamics. Well-proven algorithms for generating feedforward control signals based on kinematic or linear vehicle models leading to a sufficient control performance are widely available [30]. They show descent performance even for the nonlinear range. Nevertheless, a detailed inspection of the lateral control errors in figure 9 reveals that there are still repeating patterns in each consecutive lap. This is a clear indication that it is possible to further improve the feedforward control.

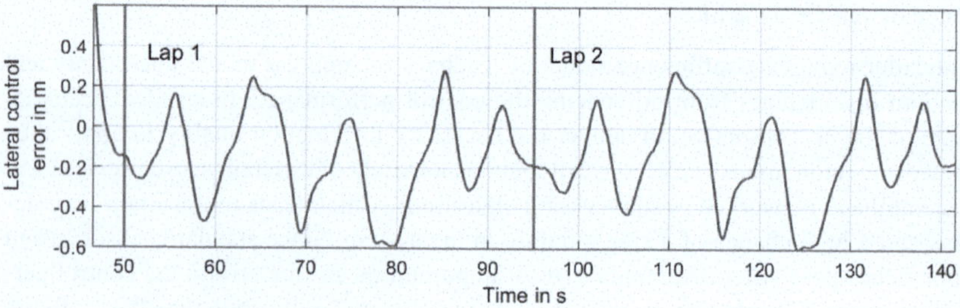

Figure 9: Lateral Control error over different laps

Several authors targeted this issue by using nonlinear vehicle models which are tailored exactly to the vehicle used in the current problem setting [65, 77]. However, this method fails to scale for the numbers of different vehicles and driving conditions available in real life. A promising approach comes from machine learning theory and adaptive control. The first, *reinforcement learning*, updates the vehicle model used for feedforward control generation (or directly updates the inverse system) based on past measurement values and directly uses the real world system in closed loop to optimize a cost functional [78]. The second, *iterative learning control*, directly minimizes the resulting control error [79]. Both methods have shown difficulties (unpredictable transient responses and convergence problems for complex systems) in application which are not tolerable for autonomous vehicles. Most severe, it is difficult to guarantee safety during the learning period. Ideas to design the learning process in further detail are found in the area of *bayesian statisics*. Partial model information can be used to determine control actions which are considered to be safe. It is an open research question, how these results can be adapted to autonomous vehicle to safely and efficiently explore the space of possible feedforward control generators. Further, the learning process might be carried out on less data and made more robust by introduction of system structure, e.g. the left turn/right turn symmetry of the differential equations could be introduced in a nonlinear learning setting. The usage of learning controllers could significantly reduce the time-to-production in series product development, while ensuring that the car adapts to slowly changing conditions, such as tire wear.

3.6 Safeguarding

An additional aim of testing autonomous software in a motorsport vehicle is the complete evaluation of the newly developed functions. The focus in this implementation is primarily on evaluating the real-time capability, performance and reliability of the functions. According to current studies, more than 100 million kilometers driven are necessary to ensure that autonomous cars are at least as safe as manually controlled cars [80]. To meet these challenges, classic test concepts for safeguarding autonomous driving functions must be combined with new innovative concepts. The following cornerstones and extensions of the existing methods are emerging in a future hedging strategy:

- Optimization of classic test concepts: All test levels (model, software, hardware and vehicle in the loop) must be used for the overall evaluation and optimally coordinated with each other. The test focus will continue to shift towards Model in the Loop (MIL) and Software in the Loop (SiL) (virtualization).
- Consistency in testing: For the efficient use of all test levels, consistency in testing is necessary, i.e. once defined or generated test scenarios must be able to be used at all levels.
- Virtualization and automation: With the increasing virtualization of security, the degree of automation can be further increased. The variation of parameters such as driving situations (e.g. inner-city traffic, motorways), weather conditions, traffic situations, etc. results in a multitude of test scenarios that cannot be mastered in reality despite virtualization and automation. Therefore, the intelligent selection of test scenarios is of decisive importance.
- Intelligent evaluation of all available data (offline testing): The consistent evaluation of trace recordings from fleet tests etc. leads to an improvement in test coverage without additional test runs. The application of statistical methods can also ensure that all relevant scenarios (e.g. with regard to Euro NCAP) are covered.
- Best practice approach: The use of state-of-the-art methods such as CI (Continous Integration) enables the efficient implementation of the development and hedging process in practice. Only then can the safety verification according to ISO 26262 be carried out continuously (traceability).

3.7 Organization & Development Process

Despite the technical problems, the motorsport environment poses additional challenges on the development workflow applied to the autonomous driving software. These can be seen a testing ground to evaluate different development methodologies for future automotive companies. First of all, limited test time and "on the point" performance requires a high level of software maturity when approaching the real car. Right now, this is difficult to achieve with classic vehicle development and software engineering processes. While the former fail to implement continuous integration and testing procedures, the latter lacks efficient methods to judge dynamics systems with several information loops closed due to algorithm interaction. It is therefore of great research interest, to design a methodology to select unit- and integration tests which allow to cover a large range of the critical scenarios for an autonomous vehicle while maintaining the set of test cases limited, to cope with limited amount of computational load and limited on-track testing time. Further, the increasing number of parameters during the use of nonlinear and complex system designs pose severe challenges in the design process, since the resulting interactions are difficult to predict.

4 Summary and Outlook

The Robocar is an electric autonomous level 5 vehicle, which is runs on the race track. From a technical point of view, the vehicle thus covers all possible extreme situations of a normal production vehicle with regard to speed, longitudinal and lateral forces as well as functionality of the algorithms. In addition, due to the repetition of the track sections, the race track offers the certainty that the developed algorithms are fully functional due to repeatability. With the start of the project, additional attention will be paid to the transferability of the developed functions to production vehicles; all knowledge and results gained can be used for the development of autonomous production vehicles.

For summing up everything, the main points why it is reasonable to do research in the field of autonomous motorsports e. g. taking part at Roborace are:

1. Autonomous level 5 vehicle: One difference to the state of the art is that in this project a level 5 vehicle is used. Thus, all the results obtained represent a strong innovative leap for vehicles that will not be on the road for 5-10 years. The knowledge generated in this project thus serves for future autonomous vehicles.

2. Sensors and hardware: The sensors used in the Robocar are similar to the hardware used in production vehicles. Future road vehicles will be having GPU based ECUs similar to the Nvidia Drive PX2. Thus, transfer, scaling and parameterization of the developed software for the Robocar is possible for future road vehicles.

3. Robustness of the algorithms: The use in racing cars involves high speeds, which means that the speed, real-time capability and robustness of the algorithms must be checked. It is to be expected that the algorithms developed will set a milestone in future software development for road vehicles. The protection at high speeds guarantees a function in city traffic at low speeds.
4. Electric powertrain: The Robocar has, as already existing and planned for future series production vehicles, a purely electric powertrain consisting of traction battery and electric motors. The new combination of potential methods for energy saving and energy recovery developed in this project enables progress beyond established technologies and can be possibly transferred to electric drive trains in passenger cars.
5. Planning of the trajectory: In all autonomous passenger cars a safe planning of the trajectory must take place. The decision algorithm for the choice of trajectory, which considers not only the primary goal of collision-free driving but also secondary and possibly temporary targets, such as maximizing the range, the fastest possible completion of a section of the track or overtaking maneuver, can be incorporated directly into a production vehicle.

Friction potential: The function developed for detecting and predicting the friction potential is a function for increasing safety in the vehicle. If the corresponding sensors are installed in the vehicle, this function can be transferred to a production vehicle.

Contributions

Johannes Betz initiated the idea of this paper and contributed essentially to the structure of this paper and mainly to the state of the art, the AI-development, perception and safeguarding part of this paper. Alexander Wischnewski contributed essentially to the control and planning part of this paper. Alexander Heilmeier, Felix Nobis, Tim Stahl, Leonhard Hermansdorfer and Boris Lohmann contributed equally to different parts of this paper. Markus Lienkamp made an essential contribution to the conception of the research project. He revised the paper critically for important intellectual content. He gave final approval of the version to be published and agrees to all aspects of the work. As a guarantor, he accepts the responsibility for the overall integrity of the paper.

Acknowledgements

First of all we want to thank you Roborace and the complete Roborace team for giving us the opportunity to work with them and using their Devbot vehicles for our research. We would like to thank the *Bayrische Forschungsstifung* for funding us in the research project *rAIcing*. We would like to thank *Conti Engineering Service* and the *TÜV Süd* for funding us with research projects. The work described in this paper was also conducted with basic research fund of the Institute of Automotive Technology from the Technical University of Munich. In addition, we want to thank the company *in-tech* for helping us with the speedgoat software development.

References

[1] Roborace, *Robocar.* [Online] Available: http://roborace.com/. Accessed on: May 07 2018.

[2] e-formel.de, *Roborace – die Rahmenserie der Formel E.* [Online] Available: https://www.e-formel.de/roborace.html. Accessed on: May 07 2018.

[3] Verband der Automobilindustrie (VDA), *Automatisiertes Fahren.* [Online] Available: https://www.vda.de/de/themen/innovation-und-technik/automatisiertes-fahren/automatisiertes-fahren.html. Accessed on: Sep. 11 2017.

[4] R. Zakharenko, "Self-driving cars will change cities," *Regional Science and Urban Economics*, vol. 61, pp. 26–37, 2016.

[5] SAE International, "Automated Driving – Levels of Driving Automation," [Online] Available: https://www.sae.org/misc/pdfs/automated_driving.pdf.

[6] Mercedes Benz, "Aktiver Park-Assistent," 2017. [Online] Available: https://m.mercedes-benz.de/techcenter/de_DE/active_parking/detail.html. Accessed on: May 07 2018.

[7] Volkswagen AG, "Spurhalteassistent "Lane Assist"," 2017. [Online] Available: http://de.volkswagen.com/de/innovation-technik/technik-lexikon/spurhalteassistentlaneassist.html. Accessed on: May 07 2018.

[8] Tesla, "Autopilot," 2017. [Online] Available: https://www.tesla.com/de_DE/autopilot. Accessed on: May 07 2018.

[9] t3n – Digital Pioniers, *Audi A8 soll „erstes Serienautomobil der Welt" mit Level-3-Automation sein.* [Online] Available: http://t3n.de/news/audi-a8-level-3-autonomes-fahren-837581/. Accessed on: May 07 2018.

[10] Audi, *So funktioniert der Audi AI Staupilot.* [Online] Available: http://blog.audi.de/2017/07/17/audi-ai-staupilot/. Accessed on: May 07 2018.

[11] Mercedes Benz, *S-Klasse Limousine – Mercedes-Benz Intelligent Drive.* [Online] Available: https://www.mercedes-benz.de/content/germany/mpc/mpc_germany_website/de/home_mpc/passengercars/home/new_cars/models/s-class/w222/facts_/comfort/intelligentdrive.html. Accessed on: May 07 2018.

[12] Audi, *Pilotiertes Fahren.* [Online] Available: https://www.audi-mediacenter.com/de/pilotiertes-fahren-3651. Accessed on: May 07 2018.

[13] Audi, *Audi RS7 piloted driving concept fährt selbständig Rekordzeit auf spanischer Rennstrecke.* [Online] Available: https://www.audi-mediacenter.com/de/pressemitteilungen/audi-rs-7-piloted-driving-concept-faehrt-selbstaendig-rekordzeit-auf-spanischer-rennstrecke-5161. Accessed on: May 07 2018.

[14] DARPA, *Urban Challenge.* [Online] Available: http://archive.darpa.mil/grandchallenge/. Accessed on: May 07 2018.

[15] Waymo, *Waymo – Self Driving Car.* [Online] Available: https://waymo.com/. Accessed on: May 07 2018.

[16] NVIDIA, *Driving Innovation – NVIDIA Automotive.* [Online] Available: http://www.nvidia.com/object/drive-automotive-technology.html. Accessed on: May 07 2018.

[17] M. Bojarski *et al.,* "End to End Learning for Self-Driving Cars," [Online] Available: http://arxiv.org/pdf/1604.07316v1. Accessed on: May 07 2019.

[18] drive.ai, *drive.ai.* [Online] Available: https://www.drive.ai/. Accessed on: May 07 2018.

[19] Stanford University, *Center for Automotive Research at Stanford.* [Online] Available: https://cars.stanford.edu/. Accessed on: May 07 2018.

[20] D. Held, J. Levinson, and S. Thrun, "Precision tracking with sparse 3D and dense color 2D data," in *2013 IEEE International Conference on Robotics and Automation (ICRA)*, Karlsruhe, Germany, pp. 1138–1145.

[21] D. Held, J. Levinson, S. Thrun, and S. Savarese, "Combining 3D Shape, Color, and Motion for Robust Anytime Tracking," 2014.

[22] J. Levinson and S. Thrun, "Robust Vehicle Localization in Urban Evnironments Using Probabilistc Maps," in *2010 IEEE International Conference on Robotics and Automation*, 2005.

[23] D. Yamamoto and N. Suganuma, “Localization for autonomous driving on urban road,” in *2015 International Conference on Intelligent Informatics and Biomedical Sciences (ICIIBMS)*, Okinawa, Japan, pp. 452–453.

[24] University of Leeds, *Driverless Cars.* [Online] Available: http://www.leeds.ac.uk/site/custom_scripts/spotlight/driverless-cars.php. Accessed on: May 07 2018.

[25] University of California Berkeley, *Berkeley Deep Drive.* [Online] Available: https://deepdrive.berkeley.edu/. Accessed on: May 07 2018.

[26] Deutsches Zentrum für Luft- und Raumfahrt e. V. (DLR), *Forschungsprojekt Pegasus: AUTOMATISIERTES FAHREN EFFEKTIV ABSICHERN.* [Online] Available: http://pegasus-projekt.info/de/about-PEGASUS. Accessed on: May 07 2018.

[27] Freie Universität Berlin, *AutoNOMOS Labs – Autonoums Cars from Berlin.* [Online] Available: http://autonomos-labs.com/. Accessed on: May 07 2018.

[28] University of Toronto, *aUToronto.* [Online] Available: https://www.autodrive.utoronto.ca/. Accessed on: May 07 2018.

[29] Massachussetts Institute of Technology (MIT), *Deep Learning for Self-Driving Cars.* [Online] Available: http://selfdrivingcars.mit.edu/. Accessed on: May 07 2018.

[30] S. Pendleton *et al.,* “Perception, Planning, Control, and Coordination for Autonomous Vehicles,” *Machines*, vol. 5, no. 1, p. 6, 2017.

[31] Udacity, *Udacity – Self Driving Nano Degree: Term 3 – Autonomous Vehicle Architecture.* [Online] Available: https://de.udacity.com/course/self-driving-car-engineer-nanodegree--nd013. Accessed on: May 07 2018.

[32] Z. Kim, “Robust Lane Detection and Tracking in Challenging Scenarios,” *IEEE Trans. Intell. Transport. Syst.*, vol. 9, no. 1, pp. 16–26, 2008.

[33] D.-K. Lee *et al.,* “Real-time lane detection and tracking system using simple filter and Kalman filter,” in *2017 Ninth International Conference on Ubiquitous and Future Networks (ICUFN)*, Milan, pp. 275–277.

[34] J. C. McCall and M. M. Trivedi, “Video-Based Lane Estimation and Tracking for Driver Assistance: Survey, System, and Evaluation,” *IEEE Trans. Intell. Transport. Syst.*, vol. 7, no. 1, pp. 20–37, 2006.

[35] A. Mukhtar, L. Xia, and T. B. Tang, “Vehicle Detection Techniques for Collision Avoidance Systems: A Review,” *IEEE Trans. Intell. Transport. Syst.*, vol. 16, no. 5, pp. 2318–2338, 2015.

[36] A. Milstein, "Occupancy Grid Maps for Localization and Mapping," in *Planning for Unraveling Deformable Linear Objects Based on Their Silhouette*, H. Wakamatsu, E. Arai, and S. Hirai, Eds.: INTECH Open Access Publisher, 2008, pp. 389–409.

[37] NVIDIA, *Nvidia Drive PX 2.* [Online] Available: http://www.nvidia.de/object/drive-px-de.html. Accessed on: May 07 2018.

[38] Speedgoat, *Mobile real-time target machine.* [Online] Available: https://www.speedgoat.com/products-services/real-time-target-machines/mobile. Accessed on: May 07 2018.

[39] M. Quigley *et al.,* "ROS: an open-source Robot Operating System," in *ICRA Workshop on Open Source Software*, 2009.

[40] Wikipedia, *Motorsport.* [Online] Available: https://de.wikipedia.org/wiki/Motorsport. Accessed on: May 07 2018.

[41] T. M. Mitchell, *Machine learning*. New York, NY: McGraw-Hill, 2010.

[42] A. C. Müller and S. Guido, *Introduction to machine learning with Python: A guide for data scientists*. Sebastopol, CA: O'Reilly Media, 2017.

[43] N. J. Nilsson, "Introduction to Machine Learning: An earyl draft of a proposed textbook," 1998. [Online] Available: http://robotics.stanford.edu/~nilsson/MLBOOK.pdf. Accessed on: May 07 2018.

[44] D. J. C. MacKay, *Information theory, inference, and learning algorithms*. Cambridge: Cambridge Univ. Press, 2011.

[45] The MathWorks, *What is Machine Learning?: 3 Things you need to know.* [Online] Available: https://de.mathworks.com/discovery/machine-learning.html?s_eid=PSM_da&hootPostID=9a946fbd13234ab6ade27d81799d17ba. Accessed on: May 07 2018.

[46] Krizhevsky Alex, I. Sutskever, and G. Hinton, "ImageNet Classification with Deep Convolutional Neural Networks," 2012.

[47] J. Leonard *et al.,* "A perception-driven autonomous urban vehicle," *J. Field Robotics*, vol. 25, no. 10, pp. 727–774, 2008.

[48] B. Ranft and C. Stiller, "The Role of Machine Vision for Intelligent Vehicles," *IEEE Trans. Intell. Veh.*, vol. 1, no. 1, pp. 8–19, 2016.

[49] L. A. Marina, F. D. Moldoveanu, and S. M. Grigorescu, "Environment perception in racing simulators using Deep Neural Networks," in *2017 International Conference on Optimization of Electrical and Electronic Equipment (OPTIM) & 2017 Intl Aegean Conference on Electrical Machines and Power Electronics (ACEMP)*, Brasov, Romania, pp. 1102–1107.

[50] D. M. Filatov, K. V. Ignatiev, and E. V. Serykh, "Neural network system of traffic signs recognition," in *2017 XX IEEE International Conference on Soft Computing and Measurements (SCM)*, Saint Petersburg, Russia, pp. 422–423.

[51] W. Hu, Q. Zhuo, C. Zhang, and J. Li, "Fast Branch Convolutional Neural Network for Traffic Sign Recognition," *IEEE Intell. Transport. Syst. Mag.*, vol. 9, no. 3, pp. 114–126, 2017.

[52] Q. Hu, H. Wang, T. Li, and C. Shen, "Deep CNNs With Spatially Weighted Pooling for Fine-Grained Car Recognition," *IEEE Trans. Intell. Transport. Syst.*, pp. 1–10, 2017.

[53] X. Wang *et al.,* "Capturing Car-Following Behaviors by Deep Learning," *IEEE Trans. Intell. Transport. Syst.*, pp. 1–11, 2017.

[54] F. Gonda *et al.,* "ICON: An interactive approach to train deep neural networks for segmentation of neuronal structures," in *2017 IEEE 14th International Symposium on Biomedical Imaging (ISBI 2017)*, Melbourne, Australia, pp. 327–331.

[55] J. Chen, H. M. Le, P. Carr, Y. Yue, and J. J. Little, "Learning Online Smooth Predictors for Realtime Camera Planning Using Recurrent Decision Trees," in *2016 IEEE Conference on Computer Vision and Pattern Recognition (CVPR)*, Las Vegas, NV, USA, pp. 4688–4696.

[56] R. Simmons, B. Browning, Y. Zhang, and V. Sadekar, "Learning to Predict Driver Route and Destination Intent," in *2006 IEEE Intelligent Transportation Systems Conference*, Toronto, ON, Canada, pp. 127–132.

[57] C. Szegedy *et al.,* "Going deeper with convolutions," in *2015 IEEE Conference on Computer Vision and Pattern Recognition (CVPR)*, Boston, MA, USA, pp. 1–9.

[58] S. Thrun, Ed., *Robotic Mapping: A Survey: Exploring artificial intelligence in the new millennium.* San Diego, CA: Morgan Kaufmann Publishers, 2003.

[59] J. O. Wallgrün, "Robot Mapping," in *Hierarchical Voronoi Graphs*, J. O. Wallgrün, Ed., Berlin, Heidelberg: Springer Berlin Heidelberg, 2010, pp. 11–43.

[60] R. Mazl and L. Preucil, “Building a 2D environment map from laser range-finder data,” in *IV 2000 Intelligent Vehicles Symposium*, Dearborn, MI, USA, Oct. 2000, pp. 290–295.

[61] A. R. Corregedor, J. Meyer, and F. Du Plessis, “Design principles for 2D local mapping using a laser range finder,” in *AFRICON 2011*, Victoria Falls, Livingstone, Zambia, pp. 1–6.

[62] Y. Abdelrasoul, A. B. S. H. Saman, and P. Sebastian, “A quantitative study of tuning ROS gmapping parameters and their effect on performing indoor 2D SLAM,” in *2016 2nd IEEE International Symposium on Robotics and Manufacturing Automation (ROMA)*, Ipoh, Malaysia, pp. 1–6.

[63] M. Rojas-Fernandez, D. Mujica-Vargas, M. Matuz-Cruz, and D. Lopez-Borreguero, “Performance comparison of 2D SLAM techniques available in ROS using a differential drive robot,” in *2018 International Conference on Electronics, Communications and Computers (CONIELECOMP)*, Cholula, pp. 50–58.

[64] P. Morin and C. Samson, “Motion Control of Wheeled Mobile Robots,” in *Springer Handbook of Robotics*, B. Siciliano and O. Khatib, Eds., Berlin, Heidelberg: Springer Berlin Heidelberg, 2008, pp. 799–826.

[65] M. Werling, *Ein neues Konzept für die Trajektoriengenerierung und -stabilisierung in zeitkritischen Verkehrsszenarien*. Zugl.: Karlsruhe, KIT, Diss., 2010. Hannover, Karlsruhe: Technische Informationsbibliothek u. Universitätsbibliothek; KIT Scientific Publishing, 2011.

[66] A. Katriniok and D. Abel, “LTV-MPC approach for lateral vehicle guidance by front steering at the limits of vehicle dynamics,” in *2011 50th IEEE Conference on Decision and Control and European Control Conference (CDC-ECC 2011)*, Orlando, FL, USA, pp. 6828–6833.

[67] A. Liniger, A. Domahidi, and M. Morari, “Optimization-based autonomous racing of 1: 43 scale RC cars,” *Optim. Control Appl. Meth.*, vol. 36, no. 5, pp. 628–647, 2015.

[68] S. Glaser, B. Vanholme, S. Mammar, D. Gruyer, and L. Nouveliere, “Maneuver-Based Trajectory Planning for Highly Autonomous Vehicles on Real Road With Traffic and Driver Interaction,” *IEEE Trans. Intell. Transport. Syst.*, vol. 11, no. 3, pp. 589–606, 2010.

[69] B. Schurmann *et al.*, “Ensuring drivability of planned motions using formal methods,” in *2017 IEEE 20th International Conference on Intelligent Transportation Systems (ITSC)*, Yokohama, pp. 1–8.

[70] L. Cardamone, D. Loiacono, P. L. Lanzi, and A. P. Bardelli, "Searching for the optimal racing line using genetic algorithms," in *2010 IEEE Symposium on Computational Intelligence and Games (CIG)*, Copenhagen, Denmark, pp. 388–394.

[71] R. Verschueren, S. de Bruyne, M. Zanon, J. V. Frasch, and M. Diehl, "Towards time-optimal race car driving using nonlinear MPC in real-time," in *2014 IEEE 53rd Annual Conference on Decision and Control (CDC)*, Los Angeles, CA, USA, pp. 2505–2510.

[72] S. Thrun *et al.,* "Stanley: The robot that won the DARPA Grand Challenge," *J. Field Robotics*, vol. 23, no. 9, pp. 661–692, 2006.

[73] D. Calzolari, B. Schurmann, and M. Althoff, "Comparison of trajectory tracking controllers for autonomous vehicles," in *2017 IEEE 20th International Conference on Intelligent Transportation Systems (ITSC)*, Yokohama, pp. 1–8.

[74] K. Hausman, S. Weiss, R. Brockers, L. Matthies, and G. S. Sukhatme, "Self-calibrating multi-sensor fusion with probabilistic measurement validation for seamless sensor switching on a UAV," in *2016 IEEE International Conference on Robotics and Automation (ICRA)*, Stockholm, Sweden, pp. 4289–4296.

[75] S. C. Patwardhan, S. Narasimhan, P. Jagadeesan, B. Gopaluni, and S. L. Shah, "Nonlinear Bayesian state estimation: A review of recent developments," *Control Engineering Practice*, vol. 20, no. 10, pp. 933–953, 2012.

[76] B. Sinopoli *et al.,* "Kalman Filtering With Intermittent Observations," *IEEE Trans. Automat. Contr.*, vol. 49, no. 9, pp. 1453–1464, 2004.

[77] L. König, *Ein virtueller Testfahrer für den querdynamischen Grenzbereich.* Zugl.: Stuttgart, Univ., Diss., 2009. Renningen: Expert-Verl., 2009.

[78] F. Lewis, D. Vrabie, and K. Vamvoudakis, "Reinforcement Learning and Feedback Control: Using Natural Decision Methods to Design Optimal Adaptive Controllers," *IEEE Control Syst.*, vol. 32, no. 6, pp. 76–105, 2012.

[79] D. A. Bristow, M. Tharayil, and A. G. Alleyne, "A survey of iterative learning control," *IEEE Control Syst. Mag.*, vol. 26, no. 3, pp. 96–114, 2006.

[80] H. WInner, S. Hakuli, F. Lotz, and C. Singer, Eds., *Handbuch Fahrerassistenzsysteme: Grundlagen, Komponenten und Systeme für aktive Sicherheit und Komfort,* 3rd ed. Wiesbaden: Springer Vieweg, 2015.

Assisted and autonomous driving on driving simulators

Mattia Bruschetta,
Senior Research Fellow, Department of Information Engineering,
University of Padova, Italy

Diego Minen,
Technical Director, VI-grade

P. Pfeffer (Ed.), *9th International Munich Chassis Symposium 2018*, Proceedings,
https://doi.org/10.1007/978-3-658-22050-1_13

Abstract

With the upcoming diffusion of autonomous vehicles, a substantial modification of the human role in the driving action is taking place. While most of the effort has been put on making the car capable of safely moving in a complex environment, the human in the control loop is becoming a critical problem: on the one hand, the driver attention in driving actions is necessary to guarantee safety in any conditions, on the other hand the driver comfort has to be considered to make the driving experience satisfactory. In this paper balancing between these two aspects is effectively investigated by means of dynamic driving simulators, particularly addressing the impact of vehicle dynamic setup.

The target user for these applications is a non-professional driver, which does not easily fit into the virtual environment. Proprietary Active Seat (AS) and Active Belts (AB) technologies are used in the driving simulator to reduce the gap between real and virtual environment. Advanced Multi-Sensory Motion Cueing Algorithm based on Nonlinear Model Predictive Control technique is used to coordinate somatosensory stimulation (AS/AB) with the vestibular one (Motion Cues).

Moreover, cognitive aspects have to be considered to collect information regarding driver attention and comfort, which are not retrievable by motion analysis. To this aim, an advanced biotelemetry device is used to collect driver's Heart Rate Variability and Skin Potential Response, which are recognized bio-signal for stress load, and that are processed by means of statistical tools to infer cognitive features. Finally, practical use cases will be presented analysing the effects of different vehicle setups on human perception while experiencing autonomous driving on the simulator.

Preliminary

It's known that OEM's attention in the development of autonomous vehicles is already very high now and will increase more and more in the immediate future. Sensor engineering and their fusion is just one of aspects of the entire problem. Among others and just from a general point of view, important areas of research being vehicle-environment simulation and its spatial resolution (micro-macro-mesa), V2X communication, trajectory planning and tracking, human-robot co-existence, human monitoring.

If we consider the aspects more related to vehicle engineering, autonomous driving is a much more challenging task than one could think it to be. It is likely that it's not just a matter of connecting sensors and actuators to the system in order to have it moving autonomously, but instead it will be essential to study in depth all aspects that relates humans in contact with robotized systems. In this respect, a few points are of fundamental importance.

Robotization acceptance (uncanny valley)

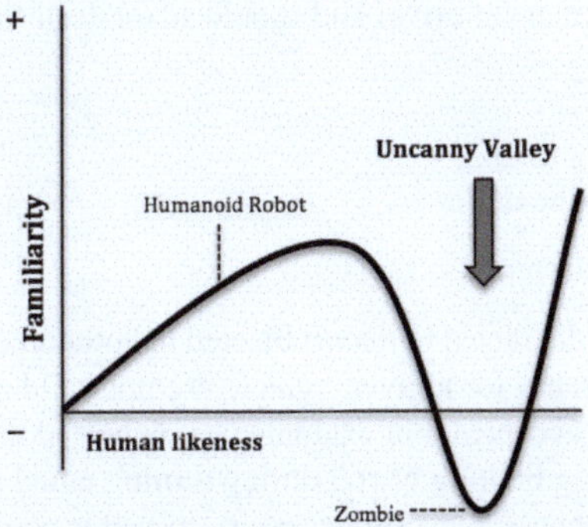

Fig 1: Uncanny valley (Picture taken from [19])

The uncanny valley principle is a known and generic problem of robotization. Basically, it states that human beings are pleased with robot actions when they are repeatable and expectable, maybe simple but consistent. Whenever a robotic action becomes weird, or the robot behaves like an (even only apparently) inconsistent human, then the level of trust to the machine decreases and the human tends to monitor the robot without trusting it blindly anymore. If we apply the principle to robotized driving, this means on one hand that the human will again need to focus visually to what happens around, therefore nullifying the efforts that are supposed to be made to make the car comfortable under unmanned control; on the other, the optimization of comfort will require a special monitoring of the human action while being driven around by the autonomous system: it is possible that an adaptive and progressive automatic setup of the vehicle dynamic characteristics is required during the familiarization phase of the robotic actions.

Human-robot transition

It is very important (especially for the non-US market of autonomous vehicles, which seems to definitely pass through a few years of human-robot co-existence) to study all the possible combinations of human robot interactions, on the one hand to make the consequences of a natural contemporary (human, robot) action manageable, on the other to study what could be accepted by the (distracted) human in terms of robot manoeuvring. Vehicle dynamics is very important in these cases: under autonomous control, the feeling of the car should be natural; under conflicting action the human should override the robot, and the feeling should be the one that the human expects to have when fully empowered in control. For example, when the human grabs the steering wheel control from the robot, special attention should be dedicated to potential dynamic overshoot due to an excessive overriding effort. Moreover, the robot should learn how the human in that particular car drives and should adapt itself to that style, with specific reference to:

- Handling actions
- Handling cues to the driver
- Ride cues to the driver

It is possible that, for a distracted human subjected to autonomous driving several rules of ideal handling/ride setup for a given vehicle are not valid as they have been traditionally and commonly accepted. For example, it is expected that under robotic driving an excessive pitch during braking or roll during turning could cause discomfort, while under human driving control they are necessary to better perceive vehicle dynamics. Similarly, it is possible that low frequency (heave, pitch, roll) compensation under robotic driving could reduce the excitation of the vestibular system of a distracted human.

Simulation and Driving Simulators

All the aspects mentioned above would require enormous resources for being studied in depth in real life, with all the risks associated having a real robotized vehicle moving in all the several needed traffic scenarios.

It is of course possible to intensively use traditional off-line vehicle simulation for assessing most of the principles for best tuning the vehicle taking into account the new requirements, for maximizing (passive) driver and passenger comfort. The metrics for deciding whether that tuning is the "best" is not completely clear, especially due to lack of experience for such robotized vehicles. One fundamental difference with respect normal human driving is that the visual control is enormously reduced (the driver/passenger watches the external scenario with a much reduced level of attention, if any). One other big difference is that the driver might not havehands on the steering wheel, and if yes , the steering (when still existing) could rotate under robotic control. As a consequence, vehicle dynamics is not any more dominantly perceived with visual cues from the scene and haptic ones from the steering, but rather from body contact to the seat and limb contact to the interior of the car, which traditionally are more cues that expert drivers use. The vestibular and somato-sensorial systems are mainly responsible to perceive the vehicle motion and are the major source of cues for the guests of a robotized car.

VI-grade and University of Padova (DEI) propose a revolutionary approach to the problem of driver/passenger comfort maximization under autonomous vehicle control:

o Application of some of the principles used for developing VI-MotionCueing (including the body contact effects as described below)as described above), to minimize the difference between motion perception in a car and on a driving simulator;

o Assessment of the driver/passenger comfort by monitoring some key psycho-pysiological parameters during driving session on a driving simulator, comparing self to robot driving in the same scenarios.

Integrated Vestibular/Somatosensory Motion Cueing

The effectiveness of driving simulators is strongly related to the quality of driver's motion perceptions, hence motion control algorithm must generate both realistic and feasible inputs to the platform. Such strategies are called Motion Cueing Algorithms (MCAs). A promising approach is that of designing MCAs based on Model Predictive Control (MPC) technique, that helps in efficiently handling platform workspace and is very effective in reproducing sense of motion, [2], [3], [4], [6], [8]. Differently from the typical usage of driving simulators, the target user for autonomous driving applications is the non-professional driver, which does not fit into the virtual environment easily [1]. Active Seat (AS) with integrated Active Belts (AB) have been introduced to reduce the gap between real and virtual environment in simulators, by providing information about sustained accelerations to the driver. These accelerations cannot be simply reproduced by the motion platform, due to its intrinsic mechanical limits. The AS/AB working principle can be seen as a somatosensory stimulation, used as a haptic feedback [13]: the pressure is interpreted by the brain as an added information on the vehicle status. The greater is the realism in the pressure reproduction and the easier should be the capability of the driver in using this information, thanks to a reduction of the perceptual conflict [14]. The same idea has been widely exploited for flight simulators [16], where AS (G-seat) is generally applied into a static environment. The typical control strategy of such tools is almost straightforward: pressure is directly associated with accelerations on the specific directions. In dynamic simulation this approach results to be misleading. As an example, let us consider a giant simulator [18], [9]: the available work-space allows to reproduce part of the sustained accelerations by means of the motion itself, and the applied pressure/tension on the AS/AB should be coordinated with that. More in general, the AS/AB has to adapt in real time to the undergoing motion behaviour.

In this paper, we propose an NMPC based Multi-Sensory Cueing Algorithm (MSCA), containing a model of a seated human body subjected to accelerations and coupled with the vestibular system. A nonlinear model is developed, taking into account the inertial body reaction, the frictions between body and seat and the damping effect. The model is then adapted to be used into an NMPC framework. An ad-hoc implementation is provided to reach real time performance. A scheme of the procedure is in Fig. 2:

1) the vehicle translational accelerations and rotational velocities {a,v} are computed by using a dynamical simulation engine;

2) signals {a,v} are then pre-processed (according to given application/performance objectives);

3) the vestibular/pressure model is used to compute the reference for the controller;

4) the platform displacements and the AS/AB pressures/tensions are computed by means of an NMPC controller, and then given as reference inputs to the motion controller.

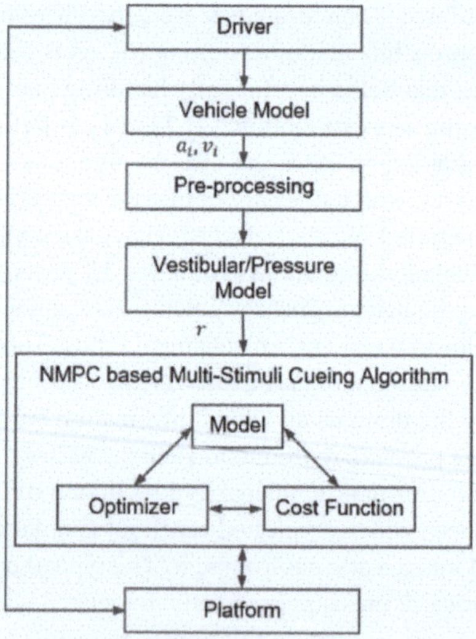

Fig. 2: Scheme of Multi-Stimuli Cueing Algorithm

The advantages of such approach are manifold:

1) the platform motion, the AS and the AB are perfectly coordinated thanks to a coupled model;

2) the usage of an optimization based controller allows to have an haptic stimulus that is as closer as possible to the real one, improving the overall realism;

3) the sickness could be reduced even for non-professional drivers, extending the class of potential users; 4) the AS/SB systems do not require to be re-tuned every time the motion strategy is modified.

Reference Motion Platforms

The algorithm has been designed to be applied to the Driver in Motion (DiM) family of simulation platforms: DiM 150 and DiM.C9/700 (see [5] for a detailed description of the DiM150). DiM150 is based on a mechanical architecture with redundant DOF: the simulator consists of a hexapodal structure mounted on a tripod frame, which moves on a flat, stiff surface sliding on airpads. The planar tripod is used to produce most of the longitudinal, lateral, and yaw sliding movements, whereas the hexapod is used for pitch, roll, vertical, and smaller longitudinal, lateral, and yaw movements. DiM.C9 is designed with the same hexapod and air-pads system, replacing the tripod with a central pulley called "discframe" driven by cables. DiM150 tripod cover a range of ±0.75[m] while the DiM.C9 specific model used for this example features a range of ±3.5[m]and is presented in fig. 3.

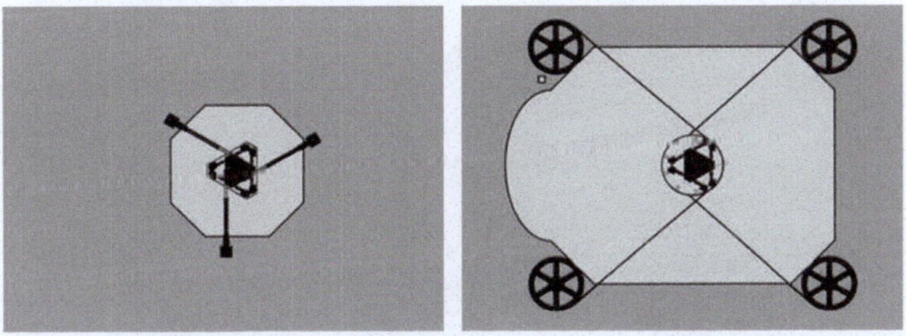

(a) **(b)**

Fig. 3: top view sketch of DiM150 and DiM.C9/700.

Active Seat and Active Belts Systems

A regular or special car passive seat is converted into active by mean of eight to ten air bladders which are inflated with compressed air via proportional valves control (see Fig. 4) properly installed in the structure of the seat.

The AB are 3, 5 or 6 points belts tensioned by means of fluidic muscles controlled by pneumatic valves. The bladders are designed to have a distributed contact area and placed to act on the body similarly to what happens in reality. Proportional valves are used to have a progressive and continuous variation of pressure, which can be controlled in the range [0-1.5] bar for the seat and [0-8] bar for the muscles.

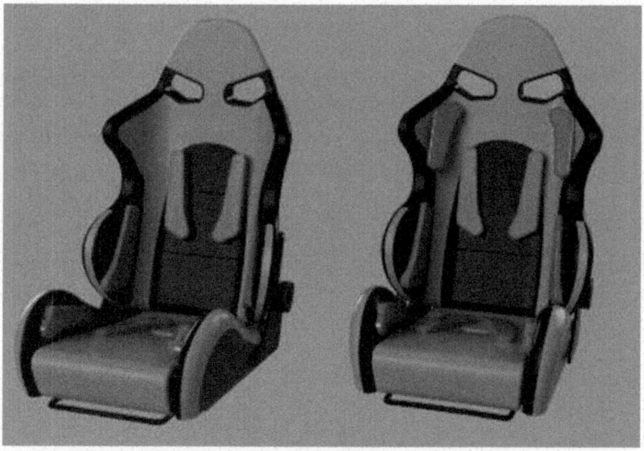

(a)

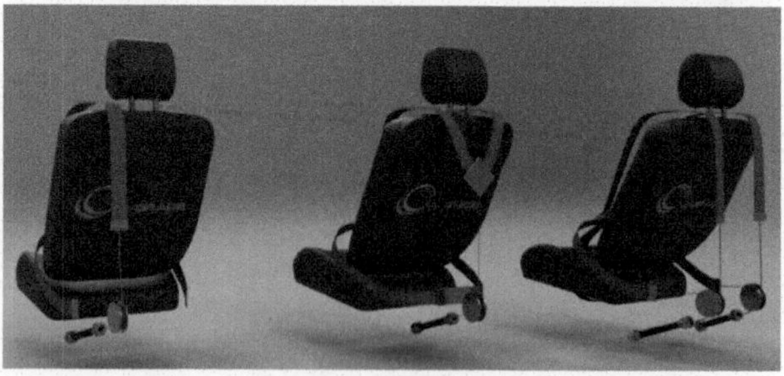

(b)

Fig. 4: Active seat and active belts system

Model for Control

The Inertial body reaction, the friction between body and seat, the seat material, the nonlinear stiffness and damping effect of the body have been considered in the model formulation. A combined vestibular/somatosensory model for control is then proposed, to be used both to compute the perceived pressure/acceleration and within an NMPC based control implementation (see scheme in Fig. 2). The proposed model neglects the pressure on legs (bladders 1 and 6) and on glutes (bladders 7 and 8), although an extension of the model to consider those elements can be derived adopting the same principles.

Lateral Pressure model

The lateral dynamic of the body is characterized by means of a mass-spring-damper model, represented by the following differential equation:

$$m\ddot{d}_y + c(d_y)\dot{d}_y + k(d_y)d_y = ma_y$$

where d_y, $\dot{d}_y$ and $\ddot{d}_y$ represent respectively the position, velocity and acceleration of the center of mass of the trunk along the lateral direction; m is the driver mass which is subjected to lateral acceleration; $c(d_y)$ is a nonlinear viscous damping coefficient; $k(d_y)$ is a nonlinear stiffness; ma_y is the external force that acts on the human body, caused by the lateral acceleration. The underlying idea is that the inertial properties of the body are considered in the mass, whereas the nonlinear stiffness/damping are used to emulate the seat elastic/viscous reaction, and $k(d_y)d_y + c(d_y)\dot{d}_y$ can be seen as the overall contact force on the trunk.

To determine the
value of m, let us define M as the mass of the driver trunk, including arms and hands, which is assumed to be about the 67% of total body weight (see [11] for details on body-mass distribution). The trunk is considered as a vertical rod of length L and center of gravity l, which rotates around the bottom edge of an angle α. According to this, the

$$I\ddot{\alpha} = F_y l$$

conservation of angular momentum law can be simplified as

The final model results to be

$$\begin{cases} \ddot{d}_y &= -\frac{c(d_y)}{m}\dot{d}_y - \frac{k(d_y)}{m}d_y - \frac{F_f(x_f)}{m} + a_y + g\phi \\ \dot{x}_f &= \dot{d}_y - \frac{|\dot{d}_y|}{h(d_y,a_x)}x_f \\ y_{p,y} &= \frac{(k(d_y)-k_0)}{A}d_y + \frac{(c(d_y)-c_0)}{A}\dot{d}_y + \bar{u}_{p,y} \end{cases}$$

where $k(d_y)$ and $c(d_y)$ have the form

$$k(d_y) = k_p d_y^p + k_0,$$
$$c(d_y) = c_p d_y^p + c_0$$

and x_f defined the friction dynamic as described in [7].

Longitudinal Pressure Model

The model that considers longitudinal acceleration describes the human body as free to move in forward direction and pressed backwards on the seat in case of positive longitudinal acceleration. In the former situation the AB system has to provide the correct tension to the driver, in the latter the task has to be accomplished by the bladders on the back. Since no significant friction phenomena between driver and seat occur, the pressure exerted by the driver's body has been determined through a linear model.

Pressure Vestibular-Coupled Model

To obtain a unique model describing both the motion and pressure perceptions, longitudinal and lateral models have to be coupled with a vestibular model. We adopt the same vestibular model used in [6] to develop a fast MPC based MCA, which is briefly reported in the following.

$$\begin{cases} \dot{\mathbf{x}}_V = A_V \mathbf{x}_V + B_V \mathbf{u}_V \\ \mathbf{y}_V = C_V \mathbf{x}_V + D_V \mathbf{u}_V \end{cases}$$

with state, input and output vectors

$$\mathbf{x}_V = \begin{bmatrix} \mathbf{x}_s^T & \mathbf{x}_{\bar{o}}^T & \mathbf{x}_I^T \end{bmatrix}^T \in \mathbb{R}^{21}$$
$$\mathbf{u}_V = \begin{bmatrix} a_x & a_y & a_z & \dot{\phi} & \dot{\theta} & \dot{\psi} \end{bmatrix}^T \in \mathbb{R}^{6}$$
$$\mathbf{y}_V = \begin{bmatrix} \mathbf{y}_s^T & \mathbf{y}_{\bar{o}}^T & \boldsymbol{\beta}^T & \mathbf{x}_I^T & \dot{\boldsymbol{\beta}}^T \end{bmatrix}^T \in \mathbb{R}^{18}$$

The state equation of the overall NMPC model results to be

$$\begin{cases} \ddot{d}_y &= -\frac{c(d_y)}{m}\dot{d}_y - \frac{k(d_y)}{m}d_y - \frac{F_f(x_f)}{m} + a_y + g\phi \\ \dot{x}_f &= \dot{d}_y - \frac{|\dot{d}_y|}{h(d_y,a_x)}d_y \\ \ddot{d}_x &= -\frac{c}{m}\dot{d}_x - \frac{k}{m}d_x + a_x + g\theta \\ \dot{\mathbf{x}}_V &= A_V\mathbf{x}_V + B_V\mathbf{u}_V \end{cases}$$

and the input and output vectors are, respectively,

$$\mathbf{u} = \begin{bmatrix}\bar{u}_{p,y} & \bar{u}_{p,x} & \mathbf{u}_V^T\end{bmatrix}^T \in \mathbb{R}^8$$
$$\mathbf{y} = \begin{bmatrix}y_{p,y} & y_{p,x} & \mathbf{y}_V^T\end{bmatrix}^T \in \mathbb{R}^{20}$$

Non linear MPC Implementation

The following optimal control problem (OCP) is solved on-line at each sampling time

$$\begin{aligned} \min_{\mathbf{x}(\cdot),\mathbf{u}(\cdot)} \quad & \int_{t_0}^{t_f} \phi(t,\mathbf{x}(t),\mathbf{u}(t))\mathrm{dt} \\ s.t. \quad & \dot{\mathbf{x}}(t) = f(\mathbf{x}(t),\mathbf{u}(t)), \\ & \underline{x} \leq \mathbf{x}(t) \leq \bar{x}, \\ & \underline{u} \leq \mathbf{u}(t) \leq \bar{u}, \end{aligned}$$

where φ is the control objective, which includes the tracking error on accelerations and displacements. Function f represents the system dynamics. The Real-Time Iteration (RTI) strategy [10] is used to satisfy the hard real-time constraint (200Hz sampling frequency or 5ms sample time). As a result, the OCP is discretized into N intervals by multiple shooting and solved by Sequential Quadratic Programming (SQP) algorithm with only one iteration. According to the common approach of formulating the problem with respect to the input and the state differences, $\mathbf{\Delta U}$ and $\mathbf{\Delta X}$, respectively, the resulting Quadratic Programming problem (QP) is

$$\begin{aligned} \min_{\Delta\mathbf{X},\Delta\mathbf{U}} \quad & \frac{1}{2}\begin{bmatrix}\Delta\mathbf{X}\\ \Delta\mathbf{U}\end{bmatrix}^T H \begin{bmatrix}\Delta\mathbf{X}\\ \Delta\mathbf{U}\end{bmatrix} + g^T\begin{bmatrix}\Delta\mathbf{X}\\ \Delta\mathbf{U}\end{bmatrix} \\ s.t. \quad & \Delta\mathbf{x}_{k+1} = A_k\Delta\mathbf{x}_k + B_k\Delta\mathbf{u}_k + c_k,\ k = 0,\ldots,N-1 \\ & \underline{x} - \mathbf{x}_k \leq \Delta\mathbf{x}_k \leq \bar{x} - \mathbf{x}_k,\ k = 0,\ldots,N-1 \\ & \underline{u} - \mathbf{u}_k \leq \Delta\mathbf{u}_k \leq \bar{u} - \mathbf{u}_k,\ k = 0,\ldots,N-1 \end{aligned}$$

where

$$\Delta\mathbf{X} = [\Delta\mathbf{x}_0^T, \dots, \Delta\mathbf{x}_N^T]^T,$$
$$\Delta\mathbf{U} = [\Delta\mathbf{u}_0^T, \dots, \Delta\mathbf{u}_{N-1}^T]^T$$

are collections of state and control variables defined in all intervals. The Hessian H is approximated using Gauss-Newton method, which requires only the first order derivative of the objective function. Matrices A_k, B_k are linearizations, computed at each time step, of system dynamics over the prediction horizon. To further accelerate the on-line computation, the adjoint sensitivity strategy [12] is used, i.e. fixed values for $A_k = A, B_k = B$ are computed off-line. In addition, the objective function φ has to be parameterized to be linearly dependent on all states and control variables. As a result, the Hessian H is time-invariant and can be also computed off-line.

The resulting QP can be solved either in the sparse form (with $(N + 1)n_x + Nn_u$ decision variables) or in the condensed form (with Nn_u decision variables) after the so-called condensing step which eliminates state variables from the OCP [17]. Since the problem has both state and control constraints, the Alternating Direction Method of Multipliers (ADMM) strategy for OCP problems [15] can be adopted to solve the sparse QP.

The hard real-time requirement is satisfied by implementing the NMPC algorithm in Matlab Executable on a PC with Intel core i7 3.60Ghz.

Simulative Results

In this section, simulative results are reported in order to analyse the AS/AB system performances. The MSCA is evaluated in the first part of the *Calabogie MotorSports Track* for the longitudinal dynamics, and in a double lane change maneuver for the lateral one. In both cases a comparison between the compact DiM 150 and the greater DiM.C is proposed.

Parameter Values

Results described in this Section are obtained with the parameter values reported in Tab. I. The nonlinear lateral pressure model $k(d_y)$ and $c(d_y)$ are described by a second degree polynomial functions. A reasonable choice for the static friction coefficient is 0.4 while the dynamic one, which is usually smaller, is set to 0.3. Stiffness and damping values and functions are set by using reasonable values and refined by an iterative tuning based on simulative results. For the application at hand, we can consider that the air cushions have size of 20 × 8 cm, hence an area of $A = 0.016$ m². Although a dedicated identification procedure would be desirable for a more precise analysis, the chosen values for parameters can be considered good enough to put in evidence advantages of the proposed algorithm.

Parameter	Value	Unit
M	50	[kg]
m	67	[kg]
$k(d_y)$	$1000(d_y)^2+1000$	[N/m]
$c(d_y)$	$200(d_y)^2+1000$	[Ns/m]
μs	0.4	
μ_d	0.3	
γ	10	[deg]
$\sigma 0$	10^4	[N/m]
$\sigma 1$	0	[Ns/m]
vs	0.005	[m/s]
kx	20000	[N/m]
cx	1500	[Ns/m]
A	0.016	[m^2]

TABLE I: Model parameters used in simulative results.

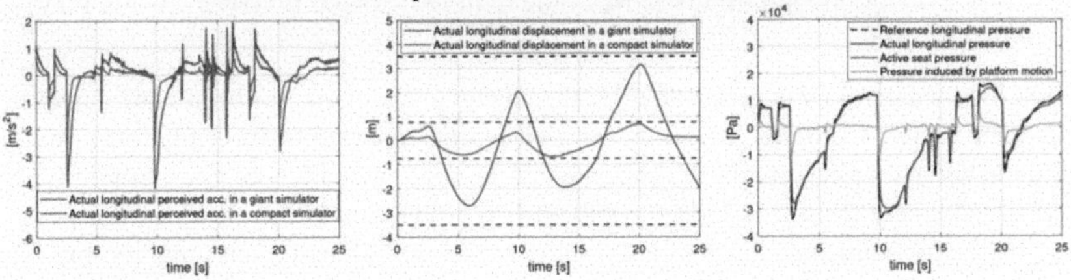

(a) Perceived longitudinal acceleration in a compact (b) Longitudinal position displacement in a com- (c) Longitudinal pressures in a compact simulator. vs giant simulator. pact vs giant simulator.

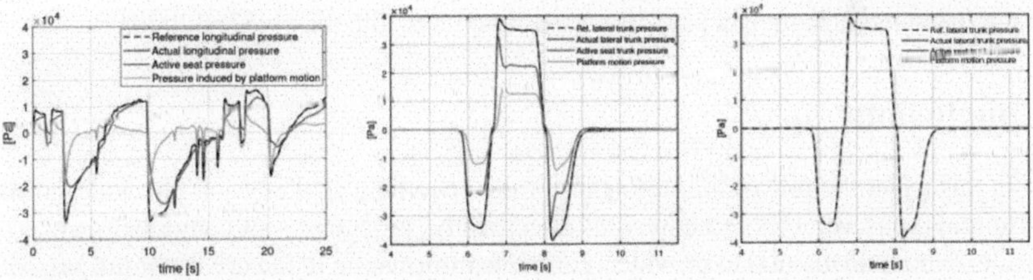

(d) Longitudinal pressures in a giant simulator. (e) Lateral pressures in a compact simulator.

(f) Lateral pressures in a giant simulator.

Fig. 5: Performance comparison: compact vs giant simulator

Regarding the longitudinal direction, the MSCA is setup so that platform working area is exploited at best, by maximizing the accelerations, using a proper scaling on the input signal. Moreover, inspired by the common practice of reproducing the braking action by means of an "impulse-like" acceleration at the beginning of the event, a high-pass filter is used in the pre-processing step (see scheme in Fig. 2). In Fig. 5a and 5b the perceived longitudinal acceleration and the displacement in the two simulators are reported. As expected, in the DiM.C a greater acceleration peak is achievable. In Fig. 5c and 5d the excellent tracking performance of AS/AB are shown, together with the pressure induced by the platform and the one added by the AS/AB system. It is interesting to note that a perfect coordination between motion and AS/AB system is obtained due to a coupled vestibular-pressure model. Moreover, the pressure induced by the simulator acceleration is significantly different in the two cases: in the compact DiM 150 it is smaller and shorter, because the force due to platform acceleration is almost 0. On the contrary, it is bigger and longer in the giant DiM.C, where the motion platform provides by itself the requested acceleration. As a consequence, the pressure request for the AS/AB is coordinated to have the same overall pressure. In both cases the need for a MSCA is evident, though, in the DiM.C, the required AS/AB pressure/tension plays a more relevant role.

As for the lateral dynamics a double lane change manoeuvre has been considered. In the compact DiM 150 the acceleration signal has been scaled to fulfil platform limits, while in the giant platform the acceleration can be reproduced 1:1. In Fig. 5e and 5f, AS performances are shown. In DiM.C a full scale reproduction of the manoeuvre is possible, hence pressure induced by motion is equal to the reference one, making the AS unnecessary. Conversely, in DiM 150, pressure peaks can be observed when reference acceleration crosses the zero value. Indeed, the AS/AB system provides the required pressure to the driver body in order to compensate the opposite sign accelerations, which are induced by the compact platform to be compliant with the physical limits.

Conclusions

In this paper the role of the simulator for the upcoming challenges in the widespread of autonomous vehicles is addressed. Specifically, an active seat/active belts system results to be fundamental to provide a complete information to the driver for the passive driving. To best handle this new technology, a novel approach for active seat and active belts systems is proposed, based on a real time nonlinear MPC implementation. A non-linear pressure model has been developed for the lateral dynamic and a linear one for the longitudinal dynamic. The two are then coupled with a vestibular one to describe the pressure and motion perception of a driver on a dynamic simulator. By means of an NMPC controller, references for AS/AB and platform displacements are generated. The advantages in using such an approach are highlighted comparing results on two different motion platforms.

References

[1] A Anund, C Ahlstrom, C Fors, and T Akerstedt. Are professional drivers less sleepy than non-professional drivers? *Scandinavian Journal of Work, Environment and Health*, 1:88–95, Jan 2018.

[2] B.D.C. Augusto and Loureiro. Motion cueing in the chalmers driving simulator: A model predictive control approach. Master's thesis, Chalmers University of technology, 2009.

[3] M. Baseggio, A. Beghi, M. Bruschetta, F. Maran, and D. Minen. An MPC approach to the design of motion cueing algorithms for driving simulators. In *Intelligent Transportation Systems (ITSC), 2011 14th International IEEE Conference on*, pages 692 –697, oct. 2011.

[4] A. Beghi, M. Bruschetta, and F. Maran. A real time implementation of mpc based motion cueing strategy for driving simulators. In *Decision and Control (CDC), 2012 51st International IEEE Conference on*, dec. 2012.

[5] M. Bruschetta, F. Maran, and A. Beghi. A nonlinear, mpc-based motion cueing algorithm for a high performance, 9-dof dynamic simulator platform. *Control Systems Technology, IEEE Transactions on*, 2016.

[6] M Bruschetta, F Maran, and A Beghi. A fast implementation of mpc-based motion cueing algorithms for mid-size road vehicle motion simulators. *Vehicle System Dynamics*, 55(6):802–826, 2017.

[7] C. Canudas de Wit, H. Olsson, K. J. Astrom, and P. Lischinsky. *A new model for control of systems with friction*. IEEE transactions on automatic control, vol. 40 no. 3, 1995.

[8] M. Dagdelen, G. Reymond, A. Kemeny, M. Bordier, and N. Ma¯ızi. Model-based predictive motion cueing strategy for vehicle driving simulators. *Control Engineering Practice*, 17(9):995–1003, 2009.

[9] J Drosdol, W. Kading, and F. Panik. The daimler-benz driving simulator. *Vehicle System Dynamics*, 14(1-3):86–90, 1985.

[10] Sebastien Gros, Mario Zanon, Rien Quirynen, Alberto Bemporad, and Moritz Diehl. From linear to nonlinear mpc: bridging the gap via the real-time′ iteration. *International Journal of Control*, pages 1–19, 2016.

[11] J. Hamill, K. M. Knutzen, and T. R. Derrick. *Biomechanical Basis of Human Movement 4 ed*. Wolters Kluwer, 2014

[12] C Kirches, L Wirsching, HG Bock, and JP Schloder. Efficient direct multiple shooting for nonlinear model predictive control on long horizons.¨ *Journal of Process Control*, 22(3):540–550, 2012.

[13] James R Lackner and Paul DiZio. Vestibular, proprioceptive, and haptic contributions to spatial orientation. *Annu. Rev. Psychol.*, 56:115–147, 2005.

[14] T Mergner, W Huber, and W Becker. Vestibular-neck interaction and transformation of sensory coordinates. *Journal of Vestibular Research*, 7(4):347, 1997.

[15] Brendan O'Donoghue, Giorgos Stathopoulos, and Stephen Boyd. A splitting method for optimal control. *IEEE Transactions on Control Systems Technology*, 21(6):2432–2442, 2013.

[16] T. W. Showalter and B. L. Parris. *The Effects of Motion and g-Seat Cues on Pilot Simulator Performance of Three Piloting Tasks*. NASA Technical Paper 1601, 1980.

[17] Milan Vukov, Alexander Domahidi, Hans Joachim Ferreau, Manfred Morari, and Moritz Diehl. Auto-generated algorithms for nonlinear model predictive control on long and on short horizons. In *Decision and Control (CDC), 2013 IEEE 52nd Annual Conference on*, pages 5113–5118. IEEE, 2013.

[18] E. Zeeb. Daimler's new full-scale, high-dynamic driving simulator - a technical overview. *Actes INRETS*, pages 157–165, 2010.

[19] Muralidharan, Laya, Ewart J. de Visser, and Raja Parasuraman. "*The effects of pitch contour and flanging on trust in speaking cognitive agents.*" CHI'14 Extended Abstracts on Human Factors in Computing Systems. ACM, 2014.

Lateral dynamics on the vehicle test bed – a steering force module as a validation tool for autonomous driving functions

Martin Förster, M.Eng. [1]

Dipl.-Ing. Rolf Hettel [2]

Dr.-Ing. Christian Schyr [2]

Prof. Dr. Peter E. Pfeffer [1]

1 Munich University of Applied Sciences, Munich, Germany

2 AVL Deutschland GmbH, Karlsruhe, Germany

P. Pfeffer (Ed.), *9th International Munich Chassis Symposium 2018*, Proceedings,
https://doi.org/10.1007/978-3-658-22050-1_14

Abstract

Autonomous driver assistant systems (ADAS) and purely autonomous driving functions (AD) such as Lane Keeping Assist, Parking Assist or Highway Pilot use the steering system to control the lateral vehicle dynamics. Testing those functions on the road is mostly time consuming and expensive due to the high number of control parameters and influences from the driving scenario. Nowadays in the most cases, they have to be tested in the complete vehicle on the road, where tests are often not reproducible and where malfunctions or fail-safe tests could lead to fatal consequences. Therefore, it would be beneficial to transfer those tests to a controlled environment where the risk for man and machine is reduced and the effectiveness is increased. Such a validation environment can use a chassis dynamometer or a powertrain test bed for complete vehicles, for mechanical stimulation in longitudinal direction.

Since the interaction with the steering system was not considered until now, the test of lateral controllers was not feasible. This paper shows an approach to also stimulate the steering system of the vehicle while keeping the interference with the vehicle at a minimum. To achieve this, the tires are mechanically decoupled from the steering system and a compact and universal steering force module is used to induce forces to the tie rod. The tires keep their static axis of rotation, only following the rotational movement of the dynos. The steering force module consists of a force controlled spindle actor and a load cell and is mounted on the underbody of the vehicle. Based on this, the vehicle can be operated under the same energetic and mechanical conditions as in the road test. Supplemented with a real time vehicle and environment simulation, the actual steering wheel angle can be received and the calculated steering force can be transferred to the tie rod, based on the virtual vehicle movement.

Besides the realistic, highly dynamic force imprint and the good reproducibility, the main benefit of this extended validation environment is the flexibility. It is possible to drive with nearly every vehicle in all possible driving scenarios, also reaching the mechanical limits. Due to the static fixation on the test bed, the risk for man and machine is reduced and the access to the vehicle while driving is improved. Measurement and calibration systems can be used with ease and more efficient. In case the research focus is on the steering (feel) evaluation, the real driver can evaluate, while not being disturbed by additional hardware in the interior. Some short examples (such as the AVL DRIVINGCUBE™) are used to demonstrate and evaluate the benefits of the steering force module, leading to the conclusion that driving lateral and longitudinal maneuvers on the chassis dyno and powertrain test bed for testing of ADAS and AD functions is now possible with high reproducibility and without risk.

Motivation

Most of the newly approved vehicles contain one or more advanced driver assistant system (ADAS) or purely autonomous driving functions (AD) such as Lane Keeping Assist, Parking Assist, Valet Parking or Highway Pilot. Some of these functions use the steering system to control the lateral vehicle dynamics and work as an essential human-machine-interface. Validating those functions on the road is mostly time consuming and expensive due to the high number of control parameters and influences from the driving scenario. A full factorial parameter variation is not possible to assure their safety and reliability. In addition, incontrollable influences from the environment, such as weather conditions, have negative effects on the test planning since they increase the overall time effort for the validation to a multiple of the pure testing time. In addition, the driver cannot rate the performance of the function efficiently, since he has to control the vehicle in parallel distracting him.

Nowadays in the most cases, those functions have to be tested in the complete vehicle (prototype or pilot production) on the public road or the proving ground, where tests are often not reproducible and where malfunctions or fail safe tests can lead to fatal consequences for operators or test equipment. Another disadvantage is the limited accessibility to the function under test (in this case the ADAS/AD function) on the road. Application of new components and doing software updates require returning to a vehicle workshop, which is time consuming and disruptive. Testing on the road also only allows three or four engineers to access the function while testing directly, making online changes nearly impossible.

A different state of the art usage for the proving ground test and the public road test is the calibration and fine-tuning of the steering system as a part of the total vehicle. For objective and reproducible assessments, the test vehicle has to be set up with automotive measuring technology as well as with robots for steering, accelerating and braking. The setup and the proving ground tests are time-consuming and expensive. Nevertheless, this effort still does not lead to optimal test conditions. Limited area on proving ground restricts the test maneuvers. Changing environmental conditions like wet, slippery or hot grounds influence the setup. This leads to less-reproducible results. Even under optimal conditions, safety plays a critical role on road tests. Due to insufficient test coverage in prior development stages, malfunctions could occur during road tests leading to unpredictable vehicle reactions [2].

Therefore, it would be beneficial to transfer tests of such ADAS/AD functions and calibration of the steering system to a controlled environment where the risk for man and machine and the overall time effort is reduced as well as the effectiveness, reproducibility and accessibility is increased.

DrivingCube Concept

The AVL DRIVINGCUBE™ is such a validation environment. It is based on a chassis dynamometer or a powertrain test bed, depending on the validation goal. Highly dynamic maneuvers may require a powertrain test bed, but better emission results can be achieved on a chassis dyno. Both environments focus on the complete vehicles and stimulate the mechanical tire-road contact (respective the longitudinal dynamics which include simulating the road resistances). A typical power train test bed for this application can be seen in Figure 2 [1, 3].

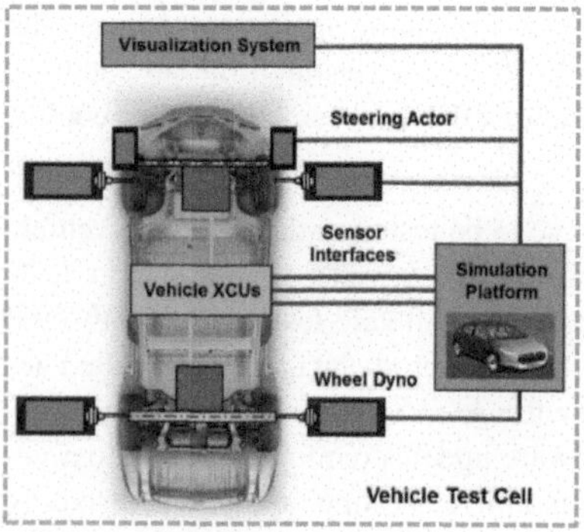

Figure 1: Mechanical layout of the validation concept with steering force module [1]

Figure 2: Example of a highly dynamic powertrain test bed

To be able to validate ADAS or AD functions in this environment, three more components have to be added in addition to the longitudinal stimulation of the tires (see also Figure 1):

- Vehicle and environment simulation
- Sensor simulation and stimulation
- Lateral stimulation for the steering system

The vehicle and environment simulation is the virtual representation of the road and contains information of the maneuvers that have to be driven to validate the system in development. It contains the virtual road, including road topology, road signs, traffic lights etc., the virtual rest vehicle model and the maneuver definition [3].

The sensor simulation and stimulation builds up the interface between the simulated environment and the physical environment (here vehicle and test bed). A simulated ideal sensor is integrated in the virtual vehicle to get ground truth information of the virtual environment and the scenarios defined. Such a sensor could be a camera providing a video stream or a radar sensor providing an object list. The ground truth data is routed through a corresponding sensor model, manipulating it with realistic effects that are caused by typical real world influences. That can be image distortion of a video stream caused by a lens system of a camera or radar radiation dampening caused by rain. Finally, the realistic sensor data is send to the physical sensor of the vehicle using a sensor stimulator. That can be for example an image projecting system for a camera or a radar receiver and transmitter. The sensor simulation and stimulation is not focus of this paper, more details can be found in [3].

The third component is the mechanical stimulation of the steering system representing the lateral movement of the vehicle on the road. Since in the validation environment the vehicle is driving on a powertrain test bed (or a chassis dyno), the axis of rotation of the tires is naturally fixed, and it is not possible to steer the vehicle while driving. Though, to be able to test functions such as highway pilot or a parking assist in the context of the complete vehicle, steering is essentially required for the vehicle. To solve this problem the steering force module is introduced. It solves the dynamic issues of classical powertrain test beds with steering function, such as low universal shaft stiffness resulting in lower dynamics (see Figure 3). On the same hand, it covers most of the benefits of a steering system test bench, such as high system integration level and good dynamics (see Figure 4).

Figure 3: State of the art powertrain test bed with steerable front wheels at the Karlsruhe Institute of Technology, Institut für Fahrzeugsystemtechnik

Figure 4: Steering system test bench for steering calibration and fine-tuning at the Munich University of Applied Sciences

Steering Force Module

The developed steering force module consists of three main components (see also Figure 5) [4]:

- The basic steering force module, which is an electro-mechanical linear actuator connected via a load cell to a carriage moving on linear guidance. The carriage carries a ball linkage that has to be connected to the tie rod of the vehicle. The steering forces in operation can be measured by the load cell.
- The vehicle adapter, which is necessary to mount, fixate and adjust the basic steering force module at the underbody of the vehicle. It is unique for vehicle or vehicle platform and connects the vehicles underbody with the standardized interface of the basic steering force module.
- The power electronics for the linear actuator and communication interface to the simulation. The interface can be used to operate the linear actuator in force-controlled or position-controlled mode, sending demand values (e.g. demand force, demand position) as well as receiving status information (e.g. error code) and actual values (e.g. steering force, actual position)

Figure 5: Components of the steering force module and mounting condition. Vehicle, test bed and environment simulation (top left), linear actuator (top right), connection between steering force module and tie rod (middle left), steering force module mounted to the vehicle underbody (bottom)

The mounting process of the steering force module is rather easy and can be accomplished in less than 90 minutes by two workers in the vehicle workshop. At first, the wheel alignment is recorded and the tie rods are disconnected from the steering rack and the wheel carriers. Than the vehicle adapter is mounted to the vehicle underbody and the basic steering force module is fixated and aligned. The steering actuator is connected afterwards by a modified tie rod on one side to the steering rack. The previously disconnected wheel is additionally supported by a connecting element to the lower wishbone or, if not possible, to the chassis of the vehicle. As a last step, the wheels are aligned to the original toe-in.

After the mechanical assembly process is accomplished, the vehicle can be transferred to the test bed and is fixated with an appropriate vehicle fixation (e.g. wheel hub fixation). Since the front wheels cannot be steered anymore after the mounting, an additional lift truck is required to move the vehicle to the test bed.

The steering force module is connected to the power supply and to a simulation environment (e.g. Vires VTD) via EtherCAT. A simulation environment is used to drive closed-loop maneuvers with the physical vehicle. To achieve this, a virtual representation of the physical vehicle on the test bed is used in the simulation to drive arbitrary maneuvers. The forces from the tie rods of the virtual vehicle are transferred to the steering force module controller and are applied to the steering rack via the linking tie rod of the physical vehicle. The resulting steering rack displacement (respectively the linear movement of the actuator) is measured and fed back to the virtual vehicle so it can steer in the simulation accordingly.

A coverage of 100% steering range for the most common vehicles (assuming 300mm of steering rack travel) can be achieved with this setup, while at the same time apply up to ±12kN of force (at the steering rack) with a maximum traverse speed of about 250mm/s. This enables the steering force module to simulate a wide variety of steering maneuvers on the test bed, and keep the high dynamic of a powertrain test bed or the realistic driving behavior of a chassis dyno, which was not achievable before. That includes maneuvers such as [2, 4]:

- Highway driving with lane changing and cornering
- Slalom and obstacle avoidance
- Euro NCAP tests for LKA (Lane Keeping Assist)
- Parking maneuvers

The steering force module is universal for different vehicle types and sizes and can be fit to most vehicles by only modifying the vehicle adapter. It also fits to the chassis dyno as well as to the powertrain test bed, whereas the chassis dyno has more restrictions in terms of assembly space (underneath the vehicle) and in terms of wheel fixation effort (the wheels axis of rotation is not fixed on the chassis dyno).

Figure 6 shows one exemplary assembly setup for the steering force module in a Sport Utility Vehicle using an advanced vehicle and environment simulation on a chassis dyno. It can be used for example to analyze the interaction between driver, steering assistance system and environment (especially the subjective perception of haptic feedback from the system to the driver).

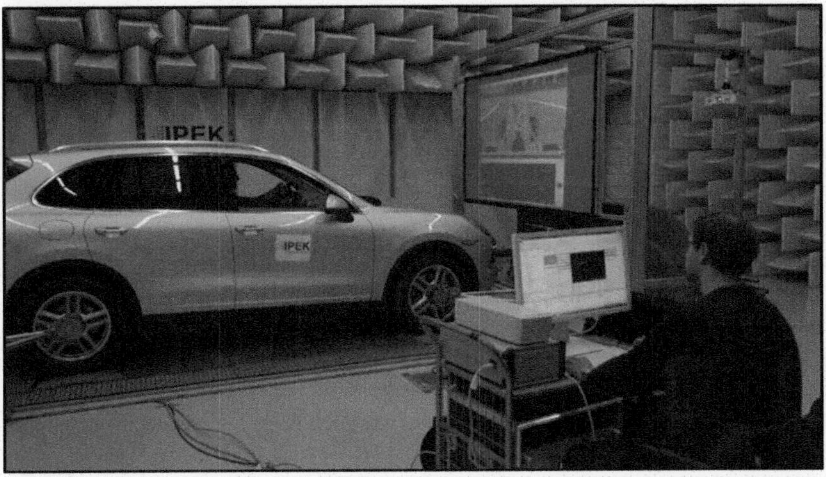

Figure 6: Steering force module mounted in a Porsche Cayenne vehicle at the chassis dyno of the Institute of Product Engineering (at the KIT), connected to a vehicle and environment simulation

Some of the key facts of the steering force module (here: the prototype shown in Figure 5) are:

- High dynamic actuator
- Traverse speed 254mm/s
- Continuous force 12kN
- Maximal travel 300mm
- Overall height 140mm
- Load cell range 20kN
- Communication interface EtherCAT and CAN
- Force- and position-demand mode

Example – Steering force module for the validation of a Parking Assistant

A typical application of the steering force module is the validation of a parking assistant on a test bed. The parking assistant is working as follows:

- Vehicle drives slowly along a road
- Driver activates parking assistant
- Ultrasonic sensors check the verge for a suitable parking lot
- After detecting a parking lot, the assistant computes a trajectory into the lot
- Assistant steers while driver has to accelerate and brake

To validate such a parking assistant on the DrivingCube, one has to do some preparation first. In a vehicle workshop, one has to equip the test vehicle with the steering force module. With the aid of lifting trucks, one can position the vehicle on the chassis dyno and fixate the wheels with wheel hub fixation. Communication between a vehicle and environment simulation and the steering force module is set up using EtherCAT. To stimulate the ultrasonic sensors, generic sensor models extended with related attributes generate data streams, which are transmitted by cables and test interfaces to an injection point behind the real ultrasonic sensor in the vehicle. Now, the setup is ready to start.

A human driver accelerates the test vehicle while the chassis dyno simulates slope and resistance of the road. If the vehicle reaches a moderate velocity, the driver activates the parking assistant. Meanwhile, ultrasonic models in the simulation environment feeds the vehicle with a scenario where parking cars with more or less distance between each other. Several parking lots could be modeled to validate whether the parking assist is able to detect suitable lots. During the whole maneuver, the steering force module send actual steering rack displacement to a vehicle dynamics simulation where a model computes the steering rack forces based on the current vehicle state (vehicle speed, road resistance, rack displacement, etc.). The steering force module induce the computed steering rack force via tie rod as a feedback for the driver and the steering system. If the parking assist has detected an appropriate parking lot, it has to ask the driver to stop the vehicle. After computing a trajectory for parking, the power assist module of the steering system should take over and maneuver the vehicle into the lot. Guided by the parking assistant and ultrasonic sensors, the driver only has to accelerate and brake. As a visual feedback for the driver, the simulation model can provide perspectives like bird's eye or drivers view, which can be projected on a screen. With this visual feedback, the driver at the test bed can also assess whether the parking assist has accomplished the task or not and rate the performance.

Steering force module as mobile steering system test bench

Another application for the steering force module is the calibration and fine-tuning of steering systems in the context of the total vehicle. It can be used for parameter variation in a controlled environment similar to a steering system test bench (see Figure 4)

In a typical setup for steering system test, the steering system is mounted on a holder which is fixed on a heavy machine bed. A shaft couples the steering wheel with a rotary motor. This drive replaces the human driver and can control angle and torque of the steering wheel. Two linear motors induces the wheels aligning forces to the tie rods. Alternatively, the linear motors can control the rack displacement. Sensors, mounted on the test benches drives, record the tie rod forces, rack displacement and the steering wheel torque and angle. To guarantee the steering system runs in the same modes as it would operate as a part of the full vehicle, a remaining bus simulation connects to the steering electronic control unit (ECU). A regenerative power supply feeds the electric power assistance system. This imitates the vehicle electrical system [5].

To stress the steering system in a physically correct manner, the steering test bench sends the rack displacement to an in parallel running vehicle dynamics simulation. Based on the specific driving maneuver and the modeled vehicle characteristics, the simulation calculates the resulting tie rod forces and the linear motors induces them into the physical steering system. This kind of test setup is called Hardware-in-the-Loop (HiL) simulation.

Engineers can fulfill a wide range of development tasks with the aid of such a steering test bench [5, 6]. During a set of open and closed loop maneuvers, the characteristic properties of a steering system can be extracted out of the measurement data. Developers can use the data to parametrize steering models. Such models can then be used e.g. to design and validate software functions in a virtual environment. Moreover, the steering test bench is a suitable tool for developing, calibrating and validating assistance functions implemented on the steering ECU. Calibration of a multi-dimensional boost curve of the power assistance and validation of an active return feature are just two of numerous tasks.

After an objective optimization of a software function with satisfied functional requirements, a human driver can replace the rotary drive stimulating the steering wheel. Now, the engineer can focus on the haptic properties and evaluate its power assistance behavior. For example he can rate if the steering feels harmonic or synthetic or if the damping on the system to high. This setup makes steering behavior noticeable for human and supports the reproducibility. If a detailed and valid vehicle dynamic simulation model was used, the steering system test bench calibration result on the test bench is a good starting point for further calibration in the physical vehicle.

Using the steering force module as a mobile and flexible steering system test bench directly in the physical vehicle brings some benefits. For example the setup of the HiL test is simplified, since the steering system is already integrated in the vehicle and therefore already uses all required interfaces. Power supply and rest bus simulation is naturally handled by the electric system of the vehicle and has not to be replicated.

Conclusion

Shown in this paper is a sophisticated approach for validation of advanced driver assistance systems and automated driving functions that require the usage of the steering system. The test of these systems can be challenging on the test track and the public road, since the testing conditions are not reproducible and the operation could harm man or machine. Transferring them to a controllable and accessible environment has many benefits and decreases the costs and overall time effort of the tests.

The demonstrated steering force module is an extension of the DrivingCube validation environment and enables the testing of lateral vehicle dynamics on a common chassis dyno and powertrain test bed. The solution convinces with mechanical simplicity as well as high coverage of testing scenarios and vehicle types. The steering force module is available as a prototype and will be developed into a series product in various sizes to be even more flexible in terms of testing scenarios and assembly space. With this setup, ADAS/AD function can be tested more efficient and reproducible in a safe environment in the future.

Bibliography

[1] Schyr C., Hakuli S., Schick B.: „Ganzheitliche Fahrzeugbewertung mittels X-in-the-Loop im Entwicklungsprozess am Beispiel Powertrain-in-the-Loop“, MTZ-Fachtagung Virtual Powertrain, 2013

[2] Specka F., Contini A., Hakuli S., Schick, B., Pfeffer, P.: „Evaluation and optimization of steering behavior in overall vehicle context at the Steering-in-the-Loop test bench”, VDI Tire-Chassis-Road Congress, 2013

[3] Schyr C., Brisard A.: „DrivingCube – A novel concept for validation of powertrain and steering systems with automated driving“, 13th international Symposium on Advanced Vehicle Control, München, 2016

[4] Schyr C., Düser T., Weck T., Schick B., WO002018046609A1, “Modular test stand for whole vehicles which are ready for driving”, PCT Patent, 2017

[5] Schimpf R., „Charakterisierung von Lenksystemen mit Hilfe eines Lenksystemprüfstands“, Dissertation, Wien, 2016

[6] Contini A., Schöning T., Pfeffer P. E., Keth N., Koegeler H.-M., „Modelbasierte Applikation der Lenkunterstützung für unterschiedliche Fahrzeuggeschwindigkeiten auf einem Lenksystemprüfstand”, 7. Tagung Simulation und Test für die Fahrzeugtechnik, Berlin, 2016

Vehicle motion control layer – a modular abstraction layer to decouple ADAS from chassis actuators

Dr.-Ing. Eckehard Münch

Harald Bestmann

Dr.-Ing. Caspar Lovell

Stephan Pollmeyer

Mehrdad Salari Khaniki

Dirk Schulte

P. Pfeffer (Ed.), *9th International Munich Chassis Symposium 2018*, Proceedings,
https://doi.org/10.1007/978-3-658-22050-1_15

Motivation

We are currently experiencing a substantial increase in both the number of mechatronic chassis actuators and driver assistance systems. For some time now the advanced driver assistance systems (ADAS) are no longer constrained by the use of a single actuator, such as electronic stability control (ESC), but usually operate several chassis actuators to direct the vehicle longitudinally and laterally (e.g. Park Assist, adaptive cruise control (ACC), Lane Centering). Similarly, it will be necessary to alternate between manual and (partly-) autonomous driving (AD) on the way to the fully autonomous vehicle. Depending on the operating mode, this has a considerable impact on the control of the mechatronic chassis actuators. This is compounded by the numerous choices when ordering a new car resulting in a substantial number of combinations of ADAS and actuators in the vehicle.

The challenge is then to counter the increasing complexity in the dependencies between the various ADAS or AD systems and the actuators. At the same time the greatest possible networking of systems must be assured.

To address this problem we shall present an approach, the objective of which is to achieve a decoupling of ADAS and actuators through the introduction of an additional Vehicle Motion Control Layer and to enable a highly modular flexible software architecture.

In addition to solving the afore-mentioned problem such an architecture has further advantages for OEMs and suppliers. By pursuing the modular approach all software modules, independent of each other, can be developed and tested both by OEMs and suppliers. The result is major time saving in the vehicle development process, resulting in a reduction of the "Time to Market" of the vehicle or vehicle family in question. Furthermore, the modular software building blocks enable customer specific combinations of ADAS and mechatronic chassis actuators to be realized with minimal effort.

Hence the introduction of an additional Vehicle Motion Control Layer offers considerable improvement potential and is described in greater detail in the following chapters.

Abstraction layer for controlling the vehicle motion

The controls architecture, which affects the vehicle motion in longitudinal, lateral and vertical direction, can be divided into the ADAS layer and the layer of chassis and drivetrain actuators. This paper deals with the introduction of an additional layer between both of them to decouple their dependencies. This additional layer is named *Motion Control Layer* in the following.

Using the layer model for structuring software architectures is an established technique in the field of mechatronic systems (see [1] [2] [3]). This is apparent from the VDI 2206 "Design methodology for mechatronic systems" guideline, which recommends the organizing principles of modularization and hierarchization for structuring complex mechatronic systems. The modules are built by the decomposition of the control tasks in individual and independent functions that are hierarchically linked to the superordinate functions. Those hierarchical levels can be seen as control layers in context of the whole software architecture.

Layer models are commonly used for structuring software architectures as well. The particular layers undertake clearly defined functions whereby the individual layer only communicates with the neighbored layers. Through the definition of abstract and generalized interfaces, an encapsulation and information hiding of the different layers is ensured. This causes an improvement of the system in clarity and serviceability. In addition to this, software components can be changed or modified easily without changing the whole system.

Figure 1 shows the three layers of this control structure. The bottom layer is on the actuator level. It contains mechatronic components like steering, brake, drivetrain or vertical actuators. All these actuators consist of their own control software, sensors and mechanics.

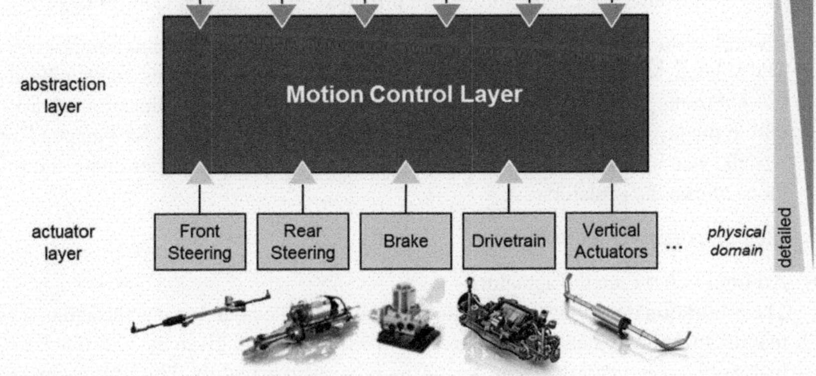

Figure 1: Layers of the control architecture

The top layer contains the ADAS, which guide the vehicle. These are equipped with comprehensive sensors that monitor the environment and generate an environmental model out of the captured data. This model represents the base for the planning of the prospective vehicle motion. The outcome of this planning is a path or a trajectory that describes the upcoming position or motion of the vehicle.

The Motion Control Layer is located between the ADAS and the actuator layer. The purpose of this layer is the abstraction and generalization of the driving task from a detailed physical view at the actuator layer to an abstract kinematic view at the ADAS layer. The actuators of the bottom layer have to deal with detailed physical processes. The actuator and measurement values contain typical parameters like angles, forces, torques, pressures or currents. Hence the control of actuators requires detailed knowledge about the physical correlations within them. In contrast to this, the ADAS provides the guidance of the vehicle along a path or trajectory. It can be described kinematically by means of the position, velocity and acceleration.

The decoupling of the ADAS and actuator layer requires a strict usage of the encapsulation principle. This means that, at the ADAS layer signals, parameters as well as models have to be described only within the kinematic domain. Or vice versa: due to this encapsulation and information hiding kinematic signals, parameters as well as kinematic models are sufficient.

The Motion Control Layer fulfills the following tasks:

- Execution of the planned paths or trajectories and determination of actuator specific set points:
  The interface between ADAS and Motion Control Layer is specified in kinematic quantities which comprises curvature, vehicle speed etc.; the actuator set points, like steering angle or engine torque, are provided by the Motion Control Layer.
- Control of the vehicle motion in longitudinal, lateral and vertical direction:
  The Vehicle Motion Control Layer merely fulfills control functions that can be allocated to the motion of the entire vehicle. This ensures a clear allocation of tasks within the layer model. The control of steering angle, slip controls for independent wheels, gear shifts or engine torque, will be handled locally as subordinate control loops within the actuators.
- Handling of driver interactions
- Arbitration between the actuators:
  The arbitration describes the allocation of the control demand to the particular actuators, meaning to handle the degrees of freedom that are given by additional actuators like rear steering, torque vectoring or vertical actuators. The arbitration has to consider the current actuator limits.

- Estimation of trajectory constraints:
  For the planning of the future vehicle motion, the ADAS needs essential information about the driving dynamics. This information is determined in the Motion Control Layer and subsequently transferred to the ADAS. This procedure maintains the abstraction principle that was introduced before.

To achieve a high flexibility the Motion Control Layer is built into a highly modular concept.

This allows the development of different replaceable instances of the particular modules. By combining appropriate instances, different requirements on the driving functions can be fulfilled. This also enables the integration of black-box-modules.

Figure 2 shows the functional architecture of the Motion Control Layer. The structure demonstrates four control paths in longitudinal, lateral and vertical direction as well as the stability controller and the modules *Signal System Consolidation* and the so called *Constraint Handler*.

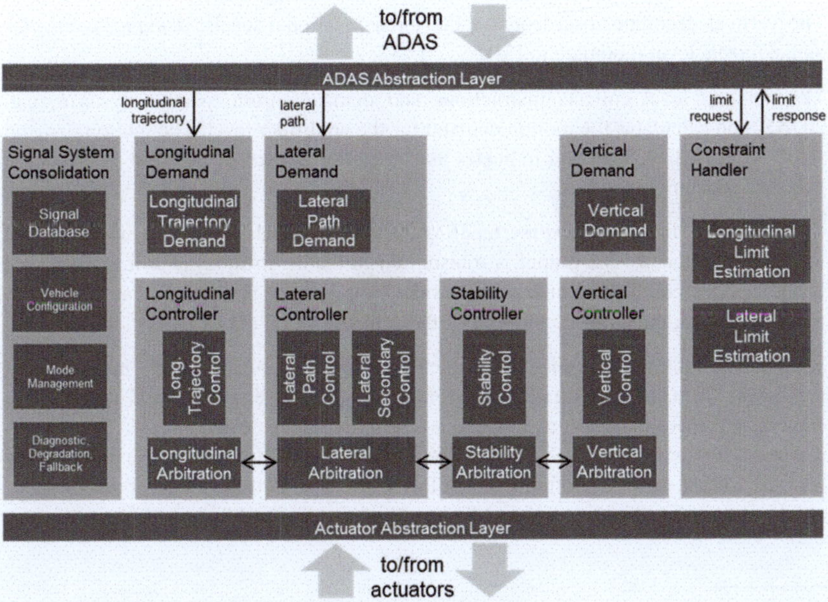

Figure 2: Functional architecture of the Motion Control Layer

The abstraction layers *ADAS Abstraction Layer* and *Actuator Abstraction Layer* allow an adaption to the properties and interfaces of the overlaying ADAS and the underlying actuators, respectively. Here various interfaces can be customized and missing functions can be implemented.

The description of these particular modules is to be found in the following chapters.

Controller

The horizontal decomposition of the controllers shown in figure 2 allows an independent implementation and configuration of the particular control paths. The drawback compared to an integral control approach consists in the potentially reduced performance especially driving at the vehicle dynamic limits. According to this the *lateral*, *longitudinal* and *vertical controller* are responsible for the comfortable vehicle guidance. The *stability controller* enables the highly prioritized fallback level when driving at the limits. ESC functions are implemented here. A separation of the motion directions does not seem reasonable, if the stabilization controller affects all motion degree of freedoms simultaneously.

The vertical decomposition leads to the three functional levels: *demand*, *controller* and *arbitrate* level (compare to [3]).

The *demand* level collects instructions and demands from external systems, e.g. ADAS, and calculates the motion demand for the controller level. The motion demand refers to the full vehicle and indicates as the motion degree of freedom of the entire vehicle for example.

The *controller* level contains the feedforward and feedback controllers which realize the motion demand. The output is transmitted to the *arbitration* level in the form of generalized forces. Taking under account the limitations of the actuators as well as the tire force limits, the arbitrators calculate the actuator set points.

The vehicle motion controller only relies on information from inertia measurement sensors or from the subordinate actuators. The usage of environmental sensors like camera, LIDAR, radar or even GPS is currently not required in this framework. Therefore the current position or orientation of the vehicle, e.g. on the road, is not known[1]. Hence this the Motion Control Layer operates in the domain of velocities or accelerations.

---

[1] If information of the actual position is available, the controller could be extended by a positon control loop.

Trajectory Demand Interface

The modules *Longitudinal Demand* and *Lateral Demand* (figure 2) process the driving demand from ADAS. The separation of the longitudinal controller and the lateral controller is reflected by the interface definition.

The interface assumes a planning algorithm located in the ADAS, which calculates the forthcoming trajectory in longitudinal and/or lateral direction. Typically the ADAS runs at lower sample rate as the underlying Motion Control Layer. The trajectory can be described as smooth piecewise polynomial curves, e.g. a spline or Bézier curves. Beside this smoothness, these curves can be interpolated to match the higher sample rate of the controllers.

The targets in longitudinal and lateral direction are transmitted as polynomial coefficients, for example of a quadratic function:

$$\kappa(s) = c_0 + c_1 s + c_2 s^2 \tag{1}$$

$$v(t) = v_0 + v_1 t + v_2 t^2 \tag{2}$$

The polynomial coefficients of the upcoming curve section are transmitted to the demand blocks of the motion control layer. If it is assumed that consecutive trajectory updates of the planning algorithm do not jump, discontinuities at the controller inputs are avoided due to the smoothness property of the polynomial.

The function $\kappa(s)$ describes the curvature over the covered distance and the function $v(t)$ is the vehicle velocity over time, starting at the current position or time respectively. The curvature can be considered as the path. Therefore it does not contain any temporal information. This has the advantage that even at standstill the path is well defined. The velocity polynomial adds time information to the target definition.

Driver interaction

The separation of the lateral controllers into the *lateral path controller* and *lateral secondary controller* is a cause of the requirement to allow driver interactions during assisted driving. The underlying assumption is that rigid steering columns will be the dominating design type for the foreseeable future, not least because of safety requirements. The rigid steering column allows the driver to control the vehicle course in case of a malfunction of the steering system. Additionally, the driver should be able to temporarily adjust the direction of travel or to provide information about a desired transversal deviation to the ADAS.

The task of the *lateral path controller* is to join the two commands from ADAS and driver and to realize the lateral guidance of the vehicle under normal, comfortable driving conditions. The controller only acts on the front steering. At the same time, the driver is able to directly introduce torques to the steering column. From a control theoretical point of view this mechanism has the same characteristics as a disturbance. Under the assumption of a reasonable driver action, this torque should be approved and be treated as a reference input. The interaction between driver torque and lateral demand yields into the actual steering angle. This angle can be interpreted as the combined driving demand.

The controls of optional actuators like rear steering systems or torque vectoring are places in the *lateral secondary controller*. The combined driving demand is used a reference signal. For a comfortable driving experience, especially when driving manually, often a feedforward control of these actuators is applied, mostly on basis of the current velocity and steering angle. Due to the dependency to the more dynamic vehicle states, a feedback control may yield to an undefined subjective driving experience.

Constraint Handler

One of the central tasks of an ADAS is the calculation of a target trajectory, which has to be executed by the Motion Control Layer and the actuators. In this case, the trajectory is the path that the vehicle follows through space as a function of time.

Among other characteristics, the curvature is a property of a trajectory:

$$\kappa = \frac{1}{r} \tag{3}$$

with radius r as the distance between the vehicle positon and the center of the driven curve. In case of steady-state driving, the curvature is

$$\kappa = \frac{a_y}{v^2} \tag{4}$$

with the lateral acceleration a_y and longitudinal velocity v. As it is known, that the lateral acceleration is limited (e.g. by the maximum lateral tire force), it is clear, that the curvature is limited and that this limit depends on the current velocity. This means the target trajectory can only be driven by the vehicle, if its curvature does not exceed this limit.

Considering transient maneuvers, the maximum lateral tire force depends on the longitudinal tire force. Hence, the maximum of lateral acceleration and - as a consequence - the maximum curvature depends on the longitudinal acceleration (see figure 3).

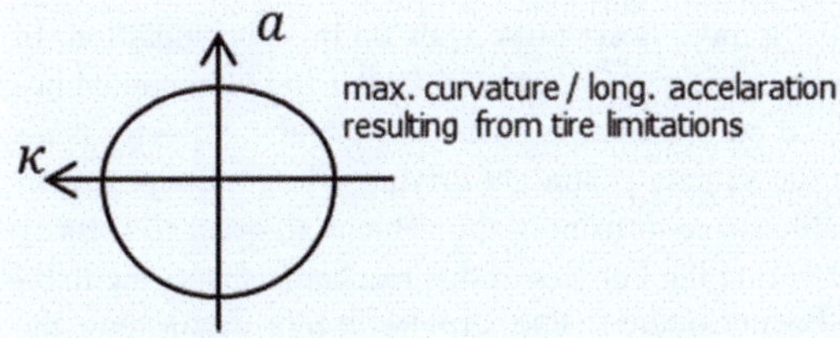

Figure 3: Correlation of maximal curvature and long acceleration resulting from tire limitations

Furthermore, the maximal curvature gradient

$$\dot{\kappa} = \frac{d}{dt}\kappa \tag{5}$$

is limited, e.g. resulting from a yaw torque limitation according to maximal tire forces. And obviously, the longitudinal velocity is limited as well (e.g. by drive characteristics).

Bringing all mentioned aspects together, a drivable trajectory is limited by the maxima of curvature, curvature gradient, longitudinal velocity and longitudinal acceleration. Further on, these limitations are called trajectory constraints.

In addition these limits affect each other. Hence, these trajectory constraints can be represented by a four-dimensional, illustrated in three-dimensional-projection in figure 4.

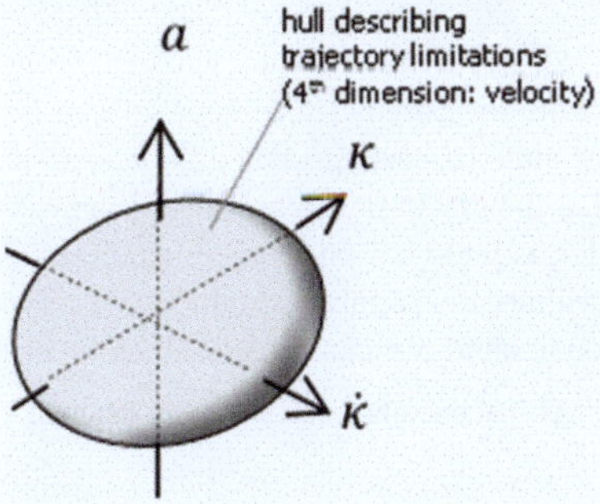

Figure 4: Hull describing trajectory limitations

This means that a target trajectory can only be driven, if the combinations of curvature, curvature gradient, longitudinal velocity and longitudinal acceleration covered by the trajectory lie within this hull.

The following example tries to give a more descriptive view on this consideration. In this example, a trajectory for a Lane Centering Control (LCC) has to be computed under the assumption, that longitudinal acceleration is small and thereby, velocity is almost constant. The initial state of the vehicle is straight driving. The limitations mentioned above constraint the reachable space in front of the vehicle. In order to identify the two extreme trajectories representing the borders of this reachable space, the maximum curvature gradient has to be considered. Furthermore, while performing the maximum curvature gradient, the curvature increases until the maximum curvature is achieved. Figure 5 illustrates these two trajectories. If the target trajectory does not exceed the limits of curvature and curvature gradient, this trajectory is in-between these two borderlines and is drivable.

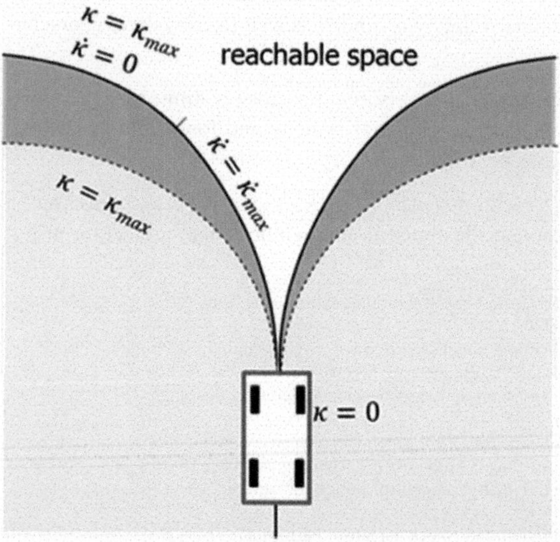

Figure 5: visualization of reachable space considering maximal curvature and curvature gradient

Beside tire characteristics, trajectory constraints could be determined by actuator limitations and attributes of the environment. E.g. the maximum longitudinal acceleration could be limited by the maximum torque of the drive system and the maximum curvature could be limited by the maximum rack force of the steering system. Furthermore, slope and bank angle as well as the road friction value influence the trajectory constraints. Obviously, these parameters can only be considered if they are known.

At this point, it is clear, that a lot of information is needed for the calculation of a drivable trajectory. If ADAS should consider all these limiting factors, the actuator limitations have to be transmitted to the ADAS and other vehicle parameters (e.g. chassis kinematics, current vehicle mass or tire characteristics) have to be known by the algorithms.

Beside mentioned physical limitations, it is conceivable to consider safety and comfort constraints, as well. E.g. as part of the safety concept, the steering force could be constraint, so the driver is capable to overrule the ADAS. For comfort reasons, it could be useful to define limits for acceleration and jerk.

As described in previous chapters, the essential idea behind the Motion Control Layer is to decouple Driver Assistance Systems and actuators. Regarding the calculation of a drivable target trajectory, this principle can be used as well. The proposal here is to implement an algorithm, which calculates the trajectory constraints, in the Motion Control Layer. The module comprising this algorithm is called Constraint Handler and transmits the calculated trajectory constraint to the ADAS. Hence, ADAS is able to compute a drivable trajectory just by considering the trajectory constraints, but without knowing properties of the vehicle or actuators.

Considering the fact that most information needed to calculate the trajectory constraints are known by the Motion Control Layer (e.g. for control tasks), the effort for implementing an ADAS in a vehicle is reduced. Furthermore - taking into account, that an increasing number of ADAS are implemented in modern vehicles - the installation of such a central service in the Motion Control Layer leads to a significant reduction of complexity.

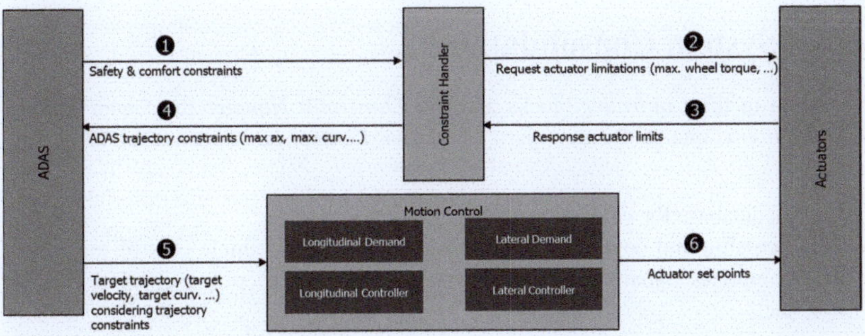

Figure 6: Proposed sequence of action ensuring decoupling of ADAS and actuators

Sequence of actions

In previous chapters, the modules of the Motion Control Layer are introduced. In this chapter, the interaction of these modules with the ADAS and the actuators is described. This process is shown in figure 6 (also see [4]):

1. ADAS transmits safety and comfort constraints to Constraint Handler.
2. Constraint Handler requests the actuator limitations.
3. The actuators provide their limitations (if necessary, considering safety and comfort constraints).
4. Constraint Handler calculates trajectory constraints (considering actuator limitations, safety & comfort constraints and vehicle parameters) and transmits the result to ADAS.
5. ADAS computes and provides the target trajectory considering the current trajectory constraints.
6. Motion Controller calculates the actuator set points according to the target trajectory.

Following this sequence of actions ensures that the target trajectory does exceed neither limits of the vehicle (including actuators) nor the safety and comfort constraints.

Contrary to this proposed sequence of actions, it would be possible to check the drivability of the target trajectory afterwards. But obviously, the result could be negative and, further on, could lead to a never ending sequence of answers and questions between ADAS and the Motion Control Layer.

Signal System Consolidation

In addition to the *Controller* blocks and the *Constraint Handler* the *Signal System Consolidation* is one of the main parts of *Motion Control Layer*. It has the following tasks:

- Central database for vehicle dynamic signals:
  Preprocessing and conditioning of sensor data for the vehicle control functions. This comprises signals like: yaw rate, acceleration and friction coefficient estimation, etc.
- Vehicle configuration:
  Handling and management of vehicle specific configurations and parameters.

- Control mode management:
  Activation/deactivation of controller considering the stability priority mode management, e.g. synchronous switching of control modes, initialization and restart routines, activation and deactivation of actuators

- Diagnostic, Degradation and Fallback Strategy:
  Provides fallback settings of the control strategy in case of (sub-)system failure, e.g. actuators, sensors; validation, logical checks and boundaries check of controller functions and their interactions as well as handling of diagnostic.

Conclusion and Outlook

To counter the increasing complexity of the dependencies between mechatronic chassis actuators and driver assistance systems an architecture concept for a vehicle motion control layer was introduced.

This control layer decouples the ADAS layer from the actuator layer. The key principle is to ensure a strict encapsulation and information hiding between the layers: detailed models in the specific physical domain on the actuator layer and kinematic models describing the motion of the entire vehicle on the ADAS layer.

Usually, the predictive calculation of the upcoming trajectory requires an entire vehicle model which describes all relevant dynamics and limitations. This conflicts with the above mentioned principle of information hiding, because limitations of the actuators have to be involved in these calculations. To solve this conflict, the approach of a constraint handler was introduced. The physical constraints of the actuator layer and the tires are transformed to a kinematic representation. Therefore the need of physical models at the ADAS layer is considerable reduced.

The next steps will be the prototypical implementation and the proof of functionality on a test vehicle. The major focus will be on the interaction between path planning, constraint handler and controller. Here one important issue will be the behavior of the system in case of actuator failures or degradations. The idea of the constraint handler allows an immediate consideration of the degraded actuator limits in the trajectory constraints and thus in the planning algorithm.

Literatur

[1] VDI 2206 „Entwicklungsmethodik für mechatronische Systeme“, VDI-Richtlinie, 2004.

[2] *A. Trächtler*: „Integrierte Fahrdynamikregelung mit ESP, aktiver Lenkung und aktivem Fahrwerk“, at – Automatisierungstechnik 53 (2005) 1, S. 11-19.

[3] *P. E. Rieth, T. Raste*: „Integrationskonzepte der Zukunft“ in „Handbuch Fahrerassistenzsysteme, Grundlagen, Komponenten und Systeme für aktive Sicherheit und Komfort“, 3. Auflage, Springer Vieweg, 2015, S. 1083-1092.

[4] *V. Held, A. Heitmann*: „Modularization of vehicle control systems based on the application of object-oriented design principles“, 8th International Munich Chassis Symposium 2017, S. 49-64.

PARALLEL STRAND II

RIDE COMFORT

Continental Air Supply – a scalable closed loop system for efficient air suspension solutions

Dr.-Ing. Uwe Folchert,
Continental Chassis & Safety,
Business Unit Vehicle Dynamics,
Segment Electronic Suspension Systems

P. Pfeffer (Ed.), *9th International Munich Chassis Symposium 2018*, Proceedings,
https://doi.org/10.1007/978-3-658-22050-1_16

Introduction

Electronic controlled air suspension systems (picture 1) provide high comfort and are effectively able to support best vehicle dynamics. Continental developed the highly energy efficient closed loop air supply system further to a high integrated compact unit. This integrated solution called CAirS® (Continental Air Supply) is scalable from basic level control of front and rear air springs up to the complete control of an adaptive air suspension chassis.

Growing application spectrum for air

Electronic controlled air suspension systems provide an opportunity to the driver to adapt the suspension/damping behaviour of the vehicle to various requirements and driving profiles. Additionally sensors can recognize the current driving state and tune the suspension automatically in an optimized manner between comfort and vehicle dynamics.

Here the transition between comfort functions and functions related to vehicle dynamics is smooth. Therefor the lowering of the centre of gravity supports a sportive manner of driving with higher curve speed and accordingly a higher cruising speed. However the lowering of the vehicle also reduces the air resistance and thereby the energy consumption considerably.

The compensation of the vehicle level at different load conditions ensures always the availability of the full suspension travel. This improves both the driving safety and the driving comfort. By adapting the stiffness of the air springs a better roll control takes place. Thus the cases of application are various.

For open loop air supply systems the supply of the needed air pressures and flow rate requires a considerable amount of energy. The use of vehicle lowering for saving energy and then the lifting according to vehicle speed and road condition requires furthermore a high system availability.

Therefor the energy efficient closed loop air supply system offers particularly big advantages. This closed loop concept has been successful in the field for many years. A few years ago Continental started the further development of the closed loop air supply up to a high integrated unit with obvious advantages concerning weight and installation space.

This resulted in the innovative concept called CAirS® - Continental Air Supply. CAirS® provides a scalable solution for controlled air suspension chassis with small dimensions and considerably reduced integration respectively assembly effort.

System features

Picture 2 shows the degree of integration of the new air supply concept. While a conventional solution consists of many single components, with CAirS® all the functions and components are integrated in one unit. Instead of the components compressor, valve block, Electronic Control Unit (ECU), motor control/relay, internal tubes and wiring, temperature and pressure sensor and multiple connectors there is now a ready-to-install unit with one central connector. Picture 3 shows clearly also an analogy to the world wide used electronic brake system (EBS) MK100. This obvious relationship is well-founded in the technology transfer within Continental, which also accelerated the development to get ready for series production. Meanwhile CAirS® is in the ramp up phase at several customers.

Compared to other discrete designed air supply systems this integrated solution shows a space reduction of up to 25%, and with a weight of 3,6 kg (without bracket) it saves about 1 kg up to 1,5 kg within the system compared to other systems (open loop) in the market.

Structure of the CAirS® unit

Many technical details of the design are derived directly from the highly integrated EBS-technology. The motor for the 2-piston-compressor is a derivate of the EBS-version. The valve concept is a carry over from the series proved MK100, but adapted due to the medium air. The coils of the solenoid valves are also carry over from MK100. They are integrated in the ECU housing and directly connected to the PCB. This is a well-proven design concept.

The integrated sensors for temperature and pressure are used to control the system state and to protect the components.

The integrated ECU allows control of the compressor motor directly by an electronic motor control. Thereby the mechanical relay is omitted. With this electronic motor control the motor inrush current can be limited and so reduced to low values. The motor load/overload can be recognized and the motor speed can be controlled.

One main component of the ECU is the microcontroller, called QUASAR (QUad-core microcontroller for Automotive Safety And Reliability). The Quasar-family was developed together with Freescale. Memory size and core number are corresponding to the requirements of the different project applications. The development of the base-software (BSW) for several Quasar-derivatives is based on a common Continental-SW-platform.

The SW architecture of CAirS® is AUTOSAR compliant and modular, see picture 4. SW modules from the customer, e.g. for damping control, can be integrated. For the current further development of CAirS® experts of Continental are working on encryption technologies of a microprocessor and SW to meet the future requirements for Cyber Security.

Application and operation

As the mechanical parts of the CAirS®-unit are widely uniform, the scaling of the functional content for different applications will be realized both by population of the PCB of the ECU and adaption of the SW-content. The range of functionality contains the basic air suspension level control up to the full chassis control with air spring / damping / spring rate control.

Working as a closed loop air supply system, CAirS® uses the reservoir to supply the needed air mass at a needed air pressure with low energy consumption within a closed loop between air spring and reservoir. Using a smart air mass management the reservoir size can widely be adapted to the requirements of the air suspension chassis for particular vehicles. Therefor often smaller reservoir sizes can be used as required for open loop air supply systems.

The 2-piston-compressor generates low vibrations and relatively low noise level. In combination with a very compact but effective isolation concept between CAirS® and bracket the integration in the vehicle becomes easy to apply.

Resume

With a relatively low energy consumption and a comfortable NVH-level the high integrated air supply system CAirS® provides an important component for future applications of air suspension systems, especially with regard to the new generation of electrical vehicles. Special features are low weight, low space and high availability for vehicle lowering as range extender.

Versions for air spring level control and also with additional damping control are prepared for series introduction. Further versions are available soon.

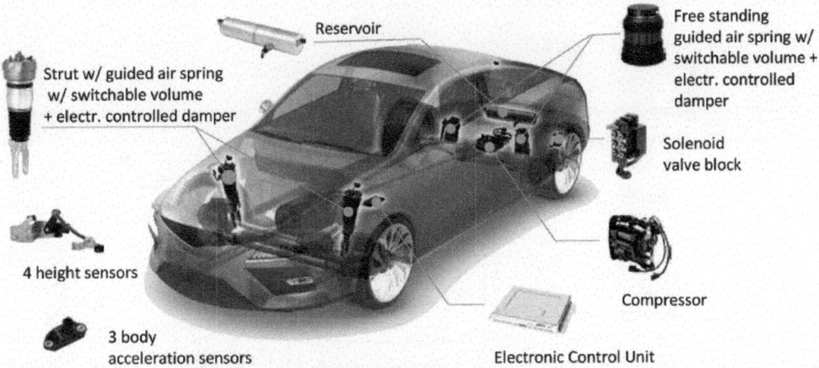

Picture 1: System components of a conventional Electronic controlled Air Suspension System

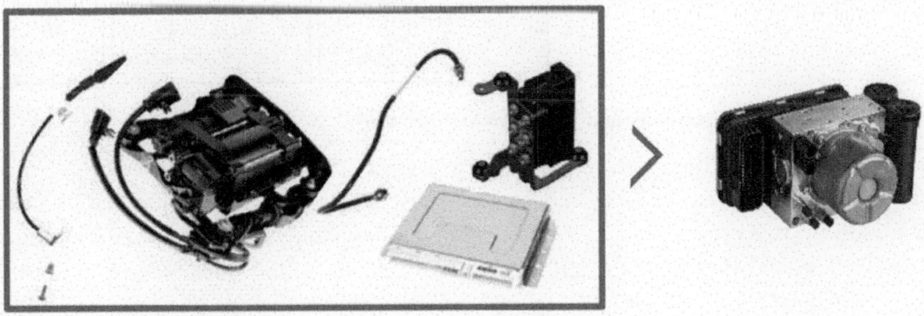

Picture 2: Comparison: single component approach versus integrated solution CAirS®

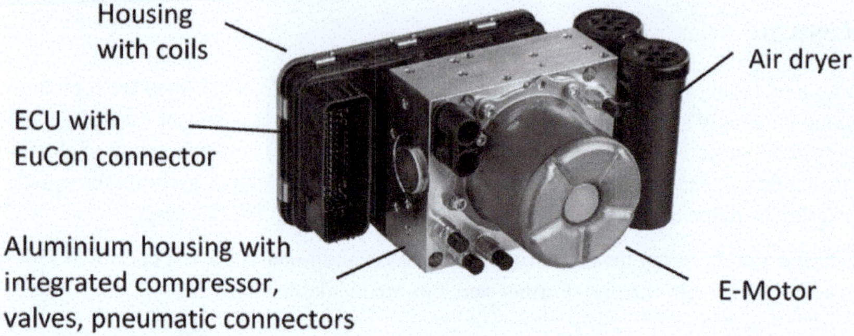

Picture 3: Continental Air Supply CAirS®

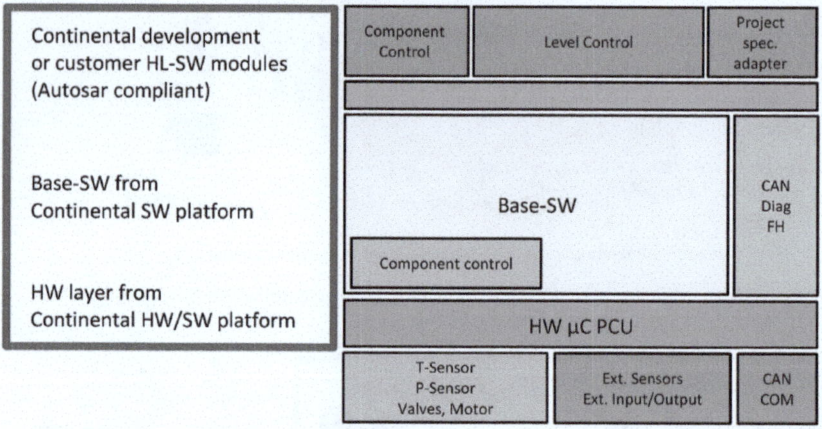

Picture 4: Schematic Electronic- and Software-Architecture

Explicit model predictive control of an active suspension system

J. Theunissen[1], A. Sorniotti[1], P. Gruber[1], S. Fallah[1],
M. Dhaens[2], K. Reybrouck[2], C. Lauwerys[2], B. Vandersmissen[2],
M. Al Sakka[2] and K. Motte[2]

1 University of Surrey, Guildford, UK
2 Tenneco Automotive Europe bvba, Sint-Truiden, BE

P. Pfeffer (Ed.), *9th International Munich Chassis Symposium 2018*, Proceedings,
https://doi.org/10.1007/978-3-658-22050-1_17

Abstract: Model predictive control (MPC) is increasingly finding its way into industrial applications, due to its superior tracking performance and ability to formally handle system constraints. However, the real-time capability problems related to the conventional implicit model predictive control (i-MPC) framework are well known, especially when targeting low-cost electronic control units (ECUs) for high bandwidth systems, such as automotive active suspensions, which are the topic of this paper. In this context, to overcome the real-time implementation issues of i-MPC, this study proposes explicit model predictive control (e-MPC), which solves the optimization problem off-line, via multi-parametric quadratic programming (mp-QP). e-MPC reduces the on-line algorithm to a function evaluation, which replaces the computationally demanding on-line solution of the quadratic programming (QP) problem. An e-MPC based suspension controller is designed and experimentally validated for a case study Sport Utility Vehicle (SUV), equipped with the active ACOCAR suspension system from the Tenneco Monroe product family. The target is to improve ride comfort in the frequency range of primary ride (< 4 Hz), without affecting the performance at higher frequencies. The proposed e-MPC implementations reduce the root mean square (RMS) value of the sprung mass acceleration by > 40% compared to the passive vehicle set-up for frequencies < 4 Hz, and by up to 19% compared to the same vehicle with a skyhook controller on the 0-100 Hz frequency range.

Keywords — Model predictive control, explicit solution, multi-parametric programming, active suspension, ride comfort

1 – Introduction

Semi-active and active suspensions with hydraulic actuators are widely used on production cars. The permanent challenge of improving ride comfort without increasing hardware costs requires the continuous enhancement of the system intelligence.

The skyhook algorithm is frequently used for primary ride improvement [1]. It is based on the introduction of a virtual damper between the sprung mass and a fixed surface, i.e., the 'sky'. Skyhook can be actuated in full only through an active suspension system, since the vertical velocity of the sprung mass can have a different sign from the relative velocity between the sprung mass and the unsprung mass. The skyhook algorithm was extended for use on controllable dampers, by introducing conditions based on the sign of the ratio of the two speeds [2]. [3] presents an extended skyhook algorithm, in which the damper force is a linear combination of a contribution proportional to the vertical velocity of the chassis (skyhook term), and a contribution proportional to the suspension deflection rate.

While skyhook reduces the vehicle body acceleration, the groundhook algorithm improves the unsprung mass dynamics, thus decreasing the oscillations of the vertical tyre load, which are detrimental to the vehicle handling performance [2]. The hybrid skyhook-groundhook controller [4-5] reduces both the dynamic tyre force and body acceleration. [6-7] introduce the semi-active suspension balance logic, targeting a reduction of the sprung mass acceleration. [8] proposes a form of groundhook blended with the balance logic. In the acceleration-driven damping algorithm [9], the shock absorber is deactivated when the body acceleration has opposite sign with respect to the suspension speed.

Model predictive control is a promising option for controllable suspension systems. In particular, i-MPC, in which the optimisation process is run on-line, requires significant computational power within the plant, which makes the practical implementations of i-MPC for high bandwidth systems, including electronic suspensions, difficult. Moreover, the implicit solution cannot be formally analysed a-priori from the viewpoint of its shape, stability and robustness. To the knowledge of the authors, most of the studies proposing i-MPC for electronic suspension systems are limited to simulation-based validations [10-16], with rare exceptions such as [17], using a high-performance 300 MHz Alpha processor.

This paper discusses an active suspension system based on e-MPC (see [18-19] for the theory). With e-MPC the optimisation problem is solved off-line, i.e., explicitly, which reduces the on-line algorithm to a function evaluation. As a consequence, e-MPC requires limited on-line computational power compared to i-MPC, while providing similar control performance and ability to handle constraints. On the other hand, the challenges of e-MPC are the increased design complexity and random-access memory (RAM) demand. e-MPC has already been implemented and, to some extent, experimentally validated on semi-active suspensions [20-23]. In all cases the simple two-mass quarter-car model was used for control system design. However, to the knowledge of the authors, e-MPC has not been proposed for fully active suspensions so far, nor model predictive control for active suspension has ever been implemented on automotive grade microprocessors. This gap is partially covered by this contribution, discussing e-MPC implementations for active suspension systems and their experimental validation on a vehicle demonstrator, including performance comparison with a skyhook controller.

The paper is organised as follows. Section 2 describes the model for control system design. Section 3 deals with the control system and mp-QP problem formulations. Finally, the explicit control law implementation and the experimental results are presented in Section 4.

2 – Model for control system design

The quarter car model in Fig. 1 is used as a basis for control system design, i.e., as prediction model. The active suspension component is a hydraulic actuator, generating an ideal force input, $u(t)$, without delays or actuation dynamics, where t is time.

The equations of motion are:

$$\begin{cases} m_1\ddot{x}_1 + k_1(x_1 - x_2) + c_1(\dot{x}_1 - \dot{x}_2) + u = 0 \\ m_2\ddot{x}_2 + k_1(x_2 - x_1) + k_2(x_2 - w) + c_1(\dot{x}_2 - \dot{x}_1) + c_2(\dot{x}_2 - \dot{w}) - u = 0 \end{cases} \quad (1)$$

where m_1 and m_2 are the sprung and unsprung masses; k_1, k_2, c_1 and c_2 are the vertical suspension stiffness, the vertical tyre stiffness, the vertical damping coefficient associated with the passive suspension components, and the vertical tyre damping coefficient; and x_1, x_2 and w are the vertical displacement of the sprung mass, the vertical displacement of the unsprung mass, and the vertical displacement of the road profile.

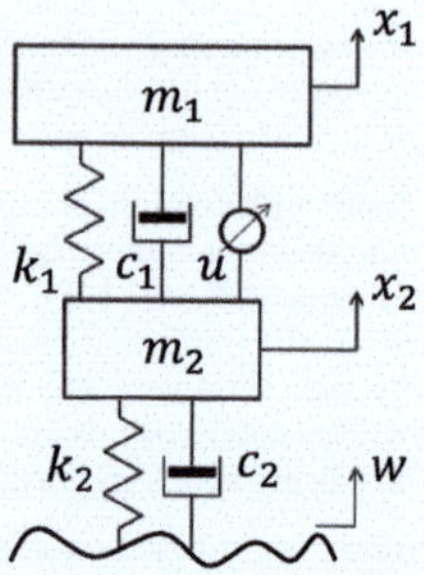

Fig. 1 – Quarter car model with active suspension system.

The system can be converted into a continuous time state-space notation:

$$\begin{cases} \dot{\boldsymbol{x}}(t) = \boldsymbol{A}\boldsymbol{x}(t) + \boldsymbol{B}u(t) + \boldsymbol{E}\boldsymbol{W}(t) \\ y(t) = \boldsymbol{C}\boldsymbol{x}(t) + \boldsymbol{D}u(t) \end{cases} \tag{2}$$

where $\boldsymbol{A}$, $\boldsymbol{B}$, $\boldsymbol{C}$ and $\boldsymbol{D}$ are the system matrices, and $\boldsymbol{E}$ is the disturbance matrix. The road input is represented by the column vector $\boldsymbol{W}(t) = [w \quad \dot{w}]^T$, where $\dot{w}$ is the vertical velocity of the road at the tyre contact point. The output, $y(t) = \ddot{x}_1$, is the acceleration of the sprung mass.

e-MPC is based on a state feedback law. Hence, the controller performance depends on the accuracy and appropriate selection of the measured or estimated states. In the specific implementation of this study, the state vector, $\boldsymbol{x}(t) = [x_1 \quad \dot{x}_1 \quad x_1 - x_2 \quad \dot{x}_1 - \dot{x}_2]^T$, contains the position and speed of the sprung mass, and the suspension displacement and deflection rate. The estimates of x_1 and $\dot{x}_1$ are obtained through the band-pass filtering and mathematical integration of the vertical acceleration measurements of the vehicle body. $x_1 - x_2$ is estimated through the direct measurement of the active suspension actuator displacement, and consideration of the suspension installation ratio. $\dot{x}_1 - \dot{x}_2$ is obtained through differentiation of $x_1 - x_2$ by using the hybrid smooth derivative method [24]. $\boldsymbol{E}$ represents the influence of the unknown disturbances w and $\dot{w}$, and is neglected during the e-MPC design.

3 – Control system formulation

3.1 – System prediction

The discrete state-space formulation of the vehicle model is:

$$\begin{cases} \boldsymbol{x}[\mathrm{k}+1] = \boldsymbol{A_d}\boldsymbol{x}[\mathrm{k}] + \boldsymbol{B_d}u[\mathrm{k}] \\ y[\mathrm{k}] = \boldsymbol{C_d}\boldsymbol{x}[k] + \boldsymbol{D_d}u[\mathrm{k}] \end{cases} \tag{3}$$

Given the initial state, $\boldsymbol{x}[\mathrm{k}]$, the initial control input, $u[\mathrm{k}]$, and the system (3), the output over the prediction horizon, $\hat{\boldsymbol{y}}$, is calculated as:

$$\hat{\boldsymbol{y}} = \begin{bmatrix} \boldsymbol{C_d A_d} \\ \boldsymbol{C_d A_d}^2 \\ \vdots \\ \boldsymbol{C_d A_d}^p \end{bmatrix}_{\mathrm{p\times1}} \boldsymbol{x}[\mathrm{k}] + \begin{bmatrix} \boldsymbol{C_d B_d} \\ \boldsymbol{C_d A_d B_d} \\ \vdots \\ \boldsymbol{C_d A_d}^{p-1}\boldsymbol{B_d} \end{bmatrix}_{\mathrm{p\times1}} u[\mathrm{k}] + \begin{bmatrix} \boldsymbol{D_d} & 0 & 0 & 0 \\ \boldsymbol{C_d B_d} & \ddots & 0 & 0 \\ \vdots & \ddots & \ddots & 0 \\ \boldsymbol{C_d A_d}^{p-2}\boldsymbol{B_d} & \cdots & \boldsymbol{C_d B_d} & \boldsymbol{D_d} \end{bmatrix}_{\mathrm{p\times n}} \hat{\boldsymbol{u}} \tag{4}$$

which, more concisely, is expressed as:

$$\hat{\boldsymbol{y}} = \boldsymbol{\Lambda x}[\mathrm{k}] + \boldsymbol{\Theta_{u0}}u[\mathrm{k}] + \boldsymbol{\Theta_u}\hat{\boldsymbol{u}} \tag{5}$$

where:

$$\hat{\boldsymbol{y}} = \begin{bmatrix} y[\mathrm{k}+1] \\ \vdots \\ y[\mathrm{k}+\mathrm{p}] \end{bmatrix}, \quad \hat{\boldsymbol{u}} = \begin{bmatrix} u[\mathrm{k}+1] \\ \vdots \\ u[\mathrm{k}+\mathrm{n}] \end{bmatrix} \tag{6}$$

n and p are the control horizon and prediction horizon, respectively. The states over the prediction horizon, $\hat{\boldsymbol{x}}$, are given by:

$$\hat{\boldsymbol{x}} = \boldsymbol{\Psi x}[\mathrm{k}] + \boldsymbol{\Omega_{u0}}u[\mathrm{k}] + \boldsymbol{\Omega_u}\hat{\boldsymbol{u}} \tag{7}$$

in which:

$$\hat{\boldsymbol{x}} = \begin{bmatrix} \boldsymbol{x}[\mathrm{k}+1] \\ \vdots \\ \boldsymbol{x}[\mathrm{k}+\mathrm{p}] \end{bmatrix} \tag{8}$$

$\hat{\boldsymbol{u}}$ is the input over the prediction horizon. The matrices $\boldsymbol{\Psi}$, $\boldsymbol{\Omega}_{u0}$ and $\boldsymbol{\Omega}_{\mathrm{u}}$ are calculated from the system model (3).

3.2 – Objective function

The general goal of suspension design is the optimisation of ride comfort, suspension rattle space and road holding. The ride comfort improvement is achieved through the reduction of the vehicle body acceleration levels, while limiting chassis motion as much as possible. Hence, this study uses a cost function penalising $\ddot{x}_1$, x_1, $x_1 - x_2$ and the control effort u. The continuous form of the performance index to be minimised, J_{e-MPC}, is:

$$J_{e-MPC} = \int_0^T \left(\frac{\rho_1}{\mathrm{n}_1} \ddot{x}_1^2 + \frac{\rho_2}{\mathrm{n}_2} (x_1 - x_2)^2 + \frac{\rho_3}{\mathrm{n}_3} x_1^2 + \frac{\rho_4}{\mathrm{n}_4} u^2 \right) dt \tag{9}$$

where ρ_i and n_i are the weighting and normalisation factors, respectively, and T is the period of observation, i.e., the duration of the prediction horizon.

3.3 – mp-QP problem formulation

J_{e-MPC} is re-arranged to be consistent with the discretised system prediction formulation of Section 3.1. J_{e-MPC} is quadratic, and is used for the following minimisation problem:

$$\min_{\hat{\boldsymbol{u}}} (\hat{\boldsymbol{y}}^T \mathbf{Q}_1 \hat{\boldsymbol{y}} + \hat{\boldsymbol{x}}^T \mathbf{Q}_2 \hat{\boldsymbol{x}} + \hat{\boldsymbol{u}}^T \mathbf{R} \hat{\boldsymbol{u}}) \tag{10}$$

where $\mathbf{Q}_1$, $\mathbf{Q}_2$ and $\mathbf{R}$ contain the factors ρ_i and n_i of (9). Through appropriate re-arrangements and simplifications, the model predictive control formulation can be represented by the following QP problem:

$$\min_{\hat{\boldsymbol{u}}} \frac{1}{2} \hat{\boldsymbol{u}}^T \mathbf{H} \hat{\boldsymbol{u}} + \boldsymbol{x}'[\mathrm{k}]^T \mathbf{F} \hat{\boldsymbol{u}} \tag{11}$$

where $\mathbf{H}$ is the Hessian matrix, and $\mathbf{F}$ includes the physical system parameters and the weighting and normalisation factors. $\boldsymbol{x}'[\mathrm{k}]$ contains the initial states of the system as well as the initial actuator force.

A conventional i-MPC would execute an on-line optimisation at each time step, Δt, for a given value of $\boldsymbol{x}'[\mathrm{k}]$, and the control law $u = u(\boldsymbol{x}')$ would be implicitly obtained by the QP solver. In the e-MPC case the optimisation is performed off-line. The QP is solved for the defined range of $\boldsymbol{x}'$, which generates the explicit solution, $u = u(\boldsymbol{x}')$. The optimisation problem becomes an mp-QP problem, generally described as follows:

$$\min_{\hat{u}} \frac{1}{2}\hat{\mathbf{u}}^T \mathbf{H}\hat{\mathbf{u}} + x'^T \mathbf{F}\hat{\mathbf{u}} + \frac{1}{2}x'^T \mathbf{Y}x' \tag{12}$$

subject to:

$$\mathbf{N}\hat{\mathbf{u}} \leq \mathbf{M_1} + \mathbf{M_2}x' \tag{13}$$

where $\mathbf{Y}$, $\mathbf{N}$, $\mathbf{M_1}$ and $\mathbf{M_2}$ are constant matrices. The constraints are typically related to actuator force and its rate. The last term in (12) is neglected, since it does not depend on $\hat{\boldsymbol{u}}$.

The solutions of the mp-QP problem is the piecewise affine function U^*, which associates the corresponding $\hat{\boldsymbol{u}}$ to each $\boldsymbol{x}'$. e-MPC uses the control input at the first time step only, i.e., $u(\boldsymbol{x}') = [\mathrm{I} \quad 0 \cdots 0]U^*$. Hence, the explicit representation of the control action is a piecewise affine state feedback law defined on a polyhedral partition of the state-space:

$$u(\boldsymbol{x}') = \begin{cases} \mathbf{L_1}\boldsymbol{x}' + \mathbf{p_1}, & \mathbf{O_1}\boldsymbol{x}' \leq \mathbf{S_1} \\ \vdots & \vdots \\ \mathbf{L_j}\boldsymbol{x}' + \mathbf{p_k}, & \mathbf{O_M}\boldsymbol{x}' \leq \mathbf{S_j} \end{cases} \tag{14}$$

where $\mathbf{L_i}$, $\mathbf{p_i}$, $\mathbf{O_i}$ and $\mathbf{S_i}$ are constant matrices.

3.4 – Distributed controller

To reduce the off-line computation time and the on-line RAM requirements of the explicit solution, four controllers based on quarter car models are used, i.e., one controller for each suspension system. A prediction model including pitch dynamics would imply a larger number of parameters per mp-QP problem, and more demanding RAM requirements.

4 – Controller implementation and experimental evaluation

4.1 – Vehicle demonstrator

The developed controller was implemented and experimentally validated on a sport utility vehicle (SUV) demonstrator (see Fig. 2) with a hydraulic active suspension system, i.e., the Tenneco Monroe intelligent suspension – ACOCAR. The sensor set and valves are identical to those of the Tenneco CVSA2 (continuously variable semi-active system with two valves) suspension technology [25]. The ACOCAR actuators are pressurised by means of a pump, which allows inputting energy into the system and actively controlling the actuation forces. The low level actuator controller, which calculates the valve currents and reference pump speed as functions of the actuator force demand and speed, already exists and is fully calibrated.

The SUV demonstrator was used to compare the experimental performance of the ACOCAR e-MPC implementations with those of:

- The passive set-up of the car. This was obtained by applying zero currents to the actuation valves, which represents the fail-safe state of the ACOCAR system, corresponding to a suspension tuning that is very close to the one of the passive version of the SUV.
- A conventional production-ready skyhook algorithm for active suspensions, configured with high gains. The adopted skyhook damping coefficients for the heave, pitch and roll motions are respectively 10,000 Ns/m, 12,000 Nms/rad and 12,000 Nms/rad.

The comparison was carried out for the excitation profile of a typical ride comfort road, i.e., the Blauwe Kei road at the Ford Lommel proving ground in Belgium, which was reproduced by means of a Schenck Instron 4-poster test rig, exciting the SUV demonstrator.

Fig. 2 – The SUV demonstrator on the 4-poster test rig.

4.2 – Controller implementation

The first step of the e-MPC implementation process was the validation of the front and rear quarter car models. This is based on experimental results obtained from measurements on the 4-poster test rig, for a set-up of the SUV demonstrator with the least possible damping.

The following variables are available in the SUV demonstrator:

- Heave, pitch and roll displacements at the centre of gravity of the vehicle body, calculated from the vertical acceleration measurements at three points of the sprung mass. Coordinate transformations are applied to obtain the velocities and displacements of the top suspension mounts.
- Suspension displacements, calculated from the suspension rotation sensors and calibration maps.

A good modelling match was achieved up to ~15 Hz, which is in line with the prediction model bandwidth. In particular, the dynamics of the vehicle body and unsprung masses at the resonance frequencies of ~1-1.5 Hz and ~10-12 Hz, respectively, are captured well.

Then the mp-QP problems for the front and rear suspensions were solved with the multi-parametric toolbox 3 (MPT3) [26], for different sets of coefficients of J_{e-MPC}. Simulations of the implemented controllers with an experimentally validated vehicle model for control system assessment were used along the Blauwe Kei profile for the identification of the coefficients of J_{e-MPC} providing the most desirable e-MPC behaviour. The performance assessment was carried out with the same RMS-based performance indicators that will be reported in the following Table I (see Section 4.4).

In the control design phase a prediction horizon $p = 5$ and a control horizon $n = 5$ were adopted, with a controller sample time $\Delta t = 10$ ms. Each actuator force was constrained to ±5,000 N. The suspension installation ratio is applied to calculate the actuator reference force, from the force demand at the wheel output by the controller, i.e., $u(t)$. In particular, at the completion of the process, two e-MPC settings, called e-MPC$_1$ and e-MPC$_2$ with objective function tunings J_{e-MPC_1} and J_{e-MPC_2}, were considered for further experimental evaluation:

- e-MPC$_1$. Compared to the skyhook, this setting reduces $\ddot{x}_1$ for frequencies < 4 Hz, without excessively increasing the acceleration levels above that frequency. The latter specification is based on the experience of Tenneco, showing that active suspensions with hydraulic actuators in parallel to the springs, as it is the case here, tend to reduce ride comfort at medium-high frequencies. This is caused by the typically limited actuator bandwidth (~8 Hz), the significant non-linearities, and the fact that the actuator force is delivered by changing the damping coefficient.
- e-MPC$_2$. With respect to the skyhook, this setting targets similar performance around the resonance frequency of the sprung mass (1-1.5 Hz), and the reduction of the acceleration levels at frequencies > 4 Hz. In comparison with the e-MPC$_1$, the e-MPC$_2$ increases the penalty on $\ddot{x}_1$ and reduces the penalty on x_1 in J_{e-MPC}.

4.3 – The explicit solution of the e-MPC$_1$

This section discusses the explicit solution of the e-MPC$_1$. The control law, $u = u(\boldsymbol{x}')$, consists of a set of functions with affine gains over 1217 polyhedral regions within the state-space. Fig. 3 shows the state-space partition, sliced at $x_1 - x_2 = 0$ and $\dot{x}_1 - \dot{x}_2 = 0$. In region 1 the control law varies as a function of x_1 and $\dot{x}_1$. In regions 2, 3, 4 and 5 the actuator force is saturated.

Fig. 3 also reports the simulation results for the front left SUV corner in the passive and e-MPC$_1$ set-ups, in the form of state trajectories for the Blauwe Kei road input. Interestingly, in the e-MPC$_1$ case the system operation is limited to one region, i.e., the first sub-partition bounded by:

$$\begin{bmatrix} 0 & 1 & 0 & 0 \\ -0.9678 & 0.2422 & 0.0682 & -0.0052 \\ -0.9690 & 0.1225 & 0.2146 & -0.0041 \\ -0.9493 & 0.0835 & 0.3030 & -0.0024 \\ -0.9402 & 0.0673 & 0.3338 & -0.0006 \\ -0.9483 & 0.0607 & 0.3116 & 0.0010 \\ \vdots & \vdots & \vdots & \vdots \\ 0.9989 & 0.0315 & -0.0331 & -0.0029 \end{bmatrix}_{25\times4} \boldsymbol{x}' \leq \begin{bmatrix} 2 \\ 0.5330 \\ 0.2914 \\ 0.2129 \\ 0.1843 \\ 0.1805 \\ \vdots \\ 0.0451 \end{bmatrix} \tag{15}$$

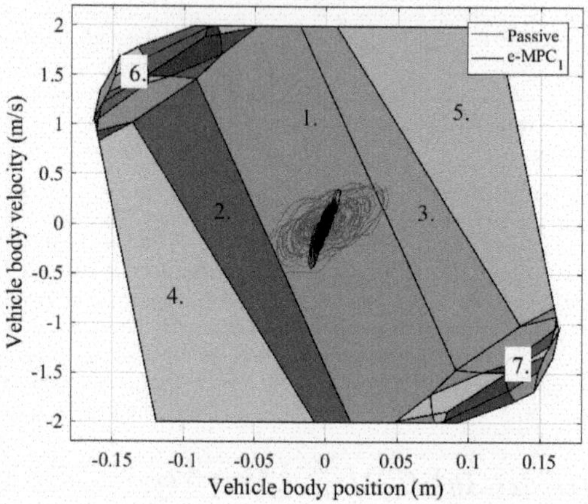

Fig. 3 – The e-MPC$_1$ state-space partition sliced at $x_1 - x_2 = 0$ and $\dot{x}_1 - \dot{x}_2 = 0$, with the simulated trajectories of the front left SUV corner with the passive set-up and e-MPC$_1$ set-up on the Blauwe Kei road.

Although Fig. 3 depicts only a two-dimensional slice of the four-dimensional state-space partition, this behaviour was verified on the four-dimensional partition. As a consequence, on the specific road the control law could be replaced by the following single affine function of the states:

$$u(\boldsymbol{x}') = [1.1086 \cdot 10^5 \quad 0.0350 \cdot 10^5 \quad -0.0368 \cdot 10^5 \quad -0.0033 \cdot 10^5]\,\boldsymbol{x}' + [9.0949 \cdot 10^{-13}] \tag{16}$$

resulting in a significant reduction of the RAM requirements. Obviously, this would not be advisable during operation on more aggressive road profiles. The potential simplification of the e-MPC control law, either through formal and systematic methods (see [27]) or the empirical observation of the most commonly used sub-partitions, will be the topic of future research.

The whole explicit e-MPC$_1$ solution from the MPT3 toolbox was uploaded on the dSPACE AutoBox rapid control prototyping unit of the SUV demonstrator.

4.4 – Experimental results and comparisons

This section reports the experimental SUV results on the 4-poster test rig along the assessed mission profile. In particular, Fig. 4 and Fig. 5 plot the frequency response characteristics of the power spectral densities (PSDs) of the heave position and heave acceleration of the centre of gravity of the sprung mass for the four considered set-ups, i.e., passive, skyhook, e-MPC$_1$ and e-MPC$_2$. Table I shows the corresponding root mean square (RMS) values for the frequency ranges below and above 4 Hz, corresponding to the so-called primary ride and secondary ride, and for the 0-100 Hz ride comfort frequency spectrum.

In the 0-4 Hz frequency range, the e-MPC$_1$ reduces the sprung mass heave acceleration by 43% with respect to the passive set-up, and by 26% compared to the skyhook, which is a major primary ride enhancement. On the other hand, above 4 Hz the skyhook strategy increases the vertical acceleration by 79% compared to the passive set-up. This phenomenon, which is well-known to the technical specialists of Tenneco, is attributed to the limited actuation dynamics of the hydraulic active suspension system. In fact, simulations of the system response with actuators with better dynamic properties did not show such a trend. This behaviour brings a deterioration of the secondary ride. In the same frequency range (> 4 Hz) the e-MPC$_1$ shows an increase in the vertical acceleration level of 65% compared to the passive set-up, which is an 8% reduction of the secondary ride problem of the skyhook. On the 0-100 Hz frequency range, the e-MPC$_1$ reduces the skyhook vibration levels by 11%.

The e-MPC$_2$ was implemented with the purpose of attenuating the secondary ride issues of the skyhook and e-MPC$_1$, while providing good primary ride performance. Fig. 5 shows that at approximately the resonance frequency of the sprung mass, i.e., at 1-1.5 Hz, the e-MPC$_2$ and the skyhook give origin to similar responses. The e-MPC$_2$ improves the skyhook acceleration performance by 22% in the 0-4 Hz frequency range, and by 19% above 4 Hz. Moreover, on the whole frequency range the e-MPC$_2$ produces lower acceleration levels than the e-MPC$_1$, which is expected given the increased penalty on $\ddot{x}_1$ in J_{e-MPC_2}. The conclusion is that for the given actuators the e-MPC$_2$ conjugates a significant enhancement of the primary ride, without an excessive penalisation of the secondary ride performance.

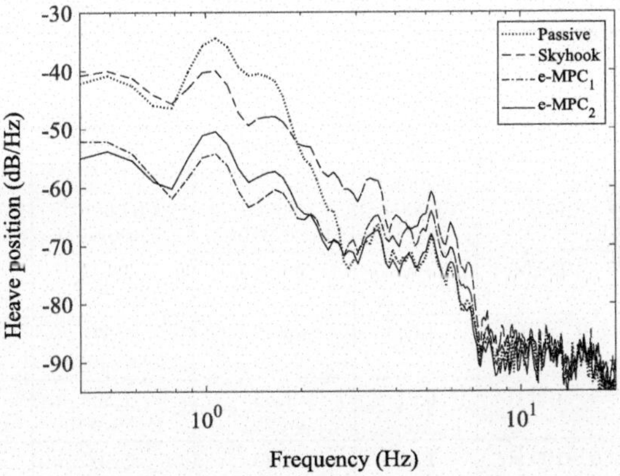

Fig. 4 – PSD of the heave displacement of the centre of gravity of the sprung mass for the passive, skyhook, e-MPC$_1$ and e-MPC$_2$ set-ups.

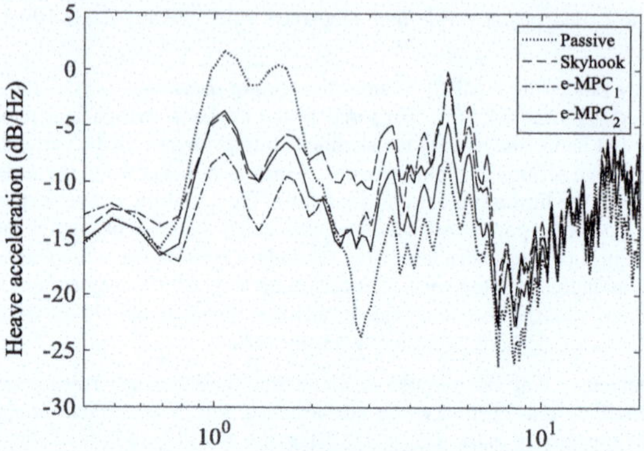

Fig. 5 – PSD of the heave acceleration of the centre of gravity of the sprung mass for the passive, skyhook, e-MPC$_1$ and e-MPC$_2$ set-ups.

Table I – RMS values for the heave position and acceleration of the centre of gravity of the sprung mass for the passive, skyhook, e-MPC$_1$ and e-MPC$_2$ set-ups.

	RMS values			
	Passive	Skyhook (wrt Passive)	e-MPC$_1$ (wrt Passive/Skyhook)	e-MPC$_2$ (wrt Passive/Skyhook)
Heave position: 0 – 4 Hz (m)	0.0132	0.0099 (-25%)	0.0024 (-82%/-75%)	0.0023 (-82%/-76%)
Heave position: 4 – 100 Hz (m)	0.0004	0.0009 (+125%)	0.0006 (+50%/-33%)	0.0004 (+0%/-55%)
Heave position: 0 – 100 Hz (m)	0.0132	0.0099 (-25%)	0.0024 (-82%/-75%)	0.0023 (-82%/-76%)
Heave acceleration: 0 – 4 Hz (m/s^2)	1.01	0.77 (-24%)	0.57 (-43%/-26%)	0.60 (-41%/-22%)
Heave acceleration: 4 – 100 Hz (m/s^2)	0.91	1.63 (+79%)	1.50 (+65%/-8%)	1.32 (+45%/-19%)
Heave acceleration: 0 – 100 Hz (m/s^2)	1.36	1.80 (+32%)	1.60 (+18%/-11%)	1.45 (+6%/-19%)

5 – Conclusion

To the knowledge of the authors, for the first time this paper discussed the application of e-MPC to an active suspension system for passenger cars, mainly targeting primary ride improvements. Multi-parametric quadratic programming was used to solve the control problem formulation, based on a quarter car model. The solution is represented by explicit control laws, based on state feedback. e-MPC brings a reduction of the computational requirements of the control system hardware with respect to i-MPC, as the on-line implementation consists of a function evaluation. The results show significant benefits of the developed controllers with respect to a pre-existing skyhook algorithm. In fact, the e-MPC_1 and e-MPC_2 implementations reduce the vehicle body acceleration levels by 26% and 22%, respectively, in the frequency range below 4 Hz, and by 8% and 19%, respectively, above 4 Hz. Future developments will focus on the systematic fine-tuning of the objective function for e-MPC design, and the assessment of the controllers on different actuation hardware.

Acknowledgement

The authors would like to thank Tenneco Automotive Europe (Sint-Truiden, Belgium) for their contribution to this research. This paper is a modified version of the conference paper [28].

References

1. Karnopp D., Crosby M.J. and Harwood RA. Vibration control using semi-active force generators, J. Eng. Ind., 1974, 96(2), 619–626.
2. Valasek M. and Novak M. Ground Hook for Semi-Active Damping of Truck's Suspension, Proceedings of the CTU Workshop 96, Engineering Mechanics, 1996, 22-24.
3. Lauwerys C. Control of active and semi-active suspension systems for passenger cars, PhD thesis, Catholic University of Leuven, 2005.
4. Ahmadian M. A hybrid semi-active control for secondary suspension applications, Proceedings of the 6th ASMA Symposium on Advanced Automotive Technologies, 1997, 16-21.
5. Goncalves F.D. and Ahmadian M. A hybrid control policy for semi-active vehicle suspension, J. Shock and Vib., 2003, 10(1), 59-69.
6. Rakheja S. and Sankar S. Vibration and Shock Isolation Performance of a Semi-Active On-Off Damper, J. Vib., Acoust., Stress, Reliab., 1985, 107(4), 398-403.
7. Stammers C.W. and Sireteanu T. Vibration control of machines by using semi-active dry friction damping, J. Snd. and Vib., 1997, 209(4), 671-684.
8. Koo J.H., Setareh M. and Murray T.M. In search of suitable control methods for semi-active tuned vibration absorbers, J. Vib. and Ctrl., 2004, 10, 163-174.
9. Savaresi S.M. and Silani E. On the Optimal Predictive Control Algorithm for Comfort-Oriented Semi-Active Suspensions, SAE Technical Paper, 2004-01-2088.
10. Nguyen M.Q., Canale M., Sename O. and Dugard L. A Model Predictive Control approach for semi-active suspension control problem of a full car, Proceedings of the IEEE 55th Conference on Decision and Control, 2016, 721-726.
11. Hu Y., Chen M.Z.Q. and Hou Z. Multiplexed model predictive control for active vehicle suspensions, Int. J. of Ctrl., 2015, 88(2), 347-363.

12. Mehra R.K., Amin J.N., Hedrick K.J., Osorio C. and Gopalasamy S. Active Suspension Using Preview Information and Model Predictive Control, Proceedings of the IEEE International Conference on Control Applications, 1997, 860-865.
13. Ahmed M. and Svaricek F. Preview Optimal Control of Vehicle Semi-active Suspension Based on Partitioning of Chassis Acceleration and Tire Load Spectra, Proceedings of the IEEE European Control Conference, 2014, 1669-1674.
14. De Bruyne S., Van der Auweraer H. and Anthonis J. Preview Control of a Constrained Hydraulic Active Suspension System, Proceedings of the 51st IEEE Conference on Decision and Control, 2015, 4400-4405.
15. Savaresi S.M., Poussot-Vassal C., Spelta C., Sename O. and Dugard L. Semi-Active Suspension Control Design for Vehicles. Chapter 5: Optimal Strategy for Semi-Active Suspensions and Benchmark, Elsevier, 2010, 91–106.
16. Göhrle C., Wagner A., Schindler A. and Sawodny O. Active Suspension Controller using MPC based on a full-car model with preview information, Proceedings of the IEEE American Control Conference, 2012, 497-502.
17. Donahue M.D. and Hedrick J.K. Implementation of an Active Suspension, Preview Controller for Improved Ride Comfort, In: Nonlinear and Hybrid Systems in Automotive Control, Springer, 2003.
18. Bemporad A., Morari M., Dua V. and Pistikopoulos E.N. The Explicit Solution of Model Predictive Control via Multiparametric Quadratic Programming, Proceedings of the IEEE American Control Conference, 2000, 2, 872-876.
19. Bemporad A., Borrelli F. and Morari M. Model Predictive Control Based on Linear Programming – The Explicit Solution, IEEE Trans. Autom. Ctrl., 2002, 47(12), 1974-1985.
20. Giorgetti N., Bemporad A. and Tseng H.E. Hybrid Model Predictive Control Application Towards Optimal Semi-Active Suspension, Proceedings of the IEEE International Symposium on Industrial Electronics, 2005, 391-398.
21. Csekő L.H., Kvasnica M. and Lantos B. Analysis of the explicit model predictive control for semi-active suspension, Polytech. Elec. Eng., 2010, 54(1-2), 41-58.
22. Canale M., Milanese M., Novara C. and Ahmad Z. Semi-Active Suspension Control Using "Fast" Model Predictive Techniques, IEEE Trans. Ctrl. Syst. Tech., 2006, 14, 1034-1046.
23. Paschedag T., Giua A. and Seatzu C. Constrained optimal control: an application to semiactive suspension systems, Int. J. of Syst. Sc., 2010, 41(7), 797-811.
24. Fallah S., Sorniotti A. and Gruber P. A Novel Robust Optimal Active Control of Vehicle Suspension Systems, IFAC Proceedings, 2014, 47(3), 11213-11218.
25. Monroe Intelligent Suspension Products, http://www.monroeintelligentsuspension.com/products/.
26. Kvasnica M. Multi-Parametric Toolbox 3, http://people.ee.ethz.ch/~mpt/3/, 2014.
27. Geyer T., Torrisi F.D. and Morari M. Optimal complexity reduction of polyhedral piecewise affine systems, Automatica, 2008, 44, 1728-1740.
28. Theunissen J., Sorniotti A., Gruber P., Fallah S., Dhaens M., Reybrouck K., Lauwerys C., Vandersmissen B., Al Sakka M., Motte K. Explicit model predictive control of active suspension systems, Proceedings of the International Conference on Advanced Vehicle Powertrains, 2017, 344-362.

Improvement of ride comfort by triple Skyhook control

Etsuo Katsuyama

Toyota Motor Corporation, Japan

P. Pfeffer (Ed.), *9th International Munich Chassis Symposium 2018*, Proceedings,
https://doi.org/10.1007/978-3-658-22050-1_18

Abstract

Various ride comfort control methods using suspension control devices have been proposed. This paper describes research into a simple control law that uses only sprung acceleration sensors to effectively reduce disturbance inputs, i.e., inputs both from the road surface and the driver, to the sprung mass of the vehicle. This control law can reduce disturbance inputs over a wide frequency range by applying skyhook control to the sprung mass using three elements instead of just the dampers: the dampers, springs, and inerters. Reflecting the number of elements, this control is called the triple skyhook control.

1 Introduction

The role of the driver is to pilot the vehicle along a target course at a desired speed by making steering, acceleration, and deceleration inputs. Since these inputs create longitudinal and lateral motions at the tire contact patches, which are far from the vehicle center of gravity, undesirable pitch and roll motions of the sprung mass may be generated. The motion of the vehicle body may also be affected by irregularities in the road surface profile. Such inputs are regarded as fundamentally undesirable and should be reduced as much as possible. In practice, body motion in response to driver inputs is suppressed at the design stage by optimizing suspension spring constants and shock absorber damping factors, as well as by optimizing the design of the suspension geometry. Suspension controls are also being researched and developed to further reduce vibration, and to achieve targets for both driver and road surface inputs.

One typical control method for ride comfort is skyhook damper control[1], which is an effective way of reducing vibration around the sprung resonance frequency. In addition, various research has been carried out into preview controls that use unsprung information and road surface profile information[3]-[7] to expand the control region to the 4 to 8 Hz frequency range, which is regarded as a source of discomfort for vehicle occupants[2], and to enhance vibration-damping performance. However, it is inevitable that the control laws for enhancing performance in this way are complex. Such controls also require additional sensors to detect the suspension stroke, state variables of the unsprung mass, and the road surface profile. Therefore, this paper proposes a methodology for reducing disturbances from both road surface and driver inputs over a wide frequency range without adding sensors and using an extremely simple control law. This paper also describes the robustness of the control with respect to changes in vehicle parameters, as well as the effects of control lag.

2 Conventional Controls

2.1 Control Effects

2.1.1 Skyhook Damper Control

Skyhook damper control (abbreviated below as "SH") is a typical ride comfort control method. It uses sensors attached to the sprung mass of the vehicle to provide velocity feedback control that damps the sprung mass without worsening inputs from the road surface. SH has been widely adopted because it is effective and simple to install. The effect of SH can be verified using the quarter-car model shown in Fig. 1. The equations of motion for the sprung and unsprung mass are expressed by Equations (1) and (2), respectively.

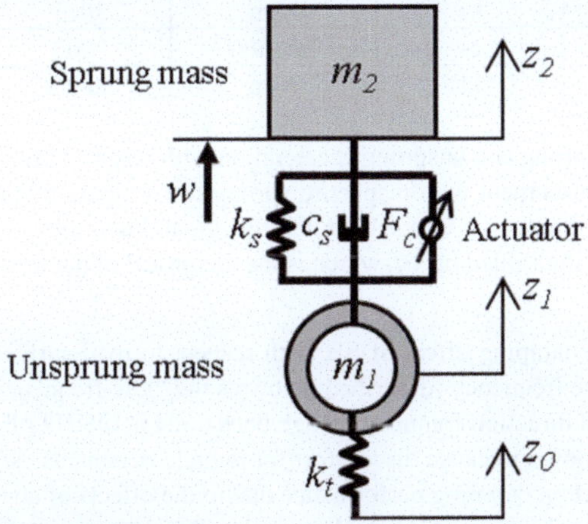

Figure 1 Quarter-Car Model

$$m_2 z_2 s^2 = (c_s s + k_s)(z_1 - z_2) + w + F_c(s) \quad (1)$$

$$m_1 z_1 s^2 = (c_s s + k_s)(z_2 - z_1) + k_t(z_0 - z_1) - F_c(s) \quad (2)$$

where, z_0, z_1, and z_2 are the vertical displacements of the road surface, unsprung mass, and sprung mass, w is the external force acting on the sprung mass, F_c is the control force of the suspension actuator, and s is the Laplace operator. The meanings and values of the other symbols are shown in Table 1. The control command value F_c (s) is applied as shown in Equation (3). c_{sh} is the control damping factor and the sign of the force is positive in the direction of repulsive force. In addition, as shown in Equation (4), $D(s)$ is the product of the first-order low pass filter (in which the cut-off frequency f_l that simulates the control response characteristics is 5.0 Hz) and the first-order high pass filter (in which the cut-off frequency f_h that assumes elimination of the integrated offset is 1.0 Hz).

$$F_c(s) = -c_{sh} z_2 s D(s) \tag{3}$$

$$D(s) = \frac{2\pi f_l}{(s + 2\pi f_l)} \frac{s}{(s + 2\pi f_h)} \tag{4}$$

Table 1 Parameters for calculation

Parameters	Symbol	Unit	Value
Sprung mass	m2	kg	500
Unsprung mass	m1	kg	50
Spring stiffness	ks	N/m	30e3
Damping coefficient	cs	N/(m/s)	2000
Tire stiffness	kt	N/m	300e3

Figure 2 shows Bode plots of the sprung and unsprung acceleration with respect to road displacement, and the sprung displacement with respect to sprung force inputs calculated using these equations. Note that $0.4 \times c_s$ was substituted for c_{sh}, and that the road displacement input amplitude was multiplied by $1/f$, which is the reciprocal of the input frequency f.

First, these plots demonstrate the damping effect of SH with respect to road surface inputs around the sprung resonance frequency (described below as the "low-frequency range"). In contrast, in the vertical direction frequency range between 4 and 8 Hz (the mid-frequency range), which is known to cause discomfort for vehicle occupants, SH actually has a slight negative effect on damping performance due to the effects of control lag. As shown by the bold line, vibration in both the low- and mid-frequency ranges can be suppressed by lowering the shock absorber damping factor c_s to counteract this effect. However, this measure causes unsprung vibration to deteriorate around the unsprung resonance frequency (the high-frequency range). Furthermore, although SH damps sprung vibration from inertial force inputs to the sprung mass caused by driver operations, this effect is also offset by lowering the shock absorber damping factor.

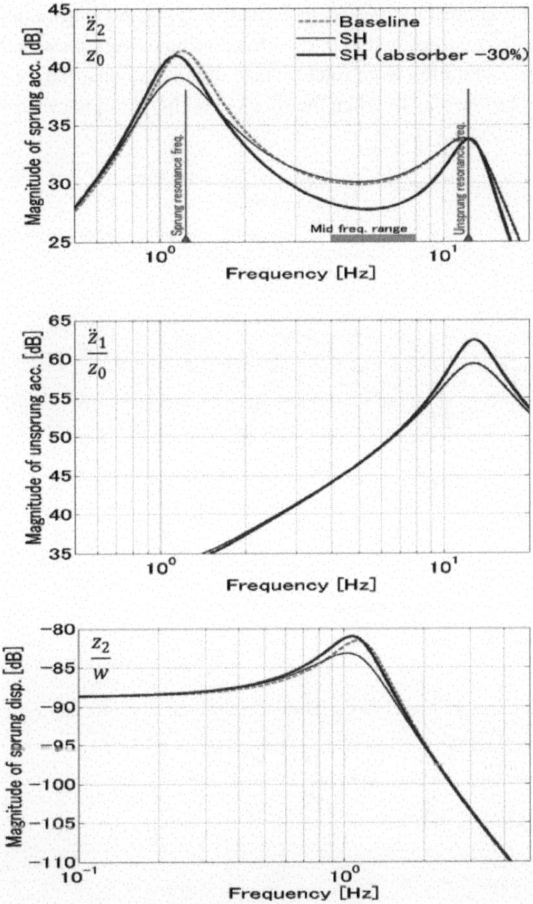

Figure 2 Effect of Skyhook Damper Control

2.1.2 Unsprung Negative Skyhook Damper Control

Mid-frequency vibration from road surface inputs can be suppressed using unsprung negative skyhook damper control (abbreviated below as "nSH"), which is a skyhook control of the unsprung mass that uses dampers with negative damping factors[8]. The control command value is applied as shown in Equation (5).

$$F_c(s) = -c_{sh} z_1 s D(s) \tag{5}$$

Figure 3 shows the effect of the control expressed in Equation (5) using the model shown in Fig. 1. In this case, c_{sh} was also defined as $0.4 \times c_s$. In addition to the substantial damping of sprung mid-frequency vibration from road surface inputs, the unsprung vibration does not deteriorate in the same way as when the shock absorber damping factor was lowered with the SH control.

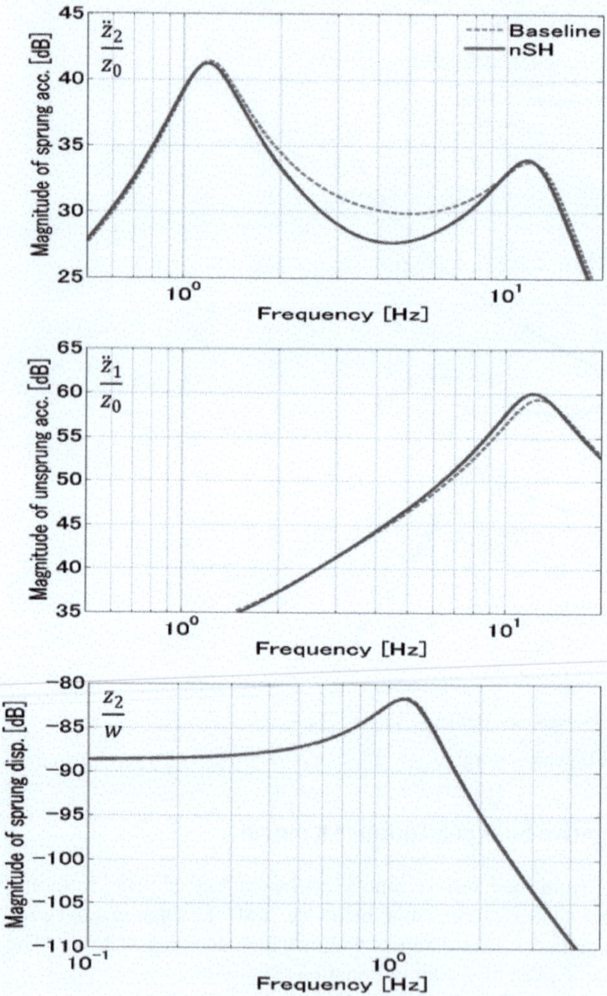

Figure 3 Effect of Unsprung Negative Skyhook Damper Control

The mechanism of this control can be described as follows. Equation (5) can be re-expressed as Equation (6). This shows that the control performs SH (with the control damping factor defined as c_{sh}) and reduces the shock absorber damping factor so that it becomes (original value – c_{sh}). The response of the control actuator becomes slower as the frequency increases and the control gain decreases. Therefore, while the damping effect described in the previous section can be obtained at low- and mid-frequencies, vibration at high-frequencies does not deteriorate because of a lack of control response. In this control, control lag is a key element that controls the performance balance between mid- and high-frequency ranges.

Furthermore, although this control does not require sensors on the sprung mass, it requires acceleration or equivalent sensors to be installed on the unsprung mass.

$$F_c(s) = -c_{sh} z_2 s D(s) - c_{sh}(z_1 - z_2) s D(s) \tag{6}$$

2.1.3 Unsprung Negative Skyhook Damper and Spring Control

As described above, nSH applies negative damping force in proportion to the unsprung velocity. In contrast, unsprung negative skyhook damper and spring control (abbreviated below as “nSH2”) can be used to simultaneously apply negative spring force in proportion to the unsprung displacement, as shown in Equation (7). In addition to the force applied by the shock absorbers, this control can also reduce the force applied by the suspension springs to the sprung mass in response to road surface inputs.

Figure 4 shows the effect of this control. The control damping factor c_{sh} and control spring constant k_{sh} are defined as $0.4 \times c_s$ and $0.4 \times k_s$, respectively. As a result, this control has a greater low-frequency sprung vibration damping effect in response to road surface inputs than nSH. Research into a control that uses road surface profile sensors and observers to obtain a damping effect at lower frequencies than the sprung resonance frequency has also been reported[7]. In practice, due to suspension stroke restrictions and the fact that roll control may not be required on some road surfaces, such as inward-facing cants on curves, the unsprung displacement required by this control may also be realistically obtained by applying a high pass filter to the integral of the unsprung velocity. In this case, although preview sensors are not required, sensors equivalent to those used by nSH remain necessary.

$$F_c(s) = -(c_{sh} s + k_{sh}) z_1 D(s) \tag{7}$$

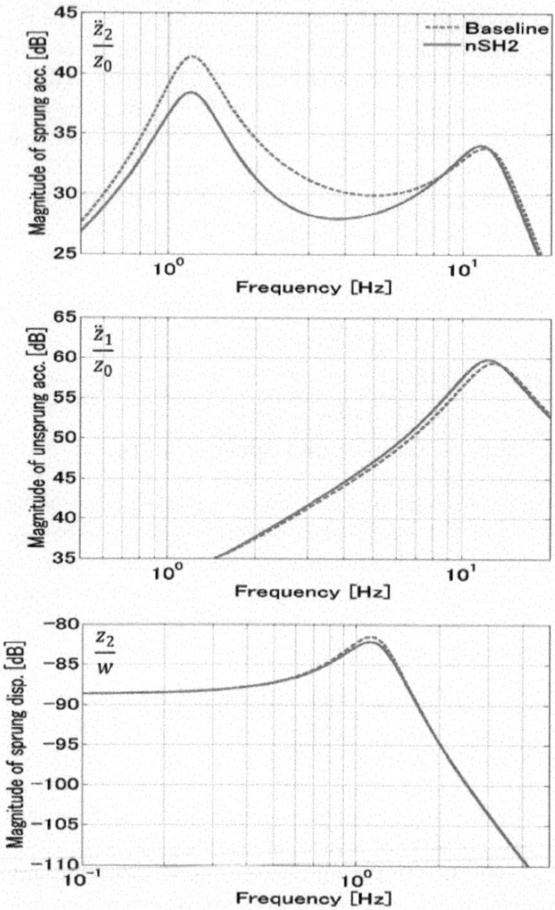

Figure 4 Effect of Unsprung Negative Skyhook Damper and Spring Control

2.2 Comparison of Controls

This section discusses these conventional controls using the quarter-car model shown in Fig. 1. Since the conventional control examples described above are all based on vertical displacement of the sprung and unsprung mass, the control force $F_c(s)$ of these controls can be defined by Equation (8). Figure 5 expresses this as a block diagram.

$$F_c(s) = H(s)z_1 + G(s)z_2 \tag{8}$$

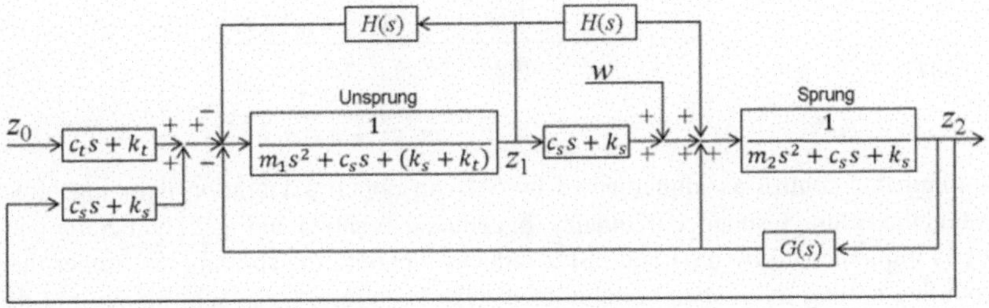

Figure 5 Block Diagram

Focusing on sprung motion, when Equation (8) is substituted into Equation (1), sprung displacement z_2 can be expressed by Equation (9). The first term on the right side is the input from the unsprung mass from the road surface, and the second term is the inertial force input from the driver.

$$z_2 = \frac{c_s s + k_s + H(s)}{m_2 s^2 + c_s s + k_s - G(s)} z_1 + \frac{1}{m_2 s^2 + c_s s + k_s - G(s)} w \qquad (9)$$

The locations of $H(s)$ and $G(s)$ in the above equation indicate that sprung information can control the denominator of the transfer function of the equation, and that unsprung information can control the numerator in response to unsprung displacement from road surface inputs and the like. Equations 10 to 12 show the control forces of SH, nSH, and nSH2 substituted into Equation (8), respectively.

SH: $$\left.\begin{array}{l} H(s) = 0 \\ G(s) = -c_{sh} s D(s) \end{array}\right\} \qquad (10)$$

nSH: $$\left.\begin{array}{l} H(s) = -c_{sh} s D(s) \\ G(s) = 0 \end{array}\right\} \qquad (11)$$

nSH2: $$\left.\begin{array}{l} H(s) = -(c_{sh} s + k_{sh}) D(s) \\ G(s) = 0 \end{array}\right\} \qquad (12)$$

Equations 13 to 15 show Equations 10 to 12 substituted into Equation (9).

SH: $$z_2 = \frac{c_s s + k_s}{m_2 s^2 + (c_s + c_{sh} D(s)) s + k_s} z_1 + \frac{1}{m_2 s^2 + (c_s + c_{sh} D(s)) s + k_s} w \qquad (13)$$

nSH: $$z_2 = \frac{(c_s - c_{sh} D(s)) s + k_s}{m_2 s^2 + c_s s + k_s} z_1 + \frac{1}{m_2 s^2 + c_s s + k_s} w \qquad (14)$$

nSH2:
$$z_2 = \frac{\left(c_s - c_{sh}D(s)\right)s + \left(k_s - k_{sh}D(s)\right)}{m_2s^2 + c_ss + k_s}z_1 + \frac{1}{m_2s^2 + c_ss + k_s}w \tag{15}$$

Denominator control as implemented by SH can obtain a vibration damping effect around the sprung resonance frequency. In contrast, as shown in Fig. 5, numerator control as implemented by nSH and nSH2 reduces inputs to the sprung mass rather than providing sprung mass state variable feedback. Therefore, these controls are not dependent on the characteristics of the sprung mass and can suppress vibration over a comparatively wider frequency range.

However, since numerator control requires information from the unsprung mass, it is necessary to add unsprung acceleration sensors, suspension stroke sensors, road surface profile preview sensors or the like, or to estimate unsprung mass state variables using an observer or other method. Consequently, to help avoid cost increases due to additional sensors, reduced reliability and durability in difficult driving conditions due to the installation of sensors close to the unsprung mass, and the effects of estimation error, the following section proposes a new control method that uses only acceleration sensors installed on the sprung mass.

3 Proposed Control

3.1 Control Law and Effect

Based on the restriction of using only sprung information, this research proposed a control law that defines the control command value as shown in Equation (16). Equation (17) can be obtained by substituting this equation into Equation (9). However, n=2.

$$\left.\begin{aligned} H(s) &= 0 \\ G(s) &= -\sum_{i=0}^{n} \alpha_i s^i D(s) \end{aligned}\right\} \tag{16}$$

$$z_2 = \frac{(c_ss + k_s)z_1 + w}{\left(m_2 + \alpha_2D(s)\right)s^2 + \left(c_s + \alpha_1D(s)\right)s + \left(k_s + \alpha_0D(s)\right)} \tag{17}$$

When the control constants α_2, α_1, and α_0 are applied as shown in Equation (18), the control force $F_c(s)$ can be expressed using Equation (19) based on Equations (8) and (16). In this case, Equation (17) becomes Equation (20).

$$[\alpha_2 \quad \alpha_1 \quad \alpha_0]^T = \alpha[m_2 \quad c_s \quad k_s]^T \tag{18}$$

$$F_c(s) = -\alpha(m_2s^2 + c_ss + k_s)z_2D(s) \tag{19}$$

$$z_2 = \frac{1}{1 + \alpha D(s)}\left(\frac{c_ss + k_s}{m_2s^2 + c_ss + k_s}z_1 + \frac{1}{m_2s^2 + c_ss + k_s}w\right) \tag{20}$$

When control lag is disregarded and $D(s)=1$, Equation (20) shows that the motion of the sprung mass in response to road surface and driver inputs is magnified by $1/(1+\alpha)$. Therefore, the proposed control can be regarded as a steady-state gain control rather than as dynamic control of the numerator or denominator. A simple way of considering the effect of this control is to envision driving on a smooth road with few surface profile irregularities in a vehicle with a lower center of gravity. nSH2 can also function as a steady-state control if c_{sh} and k_{sh} are applied so that the factors of the numerators are magnified equally. However, the proposed control may be regarded as superior since nSH2 is only effective against road surface inputs and because it requires unsprung information.

The control range is determined by the passing band of $D(s)$ and does not depend on vehicle characteristics. In the same way as nSH2, this band requires, for example, application of a high pass filter to eliminate the effects of changes in the vehicle pitch angle (due to the road surface gradient) and changes in the vehicle heave displacement at low frequencies. It also requires a cut-off at high frequencies using a low pass filter or through the control actuator response design to prevent deterioration of unsprung vibration.

As expressed by Equation (19), the proposed control uses force proportional to the displacement, velocity, and acceleration of the sprung mass. Therefore, it can be regarded as a skyhook control of the sprung mass using three elements: the springs, dampers, and inerters[9]. Figure 6 illustrates this control embodied as a suspension control device. Based on the number of control elements, this control method is called the triple skyhook control (tSH).

Figure 7 shows the effect of this control in the same way as the controls described above. In the control, α is applied so that the control steady-state gain $1/(1+\alpha)$ is 0.6. The proposed control obtains virtually the same performance as nSH2 in response to road surface inputs, while also enhancing performance in response to sprung force inputs. Therefore, this control achieves an equivalent or greater effect than conventional controls using a simple control law and only sprung acceleration sensors.

Control lag remains an important element that determines mid- and high-frequency performance. However, mid- and high-frequency vibration can be reduced with good balance by changing the first-order lag cut-off frequency f_l, which simulates control lag, to 3.0 Hz, which is less responsive than the 5.0 Hz frequency adopted by the conventional control. The reason for this is described in Section 3.3.

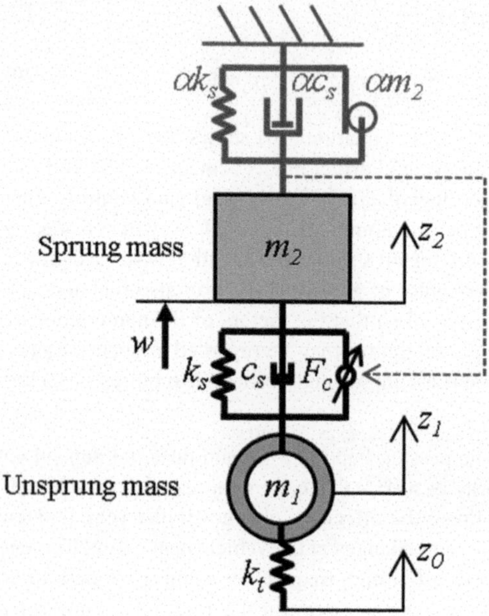

Figure 6 Quarter-Car Model for Triple Skyhook Control

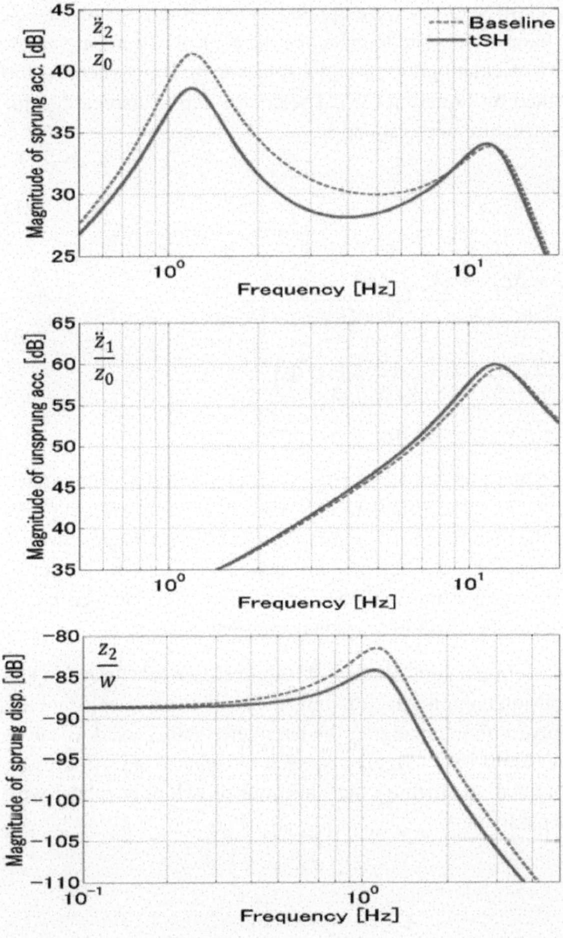

Figure 7 Effect of Triple Skyhook Control

3.2 Robustness against Vehicle Parameters

The tSH control command values are calculated using the vehicle mass m_2, the suspension damping factor c_s, and the spring constant k_s. However, in real-world conditions, the vehicle mass changes depending on the number of occupants and load, and the damping factor and spring constant may also contain errors with respect to the original design values. Therefore, the research confirmed the effects of errors in the values used for the control.

As shown in Equation (21), additive errors were applied to the vehicle values m_2, c_s, and k_s, without changing the parameters used to calculate the control command values. Equation (22) was obtained by calculating the sprung displacement. However, g_c is the control steady-state gain defined by Equation (23). The control lag is disregarded for simpler understanding of the structure of the equation.

$$\left.\begin{aligned} m_2' &= m_2 + \Delta\mathrm{m} \\ c_s' &= c_s + \Delta c \\ k_s' &= k_s + \Delta\mathrm{k} \end{aligned}\right\} \tag{21}$$

$$\begin{aligned} z_2 = g_c \Bigg(&\frac{(c_s + \Delta c)s + (k_s + \Delta k)}{(m_2 + g_c\Delta\mathrm{m})s^2 + (c_s + g_c\Delta c)s + (k_s + g_c\Delta k)} z_1 \\ &+ \frac{1}{(m_2 + g_c\Delta\mathrm{m})s^2 + (c_s + g_c\Delta c)s + (k_s + g_c\Delta k)} w \Bigg) \end{aligned} \tag{22}$$

However, $g_c = 1/(1 + \alpha)$ $(\alpha > 0)$ (23)

Equation (22) shows that the factor of the transfer function numerator, which is multiplied by the control steady-state gain, is a true value, and has the effect of suppressing road surface and driver inputs as if the control side is aware of the parameter changes. In contrast, the denominator that expresses the characteristics of the system varies by the actual amount of change multiplied by g_c. Since the applied g_c is a positive value smaller than 1, it results in motion that suppresses the effect of the actual parameter change.

Figure 8 shows the sprung acceleration control effect in response to road surface inputs when the sprung mass, suspension damping factor, and suspension spring constant are increased by 10%, respectively, without changing the parameter values used to calculate the command values. The results in this figure also factor in control lag. As a result, the targeted effect is still obtained, confirming that this control is highly robust with respect to changes in vehicle parameters.

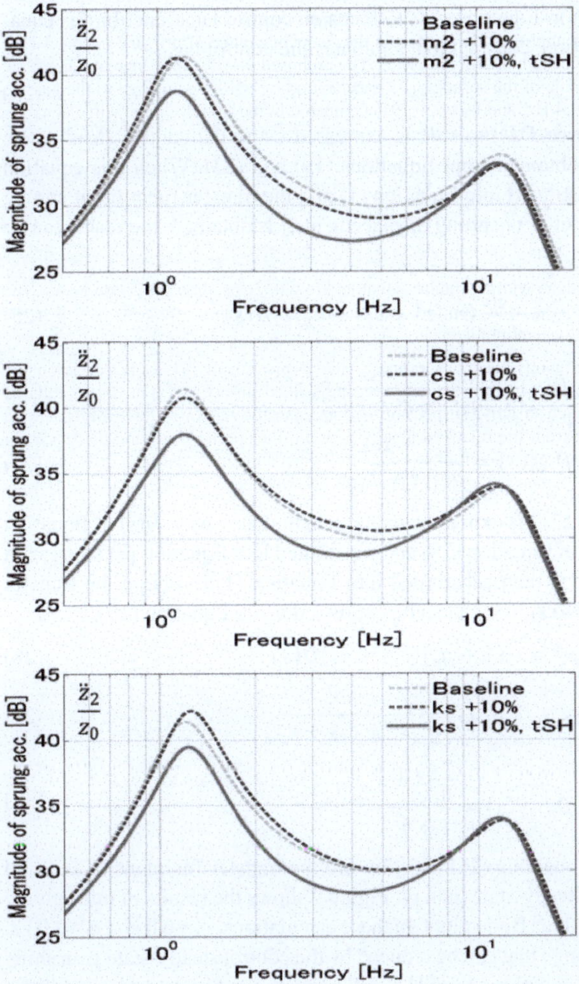

Figure 8 Robustness of Triple Skyhook Control against Changes in Vehicle Parameters

3.3 Comparison of Control Lag

Lag is unavoidably generated in the actual control. This section compares the effects of control actuator lag between the proposed tSH control and the conventional nSH2 control, both of which have been shown to effectively reduce inputs from the road surface.

$D(s)$ was refined as shown in Equation (24) to focus on control lag alone. In this equation, τ_c is a first-order lag time constant that simulates the control lag.

$$D(s) = 1/(\tau_c s + 1) \tag{24}$$

First, when the control force $F_c(s)$ of tSH is expressed using the control steady-state gain g_c, Equation (25) is obtained from Equations (19) and (23). When this equation and Equation (24) are substituted into Equation (1) and the transfer function of sprung displacement z_2 with respect to unsprung displacement z_1 is obtained, the result can be expressed using Equation (26).

$$F_c(s) = -\frac{1-g_c}{g_c}(m_2 s^2 + c_s s + k_s) z_2 D(s) \tag{25}$$

tSH:
$$\frac{z_2}{z_1} = g_c D_{tsh}(s) \frac{c_s s + k_s}{m_2 s^2 + c_s s + k_s} \tag{26}$$

However, $$D_{tsh}(s) = \frac{\tau_c s + 1}{g_c \tau_c s + 1} \tag{27}$$

Next, the control force of nSH2 is expressed using g_c in the same way. $H(s)$ of Equation (12) is rewritten as shown in Equation (28) and substituted into Equation (8) along with Equation (24). This result is then substituted into Equation (1) to obtain the transfer function of z_2 with respect to z_1, which can be expressed using Equation (29).

$$H(s) = -(1 - g_c)(c_s s + k_s) D(s) \tag{28}$$

nSH2:
$$\frac{z_2}{z_1} = g_c D_{nsh2}(s) \frac{c_s s + k_s}{m_2 s^2 + c_s s + k_s} \tag{29}$$

However, $$D_{nsh2}(s) = \frac{\frac{\tau_c}{g_c} s + 1}{\tau_c s + 1} \tag{30}$$

The difference between Equations (26) and (29) are the transfer functions $D_{tsh}(s)$ and $D_{nsh2}(s)$ multiplied by the steady-state gain g_c. Figure 9 shows the results of multiplying g_c to Equations (27) and (30) as Bode plots. In these calculations, g_c =0.6 and τ_c =0.03 s. With tSH, the frequency range that can be reduced by the control steady-state gain shifts to the high-frequency side compared to nSH2, indicating that it has control capabilities up to a higher frequency range. In other words, tSH can be used with less responsive actuators than nSH2. In addition, a comparison of Equations (27) and (30) shows that the result of multiplying the lag time constant τ_c in Equation (30) by the control steady-state gain is equal to Equation (27). The reason why the cut-off frequency is magnified by g_c can be explained by this point.

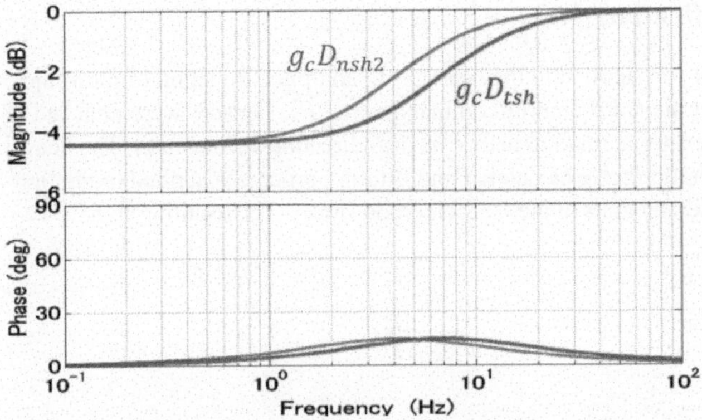

Figure 9 Comparison of Control Lag (tSH versus nSH2)

4 Actual Vehicle Evaluation

4.1 Application to Vehicle Suspension Control

The proposed control was applied to motion with three degrees of freedom: sprung heave, roll, and pitch using a vehicle installed with an active suspension system at all four wheels. Since the vehicle actuators had four degrees of freedom, it was necessary to add a further system request. For this reason, warp input, which can be controlled by the suspension actuators, was added as a command value. Equations (31) and (32) define the control vertical force $\boldsymbol{F}$ at the front-left, front-right, rear-left, and rear-right wheels, and the control command values $\boldsymbol{u}$ for roll moment, pitch moment, heave force, and warp force, respectively. The relationship between these values is expressed in Equation (33). In this equation, t_f and t_r are the track of the front and rear wheels, and l_f and l_r are the distances from the center of gravity to the front and rear axles. Commands were issued to the actuators at each wheel using this equation.

$$\boldsymbol{F} = [F_{zfl} \quad F_{zfr} \quad F_{zrl} \quad F_{zrr}]^T \tag{31}$$

$$\boldsymbol{u} = [M_x \quad M_y \quad F_z \quad F_w]^T \tag{32}$$

$$\boldsymbol{F} = \boldsymbol{C}^{-1}\boldsymbol{u} \tag{33}$$

$$\text{However,} \quad \boldsymbol{C} = \begin{bmatrix} \frac{t_f}{2} & -\frac{t_f}{2} & \frac{t_r}{2} & -\frac{t_r}{2} \\ -l_f & -l_f & l_r & l_r \\ 1 & 1 & 1 & 1 \\ 1 & -1 & -1 & 1 \end{bmatrix} \tag{34}$$

4.2 Effect Verification

The proposed control was applied in the heave, roll, and pitch direction of the vehicle and 0 was applied to the warp control command value F_w. Sensors were attached in three locations on the sprung mass to detect vertical direction acceleration. The heave, roll, and pitch acceleration were calculated and filtering processes applied to the first-order and second-order integrals of these acceleration values. The resulting signals were used for the control.

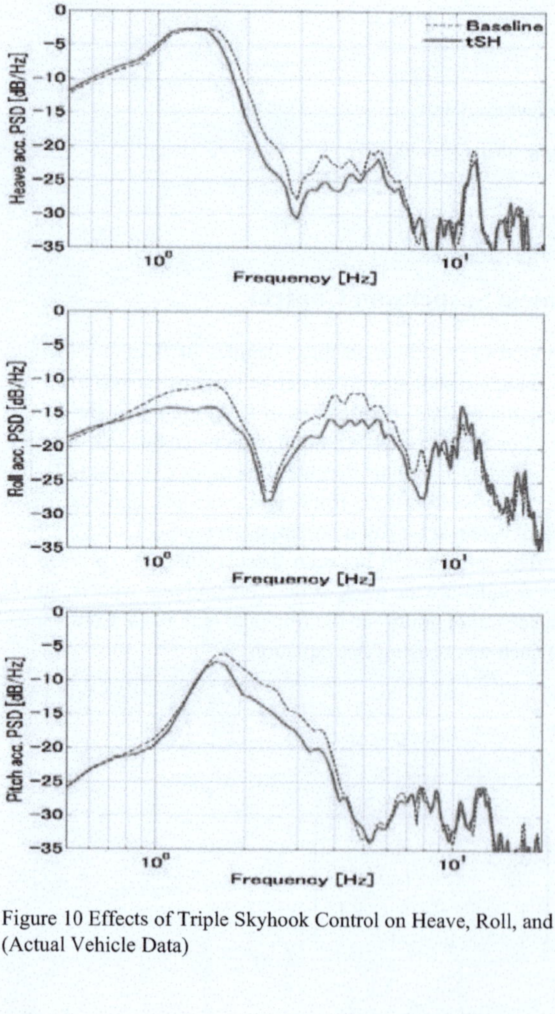

Figure 10 Effects of Triple Skyhook Control on Heave, Roll, and Pitch Acceleration PSD (Actual Vehicle Data)

Figure 10 shows the power spectral density (PSD) of the sprung acceleration when the vehicle was driven on an irregular road surface at a constant speed of 60 km/h. Vibration was suppressed from around the sprung resonance frequency to the mid-frequency range, thereby confirming the targeted control effect in an actual vehicle. This verification involved tSH by itself and it should be possible to maximize the vibration damping effect under actual conditions by combining tSH and SH, and by optimizing the shock absorber damping factor.

5 Conclusions

This paper proposed the triple skyhook control, which is capable of reducing the force applied to the sprung mass by road surface disturbances and driver inputs through feedback control of force proportional to the sprung acceleration, velocity, and displacement. The distinguishing features of this control are as follows.

- It is capable of damping vibration over a wide frequency range using a simple control law and only sprung acceleration sensors.
- Since this control uses sprung acceleration sensors, it can also reduce inertial force inputs acting on the sprung mass due to driver steering, acceleration, and deceleration operations, as well as road surface disturbance inputs.
- The proposed control is not adversely affected by changes in vehicle parameters. The control has excellent robustness and maintains its effect even when these parameters change.
- The proposed control can be used with less responsive actuators than conventional controls adopting unsprung acceleration sensors or the like.

6 References

[1] Karnopp, D. C.: Active Damping in Road Vehicle Suspension System, Vehicle System Dynamics, Vol. 12, No. 6, pp. 291-312 (1983)

[2] ISO Standard 2631-1: Mechanical vibration and shock –Evaluation of human exposure to whole-body vibration – part 1 (1997)

[3] Foag, Q., Gruebel, G.: Multi-Criteria Control Design for Preview Vehicle-Suspension Systems, IFAC, Vol. 20, Issue 5, pp. 189-195 (1987)

[4 Nagiri, S., Doi, S., Shoh-no, S., and Hiraiwa, N.: Improvement of Ride Comfort by Preview Vehicle-Suspension System, SAE Technical Paper 920277, 1992, https://doi.org/10.4271/920277

[5] Moran, A., Nagai, M. and Hayase, M.: Design of Active Suspensions with H_∞ Preview Control, Advanced Vehicle Control, pp. 215-232 (1996)

[6] Thompson, A. G., Davis, B. R.: RMS Values of Force, Stroke and Tyre Deflection in a Half-Car Model with Preview Controlled Active Suspension, Vehicle System Dynamics, Vol. 39, No. 3, pp. 245-253 (2003)

[7] Sugai, H., Buma, S., Kanda, R., Yoshioka, K., Hasegawa, M.: Preview Ride Comfort Control for Electric Active Suspension, Proceedings of the FISITA 2012 World Automotive Congress, DOI: 10.1007/978-3-642-33795-6_13 (2012)

[8] Katsuyama, E. and Omae, A.: Improvement of Ride Comfort by Unsprung Negative Skyhook Damper Control Using In-Wheel Motors, SAE Int. J. Alt. Power. 5(1):214-221, 2016, https://doi.org/10.4271/2016-01-1678.

[9] M.C. Smith, Synthesis of Mechanical Networks: The Inerter, IEEE Transaction on Automatic Control, Vol. 47, No. 10, pp. 1648-1662 (2002)

CHASSIS CONTROL SYSTEMS

Development of GVC Moment Plus Control for mass production

Author:
Daisuke Umetsu
Mazda Mortor Corporation

Co-author:
Yasunori Takahara, Osamu Sunahara, Fuminori Kato,
Mazda Motor Corporation, Japan;
Prof. Masato Abe, Prof. Makoto Yamakado, Yoshio Kano,
Kanagawa Institute of Technology, Japan;
Junya Takahashi, Hitachi Ltd., Japan

P. Pfeffer (Ed.), *9th International Munich Chassis Symposium 2018*, Proceedings,
https://doi.org/10.1007/978-3-658-22050-1_19

Abstract

GVC Moment Plus control which is integrated control system of the longitudinal acceleration control and the direct yaw moment control based on lateral jerk information has been developed for mass production. This integrated control using engine unit and brake unit improves the smooth transition between yaw, roll and pitch attitude of vehicle during turning. This results in more stable driver's steering operation and enhances the vehicle limit handling performance.

1. Introduction

When we evaluate vehicle dynamics performance, it is very important to consider closer connection and transition of all driving manoeuvres such as braking, turning and acceleration. Authors have made several researches and development on G-Vectoring Control (GVC) which enhances the closer connection between longitudinal and lateral dynamics by using brake, driving motor or high response combustion engine torque control [1][2][3].

Control principle of GVC is shown as below:

$$G_{xc} = -sgn(G_y \cdot \dot{G_y}) \frac{C_{xy}}{1+T_s} |\dot{G}_y| \qquad (1)$$

G_{xc} is longitudinal deceleration/acceleration control target which is calculated by the product of control gain C_{xy} and vehicle lateral jerk $\dot{G}_y$. And in addition, T_s is time constant of 1st order delay.

In case of high response combustion engine, GVC applies slight longitudinal deceleration to improve the vehicle's smooth transition between longitudinal and lateral acceleration during turn-in. This vehicle behaviour improvement results in the reduction of driver's steering effort not only at limit handling but also at normal driving situation.

And authors have also studied the GVC Moment Plus Control (M+ control) which is based on the same control principle as GVC. M+ control combines GVC's longitudinal acceleration control and the Electronic Stability Control (ESC)'s direct yaw moment control to further improve vehicle limit handling performance [4][5].

In this paper, we firstly explain the concept of GVC Moment Plus Control and the system configuration which is implemented in mass production vehicle. After that, the benefits of M+ control are shown in human-vehicle closed loop evaluation perspective and the vehicle's pure mechanical characteristic perspective. Throughout these evaluation results, we describe some consideration on the mechanism why driver changes driving manoeuver by M+ control.

2. Concept of GVC Moment Plus Control

Based on GVC control principle of equation (1), M+ Control exchanges the acceleration part of GVC control (during turn-out situation) into the direct yaw moment control.

M+ control principle is shown as below:

$$\begin{cases} M_{zc} = -sgn(G_y \cdot \dot{G_y})\frac{C_m}{1+T_s}|\dot{G_y}| \quad (-sgn(G_y \cdot \dot{G_y}) > 0) \\ M_{zc} = 0 \quad (-sgn(G_y \cdot \dot{G_y}) \leq 0) \end{cases} \tag{2}$$

$M_{zc}$is direct yaw moment control target which is calculated by the product of control gain C_m and vehicle lateral jerk $\dot{G_y}$. And in addition, T_s is time constant of 1st order delay.

Control concept of GVC Moment Plus Control is shown as Fig 1.

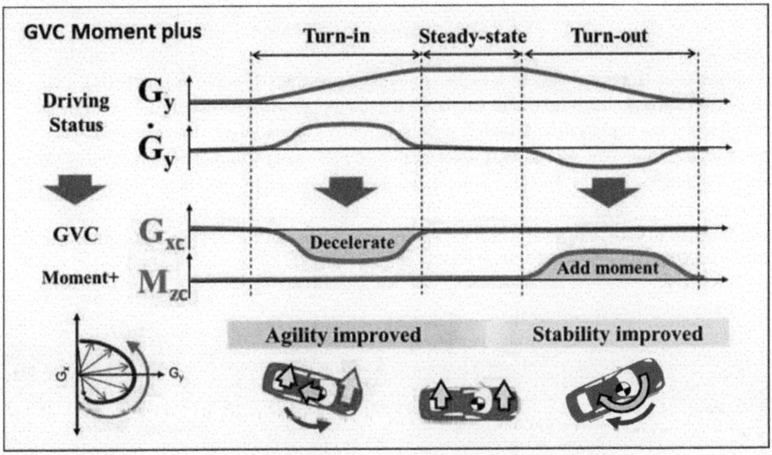

Fig.1 Control concept of GVC Moment Plus control

Driving status such as turn-in, steady-state or turn-out is calculated by lateral acceleration $G_y$and lateral jerk $\dot{G_y}$.

Based on these state information, GVC applies longitudinal deceleration G_{xc} during turn-in to improve vehicle agility, and M+ control gives additional yaw moment M_{zc} during turn-out to improve vehicle stability.

3. System Configuration

Above mentioned control concept is implemented to mass production vehicle. The system configuration is shown as Fig.2.

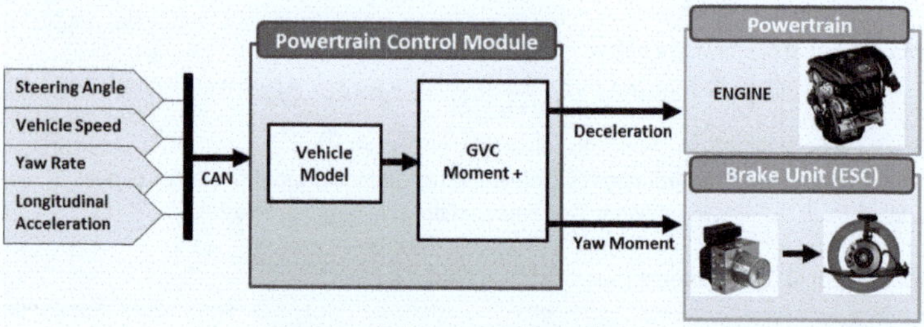

Fig.2 Control System configuration

GVC Moment Plus Control algorithm is implemented in Powertrain Control Module (PCM). PCM calculates two control requests of longitudinal deceleration G_{xc} and direct yaw moment M_{zc} based on the sensor signals from CAN bus such as Steering Wheel Angle, Vehicle speed etc. Regarding the actuation of control, GVC deceleration is achieved by torque down control of high response combustion engine, and M+ control uses direct yaw moment control of brake unit (ESC).

For one example of control flags and request signals which are calculated in PCM controller, vehicle test data (Lane change 80kph,Right direction) is shown in Fig3.

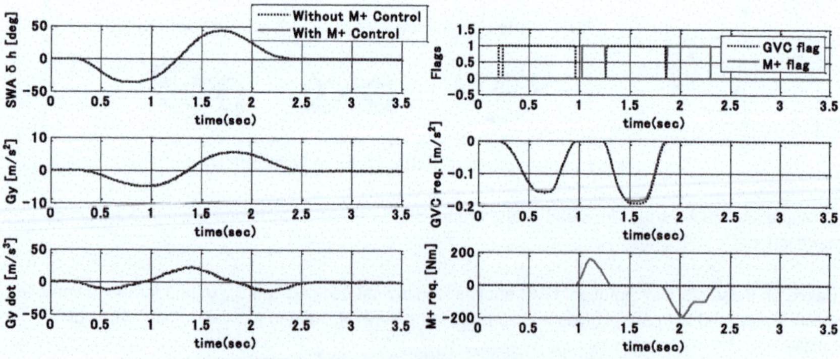

Fig.3 GVC and M+ Control Flags and Request Signals

During turn-in phase, GVC calculates longitudinal deceleration request about -0.2 m/s^2. And during turn-out phase, M+ control calculates direct yaw moment request about +/-200Nm. And both of the controls are closely cooperated and smoothly changing the control requests from one to the other.

Based on the M+ control request from PCM controller, brake unit applies the actual brake pressure on each wheel. (Fig.4)

For this test vehicle, brake unit applies pressure on front outer wheel to achieve requested yaw moment as expected. And actual brake pressure response has about 100ms delay from request

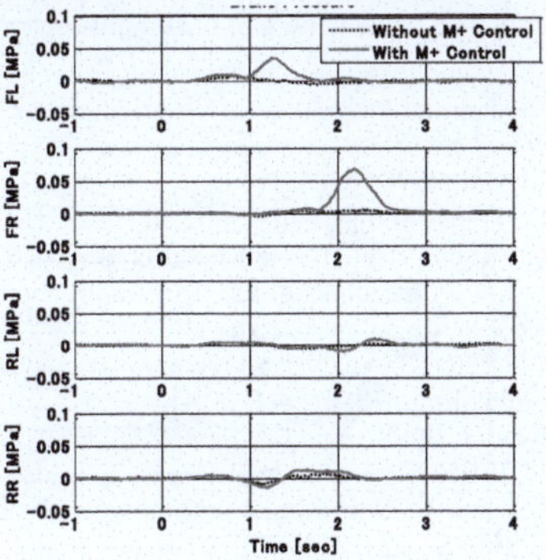

Fig.4 Yaw Moment Control Actuation by Brake Unit

4. Human-Vehicle Closed Loop Evaluation Results

4.1 Test Vehicle Information

To clarify the benefit of above mentioned control system, we firstly performed the human-vehicle closed loop evaluation test using two test vehicles. (Fig.5, Table.1)

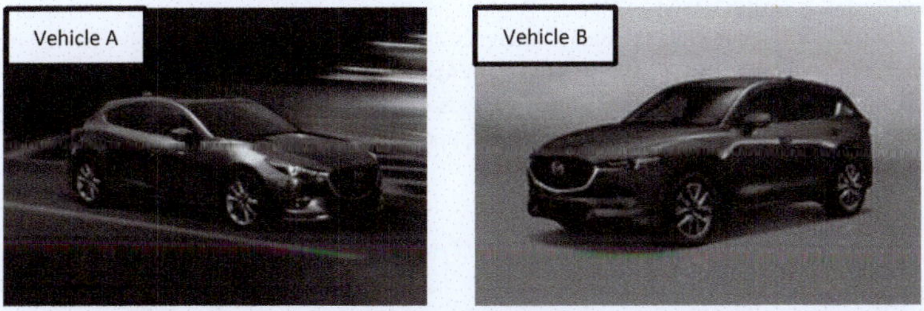

Fig.5 Test Vehicles

Table.1 Test Vehicle Specification

Dimension	unit	Vehicle A	Vehicle B
Length	mm	4470	4495
Width	mm	1795	1840
Wheelbase	mm	2700	2700
Tread (Fr/Rr)	mm	1555/1560	1595/1595
Weight(Fr/Rr)	kg	803/521	874/648
CG height	mm	561	668

4.2 Single Lane Change Evaluation

Single lane change test is performed using Vehicle A at the test course which is shown in Fig.6. Vehicle speed is kept 80km/h by cruise control function. Lateral displacement of lane is 3.0m and longitudinal displacement is 37m. LED indicator was set at the middle of lanes to notify the left/right direction to driver when vehicle front axle passes the end of straight section.

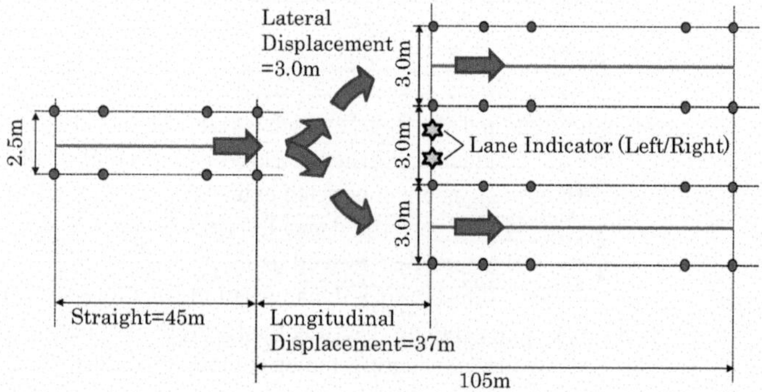

Fig.6 Single Lane Change Test Course

Two control settings are evaluated with/without M+ control (direct yaw moment control) while GVC deceleration control is always active.

Time series data of driver's Steering Wheel Angle (SWA) and vehicle's lane change path data from GPS are shown in Fig.7. Although vehicle's driving path is similar between with/without M+ control but the amount of driver's Steering Wheel Angle is reduced about 20% at latter half of lane change manoeuvre with M+ control.

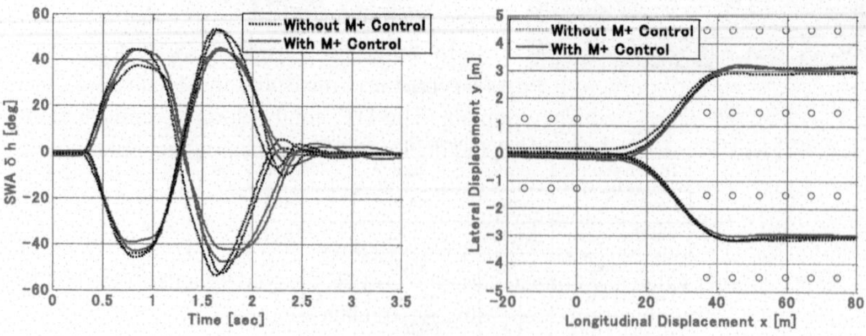

Fig.7 Lane Change Test Result

To understand control effect from driver-vehicle closed loop perspective, we identify the driver parameter τ_L which is defined by driver-vehicle model shown in Fig.8.

In previous research, meaning of τ_L is known as follows:

If identified τ_L value is greater, driver is able to handle vehicle easier with relaxed feeling.
If identified τ_L value is smaller, driver put more effort to trace target path.

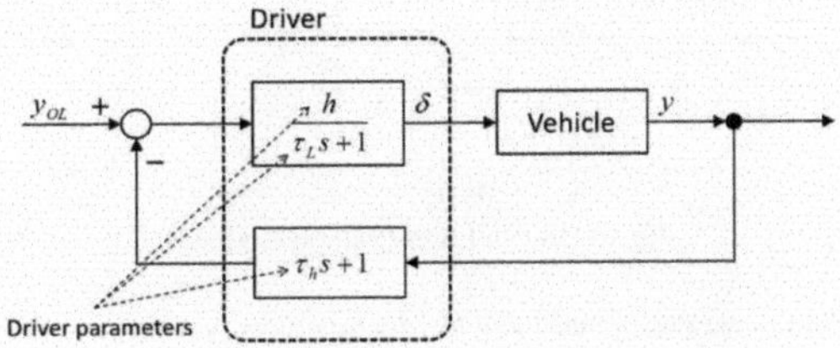

Fig.8 Driver-Vehicle Closed Loop Model

In this model, h is driver gain, $\tau_L$is driver response time, $\tau_s$is driver preview time, $y_{0L}$is target path (lateral displacement), y is vehicle's actual path (lateral), and δ_h is driver's steering angle. Driver parameters h, τ_L, τ_s are identified by minimizing the error of measured value and calculated value of y and δ_h.

Identified driver parameter τ_L is shown in Fig.9. With M+ control, $\tau_L$value is about0.012 sec greater than without M+ control. This means that the driver can handle M+ control equipped vehicle easier.

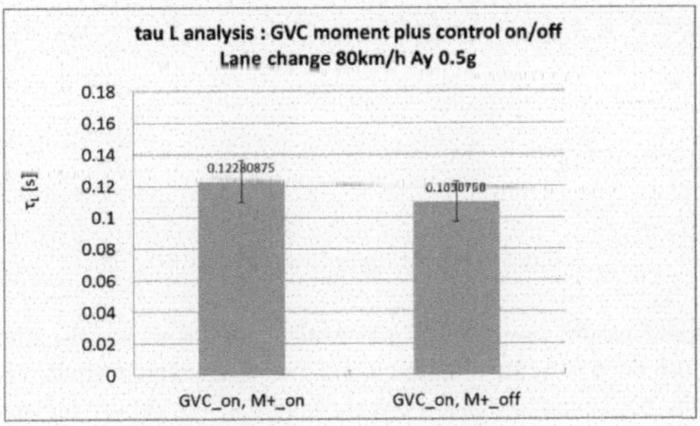

Fig.8 Driver-Vehicle Closed Loop Model

Based on driver's subjective feeling, M+ control vehicle has smoother vehicle behaviour and more stable feeling at the latter part of steering. This driver's comment is aligned with $\tau_L$analysis result.

4.3 Double Lane Change Limit Handling Performance Evaluation

Double lane change test is performed with Vehicle B to evaluate limit handling performance. Test course is shown in Fig.10. In this test, maximum vehicle speed (average speed of whole lane change period) is evaluated with/without M+ control.

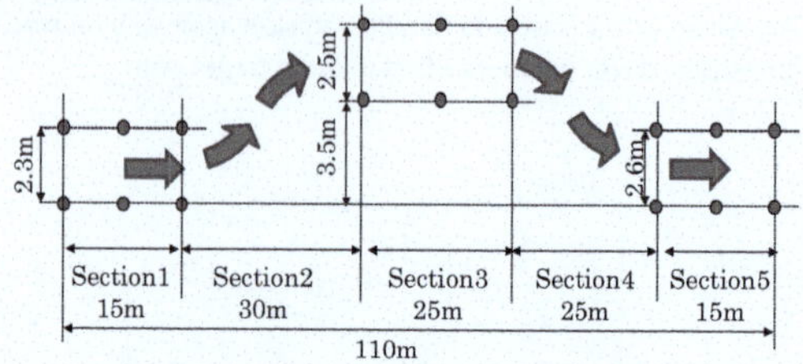

Fig.10 Double Lane Change Course

Two control settings are evaluated with/without M+ control (direct yaw moment control) while GVC deceleration control and Electric Stability Control (ESC) are always active.

Time series data of driver's Steering Wheel Angle (SWA), M+ control flag and request signal, Yaw rate and Lateral Acceleration (Gy) are shown in Fig.11.

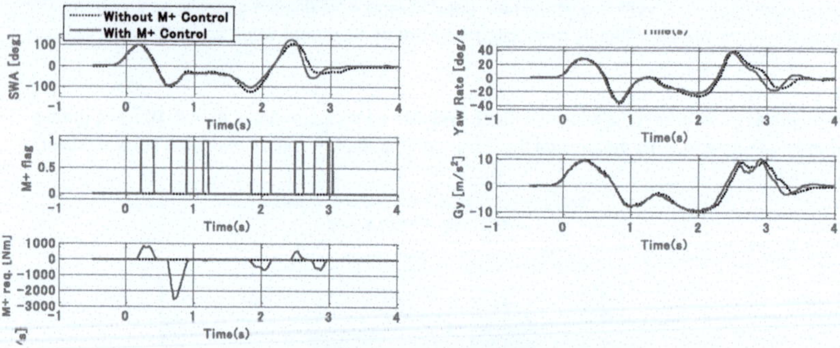

Fig.11 Double Lane Change Test Result

Since M+ control applies slight yaw moment intervention in addition to ESC stabilization control, vehicle stability performance is further improved and maximum vehicle speed increases 4.45 km/h. Vehicle motion at the beginning of Section 3 is shown in Fig.12. Vehicle movement is stabilized slightly earlier by M+ control

Fig.12 Vehicle Motion at Section 3

5. Vehicle Characteristics Evaluation Result (Open Loop Test)

In this section, we focus on vehicle characteristics by eliminating driver operation to understand what the key vehicle dynamics index is. For this evaluation we use Vehicle A and Single Lane Change Manoeuvre (same as Section 4.2).

5.1 Inputs to Vehicle

To eliminate the variation of input by driver, we use steering robot and constant throttle pedal control. Steering input wave-form for the robot is set as Fig 13 which is based on driver's lane change operation. And vehicle speed is set at 81km/h initially and the throttle pedal is set to constant value during the whole lane change test.

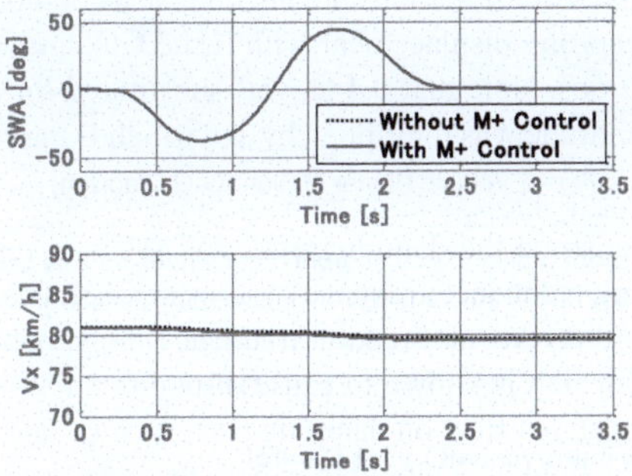

Fig.13 Vehicle Input for Open Loop Test

5.2 Vehicle Planner Motion Improvement by M+ control

To analyse vehicle planner motion improvement by M+ control, the Lissajous diagram of SWA and yaw rate is shown in Fig.14. Yaw rate response during the steer back operation is improved as expected from M+ control concept.

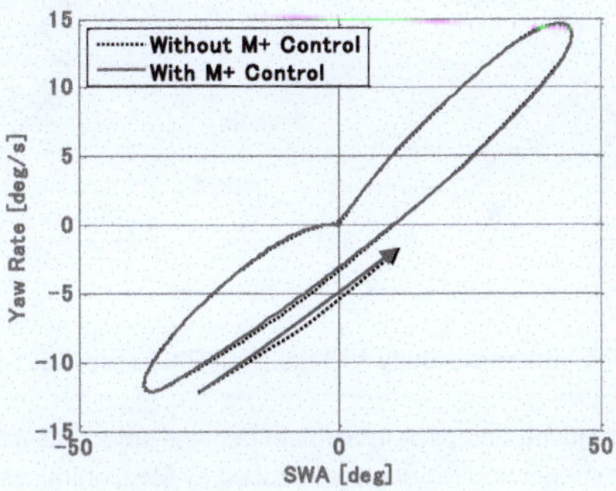

Fig.14 Vehicle Planar Motion (Yaw Rate Response)

Time series data and the Lissajous diagram of lateral and longitudinal acceleration (G-G diagram) is shown in Fig.15. Positive value of longitudinal acceleration is accelerating direction.

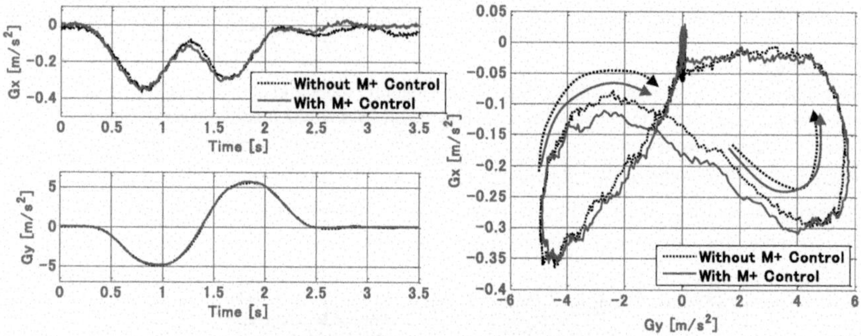

Fig.15 Vehicle Planar Motion (G-G Diagram)

Since M+ Control applies slight braking force on front outer wheel, vehicle longitudinal deceleration happens about 0.05 m/s^2. It reduces the variation of vehicle body deceleration during lane change. This means that the smoother longitudinal acceleration change during steer-back operation is achieved by M+ Control while GVC achieves the same improvement during steer-in operation.

5.3 Vehicle Body Attitude Improvement by M+ Control

Figure 16 shows the time series data and the Lissajous diagram of roll angle and pitch angle. Positive value of pitch angle is nose dive direction.

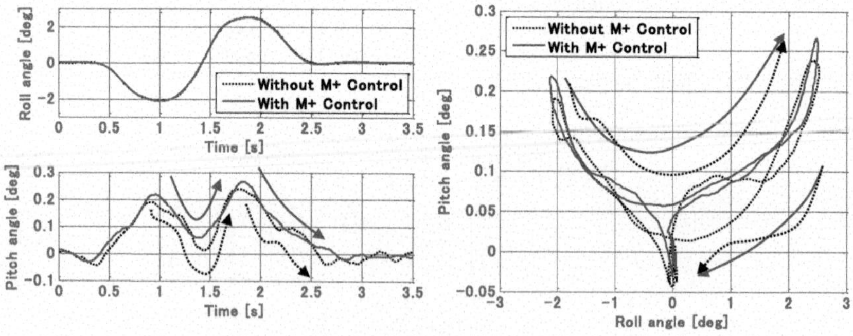

Fig.16 Vehicle Body Attitude (Roll-Pitch response)

Variation of pitch angle is reduced and pitch movement became smoother with M+ Control. And since the phase lag between pitch angle and roll angle is reduced by M+ Control, closer connection of roll movement and pitch movement is achieved.

To understand more detail behaviour of the vehicle body attitude during lane change, body height information relative to the ground at 4 corners of vehicle is shown in Figure 17.

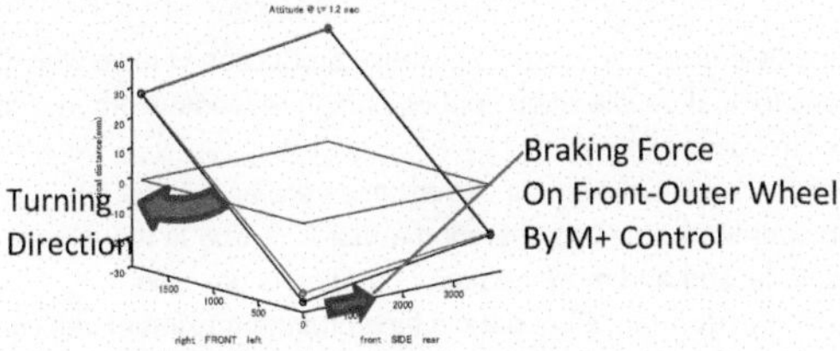

Fig.17 Vehicle Body Attitude at 1.2 sec (Right turn)

Based on the brake intervention information of Figure.4, M+ control activated period of time from 1.3 sec to 1.5 sec is continuously visualized in Figure 18. The left side is with M+ Control and the right side is without. Time is passing from top to bottom by 0.1 sec sampling.

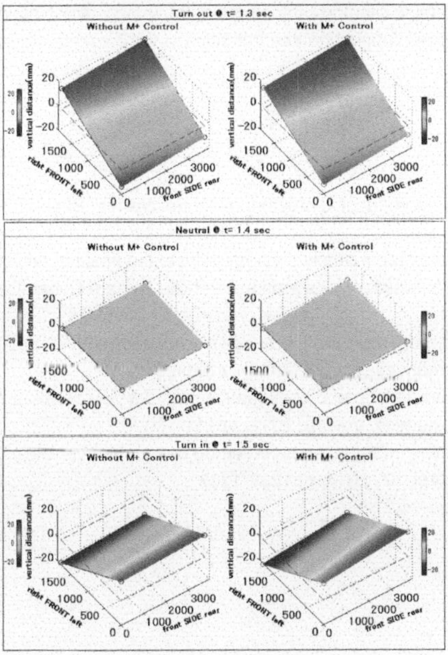

Fig.18 Vehicle Body Attitude (Turn out and Turn in)

With M+ control, the front outer side of vehicle body is returning earlier at the end of turn-out (1.3 sec) and the pitch attitude is kept as slight nose dive even at the zero roll angle timing (1.4 sec). And at the next turn-in timing (1.5 sec), vehicle pitch angle nose dive movement is earlier with M+ control.

This body attitude results in the consistent diagonal roll attitude during whole lane change manoeuvre and it might bring more confident feeling to the driver in such a high dynamic handling situation.

6. Conclusion

- GVC Moment Plus Control is developed for mass production vehicle which is the integrated control of the longitudinal deceleration control by high response combustion engine and the direct yaw moment control by brake unit.
- Customer benefit on vehicle dynamics performance are as follows:
  - Reduction of driver's steering operation about 20% due to the vehicle stability improvement. It brings safe feeling and more relaxed operation is possible
  - Limit handling performance i.e. double lane change maximum vehicle passing speed increases about 4 km/h.
- Vehicle performance improvements are as follows:
  - Vehicle planner motion: Yaw response delay during steer back operation is reduced.
  - Vehicle planner motion: Transition of longitudinal and lateral acceleration gets smoother.
  - Vehicle body attitude: Pitch angle and roll angle is consistently kept as diagonal roll attitude.
- Based on the human-vehicle closed-loop evaluation and the vehicle open loop evaluation of the M+ control effect, we think it is important to improve the quality of vehicle dynamics performance in detail such as very small connection and transition of the 3 axis 6degrees of freedom movement.

7. References

[1] Yoshioka, T. et al.," Development of G-Vectoring Control System Based on Engine Torque Control," AVEC'16, 2016.

[2] Yamakado, M. et al., "An experimentally confirmed driver longitudinal acceleration control model combined with vehicle lateral motion," Vehicle System Dynamics, Vol.46, 2008, pp. 129-149.

[3] Yamakado, M. et al., "Improvement in vehicle agility and stability by G-Vectoring control," Vehicle System Dynamics, Vol. 48, 2010, pp231-254.

[4] Yamakado, M. et al., "Comparison and combination of direct yaw-moment control and G-Vectoring control", Vehicle System Dynamics, Vol. 50, 2012, pp111-130

[5] Yamakado, M. et al., "A yaw-moment control method based on a vehicle's lateral jerk information", Vehicle System Dynamics, Vol. 52, 2014, pp1233-1253

Development of a real-time friction estimation procedure

Dr.-Ing. Gerd Müller, Technische Universität Berlin

Vincent Gregull, M.Sc., Technische Universität Berlin

Claudia Bräsemann, B.Sc., Technische Universität Berlin

Prof. Dr.-Ing. Steffen Müller, Technische Universität Berlin

Keywords: Vehicle Safety; Method for Estimation; Friction Potential; Logistic Regression

P. Pfeffer (Ed.), *9th International Munich Chassis Symposium 2018*, Proceedings,
https://doi.org/10.1007/978-3-658-22050-1_20

Abstract

Precise knowledge of the friction potential is of great importance for safe longitudinal and lateral control of a car. This potential is influenced by many parameters like weather conditions, road surface and tyres. While today it is the driver who assesses friction values, it will be necessary for future highly automated vehicles to independently obtain information on environmental conditions.

A cause-based estimation procedure for estimating the maximum friction coefficient has been developed which relies solely on information that is available without additional vehicle sensors. This information consists of data which is present in the vehicle itself, such as outside temperature, vehicle speed or rain intensity and on data provided by the surrounding infrastructure. This includes weather data from weather stations or information on road conditions obtained from road weather information systems. By combining and integrating these fields of information, the range of the maximum coefficient of friction is established using the estimation procedure developed in this project.

The result of a huge number of test brakes is a comprehensive database with more than 5,000 data sets which includes for each full braking manoeuvre more than 80 parameter information about weather, road state etc.

Based on these data the mentioned friction estimation algorithm was developed. It is based on the method of logistic regression, which is a mathematical method of statistical estimation.

In this paper the results of the brake tests and the functionality of the estimation algorithm will be shown. Furthermore, results, potentials and next development steps will be presented.

1 Introduction

Precise knowledge of the friction potential is of great importance for safe longitudinal and lateral control of a car. While today it is mostly the driver who assesses friction values and adapts his driving style accordingly, it will be necessary for future highly automated vehicles to independently obtain information on environmental conditions. Analyses of accident records show that at least 3.6 % of road deaths are due to icy road conditions. However, this number is likely to be significantly higher, since the number of accidents in Germany occurring under icy road conditions without these conditions being identified as primary causes of the accidents, is around 20 % of the total number of accidents [1].

The coefficient of friction μ is defined as the normalised resulting horizontal force which acts between tires and road:

$$\mu = \frac{\sqrt{F_X^2 + F_Y^2}}{N_Z} \tag{1}$$

F_X and F_Y act on the tire as circumferential and lateral forces and N_z is the normal or contact force. The maximum transferable friction μ_{max}, or the friction potential, is the maximum that μ can reach under the specified conditions.

The friction potential is influenced by many factors, such as the tire condition, the type of tire, or the quality of the layer between road and tire, which describes the road condition. Typical states are dry, moist, wet or snowy/icy. Estimating the friction potential is a challenge that numerous research projects have taken on [2–10]. Basically two approaches can be distinguished. The effect-based approach attempts to measure effects that result from different coefficients of friction on the tire. A prediction of the maximum friction coefficient is then issued based on these measurements. As an example, sensors have been developed which are integrated into the tire surface and deform depending on the current friction [11]. A disadvantage of effect-based methods is that they require a sufficient level of slip depending on the estimation method. With the cause-based approach, variables are measured that affect the friction potential. With the help of the measured parameters and an appropriate estimation procedure, the maximum friction coefficient is then estimated. Major disadvantages of this method are that additional sensors are necessary and an elaborate training of the estimation algorithm is required.

As part of a research project at the Technical University of Berlin that has been financed and given advisory support by Working Group 20 of the Research Association for Automotive Technology (FAT), a cause-based estimation procedure for estimating the maximum coefficient of friction has been developed which relies solely on information that is available without additional vehicle sensors. This information consists of data which is present in the vehicle itself, such as outside temperature, vehicle speed or rain intensity. On the other hand, the procedure draws on data provided by the surrounding infrastructure. This includes weather data from weather stations or information on road conditions obtained from road weather information systems. By combining and integrating these fields of information, the range of the maximum coefficient of friction is established using the estimation procedure developed in this project.

For the development of such estimation procedures it is first necessary to obtain detailed knowledge of the influence of the described information on the maximum coefficient of friction [12–15]. To this end, extensive measurement runs have been performed over a period of 30 months on a predefined route through urban and rural areas and the outskirts of Berlin. Here, the range of the friction coefficient was ascertained in real-world environments using test braking to establish the coefficient's position under varying conditions

2 Measurement of the Maximum Friction Coefficient and Results

In order to perform friction potential measurements, 32 brake points were set along a measurement course. These were positioned in town, rural areas and on highways; the driver braked on the surfaces of asphalt, concrete and cobblestones. Close proximity of the brake points to weather stations (WS) and road weather information systems (GMA)[1] was ensured. Furthermore, relevant structural features such as bridges, as well as the feasibility of brake tests in everyday traffic were taken into account. For the test runs a route in the southwest of Berlin was chosen that passes through Berlin and Brandenburg and runs further along the motorways A115 and A10. The route chosen is in proximity to the GMA Fahlhorst. Also, all the points along the route were within a distance of less than 10 km from one of the weather stations.

At each of the 32 brake points defined for this route the driver braked once in the course of each test run. At initial speeds between 30 and 120 km/h the brake pedal was depressed in such a way that in the master cylinder a minimum pressure of 175 bar built up and the braking system was reliably taken to the ABS control range. This ensured that the vehicle reaches the maximum possible deceleration. This vehicle deceleration was measured by a servo-accelerometer over a period of 0.5 s, and then averaged. From this value the average maximum possible coefficient of friction was obtained using

$$\bar{\mu}_{max} = \left| \frac{\bar{a}}{g} \right| \tag{2}$$

Since the test runs were performed over a period of 30 months, a wide range of nearly all weather conditions could be taken into account. These include various rain intensities and the resulting different water heights, measurements on closed snow cover or slush, as well as different surface temperatures under dry conditions. Summer and winter tires were exchanged regularly over the whole measurement period so that results are available for both types of tires for the weather conditions referred to above. Further tests were conducted in winter times in Sweden, where a reliable snow and ice conditions could be found.

For the evaluation of the brake tests round about 5,000 brake measurements and the associated data sets are available. Each of these sets contains 80 parameters that describe all significant variables which affect the coefficient of friction. These are weather information, vehicle-specific data and information on road surfaces. Figure 1 shows an

1 German: Glättemeldeanlage (GMA)

overview of the measurements of the maximum friction coefficients on dry roads with surfaces of asphalt, concrete and cobblestones as a function of velocity.

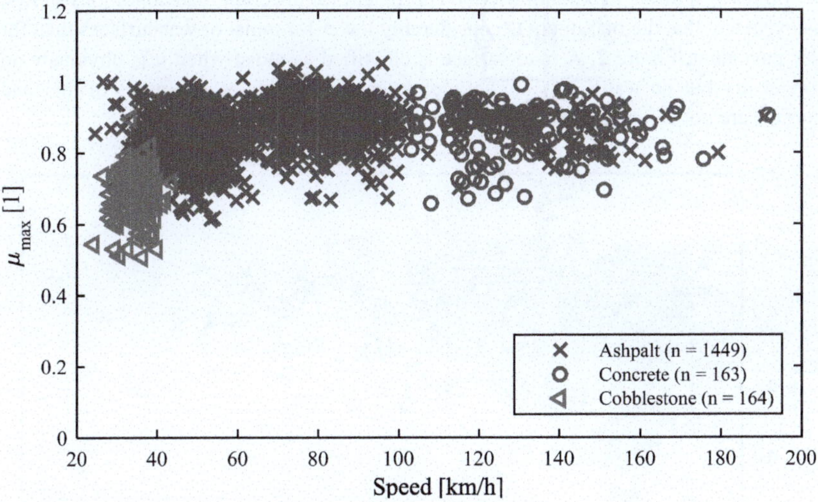

Figure 1: Maximum coefficient of friction for dry road surfaces as a function of vehicle velocity (© KFZB)

It is apparent that under dry conditions the maximum friction value is higher than $\mu = 0.5$ for all three road surfaces and all velocities. The measurements on cobblestone pavements were taken within a speed range around 40 km/h. The measured friction coefficients vary considerably for this surface and are found within a range of values from *0.53* to *0.85*. The values measured for asphalt, of which there are a lot more due to the brake point distribution, were determined at initial speeds between 30 and 190 km/h. The range of values here is *0.66* to *1.05*. The large scatter is due to the fact that along the measurement path, different varieties of asphalt were driven on. For an as much as possible exact estimation of the friction potential a small range of values is necessary. Up to now a lot of different possible parameters, which seem to influence the friction coefficient for example different asphalt surfaces, the pollen intensity, general air pollution etc. have been investigated. Obvious contexts could not be identified. For further research activities a detailed investigation of these influencing parameters is necessary.

The values measured for concrete were recorded at higher initial speeds, since this type of surface was found only on the highway part of the test section. Here the range of values for the maximum friction coefficient runs from *0.67* to *0.99*. None of the surfaces under consideration shows any significant speed dependency.

A comparison of the maximum friction coefficient for different road conditions shows, as expected, that the maximum is considerably lower for moist or wet surfaces than for dry pavement, Figure 2. A road surface is classified as moist when it is obviously no longer dry but no water is being sprayed by moving vehicles, the pores of the road surface are not closed by water, and no reflective surface has formed yet.

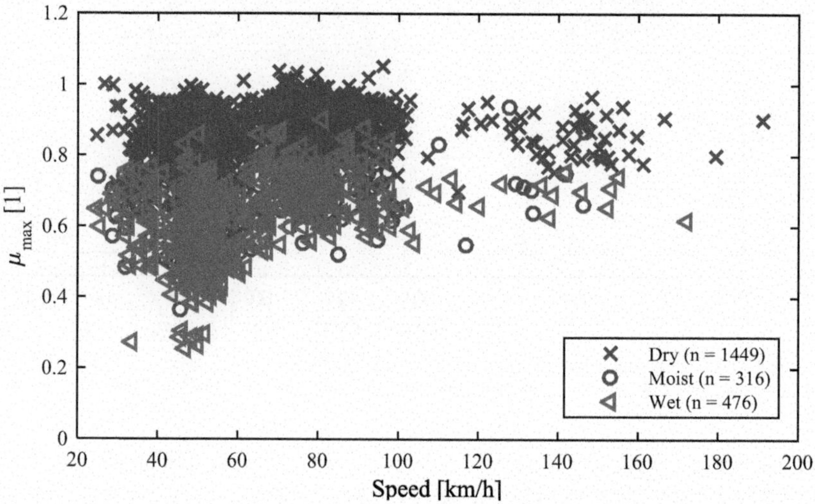

Figure 2: Maximum coefficient of friction on asphalt for different road conditions (© KFZB)

The database of maximum values of friction under different conditions, which was built up in the course of the test runs, provides a solid foundation for the development of an assessment of the friction coefficients as described below. Figure 1 and Figure 2 are examples of a variety of possible comparisons offered by this database [1].

3 Logistic Regression for the Estimation of Friction Potential

For an accurate estimate of the current coefficient of friction it is necessary to know the current status of the main influencing parameters. Among these are road surfaces and tire types. In the research project described here, the road surface was determined by a digital map, which was created for the test track. It is easy to imagine that future digital maps will provide such information for all roads. The selected type of tire was also digitally stored as information. It is conceivable that this can be done automatically in the future through corresponding coding of the tires. Communication between vehicle and tires is already standard today.

Far more challenging is an assessment of the exact road condition (dry, wet, etc.), that is to say the layer between vehicle tires and road surface. According to the requirements described at the beginning, namely that no additional measurement technology ought to be used in the vehicle other than already existing sensors, the state of this interlayer cannot be measured directly. Instead, the road condition needs to be identified with the help of other parameters. The parameters to be used here include, for example, outside temperature, rain intensity, humidity, or dew point temperature. While the outside temperature is measured directly on the vehicle and can be queried via an interface by the CAN bus, the other data has to be obtained through other means. In our project the data servers of the German Weather Service were used. The relevant information provided by weather stations along the measurement path was accessed via the Internet and transmitted directly to the vehicle via mobile communication. Additional information, such as road surface temperatures or the dew point data, was delivered by the road weather information system.

Based on this data the state of the layer between tires and road was estimated using logistic regression. This is a method of statistical analysis for dichotomous[2] distributions. For a property, there are exactly two distinct states, for example yes/no, on/off, 0/1, wet/not wet. Using this method, it is possible to describe the relationship between individual weather factors and corresponding road conditions. Based on this description, the probability of a certain state of the interlayer between tire and road can then be calculated.

To determine the road condition by means of logistic regression, the possible surfaces are defined as: dry, moist, wet or icy/snowy. Based on the database obtained in the test runs, the road conditions presented above were determined for every weather parameter and at each brake point. This results, for example, in the following distribution for the relationship between relative humidity and the state dry, Figure 3 (blue dots). Based on

2 Dichotomous (Greek): split into two parts

this distribution, the functional equation of the logistic regression is optimised. The equation has the following form:

$$P(y(x)) = \frac{c}{1+ae^{-bx}} \tag{3}$$

where P is the probability of occurrence of a particular road condition y, dry in the example shown, which in turn depends on the value of x, here the relative humidity. The free parameters a, b, and c are optimised such that the squared error between P and y is minimised. Figure 3 shows an optimised curve of the regression line (red). According to [17], the conditional variance for the function curve can be calculated, which indicates the predictive uncertainty. Its range is between 0 and 0.25. If this value is normalised and the reciprocal formed, the result is a functional assessment of the quality G of the calculated probability (orange), with

$$G(y(x)) = \frac{1}{4P(1-P)} \tag{4}$$

The same procedure is applied for all states of road conditions and for all parameters considered, so that a total yield of 25 optimised regression functions and their corresponding quality function results from five different states and five different inputs. In order to obtain an assessment of the probability of occurrence of a certain road condition as a function of the input variables, all the individual probabilities are calculated and weighted by their respective associated quality:

$$\overline{P}_{k,j} = \frac{\sum_{i=1}^{n} P_{k,j}(y_j(x_i)) \cdot G_{k,j}(y_j(x_i))}{\sum_{i=1}^{n} G_{k,j}(y_j(x_i))} \tag{5}$$

n is the number of input variables taken into account, x_i corresponds to each input variable, y_j is the road condition and k is the number of the respective brake point.

The state y_j with the highest probability calculated is the estimated road condition. The greater $P_{k,j}$, the more reliable is the estimate of the road condition. Overall, the method of logistic regression is characterised by great clarity of the structural process and ease of identification. The impact of changes in the input variables can be traced and read directly.

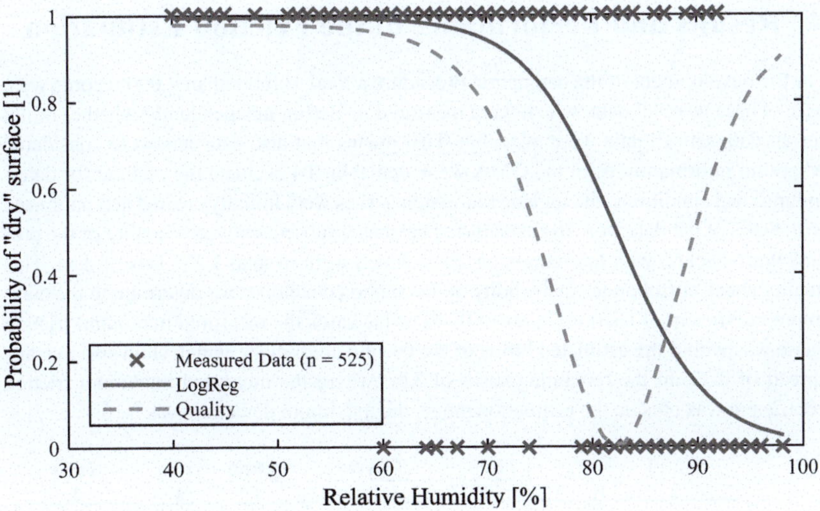

Figure 3: Logistic regression of the relationship between humidity and road condition dry (© KFZB)

The final structure of the estimation algorithm was defined in several optimisation steps. The target was to find out with which structure the best results could be achieved. One main result of this optimisation was that much better results arise, when different types of interlayer are identified in different steps. That means, that in the first level of the algorithm it is only distinguished between "dry" and "other" interlayers. For this first level the single logit function described above, were adapted and optimised. Based on the "other"-cases from the first level in the second level the algorithm distinguishes between "moist" and "other" interlayers. In the third level the algorithm differs between wet and snowy/icy. Additionally, a switch is integrated. Whenever the outside temperature is above 4°C the algorithm decides to wet, if it is below -2°C it decides for snowy/icy. That means the method of logistic regression in the third level is used for conditions between -2 and +4°C only.

Based on the estimated road condition, the actual friction potential is predicted as a next step. For this procedure the database obtained from the test drives is used. Depending on the detected road condition, a lower and an upper limit of the coefficient of friction can be established using data on the road surface, the type of tires, and the driving speed. The assessment is refined through increasing awareness of the different key parameters.

4 Results and Performance of the Friction Estimation

For the development of the estimation procedure a total of more than 4,000 records were used. These were all data, which were measured in Berlin, independently whether or not for all data sets all data were available. That means that also data sets were considered where no information from the GMA were available. To evaluate the method for determining road conditions, the logistic regression curves were initially created and optimised with 80 % of the data from the established database; subsequently, the quality of the process was checked with the remaining 20 % which were in total 1,471 sets of data. This quality check corroborates the validity of the method developed for determining the maximum coefficient of friction. In about 97 % of the cases the measured maximum friction value was within the estimated limits of the friction coefficient, which shows an average spread of 0.33. In the remaining cases of 3 %, where the measured maximum friction coefficient was outside the estimation range, the maximum deviation was 0.16.

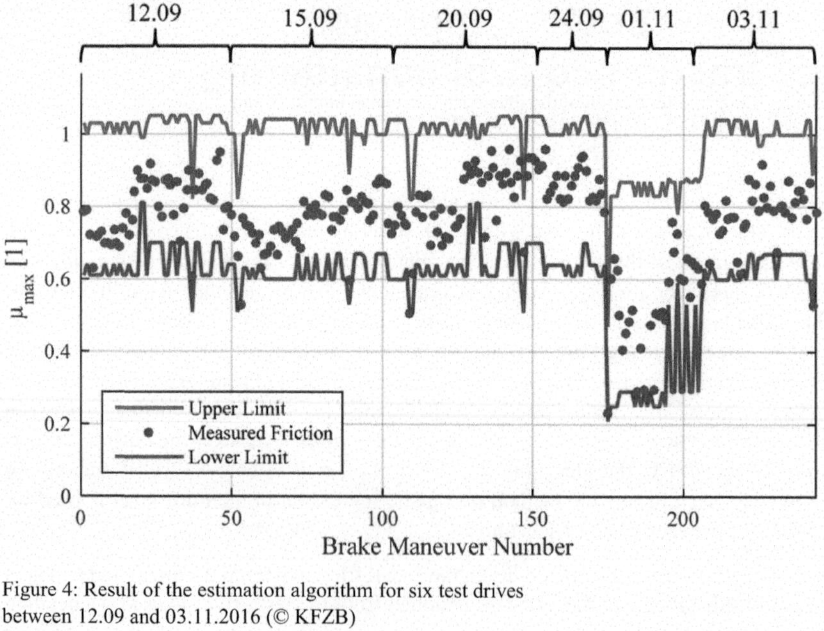

Figure 4: Result of the estimation algorithm for six test drives between 12.09 and 03.11.2016 (© KFZB)

An example of six test drives with a real time friction estimation is shown in Figure 4. The test drives were conducted between 12.09 and 03.11.2016. During that period the friction potential was estimated permanently. The green line shows the upper limit of the estimation, the red line the lower limit. The blue dots show the measured friction potential of real braking tests, which means: If the blue dots are between the green and the red line, the estimation algorithm has calculated the correct friction range. If it is outside, the algorithm was wrong. It can be seen, that for the change of weather condition, for example between brake manoeuver number 175 and 210, the friction values are much lower than in situations with dryer conditions. This was also recognised by the real time algorithm. For these test drives 241 out of 244 friction values (98.8 %) were within the estimated borders. For three brake tests the measured friction potential was below the estimated lower limit, however the maximum difference between the lower limit and the actual measured value was below 0.05.

The reduction of the average spread of 0.33 between the upper and lower limit of the friction estimation is still challenging. To achieve this goal, it is not only necessary to improve the estimation algorithm but more to get a better understanding of the parameters, which influences the friction potential. Test results of the last 3.5 years show that there are influencing parameters, which could be not yet described in total. Looking at the measured maximum friction of the 32 brake points of the test track it could be seen, that there is a significantly high spread at certain points, Figure 5. For example, at point 10 there is a difference between to lowest value of 0.64 and the highest value of 0.93 of 0.29, although it was always braked under the same weather conditions and exactly in the same place. It is remarkable as well, that this spread is clearly smaller for brake point 25.

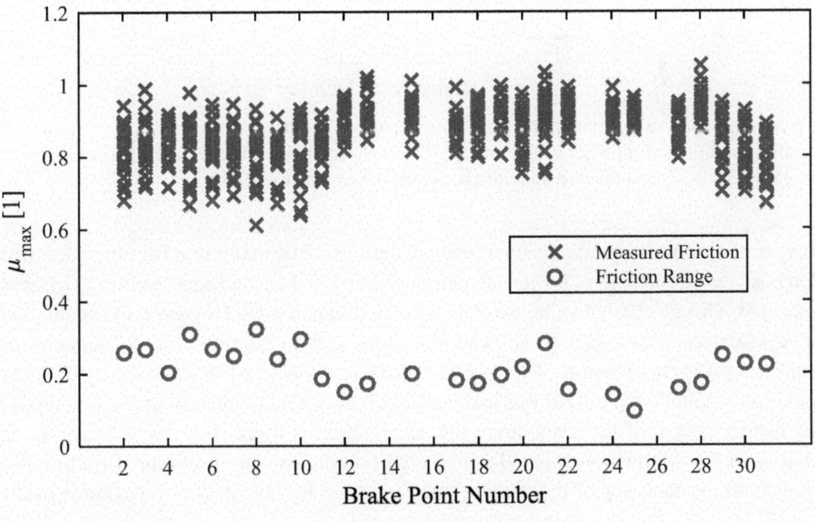

Figure 5: Asphalt-Analysis (summer tires & dry surface) (© KFZB)

One possible reason for that behaviour might be the procedure of the braking tests. Therefore, it was investigated in detail from different perspectives. First, brake tests at the same point were repeated after a short period of time. This was done for several times with a result of only minor spread. Then the braking tests along the test track were repeated on two consecutive days with almost identical weather conditions. Again, the results showed a good repeatability with an average spread of 0.09, Figure 6. This result shows, that in general the tests procedure is robust and repeatable.

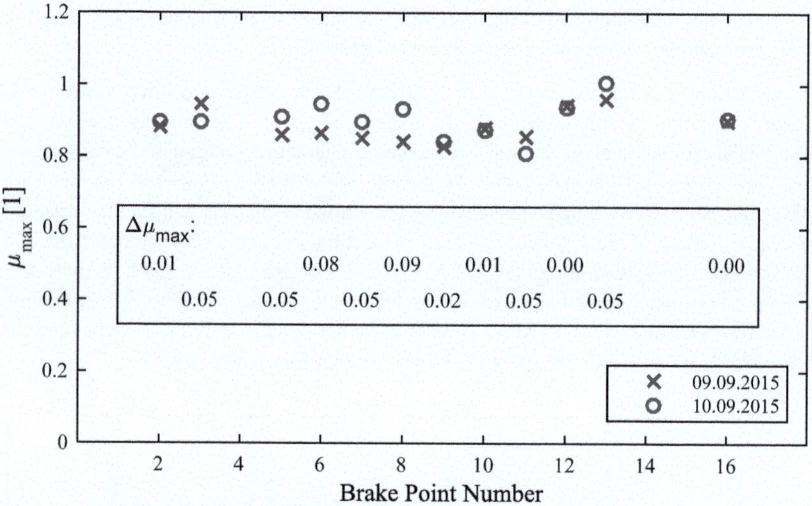

Figure 6: Analysis of the maximum friction values of two test drives with similar weather conditions (following conditions apply to both test drives: summer tires, temperature between 15 – 30°C and relative humidity between 43 and 60%)(© KFZB)

In a second step a deeper analysis of the calculation of the maximum friction value was done. As described above, for its calculation the first 0.5 second acceleration value was averaged. This calculated value was taken as friction potential. However, the evaluation of several acceleration signals showed, that especially in the first braking segment for some tests a high variation of measured values was observed. It seems that for some braking manoeuvres the ABS needs more time for a stable regulation of the tire slip. In the further course of the manoeuvre the acceleration is more stable on a lower level. That is why it is now investigated which effect on the spread of measured friction potential and the accuracy of the friction estimation can be expected, if the average of the last 0.5 second of the average acceleration is used.

5 Summary and Outlook

The evaluation of the results of the estimate of the friction coefficient clearly shows that for dry, moist, wet and snowy/icy road surface a reliable prediction of the maximum of this coefficient is possible without additional sensors. The developed estimation procedure shows a high degree of correct predictions. The acceptance of false predictions should be considered for the specific application case.

Still challenging is to reduce the range of the estimation limits. Here a better understanding for the influencing parameters of the maximum friction coefficient is needed. Detailed analysis of the data sets of single braking points should help to understand under which circumstances the friction coefficient changes.

The evaluation of the estimation algorithm showed, that it provides satisfactory results even if some information (like the GMA) are not available.

For the future it is imaginable that numerical weather simulation models, which give detailed weather information and which also consider local effects of the vegetation can improve the friction estimation a lot. Additionally, the use of vehicle dynamic parameters like wheel speed and interventions of ABS and ESP might improve the estimation algorithm.

Sources

[1] Müller, G. und S. Müller: Messung von Reibwerten unter Realbedingungen zur Erhöhung der Fahrzeugsicherheit: Proceedings of 10th VDI-Tagung Fahrzeugsicherheit, Berlin, 25-26 November 2015

[2] Müller, S., Uchanski, M. and K. Hedrick: Estimation of the maximum tire-road friction coefficient, JDSMC, 125(4), pp. 607-618, 2003.

[3] Breuer, B., Eichhorn, U. and J. Roth: Measurement of tyre/road friction ahead of the car and inside the tyre. Proceedings of AVEC'92 (International Symposium on Advanced Vehicle Control), pages 347-353, 1992.

[4] Eichhorn, U. and J. Roth: Prediction and monitoring of tyre/road Friction. XXIV FISITA Congress, London, GB, 2:67-74, June 7-11 1992. "Safety of the Vehicle and the Road".

[5] Pasterkamp, W. and Pacejka, H.: Application of Neural Networks in the Estimation of Tire/Road Friction Using the Tire as Sensor. SAE Technical Paper 971122, 1997, DOI: 10.4271/971122

[6] Pasterkamp, W. R. and Pacejka, H. B.: The Tyre as a Sensor to Estimate Friction. In: Vehicle System Dynamics, 1997, Vol. 27, No. 5-6, pp. 409-422, DOI: 10.1080/00423119708969339

[7] Becherer, Th. et al.: Der Seitenwandtorsionssensor SWT. In: ATZ Automobiltechnische Zeitschrift 102 (2000), 11, S. 946

[8] Gustafsson, F.: Slip-based tire-road friction estimation. Automatica, 33(6):1087-1099, June 1997.

[9] Kiencke, U. and A. Daiß: Estimation of tyre friction for enhanced ABS systems. Proceedings of AVEC'94, 1994

[10] Gnadler, R. und Marwitz, H.: Neues Sytem zur Ermittlung des Kraftschlusspotentials im Fahrbetrieb. In: ATZ 106 (2004), 5, S. 458-467

[11] Breuer, B., Bartz, M., Karlheinz, B., Gruber S., Semsch, M., Strothjohann, T. and C. Xie: The mechatronic vehicle corner of Darmstadt University of Technology – Interaction and cooperation of a sensor tire, new low-energy disc brake and smart wheel suspension. Proceedings of FISITA 2000, Seoul, Korea, June 12-15 2000.

[12] Reimpell, J. und Sponagel, P. (Hrsg.) (1986): Fahrwerktechnik: Reifen und Räder. 1. Auflage. Würzburg: Vogel Verlag.

[13] Thomas Bachmann (1996): Literaturrecherche zum Reibwert zwischen Reifen und Fahrbahn. Fortschr.-Ber. VDI Reihe 12 Nr. 286. Düsseldorf: VDI Verlag.

[14] J. Roth (1993): Untersuchungen zur Kraftübertragung zwischen PKW-Reifen und Fahrbahn unter besonderer Berücksichtigung der Kraftschlusserkennung im rotierenden Rad. Fortschr.-Ber. VDI Reihe 12 Nr. 195. Düsseldorf: VDI-Verlag.

[15] Bachmann, T.: Wechselwirkungen im Prozess der Reibung zwischen Reifen und Fahrbahn. Reihe 12 360, Fortschritt-Berichte CDI, 1998.

[16] http://www.geo.fu-berlin.de/met/service/pollenflugkalender/; Tagesaktuelle Polleninformationen für Berlin; accessed: 22. Oktober 2015

[17] Eid, M., Gollwitzer, M., und M. Schmitt.: Statistik und Forschungsmethoden. Lehrbuch mit Online-Materialien, 1. Auflage, Weinheim, Basel, Beltz, 2010.

Dr.-Ing. Gerd Müller,
Technische Universität Berlin, FG Kfz, TIB 13,
Gustav-Meyer-Allee 25, 13355 Berlin, Germany, +49 (0) 30 314 72 996,
gerd.mueller@tu-berlin.de

Vincent Gregull, M.Sc.,
Technische Universität Berlin, FG Kfz, TIB 13,
Gustav-Meyer-Allee 25, 13355 Berlin, Germany, +49 (0) 30 314 72 917,
vincent.gregull@campus.tu-berlin.de

Claudia Bräsemann, B.Sc.,
Technische Universität Berlin, FG Kfz, TIB 13,
Gustav-Meyer-Allee 25, 13355 Berlin, Germany, +49 (0) 30 314 72 996,
claudia.braesemann@tu-berlin.de

Prof. Dr.-Ing. Steffen Müller,
Technische Universität Berlin, FG Kfz, TIB 13,
Gustav-Meyer-Allee 25, 13355 Berlin, Germany, +49 (0) 30 314 72 970,
steffen.mueller@tu-berlin.de

ITC – model-based feed forward traction control

Dr. Lars König, Frieder Schindele, Dr. Jyotishman Ghosh

Bosch Engineering GmbH

P. Pfeffer (Ed.), *9th International Munich Chassis Symposium 2018*, Proceedings,
https://doi.org/10.1007/978-3-658-22050-1_21

Abstract

Due to the great success of Bosch Engineering's nonlinear model based lateral dynamics controller IVC, the concept has now been expanded to traction control systems. Based on the method of exact linearization, a feedforward algorithm is presented. This is done by taking the nonlinearities of tire behavior with combined lateral and longitudinal slip conditions into account. Within the target value derivation, the relationship of lateral and longitudinal slip with the aspect to the vehicle trajectory is analyzed. Asymptotic stability of the closed loop is established using a linear feedback controller, enhanced by gain scheduling and anti-windup algorithms. Since communication delay is an essential component of the control plant, stability and robustness properties are investigated by applying frequency domain based methods. The performance of the proposed control algorithm is demonstrated by means of road tests that are carried out with a rear wheel driven sports car under low-μ conditions.

1 Introduction

Traction control systems (TCS) were firstly introduced in series production cars in 1987 [1], primarily as a safety feature. In the meantime, TCS has become an essential element in order to accomplish excellent vehicle dynamics in high performance sports cars. When trying to achieve new lap records, an optimized traction control can make the difference, even for an experienced driver. Beside pure performance, drivability and driving pleasure are getting more and more into the focus of customers and manufacturers. For example, TCS driving pleasure functionalities are easing vehicle handling in drift situations [2] or enabling the driver to continuously adjust the system to his preferences [3]. However, the safety aspect is still one of the major requirements for any traction control system and has to be regarded within the design and calibration procedure.

Another upcoming trend is that original equipment manufacturers (OEMs) are requesting a time and cost reduction in tuning of vehicle dynamics controllers. At the same time, the complexity in drivetrain architectures has increased through electrification and hybridization. In order to meet all these demands and to utilize new potentials provided by electrified drivetrain configurations, Bosch Engineering GmbH is introducing ITC (Integrated Traction Control) – an innovative, model based traction control algorithm. The theory behind ITC is presented in this paper which is organized as follows. Firstly, the plant model is derived and the architecture of the control loop as well as the design objectives are described. Then, the design of the controller with separate feedforward and feedback portions is formulated and the derivation of the controller target is presented. Finally, the controller performance is investigated by means of road test results.

2 Controller design

Similar to the concept presented in [4] by Bosch Engineering GmbH involving lateral vehicle dynamics control, here a traction slip controller is developed. A model based nonlinear controller design approach, based on the principle of exact linearization [5] is utilized, as illustrated in **figure 2.1**.

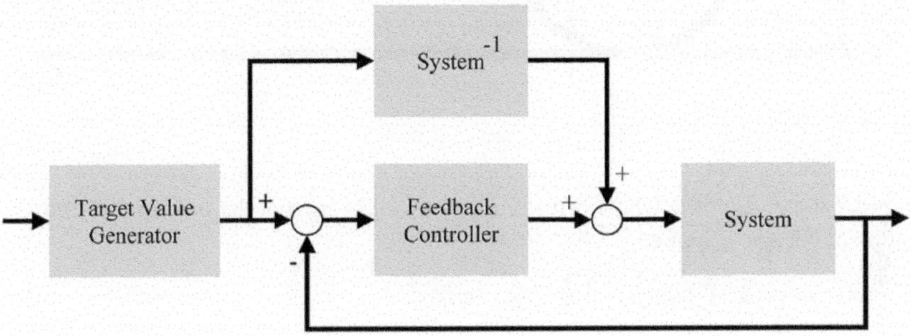

Figure 2.1: Control loop with feedforward control

The controller consists of three major elements: an inverted model of the controlled system (feedforward portion), a feedback controller and a generator of the target value.

2.1 Derivation of plant model and feedforward control

One major challenge in plant modelling for nonlinear controller design applications is finding a compromise between including all relevant effects on one hand and limiting the model complexity to a level, which can be handled in terms of model inversion and building time derivatives on the other. In **figure 2.2**, a "longitudinal quarter vehicle model" is presented. It mainly consists of one spinning wheel, which is connected to the body by a non-holonomic constraint. The differential equation of the quarter vehicle model with the moment of inertia of the drivetrain J_T, the longitudinal tire force of the rear axle F_{LR}, the drivetrain torque M_P, the angular acceleration $\dot{\omega}$ and the wheel radius r_{dyn} is denoted as

$$\dot{\omega} = \frac{1}{J_T} \cdot \left(M_P - F_{LR} \cdot r_{dyn} \right). \tag{2.1}$$

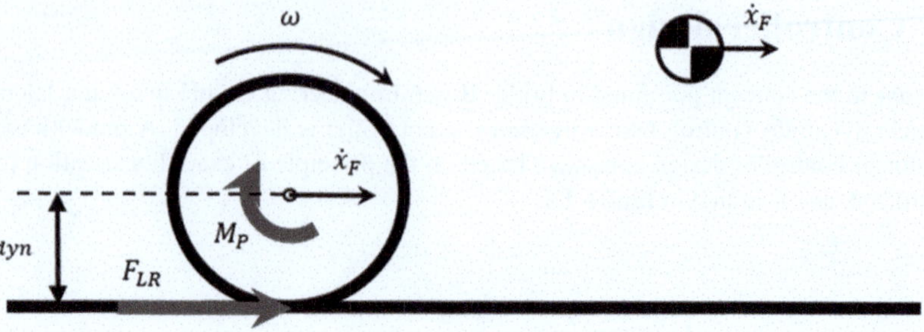

Figure 2.2: Quarter vehicle model

In order to consider the whole drivetrain's moment of inertia, the principle of conservation of energy is applied

$$\frac{1}{2} \cdot J_T \cdot \omega^2 = 2 \cdot \frac{1}{2} \cdot J_W \cdot \omega^2 + \frac{1}{2} \cdot J_{Eng} \cdot \omega_{Mot}^2 , \tag{2.2}$$

where J_W describes the moment of inertia of all drivetrain elements rotating at wheelspeed, J_{Eng} represents the moment of inertia of all drivetrain elements rotating at engine speed and ω_{Mot} represents the rotational velocity of the engine. By considering the kinematic transmission ratio $i_G = \omega_{Mot}/\omega$, (2.2)-(2.4) yield the drivetrain's moment of inertia as,

$$J_T = 2 \cdot J_W + J_{Eng} \cdot i_G^2 . \tag{2.3}$$

The friction force, F_{LR}, mainly depends on the longitudinal slip

$$\lambda_R = \frac{\dot{x}_F - \omega \cdot r_{dyn}}{\dot{x}_F} , \tag{2.4}$$

which is defined here according to [6], the sideslip angle α_R and a set of tire parameters $\boldsymbol{p}_{TR}$. The variable $\dot{x}_F$ represents the longitudinal portion of the vehicle's groundspeed at the mounting position of the wheel. By combining (2.1)-(2.4) with a tire model according to (3.6), the longitudinal quarter vehicle model's equation of motion is obtained as

$$\dot{\omega} = f_{QV}\left(M_P, \alpha_R, \lambda_R, J_T, r_{dyn}, \boldsymbol{p}_{TR}\right) . \tag{2.5}$$

The controller's feedforward portion

$$M_P^{FF} = f_{QV}^{-1}\left(\alpha_R, \lambda_{RTar}, J_T, r_{dyn}, \boldsymbol{p}_{TR}\right) + J \cdot \dot{\omega}_{tar}(\dot{x}_F, \ddot{x}_F, \lambda_{RTar}, \dot{\lambda}_{RTar}) \tag{2.6}$$

is derived by inverting (2.5), setting the wheel angular acceleration equal to its target value and splitting the actuating variable M_P into two portions: a feedforward portion M_P^{FF} and a feedback portion M_P^{FB}. The desired angular velocity can be calculated as

$$\omega_{Tar} = \frac{\dot{x}_F - \lambda_{RTar} \cdot \dot{x}_F}{r_{dyn}}. \tag{2.7}$$

It depends on the desired longitudinal slip λ_{RTar}, which is referred to in chapter 3. Here the model inversion f_{QV}^{-1} is not conducted exlusively with respect to the current system state and the inserted dynamics $\dot{\omega}_{Tar}$ is not conducted exlusively relative to the desired system state, as proposed in [5] and performed in [4, 7]. In order to reduce model complexity and sensor noise influences f_{QV}^{-1} as well as $\dot{\omega}_{Tar}$ depend on a mixture of actual and desired state variables. Thus the error dynamics of the feedforward controlled system becomes operation point dependent. Inserting the feedforward control law (2.6) into the system model (2.5) while considering major sources of model uncertainties yields the error dynamics

$$\dot{\omega}_{Tar} - \dot{\omega} = -M_P^{FB} + \xi_{c_{RL}} \cdot \omega - \xi_{ext}. \tag{2.8}$$

From a system dynamics point of view, (2.8) represents a PT1-system with an additional input ξ_{ext}. The natural frequency $\xi_{c_{RL}}$ mainly depends on the current longitudinal tire stiffness $c_{RL} = dF_L/d\lambda$.

2.2 Derivation of feedback control

Stabilizing feedback controllers with freely choosable error dynamics poles can be designed directly based upon (2.8). However, communication delay is a major factor in today's traction control system layouts. In order to consider this effect without simplifications and to apply well known robustness measures the feedback control portion is designed in the frequency domain. The ITC control loop is represented by **figure 2.3**. The plant transfer function $G_S(s)$ does not contain the nonlinear plant model (2.5), but only the remaining dynamics of the exact linearized system, which can be derived directly from (2.8) and then further extended by a time delay $e^{-s \cdot T_t}$. The feedforward transfer function $G_{FF}(s)$ can be derived directly from (2.8) as well. In order to synchronize target value and control variable, ω_{Tar} is delayed by a period of time T_t equivalent to the communication delay using the transfer function $G_T(s) = e^{-s \cdot T_t}$. The input signal ξ_{Noise} represents measurement noise as well as drivetrain oscillations, which are typically featuring a frequency range of 10-15 Hz. Due to the communication delay of $T_t \approx 0.1$ s, these oscillations cannot be damped by feedback control. However, a Chebyshev low pass filter $G_F(s)$ [8] is inserted into the control loop to prevent the controller from being stimulated by drivetrain oscillations.

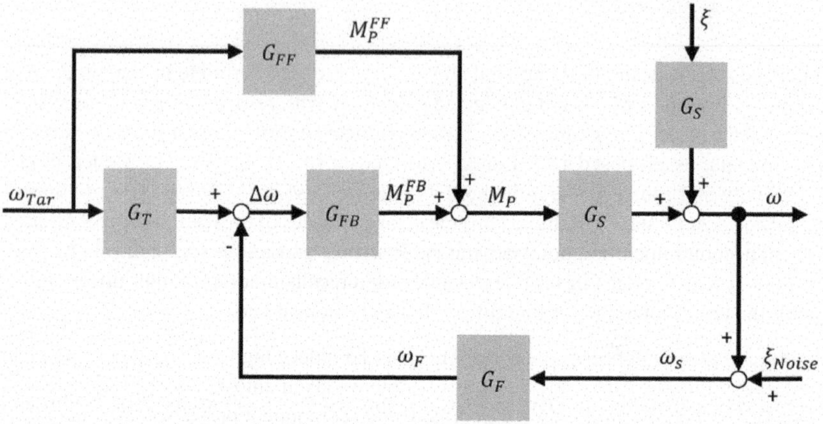

Figure 2.3: ITC control loop

The feedback controller

$$G_{FB}(s) = \frac{M_P^{FB}(s)}{\Delta\omega(s)} = \frac{1}{s} \cdot \frac{b_m \cdot s^m + b_{m-1} \cdot s^{m-1} + b_1 \cdot s + b_0}{s^n + a_{n-1} \cdot s^{n-1} + a_1 \cdot s + a_0} \tag{2.9}$$

is designed as a linear transfer function with an integrating portion, extended by an anti-windup function [9]. The control loop's transfer function

$$\omega(s) = G_{CL} \cdot \omega_{Tar} + G_{Error} \cdot \xi + G_{Noise} \cdot \xi_{Noise} \tag{2.10}$$

can be directly derived from **figure 2.3** and is a combination of the reference transfer function

$$G_{CL}(s) = \frac{\omega(s)}{\omega_{Tar}(s)} = \frac{(G_T \cdot G_{FB} + G_{FF}) \cdot G_S}{1 + (G_{FB} + G_{FF}) \cdot G_S \cdot G_F}, \tag{2.11}$$

the disturbance transfer function

$$G_{Error}(s) = \frac{\omega(s)}{\xi(s)} = \frac{G_S}{1 + (G_{FB} + G_{FF}) \cdot G_S \cdot G_F} \tag{2.12}$$

and noise transfer function

$$G_{Noise}(s) = \frac{\omega(s)}{\xi_{noise}(s)} = \frac{-(G_{FB} + G_{FF}) \cdot G_S \cdot G_F}{1 + (G_{FB} + G_{FF}) \cdot G_S \cdot G_F}. \tag{2.13}$$

As the reference action is mainly conducted by the feedforward portion, the feedback controller can be designed with major focus on disturbance and noise attenuation. In order to assess robustness properties, the Nyqusit criteria (see [10]) based on the transfer function of the open loop

$$G_O(s) = G_F \cdot G_{FB} \cdot G_S \tag{2.14}$$

is applied. A Nyquist Diagram of (2.14) is illustrated in **figure 2.4**. Therefore vehicle and controller parameters corresponding to the road tests referred to in chapter 4 have been implemented.

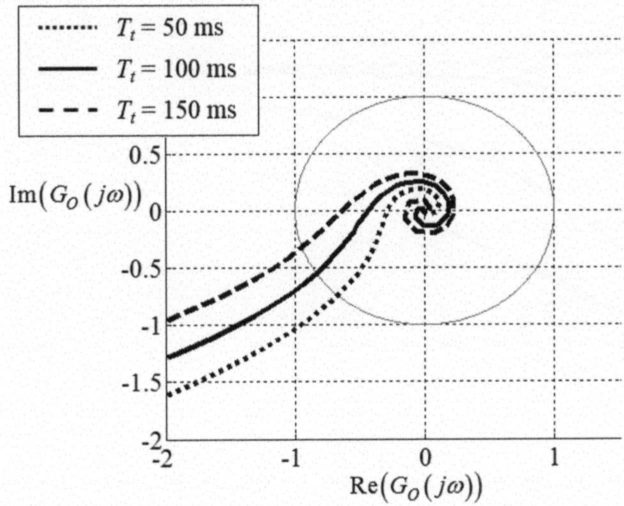

Figure 2.4: Nyquist Diagram of ITC control loop

As seen in figure 2.4, communication delay has significant influence on the robustness properties of the control loop. The phase margin varies from 18° to 48° and the gain margin from 1.6 to 3.5 when time delays from $T_t = 0.15$ s to $T_t = 0.05$ s are inserted respectively. Road tests with various vehicles have shown that there should be a phase margin of at least 30° and a gain margin of at least 2 in order to achieve an adequate controller performance for any appropriate tire choice, track conditions and operating point. As T_t is constant for a certain vehicle setup, well measurable and modelled without simplifications, it can be easily considered during the design procedure.

In contrast to the communication delay, the longitudinal tire stiffness dependent natural frequency $\xi_{C_{RL}}$ of the plant model (2.8) varies significantly during controller operation. Tire force characteristics derived from a nonlinear model including combined slip conditions as described in (3.6) are illustrated in **figure 2.5**. When the sideslip angle is small and the operating point lies within the linear range of the longitudinal force characteristics, $\xi_{C_{RL}}$ becomes large and (2.8) represents a stiff PT1-system. However, when the operating point lies within the saturation range of the longitudinal force characteristics, $\xi_{C_{RL}}$ becomes small or even negative and (2.8) represents an integrator or even an unstable PT1-system. Furthermore, figure 2.5 illustrates that there can be also a significant change in longitudinal stiffness for a constant longitudinal slip when the sideslip angle varies.

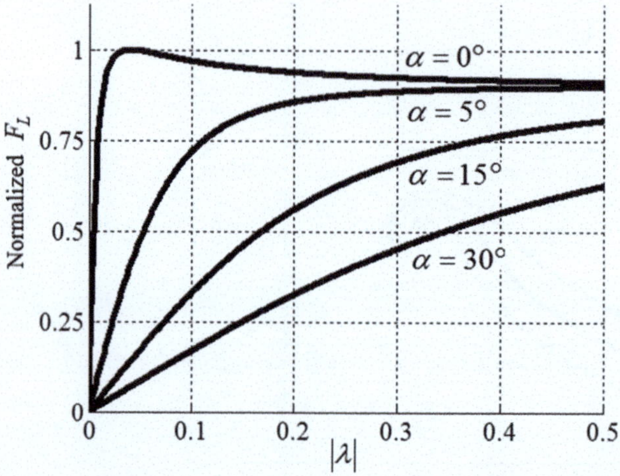

Figure 2.5: Longitudinal tire force dependent on traction slip λ and sideslip angle α

In order to obtain reasonable controller performance under any operating conditions – from an extremely stable TCS (small target slip) setup to a "drift mode" (large target slip) – the variation of $\xi_{C_{RL}}$ is considered by a gain scheduling technique. The feedback controller design is repeated serveral times with respect to various $\xi_{C_{RL}}$ operating points. Dependening on the actual value of $\xi_{C_{RL}}$, that is estimated online, the controller gains are continuously adapted during operation.

3 Derivation of the target value

Regarding straight accelerating maneuvers, a performance optimizing strategy is trivially achieved by choosing the longitudinal slip corresponding to the maximum longitudinal force as target value. However, when accelerating and cornering simultaneously, tire and vehicle properties related to combined slip conditions have to be evaluated within the derivation of the target value (compare figure 2.5). In this chapter the relationship between sideslip angle, traction slip, longitudinal and lateral acceleration is investigated, focusing on performance optimization and drift control for rear wheel driven (RWD) vehicles.

Figure 3.1 illustrates a bicycle model (compare [11]) of a vehicle with RWD layout featuring trajectory-fixed and body-fixed coordinate systems. According to body fixed coordinates Newton's equations are represented by

$$m \cdot \begin{bmatrix} \ddot{x}_F \\ \ddot{y}_F \end{bmatrix} = \begin{bmatrix} -F_{SF} \cdot \sin(\delta) + F_{LR} - F_{air} \cdot \cos(\beta) - F_{rollFA} \cdot \cos(\delta) - F_{rollRA} \\ F_{SF} \cdot \cos(\delta) + F_{SH} - F_{air} \cdot \sin(\beta) - F_{rollFA} \cdot \sin(\delta) \end{bmatrix}, \tag{3.1}$$

with m denoting the vehicle mass, $\ddot{x}_F, \ddot{y}_F$ longitudinal and lateral accelerations according to vehicle fixed coordinates, δ the front axle steering angle, the axle-wise tire forces in lateral direction F_{SF}, F_{SH} and the rear axle longitudinal force, F_{LR}. The rolling resistance forces F_{rollFA}, F_{rollRF} mainly depend on the axlewise normal forces, while the aerodynamic force $F_{air}(v, \beta)$ is basically a function of the vehicle groundspeed v and the sideslip angle β, see e.g. [11].

Transforming (3.1) to trajectory fixed coordinates yields

$$\begin{bmatrix} \ddot{x}_F \\ \ddot{y}_F \end{bmatrix} = \begin{bmatrix} \cos(\beta) & -\sin(\beta) \\ \sin(\beta) & \cos(\beta) \end{bmatrix} \cdot \begin{bmatrix} \ddot{x}_T \\ \ddot{y}_T \end{bmatrix}. \tag{3.2}$$

The accelerations $\ddot{x}_T, \ddot{y}_T$ can be rewritten dependent on vehicle groundspeed v and sideslip angle β by calculating the groundspeed's derivative with respect to inertial coordinates and then transforming back to the trajectory fixed system:

$$\begin{bmatrix} \ddot{x}_T \\ \ddot{y}_T \end{bmatrix} = \begin{bmatrix} \cos(\psi + \beta) & \sin(\psi + \beta) \\ -\sin(\psi + \beta) & \cos(\psi + \beta) \end{bmatrix} \cdot \frac{d}{dt} \begin{bmatrix} v \cdot \cos(\psi + \beta) \\ v \cdot \sin(\psi + \beta) \end{bmatrix} = \begin{bmatrix} \dot{v} \\ v \cdot (\dot{\psi} + \dot{\beta}) \end{bmatrix}. \tag{3.3}$$

Introducing the vehicle's moment of inertia J relative to its vertical axle, the longitudinal position of the center of gravity l_F, l_R and the yaw acceleration $\ddot{\psi}$, the equations of motion corresponding to the vehicle model represented by figure 3.1 are completed by Euler's equation

$$J \cdot \ddot{\psi} = F_{SF} \cdot \cos(\delta) \cdot l_F - F_{SR} \cdot l_R . \tag{3.4}$$

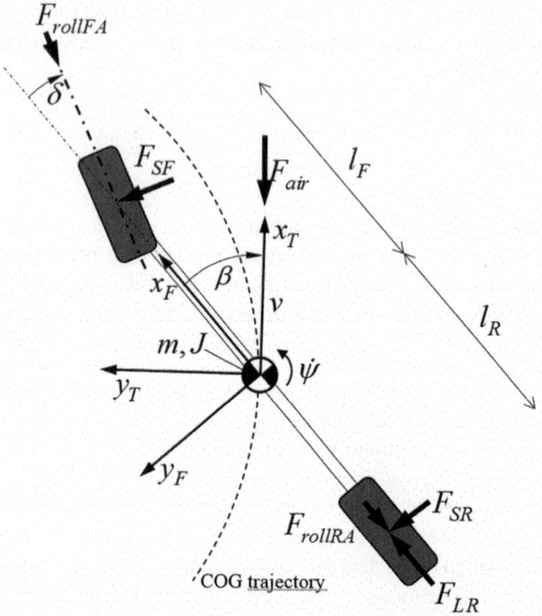

Figure 3.1: Bicycle model with trajectory-fixed and body-fixed coordinate system

By introducing the sideslip angles (see [11]) of the front and the rear axles

$$\alpha_F = \delta - \text{atan}\left(\frac{v \cdot \sin(\beta) + l_F \cdot \dot{\psi}}{v \cdot \cos(\beta)}\right), \qquad \alpha_R = -\text{atan}\left(\frac{v \cdot \sin(\beta) - l_R \cdot \dot{\psi}}{v \cdot \cos(\beta)}\right) \tag{3.5}$$

the tire forces can be written as

$$\begin{bmatrix} F_{LR} \\ F_{SR} \end{bmatrix} = \boldsymbol{f}_{TC}\left(\alpha_R, \lambda_R, \boldsymbol{p}_{TR}\right), \qquad F_{SF} = f_T\left(\alpha_F, \boldsymbol{p}_{TF}\right). \tag{3.6}$$

A common tire model $\boldsymbol{f}_{TC}$, f_T is the so called "Magic Formula" [12] where the parameter sets $\boldsymbol{p}_{TR}$, $\boldsymbol{p}_{TF}$ represent the track conditions and tire properties of the vehicle's front and rear axles. In order to obtain a solvable set of equations for the calculation of the target value, two additional constraints have to be defined. Assuming a constant sideslip angle ($\dot{\beta} = 0$), the yaw rate can be derived from (3.3) as

$$\dot{\psi} = \frac{\ddot{y}_T}{v}. \tag{3.7}$$

Furthermore, differentiating (3.7) while assuming constant lateral acceleration relative to the trajectory ($\dddot{y}_T = 0$) yields

$$\ddot{\psi} = \frac{-\ddot{y}_T \cdot \dot{v}}{v^2}. \tag{3.8}$$

By merging (3.1)-(3.8), the dependency of steering angle, ground speed derivative, trajectory lateral acceleration, rear axle wheel slip and rear axle sideslip angle denotes a nonlinear multidimensional unconstrained zero crossing detection task:

$$f_{GD}\left(\delta, v, \ddot{y}_T, \dot{v}, \alpha_R, \lambda_R\right) = \begin{bmatrix} 0 & 0 & 0 \end{bmatrix}^T \tag{3.9}$$

A direct search strategy [13] is chosen as a suitable method to efficiently find the numerical solution of (3.9). As there are six unknown variables and three equations, three variables have to be set as optimizer inputs. E.g. steering angle, cornering radius (which is in inverse proportion to $\ddot{y}_T$) and ground speed change can be calculated for a given combination of rear axle sideslip angle, wheel slip and ground speed. Or the resulting cornering radius, required (counter) steering angle and wheel slip to achieve a certain change in ground speed at a given rear axle sideslip angle and ground speed can be determined.

Solutions of (3.9) while applying a parameter set corresponding to a vehicle referred to in chapter 4 are visualized in **figure 3.2**. The track conditions are assumed to be low-μ (packed snow) and the velocity is set to 50 kph. A typical safety oriented Electronic Stability Control (ESC) setting means, that the magnitude of desired traction slip is saturated to a maximum value (under low-μ conditions approximately 4% at 50 kph) when driving straight ahead and is continuously reduced when approaching the lateral dynamics stability limit. In a rear wheel driven vehicle, this strategy prevents the driver from counter steering. However the longitudinal dynamics performance is significantly limited, as it is not possible to maintain speed or even accelerate when the rear axle sideslip angle becomes large.

Setting $\dot{v} = 0$ yields about 10% rear axle traction slip at 15° of rear axle sideslip angle and 45% rear axle traction slip at 30° of rear axle sideslip angle. This strategy allows TCS controlled steady state drifting at any sideslip angle. It relieves the driver from throttle modulation, so that he can concentrate on controlling the sideslip angle by his steering action. Road tests with various drivers and vehicles have shown, that this is a suitable strategy for enabling even unexperienced drivers to perform drift maneuvers with (super) sports cars.

In order to achieve an optimal TCS performance in terms of lap time minimization, setting $\dot{v} = 0$ is not a constructive strategy. **Figure 3.2** shows that by assuming a constant trajectory lateral acceleration below the adhesion limit (in this example 85% of the adhesion limit) solving (3.9) yields a maximum value of $\dot{v} \approx 0.9$ m/s² for small rear axle side slip angles and $\dot{v} \approx 0.5$ m/s² when α_R reaches 30°. It can also be seen, that the longitudinal slip demand is about 1.5 times of the slip demand for steady state drifting. However, regarding TCS race track applications the evaluation of (3.9) is not sufficient as it is neglecting the track layout.

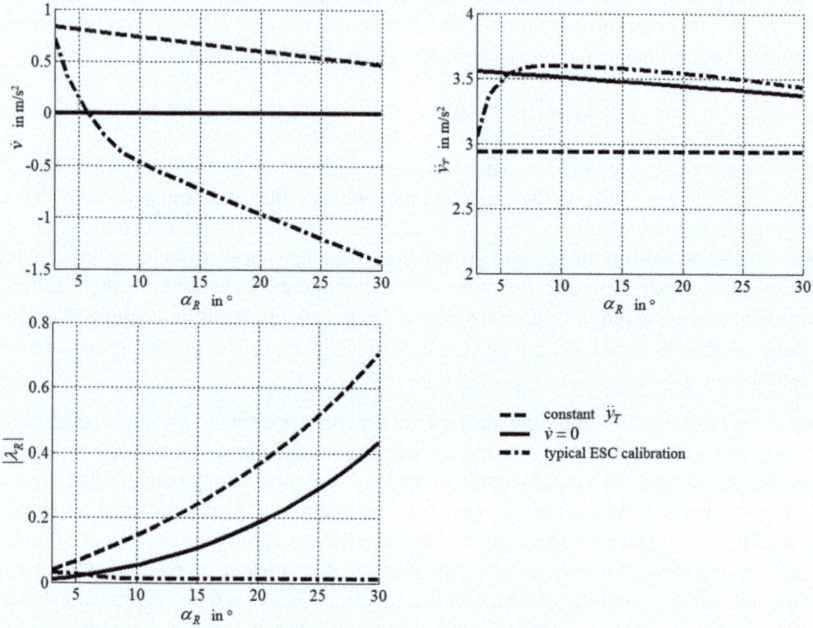

Figure 3.2: Dependency of rear axle sideslip angle, traction slip and accelerations with respect to trajectory fixed coordinates. RWD vehicle, 50 kph, low-μ-conditions

This phenomenon is illustrated in **figure 3.3**. Imagine a race driver on a flying lap with a rear wheel driven vehicle and TCS deactivated, who has reached the apex with a slight amount of oversteering. In order to avoid losing time, he will take the corner exit under slight power oversteer, controlling the sideslip angle by his steering input and simultaneously the trajectory by throttle application.

Now imagine the same driver approaching the apex with TCS switched on. He is still in control of the sideslip angle by steering wheel application, but TCS will now control the trajectory. Assuming an apex speed of v_{start}=30 kph, a rear axle sideslip angle of 10° and setting $\ddot{x}_T$ to 90% of the adhesion limit, (3.9) yields λ_R=-11% and $\dot{v}$=0.56 m/s².

The driving trajectory as well as the corner exit velocity v_{end} can be calculated by transforming (3.3) to inertial coordinates and performing numerical integration. For a tight corner exit, see figure 3.3 a), -11% of traction slip is a performance optimizing strategy, as the complete track width is utilized. However, when the corner exit is wide, see figure 3.3 b), with the same TCS settings, the vehicle would now end up in the middle of the track and the driver would typically complain about traction control "holding back" the car.

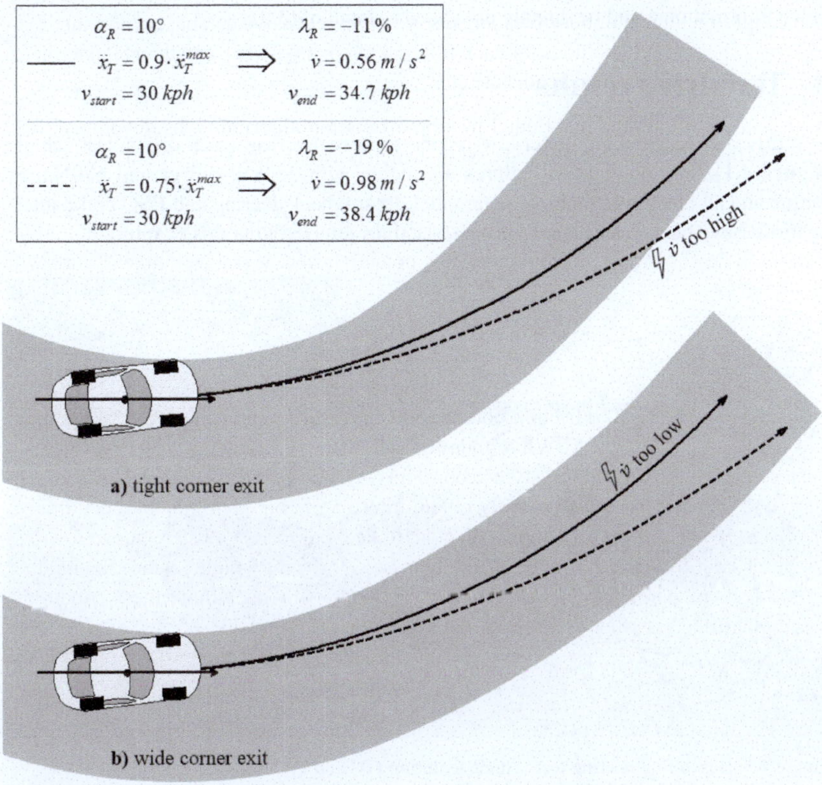

Figure 3.3: Track layout dependency for the calculation of the target value

Setting $\ddot{x}_T$ to 75% of the adhesion limit, (3.9) yields λ_R=-19%, $\dot{v}$=0.98 m/s^2 and therefore an ideal trajectory for a wide corner exit, see figure 3.3 b). However, assuming a tight corner exit the driver would now complain about "too less traction control". Apparently the track layout and the vehicle's position on the track have to be taken into account in order to generate an optimal traction slip as a target value for race track applications. Handing over a certain amount of today's TCS responsibility to the driver by introducing a throttle position dependent target value, resulted in discordant feedback of various test drivers. In further investigations it is planned to extend the calculation of the target value algorithm by video sensor and image processing based information coming from the domain of autonomous driving. By applying the well know relationship between cornering radius R, velocity and lateral acceleration ($\ddot{x}_T = v^2/R$), the current radius, sensed by video information could be directly processed as input of (3.9).

4 Test drive results

Test drives are carried out under low-μ conditions with a rear wheel driven test vehicle (**figure 4.1**). The powertrain features an engine with 384kW power and 670Nm of torque and an electronic locking differential. Rear wheel steering and ESC brake interventions have been deactivated for the test drives referred to in this chapter.

Figure 4.1: Bosch Engineering test vehicle featuring ITC and IVC

For evaluation of the driving results, an average percentage quotient for the feedforward portion is introduced as

$$\overline{FF}_{\%} = \frac{\int(|M_{FF}|)\cdot dt}{\int(|M_{FF}|+|M_{FB}|)\cdot dt} \cdot 100\% \,. \tag{4.1}$$

This quotient gives an indication about the ratio between feedforward and feedback portion.

Figure 4.2 shows results from a slalom maneuver with a velocity of ~50kph and an amplitude of the side slip angle, β of ~25°. The longitudinal target slip for the rear axle is constant at -25%. In this maneuver, the feedforward portion M_{FF} modulates the drive torque mainly depending on the side slip angle of the rear axle. This is due to the fact, that for constant target slip λ_{RTar}, the tire longitudinal force reduces with an increase in the tire side slip angle, compare figure 2.5. The feedback controller compensates only for the inaccuracy of the plant model and the disturbances coming from the inhomogeneous surface of the test track. In this maneuver, the quotient $\overline{FF_{\%}}$is about 81% and the maximum absolute control deviation $\Delta\lambda_{Rmax} = |\lambda_{RTar} - \lambda_R|$ is 6%.

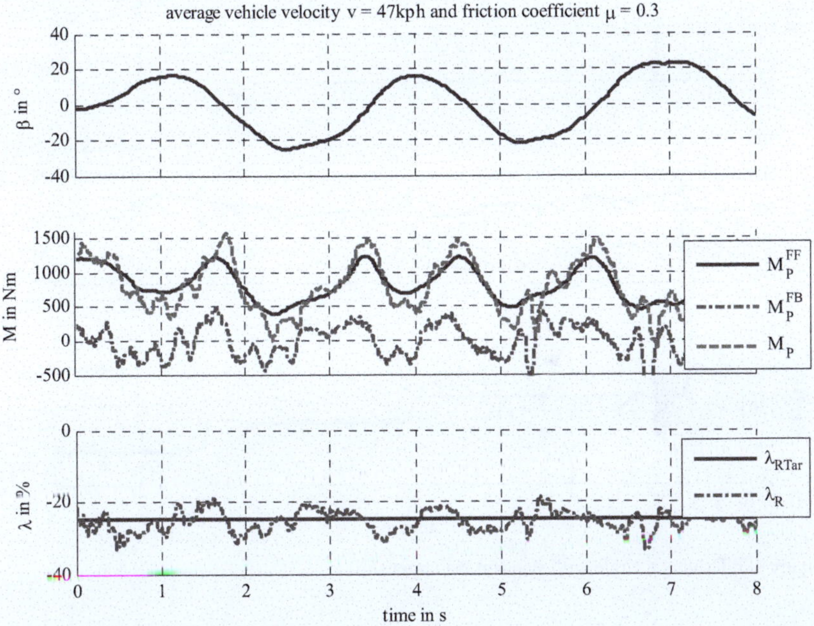

Figure 4.2: Slalom driving with constant target slip

The outcome of a circular driving maneuver where the absolute value of the target slip is increased stepwise from 20% to 41% is presented in **figure 4.3**. The feedforward portion together with the feedback portion leads to a fast reaction and prevents overshoots of the actual slip. The first crossover between target and actual slip happens ~150ms after the step request.

Within the time period 1.2-2.3 s, the friction coefficient changes to a lower value. In this situation, the feedback controller reacts on this disturbance by reducing the target torque until the friction estimation of the control system recognizes this change. In this maneuver, the feedforward quotient $\overline{FF}_{\%}$ is 83%. The evaluation of the maximal control failure of $\Delta\lambda_{Rmax}$ = 6% is here defined for the period after λ_{RTar} is constant for 100ms, due to the presence of the time delay $\dot{T}_t$.

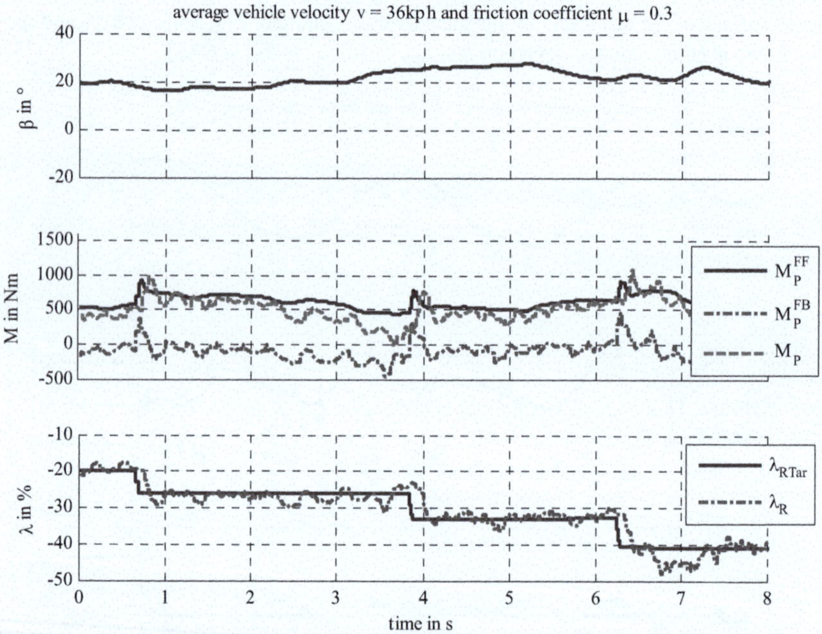

Figure 4.3: Driving a circle with steps on the target slip

The results from a drift manuever during a corner exit are presented in **figure 4.4**. In this maneuver, the target slip gets adapted depending on the throttle position. Thereby, the driver has the possibility to control the drift angle directly by throttle application. He is released from controlling the wheel slip and so he can focus on the (significantly slower) yaw dynamics. This strategy allows even nonprofessional drivers to have a command over the vehicle during driving situations featuring large sideslip angles.

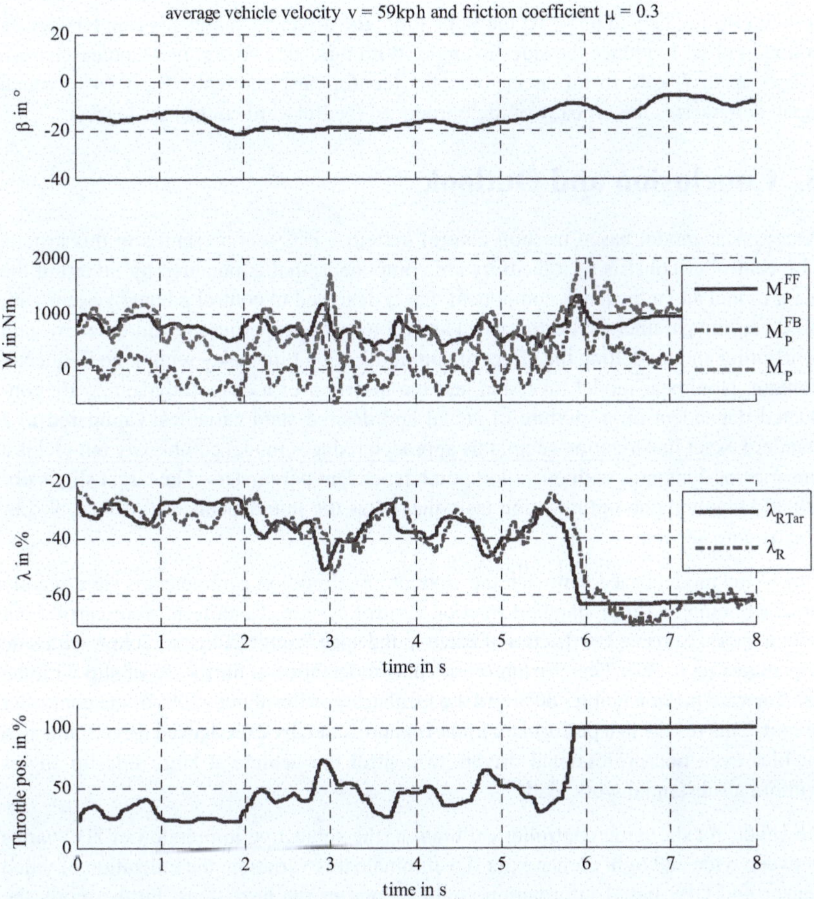

Figure 4.4: Acceleration out of a curve with a high side slip angle

Another motivation for throttle dependent target slip adaption, as mentioned in chapter 3, is enabling the driver to define the trajectory of the car with the throttle. In the beginning of the maneuver, the vehicle is about to reach the apex with a velocity of ~50kph and a sideslip angle of about 15°. Until ~5.8s, the driver adjusts the throttle position to control the trajectory while limiting the sideslip angle by steering action. At ~5.8s, the vehicle reaches the corner exit and the driver demands maximum acceleration. Therefore, he pushes the throttle to 100% and the absolute value of the target slip rises up from ~33% to 63%. At the same time, the driver increases his countersteering action in order to reduce the sideslip angle. With this, the vehicle speed increases from 57kph up to 70kph. In this maneuver, the feedforward quotient $\overline{FF_{\%}}$ is 74%, which again emphasizes the efficacy of the introduced feedforward control strategy.

5 Conclusion and Outlook

A nonlinear model based traction control system – ITC – is presented in this article. The controller consists of a feedforward component, that is obtained by inverting the plant model and a feedback component that is designed to control a linear system with time-varying parameters. Performance and robustness properties are addressed by gain scheduling and modelling the effect of communication time delay within the frequency domain. Nonlinear model inversion and the insertion of a desired dynamics are conducted dependent on a mixture of actual and desired state variables. Compared to a standard exact linearization setup, this approach reduces model complexity but yields a linear error dynamics with operation point dependent parameters. The target slip is obtained by solving an optimization task, that takes the interdependence of $\dot{v}$, v, $\ddot{x}_T$, λ_R and α_R into account.

Due to the model based approach the amount of calibration parameters is significantly reduced compared to a standard traction control system. Road tests were carried out with a sports car on a low friction surface in the speed range 30kph to 70kph with side slip angles up to 30°. Thereby the mean control deviation is only ~2% of slip with the feedforward portion being ~80% and the feedback portion about ~20% of the controller output. This results in a performance that is appreciated by experienced drivers and also enables even non-professional drivers to control the vehicle at large sideslip angles without a substantial work load.

However, for the same controller calibration, the subjective impression of ITC varies between wide and tight corner exits. To eliminate this variation, the utilization of video sensor and GPS based information could be one of the next steps. Furthermore, the potential of reducing communication delay to increase the performance of the feedback controller is under investigation. An appropriate solution could be to shift the feedback portion from the vehicle dynamics ECU to the engine ECU. Finally it is planned to extend the ITC concept to an integrated approach for more complex powertrain layouts, featuring center coupling clutches and wheel individual electric motors.

References

[1] Schöpf, H.-J.; Paul,J.: ASR Acceleration Skid Control – A Further Contribution Towards Increasing The Active Safety Of Daimler-Benz Vehicles, 22nd FISITA Congress, Dearborn MI USA, 1988

[2] Peters, M.: Fahrbericht Ferrari 458 Speciale – Spezialist für schwierige Fälle, auto motor und sport 25/2013

[3] Gebhardt, C.: Mercedes-AMG GT-R, sport auto 01/2017

[4] König, L.; Walter, T.; Gutmayer, B.; Merlein, D.: Integrated Vehicle Dynamics Control – an optimized approach for linking multiple chassis actuators, 14th Stuttgart International Symposium for Automotive and Engine Technology, Stuttgart Germany, 2014

[5] Isidori, A.: Nonlinear Control Systems. Berlin, Springer Verlag, 1995

[6] Robert Bosch GmbH: Sicherheits- und Komfortsysteme, Vieweg Verlag 2004

[7] König, L.: Ein virtueller Testfahrer für den querdynamischen Grenzbereich. Dissertation, Stuttgart University, 2009

[8] Meyer, M.: Signalverarbeitung, Springer Verlag 2017

[9] Adamy, J.: Nichtlineare Systeme und Regelungen, Springer Verlag 2018

[10] Lunze, J.: Regelungstechnik 1, Springer Verlag 2016

[11] Mitschke, M.; Wallentowitz, H.: Dynamik der Kraftfahrzeuge, Springer Verlag 2014

[12] Pacejka, H.; Bakker, E.: The Magic Formula Tire Model. Proceedings of the 1st International Colloquium on Tire Models for Vehicle Dynamics Analysis, Amsterdam Netherlands, 1993

[13] Lagarias, J.C.; Reeds J. A.; Wright, M. H.; Wright, P. E.: Convergence Properties of the Nelder-Mead Simplex Method in Low Dimensions, SIAM Journal of Optimization, Vol. 9 Number 1, pp. 112-147, 1998

Analysis of the potential of a new control approach for traction control considering a P2-Hybrid drivetrain

M. Sc. Alexander Zech, BMW Group

Dr.-Ing. Thomas Eberl, BMW Group

M. Sc. Carsten Marx, BMW Group

Prof. Dr.-Ing. Steffen Müller, Technical University of Berlin

P. Pfeffer (Ed.), *9th International Munich Chassis Symposium 2018*, Proceedings,
https://doi.org/10.1007/978-3-658-22050-1_22

Abstract

New developments in traction control consider locating the controller on the electronic control unit (ECU) of the engine. This is trivial for single engine drives operated by one ECU. Parallel hybrid powertrains on the other hand make use of two engines, an internal combustion engine (ICE) and an electric motor (EM). The two types of motors feature distinct response dynamics upon input changes. This makes a highly dynamic control problem such as traction control challenging. Both engines are controlled independently by dedicated ECUs generating a degree of freedom in providing the demanded driving torque compared to a single engine drive.

This paper investigates the given problem of maximizing the acceleration during traction control. Maintaining stability a new controller structure applied to a parallel hybrid electric vehicle is considered. First, a detailed model of the communication structure as well as engine, powertrain and driving dynamics of the considered vehicle is presented. Then a suitable allocation algorithm is applied to address the degree of freedom in the torque supply.

Using the dynamic control allocation (DCA) algorithm and locating the controller on the ECU of the EM to minimize latencies the approach is able to gain up to 6.6% acceleration compared to the state of the art.

1 Introduction

The first industrially implemented electronically controlled traction control system was introduced in 1986. Since then its development can be seen as an evolutionary process contrary to the development of powertrains which nowadays stretch from ICEs to EMs and various combinations of the two.

The state of the art structure of industrialized traction control systems consists of a dedicated driving dynamics electronic control unit (DECU), where the detection of wheel slip as well as the controller for limiting the driving torque is located [1]. This information is communicated to the electronic control unit (ECU) of the motor (Combustion-Engine-ECU: CECU and Electric-Motor-ECU: EECU) where it is simply passed through to the engine. This structure does not provide the required level of performance due to high control dynamics and latencies in the communication especially for high dynamic motors.

In [4 - 6] a new control structure was proposed in which the detection of wheel slip to generate a set point s_{max} (max. desired slip) and ω_{max} (max. desired engine speed) is computed in the DCU and communicated to the CECU. Here the control deviation is evaluated and the control algorithm is located (see figure 1). Due to communicating a signal of low dynamics via the vehicle BUS and computing the high dynamic torque demand on the ECU, this structural change enables much faster and more precise control of the wheel slip compared to the state of the art.

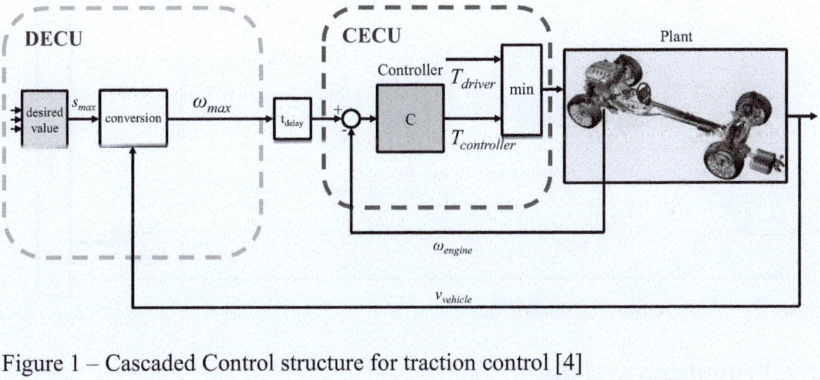

Figure 1 – Cascaded Control structure for traction control [4]

This new structure can be implemented easily into a powertrain driven by a single engine. Considering a P2 hybrid electric vehicle a new degree of freedom arises. Torque to accelerate the vehicle can now be generated by two actuators. Classic approaches to divide the requested torque between both actuators underlie mainly energetic considerations [2]. Actuator dynamics are only taken into account secondarily. However in driving dynamics the actuator response is of high interest [3] which makes this consideration essential for this research. Literature dealing with this subject, considering the dynamics of the combined system of ECUs and actuators in a hybrid electric vehicle for traction control, is unknown to the authors. In the following, the state of the art will be described as the new controller structure, applied to the CECU of a hybrid vehicle and using the default control allocation, see section 3 and figure 6.

The paper is structured as follows. In Chapter two the considered Simulation Model is described. Chapter three deals with the design of a control allocation algorithm suited for the given problem. Thereafter the control design is described. Chapter five depicts the simulation results and comparison to the state of the art followed by a conclusion and outlook in chapter six.

2 Vehicle Simulation Model

This chapter deals with the modelling of the drivetrain and vehicle dynamics as well as the ECUs and communication in between. The considered vehicle is a rear wheel drive BMW 7 series Plug-In-Hybrid-Electric-Vehicle (PHEV) with a P2-hybrid configuration. It consists of a turbocharged 4 cylinder gasoline ICE with a maximum torque of 400 Nm. The EM provides 250 Nm of torque and a maximum power output of 83 kW.

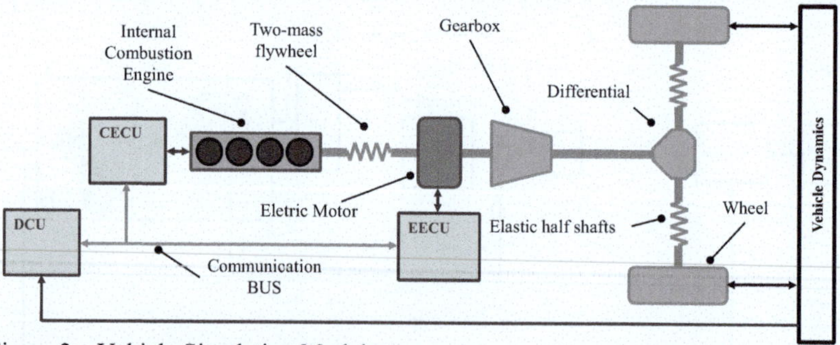

Figure 2 – Vehicle Simulation Model

2.1 Propulsion system

Most engine models used for the investigation of problems in driving dynamics simply take the maximum torque over engine speed with a constant dynamic response into account [17, 18]. Since this paper investigates the detailed differences in the dynamic behavior of the propulsion systems a more sophisticated model was developed.

2.1.1 Internal Combustion Engine

A turbocharged gasoline engine can be modeled differentiating between the fast ignition path and the slower air path. The air path itself consists of two very different dynamics: the naturally aspirated and the turbo dynamics. Figure 3 shows the general structure of the model.

Since an ICE cannot produce more torque than equivalent to the amount of air in a given cylinder, the model outputs the minimum torque of the ignition and air path. However, through variation of the ignition timing, the torque can be reduced quickly via the ignition path which is modelled as a variable first order low pass and a variable time delay. They are both driven by the speed of the engine. The time delays are set to the time between two consecutive ignitions. The change of the ignition timing can reduce torque only to a certain level. Any further reduction request is realized through cutting the torque by prohibiting ignition.

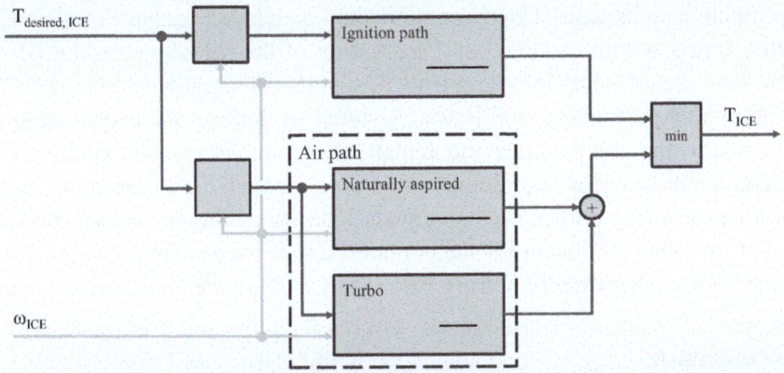

Figure 3 – Structure of the ICE-model

Modelling the response time of the ignition path the following considerations have been made. To achieve a dynamic torque model of sufficient detail a combination of Mean-Value-Engine-Model and 0-dimensional thermodynamic model has been applied. Each cylinder produces torque throughout its individual expansion stroke which lasts over 180° of crankshaft angle. The pressure build up in the cylinder to its peak takes approximately 60° of crankshaft angle [1], therefore the time constant is set to the time needed to pass 60°.

The CECU's operation frequency is set to 100 Hz. A detailed model description of the air path won't be discussed at this point since the ignition dynamics are of highest interest during traction control. The model has been validated through measurements of steps in torque demand of the test vehicle for various engine speeds.

2.1.2 Electric Motor

The EM in the analyzed vehicle is a permanent-magnet synchronous motor (PMSM). The typical behavior of a motor of such type can be modelled to a high degree of accuracy with a simple time delay and a first order low pass [2]. The time delay mainly depicts communication delays within the EECU and the low pass represents the dynamics of the current control. An intensive frequency response identification of the EM has been conducted. It resulted in values of 3 ms as the time delay and a base frequency of above 100 Hz for the low pass. The verification of the model shows that it is valid up to a frequency of half the base frequency, being much higher than the control frequency. The operating frequency of the EECU is 1000 Hz.

2.2 BUS Communication

The DCU communicates ω_{max} to both ECUs via the vehicle BUS through which both ECUs communicate as well. This communication is delayed because of several communication layers within the ECU and arbitration of the messages on the BUS. The delay times are identified by measurements of signals communicated via the BUS. The identification was performed as follows. A signal of desired value generated by the CECU is sent to the EECU. Here the actual value is measured and sent back to the CECU. The mean time for communication was evaluated. This was performed by a cross correlation of the corresponding signals for various measurements. On average values of 30 ms were evaluated for the communication latency from CECU to EECU. The latency in the other direction, from EECU to CECU, were 15 ms.

2.3 Drivetrain

The drivetrain is deduced from a complex multibody simulation model which is used to simulate acoustics having a high degree of accuracy up to frequencies beyond typical frequencies in traction control. An analysis of this complex model showed two main elastic components having eigen-frequencies in the desired range spectrum. These are the two mass fly wheel and more importantly the half shafts connecting differential and wheels. This is due to having the highest torque level in the powertrain [8 - 10]. Neglecting the degree of freedom achieved by modelling the plant with two independent wheels to depict lateral driving dynamics the resulting model is a serial three mass spring damper system. Ignoring this has no effect when only considering longitudinal dynamics having a homogeneous friction coefficient which this paper focusses on.

In order to fit the model behavior to the actual vehicle behavior seen in experimental studies, an optimization of the drivetrain parameters has been conducted. These are the elastic and damping coefficient as well as inertias. The identification method described in [7] is applied. First, measurements upon steps in the requested torque are made. These are performed on a high friction surface so that the vehicle dynamics remain in the stable region of the slip curve. The result of this identification method also applies to the drivetrain dynamics observed in the non-stable slip curve [7].

2.4 Vehicle Dynamics

To depict the nonlinear behavior of tire dynamics a common model is used: The "Magic Tire Forumula" by Pacejka [11]:

$$\mu = \mu_{max} \sin\left(C \tan^{-1}\left(Bs_x - E(Bs_x - \tan^{-1}(Bs_x))\right)\right) \tag{1}$$

The model describes the friction coefficient μ as a function of longitudinal wheel slip s_x. μ_{max} is the friction coefficient of the road surface and B, C and E are parameters that describe the trajectory of the friction coefficient.

μ multiplied with the vertical tire force F_z results in the longitudinal force F_x exerted upon the tire which also accelerates the vehicle in longitudinal direction. The vertical tire force is the sum of the static component $F_{z,stat}$ and the dynamic component $F_{z,dyn,pitch}$.

$$F_{z,stat} = \frac{m_{vehicle} g l_f}{l_r + l_f} \tag{2}$$

$$F_{z,dyn,pitch} = \frac{m_{vehicle} h_{COG} a_x}{l_r + l_f} \tag{3}$$

$m_{vehicle}$ is defined as the mass of the vehicle, g as the gravitational acceleration and a_x as the longitudinal acceleration of the vehicle. l_r and l_f are the distances between the centre of gravity (COG) and the rear and front axle respectively. h_{COG} describes the height of the COG above the road.

To avoid model instabilities at low velocities and singularities confronted with using the classic definition an alternative model of the positive wheel slip was chosen. The description and derivation can be found in [12]:

$$s_x = \frac{\omega_{wheel} r - v_x}{\overline{2} * \left(\sqrt{(\omega_{wheel} r) + \varepsilon} + \sqrt{v_x + \varepsilon} + \sqrt{\sqrt{(\omega_{wheel} r) + \varepsilon} - \sqrt{v_x + \varepsilon} + \varepsilon}\right)} \tag{4}$$

Here ω_{wheel} is the rotational speed of the driven wheel and v_x the longitudinal velocity of the wheels center point. ε is a factor which makes the expression continuously differentiable. r is the dynamic wheel radius. v_x is calculated as the time integral of the longitudinal acceleration a_x, which is defined as the air resistance F_{air} subtracted from the longitudinal force divided by the vehicle's mass:

$$a_x = \frac{F_x - F_{air}}{m_{vehicl}} \tag{5}$$

3 Control Allocation

This section deals with the design of an appropriate control allocation (CA) algorithm. First the state of the art allocation and its advantages and disadvantages are described. Then a suitable location of the controller is discussed followed by the design of an allocation algorithm meeting the requirements of the given problem.

3.1 State of the Art

Over actuated systems such as a P2 hybrid vehicle require a mechanism that defines how the additional degree of freedom will be addressed. Figure 4 shows a general control structure for a plant with two inputs. To resolve the redundancy an allocation algorithm has to be located between the controller and the system. This generates both inputs for the plant (u_1 and u_2) from the controller output v.

Since the P2 hybrid configuration was derived from conventional powertrains, the main control functions are located on the CECU which acts as master. This is where the slip controller is located in the state of the art configuration using the new cascaded structure. To coordinate the torque request between ICE and EM an allocation algorithm known as daisy chain [13 - 15] is commonly used. The principle here is to have a main actuator (ICE) to which all the torque is allocated until it reaches its static and/or dynamic limits. The excess torque demand is then shifted to the secondary actuator (EM). The EECU in the state of the art configuration mainly receives signals acting as slave. The EM therefore acts as a supporting torque source (boosting) and to shift the load point of the ICE to a more efficient point in order to generate electric power which then again can be used for electric driving. The information for this task is produced by the energetic operation strategy [2] which defines a static set point of the EM.

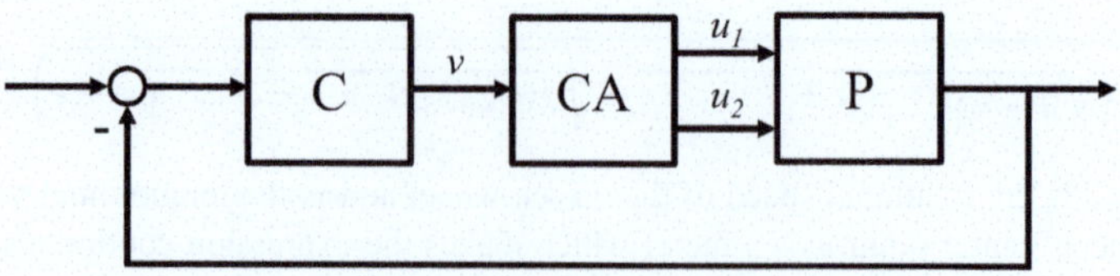

Figure 4 – Control Allocation in a feedback loop system

The state of the art approach works well realizing low dynamic and static requirements as the drivers request or a load shifting request by the operation strategy. For high dynamic torque requests though this allocation algorithm is not suited as simulations have shown. This is due to the high loss of phase delay on account of BUS communication when shifting to the slave actuator.

3.2 Location of the Cascaded Controller

Figure 5 shows the phase shift of the motor models for the two possible configurations, also taking into account the signal latency of the BUS communication. The frequency the phase shift was calculated is set to 10 Hz since this is the eigen-frequency of the powertrain for high slip values [4]. The phase shift of the ICE varies with the engine speed whereas the phase shift of the EM remains constant, see section 2.2.

The continuous line shows the phase shift for control signals generated on the CECU. The dotted line depicts the phase shift for signals generated on the EECU. For good tracking performance and disturbance rejection the fastest configuration is desired. Since the investigated hybrid vehicle configuration provides enough electric torque and power, the location of the controller for the newly proposed configuration will be chosen to be on the EECU using the EM as main actuator.

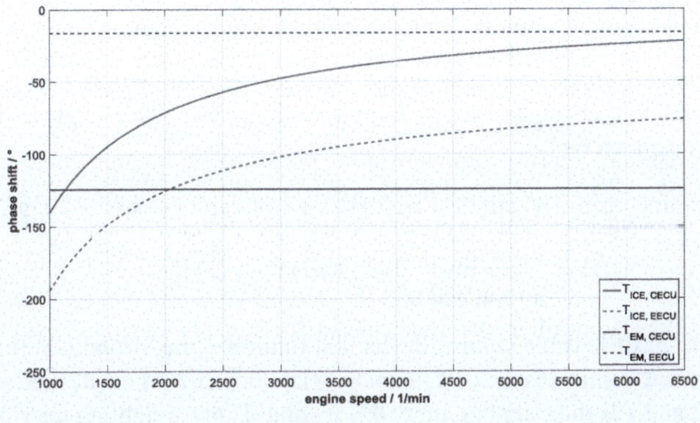

Figure 5 – Phase shift of the actuators for signals generated on different ECUs

3.3 Dynamic Control Allocation

Before designing the allocation algorithm some preliminary considerations have to be discussed. First we assume a model simplification with a rigid connection between ICE and EM. This results in both actuators acting without further dynamics upon the gearbox shaft (see figure 2).

Since there is no torque allocation function on the EECU and the state of the art approach having a bad performance for high dynamic requirements, a new algorithm will be integrated. A common approach for similar problems is the Dynamic Control Allocation (DCA) algorithm [3, 13, 16 - 18]. This is an optimization based method. The following equation describes the governing law of the allocation, where $\boldsymbol{u}$ describes the vector of input signals into the plant which has to be determined:

$$\boldsymbol{Bu}(t) = v(t) \tag{6}$$

v describes the scalar output of the controller being the virtual control input. $\boldsymbol{B}$ is defined as the diagonally occupied control effectiveness matrix. Since both actuators act upon the same shaft, both entries of $\boldsymbol{B}$ are equal to 1. $\boldsymbol{u}$ underlies further constraints. It can only take values within the boundaries of the actuator. These are determined at each sampling instant by the position and rate limits of the actuators [20]:

$$\underline{\boldsymbol{u}}(t) \leq \boldsymbol{u}(t) \leq \overline{\boldsymbol{u}}(t) \tag{7}$$

Solving equations (6) and (7) has three possible outcomes:

1. An infinite number of solutions
2. One unique solution
3. No solution

Case 2 and case 3 apply only seldom and have trivial solutions. In case 1 there is a degree of freedom that can be used to optimize an objective. This sequential optimization problem is given by

$$\boldsymbol{u}(t) = \arg \min_{\boldsymbol{u}(t) \in \Omega} \left\| W_1(\boldsymbol{u}(t) - \boldsymbol{u}_s(t)) \right\| + \left\| W_2(\boldsymbol{u}(t) - \boldsymbol{u}(t-T)) \right\| \tag{8}$$

$$\Omega = \arg \min_{\underline{\boldsymbol{u}}(t) \leq \boldsymbol{u}(t) \leq \overline{\boldsymbol{u}}(t)} \left\| W_v(\boldsymbol{Bu}(t) - v(t)) \right\| \tag{9}$$

Given Ω, the set of feasible control inputs that minimize the virtual control error, the control input that minimizes the cost function (8) is chosen. $\boldsymbol{u}_s$ is the desired steady state control input and T is the sampling time. W_1, W_2 and W_v are weighting matrices in order to define how significantly to penalize (high values) a certain term. This algorithm takes into account the steady state and dynamic boundaries of the actuators ($\underline{\boldsymbol{u}}(t) \leq \boldsymbol{u}(t) \leq \overline{\boldsymbol{u}}(t)$). Furthermore steady state control inputs are evaluated ($\boldsymbol{u}(t) - \boldsymbol{u}_s(t)$) as well as the change in the control input ($\boldsymbol{u}(t) - \boldsymbol{u}(t-T)$).

The algorithm was set to meet the goals of traction control. These are to maximize the acceleration of the vehicle on slippery road conditions and to keep the vehicle controllable during lateral maneuvers. To achieve these goals the overall actuator response upon changes in the set point and disturbances has to be as fast as possible, compare section 3.2. This results in high values of W_2 for the ICE and low values for the EM.

Since the EM has a limited torque availability, especially for high speeds, the static torque should be distributed to the ICE. This results in a steady state input vector $\boldsymbol{u_s}$ with the value of v for the ICE and 0 for the EM. W_1 in set to be an eye matrix of the values 1.

3.4 Proposed Configuration

Since an optimization algorithm is not suited for implementation on an ECU the algorithm has to be adjusted. Härkegard describes a method to convert the DCA into a linear filter. The details and derivation of this method can be found in [16]. The designed DCA was converted and implemented into the model in order to generate a concept which later can be implemented on a vehicle.

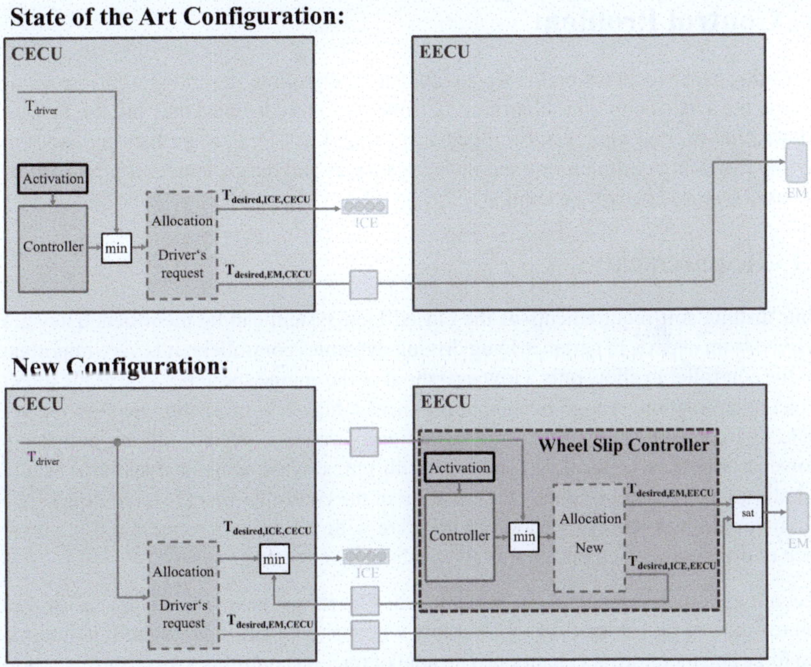

Figure 6 – State of the Art and new configuration of wheel slip controller

Both schematics in figure 6 are focused on the torque communication between the two ECUs to emphasize the distinction between the two concepts. Therefore the depiction of the speed information is neglected which is essential for the controller activation and well as the control task.

The state of the art configuration is depicted in figure 6, top. The location of the controller is on the CECU where the main controls and an existing allocation are located, see section 3.1. The configuration of the newly proposed concept is described in figure 6, bottom. It depicts the ECUs and their communication as well as the functional design of the new structure. As described in previous sections the traction controller and a new allocation algorithm will be implemented on the EECU for better tracking performance. As soon as the controller is activated the controller and new allocation take command over the torque request and distribution. The desired Torque for the ICE will then be communicated back to the CECU. The controller design will be described in the next section.

4 Control Problem

Since this paper focusses on the torque allocation algorithm described in the previous chapter the actual control problem and controller design is depicted only briefly. Further information on this topic can be found in [7] and [12]. Due to the fact that we will discuss different configurations and their advantages and disadvantages a simple linear PI based control law will be applied.

4.1 Requirements

Four distinct soft requirements of the closed loop system can be identified. They are derived from experts in parametrizing driving dynamics controllers. It is very important for the controller to allow only a low overshoot of the engine speed because of the direct effect upon traction. A high overshoot produces a high loss of traction making the vehicle less controllable resulting in a reduction of safety especially during cornering. However a perfect controller (combined with unrestricted actuator dynamics) would not allow any overshoot. Since the activation of the controller is set to a crossing of the maximum desired wheel speed to not interfere with the driver's request and the presence of drivetrain dynamics, an initial overshoot is inevitable.

Secondly a low undershoot must be achieved. The reason behind this is that a low absolute value of undershoot results in a low wheel slip which reduces the longitudinal tire force and therefore acceleration. The goal of traction control is to maximize traction which is contradicted by a high undershoot.

The third requirement is good tracking performance and a low settling time. Since a small deviation between desired and actual value results in a rather high difference in longitudinal tire force the settling time will also be considered.

A last requirement concerns the control input v. To reduce high control frequencies and amplitudes the derivative of the control input is taken into account. To achieve a smoother trajectory improving comfort as well as component protection the integral of the absolute values of the derivative of v are calculated. All requirements are depicted in figure 7.

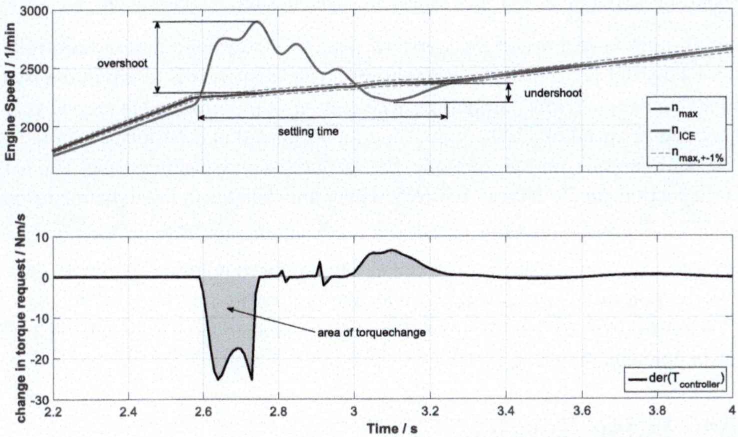

Figure 7 – Requirements of controller design

4.2 Control design

The investigated control problem consists of the inner control loop of the cascade seen in figure 1. In case of a higher torque demand than traction allows, the speed of the driven axle overshoots the maximum desired speed and the controller activates. Now the actual engine speed has to be controlled to a desired engine speed ω_{max}. Since this paper investigates the structure of the inner control loop applied to a hybrid vehicle the set point generation won't be discussed in detail. It is important though to notice that it is set to a point in the stable region of the μ-slip-curve close to its maximum.

To meet the previously defined soft requirements the control parameters were optimized according to the actual process in field tests. Another reason for this approach is that both controller settings are adjusted to the same governing law so that an unbiased comparison can be drawn.

$$J = b_1 x_{over} + b_2 x_{under} + b_3 t_{settling} + b_4 \int abs\left(\frac{dv}{dt}\right) dt \tag{10}$$

The cost function for the optimization of the controller parameters is described in (10). All requirements defined in section 4.1 are included and weighted accordingly through factors b_x. This is important to make sure that each requirement has a similar effect upon the cost. The settling time is determined through the last passing of the actual engine speed into a tube of +-1% of slip around the desired engine speed.

The optimization is performed for a critical maneuver to ensure that the controller is able to track the set point in various conditions. This maneuver is a step in the friction coefficient from high to low during high acceleration. It is performed in second gear for a medium engine speed of 3500 1/min. The first optimization is performed for the state of the art configuration for hybrid mode, the second for pure electric drive. The reason for this distinction can be seen in figure 5. There are changes in the dynamic response between the ICE and EM.

The last optimization is performed for the newly proposed configuration with the controller located on the EECU combined with the previously introduced CA. Since the EM is addressed primarily with this setup, one parameter set works equally well in hybrid and pure electric drive.

5 Simulation Results

The goal of this article is to confirm the postulation that a rearrangement of the cascaded controller structure and introducing a new allocation algorithm improves traction control performance in a P2-hybrid drivetrain. To do this both drive modes of a hybrid are evaluated: pure electric drive as well as hybrid drive. To evaluate the performance of both concepts the following test procedure was chosen. A simulation of an acceleration in 2nd gear on a low friction surface will be performed. The initial conditions prior to entering the simulated test road will remain the same for all simulations. The friction coefficient of the test road varies after a step from high to low at 0 sec between 0.18 and 0.4 and is dependent on the distance travelled. The friction corresponds to values seen on an icy road ranging from ice patches to snowy surfaces with higher grip.

5.1 Hybrid electric drive

First we will compare the simulation results of both concepts for hybrid electric drive. Figure 8 shows the actual engine speed and the desired maximum engine speed of both concepts. The desired maximum engine speed is set to the maximum slip, which is permitted dependent on driving stability limits. It is dependent on the velocity of the vehicle, see section 1. Hence the desired maximum engine speeds for both concepts progress differently.

First of all it can be seen that the initial overshoot is reduced quicker with the newly proposed configuration outperforming the state of the art. Throughout the entire course the tracking performance is significantly better making the state of the art on average accelerate less. The reason for this is that the actuator dynamic including the signal latencies differs depending on where the controller is located, see figure 5.

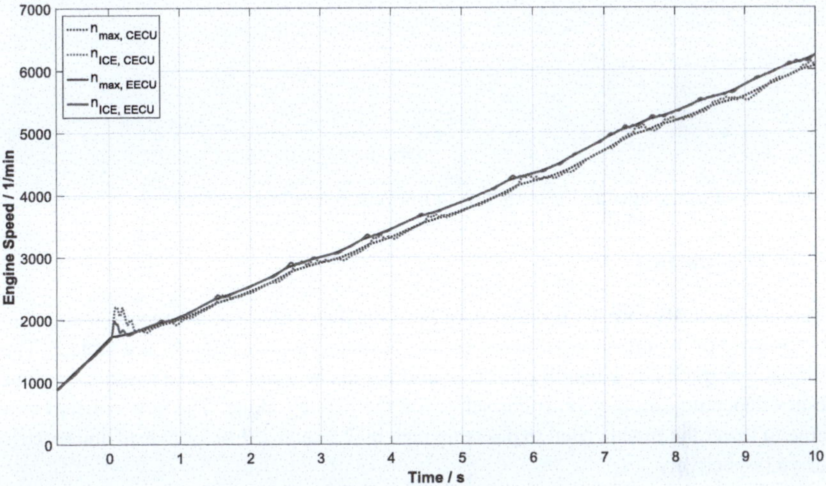

Figure 8 – Engine speeds for both configurations, hybrid drive

5.2 Pure electric drive

Next we will compare the simulation results of the two configurations for pure electric drive. Figure 9 shows the corresponding simulation results. Note that different parameters apply for the state of the art compared to hybrid electric drive, see section 4.2. The differences in control performance persist for this driving mode and deteriorate making the new concept suited even more for the control task in a hybrid vehicle.

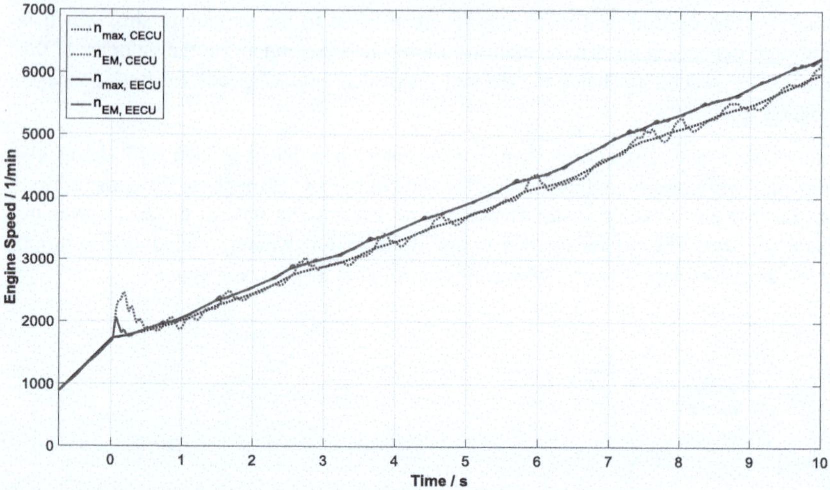

Figure 9 – Engine speeds for both configurations, pure electric drive

5.3 Discussion

This section deals with the discussion of the advantages of the new concept. Compared to the state of the art improvements in several factors were achieved, see table 1. The virtual test track has a length of 120 m and begins at 0 sec. It was passed in various times depending on the controller setup and the driving mode. The new concept was able to lower the time needed to pass the track by 2 % and 2.9 % for hybrid and electric drive respectively.

Table 1 – Improvement of new concept for a low friction track of 120 m

Driving mode	t	v_{final}	$a_{x,\,mean}$	*integral of error*
hybrid	- 2.0 %	+ 1.9 %	+ 3.9 %	- 75.7 %
electric	- 2.9 %	+ 3.5 %	+ 6.6 %	- 87.7 %

Also the final velocity was improved by the new concept. After the virtual track gains of 1.9 % and 3.5 % for hybrid and electric drive were achieved. The comparison of the mean acceleration throughout the track shows an improvement of 3.9 % for hybrid drive and 6.6 % for electric drive.

The tracking performance can be evaluated by calculating the integral of error which represents the absolute value of the surface between the desired value and actual value. Using the new configuration the integral of error was reduced by 75.7 % and 87.7 % for hybrid and electric drive respectively.

The differences in hybrid and electric mode arise from the different resulting actuator dynamics when controlling with the CECU. As seen in figure 5 the dynamics of the ICE for high speed controlled by the CECU is similar to the dynamics of the EM controlled by the EECU. This lowers the differences of the two concepts in hybrid mode. However controlling the EM via the CECU as in pure electric mode results in a high phase shift and a very low dynamic response making the new concept outperform the state of the art.

Apart from hard facts another very important benefit was achieved shifting the controller to the EECU and introducing the DCA. Just one parameter set for the controller was needed to achieve good tracking performance in both driving modes since the resulting actuator dynamics don't change. This makes the process of parameter application which is mostly done in intensive field tests easier and faster.

6 Conclusion and Outlook

6.1 Conclusion

In this contribution we proposed a new configuration for traction control of a hybrid vehicle considering a new cascaded control structure. It consists of locating the controller on the ECU of the electric motor and designing a dynamic allocation algorithm to divide the requested torque upon the two motors. The main advantage of the newly proposed allocation is that it explicitly takes the dynamics of the control signal into account. Therefore it shifts the torque demand dynamically between the two actuators.

As simulations have shown the proposed concept outperforms the state of the art. The reason for this is the minimization of latencies as well as introducing a new allocation algorithm. With this approach the gain of acceleration performance on low friction surfaces is up to 6.6 % as simulations have shown.

6.2 Outlook

It was possible to show in simulations that the newly proposed concept is able to enhance traction of a hybrid vehicle on low friction surfaces. The next step is to apply this concept to a test vehicle and confirm the simulation results in real world driving conditions.

A new approach to verify traction control systems will be discussed in [21]. Here the tested vehicle will be embedded in a test bench system to simulate vehicle dynamics. The benefit of this approach is that the influences of road uncertainties seen in field tests won't affect the results. Therefore an exact evaluation of the benefit of the new configuration proposed in this paper can be drawn.

One reason for the new concept outperforming the state of the art is that the analyzed EM is able to generate high torque and power. Further research can be directed to investigating whether this advantage still persists, if the EM-power is reduced. A use case might be mild hybrid concepts or a degraded high-voltage battery in case of of low temperature or state of charge.

References

[1] REIF, K.: Bosch Grundlagen Fahrzeug- und Motorentechnik. Vieweg + Teubner, 1. Auflage, 2011.

[2] REIF, K.: Kraftfahrzeug-Hybridantriebe: Grundlagen, Komponenten, Systeme, Anwendungen. ATZ/MTZ-Fachbuch. Vieweg + Teubner, Wiesbaden, 2012.

[3] LOOMAN, C.: Analyse, Modellierung und Entwurf einer Mehrgrößenregelung zur aktiven Schwingungsdämpfung im hybriden Antriebsstrang. Dissertation. TU zu Dresden, Dresden, 2012.

[4] ZECH, A.; EBERL, T.; MÜLLER, S.: "Analyse einer neuen kaskadierten Reglerstruktur für die Antriebsschlupfbegrenzung hochdynamischer Fahrzeugantriebe", 8th VDI Conference AUTOREG, 2017.

[5] FODOR, M.; YESTER, J.; HROVAT, D: Active Control of Vehicle Dynamics. 17th Digital Avionics Systems Conference, 1998.

[6] JAIME, R.-P.; KASTNER, F.; ERBAN, A.; KNÖDLER, K.: Regelalgorithmen für Rekuperation und Traktion bei Elektrofahrzeugen. ATZ elektronik, 02/2014.

[7] ZECH, A.; EBERL, T.; REICHENSDÖERFER, E.; ODENTHAL, D.; MÜLLER, S.: Method for developing tire slip controllers regarding a new cascaded controller structure. 14th International Symposium on Advanced Vehicle Control (AVEC), 2018.

[8] GÖTTING, G.: Dynamische Antriebsregelung von Elektrostraßenfahrzeugen unter Berücksichtigung eines schwingungsfähigen Antriebsstrangs, Dissertation, 2003.

[9] ZEMKE, S.: Analyse und modellbasierte Regelung von Ruckelschwingungen im Antriebsstrang von Kraftfahrzeugen, Dissertation, 2012.

[10] YEAP, K. Z.; MÜLLER, S.: Characterizing the interaction of individual-wheel drives with traction by linear parameter-varying model: a method for analyzing the role of traction in torsional vibrations in wheel drives and active damping. In: Vehicle System Dynamic, Volume 54, 2016 – Issue 2.

[11] PACEJKA, H. B.: "Tire and Vehicle Dynamics", 3rd Edition, Elsevir, 2012.

[12] REICHENSDÖERFER, E.; ODENTHAL, D.; WOLLHERR, D.: On the Stability of Nonlinear Wheel-Slip Zero Dynamics in Traction Control Problems. IEEE Transactions on Control Systems Technology, 2018, submitted for publication.

[13] BORDINGTON, K.A.: Constrained Control Allocation for Systems with Redundant Control Effector. Dissertation, 1996.

[14] DURHAM, W.C.; BORDINGTON, K.A.: Multiple Control Effector Rate Limiting, Journal of Guidance Control and Dynamics, 1996

[15] BUFFINGTON, J. M.; ENNS, F. D.: Lyapunov Stability Analysis of Daisy Chain Control Allocation. Journal of Guidance Control and Dynamics, 1996

[16] HÄRKEGARD, O.: Backstepping and Control Allocation with Applications to Flight Control. Dissertation, 2003.

[17] ROSENBERGER, M.; KOCH, T.; LIENKAMP, M.: Combining Regenerative Braking and Anti-Lock Braking for Enhanced Braking Performance and Efficiency, SAE International, 2012.

[18] KNOBEL, C.; PRUCKNER, A.; BÜNTE, T.: Optimized force allocation: A general approach to control and investigate the motion of over-actuated vehicles, Elsevir, IFAC Proceedings, 2006.

[19] BASSHUYSEN, R.: Handbuch Verbrennungsmotoren – Grundlagen Komponenten Systeme Perspektiven, 7. Auflage, Springer Vieweg, 2015

[20] HÄRKEGARD, O.: Efficient Active Set Algorithms for Solving Constrained Least Squares Problems in Aircraft Control Allocation, 41st IEEE Conference on Decision and Control, 2002

[21] AL-SAIDI, O.; ZECH, A.; MÜLLER S.; SCHYR, C.: Advanced Powertrain Testbed for Development and Validation of Vehicle Control Systems, 19th VDI Congress SIMVEC, 2018, submitted for publication.

CHASSIS.TECH SECTION

DEVELOPMENT METHODS

Concept study: Networking requirements for test benches which support the product release process on the system-level for Autonomous Driving (AD)*

Thomas Maur
EPS System Integration & System Qualification
ZF Groupe

* Konzeptstudie: Anforderungen an die Vernetzung von Testständen, die den systemischen Freigabeprozess von autonomen Fahrfunktionen unterstützen

P. Pfeffer (Ed.), *9th International Munich Chassis Symposium 2018*, Proceedings,
https://doi.org/10.1007/978-3-658-22050-1_23

Abstract

Focus of this presentation are the A-SPACE base practices SYS.4: System Integration Testing:

- BP1: System Integration Strategy
- BP2: System Integration Test Strategy

for autonomous driving applications.

This concept study tries to address the qualification & release challenges, which may come up, if our products [like: Steering-, Braking-, Gearbox-, Air-Bags-, ... Units] should be functionally integrated under an autonomous driving (AD) context. AD-Systems will influence our product development and release life cycles.

Furthermore, our test freedom & methods will be influenced. Test organizations should think about possible test method gaps based on the new addressed demands, which directly affect our test capabilities (e.g. new test bench interfaces). Real-Time-Testing in a Closed-Loop-Simulation approach in conjunction with "hot" connected test bench cluster, requests technical capabilities of cross functional networking.

For the near future, it is very important to develop our test organizations to face the challenges which occur under the context of autonomous driving:

- Test Organization (TO) & Test process (TP) must support a networking/shared work flow,
- Test tools & methods should support this test process,
- Test tools should base on a common standard,
- cross functional and traceable information flow of all test activities ("planning" / "preparing" / "performing" / "perfect") for each test execution should be exchanged with a common tool infrastructure: e.g. PTCI® / DOORS®,
- support mileage accumulation in a simulated a/o virtual test environment,
- execute AD test scenarios under a networking context on system level injunction with fault insertion testing (limp home mode qualification),
- Capability to execute "cyber security" test methods under AD functional context.

Exclusion:

This concept study did not include the test activities, which are requested for the AD features "sensor fusion" and "perception algorithms".

Challenge Description

Hypothesis:

AD will request the evidence of fault free mileage accumulation exceeding 30.000.000 km (*), injunction with Limp-Home-Mode qualification (Fault Insertion Testing) for product release on system level (fully integrated system).

to (*): as proposal, actual under discussion

Challenge:

Vehicle fleets will not be able to support this request during the development life cycle.

Possible Solution:

1. under real time and real environment condition (AD use case view):
   Vehicle fleet mileage accumulation: 1.00.000 km(*)
2. under real time condition (AD-Component view)
   HiL (Traffic/ vehicle, …) fleet mileage accumulation: 9.00.000 km(*)
3. under non-real-time condition (SW view)
   SiL fleet mileage accumulation: 20.000.000 km(*)

The following presentation tries to offer a possible test concept for this “hybrid” test strategy.

to (*): as proposal, relationship still under discussion

Assumptions

Following Assumption are relevant for this concept study

- AD vehicles will have the same life cycle requirements as conventional vehicles.
- Safety Integrity Level on system AD level will set to ASIL-D.
- Safety Integrity Level on component level will bequeath to 1:1.
- Well known performance and non-performance (Haptic, NVH, ...) requirements on Component level (Braking, Steering, …) will be reused 1:1.
- AD Upper System will have following Components:
  - AD-Controller with AD-Algorithm
  - AD-Sensor Components
  - Gearing with AD Interface and AD Mode Controller (*)
  - Braking with AD Interface and AD Mode Controller (*)
  - Steering with AD Interface and AD Mode Controller (*)
  - OSS with AD Interface and AD Mode Controller (*)
  - …
- Each Sub-System is qualified for the following top-level AD system integration test strategy.

to (*): AD Mode Controller handles

- Ramp-Up
- Ramp-Down
- Normal running Mode
- the limp-home-modes on Sub-System Level

EPS Test Bench Architecture (closed solution)

Follow picture shows the main test bench components of our EPS Hardware-in-the-Loop Test Bench.

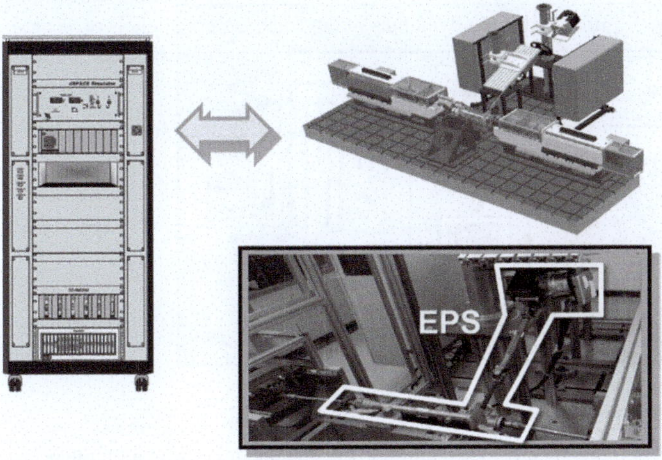

This test bench is a closed technical solution, which did not allow any further interaction which other test benches (like: ESP®, Gearbox, ...).

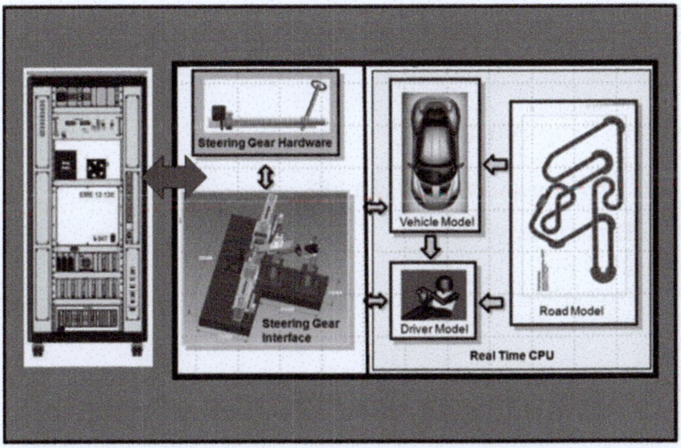

Also, the logical architecture represents a closed solution.

EPS Test Bench Architecture (Network Context)

To be well prepared for the new "networking" context, we share or test bench architecture into Master & Slave parts.

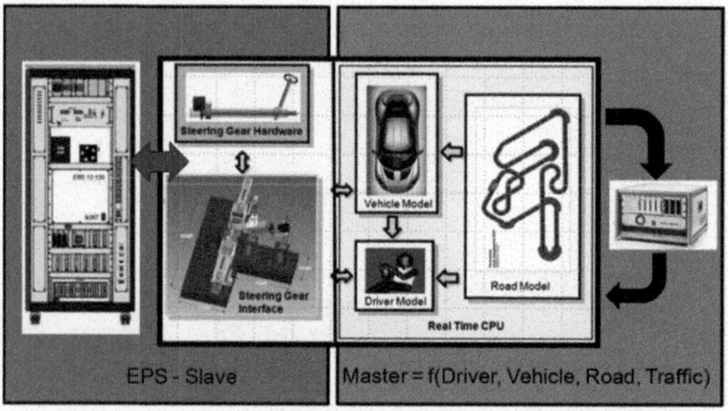

The slave part owns the EPS technology (physical & logical).

The master part represents the test bench components, which are shared with other test benches.

To setup a modular Test Bench Concept (split between Master and Slave) it makes sense to define a common definition of "Master" HiL and "Slave" HiL configuration.

Interfaces between Master/Slave should follow a standard description (Project(?), OEM (?), OEM-Consortium (?), …).

Definition of Slave (Proposal):

Standalone test facility: "open-loop-test-bench" for EPS

- Interface-1: Driver Hands to Driver Wheel
- Interface-2: Steering Tie Rod
- Test Stimulus on Driver Wheel:
  - Torque
  - Speed
  - Steering Wheel Position
- Test stimulus: on Tie Tod:
  - Position
  - Force

Independent of the EPS HiL Technology (EPS-iL, EPP-iL, ECU-iL, vECU-iL) this interface should be supported. Physical gaps should be closed via adequate simulation models.

Definition of Master (Proposal):

From EPS point of view, the stand-alone test facility for the Master includes:

- Driving Track Simulation
- Traffic Simulation
- Environment Simulation
- Vehicle Simulation
- Driver Simulation
  - Driver Hands (steer & shift)
  - Driver Feet (left / right)
  - Driver Eyes
- Master/Slave Configuration
- Test-Stimulus-Database & corresponding Expected-Behaviour-Description
  - Driver-Use-Case
  - Driving-Manoeuvre
  - Traffic Conditions
  - Limp-Home-Mode Stimulus Table
  - Diagnostic Service Routine Description
  - …

Definition of shared Slaves (for other AD-Sub-Components)

"Shared Slave" functionalities are:

- Physical test benches with physical AD-Sub-Component,
- Simulation Model of test benches which represent the corresponding AD-Sub-Component

Critical Discussion:

How is the owner of "Driver Arm" and/or "Driver Feet" Simulation?

There are good arguments to align these simulation aspects to the "Master" but also to the "Slave" functionality (e.g.: Driver Arm: EPS / Driver Feet: ESP®).

Possible solution:

Tests which request e.g. a "Driver Arm Simulation" on sub-component level, should be excluded from the AD Integration Test Strategy (AD Test Plan).

Types of Test Bench Cluster (AD Context)

Based on the previous definition, we can develop variants of AD-Test-Bench-Cluster types. Each cluster type has different advantages regarding:

- System Integration Test Strategy: stepwise system integration
- Test freedom: real-time / non-real-time
- Flexibility: OEM view / Supplier view
- Availability: virtual qualification
- Protection of Intellectual Property
- Cost sensitive (reuse of existing test technologies)

Master / Slave HiL Concept for AD (full)

This AD-test-bench-cluster type represent the complete availability of physical AD-Components.

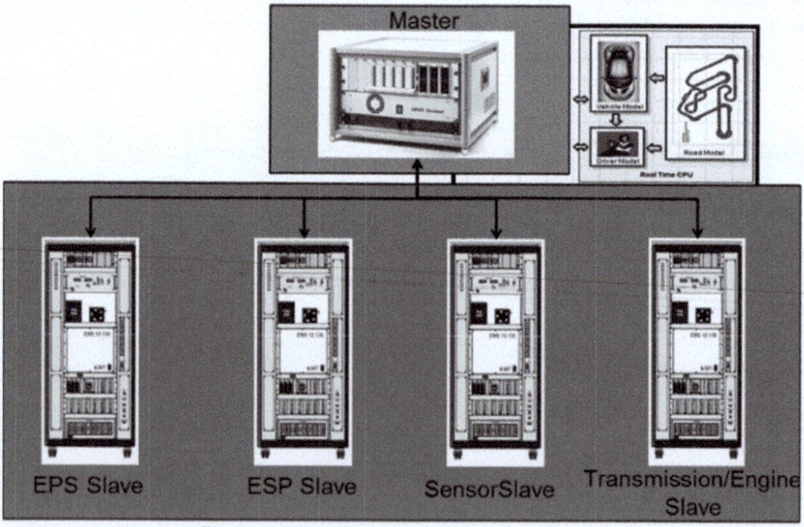

Master / Slave HiL Concept for AD (Hybrid)

This test-bench-cluster type helps in case of incomplete delivery of the devices-under-test (DuT) of missing test benches on sub-Component level.

Missing parts are replaced by the related simulation box (behavior models of the sub-component).

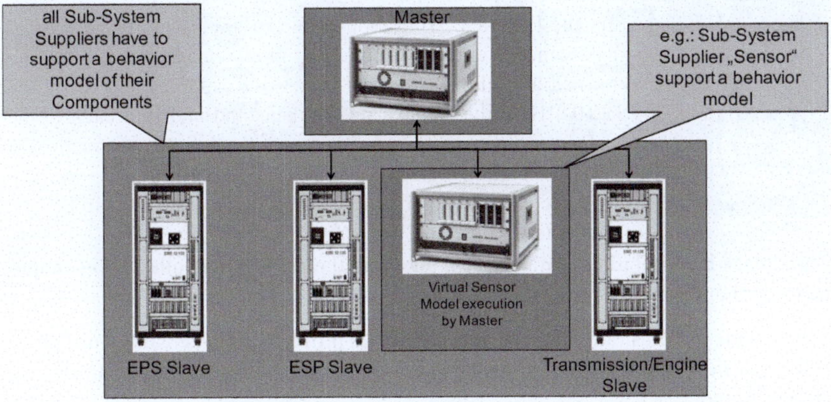

Master/Slave HiL Concept for AD (Type: Hybrid(max))

The next variant of AD-test-bench-cluster type can be used for the AD-Sub-Component suppliers. Based on this setup the supplier can qualify his component before the last step of complete AD-system integration takes place.

This picture shows for each simulated Sub-Component a separate Box "virtual Slave".

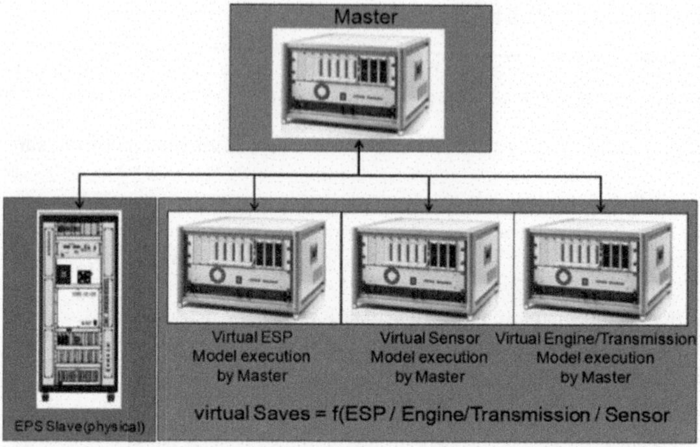

This "black box" solution guarantees the intellectual property of all other Sub-Component suppliers.

The next picture illustrates the case that intellectual property is not demanded on "physical-black-box" level. Virtual functions are included as "sFunction /DLL" (protected closed source).

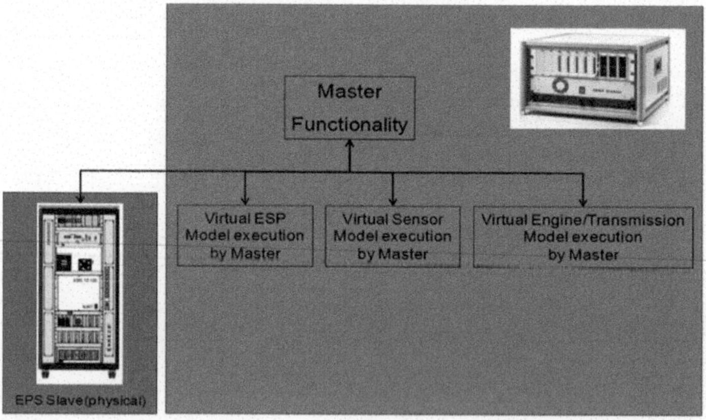

Master / Slave SiL Concept for AD (Master)

Thinking forward, the next setup of AD-test-bench-cluster abstraction takes the transfer from HiL to SiL (Hardware-in-the-Loop to Software-in-the-Loop).

Here we have following types of AD M/S SiL concepts:

1. With "physical-simulation-units" for each Sub-Component
   - Develop and review the real-time communication of the test benches without physical components
   - Development of Test Cases and virtual integration testing under real-time conditions

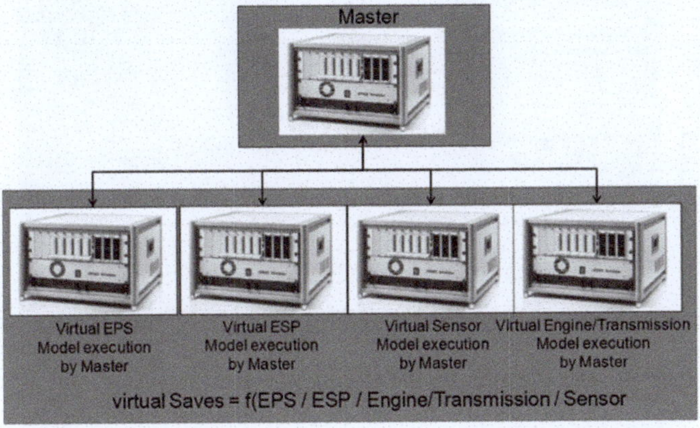

2. With "behaviour models" (only) for each Sub-Component
   - Develop and review the communication of the test benches without physical components und none-real-time (faster than) conditions.
   - Development of Test Cases and virtual integration testing under non-real-time (faster than) condition
   - After formal verification of the test bench cluster simulation (SiL) solution:
     - AD-System Integration Test Execution
     - AD-System Qualification Support
     - Faster than real-time

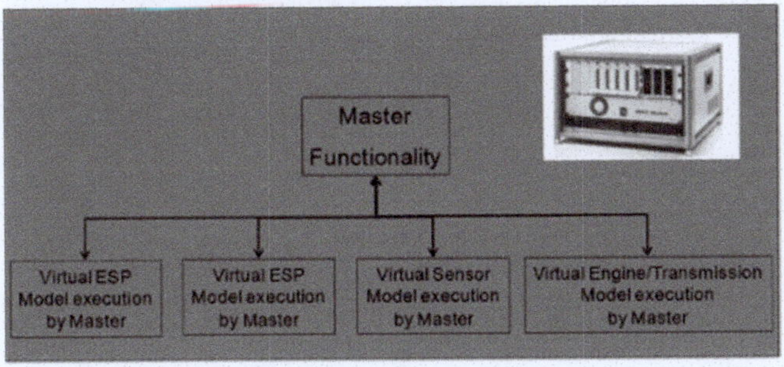

EPS HiL Test Bench Technologies (as Slave Definition)

To setup a modular Test Bench Concept (split between Master and Slave) is makes sense to define a common definition of "Master" HiL and "Slave" HiL configuration.

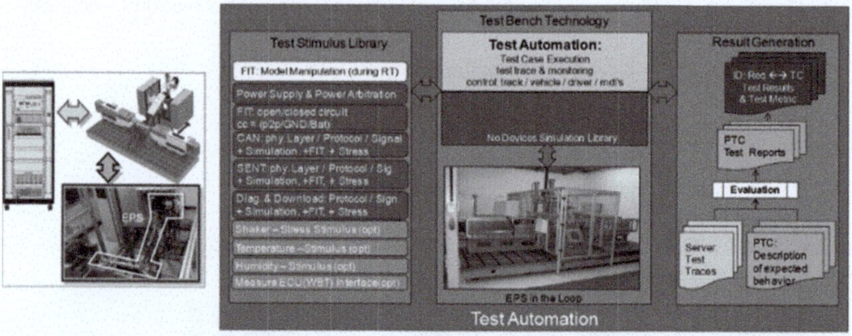

EPS-iL embedded all EPS components as requested by a vehicle setup, under real-time conditions

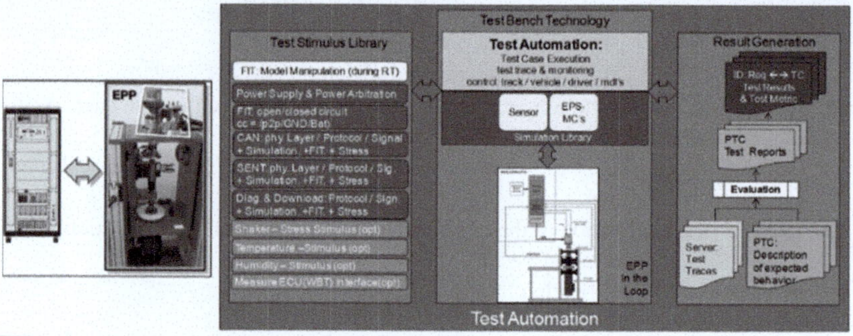

EPP-iL represent the physical EPS Components: SW / ECU / Motor mechanical EPS Components are simulated under real-time conditions

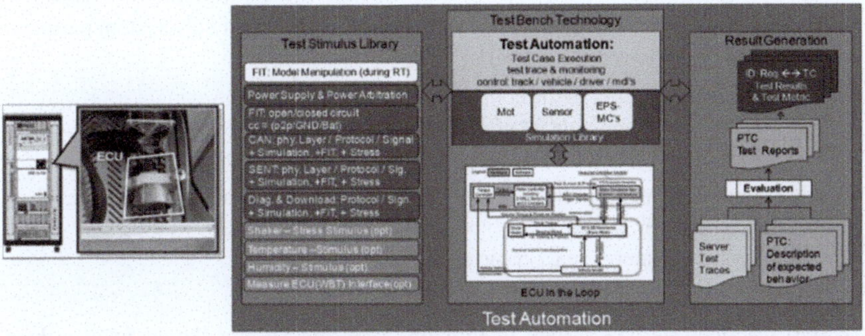

ECU-iL represent the physical EPS Components: SW / ECU EPS Motor & Motor Sensors and mechanical EPS Components are simulated under real-time conditions

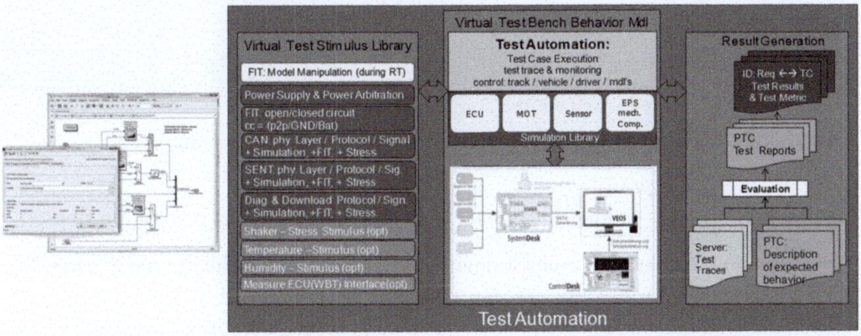

vECU-iL did not represent any physical EPS Components µP / ECU / (EPS Motor & Mot-Sensors) and mechanical EPS Components are simulated **faster** real-time conditions

Possible Top-Level Solution:

Result of this Concept Paper:

- every Sub-System supplier supports AD development with their own Technologies,
- every Sub-System owner should qualify and release their AD Interfaces (static & dynamic),
- the release activities should use verified simulation elements (including test stimulus library),
- the AD relevant simulation elements, should be designed, developed & qualified under the control of the OEM or OEM-Consortium (AD-Consortium),

- this simulation elements should be shared with all System and Sub-System partners, which belong to the AD-Consortium,
- Common understanding of:
  - SW development and Release Processes,
  - SW delivery milestones and release categories,
- Sub-System Architecture should show a maximum of independency (freedom of interference) between:
  - Sub-System Features (e.g.: basic steering functions),
  - AD-System Feature (AD-Status-Modi, Interface, ...),

To start the stepwise AD-System Integration, each Sub-System should finalize their qualification activity (see AD System Integration Strategy):

1. Integration of [AD-SW] into [AD-ECU-Host] on system level:

   supported by a complete simulated Sub-System environment,

2. Next integration steps,

   [AD-SW / ECU] with [BRAKING]: rest will be simulated,
   [AD-SW / ECU / BRAKING] with [STEERING]: rest will be simulated,
   ... .

Test integration order should be exchangeable to work on parallel integration steps(!).

Integration on target system (vehicle-level) will be started after the last integration step was performed.

Possible Consequences

The "AD-Consortium" is responsible for defining and supporting the AD specific top-level test stimuli and share this information to all involved parties:

- Standard AD driving situation,
- Limp home test scenarios,
- the boundaries of the "noise factors".

Consequences for the Sub-System Owner:

- Should adapt his test tools for the described inter acting opportunity,
- Should change his test tool roadmap (partly or completely),
- Should support a behaviour model of his Sub-System, which is capable to run on real-time conditions,
- Product-Life-Cycle and Product Release Process should be adapted,
- Architecture of the Sub-System should be modified (see freedom of interference),
- Support of real time interacting of the involved test benches.

Note:

Many thanks to dSPACE GmbH, Rathenaustraße 26, 33102 Paderborn, for supporting us with the pictures, which we are using in this article.

Other technical solutions from different vendors are also available.

Top-down development of controllers for highly automated driving using solution spaces

Jan-Dominik Korus, BMW Group

Pilar Garcia Ramos, BMW Group

Christoph Schütz, BMW Group

Markus Zimmermann, Technical University of Munich

Steffen Müller, Technical University of Berlin

P. Pfeffer (Ed.), *9th International Munich Chassis Symposium 2018*, Proceedings,
https://doi.org/10.1007/978-3-658-22050-1_24

Abstract

The complexity of highly automated driving functions and the large number of testing scenarios require the application of virtual design and testing methods. A common method to develop robust controllers is based on Γ-, B and/or Θ- stability which need highly simplified vehicle models. Most of such models are linear and neglect relevant vehicle dynamics phenomena. Thus, it cannot be guaranteed that the closed-loop system including these simplified models covers all significant effects. This paper adopts a design method that computes solution spaces for controller parameters of arbitrary black box systems and models with no restrictions regarding linearity. This way the modelling represents the real system and disturbances more accurately. Parameter space approach and solution space approach are applied to a simple linear physical model of a steering system. Then, a more detailed model demonstrates the capabilities of the design method based on solution spaces. Requirements on the closed-loop system expressed as boundaries of the stability regions are derived to ensure that the vehicle stays within a defined safety corridor next to the target trajectory. Additionally, a solution space for feasible controller parameters is specified. These requirements take nonlinear effects into account, thus increasing the validity of virtual testing. The effectiveness of this approach is demonstrated by designing the lateral controller of a highly automated vehicle.

1 Introduction

Controller systems for automated driving consist of multiple components such as controllers, disturbance observers, signal processing, safety measurements, etc. Usually they are implemented for a broad range of different vehicle types. Each vehicle can change its properties over time, when i.e. the customer changes the tires or adds different loads (sudden changes) or wearing and aging effects of the material (long term changes). Facing these demanding circumstances, the design and testing of automated driving functions become even more challenging [1]. For guaranteeing the functionality of the controller systems robust control methods are applied. A common method for robust design is the parameter space approach [2],[3]. Its basic idea is to map requirements for the closed-loop system into the space of two controller parameters. This mapping causes high computational costs and requires analytical models of plant and controllers. If the design space exceeds two dimensions an invariant plane may be found [4] to reduce the problem to a two dimensional problem. Another approach is to apply parameter sampling [5]. Each of the component provides design variables yielding additional dimensions to the design space resulting in a large scale design problem. A common method for development of mechanical systems is a cooperative design method using solution spaces [6]. First approaches to transfer the method for designing the logic and dimensions of a rear-wheel steering are shown in [7]. In this paper we apply the solution space method to controller design problems and show its advantages compared to the parameter space approach.

First, chapter 2 derives a feasible set of PID controller parameters for an exemplary system using the parameter space approach. In chapter 3 the solution space approach for the control problem is introduced and compared to the parameter space approach. In chapter 4 we show the application of the solution space approach to the design of a lateral control system of an autonomous vehicle including full nonlinear vehicle dynamics simulation.

2 Classical robust controller design

A model approximates the real system only for a limited operating range. In order to cope with a broader range, a common approach is to apply robust control design methods. Additionally, system parameters may vary over time. A classical robust control design method is the parameter space approach which is demonstrated in the following.

2.1 Plant and control logic

Both design methods are compared using a single-mass oscillator as shown in figure 1, where m is the mass, c the stiffness of the spring, d the damping, $u(t)$ an applied force and $x(t)$ the position of center point of the mass. For the control logic a PID feedback controller in parallel form with multiple design variables n>2 is chosen. Additionally a stationary feed forward controller is implemented. The PID controller contains the gains K_P, K_I and K_D. The stationary feedforward controller consists of the gain $F(0)$. Hence, the closed-loop system contains three design variables for the PID controller.

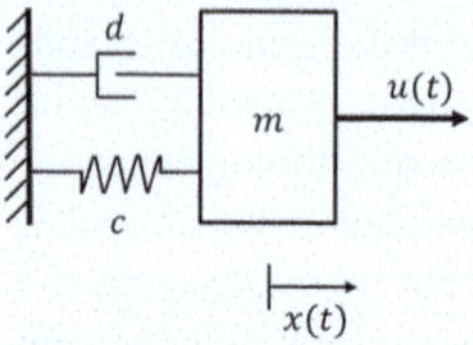

Fig. 1: Single-mass oscillator

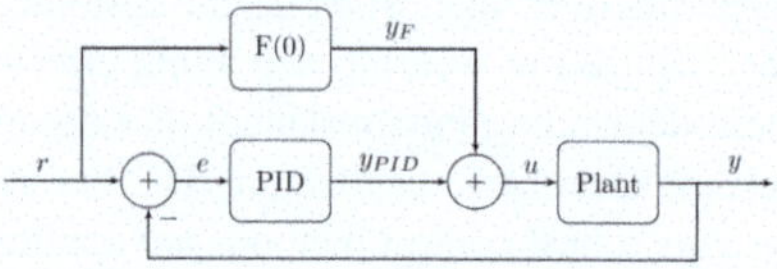

Fig. 2: Closed-loop system containing stationary feedforward controller, PID controller and PT2-Plant

The resulting closed-loop is shown in figure 2 containing the plant, PID controller and a stationary feedforward controller $F(O)$ with reference signal r, output of the feedforward controller y_F, output of the PID controller y_{PID}, plant input $u = y_F + y_{PID}$, plant output y and error $e = y - r$.

2.2 Closed-loop requirements

The basic design criterion for a control system is stability. The domain of automated driving, in particular, introduces further and more specific requirements. E.g. considering the lateral position of a car, it is necessary to restrict the maximal deviation of the trajectory to avoid a collision. Therefore we define a maximum feasible peak overshoot of the step response. Additionally, the actual position shall not change too slowly. This can be addressed by limiting the peak time of the overshoot. Those two properties can be set directly into relation with the damping and natural frequency of the closed-loop system and transferred into requirements for the position of the closed-loop systems poles [8]. Furthermore, the designed control system shall be able to handle deviations of the plants properties. Therefore robustness in terms of stability and performance is demanded. The design goal is to choose the design variables from table 1 in such a way that the closed-loop behavior with the design variables of plant parameters from table 2 satisfies the performance measures from table 3.

Tab. 1: Design variables of the control system

Controller parameters	
Proportional gain	K_{P}
Integral gain	K_{I}
Derivative gain	K_{D}

Tab. 2: Design variables of plant parameters

Plant parameters		**Nominal values**
Mass	m	$1400\ kg$
Stiffness	c	$10\,000\ kg/s^2$
Damping	d	$4\,000\ kg/s$

Tab. 3: Performance measures of the closed loop system

Performance measures		**Lower bound**	**Upper bound**
Natural frequency	ω_0	$0.25\ 1/s$	
Damping	D	$0.5\ 1/s^2$	$1\ 1/s^2$

The two performance measures for damping and natural frequency are transferred into requirements for the closed-loop poles positions in the complex plane.

2.3 Parameter space approach

The parameter space approach takes three stability definitions into account which do not only specify the systems stability but also its dynamic properties. Being defined in the complex plane the position of the systems poles mainly give information about its dynamic performance. Therefore the so called Γ-stability is analyzed [2]. For B- and Θ-stability refer to [9] and [10].

The pole condition is approximated by the hyperbola in figure 4. Every pole composition is considered permissible if it lies on the left hand side of the hyperbola. The proportional gain is chosen to $K_{\text{P}} = 10$. Due to superposition the stationary feedforward controller gain $F(0)$ is computed separately by calculating the inverse of the plants stationary gain. Therefore the two controller parameters K_{I} and K_{D} are left to determine. Using the parameter space approach the hyperbola is mapped into the $K_{\text{I}}/K_{\text{D}}$-plane. The resulting curves are shown in figure 4. The set inside the curves contain parameterizations for K_{I} and K_{D}, which satisfy the pole conditions. For evaluating the influence of K_P a sampling can be applied. For each change in parametrization the curves must be computed again.

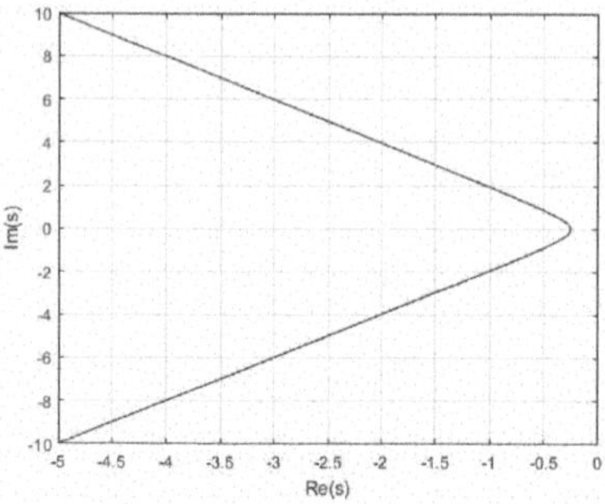

Fig. 3: Requirement – Poles have to lie on the hyperbolas left hand side

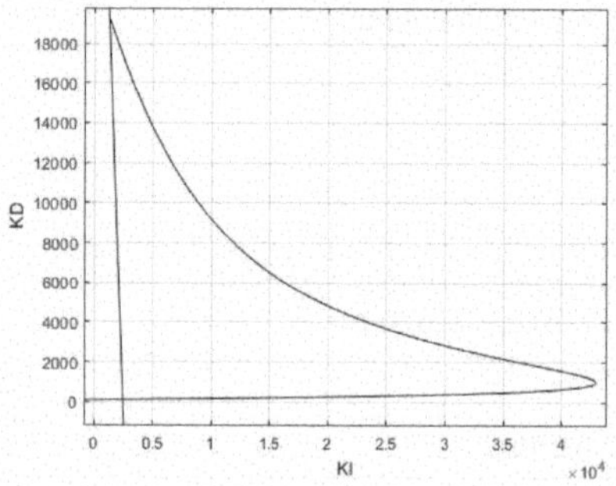

Fig. 4: Mapped Requirements into parameter space

In order to guarantee robustness w.r.t. the variation of plant properties, the mapping of the requirements is computed for each corner point of the set by variation of the plants parameters m, c and d. The overlaying intersection represents the feasible solution.

3 Introduction to controller design using solution spaces

In this chapter a cooperative design method is adapted for controller design of a single-mass oscillator. The method originates from the development of mechanical components in early development stage.

3.1 Cooperative design

The cooperative design method using solution spaces proposes a problem formulation using dependency graphs. Those give an overview about the interacting design variables, components and performance measures. For evaluation of the dependencies a bottom-up mapping between the layers of the dependency graph is needed. In a next step, the performance measures are mapped top-down into the space spanned by the design variables. The three steps for the problem formulation are shown in figure 5. The generic approach has no restrictions w.r.t. the selection of the design variables x_i. They can either be physical parameters, e.g. parameters of a mechanical component, or non-physical parameters, e.g. control parameters. In analogy the subsystems y_j can be a controller or a plant. The combination of the subsystems yield the closed-loop system for which we defined the requirements. The corresponding dependency graph for the single-mass oscillator will be shown in the next chapter.

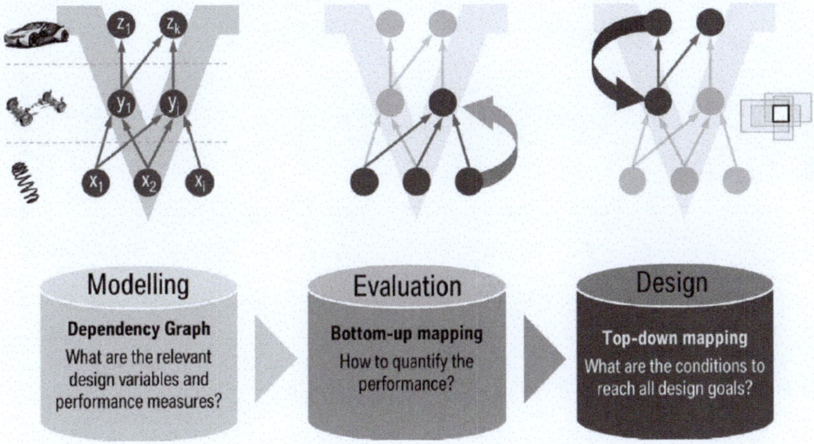

Fig. 5: Three enablers for cooperative design: (1) dependency graph, (2) bottom-up mapping, (3) top-down mapping [11]

3.2 Problem statement for single-mass oscillator

The design variables and performance measures are defined in table 1, 2 and 3. The dependency graph is shown in figure 6. In total the dependency graph consists of four layers. The bottom-up mapping of the design variables in table 1 results in the poles and zero of the PID controller while the mapping of the plant parameters results in the poles of the single-mass oscillator equivalent to a PT_2-plant. Controller and plant form the closed-loop system, which yield poles and zeros of the closed-loop system. High level requirements e.g. stability, natural frequency, damping, etc. are arranged in the top layer and have to be mapped top-down in a last step. The top level requirements are mapped into requirements for poles and zeros of the closed-loop system. The requirements of the closed-loop are then mapped down directly into the space of design variables. The plants parameters are considered arbitrary but fixed in a certain interval. For example, loads do vary or tires may be changed but are constant during driving. Loads and tires are restricted by vehicle specification such that they will only vary in a defined parameter range. In a last step a feasible set of controller parameters for the PID controller design are derived. The parameter space approach only addresses a sub problem of the problem for the cooperative design using solution spaces. The problem for the parameter space approach is shown in figure 6 by the light gray, dashed ellipses. In the case of the parameter space approach, the requirements are directly mapped from the complex plane into the K_I/K_D-plane.

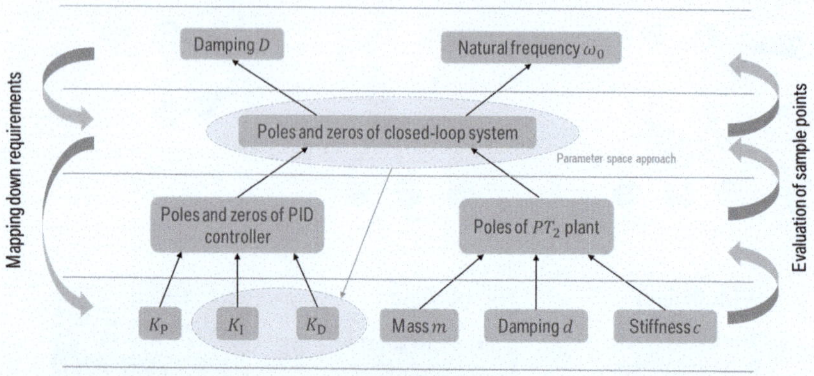

Fig. 6: Dependency graph for designing PID controller of PT_2 plant

3.3 Resulting solution spaces for single-mass oscillator

For calculation of the solution spaces the intervals of the design space must be specified in a first step. In a following step, Monte Carlo sampling is applied within the space of the design variables. Starting from the bottom, each layer can be mapped into its neighboring layer. In the case of the single-mass oscillator the poles of the resulting transfer functions are calculated. If a system is analyzed where the determination of its transfer function is difficult, an arbitrary mapping can be applied. This could be for example a complex physical model which contains detailed friction modelling, a nonphysical model like a fitted artificial neural network or a black box. The mapping must be able to generate corresponding sample points in the next layer of the dependency graph. The mapping of the single-mass oscillator generates for each sample point in the space of the design variables a set of poles in the layers above. After evaluation of the sampling points, the requirements for damping and frequency are mapped downwards back in the space of design variables and result in an area with feasible parameter combinations. In figure 7 the sampling points and the mapped sampling points are shown and colorized green when they belong to a feasible parameter combination. The first plot shows the poles of the closed-loop system. The green poles belong to a set of feasible poles. The dashed line marks the hyperbola which separates the feasible poles from the infeasible ones. Black poles belong to a group of poles which violate the damping condition. Blue poles belong to a group of poles which violate the criterion of the natural frequency. The second and third plot show the zeros of the PID controller and the poles of the plant. The sampling in the space of design variables span a six dimensional space. Therefore the projections of the controller parameters are shown in the last 3 plots. The dashed black lines mark the boxes which include feasible solutions, determined by an optimization routine using differential evolution [6].

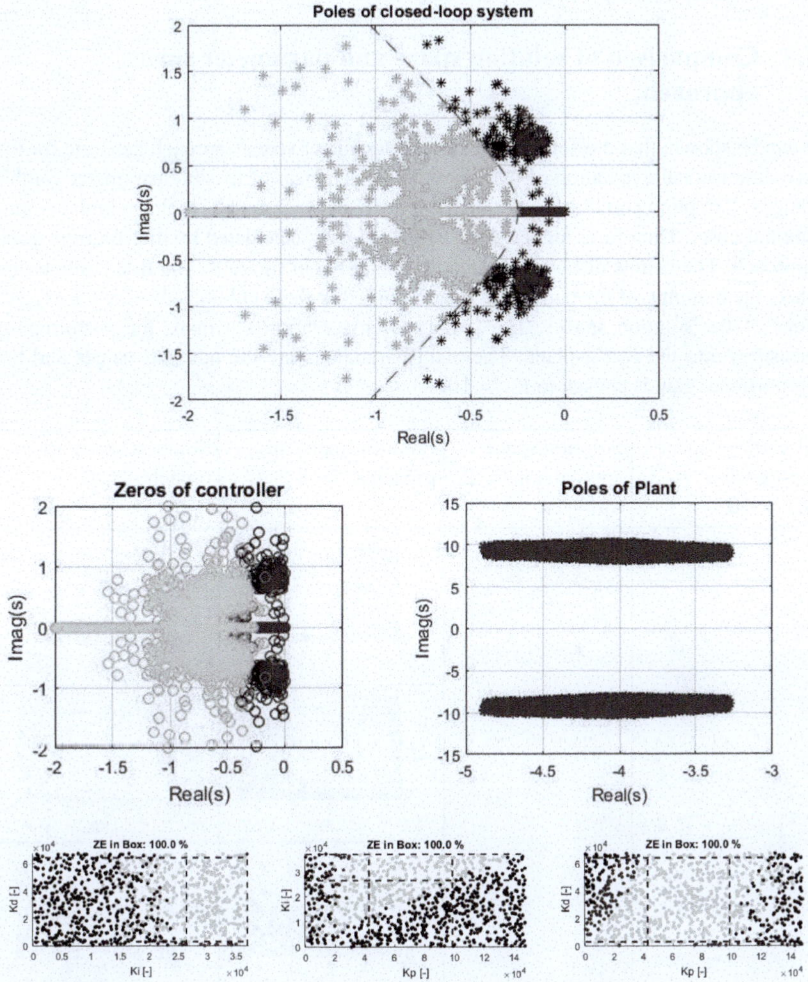

Fig. 7: Results of solution space approach corresponding to dependency graph

The edges of the hypercube provide the admissible intervals. Due to orthogonality of the edges, resulting parameter intervals can be chosen independently which allows independent development for vehicle components and control systems. Simultaneous development of vehicle components and control systems can be speed up by the calculation of independent parameter combinations.

3.4 Comparison of solution space and parameter space approach.

It can be shown, that the solution space and parameter space approach generate for the two dimensional evaluation of our example the same set of feasible parameter combinations. Sampling combined with the parameter space approach results in high computational costs. Therefore, higher dimensions are only calculated by the solution space approach. The results of both approaches are shown in figure 8. The first column displays the mapping of the parameter space. The other three columns display the projections of the solution spaces. The parameter space approach maps the performance measures into the K_I/K_D-plane. The plant parameters are the nominal values and the proportional gain is chosen to $K_P = 10$.

Parameter space approach $n = 2$, $K_P = 10$	Solution space approach $n = 2$, $K_P = 10$	Solution space approach $n = 3$	Solution space approach $n = 6$

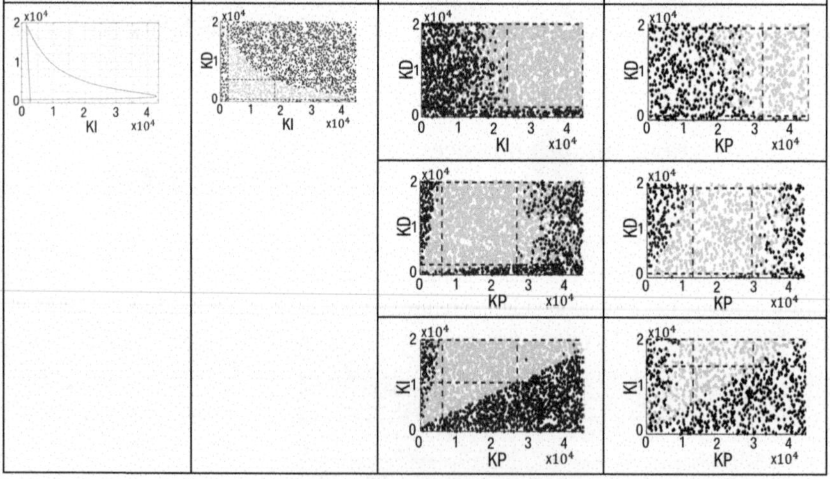

Fig. 8: Permissible ranges for control parameters resulting from parameter space approach and solution space approach with increasing number of design variables

The solution space approach solves the same problem as the parameter space approach. The feasible regions are identical. Additionally we can see the largest box for independent parameters. In the two dimensional case, the projection is the same as the solution space. In the third column K_P is varied additionally to K_I and K_D. With variation of K_P further options for parametrization are generated. If K_P stays in the given interval the feasible solutions of K_I and K_D shift in the upper right corner. The last column shows the solution spaces if not only the controller parameters but also the plant parameters are varied. Solutions that have been feasible before, now violate the performance measures, which results in smaller solution spaces. Minimizing K_P in the high dimensional mapping could produce a solution space that is similar to the parameter space approach. It would result in a smaller solution space since the optimization algorithm searches for the largest hypercube. The area of the solution space can be influenced by the design space in which the parameters are varied or by additional performance measures that adds further restrictions to the optimization problem.

4 Solution space based lateral vehicle control

Automated driving functions will be introduced stepwise [12]. For this work it is assumed that the vehicle will only be operated below a velocity of 130km/h and a lateral acceleration of 4m/s^2 [13],[14]. During operation vehicle parameters vary, e.g. a different mass due to additional loads or variation of the tire properties due to wearing or change of tires. When describing vehicle dynamics by a linear single track model simplifying assumptions have to be made. Hence, important effects such as elasto-kinematics, friction or roll torque are neglected which cause a different steering behavior than assumed for the single track model. Furthermore, internal disturbances may occur such as sensor errors or external disturbances such as different friction of the road, inclination of the road and crosswind.

4.1 Modelling and requirements

For modelling of vehicle dynamics a highly complex simulation environment based on MATLAB and Simulink [15] is being used. Furthermore a steering system is approximated by a PT_2 transfer function. The control system consist of two subsystems and is based on [16]. The combination of steering system and lateral system is regulated by two controllers as shown in figure 9.

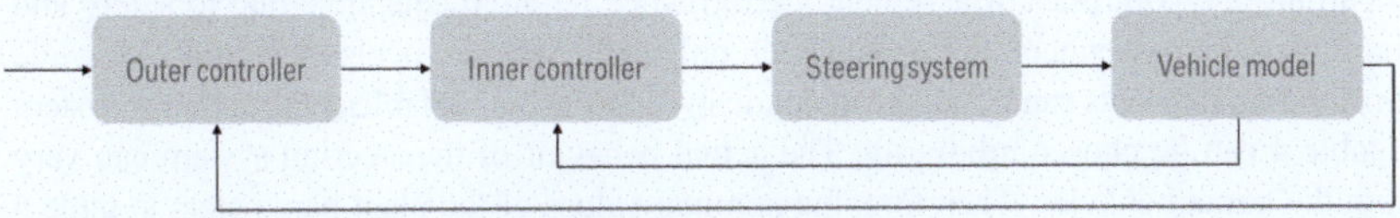

Fig. 9: Simplified structure of simulation environment

Each component interferes with the others and those interferences have to be coped with in the design process. In this example, the outer controller regulates the lateral position of the vehicle. The objective is to stay inside a corridor around the target trajectory. Therefore the peak overshoot and peak time of a step response is analyzed. The step of the lateral position can be interpreted as a lane change maneuver. Restriction of peak overshoot defines a maximum deviation and the restriction of the peak time allows only lane changes that are fast enough. The dependency graph for the solution space based design is shown in figure 10. The design variables of each component are in the bottom level. The parameters P_{C1}, P_{C2} are design variables of the outer controller and parameters P_{C3}, P_{C4} are design variables of the inner controller. The parameters K, T and D define the dynamics of the approximated steering system. The vehicle parameters are mentioned to give an overview of the whole system, but will not be varied for the design process. We assume that each component is developed by a different team. team D has already finished its design process. Team C is still developing its component, but has given a target behavior for the steering system. The remaining teams A and B are developing a controller each. The bottom-up mapping is applied and the design variables result in the systems dynamics of the components and form the closed loop system. Thus, the closed-loop system generates the step response in the layer above. The step response signal then is analyzed for peak overshoot and peak time. These performance measures are then mapped top-down into the design variables space. The goal is to determine permissible sets of parameters for team A and team B.

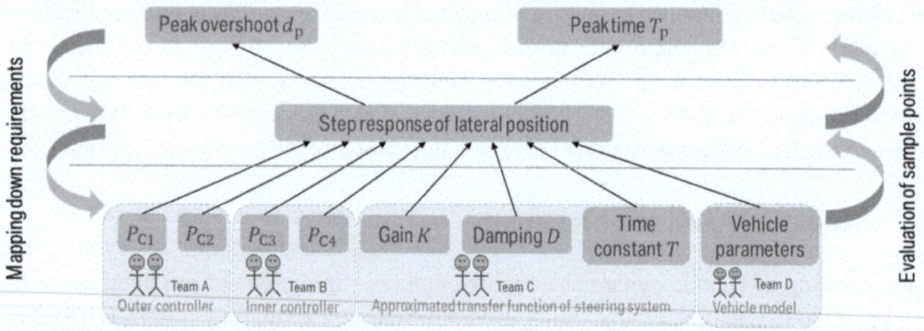

Fig. 10: Dependency graph for design of lateral control system

Regarding the controller of a vehicle, performance requirements arise due to safety and comfort of the passengers. In this approach only safety requirements are addressed, while comfort requirements can be added analogously. The design variables of the control system in table 4 can be chosen arbitrarily. The actual behavior of the steering system can vary from the target behavior. Therefore, the nominal values of the plant parameters in table 5 are varied by $\pm$ 10 percent. The values for the performance measures are shown in table 6.

Tab. 4: Design variables of the control systems

Controller parameters	
First parameter outer control cascade	P_{C1}
Second parameter outer control cascade	P_{C2}
First parameter inner control cascade	P_{C3}
Second parameter inner control cascade	P_{C4}

Tab. 5: Design variables of steering systems approximation

Plant parameters		**Nominal values**
Gain	K	1
Time constant	T	$0.5\ s$
Damping	D	$1\ 1/s^2$

Tab. 6: Performance measures of closed loop system

Performance measures		**Lower bound**	**Upper bound**
Overshoot of step response	$d_{\mathrm{p,max}}$	$0\ m$	$0.2\ m$
Peak time of overshoot	$T_{\mathrm{p,max}}$	$0\ s$	$5\ s$

4.2 Simulation results

For evaluation a seven-dimensional set is generated by Monte Carlo sampling [17]. The step responses of the closed-loop system are shown in figure 11. After evaluation the performance measures are applied. The black signals violate the overshoot criterion or become numerically unstable. The blue signals rise too slowly to the target value and the green signals fulfil both performance measures. The requirements are then mapped back in the space of design variables. The results are shown in figure 12. Blue dots show a parameter combination that leads to insufficient rising time. The red dots belong to parameter combinations causing an unstable simulation. Black dots show parameter combinations violating the overshoot criterion. The green dots mark the area with feasible parameter combinations. The dashed line shows the intervals for independent parametrization of the controllers. Nevertheless, inside the box are only feasible sample points. Parameter combinations that lie inside the hypercube but do not belong to the sample points are tested by simulation.

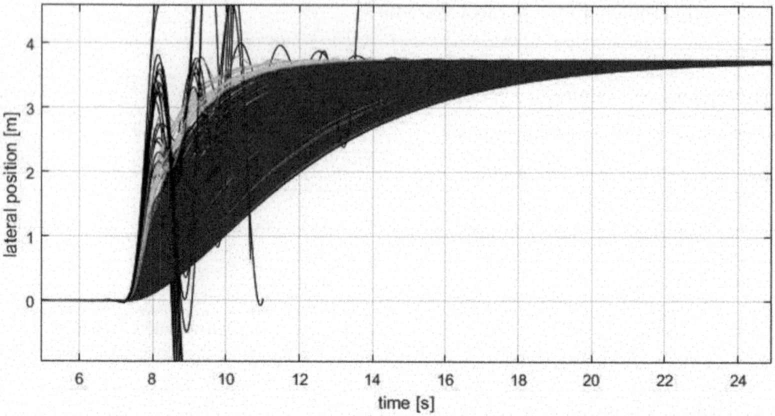

Fig. 11: Step response of lateral position of closed loop system: permissible responses (green) insufficient rising time (blue), numerical unstable and exceeding overshoot (black)

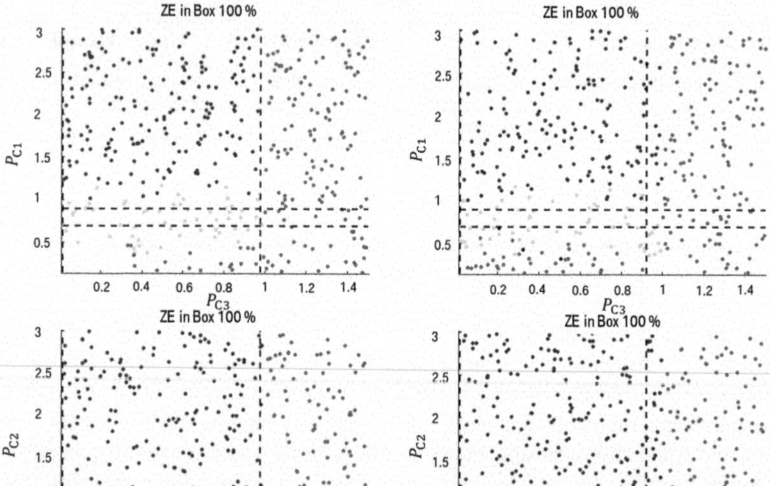

Fig. 12: Solution spaces to the lateral control problem: permissible parameters (green), insufficient rising time (blue), numerical unstable (red), exceeding maximum overshoot (black)

Therefore, four randomly chosen controller parameters are taken from the hypercube. The values are shown in table 7. As input of the controlled system a lateral trajectory is chosen. A combination of trapezoidal and sweep signal as lateral target trajectory is applied to the parametrized controller. Target (dashed line) and actual signals (solid lines) are shown in figure 13.

Tab. 7: Testing parametrization

Sets:	**Set 1**	**Set 2**	**Set 3**	**Set 4**
P_{C1}	0.8	0.8	0.75	0.8
P_{C2}	0.4	0.5	0.6	0.4
P_{C3}	0.1	0.5	0.8	0.1
P_{C4}	0.2	0.4	0.8	0.8

The actual lateral position reaches the target position within a corridor of 20 cm in about 6 seconds. At the beginning the lateral acceleration stays below 1 m/s^2 resulting in comfortable lane change maneuvers. With increasing inclination of the target signal the lateral acceleration increases up to 3.2 m/s^2 for evasive maneuvers.

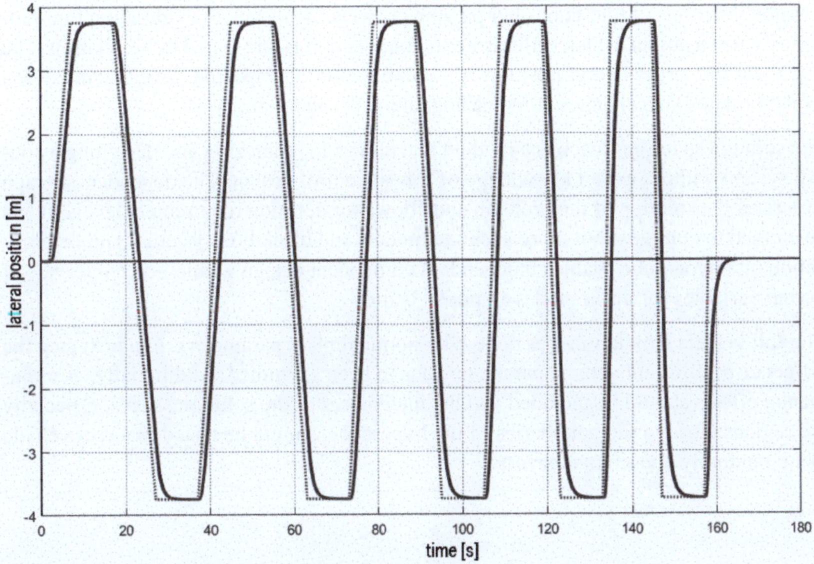

Fig. 13: Simulated lateral vehicle position with feasible controller parameters

5 Conclusions

In this paper the solution space approach is proposed as an alternative design and analyzation method to the parameter space approach. A short introduction to the principles of the solution space approach is given in chapter 2. It is used to derive feasible solutions of a PID controller that regulates a single-mass oscillator. It delivers exact results for two design variables. For a design problem with more than two design variables sampling in combination with analytical computation is typically applied. The solution space approach computes a set of feasible solutions faster due to Monte Carlo sampling. The sampling does not guarantee a valid solution in between the feasible sample points. Therefore both, the solution space approach and the parameter space approach are compared for a two dimensional design problem. In chapter 3 it is shown that both parameter space and solution space approach generate the same feasible set of design variables for an exemplary system. The solution space approach can be easily adjusted to calculate a problem with more than two design variables. In a next step the design problem of the example system was then expanded to higher dimensional problems. The results using the solution space showed a shift and magnification of the feasible solutions due to the influence of additional parameters. In chapter 4 the solution space approach is applied to a lateral vehicle control problem. Additionally, the application of a dependency graph is shown. The performance measures include overshoot and peak time of the lateral position. Intervals of four controller parameters were identified. For validation four randomly chosen points, which differ from the sample points, are tested by simulation. The signal for the target lateral position is a combination of trapezoidal signal and sweep which imitates lane changes at increasing lateral acceleration.

The solution space approach is a method that allows for analyzing and designing a control system without expert knowledge of robust control design. The dependency graph structures the solution of the problem and allows for defining requirement levels. It can be applied to complex, but more accurate models and black box systems. The determination of a hypercube enables independent component design within given intervals of design variables for different development teams.

The full vehicle simulation was done with nominal plant parameters. In a next step the influence of different vehicle parameters has to been examined. Additionally, environmental effects should be included and the interspaces of the solution spaces, especially the regions close to the boundaries of the hypercube, should be tested in a real vehicle using randomly chosen maneuvers.

References

[1] W. Wachenfeld and H. Winner, "Virtual assessment of automation in field operation a new runtime validation method," in 10. Workshop Fahrerassistenzsysteme, 2015, p. 161.

[2] J. Ackermann, P. Blue, T. B¨unte, L. G¨uvenc, D. Kaesbauer, M. Kordt, M. Muhler, and D. Odenthal, Robust Control: The Parameter Space Approach, 2nd ed. London: Springer London, 2002.

[3] J. Ackermann, J. Guldner, W. Sienel, R. Steinhauser, and V. I. Utkin, "Linear and nonlinear controller design for robust automatic steering," IEEE Transactions on Control Systems Technology, vol. 3, no. 1, pp. 132–143, 1995.

[4] J. Guldner, W. Sienal, H.-S. Tan, J. Ackermann, S. Patwardhan, T. Bunte, "Robust automatic steering control for look-down reference systems with front and rear sensors," IEEE Trans. Contr. Syst, Technol. 7(1), pp. 2-11, 1999.

[5] L. Pyta, F. Schrödel and D. Abel, "parameter space approach for large scale systems," 25th Mediterranean conference on control and automation (MED), pp. 1281-1286, 2017.

[6] M. Zimmermann and J. E. Hoessle, "Computing solution spaces for robust design," International Journal for Numerical Methods in Engineering, vol. 94, no. 3, pp. 290–307, 2013.

[7] A. Zare, K. Michels, L. Rath-Maia, M. Zimmermann, "On the design of actuators and control systems in early development stages", 8th International Munich Chassis Symposium, pp. 337-352, 2017.

[8] J. Lunze, "Regelunstechnik 1 – Systemtheoretische Gunrlagen, Analyse und Entwurf einschleifiger Regelungen, Springer-Lehrbuch, 2010.

[9] D. Odenthal, "Ein robustes fahrdynamik-regelungskonzept für die kippvermeidung von kraftfahrzeugen," Dissertation TU München, Fortschr.-Ber. VDI Reihe 12 Nr. 505. Düsseldorf: VDI Verlag, 2003.

[10] T. Bünte, „Mapping of nyquist/popov theta-stability margins into parameter space, IFAC proceedings volumes, vol. 33, issue 14, pp. 519-524, 2000.

[11] M. Zimmermann, S. Königs, C. Niemeyer, J. Fender, C. Zeherbauer, R. Vitale and M. Wahle, "On the design of large systems subject to uncertainty," Journal of Engineering Design, 28:4, 233-254, 2017.

[12] SAE International, “taxonomy and definitions for terms related to on-road motor vehicle automated driving systems“ standard SAE J3016_201401, On-Rouad Automated Driving Committee, pp 12, 2014.

[13] M. Maurer, J. C. Gerdes, B. Lenz, and H. Winner, Eds., Autonomes Fahren. Berlin, Heidelberg: Springer Berlin Heidelberg, 2015.

[14] M. Aeberhard, S. Rauch, M. Bahram, G. Tanzmeister, J. Thomas, Y. Pilat, F. Homm, W. Huber, and N. Kaempchen, “Experience, results and lessons learned from automated driving on germany’s highways,” IEEE Intelligent transportation systems magazine, vol. 7, no. 1, pp. 42–57, 2015.

[15] P. Kvasnicka, G. Prokop, M. D   orle, A. Rettinger, and S. H., “Durchgängige simulationsumgebung zur entwicklung und absicherung von fahrdynamischen regelsystemen,“ VDI 13. Internationaler Kongress, Berechnung und Simulation im Fahrzeugbau, 2006

[16] C. Rathgeber, F. Winkler, N. Nitzsche, D. Odenthal, S. Müller, “Disturbance observer for lateral trajectory tracking control for autonomous and cooperative driving,“ International Journal of mechanical, aerospace, industrial, mechatronic and manufacturing engineering, vol. 9, no. 6, pp. 886-893, 2015.

[17] N. Metropolis and S. Ulam, ”The Monte Carlo Method.,” Journal of the American Statistical Association 44:355, 1949.

Virtual chassis validation of commercial vehicles at MAN Truck & Bus

Dipl.-Ing. Manuel Armbrüster, MAN Truck & Bus AG, München

Dipl.-Ing. (FH) Alexander Schmid, MAN Truck & Bus AG, München

P. Pfeffer (Ed.), *9th International Munich Chassis Symposium 2018*, Proceedings,
https://doi.org/10.1007/978-3-658-22050-1_25

Abstract:

Using FEA (Finite Element Analysis) and MBS (Multi Body Simulation) methods or its combination is quite common in the automotive industry. The so-called virtual prototypes are a crucial element in the development processes. Due to technical innovation, these highly adapted processes must be continuously refined.

The virtual chassis validation process at MAN Truck & Bus AG is currently facing such a situation owing to the introduction of e-mobility and autonomous systems. These new technologies are making it necessary for different chassis concepts to meet package and customer requirements.

This paper describes the established virtual validation process for the chassis at MAN Truck & Bus AG. The current changes to the process, which are based on this virtual validation process, are discussed. The first part of the paper focuses on the automatic solution of the current process; the second part describes the necessary change within the CAE process. A roadmap for a closer collaboration between FEA and MBS is also presented.

1 Introduction

Increasing customer requirements in terms of cost-effectiveness, driving behaviour and comfort, as well as production and manufacturing related constraints increase the complexity of the development process of commercial vehicles. Particularly, an implementation of lightweight design leads to extended requirements for chassis development with regard to fatigue and structural dynamics.

A particularly challenging aspect of commercial vehicles is the modular build-up of the chassis in order to meet various customer demands. Using a modular system raises the efficiency in the production by using common parts. As a result, the whole truck portfolio, which consists of many thousands of basic vehicle types, can be built by using just 40 different cross members. These vehicles differ with regard to the number of axles, weight classes, wheelbases, build-up heights, and overhangs. The complexity is shown in the following image.

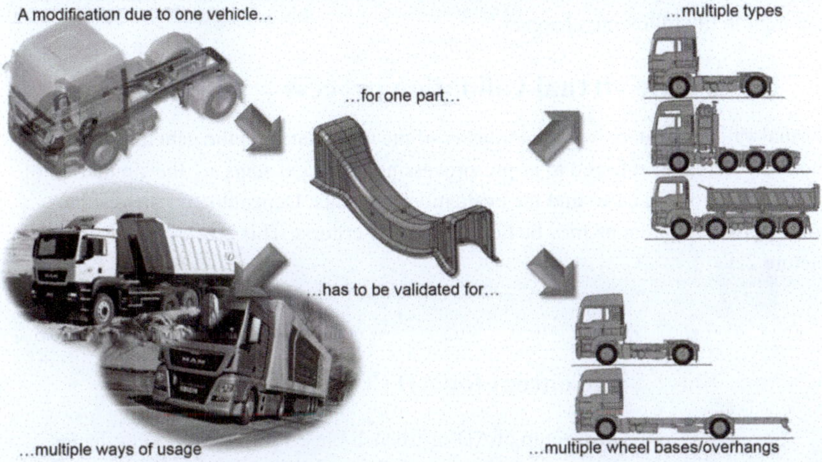

Picture 1: Variety of trucks

Besides the different design of the vehicles, the deployment of trucks shows a wide usage – from a typical long-haul tractor to a heavy-duty construction truck.

As a consequence, one cross member can be included in hundreds of different vehicle types. The number of sales of commercial vehicles is significantly less than with passenger cars – round about 100,000 units a year at MAN Truck & Bus AG compared to millions of passenger cars.

The validation of the wide truck portfolio results in special challenges for the virtual validation. Each component has to be validated for a variety of vehicles. A huge number of models has to be constructed and simulations must be performed. Nevertheless, the simulation must be efficient. The implementation of a highly automated simulation process is the only way to fulfil all these requirements. By using automation, it is even possible to reduce simulation time while raising the number of simulations and the quality. This automated process is used successfully in virtual validation and is a standard in the series development process at the present day.

In the near future, there will be changes in the design and loads due to electrification and autonomous driving. New chassis variants will be developed to enable huge battery packages and/or new interfaces between chassis and autonomous systems instead of human drivers. The impact to our virtual validation process will become apparent in the following chapters.

2 Up-to-date virtual validation process

Virtual validation starts with the transfer of the CAD data and the generation of FEA models – hereafter referred to as pre-processing. The next steps are the simulation of different static load cases and the evaluation of results. Depending on the results, some optimization loops may be needed to fulfil criteria. This process is illustrated in picture 2.

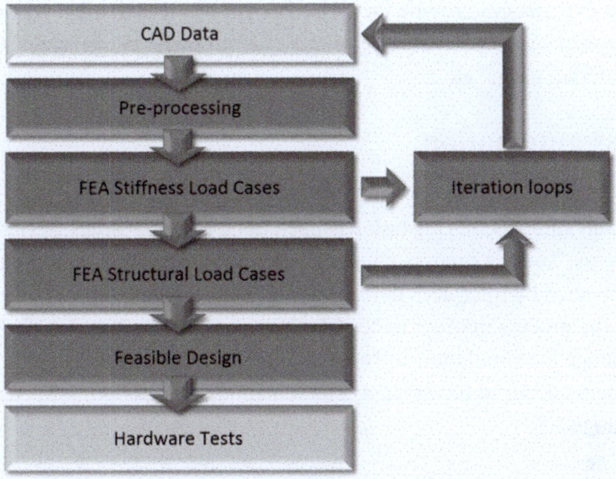

Picture 2: Up-to-date virtual validation process

Applied static load cases are documented in simulation specifications. These load cases represent all design-relevant driving situations. The translation into standard static load cases is derived from measurements, long-term comparisons between simulations and test results, the expertise of FEA simulations, and commercial specific knowledge. Picture 3 shows the evolution of virtual validation quality.

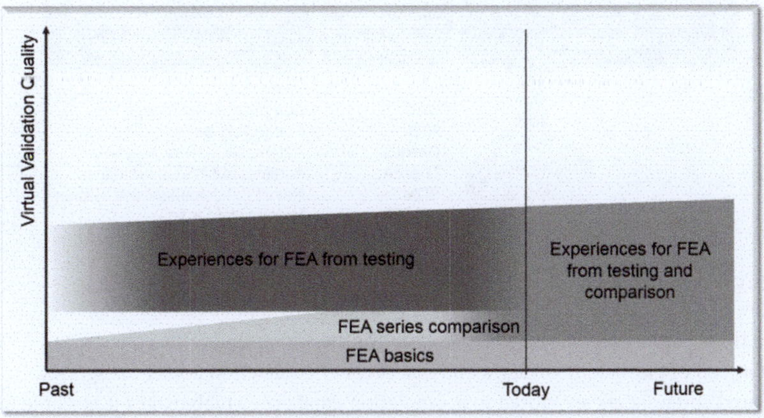

Picture 3: Origin of virtual validation quality

These coupled experiences provide the basis for virtual validation in every company. This principle works very well, as long as there are no major design changes (or changes in material, usage, etc.).

3 Process automation

For economic reasons, this process for simulations of truck frames must be standardized and automated. This task has been done at MAN Truck & Bus AG for the last five years. Today's process consists of five sections - shown in picture 4 - which are connected by interfaces defined precisely. This allows a very high flexibility for the process in order to be able to react to new tasks. For example, it is possible to change the simulation software in the process. In this case, only a small part of the process needs to be updated. The remaining process is unaffected and can be used unchanged.

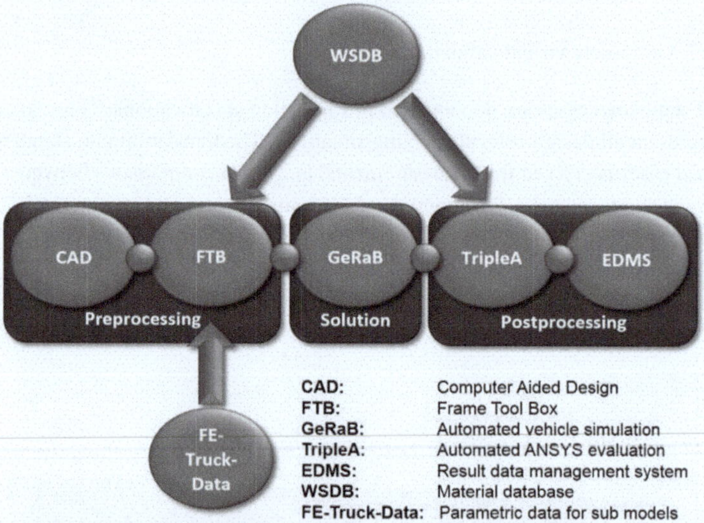

Picture 4: Automated process

CAD truck frame parts have to be prepared for the simulation. This concerns all derivations of part's mid-surfaces and markings of connection points for screws. Prepared components are stored in a simulation database, which makes it possible to reuse these components in several projects. The component's preparation is still a manual work, which is supported by macros. In future, this will also work fully automated.

For the model creation, the FTB (Frame Tool Box) has been developing in collaboration at MAN Truck & Bus AG. The FTB uses the prepared CAD data in the database and mesh them automatically. In the next step, the FTB imports and positions all meshed parts to one assembly. In the last step, connections between the parts are realized. In particular, an automated connection creation has a major impact on the reduction of simulation time, since there are around 300 connections in the truck chassis.

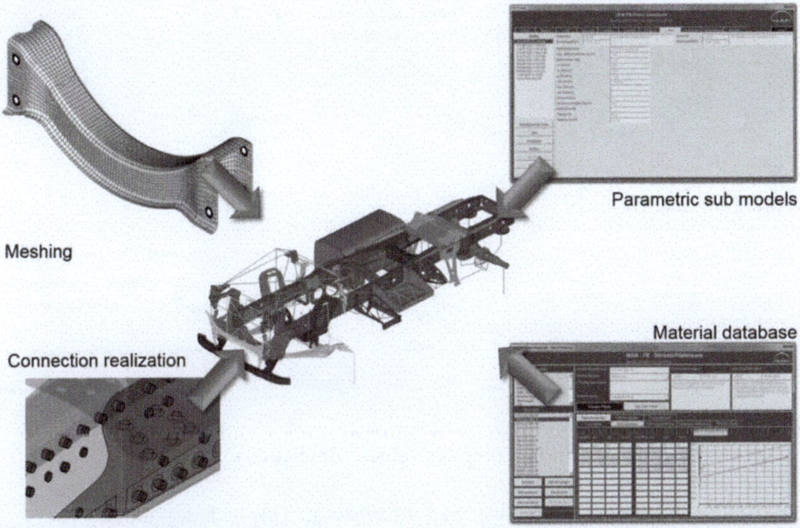

Picture 5: Assembly and realization of connections

The FTB supports different meshing styles to meet different requirements in the validation process. Even connections between parts can be realized by simple beams up to complex screw models including pretension.

The FTB accesses a material database (WSDB) with standardized material data for creating material cards. The last step is to include substitute models like models for axles, engine, or cabin. These models are realized as simple beam and mass models to reduce simulation efforts. The FE-LKW-data base creates these models. The use of a single database ensures the comparability of all models – independently from simulation engineers who created them. At the end, this highly automated process results in a simulation-ready vehicle model.

The finished model – checked by the user – is transferred to GeRaB (program for complete vehicle simulation). In addition to the model, further information regarding components, which are evaluated, and load cases, which are used, must be transferred to GeRaB. The data is provided as text files, which are also used for documentation issues because they contain all simulation relevant information.

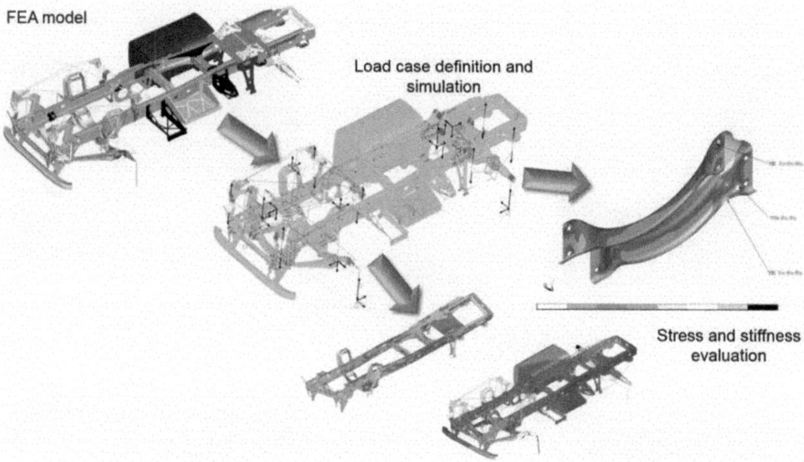

Picture 6: GeRaB (program for complete vehicle simulation)

GeRaB performs the simulation using all information. This includes not only the actual simulation of load cases but also the final preparation of the model and the derivation of all loads. The final setup of the model includes the adjustment of the payload and the pre-calculation of air spring characteristics, which correspond to the current vehicle weight. In addition, some load cases require numerical iterations to calculate final loads such as cornering and breaking. Finally, GeRaB performs the simulation run to obtain evaluation results.

After the solution, GeRaB transfers results to TripleA (program for the automated evaluation) for the post-processing. A standard chassis simulation consists of 15 load cases and 30 evaluation parts. These combinations lead to 450 part results. Due to the huge number of results, the post-processing has been automated by TripleA.

The automated evaluation contains following items:

- Animations of all load cases
- Stress evaluation for every load case and every part
- Stiffness curves of the chassis
- Preparation of screw-evaluation data

All data is stored in a presentation for documentation and visualization purposes. The presentation provides simulation engineers with a good overview of results and enables them to take a targeted look at relevant parts in post-processing.

Due to the automation, a huge number of results is generated – as mentioned above – there are some 450 results. Additionally, many complete vehicle evaluation data exists. As described above, all this data may be of interest for future projects. However, due to the large amount of data, the file-based storage makes no sense anymore. To solve this issue, a data management system was developed for the simulation results. It is called EDMS (result data management system) and deals with the huge number of data provided by GeRaB and TripleA. This system enables engineers to use all required data from previous simulations.

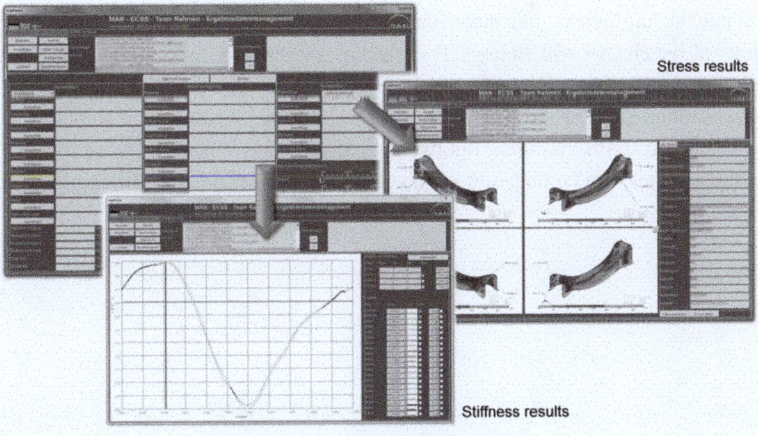

Picture 7: Using EDMS (result data management system)

Picture 7 shows some possibilities of EDMS such as finding part results or comparing stiffness curves.

EDMS can import TripleA results directly. All relevant information like loads, part numbers, materials, or evaluation values are provided directly by TripleA. Consequently, the manual work is reduced to a necessary minimum in order to add data to EDMS.

All in all, the engineering time for one simulation cycle was reduced to ca. 50% by using the fully automated process. However, the automated process has another advantage, especially in the last step. Due to automation, all results have the same mesh-, simulation-, and evaluation-quality – independently of the executing simulation engineer. Hence, EDMS results are fully comparable and can be used for the validation of different projects. The today's results are reference values for tomorrow. For this task, EDMS is most effective because it allows to recover previous results with all needed details.

4 New Challenges

Since the beginning of the truck production, the ladder frame has been used as the basis for the chassis. Up to now, load cases have been based on long time experiences with this well-known chassis concept as explained above. Based on these experiences, current load cases almost represent the real load of chassis.

Due to new technologies – like electrification and autonomous driving/ADAS – the topology of the chassis will change. This change is induced by packaging constrains to include large batteries or interface changes between chassis and cabin resulting from autonomous driving. Additionally, this design change will be used to reduce the vehicle weight to fulfil the future lightweight requirements. Figure 8 shows such a new chassis concept in comparison with a current ladder frame to illustrate big changes.

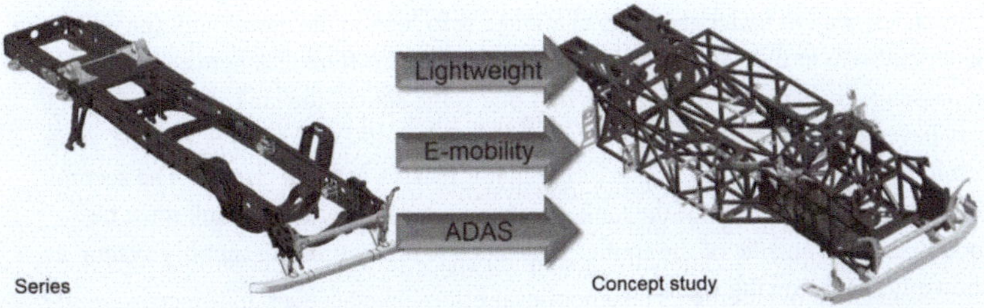

Picture 8: Series chassis versus concept study

Comparing the traditional design with the studies for possible future designs, it is obvious that vibration characteristics will change. This includes the vehicle-specific frequency band, amplitudes, and resulting vibration effects, as well as sensitivities. Consequently, the current frequency map will change. The understanding of the new vibrational phenomena plays a decisive role for the virtual validation process. Additionally, load collectives could also change. For example, an autonomous truck may have lower braking and acceleration loads than a human driver but the frequency of occurrence may increase.

The loss of experience due to new chassis concepts and the associated loss of a virtual validation quality are shown in figure 9.

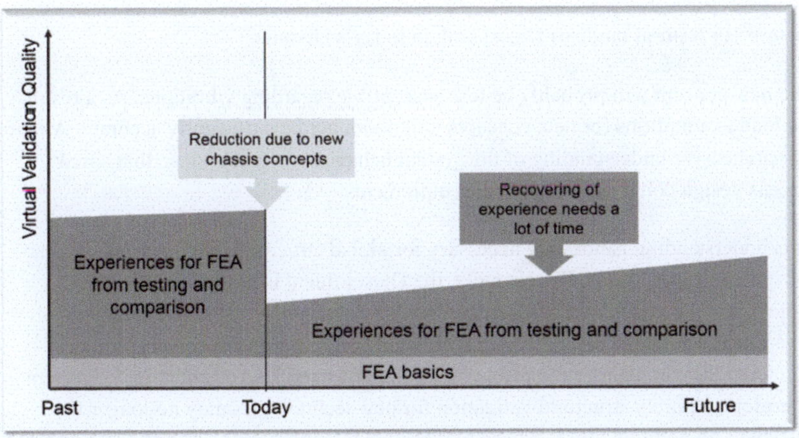

Picture 9: Challenges in virtual validation quality due to design innovations

The easiest way to understand this challenge is to look at the diesel tank (or rather the battery system in future). Today, there are two major differences between Diesel tanks in passenger cars and in trucks. The first one is the size of the tank. While passenger cars have some 50 litres, the standard truck tank has 400 to 600 litres volume. As a result, the mass of the tank is one of the major masses in truck chassis. The second difference is the position of the tank. By using a ladder frame, the tank must be mounted at the outside of the frame. This leads to a vibration susceptibly system as shown in the following figure.

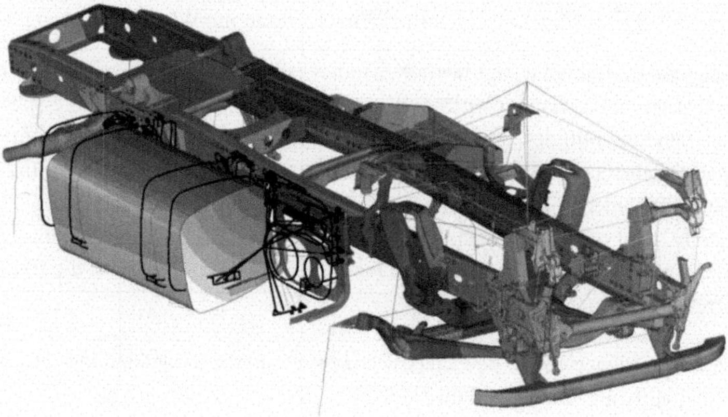

Picture 10: Natural mode of Diesel tank in today's chassis

The new concept will probably be less susceptible regarding vibrations. As a result, the load assumptions for new concepts will no longer be sufficiently accurate. A more comprehensive understanding of the system behaviour is required, so that a new chassis design could fulfil customer requirements.

This understanding is not only necessary for global structural requirements but also for specific sub-systems. For example, the Diesel fluent is not vulnerable to vibrations. The swashing Diesel fluent induces accelerations at the structural mountings, which are considered in structural load cases. In the case of batteries, accelerations can damage the inside of the batteries and lead to errors in the system. Therefore, a purely structural validation for new technologies may no longer be sufficient. In addition, the response to vibrational loading has to be validated and compared to requirements of electrical components.

Because of the brief timing for bringing electrical trucks to the market, it is not enough time in the development to gather same experiences as for conventional trucks; so, a new virtual validation process is necessary for meeting new challenges in the truck industry. This will be discussed in following pages.

5 Aspects for process optimization based on simulation

The mentioned challenge, which is presented in picture 9, shall be compensated by a complete vehicle approach using the MBS method. The hybrid approach has also the potential to increase the virtual validation quality above the current level.

The well-established MBS method at the MAN Truck & Bus AG series development is technically divided into four sections, as shown in picture 11.

Professional groups	Focus - examples
Vehicle dynamics/kinematics	kineamtics and compliance (K&C) full-vehicle dynamics
example	brake error, roll stabilization, elasto kinematics (leafspring and elastic axle beam) steady-state circular test, lane change
Vibrations	vibration phenomena - focus driving comfort road excitated (comfort road, single obstacles - edges, manholes) structural dynamics
example	configuration cab-suspension, vertical dynamics of leaf- and airsuspended chassis
Durability	vibration phenomena - focus load determination road excitation (rough road pakour, railway crossings) virtual structural durability
example	loads on components (engine mounts, chassis components), frame- or bodystructures
Powertrain and Vehicle-perfromance simulation	torsional vibration phenomena - focus isolation a. load determination powertrain excitation (engine, propeller shaft) longitudinal dynamics
example	engine/gearbox excitation by propeller shaft, rear axle excitation by ignition order

Picture 11: MBS focuses at MAN Truck & Bus AG

The basics of the integrated MBS process chain shall only be mentioned for the sake of completeness, since it corresponds to state of the art in the automotive sector.

Concept sub-models and complete vehicle models are used at MAN Truck & Bus AG. The shown sections in picture 11 describe MBS technical fields, which are relevant for commercial vehicles. The catalogues of vibration phenomena document physical mechanisms, vibration frequencies, design criteria, and the substitution tests.

The MBS toolkit underlays a continuous improvement process in terms of model topology and model validation. For this task, various MBS models are currently available for simulation.

The present paper focuses on the virtual validation process. However, this change has to consider also the collaboration with the testing department because a new virtual validation process needs the input of hardware tests.

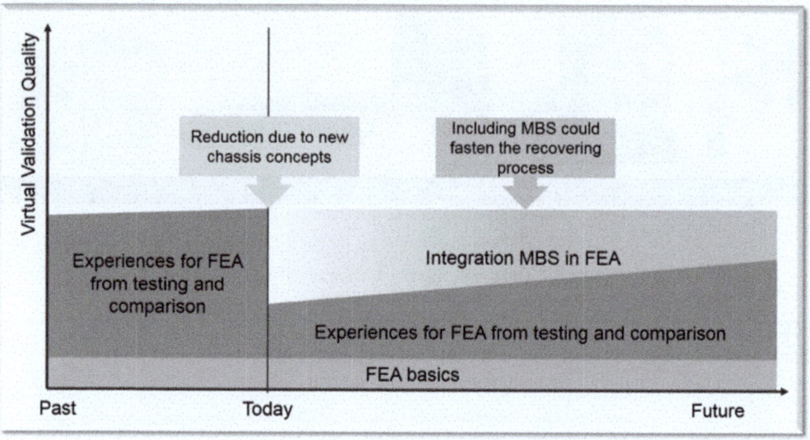

Picture 12: Implementing MBS methods fastens the recovery process

The current virtual validation process – as described in chapter 2 – has the advantage of a very high automation level, which results in a very time efficient simulation. The usage of this process will change. It will be rededicated to the main design tool using in the early stage. The new process will have more steps in the virtual validation because of discussed challenges. The implementation of the new process will advance in some steps. These steps will be discussed in the next chapters.

5.1 Redefining existing static FEA load cases using MBS input

As mentioned above, new chassis concepts will lead to different load collectives. For a successful virtual validation, today's static load cases have to be redefined for new chassis types. Due to the fact, that hardware testing will not be available to the highest possible extent in early development stages, load data from MBS can be used to recalculate static load cases. For many vehicles and chassis design types, specific load estimations can be derived and generated a sufficient database for all reference vehicles.

The advantage of this approach is the simple application and the fast response time for early optimization loops.

5.2 Dynamic analyses using FE

In a further step, the explained static load cases (chapter 5.1) can be extended by dynamic FE simulations; starting with simple modal analyses to more complex harmonic frequency response analyses. An important advantage of these analyses are short processing times because the whole system can be simulated in FEA without interaction with MBS tools.

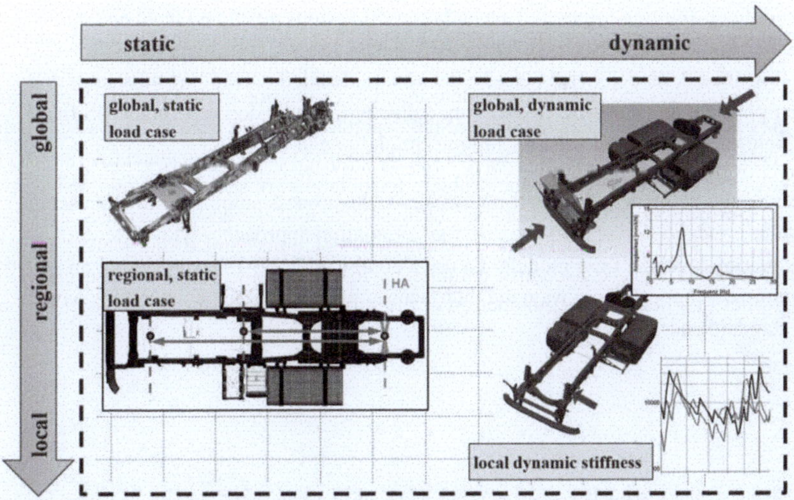

Picture 13: Fields of analysis for chassis (static/dynamic vs. global/regional/local)

Performing the analysis of the natural modes with defined boundary conditions, the linear transmission behaviour can be analysed. In this context is also included the monitoring of the local dynamic stiffness (LDS).

The harmonic response analysis – the virtual form of a "shaker" – could also be used to get information concerning the qualitative stress distribution, which is induced by vibrations.

5.3 Applying MBS loads on FE-Models

In contrast to the previous sections, the following methods deal with the coupling of MBS and FEA. This means that a simulated MBS dynamic maneuver is used as input to FEA simulations. Both linear and nonlinear analysis are applied.

In the linear dynamic transfer behaviour, the chassis is applied with forces and moments at the connection points, e.g. at steering arms. These forces, which result from harmonic excitation, are derived by means of a nonlinear full vehicle simulation in MBS. As a result, the contribution of every single structure transmission path can be analyzed, e.g. determination of relevant excitation mechanisms for the cabin.

With the usage of a virtual rough road track in MBS simulations, it is possible to obtain loads (cutting forces and accelerations at various positions) in an analogous manner to the durability tests used in FEA. These loads can be mapped to the nonlinear FEA model. MBS helps to get a more detailed view of the vibrations and their physical mechanisms. In addition, collectives can be derived from MBS full vehicle simulations and used for fatigue simulations in the FEA.

Besides the known challenges for MBS tasks, the process automation will present a further issue. MBS and FEA are different simulation approaches with different abstraction levels of their models. To use a combined MBS-FEA-process in an economically appropriate way, the interfaces and the modelling conventions have to be standardised.

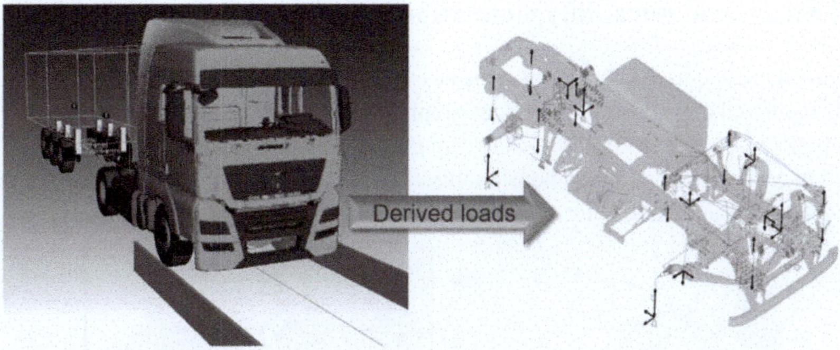

Picture 14: Deriving and using MBS loads for FEA fatigue analysis

The advantage of combining both simulation methods (FEA and MBS) offers the possibility to consider all the relevant nonlinearities (contacts and screw pretentions in FEA; friction, rubber-, and damping forces in MBS), which finally affects stress distribution in the whole chassis structure. These nonlinearities can be considered quite easily in the FEA at this point. So far, for the current FEA/MBS tool chain elastic bodies in MBS simulation are implemented as linearized (expect geometrical nonlinearities) FE-models. This topic is focused in two ways; from the software and from the engineering side in terms of the evaluation of the sensitivities.

The mechanical principle using cutting forces/moments from MBS to excite FE-structures can be exactly applied from linear harmonic excited systems and quasi-static manoeuvres in time domain (e.g. constant radius cornering or frequency sweeps). For stochastic excitations in time domain (e.g. rough road track), the transfer of the continuous MBS loads to abstracted FEA load cases is not biunique. For this reason, assumptions (analogous to durability tests) have to be made, which may very likely lead to a loss of accuracy.

This sequential process – MBS delivering input to FEA – currently results in an increased cycle time. This is, why at MAN Truck & Bus AG, this method is mainly used for detailed optimization of design concepts at an advanced stage. The earlier design steps are carried out using the methods described in 5.1 and 5.2 respectively.

5.4 Structural durability simulation based on MBS

As an alternative to 5.3, there is also the possibility to perform fatigue simulation directly, based on MBS results without simplifications when extracting the loads. The continuous deformation of the truck by simulating the virtual rough road track will be used to perform fatigue simulations. In this case, all information of MBS will be used for the fatigue simulation, not only selected and average loads and collectives.

Picture 15: Structural durability simulation based on MBS (MANLIFE)

This simulation process at MAN Truck & Bus AG (MANLIFE: **MAN** **L**oad collect**i**ve **f**atigu**e**) is used for predicting the durability of truck frames, their attachment parts, and bus chassis. The basis for the input in MBS is the rough road simulations mentioned in 5.3. The analysis tool for calculating the durability is FEMFAT.

A very important advantage to this process is the consideration of complete time dependent loads and collectives. The rough road testing as the basis for the evaluation of different concepts can be illustrated virtually. As already mentioned, it is not currently possible to include all nonlinearities of FEA models within the hybrid MBS simulation, such as nonlinear contacts or screw pretensions.

5.5 General comments on process optimization

The influence of tolerances and uncertainties that could occur in simulations, starting with idealization of the design geometry to simplified material models for FEA simulations should be kept in mind. Another point is the prediction of material characteristics influenced by the production processes like casting or bending for sheet metal parts. For the MBS part, an important issue is the measurement of the tire model parameters due to the high loads; there are no standardised measurement campaigns as there are for the passenger cars.

The quality of a complete virtual validation is defined by the greatest inaccuracy in the simulation process chain. The sensitivity of all these individual physical values and simulation steps have to be rated.

6 Conceptual design of new virtual validation processes

6.1 New virtual validation processes – general structure

The sketch below outlines a scheme of the virtual validation process for future chassis designs for series development at the MAN Truck & Bus AG. This new process will not only be applied to electric or autonomous vehicles; it will also be used for the whole portfolio to be developed in the future. The electrification of vehicles was the "enabling project" for developing and implementing the new processes (picture 16).

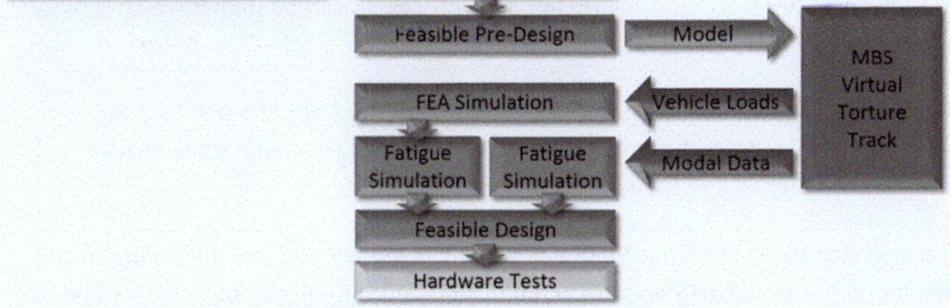

Picture 16: Structure chart of the new virtual validation process

The first steps will be adopted from the previous process (picture 2). Structural load cases will be refined and adapted by MBS simulated manoeuvers. An uprating is the implementation of dynamic load cases in FEA simulations, for example harmonic response analysis. These two FEM internal analyses are carried out in the early project phase with the goal of achieving a feasible basis design.

Based on this pre-validated design, a decision for more detailed validation steps must be considered. Should the design be approved, a transfer of the FEA models (e.g. truck chassis) to the MBS full vehicle model must be completed. Within MBS simulation, the suspension forces can be simulated by dynamic driving manoeuvres, e.g. rough road simulation on a virtual rough road track.

The results of the hybrid MBS simulation can be transferred to subsequent FE simulations (see chapter 5.3) or used as a basis for durability simulations (see chapter 5.4). If the simulation results fulfil the assessment criteria, simulation engineers will approve the developed design. This approval is necessary in order to go into hardware with its final tests; otherwise, the process must be started again with a redesigned geometry.

This new process is divided into two fields: an abbreviated simulation process in the beginning of the previous development phase, used for evaluation of many different design variants, and a sophisticated process with interaction between FEA and MBS. This last mentioned process will always be used after the first investigation of the concepts with higher maturity level.

6.2 New virtual validation process – timeline

Such described changes in the validation process take some time. Time is not only needed for developing new process steps but also for validation of the new steps through hardware tests. For this reason, new processes will be implemented step by step.

This systematic introduction is visualized in picture 17. With the first two steps: redefinition of loads, and introduction of dynamic load cases, the same virtual validation quality as it currently exists will be achieved.

The next two steps: applying MBS loads on FEA models and structural durability simulation based on MBS are more complex tasks. Using these steps, we will be able to achieve an even higher grade of virtual validation quality than today. Parallel to the development of these four steps, we have been working on methods for reducing the inaccuracies mentioned in 5.5; otherwise, increasing the quality above today's level will simply not be possible.

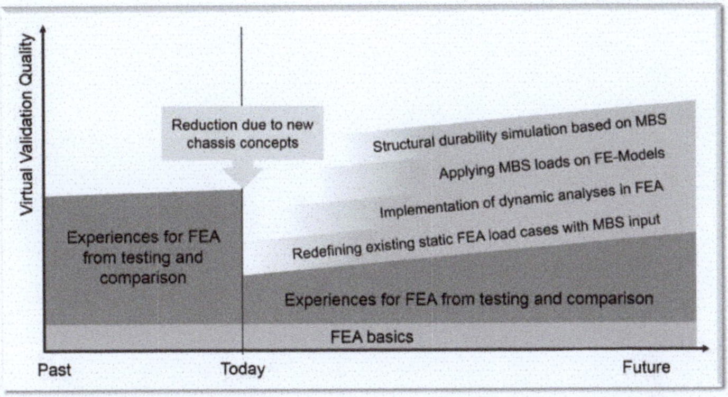

Picture 17: Chronological order

7 Conclusion and perspective

As shown in this paper, a continuous evolution of the virtual validation process is necessary. External influences, like larger design changes or usage of new material types, lead to a revision of the process. With the close collaboration between FEA, MBS, and testing, new methods can be implemented to face these challenges. By using MBS methods, it is not only possible to speed up the development of trucks, it will even be possible to increase the level of virtual validation quality in the near future.

An up-to-date validation process is needed for an economic and expedient development. For this reason, new effects have to be identified and evaluated concerning sensitivity and influence on the virtual validation process. Implementations of new physical mechanisms only lead to the desired results if their improvements are essential in comparison to current knowledge.

Finally, every change in the virtual validation process must be automated; otherwise, needed time for simulations will probably increase with every step. This would not be acceptably.

The virtual validation at MAN Truck & Bus AG is currently at a turning point. Our simulation team is well prepared for the challenges that have been discussed in this paper.

ELASTOMERIC BUSHING AND WHEEL SUSPENSION

Potential of elastodynamic analysis for robust suspension design in the early development stage

Stefan Buechner, M. Sc.
Institute of Automotive Technology, Technical University of Munich

Patrick Streubel, B. Sc. BMW Group

Dipl.-Ing. Norbert Deixler BMW Group

Dr. Ralf Stroph BMW Group

Dipl.-Ing. Udo Ochner BMW Group

Prof. Dr.-Ing. Markus Lienkamp
Institute of Automotive Technology, Technical University of Munich

P. Pfeffer (Ed.), *9th International Munich Chassis Symposium 2018*, Proceedings,
https://doi.org/10.1007/978-3-658-22050-1_26

Abstract

In the chassis development process, requirements for each subsystem are deduced from complete vehicle targets regarding safety, ride comfort and driving dynamics. After the design of individual components, they are integrated into respective subsystems resulting in a complete vehicle.
Late modifications in the development process after vehicle testing resulting from insufficient quality of simulations (e.g. because of disregarded effects or parameter uncertainties) and unnecessary target conflicts (e.g. regarding package) due to disregarded solution spaces lead to increasing development costs and time.

Chassis development is focused on the design and optimization of the suspension system. Characteristic values calculated from kinematics and compliance are in the center of interest and play a decisive role in evaluating chassis performance. Interactions with the steering system are partially modeled with reduced complexity or not considered at all.

This paper deals with a newly developed approach to evaluate suspension systems considering steering properties without a complete vehicle. For this purpose, analysis of a complete front suspension system including steering system is carried out using multi-body simulation. The suspension model comprehends kinematic and compliant properties. The steering system is modeled with relevant elasticities as well as speed-dependent steering assistance.

Stiffness properties of the suspension system, including steering system, are studied using the compliance matrix of the suspension system. The focus is on the analysis of the compliant steering axis by means of the compliance matrix method in comparison to the kinematic steering axis. The proposed method is also used to calculate corresponding characteristic values for the suspension system. These are analyzed for relevant load cases.

The objective is to gain further insights into suspension design to characterize and eventually optimize suspension performance in the early development stage. Therefore, respective interactions between the individual subsystems (suspension and steering system) and their components are analyzed. Their corresponding influences on the overall system are quantified. The advantage of considering compliant effects and the mentioned interactions for robust suspension design is shown.

1 Introduction

1.1 Motivation

In the chassis development process, the suspension system and its design play a decisive role, as they significantly influence vehicle targets regarding safety, ride comfort and driving dynamics. Currently, chassis development for new models is mainly based on the improvement of previous vehicle models. Working suspension concepts have been established and are adapted to respective vehicle targets while meeting defined requirements.

The automotive industry, however, is currently in change and subjected to several considerable trends, which have an impact on chassis development as well [1]. Modularization and standardization are focal points for reducing product complexity and variants without affecting chassis characteristics. The development of electric vehicles with various drivetrain topologies may result in new challenges and problems for suspension design, as it is influenced by the respective drivetrain and therefore dependent on it. Ride comfort has consistently become more and more relevant. With the rise of automated driving, it continues to gain in importance, whereas driving dynamics is expected to become less important. The presented trends affect the requirements defined for the chassis. The changing requirements influence both the selection of the suspension concept and its design. Increasing uncertainty during the development process is the result and leads to more development loops due to modifications causing additional costs.

Furthermore, early determination of product properties is a key factor during product development to reduce costs due to modifications [2]. Late modifications in the development process lead to increasing development costs and time.

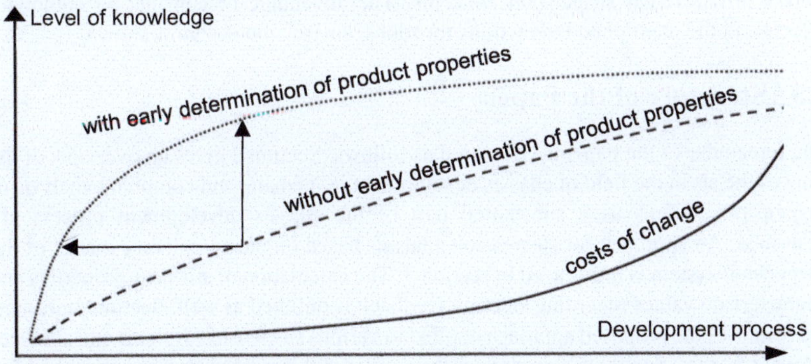

Figure 1: Level of knowledge and costs of change during product development based on [2]

By improving quality and validity of simulations due to new insights into suspension design, earlier and more detailed determination of suspension characteristics is possible. Either the same level of knowledge is gained at an earlier time or a higher level of knowledge is gained at the same time. Therefore, earlier modifications in the chassis development process are possible or they can be avoided entirely. As a result, development loops can be reduced saving cost and time.

1.2 Challenges and Goals

This paper aims at improving suspension design in the early development stage by means of multi-body simulation. Chassis development is already a well-established process and much research has been done on suspension design. Nevertheless, it is still a highly iterative process and poor chassis performance often goes unnoticed before vehicle testing. Proven characteristic values of the suspension system are in the center of interest during suspension design. It is state of the art to take them into account and aim at appropriate target areas, which are based on long-time experience. On the one hand, however, some of the characteristic values are calculated from suspension kinematics while ignoring its compliance. On the other hand, some of them only include the suspension system, while the steering system is not considered. This paper presents an approach for extended suspension analysis based on the compliance matrix of the suspension system. It allows a more detailed and extensive analysis of the suspension characteristics. The objective is to gain further insights into suspension design to characterize and eventually optimize suspension performance in the early development stage. Therefore, respective interactions between the suspension system and the steering system are analyzed. Their corresponding influences on the overall system are quantified. Furthermore, the compliant steering axis, which is dependent on the respective load case, is compared to the kinematic steering axis. Their properties are studied for different suspension concepts. The advantage of considering compliant effects and the mentioned interactions for robust suspension design is shown.

1.3 Structure of the Paper

The remainder of the paper is structured as follows. Section 2 gives an overview of the state of the art in the field of chassis development. Kinematic and compliant analysis of suspension systems as a substantial part of the chassis development process are described. An approach for suspension analysis based on the compliance matrix of the suspension system is introduced in section 3. The calculation of stiffness properties and characteristic values regarding steering feedback is outlined as well. Section 4 presents the results of the proposed approach applied to various suspension systems and analyzes relevant factors. Section 5 summarizes the results of this work and gives an outlook on further research.

2 State of the Art

Current research in the field of chassis development is focused on the chassis development process itself and new approaches to make it more efficient. Furthermore, much effort has been devoted to the analysis and design of suspension systems to improve its characteristics.

2.1 Chassis Development in the Vehicle Development Process

In vehicle development as well as in chassis development, the V-model has become established [1],[3],[4]. On the specification branch, full vehicle targets are defined from customer preferences. Requirements on the system level (chassis, drivetrain, etc.) and subsystem level (suspension system, steering system, tires, etc.) are subsequently derived from them top-down. This process is also called target cascading. It is followed by the design and analysis of each subsystem trying to meet the defined requirements. However, the development of individual subsystems is mostly independent of each other and interactions between the respective subsystems are not considered to some extent. To solve this problem, [5] has proposed an approach for the simultaneous design of axles and tires based on the computation of solution spaces. For the suspension system, kinematic and compliant properties are in the center of interest, as is shown in the next subsection. Typically, multi-body simulation is used for suspension analysis and design. A focus of recent research has been on sensitivity analysis and optimization [6],[7],[8]. The next step is the design of individual components.

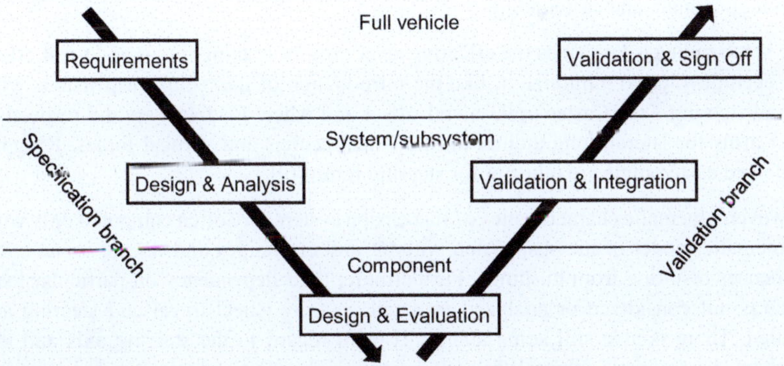

Figure 2: V-model of the chassis development process based on [1],[3],[4]

On the validation branch, components are assembled bottom-up into subsystems and systems again, which are validated against their specifications. In the beginning, validation is mostly done with simulations. This is followed by hardware testing, for example Kinematics & Compliance testing, or full vehicle testing. Especially for the steering system, much fine-tuning regarding steering feel is done during complete vehicle testing. The V-model is an iterative process until the defined requirements are met. For this reason, a robust and concurrent design of the subsystems, for which accurate simulation results are essential, is important in order to minimize unnecessary iterations due to required modifications.

2.2 Suspension Analysis and Design

Chassis development is focused on the design of the suspension system with kinematic and compliant characteristics being in the center of interest. In recent years, new methods for modeling, analysis and optimization of suspension systems have been proposed and studied, whereas the established characteristic values have not changed.

Kinematic analysis deals with the analysis of the wheel movement during wheel travel and steering, whereas external loads are not taken into account. In addition to wheel alignment characteristics, like toe and camber, and instant centers, the steering axis plays a decisive role for front suspensions [1],[9]. The characteristic values resulting from the steering axis are relevant for steering feedback and include:

- Scrub radius and caster trail (both at contact patch)
- Lateral and longitudinal offset (both at wheel center)
- Wheel load arm
- Kingpin angle and caster angle

The scrub radius is known as the effective lever arm for braking forces and great effort has been devoted to minimize it, like the introduction of multi-link suspensions with virtual steering axis. Caster trail, lateral offset and wheel load arm are the respective lever arms for lateral, longitudinal (at the wheel center) and vertical forces. Kingpin angle and caster angle are relevant for steering return [9].

However, kinematic characteristics of suspension systems are often computed only with a kinematic model of the suspension without any elasticities and do not include the influences resulting from bushings. Furthermore, their dependency on particular load cases is not considered or analyzed in detail, as only wheel travel and steering are studied. There is also still some ambiguity with regard to the steering axis and the resultant characteristic values. While [9] defines scrub radius and caster trail as effective lever arms for braking and lateral forces with regard to steering feedback, [10] relates them to wheel movement, more precisely to toe behavior.

In contrast, compliance analysis studies the wheel movement under external loads, i.e. braking forces, longitudinal forces at the wheel center, lateral forces as well as wheel loads. Like in kinematics, wheel alignment characteristics, especially toe behavior under braking and lateral forces, are studied. Besides, stiffness characteristics are in the focus, as they are relevant for both comfort and handling. Suspension stiffness has decreased more and more in the past years due to the enhanced effort devoted to improve comfort, especially in the premium segment. According to [1], the most relevant stiffness characteristics are:

- Longitudinal and lateral stiffness (both at contact patch and wheel center)
- Wheel rate
- Camber and toe stiffness (both regarding longitudinal and lateral forces)

To some extent, the influences on the mentioned stiffness characteristics are not studied in detail though, in spite of their dependency on particular load cases and the steering system. As pointed out in [11], the elasticity resulting from the steering system has an effect on vehicle behavior. Moreover, [12] and [13] mention the influence of steering elasticity on steering precision. Nevertheless, in compliance analysis, it is often not considered in detail or at all.

3 Approach

This paper presents an approach to evaluate suspension systems considering steering and tire properties without a complete vehicle. The overall concept is shown in Figure 3.

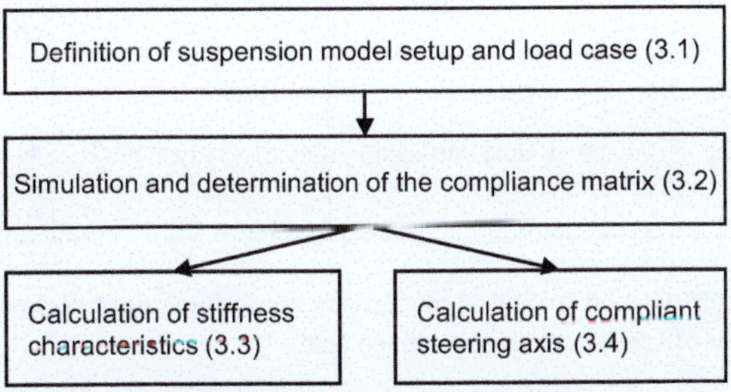

Figure 3: Overall concept of the approach proposed in this paper

For this purpose, multi-body simulation is performed using the simulation tool MSC Adams/Car®. First of all, an appropriate model of the suspension system and a specific load case are defined. Afterwards, the compliance matrix of the suspension system is determined, while simulation is carried out. Finally, stiffness characteristics and characteristic values for steering feedback, which result in the compliant steering axis, are calculated from the compliance matrix. While the former are analyzed regarding influences resulting from the steering system, the latter are compared to their kinematic equivalents.

3.1 Suspension Models and Multi-Body Simulation

Two different suspension concepts are studied in this paper, a double wishbone suspension and a multi-link suspension. The suspension model comprehends kinematic and compliant properties. Bushings, springs as well as bump and rebound stops are modeled with corresponding stiffness characteristics. The steering system also considers relevant elasticities. Hardy disk and torsion bar are modeled with linear stiffness properties. Steering assistance is dependent on different speed characteristics according to [14]. Therefore, the resulting steering elasticity is also assumed to be speed-dependent [13]. It is possible to convert the compliant model into a kinematic one by replacing bushings with joints and removing force elements. The system boundary is varied by setting the steering actuator to either rack or steering wheel. If it is set to rack, steering elasticities are not considered, as the rack is fixed. An overview of various suspension model setups, which are studied in this paper, is given in Table 1. Tire behavior is considered by using pneumatic trail and scrub as the point of force application (in this contribution 25 mm and 0 mm respectively).

Table 1: Overview of suspension model setups for this contribution

Suspension type	Model complexity	Steering actuator	Steering assist characteristics
• Double wishbone • Multi-link	• Kinematic • Compliant	• Rack • Steering wheel	• 0 km/h • 80 km/h • 200 km/h

Specific load cases derived from real maneuvers are defined next. Relevant load cases for the evaluation of suspension performance are:

- Braking forces (applied at contact patch)
- Traction forces and disturbance forces (applied at wheel center)
- Lateral forces (applied at contact patch, displaced by pneumatic trail)
- Wheel travel/wheel loads

After the suspension model has been configured and the load case has been defined, simulation is carried out.

3.2 Compliance Matrix of the Suspension System

The approach for suspension analysis presented in this paper is based on the compliance matrix of the suspension system. The suspension system is simply an elastic system consisting of (flexible) linkages as well as bushings and springs with corresponding stiffness characteristics, which connects the wheel to the vehicle body. Therefore, its properties for a specific state can be expressed by the elastic system's stiffness matrix K, as outlined in [15],[16],[17]. The compliant characteristics resulting from the stiffness matrix and from the original system are the same in this state.

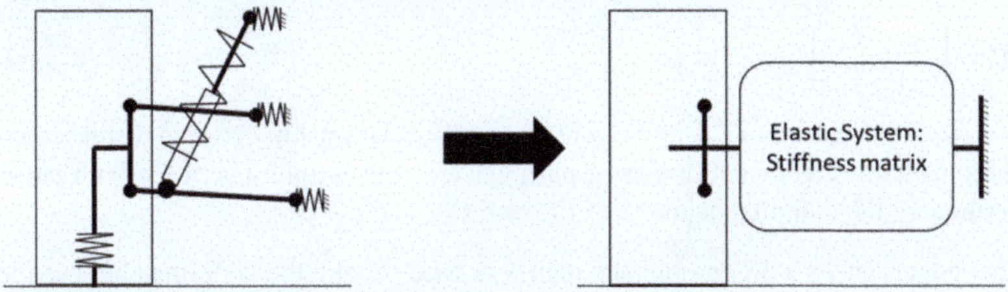

Figure 4: Suspension system as an elastic system represented by its stiffness matrix based on [16]

The stiffness matrix K is a 6 x 6 matrix and describes the resulting force vector $\vec{F}$ for a given displacement vector $\vec{d}$.

$$\vec{F} = K\vec{d} \tag{1}$$

The force vector $\vec{F}$ can be subdivided into a force $\vec{F}_t$ and a torque $\vec{F}_r$.

$$\vec{F} = \begin{pmatrix} \vec{F}_t \\ \vec{F}_r \end{pmatrix} = (F_x, F_y, F_z, M_x, M_y, M_z)^T \tag{2}$$

The displacement vector $\vec{d}$ can also be subdivided into a translational and a rotational part.

$$\vec{d} = \begin{pmatrix} \vec{d}_t \\ \vec{d}_r \end{pmatrix} = (d_x, d_y, d_z, \theta_x, \theta_y, \theta_z)^T \tag{3}$$

The compliance matrix C is defined as the inverse of the stiffness matrix K.

$$C = K^{-1} \tag{4}$$

Each matrix element represents the respective compliance for any degree-of-freedom due to forces in any degree-of-freedom.

$$C = \begin{bmatrix} c_{11} & \cdots & c_{16} \\ \vdots & \ddots & \vdots \\ c_{61} & \cdots & c_{66} \end{bmatrix} \tag{5}$$

Now we can derive the resulting displacement vector $\vec{d}$ for a given force vector $\vec{F}$.

$$\vec{d} = C\vec{F} \tag{6}$$

The coupling between different parts is considered by extending the compliance matrix C to a 12 x 12 matrix. For two parts, the following equation is obtained.

$$\begin{pmatrix} \vec{d}_1 \\ \vec{d}_2 \end{pmatrix} = C \begin{pmatrix} \vec{F}_1 \\ \vec{F}_2 \end{pmatrix} \tag{7}$$

where

$$C = \begin{bmatrix} C_{11} & C_{12} \\ C_{21} & C_{22} \end{bmatrix} \tag{8}$$

The compliance matrices C_{11} and C_{22} represent the relationship between displacements and forces with reference to the same part, whereas the coupling between both parts is described by the compliance matrices C_{12} and C_{21}.

In this paper, an 18 x 18 compliance matrix is used to take the couplings between left wheel, right wheel and rack into account. A user-written subroutine calling a utility subroutine is implemented in Adams/Car®, as it turned out to be an easy and fast way to obtain the system's compliance matrix. As a result, during the simulation, the compliance matrix is returned for each simulation step. Suspension characteristics are then calculated from the compliance matrix as described in the following subsections.

3.3 Suspension Stiffness Characteristics

This section deals with the calculation of suspension stiffness characteristics from the obtained compliance matrix, as already shown in [4] and [16]. The approach is outlined for the suspension of the left wheel here. Hence, the compliance matrix is just a 6 x 6 matrix. For brevity, only the calculation of the longitudinal stiffness characteristics is explained in detail. Lateral stiffness and wheel rate as well as toe, camber and wind-up stiffness are obtained analogously. As mentioned before, couplings between left wheel, right wheel and rack are considered by extending the compliance matrix to an 18 x 18 matrix. This is necessary, for example, to incorporate anti-roll bar properties.

In this paper, the compliance matrix is obtained with reference to the wheel center. Therefore, longitudinal stiffness at the wheel center is simply calculated from the compliance matrix element c_{11}, which describes the relationship between a longitudinal force and a translation along the x-axis.

$$K_{x,wc} = \frac{1}{c_{11}} \tag{9}$$

Forces $\vec{F}_t$, which are not applied at the wheel center, generate a torque $\vec{F}_r$.

$$\vec{F}_r = \vec{r} \times \vec{F}_t \tag{10}$$

The resulting torque $\vec{F}_r$ is dependent on the point of application $\vec{r}$ in the yz-plane.

$$\vec{r} = (0, y, z)^T \tag{11}$$

Only a longitudinal unit force is applied here, as the longitudinal stiffness is calculated. A unit force can be used, as the system is assumed to be linear in the current state of the system. Therefore, the force $\vec{F}_t$ can be written as

$$\vec{F}_t = (1, 0, 0)^T \tag{12}$$

Equations (10) and (12) yield the force vector $\vec{F}$.

$$\vec{F} = \begin{pmatrix} \vec{F}_t \\ \vec{F}_r \end{pmatrix} \tag{13}$$

Next, the displacement vector $\vec{d}$ is calculated.

$$\vec{d} = C\vec{F} \tag{14}$$

Now we can derive the resulting displacement $\vec{d}_{res}$ at the point of force application.

$$\vec{d}_{res} = \vec{d}_t + \vec{d}_r \times \vec{r} \tag{15}$$

Longitudinal stiffness K_x at the point of force application is then given by

$$K_x = \frac{F_x}{d_{res,x}} \tag{16}$$

Making use of equation (12), equation (16) is simplified to

$$K_x = \frac{1}{d_{res,x}} \tag{17}$$

After calculating the longitudinal stiffness for each point in the yz-plane, the suspension system's stiffness characteristics can be visualized.

3.4 Characteristic Values for Steering Feedback

In addition to stiffness characteristics, the obtained compliance matrix is also used to calculate characteristic values regarding steering feedback, which add up to the compliant steering axis. For space reasons, only the calculation of the scrub radius is explained here in detail. Castor trail as well as longitudinal and lateral offset at the wheel center are determined analogously. The extended 18 x 18 compliance matrix is necessary, as the couplings between wheels and rack need to be considered, when the relationship between forces applied to the wheels and resulting steering feedback is analyzed.

The force vector $\vec{F}$ can now be subdivided into a force applied to the left and right wheel as well as to the rack. A unit braking force is applied to the left wheel at contact patch with an offset y.

$$\vec{F} = \begin{pmatrix} \vec{F}_{wheel,left} \\ \vec{F}_{wheel,right} \\ \vec{F}_{rack} \end{pmatrix} \tag{18}$$

where

$$\vec{F}_{wheel,left} = (1,0,0,0,-1 \cdot R, -1 \cdot y)^T \tag{19}$$

$$\vec{F}_{wheel,right} = (0,0,0,0,0,0)^T \tag{20}$$

$$\vec{F}_{rack} = (0,0,0,0,0,0)^T \tag{21}$$

The displacement vector $\vec{d}$ can also be subdivided into a displacement with reference to the left and right wheel as well as to the rack.

$$\vec{d} = \begin{pmatrix} \vec{d}_{wheel,left} \\ \vec{d}_{wheel,right} \\ \vec{d}_{rack} \end{pmatrix} \tag{22}$$

Next, the displacement vector $\vec{d}$ is calculated with the compliance matrix C.

$$\vec{d} = C\vec{F} \tag{23}$$

The 14th row of equation (23) describes the translation of the rack due to the braking force.

$$d_{rack,y} = c_{14,1} - c_{14,5}R - c_{14,6}y \tag{24}$$

As mentioned before, the scrub radius is known as the effective lever arm for braking forces. Therefore, a braking force, which is displaced by the scrub radius r_k, has no effect on the steering system. Under the condition that it has no effect, the translation of the rack due to this braking force has to be zero.

$$d_{rack,y} = 0 \tag{25}$$

Now we obtain the scrub radius r_k from equation (24).

$$r_k = \frac{c_{14,1} - c_{14,5}R}{c_{14,6}} \tag{26}$$

Scrub radius, caster trail as well as longitudinal and lateral offset add up to the compliant steering axis. Kingpin and caster angle as well as the wheel load arm can then be derived from the compliant steering axis.

4 Results

In this section, the application of the proposed approach to various suspension systems is demonstrated. The first two subsections deal with the suspension stiffness characteristics and relevant influences of the steering system. The last two subsections focus on the comparison of the compliant steering axis with the kinematic one as well as on its load case dependency.

4.1 Suspension Stiffness Characteristics

As already pointed out in subsection 3.3, the stiffness characteristics of the suspension system are visualized after being calculated. This can be used to evaluate the effects of parameter modifications more easily, as the overall stiffness characteristics are visible at a glance. Next to the maximum stiffness, relevant characteristic stiffness values at wheel center and contact patch can also be determined. The multi-link suspension with consideration of its compliance is used here. The steering actuator is set to the rack, so steering stiffness has no influence on the results. Therefore, the selected steering assistance characteristic is not relevant here. The results here are obtained for the design position. Figure 5 shows the longitudinal and lateral stiffness characteristics of the suspension system dependent on the location of additional forces.

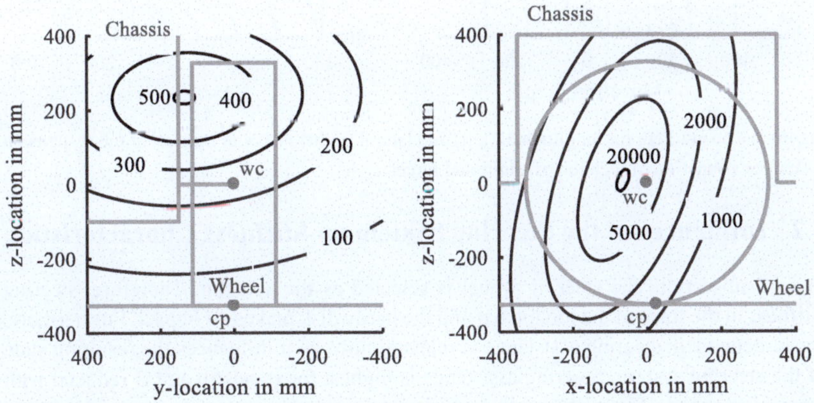

Figure 5: Longitudinal (left) and lateral (right) stiffness in N/mm of the suspension system dependent on the location of additional forces

Longitudinal stiffness at wheel center $K_{x,wc}$ and at contact patch $K_{x,cp}$ are approximately 0.2 kN/mm and 0.1 kN/mm respectively. The maximum value of the longitudinal stiffness is 0.5 kN/mm and is located close to the upper wishbone. Lateral stiffness at wheel center $K_{y,wc}$ and at contact patch $K_{y,cp}$ are 9.4 kN/mm and 1.2 kN/mm respectively. The maximum value of the lateral stiffness is 21 kN/mm and is located close to the position of the tie rod.

Figure 6 shows the toe gradient of the suspension system for longitudinal and lateral forces dependent on the location of additional forces. As already mentioned in [16], this plot can be used to study the robustness of the suspension system to changing points of force application due to tire behavior. The toe stiffness is then calculated as the inverse of the toe gradient. Toe stiffness regarding longitudinal forces at wheel center $K_{toe,Fx,wc}$ and at contact patch $K_{toe,Fx,cp}$ are 0.15 kN/min and 0.13 kN/min respectively. Toe stiffness regarding lateral forces at wheel center $K_{toe,Fy,wc}$ and at contact patch $K_{toe,Fy,cp}$ are 0.24 kN/mm and 0.12 kN/mm respectively.

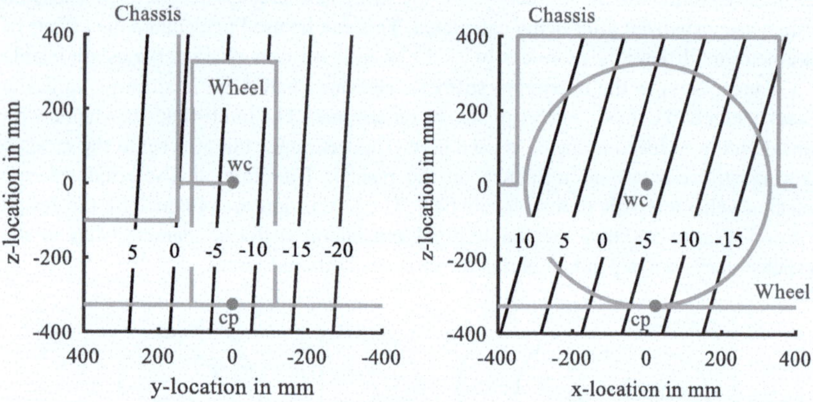

Figure 6: Toe gradient for longitudinal (left) and lateral (right) forces in min/kN of the suspension system as a function of location of additional forces

4.2 Influences of the Steering System on Stiffness Characteristics

In this subsection, the steering wheel is selected as the actuator. Therefore, steering stiffness in the form of the torsion bar and the hardy disk makes an impact. Furthermore, speed-dependent steering assistance is relevant now, as it influences the resulting state of the steering system. Steering assistance is highest for standstill and is reduced with increasing speed. The stiffness characteristics from subsection 4.1 are now obtained with consideration of steering stiffness and different steering assistance characteristics. The results are shown in Table 2.

Table 2: Overview of evaluated suspension model setups

Steering actuator		Rack	Steering Wheel	Steering Wheel	Steering Wheel
Steering assist		-	0 km/h	80 km/h	220 km/h
$K_{x,wc}$	N/mm	237	233	231	230
$K_{x,cp}$	N/mm	74	74	74	74
$K_{y,wc}$	N/mm	9406	9094	8927	8813
$K_{y,cp}$	N/mm	1178	1124	1095	1076
$K_{toe,Fx,wc}$	N/min	149	97	81	72
$K_{toe,Fx,cp}$	N/min	132	114	106	101
$K_{toe,Fy,cp}$	N/min	120	90	78	72

While the steering system has no influence on the resulting longitudinal stiffness, the resulting lateral stiffness is slightly reduced by adding the steering stiffness. The biggest influence, however, is on the toe stiffness, which decreases considerably. Furthermore, the steering assist characteristic is relevant when studying the toe stiffness, as it is influenced by the amount of assistance.

4.3 Kinematic and Compliant Steering Axis

In this subsection, the kinematic steering axis and the characteristic values for steering feedback are compared to the compliant ones. While the kinematic characteristic values and steering axis are determined according to [9], the compliant ones are calculated from the suspension system's compliance matrix as described in subsection 3.4. The rack is selected as the actuator again, so the steering stiffness has no influence. The results here are obtained both for the multi-link and for the double wishbone suspension, at first for the design position. Figure 7 shows the results for the multi-link suspension. While a distinct difference is visible in the front view, especially for the scrub radius, no difference is visible in the side view due to the higher lateral stiffness. The results for the double wishbone suspension are shown in figure 8. Kinematic and compliant steering axes are located close to each other both for front and side view. Kinematic and compliant characteristic values for steering feedback resulting from the steering axis are summarized in Table 3. For the multi-link suspension, there are considerable differences between kinematic and compliant scrub radius as well as lateral offset. The compliant scrub radius is more than three times the kinematic one.

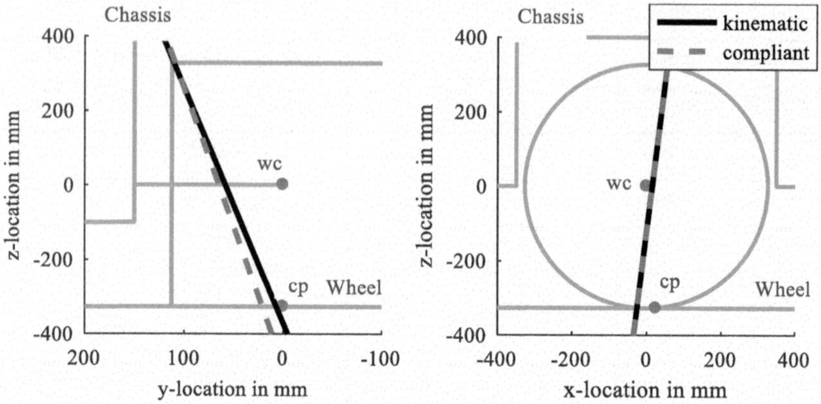

Figure 7: Kinematic and compliant steering axis in front view (left) and side view (right) for multi-link suspension

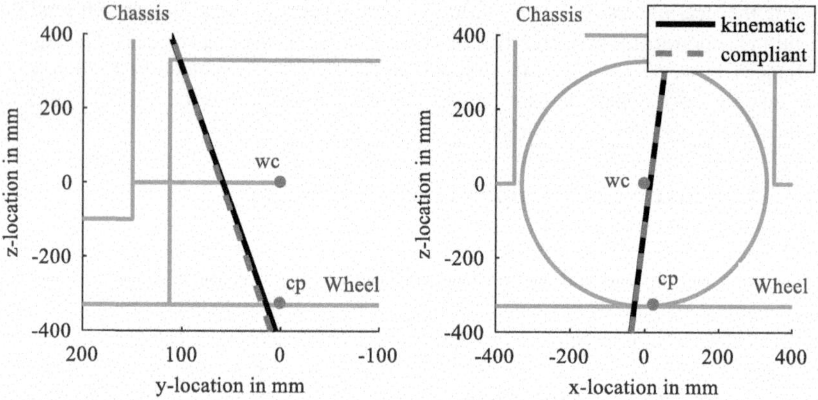

Figure 8: Kinematic and compliant steering axis in front view (left) and side view (right) for double wishbone suspension

Remarkably, this is only valid for the multi-link suspension. Kinematic and compliant scrub radius as well as lateral offset are close to each other for the double wishbone suspension. Thus, in total, the compliant scrub radius of the multi-link suspension is the same as that of the double wishbone suspension, though the kinematic ones vary significantly from each other. For caster trail and longitudinal offset, however, no significant differences between the kinematic and compliant results were found both for the multi-link and for the double wishbone suspension.

Table 3: Comparison of kinematic and compliant characteristic values for steering feedback for multi-link and double wishbone suspension

		Multi-link		Double wishbone	
		Kinematic	Compliant	Kinematic	Compliant
Scrub radius	mm	7	22	17	22
Caster trail	mm	26	25	27	27
Lateral offset	mm	59	65	59	61
Long. offset	mm	15	15	13	14

The characteristic values have previously only been obtained for the design position. The compliant ones, however, are dependent on the respective load case. For this reason, a closer look is taken at how they behave for different load cases. Due to restrictions of the length of this publication, only scrub radius and caster trail in dependency of braking forces for the multi-link suspension are presented here (see Figure 9). As already mentioned, the compliant scrub radius for the design position is more than three times the kinematic one. However, under braking forces, it approximates the kinematic one. The compliant caster trail for the design position is close to the kinematic one again. It decreases to less than half of it under braking forces. Consequently, when using the characteristic values for steering feedback for the evaluation of suspension systems, their behavior in terms of relevant load cases should be considered.

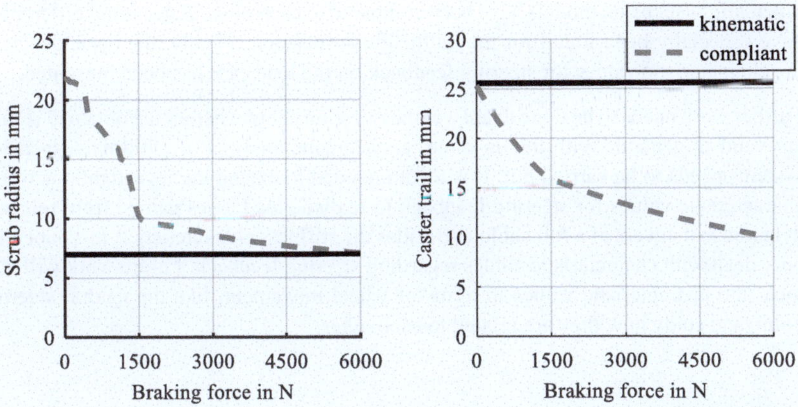

Figure 9: Scrub radius (left) and caster trail (right) under braking forces for multi-link suspension

5 Summary and Outlook

In summary, this paper has presented an approach for the analysis and evaluation of suspension systems based on their compliance matrix. The goal was to gain further insights into suspension design, which can be used to improve the development process and optimize suspension performance.

The literature on suspension design shows that stiffness characteristics as well as the steering axis and resulting characteristic values for steering feedback play a decisive role in the development process. Several publications, however, focus only on a kinematic model of the suspension system and do not consider compliance. Furthermore, the steering system with its stiffness and speed-dependent steering assistance are often not investigated in detail.

Our method for suspension analysis is based on the suspension system's compliance matrix taking kinematics as well as compliance into account. After definition of the suspension model setup and load case, simulation is carried out. During the simulation, the compliance matrix is obtained for each simulation step. It is then used for the calculation of stiffness characteristics and characteristic values for steering feedback. This paper has highlighted the suitability of the proposed method to calculate various characteristic values of suspension systems easily and fast. Calculation as well as results of longitudinal, lateral and toe stiffness characteristics have been presented. The results of this study showed that steering stiffness and steering assistance had almost no influence on longitudinal and lateral stiffness. Their effect on toe stiffness, however, has been clearly demonstrated. Moreover, the calculation of characteristic values regarding steering feedback has been described. Results for the compliant and kinematic suspension models have been compared to each other. Substantial differences have especially been found for the multi-link suspension. Finally, the dependency of the characteristic values for steering feedback on the load case has been presented.

Further work needs to be done to answer the question about whether the obtained results are valid in general. With this in mind, a systematic analysis of multiple suspension systems needs to be carried out. It is recommended to extend the investigations of the characteristic values for steering feedback to realistic load cases gained from relevant dynamic maneuvers of a full vehicle. Besides the stiffness characteristics, this paper has only dealt with characteristic values regarding steering feedback. Future work will also look into characteristic values in terms of wheel movement, like the neutral steering point, and study how they are related to each other.

Acknowledgement

This research is funded by the BMW Group. The author would like to thank them for their support.

Contributions

Stefan Buechner (corresponding author) initiated and implemented this paper. Patrick Streubel made essential contributions to the implementation of the proposed methodology as part of his master thesis. Norbert Deixler contributed to a critical discussion of the proposed methodology. Ralf Stroph and Udo Ochner revised the manuscript critically for important intellectual content. Markus Lienkamp made an essential contribution to the conception of the research project. He revised the paper critically for important intellectual content. Mr. Lienkamp gave final approval of the version to be published and agrees to all aspects of the work. As a guarantor, he accepts responsibility for the overall integrity of the paper.

References

[1] Ersoy, M.; Gies, S.: *Fahrwerkhandbuch*. Wiesbaden: Springer Fachmedien Wiesbaden, 2017.

[2] Lindemann, U: *Methodische Entwicklung technischer Produkte: Methoden flexibel und situationsgerecht anwenden*. Berlin, Heidelberg: Springer Berlin Heidelberg, 2009.

[3] Weber, J.: *Automotive Development Processes*. Berlin, Heidelberg: Springer Berlin Heidelberg, 2009.

[4] Blundell, M.; Harty D.: *The Multibody Systems Approach to Vehicle Dynamics*. Oxford, England, Waltham, Massachusetts: Butterworth-Heinemann, 2015.

[5] Wimmler J.; Schramm D.; Wahle M.; Zimmermann M.: *Concurrent design of vehicle tires and axles*. In: Pfeffer P. (eds) 6th International Munich Chassis Symposium 2015, pp. 839–851. Wiesbaden: Springer Fachmedien Wiesbaden, 2016.

[6] Kang, D.O.; Heo, S.J.; Kim, M.S.: *Robust design optimization of the McPherson suspension system with consideration of a bush compliance uncertainty*. In: Proceedings of the Institution of Mechanical Engineers, Part D: Journal of Automobile Engineering, vol. 224, no. 6, pp. 705–716, 2010.

[7] Park, K., Heo, S.J., Kang, D.O. et al.: *Robust design optimization of suspension system considering steering pull reduction*. In: Int. J. Automot. Technol., vol. 14, no. 6, pp. 927–933, 2013.

[8] Klose, G.; Kopp, C.: *Suspension Design based on DoE Studies Using a Virtual Roller Test Rig*. In: 2016 Vehicle Dynamics Conference. München, 2016.

[9] Matschinsky, W.: *Radführungen der Straßenfahrzeuge. Kinematik, Elasto-Kinematik und Konstruktion*. Berlin, Heidelberg: Springer-Verlag Berlin Heidelberg, 2007.

[10] Schultz, E.: *Toe-Correcting-Twistbeam (BTCT) – extended usage of twistbeam axles*. In: Pfeffer P. (eds) 7th International Munich Chassis Symposium 2016, pp. 159-177. Wiesbaden: Springer Fachmedien Wiesbaden, 2017.

[11] Van Ende, K.; Kallmeyer, F.; Nippold, C. et al.: *Analysis of steering system elasticities and their impact on on-centre handling*. In: Int. J. Vehicle Design, vol. 70, no. 3, pp. 211–233, 2016.

[12] Heißing, B.; Brandl, H. J.: *Subjektive Beurteilung des Fahrverhaltens*. Würzburg: Vogel, 2002.

[13] Buschardt, B.: *Synthetische Lenkmomente*. Dissertation. ZMMS-Spektrum 16. Düsseldorf: VDI-Verlag, 2003.

[14] Pfeffer, P.; Harrer, M.: *Lenkungshandbuch*. Wiesbaden: Springer Fachmedien Wiesbaden, 2013.

[15] Yao, G; Hou, J; Zhao, P.: *A new methodology to calculate the equivalent stiffness matrix of the suspension structure with flexible linkages*. In: Advances in Mechanical Engineering, vol. 9, no. 7, pp. 1–8, 2017.

[16] Gerrard, M.: *The Equivalent Elastic Mechanism a Tool for the Analysis and the Design of Compliant Suspension Linkages*. In: SAE World Congress & Exhibition. SAE Technical Paper Series. Warrendale, PA: SAE International, 2005.

[17] Nishimura, K.; Nozawa, T.: *Development of Suspension Design Technology Applying Principal Elastic Axes*. In: SAE World Congress & Exhibition. SAE Technical Paper Series. Warrendale, PA: SAE International, 2007.

The behavior of elastomer components in chassis systems under operating and special loads in real operating conditions and in the computational determination of sectional loads

Frieder Riedel

P. Pfeffer (Ed.), *9th International Munich Chassis Symposium 2018*, Proceedings,
https://doi.org/10.1007/978-3-658-22050-1_27

Abstract

For the digital support of the durability assurance virtual methods are used in early development phases. An effective means is the multi-body simulation. In order to improve the predictive quality of the calculations of load-time functions for operating and special loads, complex submodels of individual components have to be used. These include the chassis elastomer bushings. The elastomer bushings behave strongly non-linearly under high loads and exhibit at the reversal points an increased hysteresis widening of the force-displacement curve. A benchmark analysis shows that simple analogous models, such as the Kelvin-Voigt element, are unable to map the amplitude and frequency dependent behavior as well as the dissipative properties accordingly. Only by using more detailed friction models the real elastomer bushing behavior can be approximated in the high load range. The models according to Dzierzek and Daimler AG (GuHyLaPu) have a corresponding friction modeling. Nevertheless, both approaches have deficits. Accordingly, a new friction model has to be developed, which is suitable for the calculation of load-time functions in the medium and high load range and is also parameterizable efficiently.

1 Introduction

Virtual methods are a leading part of the current automotive development process in order to meet the shorter development cycle times and the growing variety of derivatives. For the digital support of the durability assessment of chassis, pre-optimizations of components with respect to strength capacity occur due to numerical simulations, before the actual production process is initiated. The multi body system (MBS) simulation is a proven means of calculating required sectional loads for the component design. For the virtual determination of load-time function, components are flexibly integrated and submodels for tires, shock absorbers as well as elastomer bushings are used.

In case of durability, the loads are divided in operating, special and misuse loads. The latter is part of the improper use of a vehicle and primarily serves the design of the damage chain in case of component failure. Operating and special loads, in contrast, are subject to intended use, whereby the component is tested for fatigue strength or plastic deformation. Special events, such as crossing a cross channel or braking in a manhole cover, occur with very low frequency, but introduce extremely high force amplitudes in the chassis components. These force peaks can lead to local, permanent plastic strains which may not exceed a specific material limit.

Figure 1 shows an example of the force range of a standardized measurement of a force-displacement curve of an elastomer bushing. For the shaded areas of gray there are no explicit functional data. The highly non-linear characteristic as well as the damping behavior of the elastomer bushings must therefore be estimated by the calculation engineer in the high load ranges. The importance of the elastomer bushing behavior in the high load range is shown by means of Figure 2. The upper part shows the force signal, the level-crossing counting and the damage sum from an operating load measurement. Only a few cycles exceed the force range of ±1, but these have a significant effect on the damage sum. If the force signal of a special load measurement is considered (see Fig. 2 below), an even more drastic picture appears. The majority of the vibration amplitudes are outside the measured area. Special events can lead to a loss of the tire ground contact, which makes the suspension a free-swinging system. Hence, the knowledge about the elastomer bushing, acting as an energy storage and sink, is of enormous importance for this particular purpose.

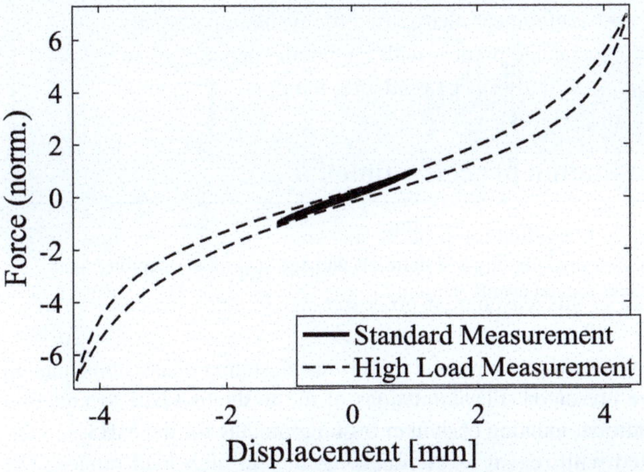

Fig. 1: Force-displacement curve of an elastomer bushing

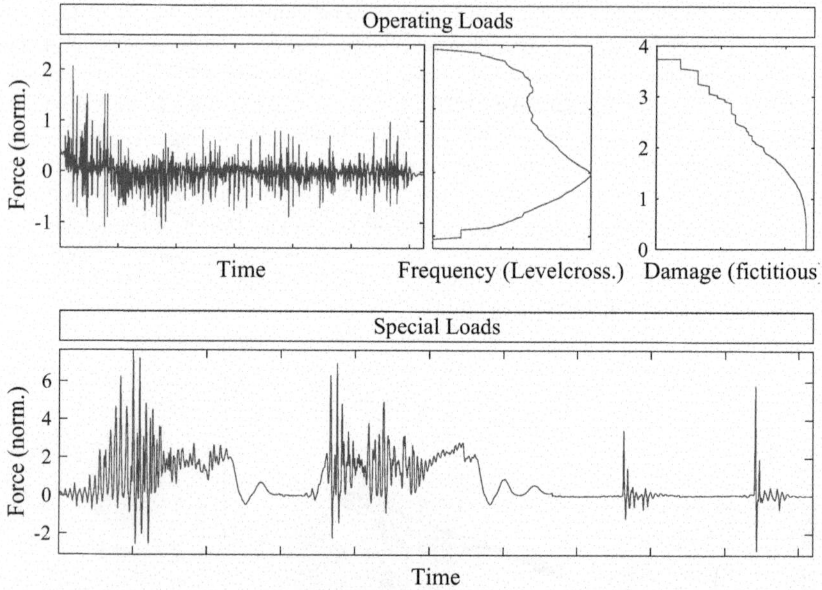

Fig. 2: Force signal, level-crossing counting and fictitious damage sum of an operating load (top); force signal of different special events (bottom)

In the following, findings from measurement of various elastomer chassis bushings in the high load range are presented. FE simulations of the bushings serve to underpin observed effects. In addition, existing elastomer bushing models for the MBS-simulation will be investigated with regard to the image quality in high load ranges. The demonstration of optimization potential completes the work.

2 Elastomer bushing behavior under high loads

Special loads are distinguished from the operating loads by a relatively narrow-band frequency spectrum. In Figure 3 above, for two special events, the force responses are shown for one in the spring link to the other in the strut rod. Using a so-called modified S-Transformation (Stockwell-Transformation [6]), it is possible to investigate the intensity of occurring frequencies of transient signals. The result of the frequency analysis for the force signals is shown in Figure 3 below. A large amount of the energy input is available between 15 and 30 Hz for braking on a bad road (left) and crossing a vertical event (right). The signals also show further frequency contents in a slightly weaker form up to 80 Hz. This observation can also be found for other special events and components.

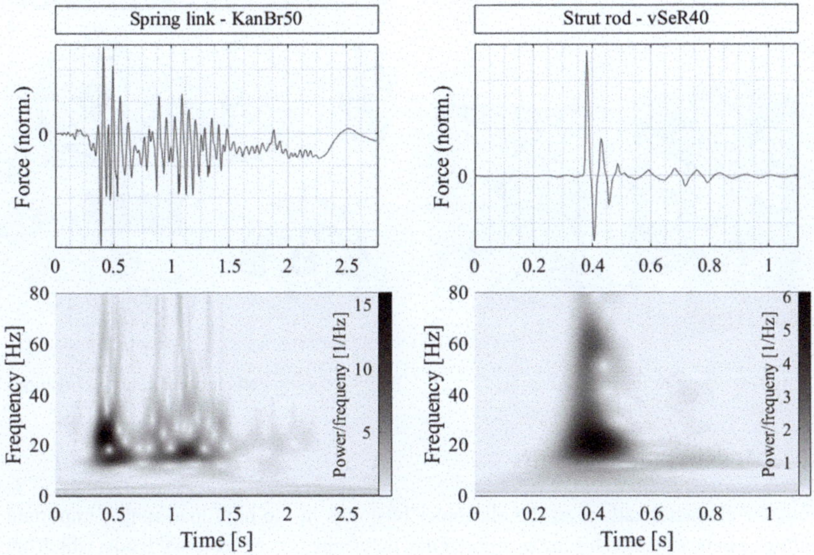

Fig. 3: Force signal and S-Transformation in the spring link for braking on a bad road (left) and in the strut rod for crossing a vertical event (right)

2.1 Set-up and scope of testing

For the static and dynamic characterization of the elastomer bushings in the desired load and dynamic range, a specially developed servo-hydraulic, uniaxial test rig is designed (see Fig. 4). The differential cylinder has a static force of 180 kN. Using a three-stage servo valve with 630 l/min and the given dimensions of the hydraulic cylinder, a maximum excitation speed of 1300 mm/s can be reached. The force measurement takes place above the adaption fork. Two linear potentiometers connected in parallel to the adaption fork and ring record the deflection.

As part of the investigation, five representative suspension bushings are selected – a conventional bushing (1), a pre-compressed solid rubber mount with inlay (2), a non-symmetrical pocket bushing (3), a pre-compressed pocket bushing (4) and a hydraulic suspension bushing (5) (see Fig. 5). The elastomer bushings are statically and dynamically measured with the maximum force amplitudes known from vehicle measurement. The achievable frequency for harmonic sinus excitations is limited by the maximum excitation speed and a maximum value of 50 Hz.

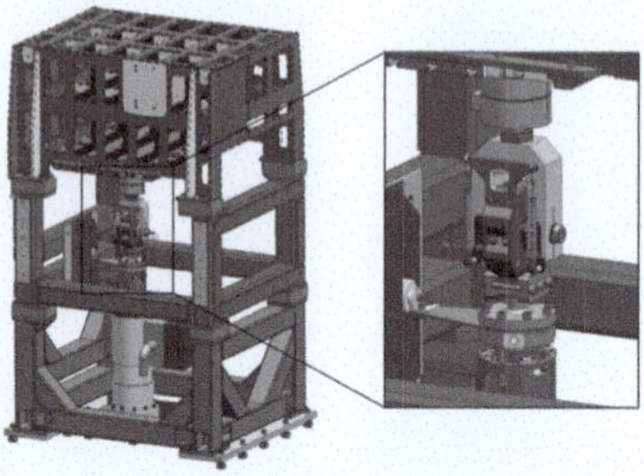

Fig. 4: Test set-up of the measurement

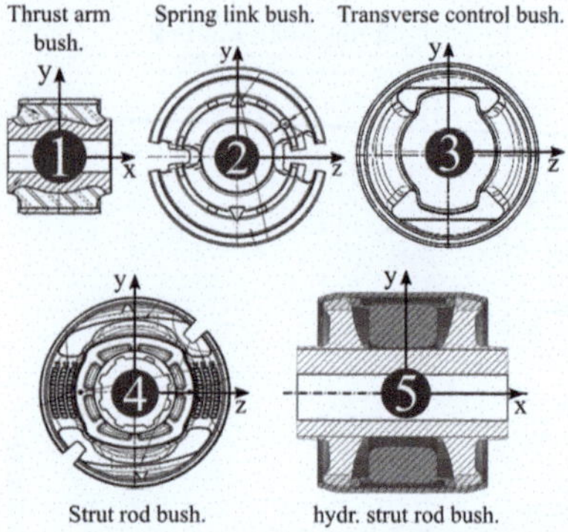

Fig. 5: Investigated elastomer chassis bushings

2.2 Measurement results

From the characterizations of the elastomer bushings in the high load range, important findings can be obtained for the computational determination of sectional loads, which are shown below by means of exemplary measurements of corresponding bushing types.

2.2.1 Softening

Elastomers have the special property to soften in the first load cycles. This phenomenon only occurs with new, previously unloaded specimens and is referred to as the Mullins effect. In addition to Mullins et. al [14], the effect is studied in numerous other papers (Harwood, Mullins, Payne (1966) [7], Ihlemann (2002) [8}, Böl (2005) [2]). Lion (2006) [10] explains the Mullins effect in the destruction of the weakest physical bond in the elastomer. The softening in the first cycles sometimes has less practical background in the virtual fatigue design. The frequently occurring maximum force amplitudes at operating loads quickly lead to a stationary state – the material is no longer softened. However, if the deformation of the elastomer in the chassis bushing is further increased, for example by force peaks caused by special events, a renewed decrease in tension respectively material softening occurs. The extent to which individual, dynamic cycles in the high load range affect the base stiffness of the elastomer bushings is the subject of the characterization. The potential of softening is fundamentally dependent on the distortions occurring in the material. Figures 6 and 7 show the first ten cycles of constant amplitude for two elastomer bushings. The heavily pre-compressed spring link bushing (SLB) loses much of its base stiffness after the first cycle. In contrast, for the Transverse control arm bushing (TCB) respectively pocket bushing no significant Mullins effect can be seen at zero crossing. Only in the progression of the force-displacement curve is a slight decrease in the maximum force amplitude visible. Through the lower free rubber surface, the pretension and the metal inlay, the elastomer is distorted severely in the SLB compared to the TCB, which in turn leads to the corresponding softening (see Sec. 3.2).

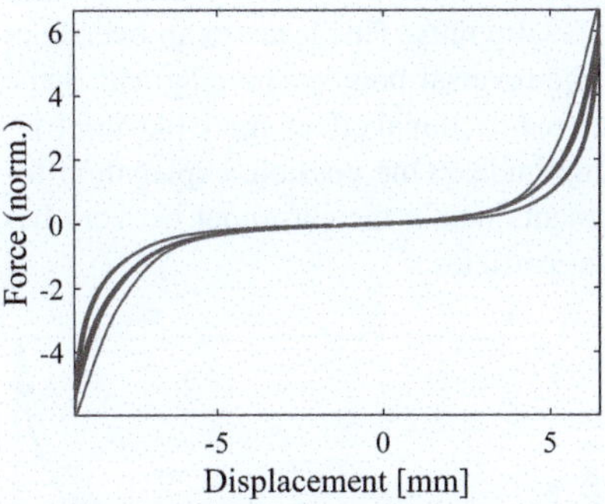

Fig. 6: TCB: Mullins effect

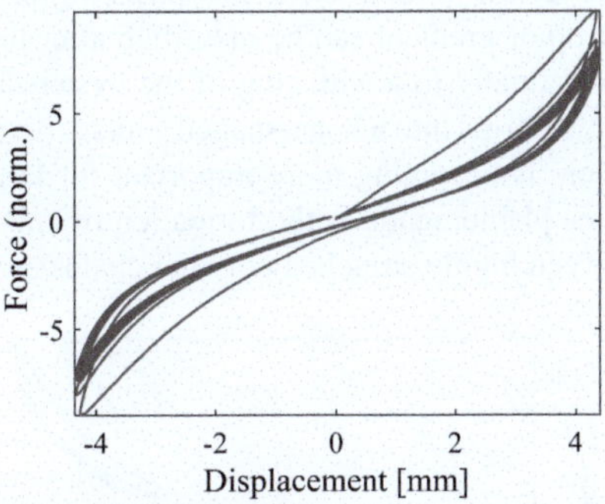

Fig. 7: SLB: Mullins effect

In the literature, the Mullins effect is referred to as a quasi-static phenomenon. However, the load levels induced by special events are initiated dynamically into the chassis (frequencies greater than 15 Hz). For this reason, the softening for different excitation velocities at constant deflection is examined. It turns out that the elastomer bushings soften slightly less as soon as the velocity amplitude of the harmonic excitation increases respectively the time-at-level in the high load ranges decreases. The base stiffness of the SLB is reduced, for example, by 41.1 % after three cycles with $\hat{F} = 75$ kN and $f = 0.1$ Hz, and by 32.8 % at an increased frequency of $f = 20$ Hz. All further investigated elastomer bushings exhibit an analogous behavior, but the effect is less noticeable.

It follows for one thing, that even a small number of cycles in the high load range can lead to a big drop in base stiffness for certain bushings. On the other hand, it is not possible to gather the stiffness conditions in the elastomer bushing after a special event from quasi-static pre-conditioning in the high load range during bushing characterization.

2.2.2 Quasi-static hysteresis

The force-displacement curve of elastomer bushings is characterized in radial direction by a pronounced progressive behavior. Figures 8 and 9 illustrate the force-displacement hysteresis for two types of bushings. In addition to the quasi-static curves, the relaxed force-displacement curves (equilibrium state) are shown, which are obtained by starting a series of hold times in one excitation cycle. The equilibrium curve is amplitude independent and is used to separate the purely elastic force response from the quasi-static hysteresis. It remains the dissipative portion, which is continuously referred to as pure hysteresis (see Figs. 10 and 11). For smaller amplitudes, the pure hysteresis assume an elliptical shape reminiscent of viscous friction behavior. The hysteresis shape of the respective larger amplitudes, in which the progression sets in, shows a clearly divergent behavior. In addition to the stiffening in the force increase path, an increase in force drop or gradient can be identified after the reversal points. This leads to an overshoot and finally to a widening of the hysteresis at the reversal points. The elliptical shape transforms into a bone-shaped course. An additional frictional effect, more like dry friction, is becoming more important for large amplitudes. One possible explanation for this phenomenon is the formation of entanglements within the elastomer network between highly stretched polymer chains and filler particles.

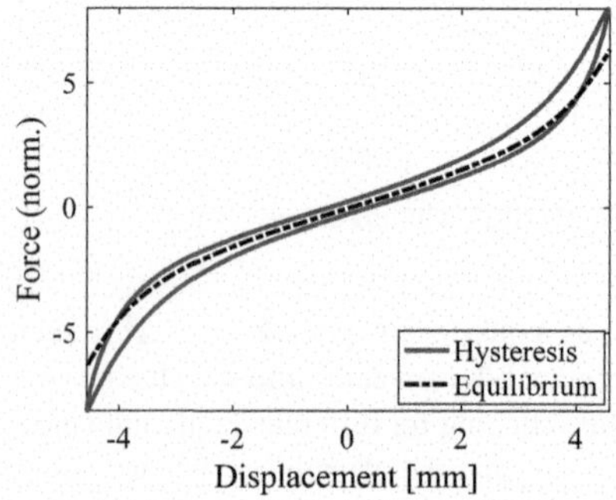

Fig. 8: SLB: hysteresis and equilibrium

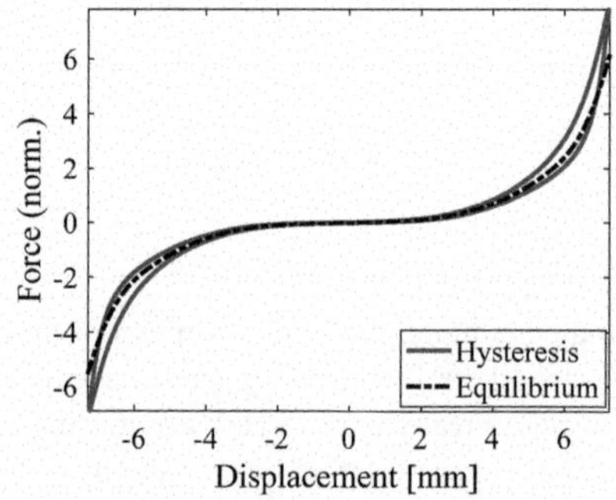

Fig. 9: SRB: hysteresis and equilibrium

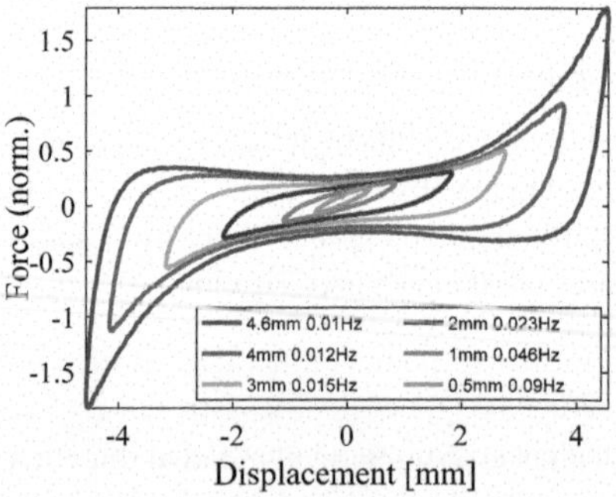

Fig. 10: SLB: pure hysteresis

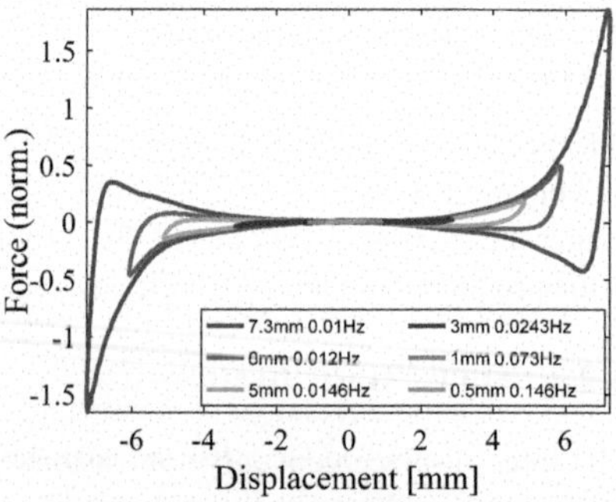

Fig. 11: SRB: pure hysteresis

2.2.3 Dynamic properties

The dynamic properties of the elastomer bushings result from the viscoelasticity of the polymer. They can be illustrated by the scalar parameters dynamic stiffness c_{dyn} and the loss angle φ. Figure 12 shows the determination of the two parameters. The dynamic stiffness is not calculated as per DIN 53513 [3] with an endpoint linearization. Rather, it is determined with a linearization around the zero position with defined sampling points according to equation 1. This procedure does not distort the dynamic stiffness due to the strong progression in the high load range.

$$c_{dyn}(f) = \frac{c_{1/3}+c_{7/9}}{2} \tag{1}$$

The loss angle can be calculated on the one hand, as described in DIN 53513, by means of the hysteresis area (applies only to elliptic hysteresis), on the other hand, by means of a Fourier transformation. The loss angle results in the second case from the difference of the phase angles of the first harmonic components of the force and the displacement to

$$\varphi(f) = \left[\tan^{-1}\left(\frac{\mathrm{Im}(\mathcal{F}(F))}{\mathrm{Re}(\mathcal{F}(F))}\right) - \tan^{-1}\left(\frac{\mathrm{Im}(\mathcal{F}(x))}{\mathrm{Re}(\mathcal{F}(x))}\right)\right]. \tag{2}$$

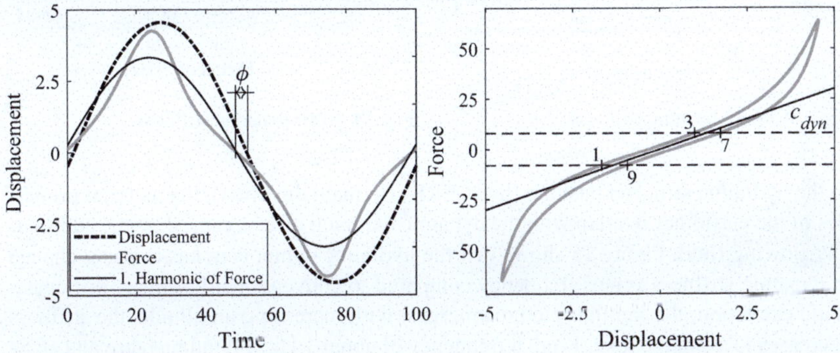

Fig. 12: Definition of dynamic stiffness and loss angle

The formation of the first harmonic loses much of the information content of the originally nonlinear force signal. For this reason, the loss angle is not used in the further course for the assessment of the damping properties, but the loss work. This corresponds to the area enclosed by the hysteresis:

$$W_{damp}(f) = \oint F\mathrm{d}x\,. \tag{3}$$

Dynamic stiffness

The dynamic stiffness of the thrust arm (TAB) and spring link bushing show qualitatively the same characteristics. Over the frequency, the stiffness increases to about 10 Hz degressive, moreover, a smaller, linear increase can be seen (see Fig. 13 and 14). The amplitude dependent behavior indicates an inversely proportional relationship over the measured amplitudes. This relationship is equated with that of pure material behavior. According to Payne [16], the shear modulus decreases significantly in an amplitude range of 0.1 to 10 % strain. He attributes the effect solely to the filler-filler network, but recent works explain the amplitude dependency with the "model of variable net bow density" [12].

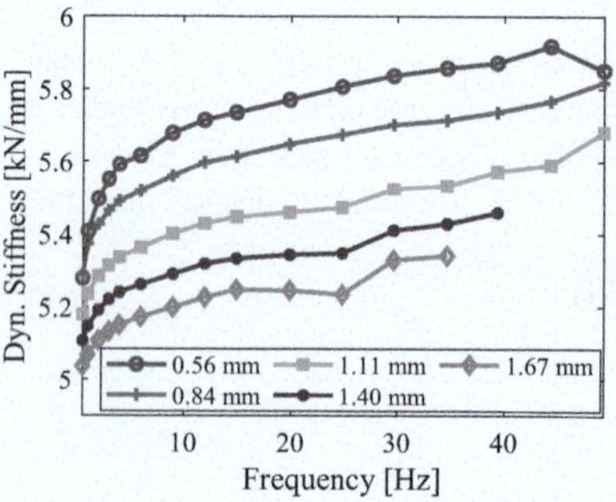

Fig. 13: TAB: dynamic stiffness

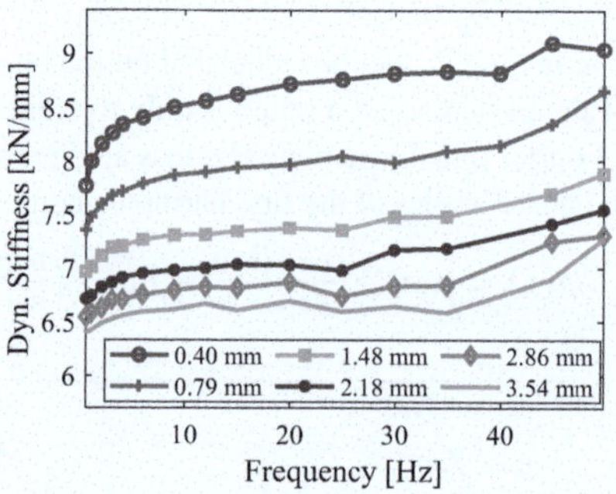

Fig. 14: SLB: dynamic stiffness

In the hydraulic strut rod bushing (hydr. SRB) (in main direction), the material properties of the elastomer overlap with the dynamic damping effects caused by the additionally flowing fluid. Figure 15 shows that the hydraulic component has a strong change in dynamic stiffness versus frequency compared to conventional bushings. For quasi-static excitation, the familiar picture emerges: with increasing amplitude, the stiffness decreases to a small extent. From a frequency of about 1 Hz, a rapid increase and an S-shaped course of the stiffness are observed, whose development decreases with increasing amplitude. The cause of this characteristic is the fluid, which is pumped back and forth between two working chambers through a narrow channel. Duc to the inertia of the fluid mass and the swelling stiffness of the working cambers a vibration absorber effect occurs. Higher frequencies (above the natural frequency of the absorber system) lead to a reduced exchange of the fluid between the chambers. The dynamic stiffness of the bushing then results from a parallel connection of the stiffness of the elastomer body and the swelling stiffness of the working chambers.

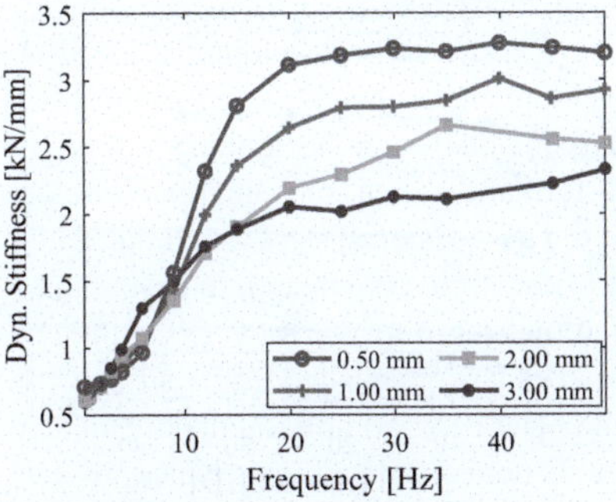

Fig. 15: hydr. SRB: dynamic stiffness

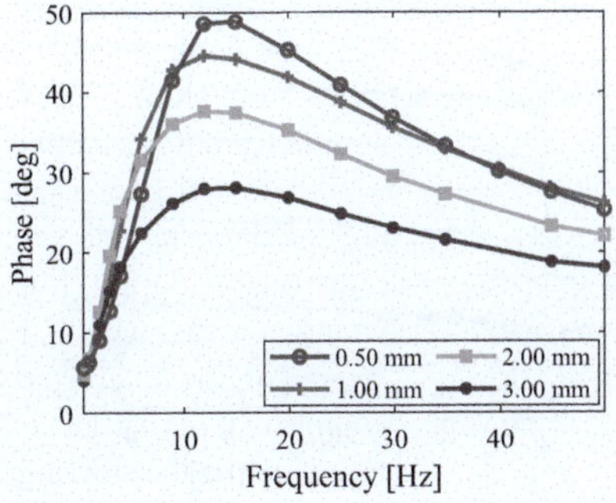

Fig. 16: hydr. SRB: loss angle

The dynamic stiffness depends to a small extent on the temperature. By a temperature increase from 25 to 80 ° C, the stiffnesses for all tested bushings, apart from the hydraulic component, progressively increase between 10 and 20 %. The decisive factor is not the entropy elasticity of the elastomer, but the increased stress due to thermal expansion (see Sec. 3.2). In the hydraulic bushing, the properties of the hydraulic system dominate under temperature influence. A rising temperature involves a slight reduction of the dynamic stiffness for amplitudes with $\hat{x} \leq 1$ mm. The cause lies in the temperature-decreasing dynamic modulus of the filled elastomer, which represents the swelling stiffness (see [9]). The characteristic of the amplitude $\hat{x} = 4$ mm exhibits a pronounced maximum at a frequency of approx. 18 Hz during external heating. It appears that the reduced damping due to viscosity drop would lead to higher amplitudes of the fluid mass in case of resonance.

Loss work

The loss work serves in the further course as the characterization of the damping. Basically, a progressive increase of the loss work over the amplitude is evident (see Fig. 17 and 18), which can be justified in the widening of the hysteresis to larger amplitudes. This behavior is particularly pronounced for pocket bushings in main direction, as comparatively more energy is dissipated by deformation of the bump stop. Over frequency, a degressive increase in loss work by about 40 to 60 % can be observed. The hydraulic strut rod bushing is in turn an exception. Similar to the transfer behavior of a Maxwell-element (see Sec. 4.2), the hydraulic bushing has a maximum loss angle or maximum loss work. The maximum of the loss angle is located in about the frequency point at which the dynamic stiffness undergoes the greatest change (see Fig. 15 and 16). An analogous characteristic applies to the loss work of the hydraulic bushing, with the difference of the maximum value at 20 Hz.

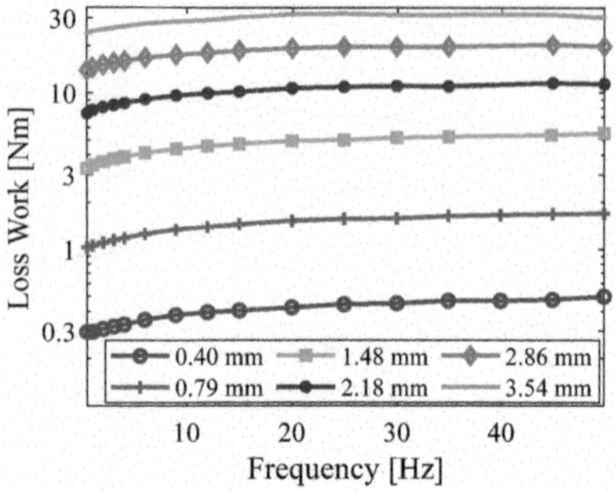

Fig. 17: SLB: loss work

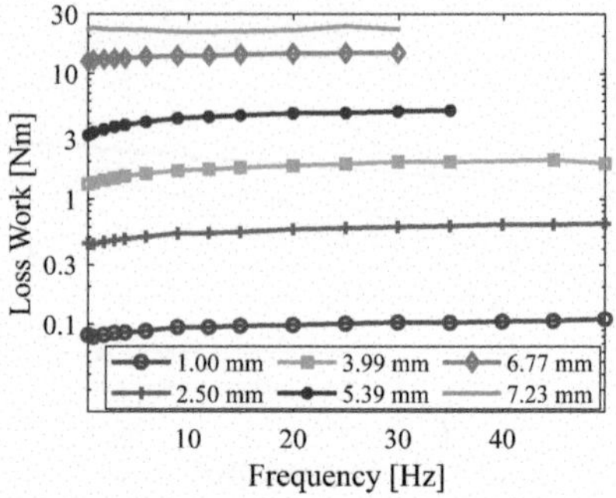

Fig. 18: TCB: loss work

The change in ambient temperature has a relatively strong effect on the loss work. When the bushings are heated from 25 to 80 °C, the loss work of all amplitudes for the pocket bushings decreases by about 30 to 40 %, for the solid rubber mounts by about 40 to 50 %. Again, the hydraulic variant behaves different. Much of the dissipated energy can be attributed to the hydraulic system. The fluid friction is fundamentally dependent on the type of flow. In laminar flow, the frictional force is proportional to viscosity (for Newtonian fluids). In turbulent flows, instead of the substance-dependent viscosity, the turbulence viscosity is taken as the basis. Due to a temperature increase, the viscosity of the fluid reduces, which means that for small excitation amplitudes and frequencies respectively flow rate (laminar flow) the loss work decreases by a small amount (see Fig. 19). For larger excitation velocity amplitudes ($\hat{x} = [2, 3]$ mm, $f > 8$ Hz) a transition from laminar to turbulent flow is apparent and the fluid friction rises above the temperature. This increase in friction is attributable to the turbulence viscosity, which increases as the temperature rises due to increasing molecular movements.

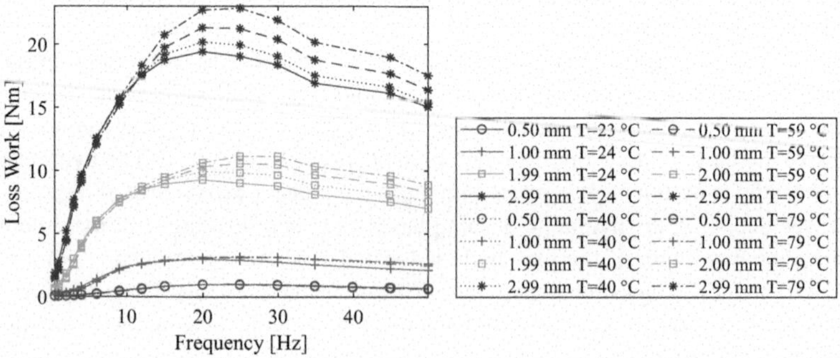

Abb. 19: hydr. ZSL: temperature behavior of loss work

The most important findings from the measurements in the high load range are summarized below:

- Significant decrease in the stiffness of the solid rubber mounts due to the Mullins effect; dependence of the softening on the time-at-level in the high load range
- Increase of the influence of inner friction for large excitation amplitudes
- Low dynamic hardening of the examined elastomer compounds
- Progressive increase in loss work over the amplitude
- Slight increase in dynamic stiffness and large decrease in loss work over temperature

3 Static FE-calculations

By means of static finite element calculations, local strain states, the formation of the progression in bump stop areas, and the effect of rubber dilation due to temperature increases at the elastomeric bushings can be determined. Since a detailed measurement of the material is not available and thus the parameters of the material model can only be estimated, the following calculation results serve for the qualitative assessment of the elastomer bushing behavior.

3.1 Implicit vs. explicit

As part of the work, the influence of the integration method is examined first. For this purpose, quasi-static implicit and explicit calculations (Abaqus) of the thrust arm bushing in radial direction with large deformations are compared. Boundary conditions and the hyperelastic material (*Yeoh*-parameters) remain unchanged. The radially symmetrical component is

- implicit: hexahedral meshing (C3D8H),
- implicit: tetrahedral meshing (C3D4H) using mesh-to-mesh solution mapping,
- explicit: hexahedral meshing 1. order (C3D8R) and
- explicit: tetrahedral meshing 2. order (C3D10M)

calculated. It can be shown that the convergence on implicit scheme can be achieved by an appropriate hexahedral meshing even for larger deformations. However, elastomer bushings also exhibit more complex, asymmetrical geometries, making hexahedral meshing not always readily feasible. The remedy is tetrahedron elements. In order to maintain their stability in the implicit scheme, an adaptive meshing technique has to be used. One is the mesh-to-mesh solution mapping, where the entire simulation history is divided into individual sequences. After each sequence, the deformed geometry is extracted and re-meshed. This is followed by the next sequence as a new calculation job with the newly meshed deformed geometry and an initial stress state, which is taken from the result of the previous sequence. This technique is used for first-order tetrahedral elements. In the explicit integration scheme, no convergence problems can occur, but attention must be paid to the wave propagation speed within the elements. For this reason, the elements of explicit meshes have to be larger in comparison to implicit ones.

Figures 20 and 21 show the force-displacement curves and their derivatives for all variants. It is evident that the explicit calculation with second-order tetrahedra is almost the same as the implicit approach with the first-order hexahedra. A somewhat lower stiffness results for the explicit first-order hexahedral meshing. In the mesh-to-mesh method (Map2Map), discontinuities of stiffness are noted for $x = 0.6$ and 1.2 mm. The re-meshing removes the strong distortions of the old mesh, resulting in a small loss of internal energy.

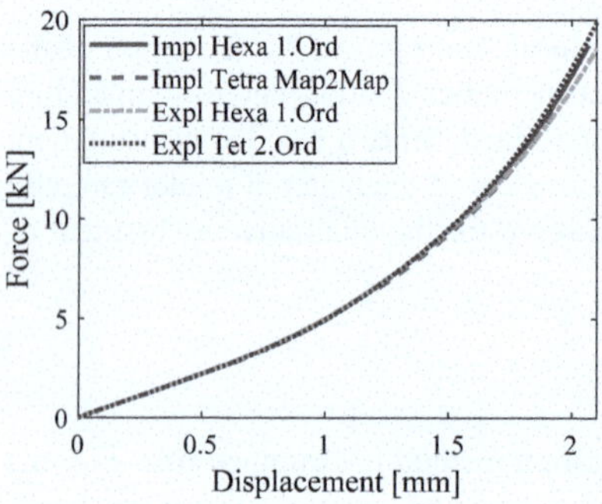

Fig. 20: FEM TAB: force-displacement curve

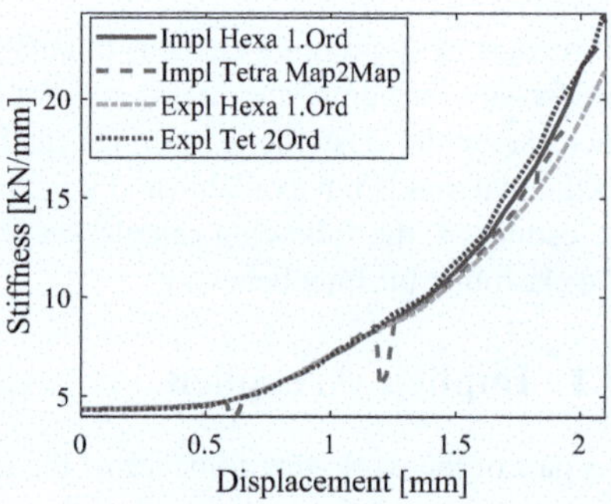

Fig. 21: FEM TAB: stiffness curve

The investigation shows that a negligible difference exists between the calculation types. In the broader context, the explicit method is preferred, as no convergence problems occur. Moreover, with tetrahedral second-order elements, it is possible to link arbitrary geometries and achieve approximately the same results as compared to the implicit.

3.2 Static curve

The calculation of the thrust arm bushing made in the previous section were based on the assumption that a stress-free state prevails in the elastomer in case of a non-deflected bushing core. In the real elastomer bushing, this does not apply. On the one hand, the cooling of the bushing after vulcanisation[1] leads to tensile stresses due to different expansion coefficients of the aluminium parts and the elastomer ($\alpha_{Rubber} \approx 6 \cdot \alpha_{Alu}$). Since these have a negative effect on the lifetime of the elastomer, on the other hand, a

1 Manufacturing process for the formation of sulfur bridges in the elastomer

calibration[2] of the bushing is carried out, which induces compressive stresses in the elastomer. Thus, the initial, real state of stress and strain can only be approximated if the cooling and calibration process is also simulated.

To evaluate the strain state, the equivalent in the form

$$\varepsilon_V = \frac{2}{\sqrt{3}} \sqrt{\frac{1}{6} \left[(\varepsilon_1 - \varepsilon_2)^2 + (\varepsilon_2 - \varepsilon_3)^2 + (\varepsilon_3 - \varepsilon_1)^2\right]} \tag{4}$$

with the principal strains ε_i ($i = 1,2,3$) can be used. In the thrust arm bushing, the deflection of the bushing core by $x = 2$ mm leads in a wide range to equivalent strains in the elastomer of $\varepsilon_V > 1$ (see Fig. 22). The almost incompressible elastomer is pressed by the high deflection on the one hand to the outside. On the other hand, the displaced volume distributed against the loading direction, but is inhibited by the binding to the bushing core and sleeve therein. In contrast, the pocket bushing has lower minimum and maximum strains, since the elastomer can deform unrestrained by the larger free surface. For comparable deflections of the bushing core ($x = 6$ mm), approximately the same strain values result in the bump stop. In the web area, however, only a maximum equivalent strain of $\varepsilon_V \approx 0.65$ is achieved. The lower distortions in the web lead to the relatively small influence of the Mullins effect, which was discussed in Section 2.2.1.

In Section 2.2.3 the influence of temperature on stiffness was discussed. As already mentioned, the effect occurring in the radial loading direction is the result of the thermal expansion. Figure 23 illustrates the change in the maximum principal stresses of the thrust arm bushing at a temperature increase from 25 to 80 °C. Only the expansion coefficient is considered (no temperature dependence of the σ/ε-relationship of the elastomer). In addition to the increased stresses, it can be seen that the cross-sectional area of the rubber contour increases and the free space between the bearing sleeve and core is closed by the additional material volume. This results in the force-displacement curve shown in Figure 24 with higher stiffness around the zero position and earlier onset of the progression.

2 Diameter reduction of the bushing sleeve

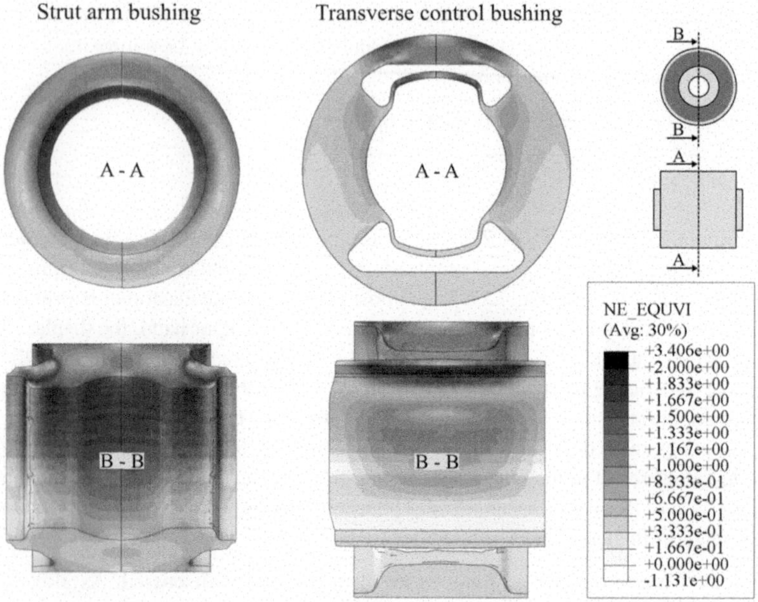

Fig. 22: FEM TAB and TCB: contour-plot (undeformed) of the equivalent strain ε_V for comparable deflections

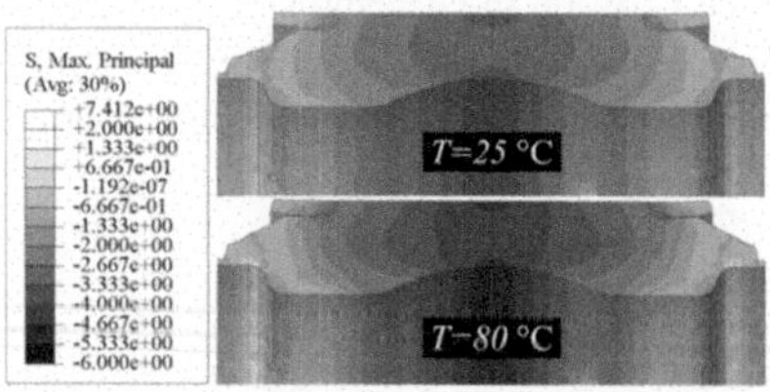

Fig. 23: FEM TAB: max. principal stresses (cross section of the rubber contour)

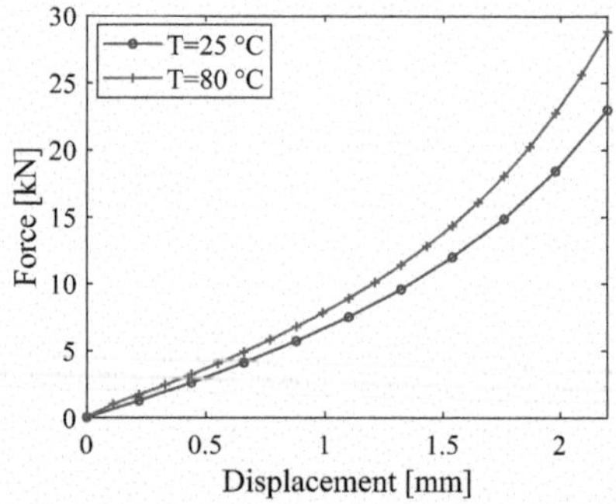

Fig. 24: FEM TAB: force-displacement curve

The pure elastomer shows a contrary behavior – with increasing temperature the stiffness decreases. However, as can be seen from the measurements of this work and, for example, that of Koch [9], the effect of thermal expansion dominates in radial direction.

4 Benchmark-analysis of elastomer bushing modeling

In the past decades, various modeling methods were developed to approximate the real elastomer bushing behavior for MBS simulation. The phenomenological, rheological model approaches represent a tried and tested remedy. In the following, the suitability of different elastomer bushing models for application in the high load range is examined by means of a benchmark analysis. The focus is on the pocket bushing as a representative of a highly nonlinear elastomer bushing.

4.1 Examined models

The benchmark compares the following models:

Tab. 1: Investigated phenomenological model approaches

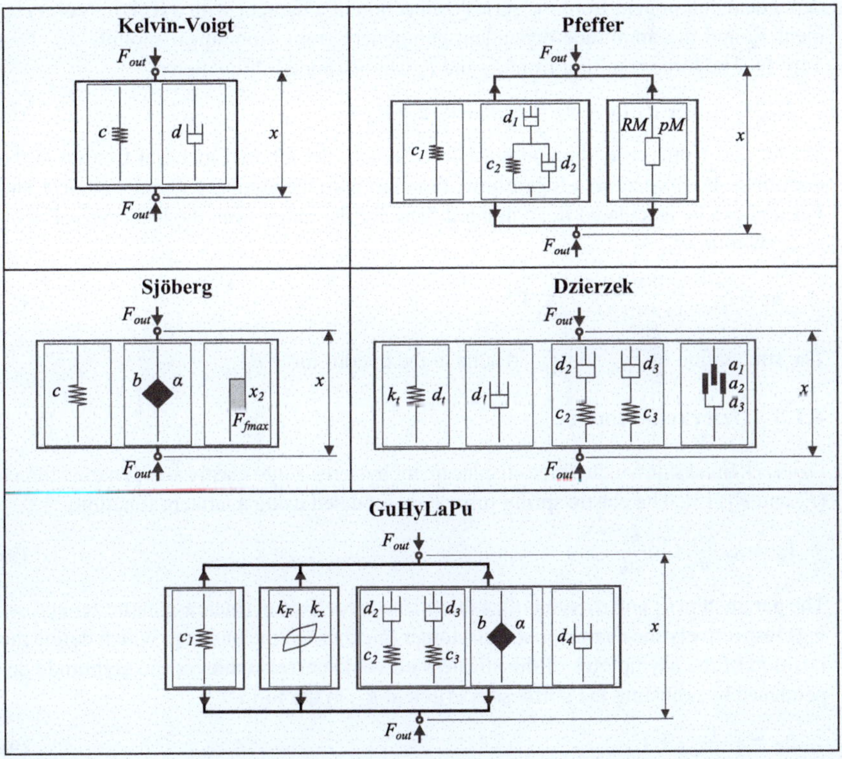

4.1.1 Kelvin-Voigt

The Kelvin-Voigt element is widely used in practical applications. The parallel combination of linear or non-linear spring and damper approximates the dynamic stiffness and loss angle as a linear relationship over the frequency. The model is not able to map the amplitude dependence. Furthermore, the loss angle starts from zero. The Kelvin-Voigt element must therefore be set to an operating point (frequency and amplitude), which provides only a rough approximation to the real system behavior.

$$F_{out} = c\,x + d\,\dot{x} \tag{5}$$

4.1.2 Pfeffer-Model

A much more complex construction is shown by the model by Pfeffer and Hofer [17]. The elastic component of the elastomer results from the spring force F_e, the frequency-dependent component from the series connection of a damper with a Kelvin-Voigt element F_z and the amplitude-dependent component from a nonlinear module F_{NL} (see Tab. 1). The force resulting from F_e and F_z can be found in Equation 6.

$$F_{lin} = c_1\,x + d_1\,(\dot{x} - \dot{z}) = c_1\,x + c_2\,z + d_2\,\dot{z} \tag{6}$$

Pfeffer and Hofer refer the nonlinear force F_{NL} to the internal material friction in the elastomer. It is based on a logarithmic function with the auxiliary parameters RM and pM and can be calculated in an integral form for the range from the last reversal point x_{Um_j} as follows:

$$F_{NL} = \int_{t=0}^{t\prime} \frac{RM\,pM\,\dot{x}(t)}{1+pM\left|x(t)-x_{Um_j}\right|}\,\mathrm{d}t\,. \tag{7}$$

The summation of F_{lin} and F_{NL} results in the overall force F_{out}.

4.1.3 Dzierzek-Model

Dzierzek breaks down the force-displacement hysteresis into elastic and dynamic forces (F_e and F_d) [5]. The elastic spring force F_e is modeled using a tangent function:

$$F_e = k_t \frac{2d_t}{\pi} \tan \frac{\pi x}{2d_t}\,. \tag{8}$$

The parameters k_t and d_t symbolize the stiffness coefficient (increase in the zero crossing) respectively the characteristic elastomer thickness of the bushing, which define the position of the asymptotes of the stiffness curve. Dzierzek achieves the amplitude dependence by adjusting the stiffness coefficient k_t, in the form

$$k_t = \frac{p}{x_{max}+q} + r\,. \tag{9}$$

The coefficient k_t is thus dependent on the maximum excitation amplitude x_{max} and three other subparameters.

The dynamic force F_d is formed by the summation of a friction force F_f and a purely frequency-dependent force component, which in turn results from a parallel connection of a linear damper and two Maxwell elements[3]. The amplitude response of a Maxwell element increases over the frequency and has a point of inflection at the position $f = 1/(2\pi\,\tau)$ with the time constant $\tau = d_{MW}/c_{MW}$. At the same frequency point, the phase response increases to a maximum value and decays again for higher frequencies. In order to reduce the phase drop and to approximate the real component behavior for a larger frequency range, two Maxwell elements with different time constants are connected in parallel. An additional linear damper also counteracts this problem.

The frictional force F_f maps the hysteresis expansion at the reversal points for radial deflections and is also dependent on the profile of excitation velocity (see Eq. 10).

$$F_f = \left[a_1\left(k_t\frac{2d_t}{\pi}\tan\frac{\pi x}{2d_t} - k_t x\right)^{a_2} + a_3\right]\frac{\dot{x}}{\sqrt{|\dot{x}^2 - x\,\ddot{x}|}} \qquad (10)$$

4.1.4 Sjöberg-Model

The modular model approach according to Sjöberg [19] includes three components: the spring force (analogous to Kelvin-Voigt), the force of an element with a fractional derivative and the frictional force. As explained in the previous section, the loss angle of the Maxwell element has a maximum. Although the parallel connection of several Maxwell elements leads to an improvement in the image quality, but also to a sharp increase in model complexity and computing time. So-called Spring Pot elements with fractional derivatives provide a way to circumvent the limiting properties of Maxwell elements. The constitutive equation for the Spring Pot module is

$$F_{SP} = b\,D^{\alpha}x(t) \qquad (11)$$

with the fractional derivative $D^{\alpha}x(t)$ of the order α $(0 \le \alpha \le 1)$ and the fractional damping parameter b. For $\alpha = 0$ an ideal elastic behavior results, for $\alpha = 1$ an ideal viscous behavior. A Spring Pot element thus represents a mixture between a spring and a damper. For detailed information on fractional derivatives, reference is made to [15] and the numerical implementation can be seen [19].

In order to depict the friction behavior and thus the Payne effect, Sjöberg uses the approach from Berg [1]. The upper hysteresis curve results from equation 12 and the lower one from equation 13.

3 Series connection of a spring and a damper

$$F_f = F_{fs} + \frac{x - x_s}{x_2(1-\mu)+(x-x_s)}(F_{fmax} - F_{fs}) \quad \text{for } \dot{x}(t) > 0 \tag{12}$$

$$F_f = F_{fs} + \frac{x - x_s}{x_2(1+\mu)-(x-x_s)}(F_{fmax} + F_{fs}) \quad \text{for } \dot{x}(t) < 0 \tag{13}$$

with $= F_{fs}/F_{fmax}$.

The parameter F_{fmax} represents the maximum frictional force and x_2 the deflection needed to reach half the maximum frictional force. F_{fs} and x_s represent the starting point for the respective hysteresis curve and must be updated each time the deflection is reversed. Figure 25 illustrates the operation of the friction module using force-displacement hysteresis.

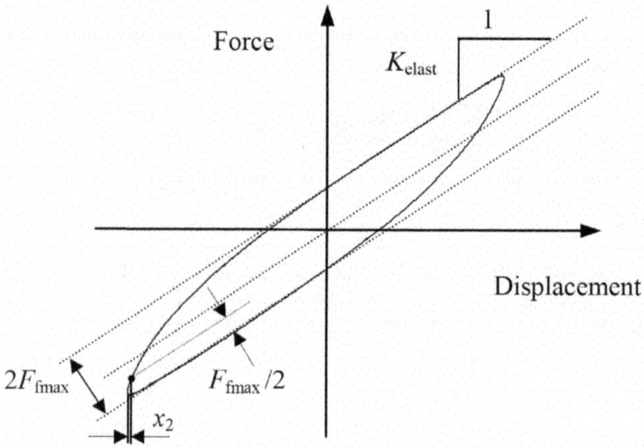

Abb. 25: Modeling of hysteresis according to Berg {Sjöberg 2000 #15}

4.1.5 GuHyLaPu

The GuHyLaPu is a phenomenological, modulary model developed by Daimler AG (see [13], [11], [4], [18]). It is divided into three branches analogous to Sjöberg: the first branch reflects the static stiffness, the second one the static friction and amplitude dependence, and the third one the frequency-dependent properties. The latter branch differs from the Sjöberg model in that you can choose between a Spring Pot or two Maxwell elements (here: spring pot activated). On the other hand, always an additional, linear damper is active. The friction model is also based on the algebraic approach of Berg, but differs as follows [13]:

$$F_f = F_{fs}^l + k_F \frac{2(x - x_s^l)}{2k_x + (x - x_s^l)} \quad \text{for } \dot{x}(t) > 0 \tag{14}$$

$$F_f = F_{fs}^u + k_F \frac{2(x - x_s^u)}{2k_x - (x - x_s^u)} \quad \text{for } \dot{x}(t) < 0 \tag{15}$$

The slope after the inversion points ($[x_s^l, F_{fs}^l]$, $[x_s^u, F_{fs}^u]$) is the ratio of the auxiliary parameters k_F to k_x. In order to map the rubber memory with the equations 14 and 15, a state machine is additionally required, which sets the last reversal points in response to the opening or closing of hysteresis loops.

The functionalities of the friction formulations according to Berg (see Eqs. 12, 13) and according to Meljnikov (see Eqs. 14, 15) are illustrated in Figure 26. Within a large hysteresis loop 1-2, a smaller one is opened by the deflection from 3 to 4. When closing hysteresis loop 3-4-3 in direction 2, differences between the friction models are now indicated. The Meljnikov model would take the dashed path with deactivated state machine, with an activated one the course from 3 to 2 and thus the course of the real elastomer (rubber memory). The dotted line illustrates the force path of Berg's model. In this case, the point 3 is not closed exactly and the old force path 3-2 is not reactivated, since on the one hand the approach function is scaled down shallower depending on the position of the reversal point and on the other hand no state machine exists.

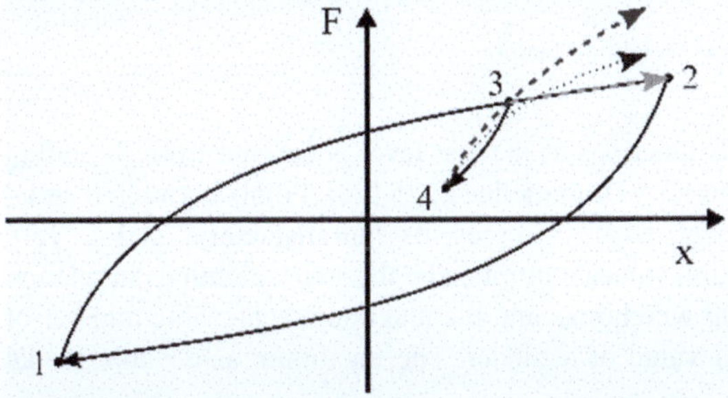

Fig. 26: Behavior of the friction models according to Berg (dotted line) and Meljnikov without and with state machine (dashed line: red respectively green)

Furthermore, the GuHyLaPu has a scaling function for the hysteresis force, which ensures the hysteresis widening at the reversal points (see Fig. 27).

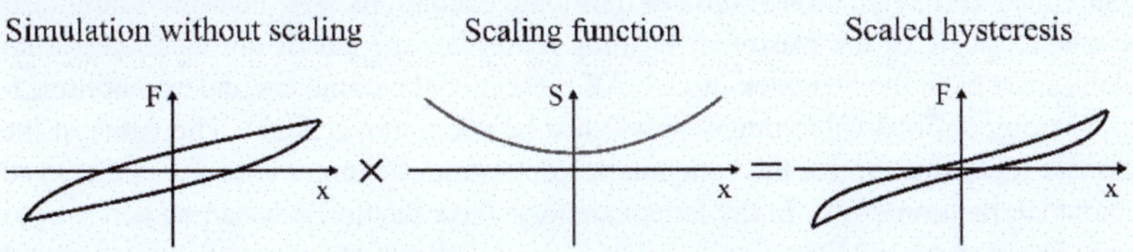

Fig. 27: Scaling function of GuHyLaPu

4.2 Utility value analysis

In the form of a utility value analysis, the suitability of the existing elastomer bushing models with respect to the use in the MBS simulation can be assessed quantitatively. The models are examined according to the criteria listed in Figure 28, whereby the value of a criterion results from a stepwise abstraction. The node weighting G_K indicates which portion of the respective subcategory is assigned to the superordinate one. The sum of these gives the value 1. By forming the product of the node weights along a path, the step weight G_S is finally determined. At one level, these add up to the value 1.

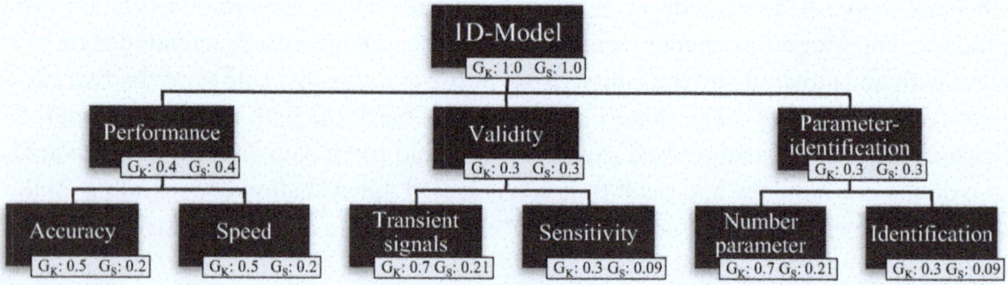

Fig. 28: Outline of the benchmark and weighting factors

A quantitative value for the sub-functions at the last level is derived from the rating scale of technical value according to VDI guideline 2225 [20]. Points between 0 and 4 are awarded for goal achievement, where 0 represents "unsatisfactory" and 4 "very good". The sum of the sub-function values multiplied by the step weightings in relation to the product of the summed up weightings and the maximum achievable number of points finally provides the total value of a model. The maximum achievable model value is thus 1.

In order to evaluate the performance and validity, the elastomer bushing model must be parameterized. In order to compare the pure modeling quality independently of the respectively recommended parameterization routines, the model parameters are determined via a global optimization method (Genetic Algorithm). The parameter identification takes place via the minimization of a relative, quadratic error deviation RQF of measurement and simulation results of harmonic excitations. The measured nonlinear base characteristic of the elastomer bushing represents the elastic spring force for all models, apart from the Dzierzek model. All other model parameters undergo optimization, whereby defined value range limits must be taken into account. The value of the parameter identification for the benchmark results from the number of parameters and the parameterization effort. In the latter category, the valuation is based on soft factors that are based on VDI 2225.

The performance is subdivided on the one hand into the accuracy, which refers primarily to the relative error deviation of the dynamic stiffness and loss work. On the other hand, the simulation time influences the value of the performance. The simulation results of the model according to Sjöberg and the GuHyLaPu model for the pocket bushing can be taken from illustrations 29 and 30 by way of example.

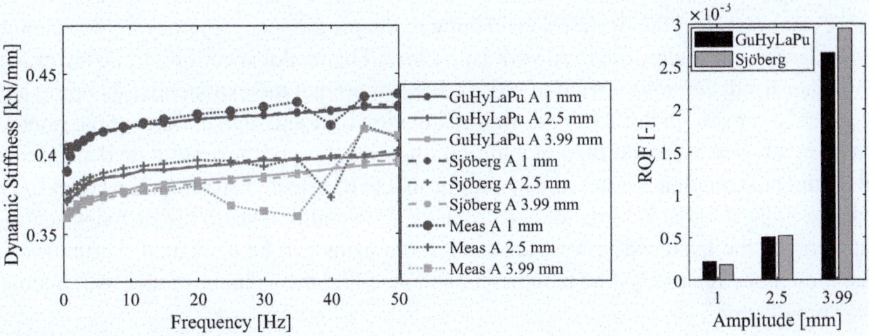

Fig. 29: Comparison of the Sjöberg and GuHyLaPu model: dyn. stiffness over frequency (left) and *RQF* over amplitude (right)

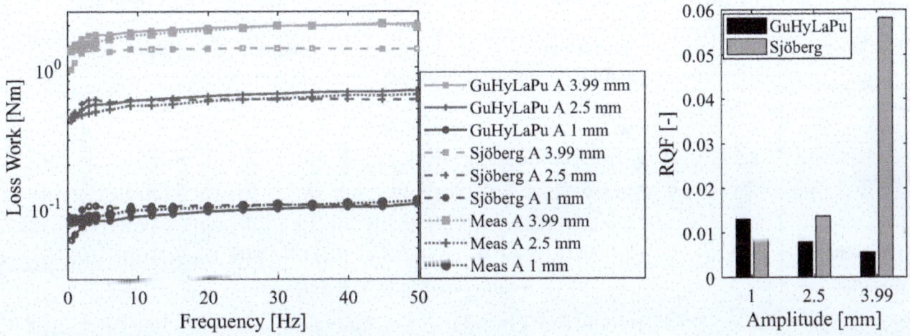

Fig. 30: Comparison of the Sjöberg and GuHyLaPu model: loss work over frequency (left) and *RQF* over amplitude (right)

The validity includes the sensitivity of the models, which is analyzed using the "one-factor-at-a-time" (OFAT) method and is also rated with soft factors. The second sub-function of validity is the simulation quality for transient excitation signals (special events).

For the transfer behavior of the described elastomer bushing models in the load range of operational and special events, the benchmark leads to the result summarized in Table 2. Among all models, the Kelvin-Voigt element has the lowest performance (apart from the simulation time) and validity. The reason for this, as mentioned in Section 4.1.1, is a linear frequency dependence and not an amplitude dependency. A positive factor is the very low parameterization effort. The Pfeffer and Sjöberg models provide good results for small to moderate excitation amplitudes. Likewise, in the user friendliness they show passable properties. In the simulation of excitation amplitudes in the high load range, however, there are deficits. The friction models of Pfeffer and Berg can not reflect the hysteresis widening in the progression, whereby at correspondingly large amplitudes, the loss work is too low. The model according to Dzierzek, on the other hand, contains a friction model that can predict the exponential damping and thus the loss work. In this case, the high calculation time and high number of parameters have an adverse effect on this. In addition, the modeling of the amplitude dependence according to equation 9 is not directly applicable to transient excitation. The GuHyLaPu has the highest value in terms of performance and validity. Due to the included scaling function of the hysteresis curve, arbitrary force paths can be generated. Furthermore, the model has relatively few parameters and acts like the Sjöberg model with a small real-time factor.

For use in MBS simulation for durability, the GuHyLaPu emerges from the elastomer bushing models tested in this work as the most performant and highest quality model.

Tab. 2: Result of the benchmark analysis for the transverse control arm bushing

Benchmark	**Performance**					**Validity**		**Param.-Ident**		
	Accuracy				Speed					
Nr.	1	2	3	4	5	6	7	8	9	Σ
Factor	RQF total	Variance Ampl.	Variance Freq.	Quality Hys.	Simulation time	Transient signal	Sensitivity	Number Param.	Expenditure	
Gewichtung	0.1	0.04	0.02	0.04	0.2	0.21	0.09	0.21	0.09	1
Model	Value according to VDI 2225									W
Kelvin Voigt	0	0	0	0	4	2	3	4	4	0.67
Pfeffer	3	4	3	1	3	2	2	3	2	0.64
Sjöberg	3	4	4	1	4	2	3	4	3	0.80
Dzierzek	2	2	2	3	2	2	3	1	2	0.48
GuHyLaPu	4	4	4	4	4	4	3	3	2	0.88

5 Outlook

The benchmark analysis shows that most of the models perform well in the small-signal range, but that the real force-displacement hysteresis in the high load range can not be adequately reflected. Only the Dzierzek and GuHyLaPu models have model approaches that predict the loss work more accurately. Due to the disadvantages of the Dzierzek model discussed in the previous section, it can not be used for the calculation of special loads without adjustments. The GuHyLaPu also has a significant disadvantage. This concerns the parameterization of the scaling function, which is only possible if a quasi-static measurement of the force-displacement curve exists in the high load range. In early and later developmental phases, however, such measurements are usually not available. For this reason, the friction models have to be adapted or further developed in order to increase the simulation quality under special load conditions and thus also the quality of the prediction of the MBS vehicle calculations.

Literature

[1] Berg, M. 1998. A Non-Linear Rubber Spring Model for Rail Vehicle Dynamics Analysis. *Vehicle System Dynamics* 30, 3-4, 197–212.

[2] Böl, M. 2005. *Numerische Simulation von Polymermetzwerken mit Hilfe der Finite-Elemente-Methode*. Zugl.: Bochum, Ruhr-Univ., Diss., 2005. Ruhr-Universität, Bochum.

[3] DIN Deutsches Institut für Normung e.V. 1990. *Bestimmung der visko-elastischen Eigenschaften von Elastomeren*. Beuth Verlag, Berlin, DIN 53 513.

[4] Dronka, S. and Rauh, J. 2007. *Modell zur Simulation von Fahrwerks-Gummilagern. Eigenschaften – Modellstruktur – Parametrierung*. VDI Berichte 2003.

[5] Dzierzek, S. 2000. Experiment-Based Modeling of Cylindrical Rubber Bushings for the Simulation of Wheel Suspension Dynamic Behavior. In . SAE Technical Paper Series. SAE International400 Commonwealth Drive, Warrendale, PA, United States.

[6] Giron-Sierra, J. M. 2017. *Digital Signal Processing with Matlab Examples, Volume 1*. Springer Singapore, Singapore.

[7] Harwood, J. A. C., Mullins, L., and Payne, A. R. 1966. Stress Softening in Natural Rubber Vulcanizates. Part II. Stress Softening Effects in Pure Gum and Filler Loaded Rubbers. *Rubber Chemistry and Technology* 39, 4, 814–822.

[8] Ihlemann, J. 2003. *Kontinuumsmechanische Nachbildung hochbelasteter technischer Gummiwerkstoffe*. Zugl.: Hannover, Univ., Diss., 2002. VDI-Verl., Düsseldorf.

[9] Koch, P., Angrick, C., Beitelschmidt, D., Prokop, G., and Knauer, P. 2015. Influence of Rubber Temperature on Transfer Functions of Bushings. *SAE Int. J. Passeng. Cars – Mech. Syst.* 8, 4.

[10] Lion, A. 2006. *Dynamische Modellierung von elastomeren Bauteilen*.

[11] Magnus Lahti. 2003. *Modelling Rubber Mounts. Using Non-linear Elastic Force, Friction and Fractional Derivatives*. Master thesis.

[12] Maier, P. and Göritz, D. 1996. Molecular interpretation of the Payne effect. *Kautschuk Gummi kunstoffe,* Vol. 49, 18–21.

[13] Meljnikov, D. 2003. *Ein praxisgerechtes 1D-Gummilagermodell für die Fahrzeugdynamiksimulation*. Presentation at the HdT-Essen-Seminar, München.

[14] Mullins, L. 1969. Softening of Rubber by Deformation. *Rubber Chemistry and Technology* 42, 1, 339–362.

[15] Oldham, K. B. and Spanier, J. 2010. *The fractional calculus. Theory and applications of differentiation and integration to arbitrary order*. Mathematics in science and engineering v. 111. Academic Press, New York.

[16] Payne, A. R. 1962. The dynamic properties of carbon black-loaded natural rubber vulcanizates. Part I. *J. Appl. Polym. Sci.* 6, 19, 57–63.

[17] Pfeffer, P. and Hofer, K. 2002. Einfaches nichtlineares Modell für Elastomer und Hydrolager zur Optimierung der Gesamtfahrzeug-Simulation. *ATZ Automobiltech Z* 104, 5, 442–451.

[18] Sedlaczek, K., Dronka, S., and Rauh, J. 2010. Advanced modular modelling of rubber bushings for vehicle simulations. *Vehicle System Dynamics* 49, 5, 741–759.

[19] Sjöberg, M. 2000. Rubber Isolators – Measurements and Modelling using Fractional Derivatives and Friction. Ph.D. thesis. In . SAE Technical Paper Series. SAE International400 Commonwealth Drive, Warrendale, PA, United States.

[20] VDI Verein Deutscher Ingenieure. 1998. *Technisch-wirtschaftliches Konstruieren* 03.100.40. Beuth Verlag, Berlin 03.100.40, VDI 2225.

Enhanced experimental suspension examination

Maximilian Georg Reisner
Chair of Automotive Engineering, Technical University of Dresden

Helmut Krome
Chassis development, AUDI AG

Prof. Dr-Ing. Günther Prokop
Chair of Automotive Engineering, Technical University of Dresden

P. Pfeffer (Ed.), *9th International Munich Chassis Symposium 2018*, Proceedings,
https://doi.org/10.1007/978-3-658-22050-1_28

1 Introduction

The fundamentals of the validation process of suspension models are primary based on experimental kinematics and compliance testing of suspension systems. However the suspension system behaviour looks like a black-box conditioned by the application of forces and the synchronous capture of the wheel motions. This happens by gathering only the input and output quantities. An examination on component level is necessary for an adequate breakdown of the subsystem characteristics.

As part of a research project experimental tests are performed with a suspension system to get an enhanced event chain knowledge between the suspension components and the whole subsystem. A five-link suspension system on a kinematics and compliance test rig (KNC) serves as test object. Besides the conventional KNC-testing the coordinates of discrete points on the suspension system are measured with the help of an optical measurement device. Several decomposed load excitations are applied on the system.

The analysis of kinematic quantities, such as position and orientation of the imaginary steering axis and the measurement of displacements in bushings and the local deformations of single metal suspension parts are possible by means of this approach. The examination enables to derive the compliance influence of different parts on the elastokinematic behaviour. In this way, flexible multibody systems (MBS) can be validated on component level and differences between the simulation and test rig results can be understood. This procedure enables an improved model validation process.

2 Conventional KNC testing

The elastokinematic test rig is one of the most important testing facility in the chassis development regarding vehicle dynamics. It is possible to gather the system characteristics of whole suspension systems. The data, which is captured with the KNC test rig, serves as an insurance for the defined suspension target values, as parameterization of characteristic line and field based models for vehicle dynamics and ride comfort, as validation of multibody systems as well as the examination of kinematics and compliance concepts of competitors.

The functional principle of a KNC test rig is based on the application of loads and deflections on suspension systems during capturing the deformations and reaction forces. The six degrees of freedom of the wheels are the primary deformation quantities. These are gathered by a mechanical based wheel motion sensor arm. Alternatively this could be superseded by an optical measurement system ([7], pp. 323 – 324). All four wheels of a whole vehicle can be simultaneously measured in most cases of test rig design. The vehicle body is usually fixed on four points of the rocker rail. In this way the local attachment stiffnesses of the bodywork are also included during the testing. If only the chassis characteristics are of interest, a rigid construction can substitute the flexible bodywork. It is also possible to separate the influence of the rim and tire by replacing the wheel through a rigid spindle adapter. This is an often used approach to distinguish the characteristics of the wheels and the suspension system. The way of excitation of the test rig is another distinctive feature. Besides the quasistatic KNC test rigs (see Figure 1 left-hand side) there are also test systems, which enable the examination of dynamic suspension characteristics (see Figure 1 right-hand side). A comparison of the two types of KNC test rigs is described in [9].

Figure 1: Quasistatic (left-hand side) [1] and dynamic (right-hand side) KNC test rig [8]

3 Enhanced KNC testing

Also internal quantities of the systems are often of prime importance apart from the global suspension characteristics. The travel range of actuators or the forces of the car chassis to the bodywork are two possible relevant quantities. In this way additional measurement equipment such as thread potentiometers, inductive displacement sensors or strain gauges can be necessary. This supplementary channels are often directly synchronized with the KNC test rig. The desire for a better knowledge about the internal suspension cause-effect relationships leads to considerations regarding an enhanced KNC measurement with the usage of an optical measurement system.

3.1 Optical measurement as a tool in the chassis development

Optical measurement techniques are nowadays an integral part of the chassis development. Different problem types are solved with the approach of photonic in almost every department: Thermography cameras are used in the tire development to identify the heating of a tire. Also for aerodynamic testing in the wheel development optical flow measurement techniques serve as tools. Mechanical strains, stresses and damages of chassis components are analysed in the department of strength. Up to the serial development and the manufacture different approaches of photogrammetry are used for shape identification to capture deviations and deflections of components, such as the subframe. Also in the department of vehicle dynamics and ride comfort serve these systems for examinations of mechanical vibrations as well as displacement and velocity measurements [5].

In the following study it should be analysed, whether the usage of optical measurement equipment is useful for the motion and deformation examination of chassis components on a KNC test rig. The digital image correlation is chosen as approach. Different patterns on surfaces of the measurement object are tracked over time within this technique. At least two optical sensors and an illumination unit are necessary for this method. The pattern could be fixed on the measurement object in the form of a stochastic point pattern, the so called speckle pattern, or also adhesive bonded measurement markers. The stochastic point pattern is used primary for surface examinations. The assignment of a single pattern to the respective picture, for example, takes place with the help of greyscale correlation. The stereo camera principle enables the detection of the coordinates in space.

3.2 Description of the test setup

A five-link rear suspension of an AUDI Q7 (model year 2018) serves as test object. This suspension system is examined separately and is integrated in a rigid bodywork construction, which is shown in Figure 2 on the left-hand side. The attachment of the wheel hub on the test rig is realized through a rigid spindle adapter. The motion of the wheel hub is metrological detected by a mechanical wheel motion sensor arm. The occurring forces and torques are gathered by a load cell, which is integrated in the excitation device, located under the wheel contact point.

Only one half of the suspension system is focused because the construction of the chassis is considered as almost symmetric regarding to the median longitudinal plane of the vehicle. Figure 2 shows on the right-hand side an optical measurement system called ARAMIS of the company GOM, which serves as measurement equipment. The measurement principle is based on the digital image correlation. At first it is looked for an optimal installation position of the system. As many as possible chassis components should be included in the examination. The measurement system is positioned under the chassis with the view from the inside between the two excitation devices of the KNC test rig. Reflective measurement markers serve as pattern, which are fixed on the test objects. Some components are hidden through others. To serve this, small metal brackets are adhesive bonded on the chassis components. Through these extended lever arms also data of the hidden components could be acquired. In addition, measurement markers are fixed on the excitation device of the test rig because the motions of the test rig actuators are of interest. The measurement markers of the rigid construction serve as reference system. At least three measurement markers at each body are necessary to gather the rigid body motion. An appropriate number of markers should be chosen for examinations regarding to component deformations.

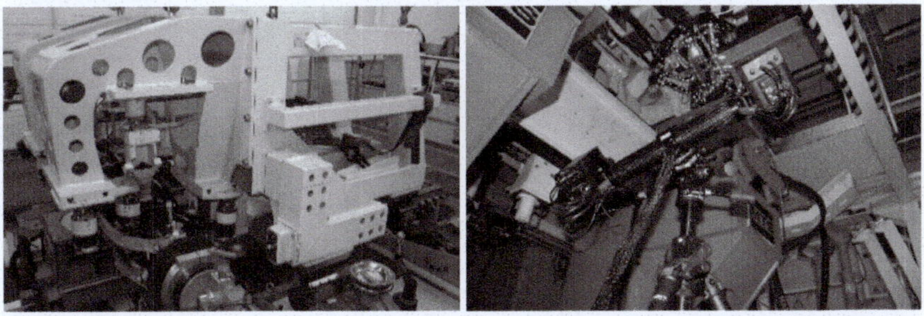

Figure 2: Chassis fixed to rigid construction (left-hand side) and optical measurement system (right-hand side)

The defined test rig manoeuvres for the examination are captured in Table 1. Force and displacement controlled vertical excitations are used for the analysis of the kinematic behaviour of the chassis. The separate application of longitudinal forces, lateral forces and toe torques provides information about the elasticity of the chassis system. Furthermore a lateral force with the pneumatic trail is reproduced through the superposition of a toe torque and a lateral force. Also opposite acting forces are applied because there is a direct coupling between the wheels through the stabilizer bar and the flexible mounted subframe. This enables the separation of interactions between wheels and single behaviour of one suspension half.

Table 1: Test rig manoeuvres

Vertical excitation	Lateral force F_y	Lateral force and toe torque $F_y - M_z$	Toe torque M_z	Longitudinal forces $-F_x$
s_z 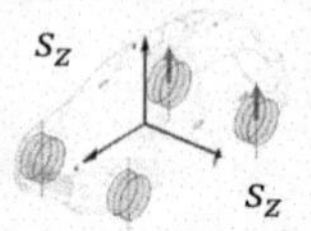 s_z	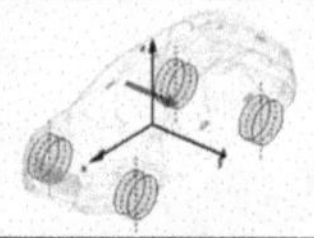	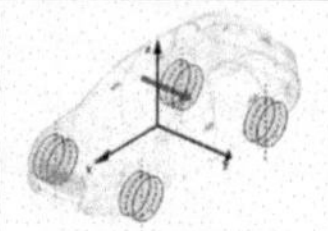	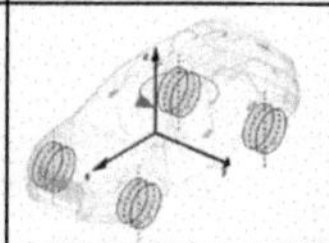	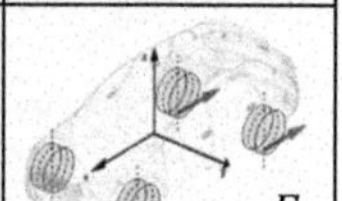 $-F_x$
F_z 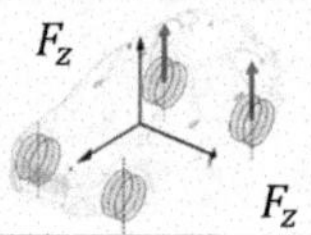 F_z	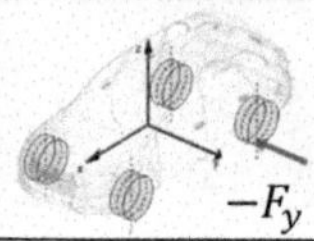 $-F_y$	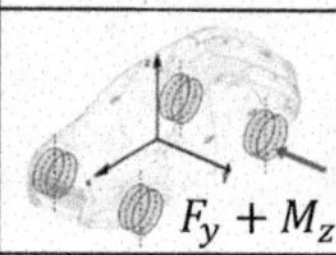 $F_y + M_z$	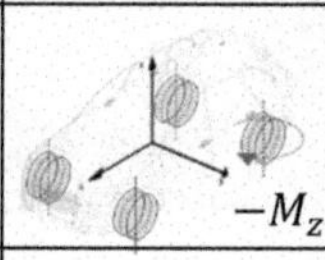 $-M_z$	F_x 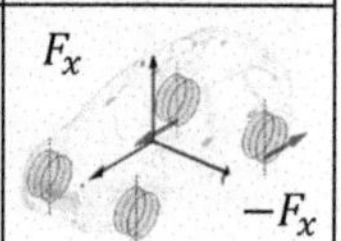 $-F_x$
F_z 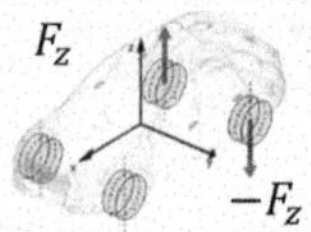 $-F_z$	F_y 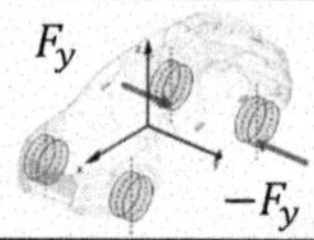 $-F_y$	$F_y - M_z$ 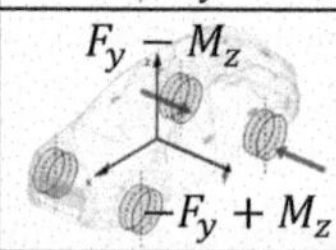 $-F_y + M_z$	M_z 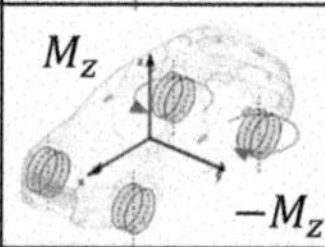 $-M_z$	F_x 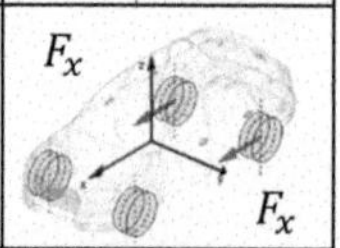 F_x
				F_x 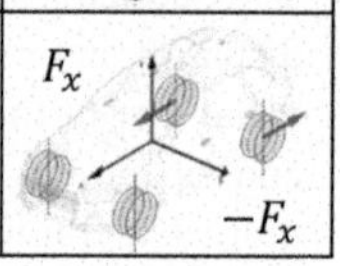 $-F_x$

3.3 Experimental procedure

A communication interface for synchronization is necessary because the test is operated with two principal different measurement systems. To serve this, selected force and displacement channels of the KNC test rig are fed in the optical measurement system. The data of the KNC test rig as well as the data of the optical measurement system are acquired with a sample rate of 10 Hz during the quasistatic test rig manoeuvres. The optical measurement system is once calibrated before the start of the measurement.

3.4 Postprocessing steps

The postprocessing of the generated test data seems to be a great effort. The size of the raw data amount of each measurement of a KNC manoeuvre rises up to several gigabytes. The respective measurement markers are calculated from the photo material in the context of computing. Further graphical postprocessing is done because the matching of a single marker from a point cloud to a chassis component seems to be very difficult without a measurement systems' picture. In the first step the identified measurement markers are matched to each single chassis component and marked with a specific colour (see Figure 3 on the left hand-side). The hardpoints of a suspension system are very important for the postprocessing of kinematics. However, it is comprehensible, that these points are not directly available through the measurement system. Therefore, the usage of an auxiliary geometry can be useful. If the components aren't available as construction data, the data can be generated by an approach of the 3D shape reconstruction, for example the strip projection method ([7], pp. 44 ff.). The CAD geometries are positioned with the help of a special algorithm. In this way, the measured coordinates of the measurement markers are located near the surfaces of the CAD parts. While doing so, the deformed chassis components make some trouble because the chassis is already preloaded in the design altitude. In this way, there are small deviations between the ideal CAD geometry and the real chassis components. As next step it is possible to affix the CAD part to the related point component. Figure 3 on the right hand-side shows the positioned CAD geometries of the five links, the stabilizer bar, the subframe, the wheel carrier and the wheel hub including the brake disk.

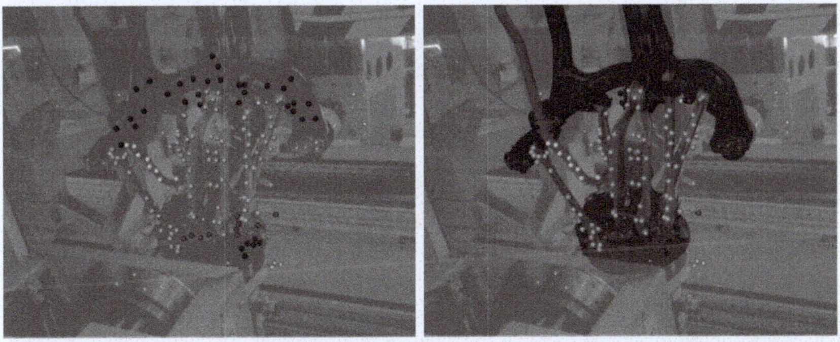

Figure 3: Assignment of measurement markers to chassis components (left-hand side) and connection to CAD parts (right-hand side)

Now it is possible to locate the initial position of the chassis hardpoints with the help of the CAD components. Furthermore, the component coordinate systems are defined analogy to a multibody system model. Coincident coordinate systems are created in the location of bearings to identify the torsional and gimbal angular as well as the translational displacements. Also, other chassis quantities, referring to ISO 8855 [2], are defined to verify the measurement with the help of miscellaneous geometry elements.

While there are no geometrical changes, respectively no changes of the measurement markers, the definition of each element has to be performed only onetime. In this way the postprocessing template can be retained for all kinematics and compliance test rig manoeuvres.

4 Evaluation of a chassis system

Hereinafter the result of the performed experimental tests are presented. At first the comparison between the results of the KNC test rig and the optical measurement system takes place. Other until now not measurable internal chassis kinematics and compliance quantities are further analysed by reference to the verified optical measurement data.

4.1 Verification of the measurement systems

The measurement channels of the KNC test rig are just a small subset of the channels of the optical device. Nevertheless, the alignment of the mechanical tracked measurement quantities and the postprocessed digital image correlation is verified. In the following, the similarity of the results is checked during great deformation manoeuvres, such as the equilateral wheel bounce.

The toe and camber angles belong to the most important wheel position quantities. Figure 4 shows the measured vertical displacements of the wheel centre point s_z above the wheel steer angle δ_{2R} (left-hand side) and the camber angle $\varepsilon_{V(2R)}$ of the rear right wheel during a equilateral bounce of the wheels. Both of the presented angles as well as the wheel travel displacements are separately gathered by the measurement systems. The wheel travel displacements over the measured camber angles expose a nearly congruent agreement. Merely the two steer angles differ to each other through small deviations regarding the gradient. The angle and displacement information of the optical system are reached through the measurement markers on the brake disk (see Figure 3). The brake disk is rigidly connected to the wheel motion sensor arm. The resolution of the angle information of the toe angle is not as high as the resolution of the KNC test rig. This is accounted for by the fact that the measurement markers on the disk are very near to each other in longitudinal direction. The comparison of the remaining KNC manoeuvres indicates a sufficient alignment of the results of both measurement systems.

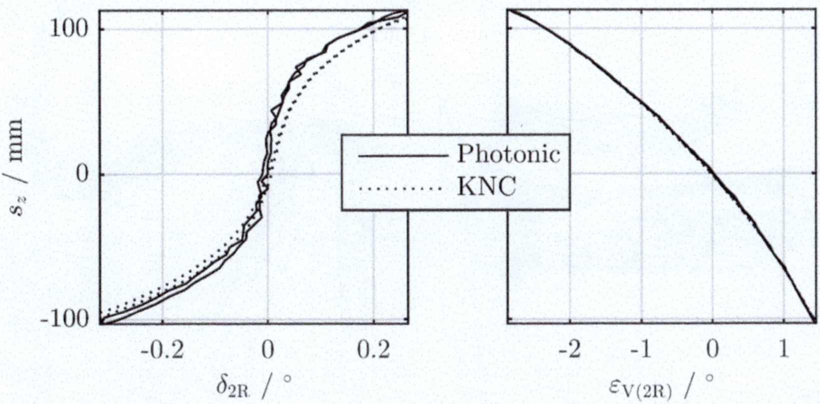

Figure 4: Verification of KNC test rig and optical measurement system

4.2 Kinematic evaluation

A number of other chassis internal kinematic quantities exist besides the global kinematic quantities, which are determinable with a conventional KNC test rig, such as the toe and camber angle in chapter 4.1. The imaginary steering axis is one of these quantities. It is obtained from the position of the upper and lower strut level ([6], p. 240 – 241). The determination of this axis in a fully resolved suspension system is difficult without the knowledge of the orientation of the links. The measured points on the chassis components enable the calculation of the imaginary steering axis for each time step of the measurement. Figure 5 on the left-hand side shows the chassis links and the calculated imaginary steering axis in the design attitude.

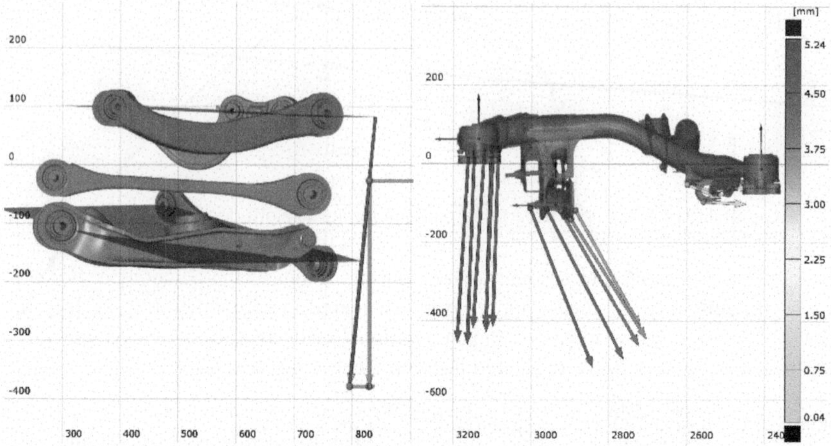

Figure 5: Imaginary steering axis (left-hand side) and global motion of subframe (right-hand side)

This axis can change the orientation as well as the position during variation of the wheel travel. Three characteristic values in the view towards the vehicle longitudinal direction are the kingpin inclination angle σ, kingpin offset at wheel center r_σ and kingpin offset at ground r_k. The test with the optical measurement system enables the determination of these characteristic values with respect to the wheel travel. The courses of these quantities are shown in Figure 6. The castor angle, castor offset at ground and castor offset at wheel centre are examined in the same manner.

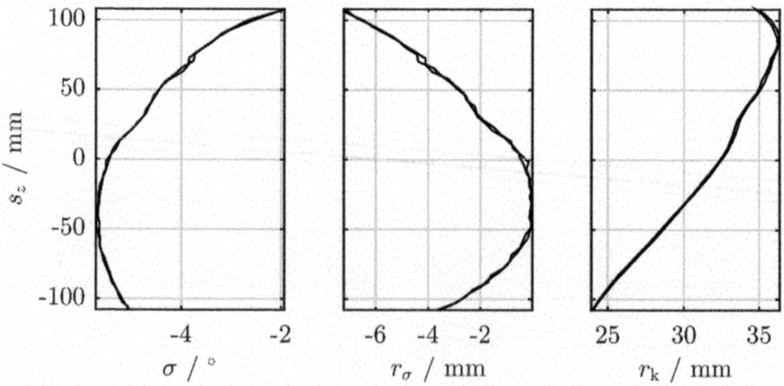

Figure 6: Kingpin inclination angle σ, kingpin offset at wheel centre r_σ and kingpin offset at ground r_k

Also global motion trajectories can be analysed within this approach. The elastic mounted subframe performs a complex motion during the bounce of the wheels. This motion is effected through the stiffnesses of the subframe bushings, the lever ratios and the spring ratio. Figure 5 on the right-hand side shows the displacement vectors of the measurement marker relatively to the design attitude during the duplication of the tire contact force compared to the curb weight of the vehicle. It is clearly recognizable that the rear end of the chassis subframe is pushed down through the spring ratio. The displacement vectors of the measurement markers on the bracket of the support arm indicate, that the subframe performs simultaneously a rotational motion. It is possible to calculate the velocity vectors, because the displacement vectors are tracked above the time during the measurement. The velocity vectors are the trajectory tangents of an instantaneous centre of rotation. Therefore, the herpolhode and the polhode of the chassis subframe can be analysed during variation of wheel travel displacement ([4], pp. 128 ff.). An obvious change of the instantaneous centre of rotation is recognizable at the time of the application of the compression bump stops.

The ratio of spring, damper and stabilizer bar over the wheel travel as well as a lot of other suspension internal quantities can be examined besides the presented kinematic quantities.

4.3 Detection of structural deformations

The compliance characteristics pose a challenge in the validation process of multibody systems within KNC measurements. Possible deviations in the wheel position quantities can usually only compensated through changes in the bushing characteristics or sometimes through increasing the level of detail. One measure of increasing the level of detail can be the consideration of chassis component stiffnesses. If there is no data for validation purposes of the suspension internal stiffnesses, these two measures can be the only available procedures for model optimization within the scope of validation.

However, the optical measurement technique, which is used in this examination, enables beside the description of the kinematics also deep insights in the deformation modes of the chassis components. Therefore, it is possible to gather the deformations of the bushings as well as the deformation of whole metal components. In this way, completely new possibilities in the validation process of multibody models of suspension systems are developed.

The deformations of bushings can be captured through the application of measurement markers on the adjacent interaction parts. Subsequently the measurement markers on each part area are defined to its own rigid body component. Therefore, the translational displacements and the gimbal and torsional angles can be reached within the two rigid bodies.

The same procedure can be transferred to the metal chassis components. Figure 7 on the left-hand side shows exemplarily the support arm from below. Respectively five measurement markers, which are near located to the hardpoints, are summarized to one rigid body component. The displacements and rotations of both coordinate systems in reference to each other describe the deformation modes of the component. A lateral force F_y applied in the wheel contact point leads to a deformation Δl of the component. The length variation of the support arm during a lateral force manoeuvre is shown in Figure 7 on the right-hand side. The stiffness of this component is during the lateral force application as expected constant with an on the lateral force based stiffness value of -35.6 kN/mm. This value was determined by a regression line. If the lateral force manoeuvre is virtual simulated with a MBS-model, it would be possible to directly determine the stiffness of the part through the simulated arm force.

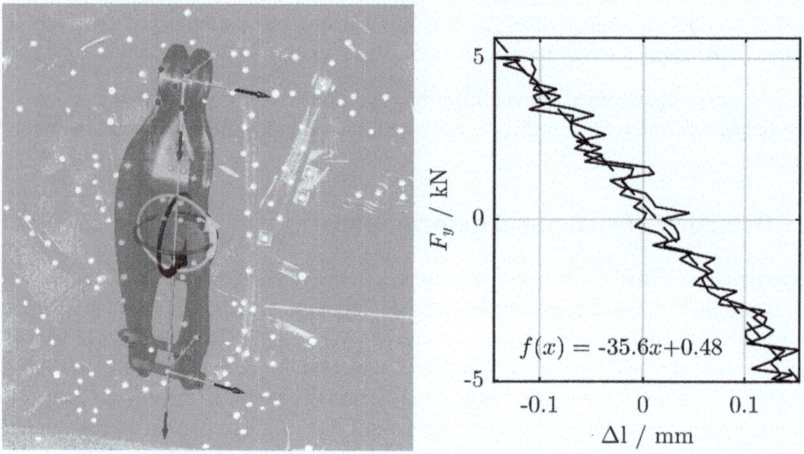

Figure 7: Deformation of the support arm during lateral force excitation

5 Approach for influence dctermination of suspension modifications

The in chapter 4 presented areas of application serve an improved understanding of the chassis internal cause-effective relationships. The influence of different component concepts is analysed in this chapter. Two new examinations are performed to serve that. The focus of these measurements is not on the whole suspension system but on only one component. This way enables a higher resolution of the measurement quantities because the measurement space is many times smaller. Furthermore a better discretisation with measurement markers of the investigating component is possible. Two conceptual different 2-point links serve as test objects. The different concepts are necessary to satisfy the requirements of the component strength ([3], pp. 334 – 336). A forged aluminium component as well as a reshaped sheet steel link are sticked all over with measurement markers, which is shown in Figure 8. The two rubber bushings with slit outer sleeves are pressed in the links. The bushing characteristics are equal for both links on each bearing position. In addition, further measurement markers are applied on the bracket of the subframe and the attachment of the wheel carrier. By doing this it is possible to determine the bushing deformations. It makes sense to increase the density of measurement markers on the bearing positions to enhance the resolution of the deformations. In reference to the already performed measurements, the known KNC-manoeuvres are repeated.

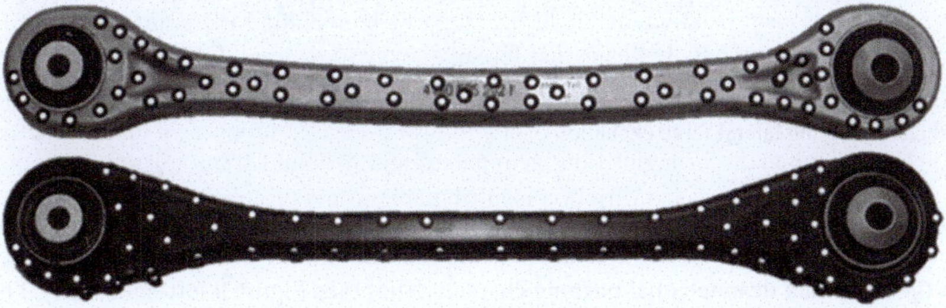

Figure 8: Forged aluminium link (above) and reshaped steel sheet link (below) as test objects

The evaluation of the deformations indicates, that even the single component modes differ to each other during a lateral force excitation. Both components perform primary a bending deformation. However the torsional deformation r_x of the sheet link is many times higher in reference to the aluminium link (see Figure 9 right-hand side). The cause for this behaviour is the open cross-section of the profile. The forged aluminium exhibits a higher torsional section modulus.

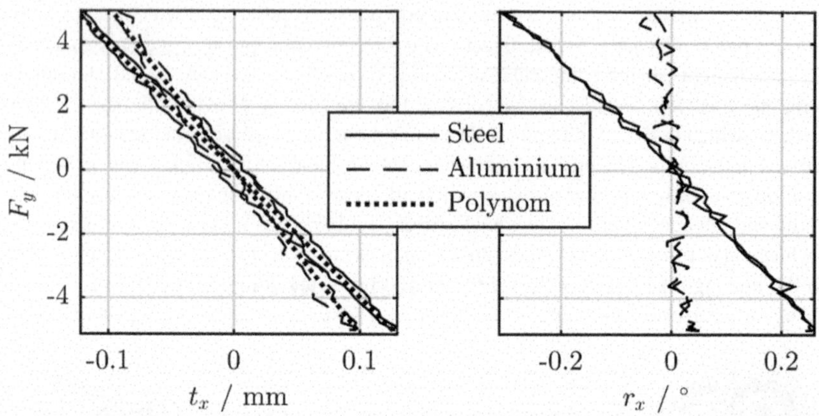

Figure 9: Different bushing deformations (left-hand side) and component torsional angles (right-hand side) during lateral force excitation

Another interesting circumstance is detected during the examination of the bushing deformations. The bushing stiffnesses of the aluminium link are higher as the stiffnesses of the sheet link despite equal bushing characteristics (see Figure 9 left-hand side). The loss of radial bushing preload, caused through the bushing attachment of the sheet link, is the reason for this behaviour. On the other hand embraces the aluminium arm the bushing about the whole width. In this way the radial preload of the bushing is ensured.

6 Enhanced validation of MBS-models

It is often very difficult to identify the cause for deviations during MBS-model validation on suspension level. Therefore, the chassis internal deformations, which are gathered by the optical measurement system, are used for an enhanced model validation.

It is useful to overtake the conventions of a MBS-model during the postprocessing of the optical measurement data. This includes the definition of local body reference systems as well as coordinate systems of force elements, such as a bushing models. Based on this procedure the foundation for an enhanced model validation of MBS-model is enabled.

Hereinafter, the experimental tests are virtually reconstructed by the usage of a flexible multibody model. The metal components itself are modelled with modal reduced finite element (FE) structures in the MBS environment. The coordinates of each single measurement marker in the local body reference coordinate system are exported to make the markers also available in the simulation model. The markers are imported as geometry instances inside the MBS environment and fixed on the FE nodes, which are located near to the marker. Figure 10 on the left-hand side shows the experimental test rig constitution and on the right-hand side the virtual simulation model including the imported measurement markers.

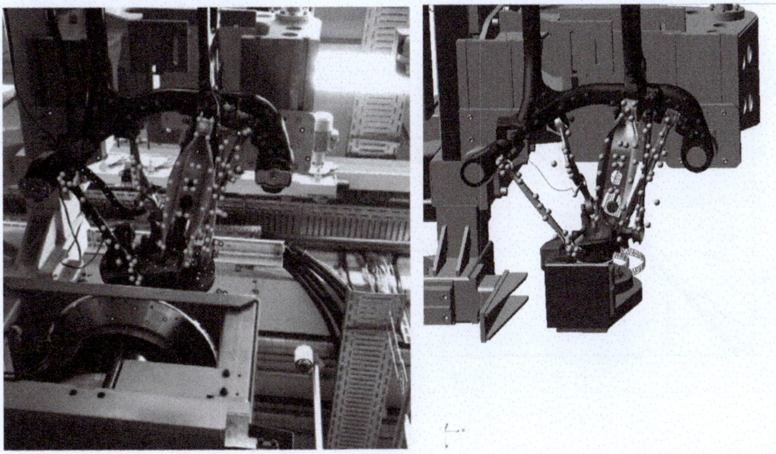

Figure 10: Real test rig (left-hand side) and MBS-model with measurement markers (right-hand side)

The measured manoeuvre 'lateral force applied in the pneumatic trail' serves to check the components' deformations. The deformations of the ten bushings and five chassis arms are evaluated with the digital image correlation as described in chapter 4.3. Following this, the test rig manoeuvre is performed on the virtual kinematics and compliance test rig within the scope of a quasistatic simulation. As next step, the deformation quantities of the components in experiment and simulation are compared to each other. In fact it is now possible to determine the forces in the links. In this way the combination of the optical measurement and the data of the MBS-model enables to deduce the stiffnesses of the components.

Figure 11 exemplarily shows on the left-hand side the radial deformation s_x of the support link bushing and on the right-hand side the variation of length Δl of the support link. The occurring deformations during a lateral force of 5000 N and a toe torque of 175 Nm are very small reasoned by the high component stiffnesses. Thus the deformations are in the range of a tenth of millimetre. The plots figure out, that the bushing model as well as the reduced FE-model are described a little bit too stiff.

The accuracy of the optical measurement system is individual for each quantity. It is defined through the position of the markers in the measurement space. Additional the amount of measurement markers and the position of these to each other define the accuracy in the case of deduced quantities. However, it is possible to improve the appearance of the quantities with the usage of a polynomial approach together with multiple time steps, as shown in Figure 11.

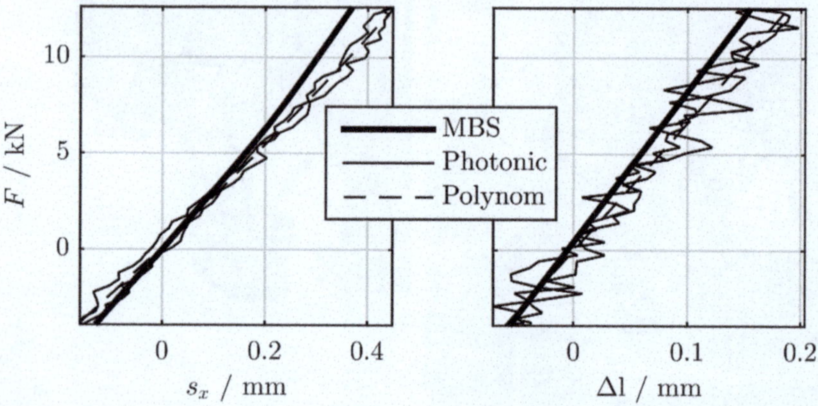

Figure 11: Bushing and component behaviour in the simulation and experiment

The direct comparison of the deformations of the markers in reference to a local body coordinate system in the moment of maximal loading through a lateral force is shown in Figure 12. The translational displacements are visualized through scaled vectors. Crucial for the visualization is the chosen system of reference. The body fixed reference system of the support link is located in the middle of the bushing. Subsequently, all deformations are described in reference to this system. It can be conducted within a comparison between the vectors of simulation and experiment, that the support link figures a higher bending deformation in the simulation. This is recognized by the vector orientations inside the X-Y-plain about the length of the link.

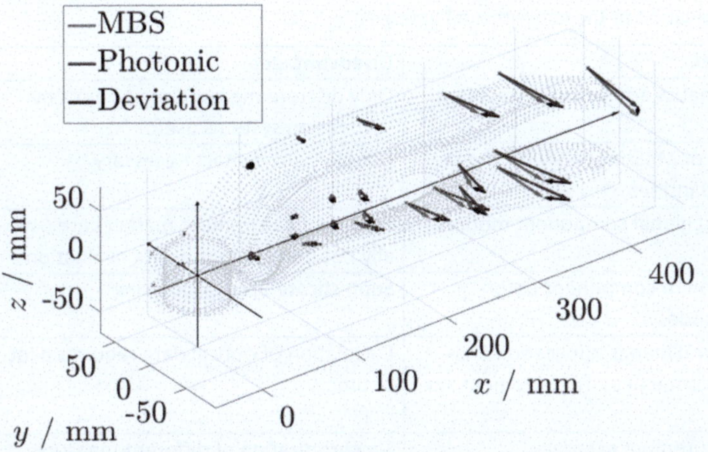

Figure 12: Displacement vectors of simulation and measurement of the support link

7 Outlook and conclusion

The presented approach enables insights in the suspension internal event chains, which was till now difficult to describe experimentally. The usage of an optical measurement system in combination with a KNC test rig shows that different quantities can be determined similarly to a multibody simulation with the effort, that the gathered data has its origin from a real system. However, the effort of this approach must not be underestimated. Especially the comparison of different test setups is affiliated to a high effort. Table 2 shows a summary of the advantages and disadvantages of the experimental method.

Table 2: Evaluation of the experimental approach

Advantages	**Disadvantages**
Real kinematics and compliance quantities	Only discrete measurement points by suspension systems useful
Validation of MBS-models with chassis internal quantities	Double sight contact necessary
Analyses of global component motions possible	Missing measurement markers are usually not detected until data evaluation
Examination of component deformations modes	Sophisticated postprocessing
Suitable for dynamic application enabled with contactless measurement system	Calculation of optical data requires a lot of time
High measurement accuracy ~ 0.015 mm	Synchronization of different measurement systems necessary

For prospective examinations it is conceivable to use a second optical measurement system from another point of view. In this way, a better description of the characteristics of the hidden chassis components is possible.

Furthermore the approach could be used for the competitor analyses beyond the conventional KNC testing. The creation and parameterization of a whole virtual kinematics and compliance model is conceivable. At first the geometry of the chassis is to be determined with a 3D shape identification method. In this way the identification of the kinematic behaviour of the suspension system is possible. The local deformations of the bushings and the components can be investigated by the application of measurement markers on the interaction parts of the chassis components. The internal loads are calculated with the MBS-model, which is generated from the 3D shape identification method. The combination of testing and simulation enables therefore to determine the stiffnesses of each single stiffness. In case of a strong coupled system of the stiffnesses, it is conceivable to describe the stiffnesses as linear parameters or polynomial terms. These parameters can be fitted to the measured deformations within the method of least squares in combination with the MBS-model.

Acknowledgements

Mr. Matthias Springer and Mr. Jürgen Gräf have accomplished an essential contribution to the success of this examination. Through their competent commitment within the scope of measurement and test rig techniques, the tests were run smoothly. The authors would like to thank them.

References

[1] Audi AG:
Ingolstadt, 2018

[2] DIN ISO 8855:2013-11:
Straßenfahrzeuge – Fahrzeugdynamik und Fahrverhalten – Begriffe (ISO 8855:2011)

[3] Heißing, Bernd; Ersoy, Metin; Gies, Stefan (Ed.):
Fahrwerkhandbuch
4th edition, Wiesbaden : Springer Fachmedien, 2013

[4] Holzmann, Günther; Meyer, Heinz; Schumpich, Georg
Technische Mechanik Kinematik und Kinetik
10th edition, Krefeld : Vieweg+Teubner, 2010

[5] Käppler, Wolf Dieter:
Smart Vehicle Handling – Test und Evaluation in der Fahrzeugtechnik
Berlin : Springer Verlag, 2015

[6] Matschinsky, Wolfgang:
Radführung der Straßenfahrzeuge – Kinematik, Elasto-Kinematik und Konstruktion
3th edition, Berlin : Springer Verlag, 2007

[7] Schmuth, Wolfgang:
Handbuch Optische Messtechnik : praktische Anwendungen für Entwicklung, Versuch, Fertigung und Qualitätssicherung
München : Hanser Verlag, 2017

[8] Technische Universität Dresden:
Professur für Kraftfahrzeugtechnik, Dresden 2018

[9] Wei, T. E.; Dorfi, H. R.:
Vehicle Suspension Measurements: Evaluation of the Benefits of Dynamic over Quasistatic Kinematics and Compliance Testing
Vol. 37, No. 1, 32-46, Tire Science and Technology, 2009

INNOVATIVE MATERIALS AND METHODS

Carbon Fiber Reinforced Plastic (CFRP) anti-roll bar and CFRP drop link in the Porsche 911 GT2 RS and 911 GT3 RS with Weissach Package

Boris Schmidt, Dr. Ing. h.c. F. Porsche AG

Eric Begenau, Dr. Ing. h.c. F. Porsche AG

P. Pfeffer (Ed.), *9th International Munich Chassis Symposium 2018*, Proceedings,
https://doi.org/10.1007/978-3-658-22050-1_29

CFRP anti-roll bar and CFRP drop link in the Porsche 911 GT2 RS and 911 GT3 RS with Weissach Package

The lightweight concept is a primary focus in the development of Porsche vehicles. Regarding lightweight construction, future vehicle projects will challenge standard materials even more. Especially the transition to electric mobility leads to an increase in vehicle weight. With regard to anti-roll bars, only small steps can be expected by using the existing and tested materials and manufacturing processes. Significant weight reductions only seem to be achievable with new materials.

Figure 1: Pioneer in CFRP[1]- anti-roll bars; Porsche 918 Spyder with Weissach Package on the Nürburgring Nordschleife

Figure 2: Porsche 911 GT2 RS with Weissach Package on the Nürburgring Nordschleife.

At Porsche, we have developed an anti-roll bar in a hybrid design for the rear axle of the Porsche 918 Spyder as a first step and thus laid the foundation for CFRP chassis components in Porsche road vehicles. It consists of a CFRP torque tube, which is positively connected to aluminium levers and offers a weight advantage of 63 percent.

For the current Porsche 911 GT2 RS and 911 GT3 RS, the further development has led towards one-piece CFRP anti-roll bars and CFRP drop links. The conflicting factors of technological standards, economic viability, quality requirements, testing scopes and shorter development periods represent a great challenge. Because components of this kind have up to now only been produced in prototype-typical quantities or not at all, a supplier environment with experience of components and materials in large volumes does not exist. Fortunately, this situation is improving so that more potential partners will be available for future projects.

1 CFRP: carbon fibre reinforced plastic

The development of the CFRP anti-roll bar and CFRP drop link focused on the following topics:

- functionality,
- weight,
- structural durability.

Functionality

The function of the anti-roll bar is to counteract the roll movement of the body. Furthermore, it is a control variable for coordinating the roll torque distribution between front and rear axle and thus the self-steering properties of the vehicle when cornering.

Steel components of a conventional design are installed as standard in the Porsche 911 GT2 RS and 911 GT3 RS. The CFRP anti-roll bars and the CFRP drop links are available in the Weissach Package special equipment. In terms of function, both versions have the same characteristics.

Porsche GT vehicles always offer the customer the option to alter the handling depending on the route and the ambient conditions. In the case of the anti-roll bars, this is enabled via varying the connection position between drop link and anti-roll bar and thus influencing the anti-roll bar rate. Adjustments are possible in three stages on the front axle and in two stages on the rear axle.

Weight

Aligned with the central objective of the Weissach Package, the emphasis is on optimum lightweight construction. Not only is the focus on the use of CFRP, but also on optimisation of the drop link spherical bearings.

All components are manufactured using the CFRP prepreg bladder moulding process. This is based on a mix of high tensile and high modulus fibres. The combination, arrangement and orientation of the fibre types are adapted to the component function for maximum durability and minimum weight.

For the application of force to the joints and bolted connections, metallic inserts were consistently laminated in. The joint movement of the drop links is realised via specially developed Unibal and patented wide-angle joints.

Compared with conventional steel components, the CFRP package consisting of two anti-roll bars and four drop links achieves a weight reduction of 51 percent. The rear-axle anti-roll bar harbours the greatest potential with a reduction of 59 percent.

Structural durability

For testing and release of the components, load and temperature data was collected on various test and race tracks worldwide.

This forms the basis for qualification of the components via preliminary component testing and subsequent RLD[2]-based rig testing under varying temperature and moisture conditions (see Fig. 4). Testing in various vehicle endurance runs rounds off the test programme.

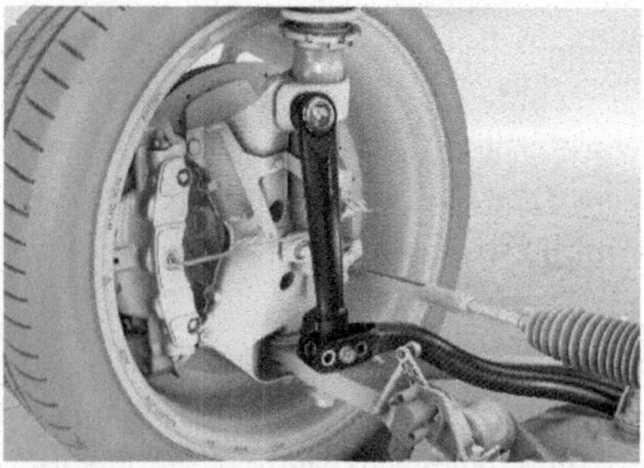

Figure 3: Current Porsche 911 GT2 RS front axle with CFRP anti-roll bar and CFRP drop link

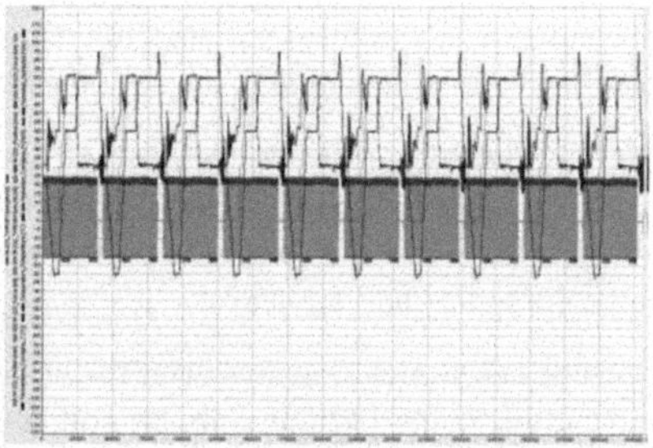

Figure 4: RLD-based rig test curves with operating loads (magenta), temperatures (red) and moisture (blue)

The difficulty at the outset was that no component test standards existed for FRP[3] chassis components. In cooperation with the Porsche Material Technology and Durability Department, we have developed a new testing strategy which takes account of the CFRP-specific material properties.

This mainly includes:

- Testing under consideration of the expected peak loads, which have a significantly higher impact on the lifetime of FRP components and on crack formation compared to metallic materials. A test with constant amplitude at a lower but equally damaging load level, which is normally used on metallic materials, delivers varying results.
- Evaluation of rigidity. Depending on load and environmental influences, structural FRP components are subject to a continuous reduction in rigidity over their lifetime. This reduction can be minimised by constructive measures with regard to fibre alignment.

2 RLD: road load data

3 FRP: fibre reinforced plastic

- Influence of temperature and moisture. For all sign-off relevant testing, preconditioned components were used and almost all tests were performed under climatic conditions. Especially the resin-dependent parameters can strongly influence the stability of the laminate and therefore the test result.
- Electrochemical corrosion and media resistance, in which thermoset resins exhibit stability with respect to the customary operating media. The combination of carbon fibres with metallic materials is usually only possible through targeted material selection or additional measures.

In the case of FRP materials, the alignment of the individual layers is of fundamental importance due to the differing properties in the fibre direction and transversely to it.

According to our philosophy, the laminate must not only form a composite, but also function as such. Here it is important that all layers are subjected to the same load. Only in this way it can be ensured that inter-laminate tensions and distensions are minimised and that the lifetime is consequently maximised with regard to material use.

This patented approach enables targeted influencing of the reserve factors in the laminate during design, the creation of soft transitions and thus the construction of extremely light and durable CFRP anti-roll bars.

In addition, the geometric freedom of design with radii and transitions appropriate for the fibres has a positive effect on the lifetime.

Generally, FRP components react more weakly under compression than under tensile load as the fibres tend to buckle under pressure. Due to a patented design derived from the Porsche 919 Hybrid, the drop links exhibit similar rigidity and strength characteristics under compression as under tensile load. Across the entire operating temperature range, the strength properties are higher than the characteristic values for steel components.

Further steps

Following seven years of development work, it can be said that the CFRP material is fundamentally suitable for a functioning structural component, weight reductions of more than 50 percent are possible and the fibre composite segment in the supplier industry is growing.

The growth in experience in component testing from which the test standards for fibre reinforced components can be derived will reduce the testing effort for new series production developments in terms of time and expense in the future.

Automation and process monitoring play an important role. Only through targeted selection of the production processes consistent quality can be ensured for all stakeholders and the risks can be evaluated.

The attractiveness for larger production series can be increased through the reduction of material costs. The use of standard fibres must be considered for this purpose. A reduction in the fibre quantities used as well as the substitution of metallic joints harbours further cost-saving potential.

Capability of glass fiber reinforced plastics for lightweight design control arms in wheel suspensions

Dr.-Ing. Stefan Kurtenbach, Automotive Center Südwestfalen
Dipl.-Ing. Georg Schöntauf, Automotive Center Südwestfalen

Prof. Dr.-Ing. Andreas Nevoigt, Fachhochschule Südwestfalen
M.Sc. Dimitri Butakov, Fachhochschule Südwestfalen

P. Pfeffer (Ed.), *9th International Munich Chassis Symposium 2018*, Proceedings,
https://doi.org/10.1007/978-3-658-22050-1_30

1 Introduction

The use of lightweight components in the chassis of passenger cars is state of the art since many years. In particular for upper-class and sports vehicles, light-alloy materials are often used instead of steel for chassis components. However, today the use of plastic components in the field of load-carrying structures is still limited to particular applications. A number of concepts have been presented for this purpose in the past (e.g. Mue15, ZF18), but so far only few standard-production applications are established, especially for leaf springs in passenger cars and light duty trucks (e.g. Mo18).

In addition to their functional properties and their potential for weight reduction, the economic efficiency, which is largely determined by the material costs and the costs from production process, is an important criterion for the evaluation of plastic components. In particular for large-scale applications the acceptance of plastic components will significantly depend on their cost-effectiveness compared to the established metal components. The prepreg process, which is used today for production of leaf springs, is not optimal in terms of its economic efficiency for very large quantities. The injection moulding technology promises a better economic efficiency of mass-produced plastic components for chassis applications.

Main task of the government-funded project IMoLA (Injection Moulded Long fibre reinforced control Arm) was research and engineering of technical requirements in order to produce long-fibre reinforced plastic parts economically. Those samples made of thermoplastics have to withstand high static and dynamic loads. As a reference a chassis part has been chosen.

The concept of using the injection-moulding technology in combination with glass-fibre-reinforced plastic material as an approach for weight reduction of chassis parts in described in this paper. A consortium consisting of plastic parts manufacturer, tool developer, research and development service provider and university has successfully implemented this concept on the example of a longitudinal control arm of a multi-link suspension. Essential tasks have been the elaboration of a material-compliant design of the component, the development of a suitable tool layout and specific improvements of the injection-moulding process. In addition to that, investigation of the impact from environmental conditions on load carrying capacity of the plastic component was in focus of the project. The results show, that temperature, moisture absorption and stone chipping must be considered by component dimensioning.

2 Basic considerations concerning manufacturing process and material

2.1 Starting Position

At the beginning of the research project with thermoplastic injection moulding, reduced fibre lengths have been state of the art. It has to be specified, if Polypropylene or Polyamide is used. Typically PP allows longer fibres with moulding. PA—material is limited due to its chemical and thermal behaviour. Nevertheless PA is the preferred substance to be used with chassis-parts. Reasons therefor are its performance with extreme temperatures (hot and cold) as well as its chemical resistance, e.g. to oil, brake fluid etc. By using PA, fibre lengths of <<1 mm have been realized so far.

KraussMaffei as a producer of injection-moulding-machines has performed some basic research with regard to fibre length [KM03]. Conclusion was that enlarging fibre-length will result in improved mechanical properties of thermoplastic material. Tensile modulus as well as stress and strain at break increase with remaining fibre dimensions inside moulding.

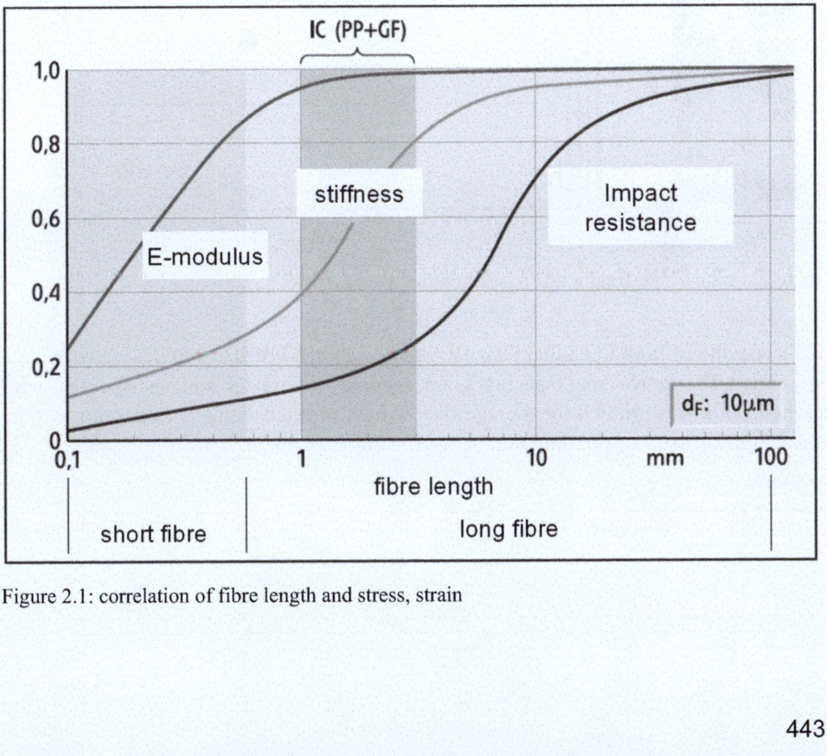

Figure 2.1: correlation of fibre length and stress, strain

This argumentation has been followed up. Different solutions have been analysed in order to enlarge fibre-length inside a complex geometry. Therefore, various basic studies have been done. For example different effects of tooling design were investigated.

Additionally different material compositions were benchmarked. Standard FRP-materials have been used in comparison to compounds, which were produced at the facilities of Automotive Center Südwestfalen (acs) during the research project.

2.2 Modification of tool-design

For initial studies, an existing benchmark-tool was used. With the help of that one, first of all simple geometries (e.g. tensile bars) have been manufactured. Those samples were analysed to identify realized fibre lengths. Results and improved material understanding were transferred into rheological simulation. FE-models were improved unless an appropriate accordance between simulation and real-world-analysis was established. Basis of those research topics was material PA6.6 with a fibre content of 30% of glass.

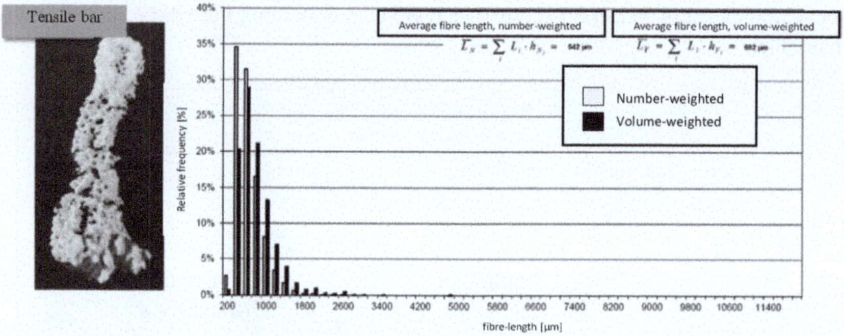

Figure 2.2: initial fibre-lengths and its distribution (tensile bar)

Acs's benchmark-tool including its gate system were modelled at Moldex3D-software. After model's improvement a very good accuracy of model and test-samples was reached. This was proofed by average fibre-length of a tensile bar of 0.53 mm with a real sample. The corresponding number of rheology was 0.542 mm.

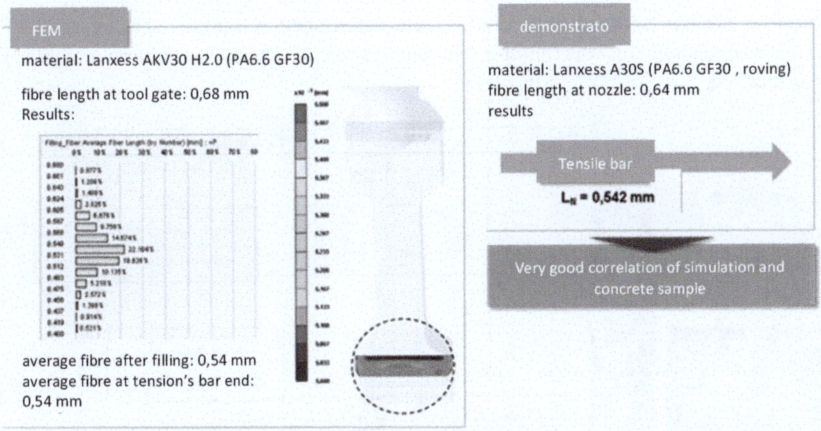

Figure 2.3: comparison of simulation's and sample's fibre length

Next step was to use these experiences in order to improve tooling design. Thus enhanced fibre-lengths were enabled.

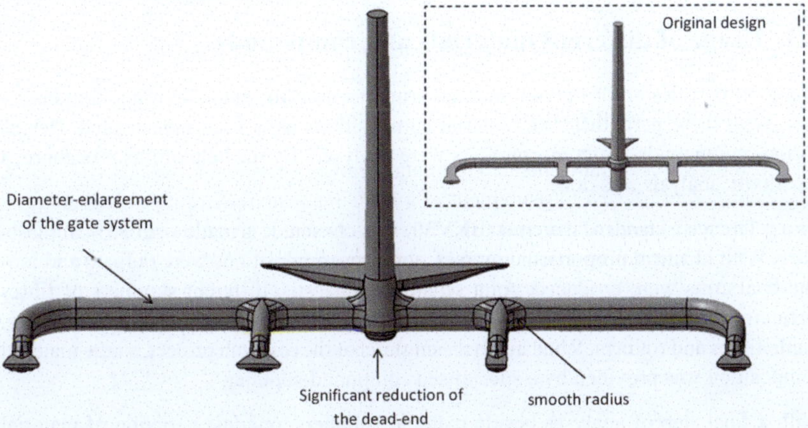

Figure 2.4: adaption of tooling-design (especially gate-system)

Modification of tooling-design was transferred into the already available benchmark-tool. Following new series of test samples was produced and analysed. The fibre lengths of these samples were compared to those of initial test series. With new samples, explicit enlarged fibre lengths were determined. Summarized an increase of average fibre length of 71% (number-weighted) respectively 240% (volume-weighted) was reached.

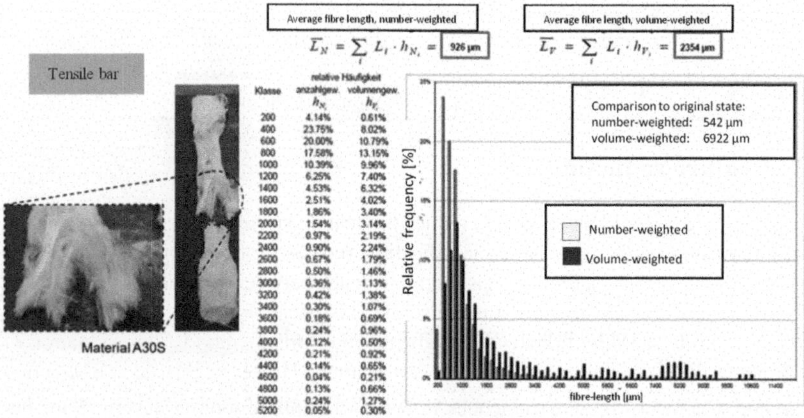

Figure 2.5: final fibre-lengths and its distribution (tensile bar)

2.3 Usage of different materials and compounds

The final part-design of the longitudinal arm was used for manufacturing demonstrators. With these parts different material compositions have been investigated. Beside various compounds, different suppliers were analysed. By the help of this procedure, a sensitivity analysis was done.

As a reference a standard material (BKV30) was chosen. It is made with 30% of glass-fibre. With identical proportional masses, acs-own compositions were defined and longitudinal arms were produced. As a sensitivity analysis, different suppliers of fibres were considered. Besides that, different types of fibres were used, such as short fibre, staple fibres and rovings. With an advanced stage of the research project, a new material composition was provided by a specialised compound-supplier.

With a final step of analyses beside material suppliers, one last variation of material compounds has been evaluated. Fibre content was increased up to 50% instead of 30% originally.

Process parameters were adapted according to the specific material combinations. As a first step, samples have been evaluated by quasi-static tests.

3 Requirement profile and component design

3.1 Considerations for component selection

The specific properties of plastic material compared to standard steel material play a decisive role in regard to the selection of a suitable component for material substitution. Basically, plastic materials are characterized by a significantly lower density compared to steel. Crucial for the lightweight construction potential, however, is the ratio of strength to density respectively rigidity to density. Figure 3.1 shows such a comparison for a polyamide compound with 50% glass fibre content compared to a cold-formable steel sheet. The figure shows (in relation to the respective values for steel) the density ρ, the ratio of modulus of elasticity and density as well as the ratio of tensile strength and density. The bar diagram clearly shows that a weight reduction for the plastic component is to be expected in the area of the load-bearing capacity. In contrast to that, with regard to stiffness the use of plastic material might be disadvantageous. As the bars show, bearing the same external load indicates a weight potential of approx. 30% for the plastic component, whereas with the same deformation, even a higher weight can be expected for a plastic part.

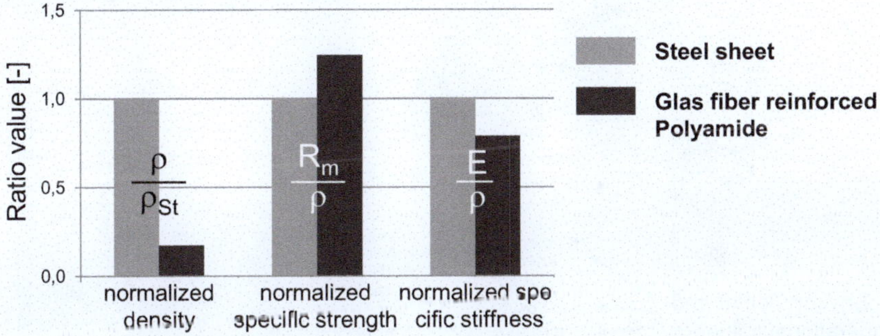

Figure 3.1: Fundamental differences in material properties

On this basis, only such components are qualified for a material substitution, in which stiffness losses can be accepted to a limited extent. With respect to this assumption, for further investigations the longitudinal control arm (sword bar) of a 4-link suspension was selected for material substitution. This rear axle concept is used in large quantities in the compact und middle class segment by various car manufacturers.

3.2 General requirements and load spectrum

Regarding the mechanical loads on the component, it has to be differentiated between normal operating loads resulting from the intended use of the vehicle and misuse load cases. In terms of the operating loads, due to its arrangement in the axle construction, the control arm essentially has the task of bearing the longitudinal forces resulting from braking (respectively acceleration in the case of a driven axle) and to stabilize the braking torque. As a result from this, the component is essentially loaded by tension/compression and bending. In the case of misuse, such as passing through a transverse channel, the longitudinal control arm is exposed to tensile forces which, however, can be 4 to 5 times higher than the longitudinal forces resulting from operational load cases. While the components must withstand the operating loads for a high number of load cycles without cracking, breakage and permanent deformation, for misuse loads a non-problematic behaviour of the control arm (in the first place avoiding brittle fracture) is desired.

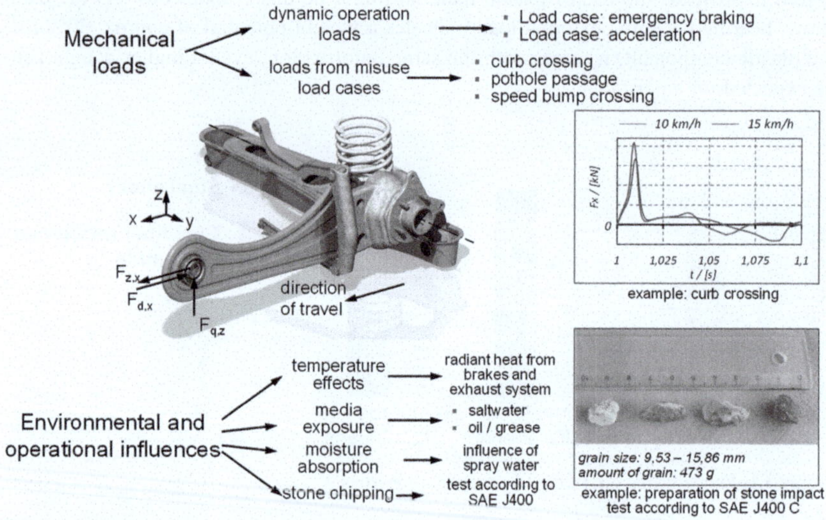

Figure 3.2: Load spectrum from operation and environmental conditions

In addition to the mechanical loads shown in Figure 3.2, the consideration of environmental influences on the component is of crucial importance. Particularly for plastic components moisture absorption, high temperature and stone-chipping have a significant influence on material properties and the resulting load-bearing capacity. For this reason, the moisture absorption and the influence of high temperature on material properties were taken into account already during design of the component. Furthermore, the influence of stone-chipping was considered in component testing. In summary, Figure 3.2 shows the mechanical operational loads as well as environmental influences, which have been considered in component design and testing.

Figure 3.3 schematically shows the influence of moisture absorption and temperature on the stress-strain behaviour of plastic material. It can be seen from the diagram (solid lines) that the curve progression is lower with increasing temperature. This means, that on the one hand the modulus of elasticity and thus the rigidity decrease with increasing temperature and on the other hand it means, that similar strain provides lower stress values with increasing temperature. Comparing the solid curves (dry material) with the dashed (material in the saturated state) indicates a similar correlation: it is evident that in the saturated state the stiffness as well as the load bearing capacity decrease with similar elongation. It is important to mention, that the differences in material behaviour between dry and saturated conditions are in particular evident in the range of room temperature.

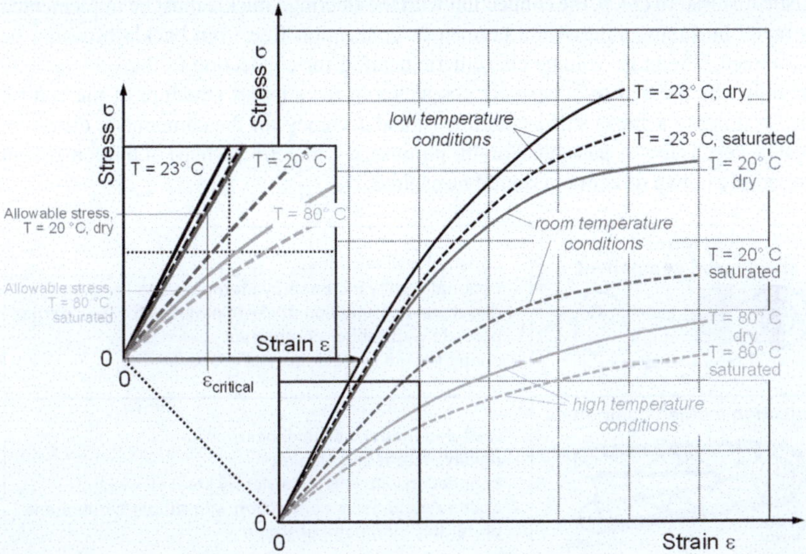

Figure 3.3: Impact of temperature and moisture on stress-strain-characteristics (according to Cam15)

For a design concept as it was used for the longitudinal control arm, with a dimensioning against maximum allowable elongation, the relations described above, have a decisive impact. Considering moisture absorption (saturated state) and temperature (80 °C) imply, that for a certain elongation (assumed to be critical) the allowable stress value is considerably lower than that for dry material at room temperature. Figure 3.3 schematically shows this correlation for the range of small strains, which is solely qualified for continuous load.

3.3 Derived component design

The design of the injection moulded plastic component requires a number of fundamental differences compared to a conventional sheet steel component. Figure 3.4 summarizes the most important differences. In order to avoid large deformation under high loads, the component should preferably be subjected to tensile or pressure loads instead of bending and torsion. This leads to a component shape with a preferably straight connection between the load introduction points. A best possible stress distribution is achieved by a wall thickness adapted to the loads. An additional stiffening of the structure is achieved by a suitable ribbing of the component. Due to the tendency of the plastic material to creep, the connection with the steering knuckle must be implemented by metal bushings, otherwise a permanent connection free from backlash cannot be guaranteed. The large-volume elastomeric bearing for connection to the vehicle body can either be pressed or adhesively connected to the support structure of the control arm. In order to achieve sufficient dimensional accuracy for the component, distortion must be kept as low as possible. For this purpose, a symmetrical shape of the component (preferably in two directions) should be aspired.

cold formed sheet metal component 	▪ constant wall thickness ▪ optional rigidity increase by stamping and flanging ▪ direct srew connection of sheet metal with wheel carrier ▪ press fit with rubber component ▪ shape limited by cold forming process
injection molded component 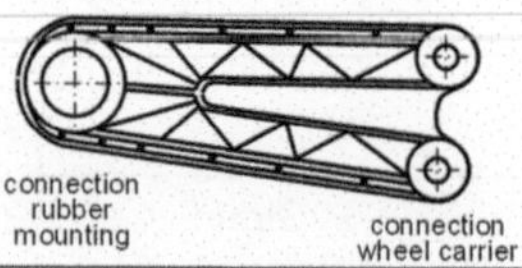	▪ load-dependent wall thickness ▪ rigidity increase by ribbing ▪ connection with wheel carrier by steel sleeves ▪ optional adheasive connection with rubber component ▪ preferably symmetrical shape

Figure 3.4: General differences in design of injection moulded plastic parts versus steel part

Considering the above guidelines for a plastic-friendly design, the control arm was finally designed in the way shown in Figure 3.5. The component has the following essential design features:

1	bearing lug	enlarged wall thickness with ribbing for connection of the elastomer bearing and load transition to the main load paths
2	connection to	holes with variable curvature for assembly of the metal the steering knuckle bushings
3	main load path	straight-line connection of the bearings preferably subject to normal stresses due to tension / compression and bending
4	outer rib	enlargement of the axial section modulus
5	main rib	
6	transverse ribs	increasing torsional rigidity and stability of the component
7	wall thickness profile	load dependent modulation of wall thickness

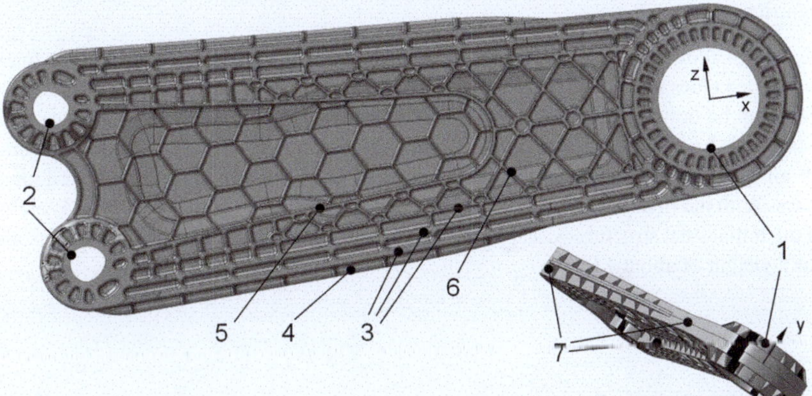

Figure 3.5: Design of the control arm for injection moulding

Moreover, the control arm was designed symmetrically in y- and z-direction in order to avoid distortion due to inhomogeneous cooling of the component. To counteract contraction cavities caused by excessive material accumulation, the tool was equipped with cores especially in the areas of the attachment points to the bearings (1 and 2). The shape of the metal bushes for the attachment with the steering knuckle was designed polygonal instead of round in order to achieve homogeneous load distribution and avoid hot spots.

The above described component design, which was developed together with a tool manufacturer and a producer of plastic components for the automotive industry, finally results in a stress distribution as shown in figure 3.6. The dimensioning was based on the values for permissible strain of the material PA6.6 with a 50 % content of glass fibre (relating to proportion of weight). At a temperature of 80 °C and 1,2 % humidity (meaning saturated state of the material) the values are as follows:

critical tension in flow direction: $\varepsilon_{II,critical} = 0,6$ %
resulting allowable stress in flow direction: $\sigma_{II,critical} = 50$ MPa
critical tension transverse to flow direction: $\varepsilon_{\perp,critical} = 0,2$ %
resulting allowable stress transverse to flow direction: $\sigma_{\perp,critical} = 18$ MPa.

The different values come from the fact, that the fibres contained in the material will preferably be oriented in flow direction. As a result from this, the material has a higher load capacity in flow direction (corresponding to orientation of the fibres) than perpendicular to flow direction. The design of the component ensures that fibre orientation (and thus direction with higher load capacity) is widely similar to the orientation of the significant loads.

Figure 3.6 shows the resulting main normal stresses for the example of a misuse load case. The values indicate that the permissible loads are not exceeded. The same conclusions can be drawn from analyzation of shear stresses as well as other load cases like usual work loads resulting from braking and acceleration. In summary, the results from FE-analysis predict a sufficient dimensioning of the control arm for all considered load cases. With the final design, a weight reduction of about 30% has been realized. Weight of the initial steel structure was approximately 1.500 gr, whereas the final plastic design has a weight of about 1.000 gr.

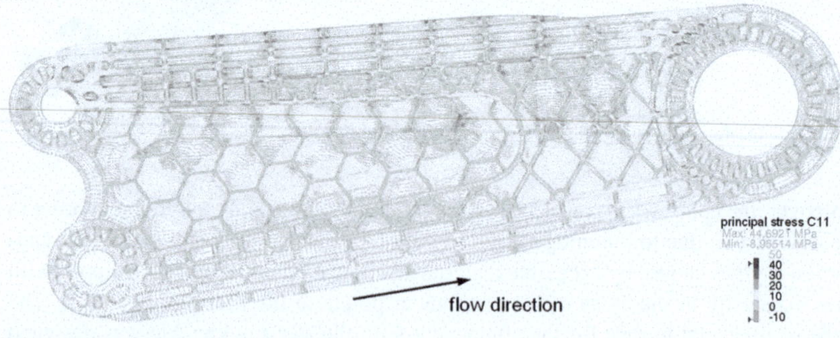

Figure 3.6: Calculation results

4 Results from production tests and experimental test results

The experimental validation of the component was carried out in two steps. In a first step the basic influences on the load carrying capacity of the plastic components have been investigated. These tests have been conducted with geometrical simple specimen for tension tests. The results from these tests helped to derive optimization measures for the manufacturing process and to find most suitable fibre-matrix combinations. Subsequently, fatigue tests with the load bearing structure of the control arm and finally tests of the complete control arm (support structure with elastomer bearing) were carried out.

4.1 Investigation for environmental influences

As an example of the impact from environmental exposure and operating conditions, figure 4.1 illustrates the influence of stone chipping on the attainable load cycles under a dynamic tensile load. In these tests, the specimens were subjected to pulsating tensile stresses with different levels of magnitude until fracture of the structure. These tests were carried out with glass-fibre reinforced samples of PA6.6 in dried state as a new part (1) and pre-damaged after performing a stone chipping test (2). The results show on the one hand the impact of the load level on the attainable number of cycles (characterized by a growing number of cycles with decreasing load level). On the other hand the results clearly indicate the impact of near-surface damages caused by stone chipping. Due to pre-damage of the parts, their load cycle values are up to 40 % below those of an undamaged sample. This can be explained by the damage on the surface which has the same effect like a score and therefore acts as starting point for cracking under cyclic dynamic load.

Due to the much larger cross-sectional area of the real component in comparison to the tensile bars tested here, the pre-damage caused by stone chipping has a significant lesser impact on the real control arm than on the tensile bar.

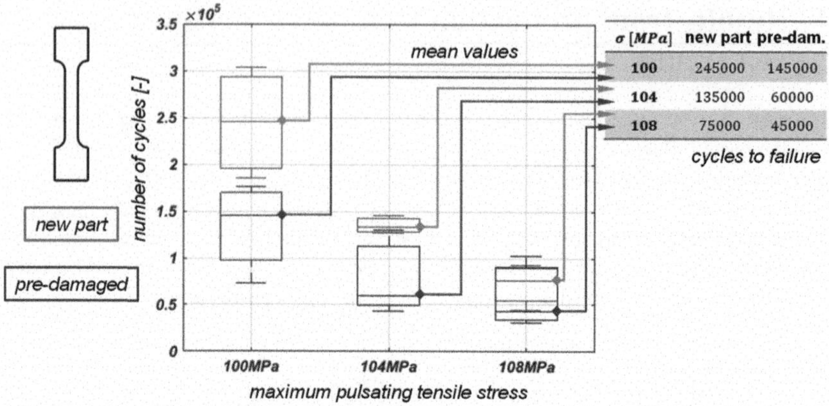

Figure 4.1: Impact of stone chipping on load cycles

In a similar way, test series were carried out to determine the influence of moisture absorption on the dynamic load capacity of tensile specimens. Again, a significant impact could be detected. With the same number of cycles to failure, the permissible stress for a saturated material decreases by approximately 15 % compared to the value for a material in the dry state. Furthermore, the experiments with cyclic loading have shown that a homogeneous distribution of the glass fibres in the component is a necessary requirement for sufficient dynamic load capacity of the samples. Local accumulations of fibres act like a notch and provide premature formation and growth of cracks. With increasing number of load cycles this mechanism leads to early failure of the samples. This makes a decisive difference to conventional tensile tests, in which the described defects do not have a noticeable negative effect on attainable values for strength and stiffness.

4.2 Results from production tests

The final part-design of the longitudinal arm was used for manufacturing demonstrators. With these parts different material compositions have been investigated. Beside various compounds, different suppliers were analysed. By the help of this procedure, a sensitivity analysis was done.

As a reference a standard material (BKV30) was chosen. It is made with 30% of glass-fibre. With identical proportional masses, acs-own compositions were defined and longitudinal arms were produced. As a sensitivity analysis different suppliers of fibres were considered. Besides that, different types of fibres were used, such as short fibre, staple fibres and rovings. With an advanced stage of the research project, a new material composition was provided by specialised compound-supplier.

With a final step of analyses beside material suppliers, one last variation of material compounds has been evaluated. Fibre content was increased up to 50% instead of 30% originally.

Processes' parameters were adapted according to the specific material combinations. As a first step, samples have been evaluated by quasi-static tests.

Demonstrators were mounted at its original fixation points. With the help of a rigid frame they were placed inside a standard tension testing machine. Sample's orientation was according to its load introduction when mounted at a car. Thus quasi-static towing tests with a representative number of repetitions were done. During these test series a basic correlation between original fibre length and sample's performance was evaluated. With increasing fibre lengths (original state) a corresponding increasing strength was recognised.

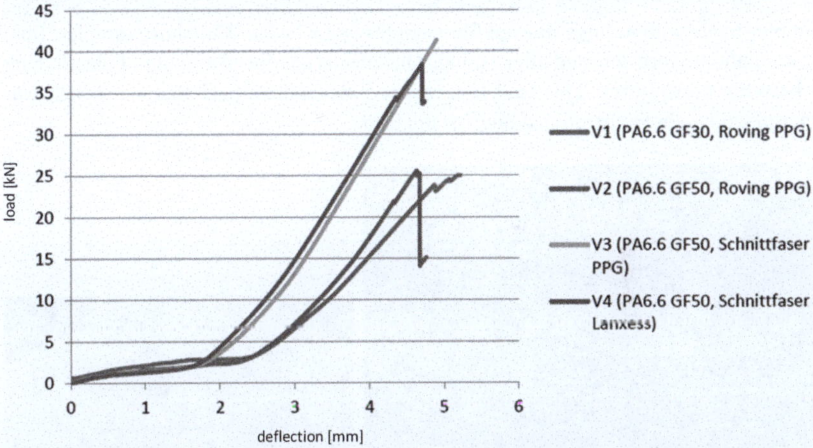

Figure 4.2: Correlation between original fibre length and part's stress performance

Later on ashings of the samples showed additional details of fibre orientation and lengths. With samples, which owned short fibres, fibre lengths of 0.3-0.5 mm were realised. After ashing a homogeneous distribution of the fibres was detected.

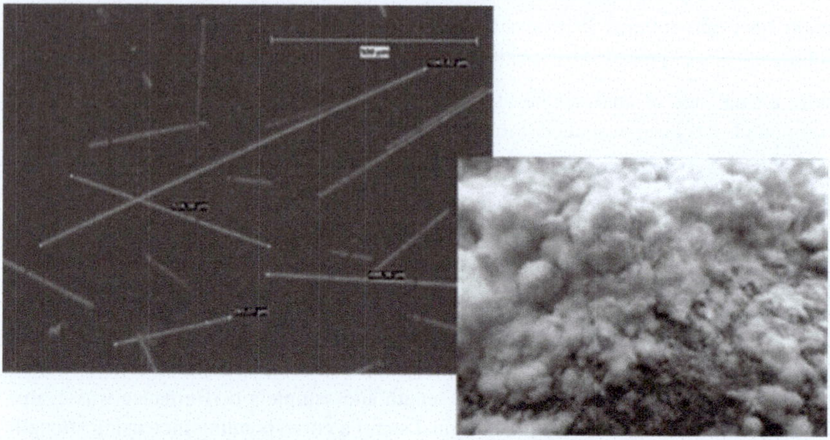

Figure 4.3: analysis of a short-fibre sample

Samples made with rovings showed fibre lengths up to 2 mm. With those samples, fibre accumulations were detected after ashing. This means a reduced amount of plastic-melt in between these fibres. This lead to a reduced mechanical performance, which was determined by above mentioned quasi-static tests.

Figure 4.4: analysis of a roving made sample

With a series of quasi-static tests all sample configurations (different compounds, different suppliers, different fibre contents) were tested. Boundary conditions of the tests were fixed. Ashing of the new material compound showed fibre lengths of 0.5-0.75 mm. This means about 60% enhanced fibre length in comparison to staple fibres. The following table shows the average results of different materials.

design	material	max. load F_{max} [kN]	elongation at F_{max} [%]	deflection at F_{max} [%]
V5	A30S GF50 (Lanxess)	64,84	1,24	4,72
V6	A30S GF50 (PPG)	60,98	1,19	4,46
V7	Domamid Compound	73,36	1,46	5,56

Figure 4.5: mechanical properties of sample tests depending on material supplier

According to those results, the breaking load as a failure criterion of the real longitudinal arm increases. An improved performance by 15 % with regard to standard staple fibres was established.

With improved analysis the process of fibre length-reduction during injection moulding was determined. With the help of this analysis it was shown, that at machine's injection nozzle a massive reduction of fibre length has already taken place in comparison to the original length. Due to that fact, inside the geometrically complex sample additionally reduced fibre length can be found.

4.3 Results from endurance tests with components

On basis of the findings from the model tests as well as using the component design described in section 3, support structures of the control arm have been produced by injection moulding. These parts were subsequently tested under the previously determined operating loads. Figure 4.6 shows the installation of the control arm on the test rig and figure 4.7 shows a section of the test program carried out with the support structure. The parts were subjected to a cyclic load on the test rig, which corresponds to the load case of full braking. In order to consider scattering of material properties as well as variation of loads, the load amplitudes were increased by a safety factor of 1.25. During the tests both tensile and bending loads occur in the component. The results show that in all test configurations, the expected number of load cycles is achieved or sometimes even significantly exceeded, so that it can be assumed that the component has a sufficient design and dimensioning.

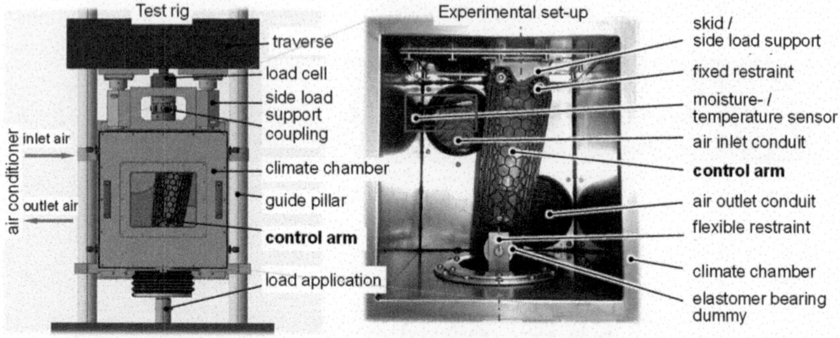

Figure 4.6: Installation of the component on the test rig

Due to the significantly larger cross-section of the control arm compared to the tensile specimen, pre-damages introduced by stone chipping do not come to the same extent of shortening of life, although the damages have specifically been introduced at those areas of the component with the highest stress. Therefore, a sufficient dimensioning of the component can be assumed for all test conditions and overall, the injection-moulded plastic component meets the dynamic operating load requirements. To determine the load limits and learn more about failure mechanisms of plastic components, it is planned to continue the test series with larger dynamic loads in the future.

test part condition	test temperature	working load factor	load cycle factor	result
dry	T = 80 °C	1,25	2,0	☺
saturated	T = 80 °C	1,25	1,0	☺
pre-damaged by stone chipping	T = 80 °C	1,25	1,5	☺

Figure 4.7: Test conditions for component validation

5 Conclusions and outlook

Publicly funded research project IMoLA has led to improvement of enlarged fibre lengths at complex sample designs. Starting with simple geometry and corresponding rheological simulation, material compounds with special regard to fibre length have been validated. These results have been transferred into a real chassis component, a longitudinal control arm of a rear axle. For analysing mechanical properties of these arms, quasi-static and dynamic tests have been performed until samples have collapsed. Within this research different material compositions (standard ones as well as acs-own ones) have been analysed and compared. Due to project's results, improvements of fibre reinforced components have been proved.

While working on project fibre-reinforced demonstrators have been manufactured. Those samples have contained increased fibre lengths in comparison to state of the art. Nevertheless, improvement has been less than initially estimated. During the research project some new aspects and challenges have been identified. Results showed a large field of improvement inside the injection machine. The improvements should be done be machines' manufacturer as well as by material and process development.

With focus on special demands of plastic material concerning material properties and production, a compatible design of the control arm has been developed, leading to a weight reduction of about 30 % compared to standard steel design. FE-calculations as well as endurance tests with the load-carrying structure of the component prove the basic functionality of the lightweight control arm. Consideration of impacts from moisture absorption and temperature already in the design state of the component and dimensioning of the component based on critical strain have proven to be a suitable method for sufficient endurance performance of the plastic component. In conclusion, using the injection moulding process together with glass fibre reinforced plastic material has proved to be an option for economic production of load-carrying lightweight structures in the chassis.

Injection moulding technology offers the possibility of a direct connection of the bearing component in the manufacturing process by using 2-component-technology. Using this option would further enhance the economy of the lightweight component.

6 Acknowledgements

The presented results were developed in the project "Injection Moulded Long fibre reinforced control arm – IMoLA". The authors would like to thank the Ministry for Economics (BMWi) of the German Federal Republic for the financial support of the project within the funding program ZIM – "Zentrales Innovationsprogramm Mittelstand".

Bundesministerium für Wirtschaft und Energie

ZIM impulse für wachstum Zentrales Innovationsprogramm Mittelstand

7 References

Cam15 CAMPUS "a material information system for the plastics industry", 2015 online: http://www.campusplastics.com, release 2015

KM03 KM03 Krauss-Maffei Kunststofftechnik GmbH, Carl Hanser Verlag (München), Zeitschrift Kunststoffe, Ausgabe 03/2003

Mo18 Fahrwerk XC90II https://www.motor-talk.de/bilder/fahrwerk-xc90ii-g69211699/ausschnitt-blattfeder-i207862571.html

Mue15 Beitrag zur Entwicklung von Achsfedern aus Glasfaser-Kunststoff-Verbund Dissertation TU Darmstadt Shaker Verlag Aachen 2015, ISBN 978-3-8440-3803-3

ZF18 Leichtbau-Federbein-Radträgermodul online: https://press.zf.com/site/press/de_de/microsites/press/list/release/media_2859.html, release 2018

Damper diagnosis by artificial intelligence

Thomas Zehelein, M. Sc.
Alexander Merk, B. Sc.
Prof. Dr.-Ing. Markus Lienkamp

Institute of Automotive Technology, Technical University of Munich

P. Pfeffer (Ed.), *9th International Munich Chassis Symposium 2018*, Proceedings,
https://doi.org/10.1007/978-3-658-22050-1_31

Abstract

Diagnosis functions are commonly used within modern vehicles. These functions are applied to subsystems as the engine, exhaust system or electric system. An important component regarding safety as well as driving characteristics is the damper. However, damper defect diagnosis is not performed in production cars. Also in literature, there are only very few approaches that address damper diagnosis during driving. This paper shows an approach to diagnosing damper defects with machine learning. A support vector machine (SVM) architecture is used to classify measurement data with different settings of the damper functionality. Therefore, test drives were conducted with a vehicle, equipped with semi-active dampers and measurement data of longitudinal and lateral accelerations as well as wheel speed signals was gathered. Several signal features are calculated to characterize the measurement data of each sensor. The most meaningful features are explained in this paper. Feature reduction is required to maximize the classification result. This is done by three different methods and the performance of each technique is studied. Finally, a linear SVM classifier is trained that allows the detection as well as the isolation of the vehicle's damper states.

1 Introduction

With the increasing importance of autonomous driving, the driver will not be part of the actual driving process anymore. This means that he will not realize if a specific driving characteristic of the vehicle is intended or if there are any chassis system damages or defects. Another trend in the automotive sector is car sharing. Herein, the driver does not drive his own car on a frequent basis but many different ones on an occasional basis. Therefore, the driver does not know the correct driving behavior of a specific vehicle model and he will not realize if the driving behavior of a specific car is based on the actual chassis system design or if it is caused by a component failure or aging. Another aspect that needs to be considered in terms of chassis system component failures and aging is driving safety. A periodic technical inspection covers only a fractional amount in terms of detecting component defects and ensuring driving safety. Failures that arise shortly after the inspection will be detected at the next inspection after up to two years and the vehicle will travel a long distance with a faulty chassis system. Therefore, an automated system for detecting chassis system failures that is functional even during normal driving with commonly available driving dynamics data such as wheel speeds as well as longitudinal and lateral acceleration (ESC sensors) is required to keep up with the current trends in the automotive sector and to increase driving safety.

In former years, the capacity of electronic control units (ECU) was limited in terms of calculation power and data storage capacity. However, calculation power of ECUs increases and it is even possible to send data to servers that can handle higher demands for calculation power for more extensive diagnostic functions. Therefore, diagnosis functions become a more relevant part in engineering applications [1, 2].

A broad field for the application of diagnosis functions are industrial applications such as rotating machineries [3] as well as bearings [4]. Railway applications are another field with a lot of research regarding diagnosis and condition monitoring [5–9]. In contrast to these applications, diagnosis of chassis system components poses a challenge as parameters of a vehicle can vary widely (mass, cornering stiffness etc.) and environmental circumstances such as road excitation are only known to a small extent. Therefore, research activities with regard to vehicle chassis system diagnosis are often applied to the detection of sensor or actuator failures [10–13]. Only a few papers deal with diagnosing actual chassis system components. Previous approaches of damper defect diagnosis often require additional sensors such as vertical accelerations of the vehicle body mass and/or the tire mass or require test bench measurements [14–19]. However, as dampers have a significant influence on driving safety [20], it is essential to focus on detecting faults of dampers during normal driving with commonly available driving dynamics data, especially in series vehicles. Only the approach by Jautze [21] works on ESC sensor data. However, as this approach did not make it to series production even though it was developed in cooperation with an original equipment manufacturer (OEM), reliability seems questionable.

New developments in the field of diagnosis applications showed, that using artificial intelligence (AI) systems generates promising results [22–24]. Another advantage of AI systems compared to diagnosis methods based on traditional control theory is the independency from any model representation. A given approach is applicable to other fields of engineering. An approach for diagnosing damper defects should therefore also be applicable to other component defects.

This paper will show an approach to diagnosing damper defects with the machine learning architecture support vector machine (SVM).

Chapter 2 shows related work concerning damper defect detection. Chapter 3 explain the general test setup. This includes a description of the realization of damper defects as well as conducted test drives. The developed fault detection and isolation (FDI) system is explained in chapter 4. After a general development overview, some meaningful features are presented as well as relevant feature reduction algorithms. Finally, classification results of measurement data for different damper settings are summarized in chapter 5 and an outlook for future research is given in chapter 6.

2 Related Work

There have been many approaches for diagnosing damper and suspension faults, most of them in automotive or railway applications. Diagnosing damper faults is often approached by modelling a state observer [25–28]. Börner [25] describes a quarter vehicle model in order to compare calculated variables with measured driving dynamic data. This requires the measurement of the damper deflection or the vertical acceleration of the wheel carrier. Damper property changes can be detected from the calculated residuals directly and in real time. Jautze [21] describes a signal model which detects damper defects using the spectral power density of wheel speed signals. The proposition is that dynamic wheel load differences with faulty dampers lead to an increased amplitude at the wheel carrier vertical natural frequency. This can be detected by evaluating the wheel speed signals. Analyses show that the amplitude overshoot depends on the road excitation, for which a normalized damper damage characteristic value (DSKW) is calculated. Furthermore, the DSKW shows a non-linear dependence on the vehicle speed, which might be a reason for problems regarding reliability.

The hybrid detection system by Fischer [29] also uses a quarter-vehicle model with state observers. A neural network for parameter estimation is implemented. Also, the system relies on additional sensors in form of acceleration sensors at the wheel carriers.

The system developed by MEI [9] detects faults in a train suspension without using measurements of wheel carrier vertical acceleration. The data sets are generated in a simulation environment, where the train chassis is only equipped with sensors for roll and pitch acceleration measurements. Cross-correlation and relative variance of these signals are used to develop a simple threshold excitation detection. A further combination of different features does not exist for this system.

Further approaches for damper fault detection have been made with machine learning methods. HARDIER [17] models the wear behavior of the damper based on a parameter estimation method via a radial basis function network. The algorithm detects observations that do not correspond with the typical map of an intact attenuator but with a hysteresis curve. Again, damper deflection is needed to realize the fault detection. The systems of PACIELLO [30] and LIU [31] use a neural network or a fuzzy logic for the detection of damper or sensor faults. Substantial research on damper fault detection and isolation systems has been done for railway applications. WEI [8] describes a system for monitoring the primary suspension of a train, where defects are detected by a consensus principle components analysis. The dataset consists of wheel carrier and body vertical acceleration time signals. The obtained principal components of the time signals are used as feature vectors for the consensus principle components analysis algorithm.

DAOCHAO [32] also realizes a SVM-based defect detection of a train chassis with a similar structure to [8] . The features used are derived from both the time and the frequency domain of the acceleration signals. In addition, the energy coefficients of a wavelet-transformation are committed to the system. The selection of the best features is done by recursive feature elimination. As well as WEI and DAOCHAO, ZHIDAN [33] uses a SVM for its train chassis monitoring system. In this case, the features consist mainly of the sample entropy of vertical acceleration signals, which allows the system to distinguish between different defect cases.

Even though, some methods for damper diagnosis already exist, there is no reliable approach, that allows diagnosing dampers based on ESC data.

3 Test Setup

The analysis in this paper is based on a research vehicle, type BMW 650i Grand Coupé. It is equipped with semi-active dampers. An engineering interface enables the application of individual constant damper currents for the rebound and compression stage of each damper.

3.1 Measurement sensors

The recorded data is obtained directly from the vehicle's chassis system CAN. Lateral and longitudinal acceleration as well as the wheel speed of each wheel represent the relevant sensors. All sensor signals are updated with a frequency of 50 Hz.

3.2 Implementation of damper defects

Figure 1 and Figure 2 show the characteristic curves of the semi-active dampers for different damper current settings. The "intact" mode has a specific constant rebound and compression stage current. Based on expert knowledge, this current setting is suited to simulate the behavior of a passive damper. The "defect" mode generates the least damping forces mirroring real world damage characteristics. A third mode generates high damping forces and leads to a "stiff" chassis system. It was chosen to cover greater differences between the considered damper settings. The vehicle's "sport" and "comfort" mode of the semi-active dampers cannot be shown, as those driving modes consist of varying damper currents.

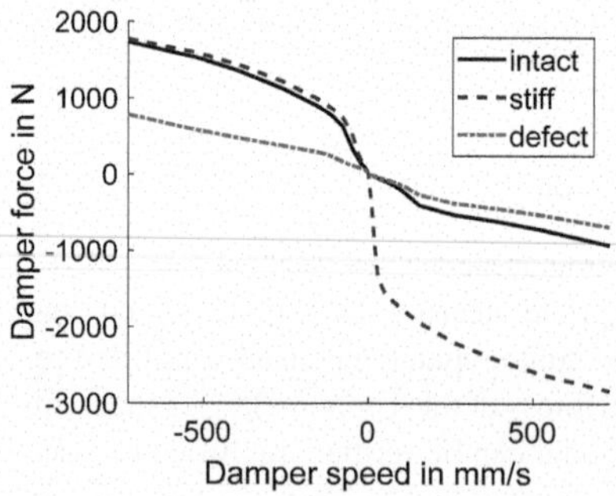

Figure 1: Characteristic curve of front axle damper

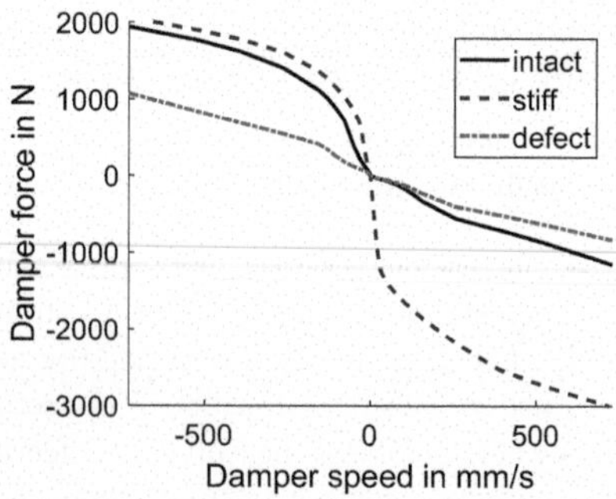

Figure 2: Characteristic curve of rear axle damper

3.3 Test roads and resulting datasets

Two different datasets were created for the analysis in this paper. The first dataset ("intact vs. defect") consists of several test drives with either all dampers set to "intact", all dampers set to "defect", or the front left or the rear right damper set to "defect" while the other dampers are in "intact" mode. Measurements of the "intact vs. defect" dataset were conducted on several roads with different track conditions. A combination of the German Autobahn A9 and A92 was driven from the connection point "Garching Nord" via the interchange "Neufahrn" to the connection point "Unterschleißheim". While the roughness of the A9 is smooth, the A92 consists of cross groves, which results in more road excitation compared to the A9. Furthermore, four different country roads were driven. Hereof, two are badly conditioned country roads, one of which with short-wavelength radiation, which results in a high vertical wheel carrier excitation, and one with long-wavelength radiation, which results in a high body mass excitation. The other two country roads are average German roads with no specific characteristics. The "intact vs. defect" dataset consists of 280 minutes of driving with speeds between 60 and 140 km/h.

The second dataset ("stiff vs. defect") evaluates the performance of the algorithm for greater differences in the damper settings. Due to the availability of the research vehicle, it is slightly different from the "intact vs. defect" dataset. It consists of test drives with either all dampers set to "stiff" or any (all or one) damper set to "defect" while the other dampers are in "stiff" mode. Measurements of the "stiff vs. defect" dataset were conducted at the same German Autobahn as for the first dataset as well as on an average country road with no specific characteristic. There are no badly conditioned roads in this dataset. The "stiff vs. defect" dataset consists of 190 minutes of driving with speeds between 60 and 140 km/h.

4 Fault Detection and Isolation System

The following chapter shows the development of the fault detection and isolation system. After a brief general overview of this process, as shown in Figure 3, more details of the relevant steps are provided in the following subchapters.

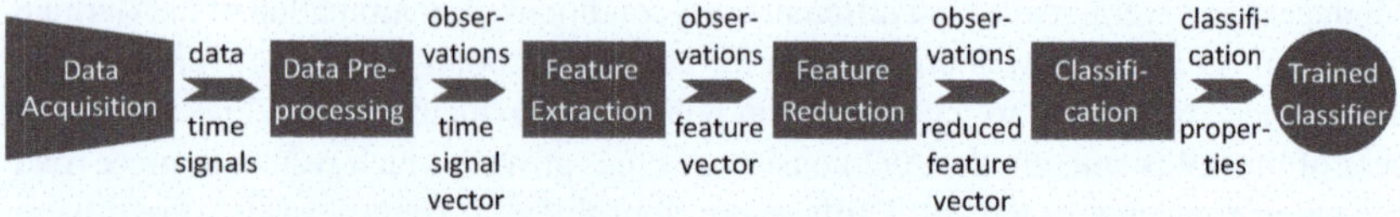

Figure 3: Overview of the fault detection and isolation system training process

Data acquisition is done according to the measurement setup explained in chapter 3. The output is raw data in form of time signals of each sensor.

At first, the raw data needs to be preprocessed. The long-time sequences are cut to short-time sequences of 5 seconds. The mean longitudinal and lateral accelerations of those short-time sequences need to be below a threshold of 1 m/s² and the velocity needs to be above 30 km/h in order to consider only situations with similar and relevant dynamic circumstances. This requirement holds for large portions of the measurement data, as customer-relevant driving is often considered as driving straight at a constant speed. Only short-time sequences meeting those requirements (further called observations) are considered for the training of the classifier.

In the next step, features are extracted for each sensor time signal of each observation. A feature is a specific characteristic value of a time signal of an observation. Features can be of different domains e.g. from the time domain with a mean, maximum or minimum signal value, or from the frequency domain with an amplitude at a given frequency or the location of an eigenfrequency. The result of the feature extraction is a feature vector for each observation.

A high amount of features does not guarantee a good classification. The better the differences of the data classes are represented by the supplied features, the more accurate the classification. The feature reduction process takes care of the quantification of this dependency, which results in a reduced feature vector.

Finally, the observations with the reduced feature vectors are classified. This can be conducted with a binary (defect present or not) or a multi-class (which damper is defect) approach. The main output of this step is a trained classifier for the classification of further data. However, there are additional properties of the classification, e.g. a trained classifier can be prone to overfitting and therefore lose its capability for generalization.

Also quantifying the confidence of each observation classification is important for the further usage of the generated knowledge as misclassification is always an issue in the automotive field.

4.1 Feature extraction

In principle, features for classification can be generated in two ways. One way is to extract features that represent physical or mathematical relationships of a system. Another way is calculating artificial features by means of neural networks. This work focuses on physical and mathematical features originating from the frequency or time domain. In the following, the most meaningful features of the fault detection and isolation system are exemplified.

4.1.1 Singular value decomposition

The interpretation of singular values is similar to the meaning of eigenvalues. They represent the character of a matrix. Singular value decomposition is the generalized formulation of eigenvalue decomposition and is used to compute the singular values of non-orthogonal matrices. By means of a transformation, the non-orthogonal matrix A of size $m \, x \, n$ is decomposed into the matrices U (size $m \, x \, r$), V (size $n \, x \, r$) and $\sum$ (size $r \, x \, r$) [34, 35] by

$$A = U \sum V^T \tag{1}$$

$\sum$ contains the singular values on the diagonal and can be found by solving the equations

$$A^T A = V \sum \sum^T V^T \tag{2}$$

$$AV = U \sum. \tag{3}$$

The SVD feature is evaluated for each observation. Matrix A consists of the six sensor time signals of an observation interval of size ($t_{\text{obs}} * f_s$ x $6 = 250$ x 6) with the sample rate f_s and the observation time t_{obs}. Hereby, the singular values characterize the relationship of the sensor signals. The calculated six singular values are normalized with respect to the largest singular value within an observation and are stored in descending order.

4.1.2 Magnitude and frequency of maximum power spectral density

Several features can be obtained by evaluating the power spectral densities especially for the wheel speed data, but also for acceleration sensor data. The power spectral density of the wheel speed signal shows a shift and increase in the maximum value of the region for the wheel carrier and vehicle body vertical natural frequency [21, pp. 42-43]. Therefore, the shift and peak values are obtained for both eigenfrequency intervals, in which filtered and unfiltered preprocessed data is used. A band-pass filter is applied to the wheel carrier vertical natural frequency to emphasize the peaks and the filtered signal's power spectral density is evaluated. The band-pass frequency range is set to 10 to 15 Hz to cover shifts of the eigenfrequency. This is also conducted for a reference interval at the eigenfrequency of the vehicle body (2.5 to 5 Hz).

Another feature is calculated by dividing the magnitude peak of the reference interval by the magnitude peak of the wheel carrier natural frequency interval.

4.1.3 Frequency band energy

This feature is calculated by evaluating the spectral power density in a specific frequency interval. According to the Nyquist-Shannon theorem, the power spectral density can be resolved with a maximum frequency 25 Hz (half of the sample frequency) [36]. Therefore, five intervals are set for calculating the frequency band energy (1 - 5 Hz, 5 - 10 Hz, 10 - 15 Hz, 15 - 20 Hz, 20 - 25 Hz) for each sensor signal. In addition to the five evaluation ranges, the frequency band energy is calculated for the whole frequency range and for an interval of ± 2 Hz from the peak of the wheel carrier's vertical natural frequency range of the power spectral density.

4.1.4 Entropy

In the information theory, entropy is a measure of the average information content of a message [37, pp. 61-95]. In particular, the repeatability or predictability of a signal can be measured by means of entropy [38, pp. 349-352]. Among others, approaches for sample entropy, approximate energy entropy, Shannon entropy and log energy entropy exist for entropy calculation [37, pp. 61-95, 39]. All mentioned approaches are used for feature generation of the fault detection and isolation system. In the following, the approach of approximate energy entropy $ApEn$ is highlighted.

To calculate the entropy value [38], a time signal $u = u(1), u(2) \dots u(N)$ is divided into $N - m + 1$ sequences $x(i) = u(i), u(i+1), \dots, u(i+m-1)$ with $i = 1, \dots, N - m + 1$ with the length of the sequences m which is often chosen in the range $2 \leq m \leq 4$ [38, pp. 349-352]. The predictability of the time signal is calculated by using the similarity of the sequences to each other. This is done by calculating the distance

$$d[x(i), x(j)] = \max_{k=0,\dots,m-1} (|u(i+k) - u(j+k)|), \tag{4}$$

which indicates the distance of a sequence $x(j)$ to the respective reference sequence $x(i)$ [40]. Using the filter stage r, all distances fulfilling the condition

$$d[x(i), x(j)] < r \tag{5}$$

are counted.

$$C_i^m(r) = \frac{\sum_{j=1}^{N-m+1} d[x(i), x(j)] < r}{N - m + 1} \tag{6}$$

gives a measure of repeatability within the signal [40]. These steps are repeated for a sequence length of $m + 1$. Finally, the approximate entropy is found by evaluating

$$\Phi^m(r) = \frac{\sum_{i=1}^{N-m+1} \log(C_i^m(r))}{N - m + 1} \tag{7}$$

$$ApEn = \Phi^m(r) - \Phi^{m+1}(r) \tag{8}$$

4.1.5 Autoregressive model (AR)

Autoregressive models are used to analyze complex stochastic processes and consist of linear difference equations. The basis of the AR-model is the assumption that the target variable y_t depends linearly on its p previous values $y_{t-1}, y_{t-2}, \dots, y_{t-\mathrm{p}}$ [39]. The regressed target variable results in

$$y_t = c + \sum_{i=1}^{p} \varphi_i \, y_{t-i} + \varepsilon_t, \tag{9}$$

in which the p model parameters φ_i can be found using the Yule-Walker Equation or the least square method. Finally, the p AR-model coefficients φ_i for every sensor signal are used as features to design the defect detection system.

The high number of AR-model parameters poses a challenge for the fault detection and isolation system, as the number of features increase the risk of overfitting. Therefore, a feature optimization is necessary to identify an optimum between the amount of AR-model coefficients (order of the model) and the categorization ability of the feature block. One way to optimize the model order p is to train and validate a classifier for different AR model orders. For this reason, a linear SVM is trained with only AR-model features of different order to subsequently classify a test dataset. In this analysis, the AR-model with the lowest classification error for predicting a binary dataset is $p = 10$.

4.1.6 Autocorrelation

The autocorrelation feature is described using past values of its own time series. In contrast to autoregression, autocorrelation builds the residuals of the signals y_t and the time series y_{t+k} delayed by k to analyze different signal characteristics.

Accordingly, the autocorrelation

$$R_k = \frac{\sum_{t=1}^{N-k}(y_t - \bar{y})(y_{t+k} - \bar{y})}{\sum_{t=1}^{N-k}(y_t - \bar{y})^2} \tag{10}$$

evaluates the relationship between the original and k-shifted time series for a certain number of delays n with $k = 0, \dots, n$. Like the auto-regression, the autocorrelation feature implies the risk of overfitting. It is therefore necessary to determine the optimal order n of the autocorrelation model to ensure the highest possible categorization accuracy of the autocorrelation feature. The minimum classification error for binary classification was found for $n = 8$.

4.1.7 Further features

In addition, various features from the time and frequency domain are generated, which further increase the separability. These are listed in Table 1.

Table 1: Additional features for damper defect detection

Feature-block	Domain	Number of features
Mean	Time	6
Standard deviation	Time	6
Mean absolute deviation	Time	6
Maximum	Time	6
Minimum	Time	6
Kurtosis	Time	6
Skewness	Time	6
Interquartile range	Time	6
Root mean square	Time	6
Range	Time	6
Crest	Time	6
Mean frequency	Frequency	6

4.2 Feature reduction

The goal of the feature reduction is to reduce the feature space in order to prevent over- and underfitting, and to improve classification results. A feature reduction can be achieved using various methods. A distinction is made between wrapper, embedded and filter-based methods. Wrapper methods use the classification accuracy of a classification algorithm to rank the features. The advantage of this method is the evaluation of the final classification accuracy with different features. However, this feature selection method is dependent on the classification method, whereby the selected features do not directly guarantee equal results with other classification methods. Embedded feature reduction methods are also dependent on a certain classification method, which classify using a direct selection of the features (e.g. decision trees). Filter-based feature reduction approaches compute an independent criterion that can be used to rank the best features. These approaches usually regard only a single feature and do not consider the dataset separability by projecting multiple features. [41]

Three different feature reduction methods were tested within the scope of this paper, which are the filter-based methods S-score and Fisher-score as well as the wrapper method random feature elimination.

The S-score calculates the information content

$$S_k = \frac{|\mu_1^k - \mu_2^k|}{(\sigma_1^k)^2 + (\sigma_2^k)^2} \tag{11}$$

of the k-th feature by evaluating the standard deviation σ_i^k and the mean μ_i^k for the respective class i (e.g. $i =$ "intact", "defect"). he S-score is a measure of the dataset separability using the k-th feature and can only be evaluated for binary classification problems. For multiclass problems, the S-score must be calculated for each binary classifier (one-vs-one or one-vs-all).

A similar filter-based feature selection method is the Fisher-score F_k. The Fisher-score

$$F_k = \frac{\sum_{i=1}^{c} n_i \, (\mu_i^k - \mu^k)^2}{\sum_{i=1}^{c} n_i \, (\sigma_i^k)^2} \tag{12}$$

consists of the mean μ_i^k and standard deviations σ_i^k of the k-th feature (k as superscript) for n_i observations of the respective class $i = 1, \dots, c$. The mean of the k-th feature of the whole dataset is denoted by μ^k. In addition to the S-score and Fisher-score, the wrapper method random feature elimination is investigated. To determine this feature ranking, a bottom-up or top-down approach can be used. The bottom-up approach trains with ever-increasing feature sets, whereby features with the greatest positive influence on the classification accuracy are selected. The top-down approach works in the opposite way, eliminating the worst features and thus ranking them. The feature selection of the developed fault detection and isolation system is conducted via top-down random feature elimination, as this approach reacts more insensitive to high-dimensional feature blocks. [42, 43]

4.3 Classification

There are numerous approaches for diagnosing a physical component. Besides regressive approaches by calculating a state of health of a system, several methods classifying datasets into binary or multiclass systems exist [44]. The SVM classifier represents one of the most broadly used classification technique because of its robustness and performance [45]. Which is why SVM classifiers are preferably used for defect detection systems [32, 33].

Considering a dataset of a two-class SVM classification problem

$$\{(x_i, y_i) | i = 1, \ldots, m; y_i \in \{-1,1\}\} \tag{13}$$

with y_i being the class affiliation of observation x_i described by n features [46]. A feature space exists which allows a linear separability of the observations of different classes. A kernel transformation $\phi(x)$ transforms the dataset into a higher dimension in order to create a linear hyperplane

$$y(x) = w^T \phi(x) + b. \tag{14}$$

The margin is given by the perpendicular distance of the hyperplane to the closest points x_i from the data set (support vectors) [44]. To find the maximum margin solution, the parameters w and bias b need to be optimized by solving

$$\arg max_{w,b} \left\{ \frac{1}{\|w\|} min_i \left[y_i \left(w^T \phi(x_i) + b \right) \right] \right\}. \tag{15}$$

The trained SVM model can than classify the observations into the respective classes by evaluating

$$y(x) = sign(w^T \phi(x) + b). \tag{16}$$

Support vector machines are non-probabilistic classifiers. Nevertheless, it is possible to evaluate the confidence level of categorizations. By evaluating margins of specific observations, a measure of classification certainty can be obtained. Dealing with multiclass-problems, SVMs can be transformed to multiple binary-classification problems. Common approaches are creating several one-vs.-all- or one-vs.-one-class SVMs. [44]

A general problem of machine learning methods is overfitting. On the one hand overfitting depends on the selected features, which specify the shape of the data set in the high-dimensional feature space. On the other hand, there is a possibility to allow a certain number of misclassifications (C-parameter) to prevent overfitting. Both parameters determine the general validity of the trained SVM classifier. An iterative adaptation of the selected features and C-Parameters provides a universal separation plane and determines an ideal number of support vectors. [44]

The classification of the datasets for damper diagnosis should cover not only the defect detection but also the defect isolation. The fault detection and isolation system should be able to distinguish a defect of a single damper from a total defect of all dampers and assign the respective defect to a specific wheel. Therefore, the system design consists of a multiclass classifier, which distinguishes the observations directly into the respective classes of which damper is defect. By a redefinition of the multiclass labels into the binary labeling convention (any defect present or not), the detection function can be performed. Hereby, observations that are "defect" in reality but are incorrectly classified according to the defect position still correspond to the correct detection class.

5 Results

The circumstances of many features with moderately high information content results in a high-dimensional feature space, which is why the SVM was chosen for this classification problem. A linear SVM classifier was selected because of its higher robustness in terms of overfitting compared to higher order SVMs.

The proposed feature reduction methods S-score, Fisher-score and RFE were evaluated for both datasets including binary and multiclass classification. S-score and Fisher-score provide similar features for the same datasets and classification type. Nevertheless, the classification accuracy remains below the level of the RFE feature selection.

Figure 4 shows the accuracy of the classification of the “stiff vs. defect” dataset for different numbers of features for the actual classification with a linear SVM. Selecting features based on the Fisher- or S-score leads to similar classification results for the same number of selected features. However, the resulting classification accuracy using features selected with the RFE algorithm shows the best results. The small differences for using all features results from the different alignment of the features within the observation matrix, which can result in different SVM parameters and therefore in small differences of the classification accuracy.

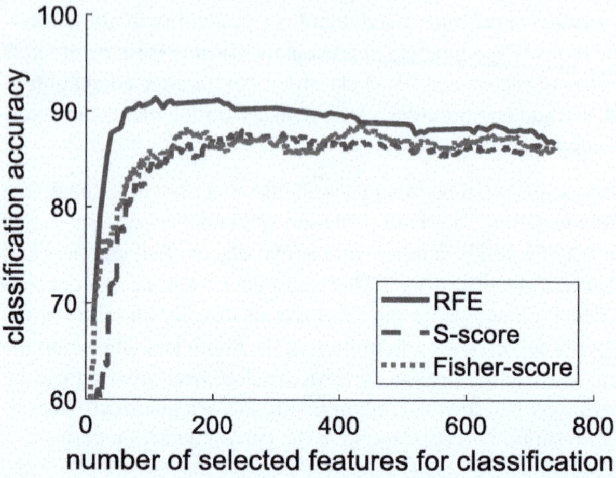

Figure 4: Classification accuracy for different number of selected features using different feature reduction methods for the "stiff vs. defect" dataset with a linear SVM

The feature space, based on the RFE feature ranking and the simultaneous validation of the classifier to avoid overfitting for the different datasets, leads to the finally used feature set. Using 239 features led to the best overall results for both datasets. The used linear SVM classifiers are trained based on 70 % of the complete dataset and tested with 30 % of the dataset. The 70 % training data is further divided to 75 % actual training data and 25 % holdout validation data to prevent overfitting. The classification accuracies for the respective test data of each dataset are listed in Table *2*.

Table 2: Correct classification rate of the test data for both datasets

Dataset	Detection	Isolation
Stiff vs. defect	94,43 %	92,96 %
Intact vs. defect	84,90 %	72,46 %

The classification results are promising especially for the "stiff vs. defect" dataset. The results of the "intact vs. defect" dataset regarding the failure isolation need to be improved further. However, the difference of the damper forces between the "intact" and "defect" damper settings are small in the damper compression stage (Figure 1 and Figure 2). Those differences were already barely noticeable by the driver. Even those small differences are detected by a classification rate of 84 % and isolated by 72 %.

Further improvements of the classification can be achieved by implementing the classification confidence level. If a critical value of the confidence level is undercut, no classification of the observation takes place. Sorting out observations, which are difficult to classify can bring up to 2 % higher classification accuracies. Furthermore, a final indication of a damper defect can be deployed only when a specific number of consecutive observations is detected as a defect.

6 Summary and Outlook

This paper discusses a novel approach for detecting damper defects based on ESC sensor readings as longitudinal and lateral acceleration as well as wheel speed sensors. The developed fault detection and isolation system based on feature classification by a support vector machine was explained. Herein, some significant features as well as different feature reduction techniques and their impact on the classification accuracy were presented. Finally the classification results of this detection system were displayed.

Extracting meaningful features is a key to generate good classification results. To prevent the step of searching for the right features manually, a future step is the automatic generation of features. Several approaches for this topic exist in literature. Another step is the direct classification of time signals by neural networks, which also prevents the necessity of extracting the right features manually.

7 Contributions

Thomas Zehelein (corresponding author) initiated this paper. He essentially contributed to the overall methodology of the proposed approach and supervised the master thesis of Alexander Merk. Alexander Merk made essential contributions to the implementation of the proposed approach. Both implemented this paper.

Markus Lienkamp made an essential contribution to the conception of the research project. He revised the paper critically for important intellectual content. Mr. Lienkamp gave final approval of the version to be published and agrees to all aspects of the work. As a guarantor, he accepts responsibility for the overall integrity of the paper.

8 Acknowledgement

This research received no specific grant from any funding agency in the public, commercial, or not-for-profit sectors. However, the author would like to thank BMW AG for supplying the research vehicle for independent research.

9 References

[1] Z. Gao, C. Cecati, and S. X. Ding, "A Survey of Fault Diagnosis and Fault-Tolerant Techniques—Part I: Fault Diagnosis With Model-Based and Signal-Based Approaches," *IEEE Trans. Ind. Electron.*, vol. 62, no. 6, pp. 3757–3767, 2015.

[2] Z. Gao, C. Cecati, and S. Ding, "A Survey of Fault Diagnosis and Fault-Tolerant Techniques Part II: Fault Diagnosis with Knowledge-Based and Hybrid/Active Approaches," *IEEE Trans. Ind. Electron.*, p. 1, 2015.

[3] R. Liu, B. Yang, E. Zio, and X. Chen, "Artificial intelligence for fault diagnosis of rotating machinery: A review," *Mechanical Systems and Signal Processing*, vol. 108, pp. 33–47, 2018.

[4] Y. Lei, F. Jia, J. Lin, S. Xing, and S. X. Ding, "An Intelligent Fault Diagnosis Method Using Unsupervised Feature Learning Towards Mechanical Big Data," *IEEE Trans. Ind. Electron.*, vol. 63, no. 5, pp. 3137–3147, 2016.

[5] Z. H. Mamar, P. Chainais, and A. Aussem, "Combining learning methods and time-scale analysis for defect diagnosis of a tramway guiding system," in *16th Mediterranean Conference on Control and Automation, 2008: MED '08 ; 25 – 27 June 2008, Ajaccio, Corsica, France ; conference proceedings*, Ajaccio, France, 2008, pp. 1780–1785.

[6] C. P. Ward *et al.*, "Condition Monitoring Opportunities Using Vehicle-Based Sensors," *Proceedings of the Institution of Mechanical Engineers, Part F: Journal of Rail and Rapid Transit*, vol. 225, no. 2, pp. 202–218, 2011.

[7] P. Li *et al.*, "Estimation of railway vehicle suspension parameters for condition monitoring," *Control Engineering Practice*, vol. 15, no. 1, pp. 43–55, 2007.

[8] X. Wei, Y. Guo, L. Jia, and H. Liu, "Fault detection of rail vehicle suspension system based on CPCA," in *Conference on Control and Fault-Tolerant Systems (SysTol), 2013: 9 – 11 Oct. 2013, Hotel Le Negresco, Nice, France*, Nice, France, 2013, pp. 700–705.

[9] T. X. Mei and X. J. Ding, "New condition monitoring techniques for vehicle suspensions," in *Railway Condition Monitoring, 2008 4th IET International Conference on: Date, 18-20 June*, Derby, UK, 2008, p. 39.

[10] J. P. N. González and P. P. Villanueva, "Vehicle Lateral Dynamics Fault Diagnosis Using an Autoassociative Neural Network and a Fuzzy System," in *Lecture Notes in Computer Science, Advances in Computational Intelligence*, D. Hutchison et al., Eds., Berlin, Heidelberg: Springer Berlin Heidelberg, 2013, pp. 236–246.

[11] Y. Jeong *et al.,* "Vehicle sensor and actuator fault detection algorithm for automated vehicles," in *2015 IEEE Intelligent Vehicles Symposium (IV): June 28, 2015 – July 1, 2015, COEX, Seoul, Korea*, Seoul, South Korea, 2015, pp. 927–932.

[12] Y. Lin, B.-C. Chen, H.-C. Tsai, and B.-C. Luan, "Model-Based Sensor Fault Detection for Advanced Driver Assistance System," in *The 13th International Conference on Automotive Engineering*, 2017.

[13] M. L. Schwall, J. C. Gerdes, and F. M. A, "Multi-modal diagnostics for vehicle fault detection," (English), *American Society of Mechanical Engineers, Dynamic Systems and Control Division (Publication) DSC*, vol. 70, 2002.

[14] D. H. Alcantara, R. Morales-Menendez, and L. Amezquita-Brooks, "Fault diagnosis for an automotive suspension using particle filters," in *2016 European Control Conference (ECC): June 29-July 1, 2016, Aalborg, Denmark*, Aalborg, Denmark, 2016, pp. 1898–1903.

[15] S. Azadi and A. Soltani, "Fault detection of vehicle suspension system using wavelet analysis," *NVSD*, vol. 47, no. 4, pp. 403–418, 2009.

[16] C. Ferreira *et al.,* "Sensing methodologies to determine automotive damper condition under vehicle normal operation," *Sensors and Actuators A: Physical*, vol. 156, no. 1, pp. 237–244, 2009.

[17] G. Hardier, "Recurrent RBF networks for suspension system modeling and wear diagnosis of a damper," in *ICNN '98 – International Conference on Neural Networks*, Anchorage, AK, USA, May. 1998, pp. 2441–2446.

[18] D. Hernández-Alcántara, J. C. Tudón-Martínez, L. Amézquita-Brooks, C. A. Vivas-López, and R. Morales-Menéndez, "Modeling, diagnosis and estimation of actuator faults in vehicle suspensions," *Control Engineering Practice*, vol. 49, pp. 173–186, 2016.

[19] J. Lozoya-Santos, J. C. T.-M. nez, R. Morales-Menendez, R. R. rez-Mendoza, and A. M. Gutierrez, "Fault Detection for an Automotive MR Damper," *IFAC Proceedings Volumes*, vol. 45, no. 6, pp. 1023–1028, 2012.

[20] M. D. Bedük, K. Çalışkan, R. Henze, and F. Küçükay, "Effects of damper failure on vehicle stability," *Proceedings of the Institution of Mechanical Engineers, Part D: Journal of Automobile Engineering*, vol. 227, no. 7, pp. 1024–1039, 2012.

[21] M. Jautze, *Ein signalmodellbasiertes Verfahren zum Erkennen von Dämpferschäden bei Kraftfahrzeugen*. Zugl.: Erlangen-Nürnberg, Univ., Diss., 2002. Düsseldorf: VDI-Verl., 2002.

[22] J. Tao, Y. Liu, and D. Yang, "Bearing Fault Diagnosis Based on Deep Belief Network and Multisensor Information Fusion," *Shock and Vibration*, vol. 2016, no. 7, pp. 1–9, 2016.

[23] T. van Tran, F. AlThobiani, and A. Ball, "An approach to fault diagnosis of reciprocating compressor valves using Teager–Kaiser energy operator and deep belief networks," *Expert Systems with Applications*, vol. 41, no. 9, pp. 4113–4122, 2014.

[24] S. Dong, Z. Zhang, G. Wen, and G. Wen, "Design and application of unsupervised convolutional neural networks integrated with deep belief networks for mechanical fault diagnosis," in *Conference proceedings: 2017 Prognostics and System Health Management Conference (PHM-Harbin)*, Harbin, China, 2017, pp. 1–7.

[25] M. Boerner, M. Zele, and R. Isermann, *Comparison of different fault detection algorithms for active body control components: ACC : June 25-27, 2001, Crystal Gateway Marriot, Arlington, VA, USA*. Evanston, Ill, Piscataway, N.J: American Automatic Control Council, 2000.

[26] D. Hernandez-Alcantara, R. Morales-Menendez, Amezquita-Brooks, L. Sename, O., and L. Dugard, *Fault Estimation Methods for Semi-Active Suspension Systems: 2015 IEEE International Autumn Meeting on Power, Electronics and Computing (ROPEC), Ixtapa, Mexico, 4-6 November 2015*. Piscataway, NJ: IEEE, 2015.

[27] H.-B. Moncada and J.-A. Marin, "Fault Detection for a Bus Suspension Model Using an Utkin Observer," in *2011 IEEE Electronics, Robotics and Automotive Mechanics Conference*, Cuernavaca, Morelos, Mexico, 2011, pp. 246–251.

[28] J. Tudón-Martinez *et al.*, *Fault Tolerant Control with Additive Compensation for faults in an Automotive Damper: 10th IEEE International Conference on Networking, Sensing and Control (ICNSC), 2013, 10 – 12 April 2013, Evry Val-d'Essonne University*. Piscataway, NJ: IEEE, 2013.

[29] D. Fischer and Eberhard, K., Isermann, R., *Fault detection for an active vehicle suspension: ACC : June 4-6, 2003, the Adams Mark Hotel, Denver, Colorado, USA*. Piscataway, N.J: IEEE, 2003.

[30] V. Paciello, P. Sommella, V. Paciello, and P. Sommella, "Smart Sensing and Smart Material for Smart Automotive Damping // Smart sensing and smart material for smart automotive damping," *IEEE Instrum. Meas. Mag.*, vol. 16, no. 5, pp. 24–30, 2013.

[31] Y.-m. Liu *et al., Eds., Suspension system fault diagnosis method based on fuzzy mathematics: ICEMI '09 ; 16 – 19 Aug. 2009, Beijing, China*. Piscataway, NJ: IEEE, 2009.

[32] T. Daochao, J. Weidong, Q. Na, and L. Hui, "Feature selection and analysis of single lateral damper fault based on SVM-RFE with correlation bias reduction," in *2016 35th Chinese Control Conference (CCC)*, Chengdu, China, 2016, pp. 3840–3845.

[33] W. Zhidan, J. Weidong, and Q. Na, "Fault feature analysis of high-speed train suspension system based on multivariate multi-scale sample entropy," in *2016 35th Chinese Control Conference (CCC)*, Chengdu, China, 2016, pp. 3913–3918.

[34] G. Golub and W. Kahan, "Calculating the Singular Values and Pseudo-Inverse of a Matrix," *Journal of the Society for Industrial and Applied Mathematics Series B Numerical Analysis*, vol. 2, no. 2, pp. 205–224, 1965.

[35] K. Modarresi, "Unsupervised Feature Extraction Using Singular Value Decomposition," *Procedia Computer Science*, vol. 51, pp. 2417–2425, 2015.

[36] C. E. Shannon, "Communication In The Presence Of Noise," *Proc. IEEE*, vol. 86, no. 2, pp. 447–457, 1998.

[37] R. M. Gray, *Entropy and Information Theory,* 2nd ed. s.l.: Springer Science + Business Media, 2011.

[38] J. M. Yentes *et al.,* "The appropriate use of approximate entropy and sample entropy with short data sets," (eng), *Annals of biomedical engineering*, vol. 41, no. 2, pp. 349–365, 2013.

[39] *Pandas feature database.* [Online] Available: http://tsfresh.readthedocs.io/en/latest/api/tsfresh.feature_extraction.html. Accessed on: Apr. 18 2018.

[40] S. M. Pincus, “Approximate entropy as a measure of system complexity,” *Proceedings of the National Academy of Sciences*, vol. 88, no. 6, pp. 2297–2301, 1991.

[41] Isabelle Guyon and Andr´e Elisseeff, “An Introduction to Variable and Feature Selection,” *Journal of Machine Learning Research*, no. 3, pp. 1157–1182, 2003.

[42] T. Daochao, J. Weidong, Q. Na, and L. Hui, “Feature selection and analysis of single lateral damper fault based on SVM-RFE with correlation bias reduction,” in *2016 35th Chinese Control Conference (CCC)*, Chengdu, China, 2016, pp. 3840–3845.

[43] Sun, Zhang, R. Sun, and X. Zhang, “Top-Down versus Bottom-Up Learning in Cognitive Skill Acquisition,” *Cognitive Systems Research*, vol. 5, no. 1, pp. 63–89, 2004.

[44] C. M. Bishop and C. M. Bishop, *Pattern Recognition and Machine Learning,* 8th ed. New York, NY: Springer, 2009.

[45] M.-C. Lee and C. To, “Comparison of Support Vector Machine and Back Propagation Neural Network in Evaluating the Enterprise Financial Distress,” *IJAIA*, vol. 1, no. 3, pp. 31–43, 2010.

[46] S.-B. Zhe, Z.-Z. Han, H.-J. Tang, and Y. Zhang, “Application of support vector machines to sensor fault diagnosis in ESP system: Machine Learning and Cybernetics, 2004. Proceedings of 2004 International Conference on Machine Learning and Cybernetics, Shanghai, 26-29 August 2004,”

STEERING.TECH SECTION

STEERING FEEL AND HUMAN-MACHINE INTERFACE (HMI)

Steering features of the new Ford Focus

Dr. Thorsten Hestermeyer

W. Bongarth,

J. Dornhege

P. Pfeffer (Ed.), *9th International Munich Chassis Symposium 2018*, Proceedings,
https://doi.org/10.1007/978-3-658-22050-1_32

Introduction

The introduction of Advanced Driver Assistance Systems (ADAS) in automobiles is growing since years and speed of technical innovation is ever increasing. Customers are becoming more demanding in use of ADAS and comfort/safety features and expectation of technical execution is rising.

As a reaction to the demand of Steering enabled or EPAS executed features since several years Ford Chassis Engineering has pulled the development of Steering Control features and functions in-house. This development approach of steering features and functions provides several advantages and enhances the ability to respond to market trends quicker.

On the one hand in-house development can reduce the complexity of software for the multiple program top-hat applications and various tier 1-suppliers for the EPS system. Via the used development and change management processes reusability of SW code and increase quality and robustness of the algorithms can be ensured.

But also consistency of the feature's subjective steer feel and functionality is ensured by sharing the in-house steering software across all carlines globally.

The all new Ford Focus is spear heading the global introduction of a wide range of Ford owned steering feature/function software which enhances steering error state mitigation as well as an extension of ADAS and comfort/safety functions.

The steering disturbances which impact the quality perception of the customer steer feel are normally considered as any kind of unintended vibrations or torque overlays while driving. The well-known and established functions in today's production vehicles are for instance Active Nibble Control and Pull Drift Compensation, which have already significantly contributed to the reduction of error states in recent years.

Powertrain and braking induced steering torque disturbances so far however were not mitigated or reduced to acceptable level in the past. This has led to the development and introduction of a function called Steering Torque Disturbances Reduction (STDR) which enables Torque Steer Compensation and Brake Pull Reduction SW functions in the new Ford Focus.

Beyond the customer transparent error state mitigation functions we respond to the increasing demand for ADAS features in modern automobiles with the introduction of various advanced steering based features. These include the well-known Lane Keeping System and Lane Departure Warning features which are extended by the Road Edge Detection functionality. The current trend and customer desire towards more lateral control features of the vehicle is realized by the introduction of a Lane Centering Aid (LCA).

This "hands on the steering wheel" SAE level 2 feature enables a Traffic Jam Assist function as well as guiding the vehicle at higher speeds within the center of the chosen lane.

A further function to enhance the safety of the vehicle in emergency situations is the Evasive Steer Assist (ESA) which is a camera and front radar enabled system to help the driver to steer around hazards or objects when breaking alone would not be sufficient to avoid an accident.

All above steering based features will be introduced and available with the new Ford Focus in its fourth generation and will push the boundary in the compact vehicle segment with regard to the steering enabled software features and functions.

Integration of Ford In-house Software into the Steering

For implementation of steering features, Ford uses an in-house process. In-house functions are integrated as obj-files into the supplier software as described in [1]. The resulting software structure is shown in Figure 1.

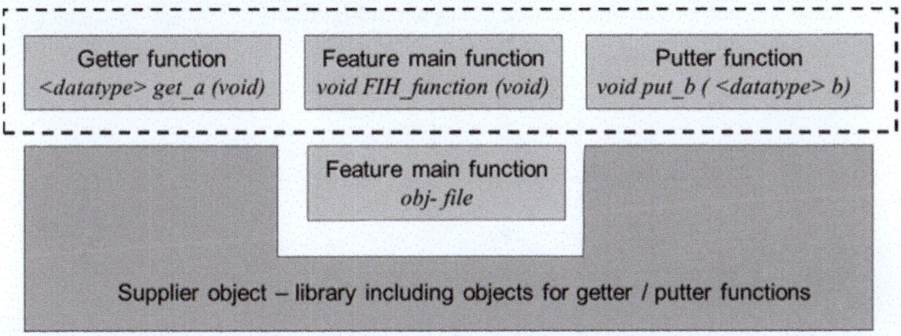

Figure 1: Integration of Ford In-House Software Functions into the Supplier Software (rf. [1])

The in-house functions generate a torque that is added to the normal EPAS torques. Ford typically differentiates between two major torque input interfaces: one input is added as a torque that adds to the pinion torque (→ pinion torque interface), the other torque is added like an additional driver hand-wheel-torque (→hand-wheel torque interface). The effect of additional driver-torque input is different from a pinion torque input, as driver torque is amplified via the basic assist functions in the steering.

In order to lower the Safety integrity required for the Ford functions, the functional safety concept forces integration via limiters, monitors and a central arbiter function (see Table 1). These concepts are described in more detail in [1].

Table 1: Functional Safety Elements for Integration of Ford In-House Functions

Restriction Safety Element	Safety Element
Torque level of features are usually significantly lower	Limiter that reduce the torque according a given shape *(Torque Shape Limiter – TSL)*
Some features are only active in special situations	Arbiter functionality that enforces operation only in a given safe condition
Some feature torques have a characteristic property that improve controllability	Limiter that diagnoses whether characteristic behavior (e.g. frequency) is kept and that reduces torque otherwise *(Monitor)*
For many driving situations, there is a characteristic connection between driver torque and steering wheel reaction	Limiter that checks for the vehicle reaction and ensures torque limitation if out of allowed reaction/torque band *(Steering Reaction Limiter – SRL)*

These ideas are put into an architecture as shown in Figure 2.

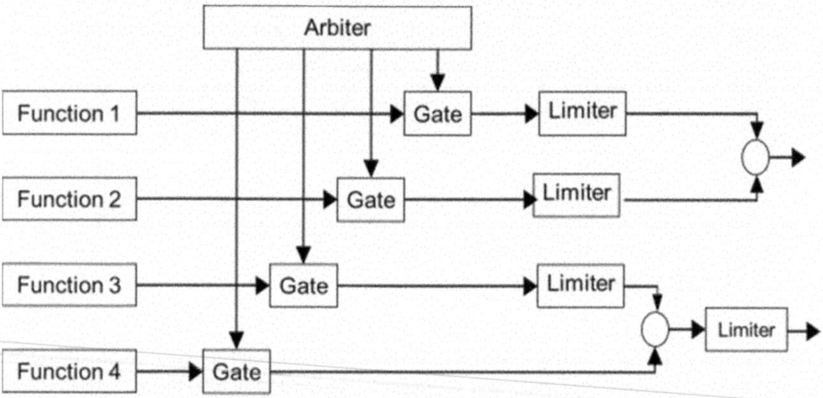

Figure 2: Monitor & Limiter structure for inclusion of Ford In-house Functions

Functional Deep Dive into the Ford Focus Functions

As described in the introduction, Ford in-house features are either used to support steer-feel by suppressing typical error-states nibble, pull or torque steer, or they deliver functionality for advanced driver function assistance systems. The Ford Active Nibble and

Pull/Drift Compensation functions are already in production in other carlines, as are the Ford Lane Keeping and Lane Departure Warning Functionality. Therefore this paper focuses on the functionality of the all new Torque Steer and Brake Pull Reduction, Evasive Steer Assist and Lateral Control.

Torque Steer and Brake Pull Reduction

Importance of Steering Torque Feedback

The steering torque felt at the hand-wheel belongs to the most important information to allow a driver controlling the vehicle path. Although most of the information to build the intended path is based on visual information, the haptic feedback of the steering wheel to tire forces allows better control on following that path, than the delayed visual impression of vehicle reaction. If the steering torque builds consistently on the lateral forces of the front suspension this helps the driver to understand upfront and intuitively, what the reaction and path of the vehicle will be. To allow this precise display a steering system with minimized internal friction is required. Especially for electrical assisted systems emphasis needs to be placed on mechanical optimization. Unfortunately, the intended superior sensing of lateral tire forces goes hand in hand with unintended sensitivity for erroneous disturbance forces. These result from road irregularities, wheel rotor imbalance (Nibble), or suspension asymmetry caused by production tolerances, wear or impacts (Pull). An additional source of disturbance are longitudinal tire forces, i.e. the steered wheels also driven by the engine (Torque Steer) or braked (Brake Pull).

The Impact of Driving and Braking Torque on Steering Feedback

While Nibble and Pull disturbances seem to be decreasing due to improved manufacturing quality, robust designs and proven compensation functions, Torque Steer and Brake Pull disturbances take precedence with the trend to stronger engines and large, low aspect tires in the automotive industry.

Although internal suspension and drive train design parameters of Torque Steer robustness are understood, a residual disturbance cannot be excluded. The geometric optimization is bounded by cost, weight and package constraints. Furthermore, uneven ground can move the center point of the tire to road contact patch laterally. This will lead to asymmetric driving or braking torque application and finally to residual torque disturbing the driver feedback at the steering wheel.

Approach

In general, the Steering Torque Disturbance Reduction (STDR) function calculates tie rod forces acting at the steering rack using several methods. One model for rack force calculation is based on overall vehicle data, depicting the driving situation, and serving as reference for the situation based ideal rack force. This will be called the estimated rack force ($F_{Rack\ est}$) in the following text. A second rack force calculation bases on steering system data, representing the current state of the system including all disturbances. This will be called the current rack force ($F_{Rack\ cur}$). Both calculation models must match the physical rack force closely in vehicle driving situations without disturbance. While in situations with steering torque disturbance this influence will be reflected in $F_{Rack\ cur}$, but will not be seen in $F_{Rack\ est}$. Therefore, in situations, where a disturbance is likely, e.g. high propulsion torque requested, the disturbance reduction algorithm will become active and request the difference of $F_{Rack\ est}$ and $F_{Rack\ cur}$ with appropriate gear ration conversion as disturbance reduction torque.

As the function is intended to work as embedded software in an automotive steering control unit, the execution time for calculations has to be reduced to the absolutely required level. Even if better approaches are known to solve partial problems, efficient solutions have to be chosen.

Figure 3 provides an overview on the control loop during active reduction.

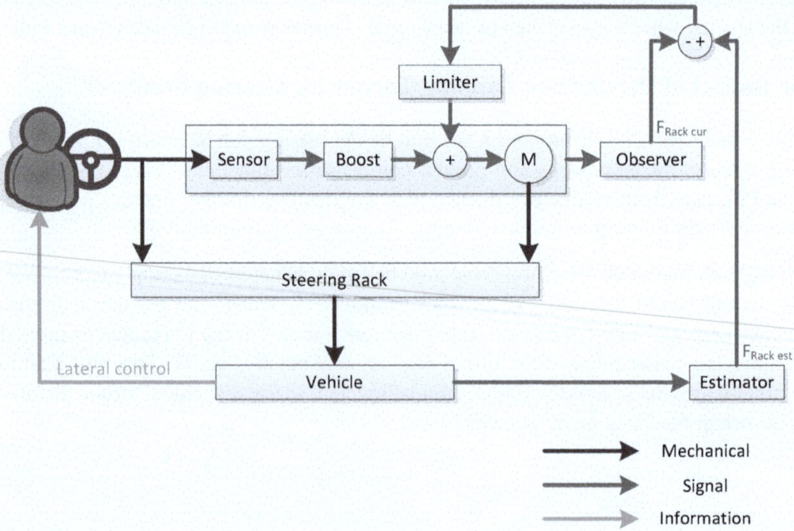

Figure 3: Steering Torque Disturbance Reduction

Estimated Rack Force

Background for the calculation of the reference rack force $F_{Rack\ est}$ is the finding that for the majority of driving situations rack force is basically linearly dependent on vehicle lateral acceleration. To avoid the significantly phase delay of sensor measured vehicle lateral acceleration to steering input a simplified vehicle model is calculated. As Torque-Steer is a phenomenon mostly recognized during mainly linear driving situations with minor steering angle change a linear vehicle representation was chosen. Some parameters of this vehicle model are continuously updated while driving based on available sensor signals. The basis of $F_{Rack\ est}$ is formed by lateral acceleration, vehicle front axle loading condition and suspension geometry.

An additional contributor to $F_{Rack\ est}$ is the tire aligning torque based on the tire side slip angle calculated in the vehicle model. As customers will replace tires with aftermarket tires of unknown specification, e.g. different size, manufacturer, summer vs. winter or all season, this tire relevant parameter is adjusted also during driving to minimize deviation of $F_{Rack\ est}$ from $F_{Rack\ cur}$. While this parameter adaptation helps to adjust the model as long as no disturbance is present, it is suppressed during active phases of the disturbance reduction algorithm to prevent wrong values.

Current Rack Force

To calculate the rack force that is physically acting at the steering system a disturbance observer has been chosen requiring an appropriate steering gear model. A complex model including linkages, rims and tire to road contact patch has been simplified to a dual mass representation with the connecting stiffness of the Torsion-Bar. As the torque in this Torsion-Bar is measured, the steering model can be further simplified to a state space system containing a single mass.

Based on this state space system a Luenberger Observer calculates the external rack force acting at the tie rods of the ideal model. Corrected by a friction model this results in $F_{Rack\ cur}$.

Situation Detection

The situation detection has to decide on the probability of Torque-Steer or Brake-Pull disturbances in the current driving situation. The activation of the disturbance reduction is connected to a set of enablers for torque steer, a set of enablers for Brake-Pull, and a set of suppressions for both. While for activation all enablers of a group must be valid, any suppression will be sufficient to ramp down the torque request. Basic input for this decision are signals received from the respective control modules via the vehicle CAN. The powertrain control module sends propulsion torque, representing the torque delivered by the combustion engine already amplified by the current gear ratio. From the brakes module the current front brake pressure is received, which is proportional to the applied brake torque. Both values are compared to tunable thresholds and form enablers to activate the disturbance reduction. As not every acceleration or deceleration of the vehicle will result in a disturbance situation, a certain deviation of $F_{Rack\ est}$ from $F_{Rack\ cur}$ is required to start disturbance reduction. Further enablers are formed by signals representing the current driving situation, e.g. vehicle speed, steering wheel angle, lateral acceleration and more. Suppressions are applied in situations where the underlying models for $F_{Rack\ est}$ and $F_{Rack\ cur}$ are expected to be invalid. One example for a model outside its valid range is the linear vehicle model during excessive under- or oversteer.

Customer Usage

The STDR function will remain unnoticed by the customer, as it activates and deactivates automatically. Using this disturbance reduction algorithm, the compromise of a steering system displaying all intended lateral tire forces while being robust against that unintended feedback from longitudinal driving forces can be resolved much better than before. Especially for high powered engines in performance vehicles annoying torque steer can be significantly reduced while maintaining a natural and transparent steer feel.

Additional details of Steering Torque Disturbance Reduction can be found in [2].

Evasive Steer Assist

Ford's Evasive Steer Assist (ESA) supports the driver in avoiding rear-end collisions. Designed as a support function, that only operates when the driver has initiated the evasive maneuver, the system generates additional steering assist to go around the obstacle and stabilizes in the chosen avoidance path. The obstacle can be a stationary target vehicle or a moving target vehicle. In future versions, the system will be extended to support not only avoiding cars, but also pedestrians and bicycles. The description here is based on [3].

Motivation of Evasive Steer Assist – Why is Ford doing this?

Statistics (ref. Figure 4) show, that in 19.3% of rear-end accidents, the driver had started to steer but still hit the target vehicle. This includes the partially overlapping crash cases which are especially dangerous. With an evasive steering assist system helping the driver to complete the scenario, this could be improved.

host driver reaction	number of collisions
unknown	130
complete lane change	38
steering	744
no steering	2935
total	3847

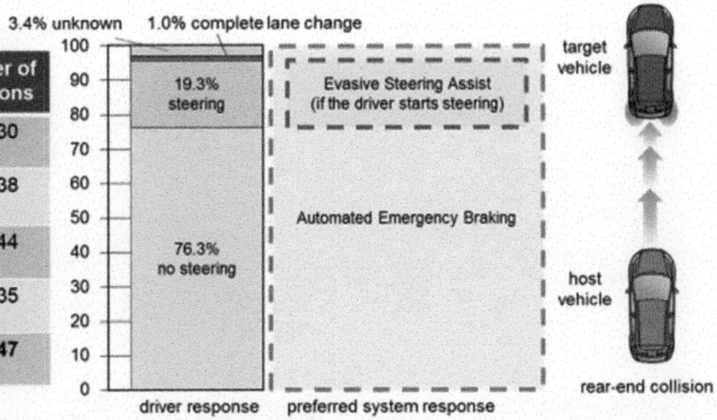

Figure 4: Accident Statistics motivating Evasive Steer Assist [4]

The question remains, whether an Evasive Steer Assist system is really required on top of existing systems that brake the car. The following comparison answers this question:

What is the distance required for braking the host vehicle before it reaches the target vehicle?

Let us assume the host vehicle moves in x-direction with speed v_{x0}, the obstacle vehicle stands still (see also Figure 5). Upon perceiving the obstacle, the host starts to brake with a constant longitudinal deceleration which cannot go beyond

$$a_x = -\mu g \tag{1}$$

The longitudinal position and speed of the host can thus be written as

$$x = v_{x0}\, t - \tfrac{1}{2}\mu g t^2 \tag{2}$$

The host vehicle stops after time t

$$t = v_{x0}/\mu g \tag{3}$$

This gives the stopping distance with (3) in (2), which growths quadratic with speed.

$$L_{stop} = x_{v=0} = \tfrac{1}{2} v_{x0}^2/\mu g \tag{4}$$

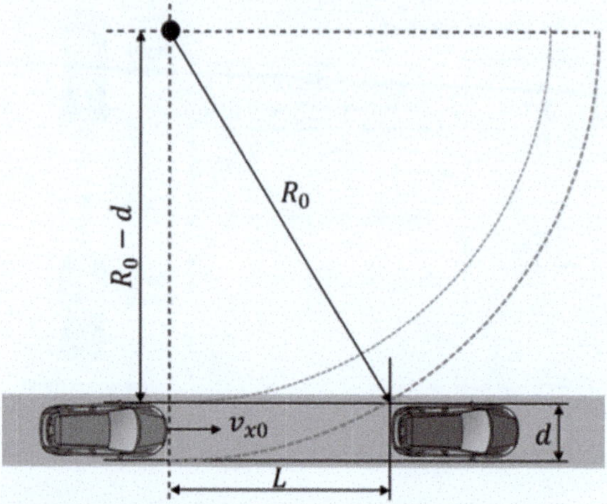

Figure 5: Setup for Comparison of Braking vs Steering in a Rear-End Collision Scenario

What is the distance required to steer the host vehicle around the target vehicle?

Again the host vehicle moves with speed v_{x0}, the obstacle vehicle stands still. The lateral distance the host has to move to avoid the accident shall be d. Assuming the host drives around the obstacle with a constant curve radius, the right front end of the host vehicle moves along a circular path of radius R_0. The triangle of host vehicle front right and obstacle vehicle left and right end yields the following geometry, assuming $d \ll R_0$:

$$R_0 = \frac{L^2}{2d} \tag{5}$$

The required lateral acceleration is – using (5) – given by

$$a_y = \frac{v_{x0}^2}{R_0} = \frac{2dv_{x0}^2}{L^2} \tag{6}$$

As the maximum lateral acceleration corresponds to the maximum longitudinal acceleration (1):

$$a_y = \mu g \tag{7}$$

The minimum distance to drive around the target growth proportional to speed

$$L_{steer} = v_{x0}\sqrt{d2/\mu g} \tag{8}$$

Comparing the two required distances (4) and (8), assuming an optimal friction coefficient of $\mu = 1$ and a typical lateral offset of $d = 2.5m$ yields Figure 6.

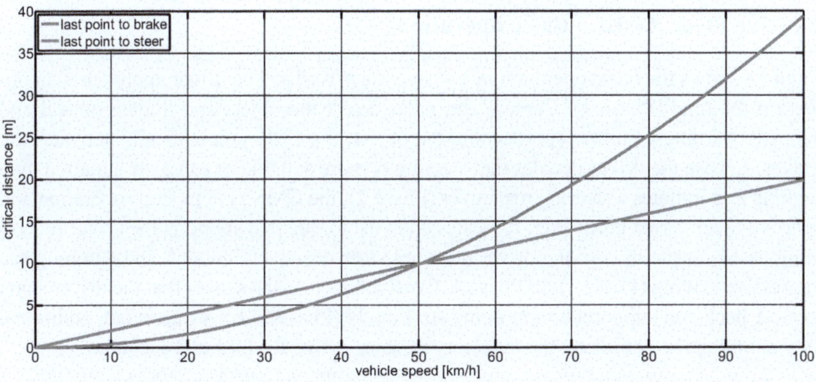

Figure 6: Comparison of required stopping distance for braking and steering

Above 50kph, it is more effective to drive around the obstacle, below 50kph, it is more effective to brake. With lower friction, this vehicle speed goes down.

Evasive Steer Assist – Basic setup

The new Evasive Steer Assist System is integrated with the Ford Collision Mitigation System (CMS), also available e.g. on the current Ford Focus. The CMS continuously checks for objects on the road, gives a warning to the driver in case an obstacle blocks the road and finally triggers the brakes. The CMS as well as the Evasive Steer Assist use forward looking camera and radar as shown in Figure 7, right.

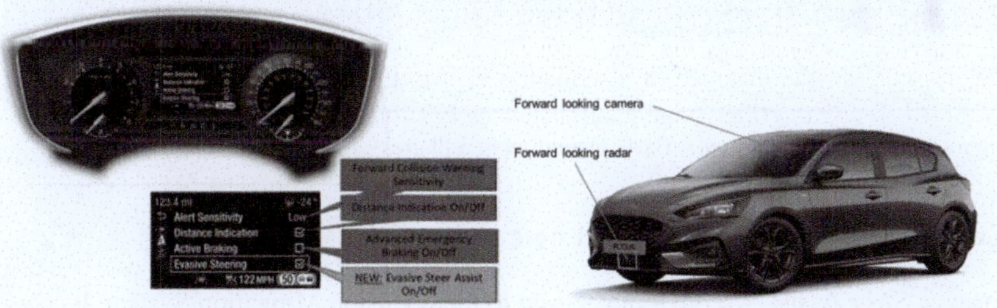

Figure 7: Cluster settings and sensors for Evasive Steer Assist

The driver can adjust his collision mitigation settings in the cluster menu. Here he also finds a setting to activate or deactivate the Evasive Steer Assist (Figure 7, left).

Evasive Steer Assist – Basic operation

Figure 8 shows the basic operation of Evasive Steer Assist. The driver approaches an object on the road (Phase 1). Camera and radar detect the object and initiate optical and acoustic warnings. Further approaching the object, the CMS automatically activates the brakes. In case the driver decides that steering is more suitable to solve the situation than braking and initiates a steering maneuver (Phase 2), the CMS system stops operation and Evasive Steer Assist takes over. It creates an assist torque that supports the driver by creating assist torque in case the driver does not steer decisively enough and damping too aggressive driving (Phase 3). In Phase 4, Evasive Steer Assist senses that the driver starts to steer back and supports him in doing so. Finally (Phase 5) ESA ramps out, additional yaw damping for stabilizing the trajectory remains active for further 2 seconds.

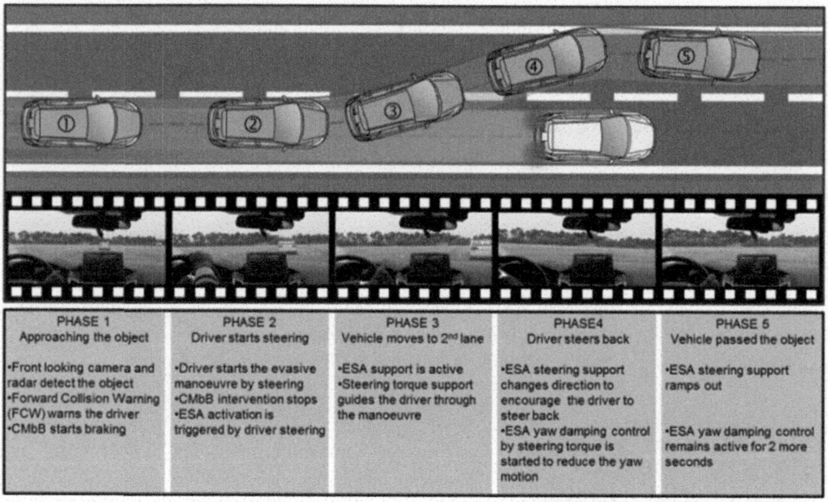

Figure 8: Evasive Steer Assist – Basic Operation in 5 Phases

Evasive Steer Assist – Control Logic

Evasive Steer Assist is setup as an add-on to the Collision Mitigation System. Its controls run completely on the Power Steering Control Module (PSCM). The controls logics are shown in Figure 9.

Within the CMS, radar and camera data are fused to create an environmental model. Out of this environmental model a threat assessment extracts the criticality of the current driving situation, forwarding the threat level to the CMS state machine. Depending on the threat level and the driver action, CMS creates warning requests to the driver information and activates braking (not shown). If the driver steering override detection recognizes that the driver wants to steer around the obstacle, CMS is stopped and further information transmitted to the Evasive Steer Assist.

The Evasive Steer Assist controller creates a path out of the provided CMS information and controls the torque generation along the path via the path control and yaw damping logic, which uses chassis sensors for the control action. The driver detection logic in combination with the Evasive Steer Assist mode manager supervises the different phases of the ESA interaction.

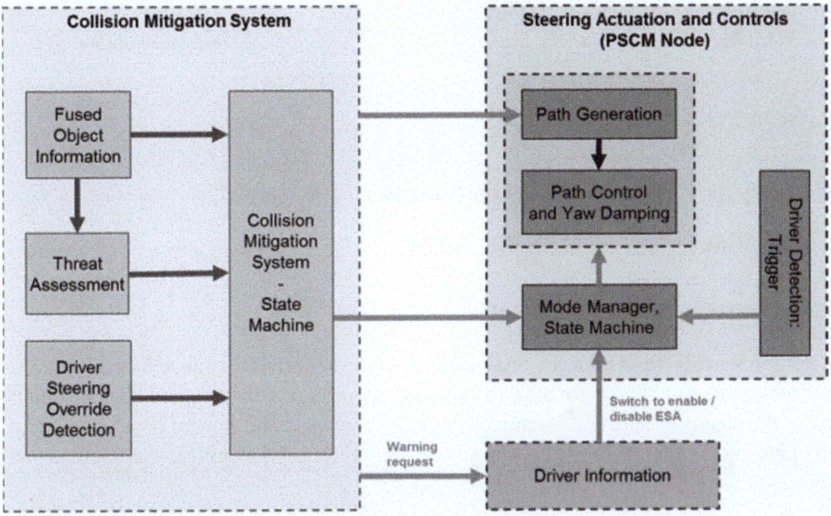

Figure 9: Evasive Steer Assist – Controls Logic

Evasive Steer Assist – Path Control

The ESA path control uses a preview technique to create the assist torques. Figure 10 illustrates the idea (assuming small yaw angles ψ):

The host vehicle moves with speed v and lateral acceleration a_y. Starting from the current position at time t (denoted by y_t), the controller computes a preview position y_{prev} by projecting the current motion along the current angle θ_t into the future time $t + \Delta t_p$ assuming constant speed and circular movement.

$$y_{prev} = y_t + v\,\psi_t \Delta t_p + \frac{1}{2} a_y\,\Delta t_p^2 = y_t + v\,\psi_t \Delta t_p + \frac{1}{2} v\,\dot{\psi}_t\,\Delta t_p^2 \quad (9)$$

The ESA path control computes a proportional control torque for the projected lateral error e_y between the preview position y_{prev} and the reference position at time $t + \Delta t_p$ denoted $y_{ref,\Delta t_p}$:

$$T_{ESA} = K_t\, e_y = K_t \left(y_{ref,\Delta t_p} - y_{prev}\right) \tag{10}$$

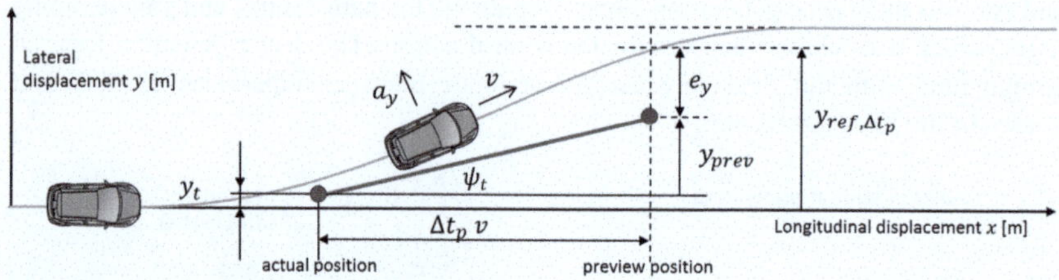

Figure 10: Evasive Steer Assist – Path Control

Using (9), the ESA-torque can be simplified to

$$T_{ESA} = K_1 \left(y_{ref,\Delta t_p} - y_t\right) + K_2\psi + K_3\dot{\psi} \tag{11}$$

Evasive Steer Assist – Realization

Fig shows a test drive with the realized ESA control system. In subjective drive tests, the factors K_i of (11) can be used to calibrate the response. Tests show, that K_1 dominates the steer-out part, K_2 dominates the steer-back part and K_3 can be used to adopt the phasing, so that 0-torque crossing between steering wheel torque and incremental torque fit.

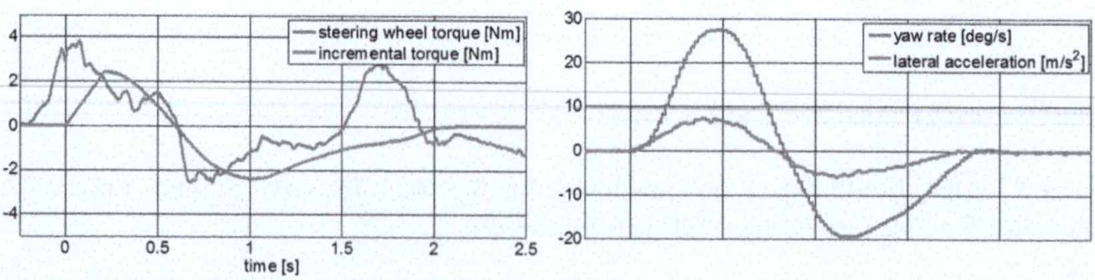

Figure 11: Evasive Steer Assist – Realization

Lane Centering Aid

In the Focus, Ford offers its new Lane Centering Aid combined with the Adaptive Cruise Control (ACC) system, which controls the longitudinal movement of the car. The Lane Centering Aid function is engineered as a level 2 autonomy system, classified as ACSF B1 function according to the new ECE79. The function offers a continuous lateral support that allows drivers on the one-hand to loosely hold the steering while being driven by the system while always monitoring the road, but also taking over and driving with hardly any impact on the steering-feel. Combined with a stringent hands-off warning strategy, this smooth integration between driving and "being driven" ensures that drivers are kept in the loop and perceive the system as an assistant for their personal driving rather than a level 3 or 4 autonomy system where they can disconnect from the driving task.

Lane Centering Aid – Basic operation and Warning Strategy

Lane Centering Aid with ACC is considered one guiding system. Therefore, a first step for Lane Centering Aid activation is the activation of the overall system via the System on/off (1) button. After selection, ACC Gap setting (4) and ACC state (5) are turned on grey to indicate stand-by. In order to choose Lane Centering Aid, the LCA On/Off Switch (2) needs to be selected. Choice of LCA On/Off is memorized over ignition cycles. After selection, LCA Status Display (6) and LCA Status Icon (7) turn on grey – stand-by. Finally, after setting the set-speed via Set or Resume (3), ACC and Lane Centering Aid are activated; (4),(5),(6),(7) turn green and the car controls longitudinal and lateral motion. In case Lane Centering Aid becomes unavailable due to e.g. missing lane markings, (6) and (7) turn grey as indication to the driver.

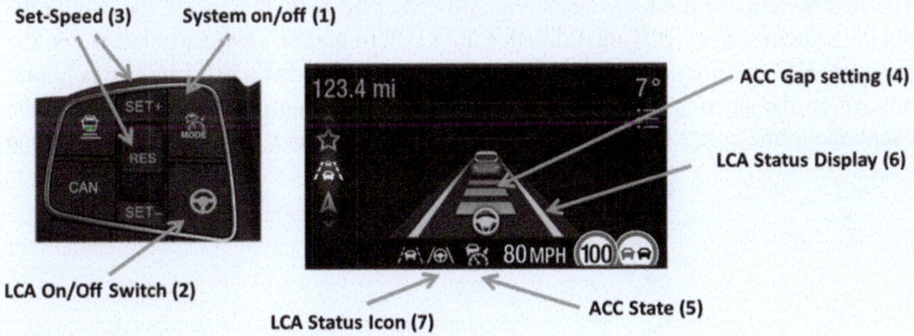

Figure 12: Lane Centering Aid Operation Elements

In order to ensure that drivers do not misuse the Lane Centering Aid, a vehicle speed dependent three-level warning strategy has been implemented. As basic input, a hands-on/off detection algorithm in the steering is used, that analyses the measured hand-wheel torque signal and returns a hands-on/off status to the main Lane Centering Aid supervision system. Depending on the vehicle speed, the escalation triggers first a visual warning if hands are not on the steering wheel, then an audible warning. If the driver still does not react, Lane Centering Aid is cancelled and a take-over strategy is started. During the take-over strategy, audible and visual warnings are active and the vehicle speed is reduced to 10kph in case of available lane information, to 0kph otherwise. Lateral support is still given as this is the minimum risk maneuver. In order to leave the take-over strategy, the hands need to be back on the steering wheel and the resume button needs to be pressed. Accelerator pedal activation will also exit the take-over strategy but stops Lateral Centering Aid.

Table 2 shows the selected timing for the escalation strategy dependent on the vehicle speed.

Table 2: Lane Centering Aid – Hands-off

Vehicle Speed	Visual Warning at	Audio-Visual Warning at	Take-over Strategy starting at
<10kph	*none/maintained*	*none/maintained*	*none/maintained*
10 kph– 60kph	15sec	30sec	60sec
>60 kph	15sec	19sec	25sec

Lane Centering Aid – System Setup and Control Logic

The functionality of the Lane Centering Aid is located on two main ECUs – while the overall system status, HMI and reference track determination are mainly located on the forward looking camera, the main path control and the hands-on/off detection is implemented on the steering ECU. Next to additional chassis sensors these ECUs form the backbone of the core LCA. Figure 13 gives an overview for the ECUs required for the total system.

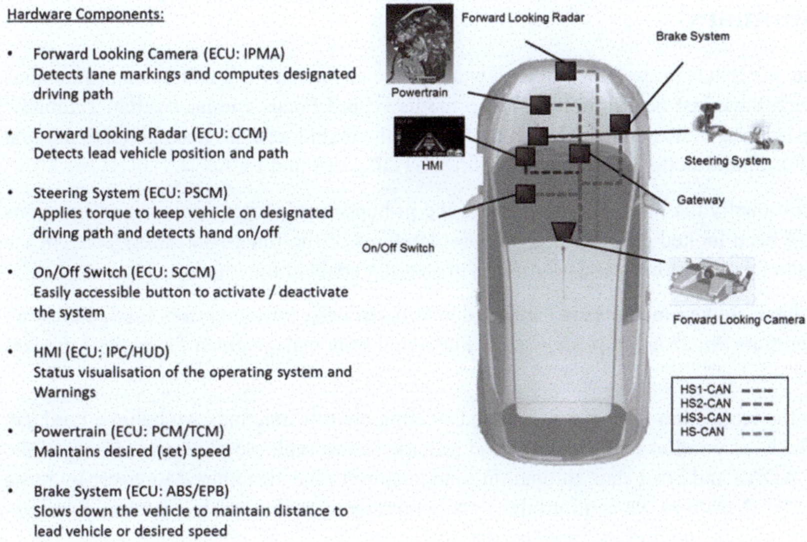

Figure 13: Lane Centering Aid System Setup

The continuous control of the Lane Centering Controller is setup as a hierarchical controller with two control loops:

The outer control loop focuses on position and orientation of the car. It uses measurement signals of these signals from the camera, which holds the fused radar/camera information. Via a none-linear "micro-path" filter, which determines a path to a target point approx. 30-150m in front of the car, a reference signal for the inner control-loop is computed.

The inner control-loop controls the curvature of the current vehicle motion. The measurement curvature is constructed from chassis sensors. The control error is minimized by a curvature controller, whose feedback-strength is scheduled by a driver-in-the-loop detection system.

As last building block, a feed-forward controller for road curvature improves curve performance.

Summary

This paper has provided an overview about the advanced Ford features and functions which have just been introduced into the new Ford Focus compact vehicle globally. Besides the well-established steering features the main focus has been given to the first in Ford introduction of the functionalities STDR, ESA and LCA.

Additionally the integration aspects of the in-house engineered features and functions have been looked at including the considerations of the functional safety concept via limitation, monitoring and arbitration of steering enabled functions.

In detail above functions are explained in terms of basic customer uses case and feature operation. An abstract of the control logic- and principals is given for each of the features.

By the realization of in-house enabled steering control functions we believe Ford can quickly respond to market trends and provide across multiple platforms a state of the art ADAS and error state mitigation functionality to the customer. Moreover by reuse of SW algorithms we can introduce very consistently and robustly customer functionality even in compressed program timings.

References

[1] Sicherheitsrelevante Inhouse-Software Entwicklung bei Ford Steering; T.Hestermeyer, W.Bongarth; chassis.tech plus, 7th International Munich Chassis Symposium; ATZ live; Munich; 14-15 June 2016

[2] Steering Torque Disturbance Rejection; J.Dornhege, S.Nolden, M.Mayer; SAE World Congress Experience 2017; Detroit; 4-6 April 2017

[3] Ford Evasive Steering Assist "steering you out of trouble"; P.Zegelaar, H.Bosch, M.Schiebahn, S.Notelaers, G.Allexi, E.Vukovic; 11th International IQPC Conference "Steering Systems"; Berlin; 14-16 November 2017

[4] Analyses and reconstruction of accidents in the greater areas of Hanover and Dresden; www.gidas.org (GIDAS = German In-Depth Accident Study); 2005 – 2016; Total number of investigated accidents: 29943

Objectification of the feedback behavior of the suspension and steering system

M.Sc. Dario Düsterloh

Dr. Ing. Anton Uselmann

B.Eng. Julian Scherhaufer

Dr. Ing. Christoph Bittner

Prof. Dr. Ing. Dr. h.c. Dieter Schramm

P. Pfeffer (Ed.), *9th International Munich Chassis Symposium 2018*, Proceedings,
https://doi.org/10.1007/978-3-658-22050-1_33

Abstract

This paper discusses the objectification of the influence of suspension and steering system (hardware and software) on the feedback from the road to the steering wheel. For the vehicle guidance, feedback regarding changes in the forces and moments at the wheels is essential. These variations can be caused by fluctuation in the wheel load, the rolling resistance, the longitudinal and lateral forces as well as the road and weather conditions. As a result, the feedback describes the transmission of information on the force and torque states at the tire and, as a consequence, on the traction potential and traction reserve. In this case, the information can be subdivided into beneficial and interfering information by means of the excitation frequency.

Within the scope of this article various possibilities for the objective determination and visualization of the feedback behavior are presented. For this purpose, the frequency-based determination of the feedback behavior with the aid of a steering test bench is discussed. This method is used at the Dr. Ing. h.c. F. Porsche AG, in order to support the steering controller and steering software development. In addition, the method allows the comparison of different steering systems and provides valuable parameters for the parameterization of simulative steering models as well as the application of the steering feel. Due to the fact that the steering test bench can only be used to consider the properties of the steering system, separated from the composite with the suspension, additional tests have been carried out on a special test setup. This test setup allows the analysis of the influence of the suspension on the feedback characteristics. Finally, driving tests are presented to objectively examine the feedback in the complete vehicle.

The Dr. Ing. h.c. F. Porsche AG develops an own steering controller including the corresponding steering functions to achieve a Porsche typical steering feel (software in house development). The effect of this development can be objectively proven both at the test stand and in the driving test. The results of a press analysis for the subjective evaluation of the feedback behavior further prove the effect.

1 Introduction and State of the Art

In the following sections the guidance behavior and feedback of the steering are defined. Subsequently, different types of steering systems and the servo assistance are described and their respective advantages and disadvantages are explained. The chapter concludes with the definition of target values for the guidance and feedback behavior.

1.1 Guidance Behavior and Feedback Behavior

The steering must allow the driver to be able to guide the vehicle safely and foreseeably in any situation. Driver, vehicle and driving environment form the driver-vehicle-control-loop shown in figure 1. In this control-loop, the driver represents the controller which responds to optical, haptic, acoustic and kinesthetic sensations. Foot pedals and steering wheel act as actuators to form the interfaces between driver and vehicle. The vehicle represents the control path and the driving course represents the control values. The difference between the target value and the current driving course is used as the feedback value. All the forces acting on the vehicle from outside (road irregularities, friction changes, weather, side wind, centrifugal forces, etc.) are considered as disturbing variables.

The vehicle guidance can be divided into an anticipatory and a compensatory part. The anticipatory vehicle guidance is based on the driver's experience and his estimates of the following road and the vehicle reaction. The compensatory vehicle guidance results from the feedback value, i.e. from the course deviation between the road and the current driving course. [Braess, 2004] [Pfeffer & Harrer, 2013] [Schindler, 2007]

The vehicle guidance in lateral direction is decisively influenced by the driver with the aid of the steering. Therefore, the steering feel is of crucial importance for the driver as the controller. It is defined in [Pfeffer & Harrer, 2013] as follows: „Steering-feel is the complete optical, kinesthetic, and haptic sensations of the driver while steering a car, resulting in a complex, subjective experience."

Regarding steering feel, the steering systems fulfills two basic functions according to figure 1. On the one hand, it serves as the driver's control element for the vehicle guidance. On the other hand, it serves as a source of information for the feedback of the road condition to the driver. In the specific definition, the steering feel describes the steering wheel angle to be applied by the driver as well as the noticeable steering wheel torque. In the encompassing definition, the reactions of the vehicle to the driver's inputs at the steering wheel are also considered.

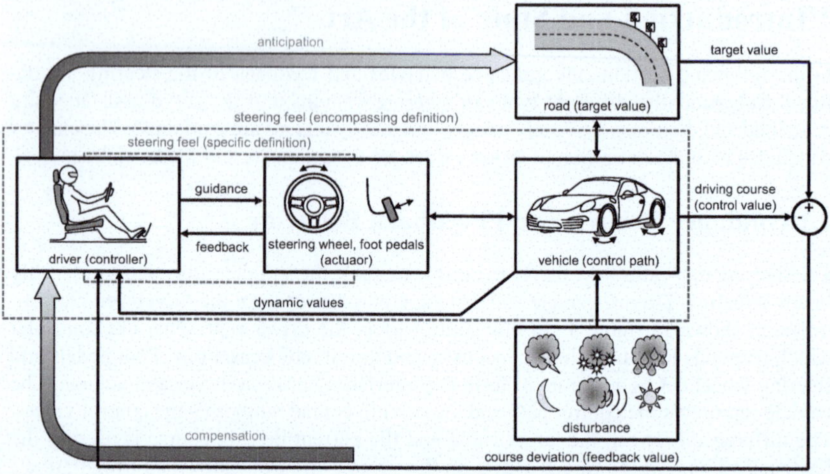

Figure 1: Driver-vehicle-control-loop according to [Pfeffer & Harrer, 2013]

The guidance behavior characterizes a vehicle with regard to its lane guidance and lane keeping. The steering commands applied by the driver must be carried out with an adequate steering afford (steering angle and steering torque requirement) for an optimal guidance behavior and be converted into a vehicle reaction with minimal time delay. The reaction of the vehicle must be foreseeably and manageable. To fulfill these requirements, there must be a continuous ratio between the steering wheel angle and the steering angle of the wheels. Variations in vehicle loading, weather or roadway conditions must not have negative effects on the guidance behavior.

For a good steering feedback, variations of the forces and torques acting on the wheels due to changes in the wheel load, the rolling resistance or the longitudinal and lateral forces should result in a change in the steering wheel torque, which is noticeable and comprehensible for the driver. The feedback behavior thus describes the transmission of information on the forces and torques at the wheels. This information can be subdivided into beneficial and interfering information. For the vehicle guidance, the useful information, such as the feedback on the adhesion limit of the wheels, is of decisive importance. Interfering information, such as brake force fluctuations or wheel imbalance variations, are often characterized by periodic excitations of the vehicle and should be suppressed via the transmission path of the steering.

The transmission paths of the guidance behavior from the steering wheel to the wheels and the feedback behavior from the wheels to the steering wheel are not invertible. From the knowledge of one transmission path, therefore, no error-free conclusions can be made to the other transmission path. The steering feel is decisively influenced by the steering system design and its parameterization as well as by the characteristics of the tires, the suspension kinematics and elastokinematics. [Braess, 2004] [Henning et al, 2015] [Pfeffer & Harrer, 2013]

For more detailed information regarding the characterization and objectification of the guidance behavior, the publications of [Pfeffer & Harrer, 2013], [Harrer, 2007] and [Zschocke, 2009] as well as [Wolf, 2009] shall be mentioned. The following discussion of this article deals with the feedback characteristics of steering systems. The different types of steering systems are briefly explained in the next section.

1.2 Types of Steering Systems

In order to meet the large number of requirements, the Dr. Ing. h.c. F. Porsche AG has strengthened the use of mechatronic systems in the vehicle, and consequently has successively replaced hydraulic power steering systems (HPS[1]) with electric power steering systems (EPS) across all model series. Figure 2 shows the basic structure of an HPS as well as an EPS with axle-parallel servomotor (apa-EPS[2]) and a rack-concentric servomotor (rc-EPS[3]). [Pfeffer&Harrer, 2013]

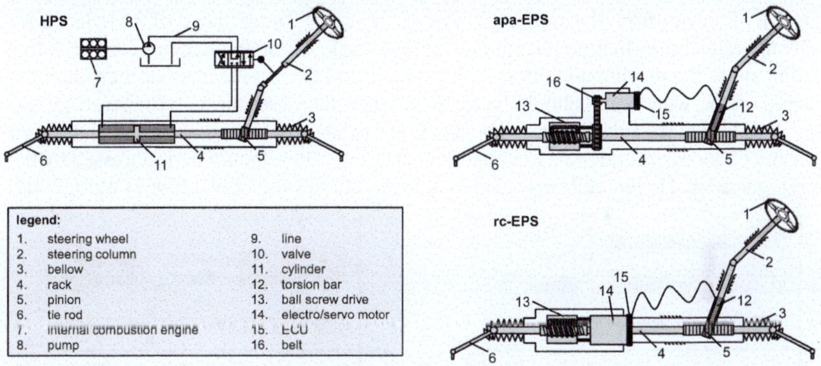

Figure 2: Types of steering systems according to [Pfeffer & Harrer, 2013]

The main advantages of the EPS compared to the HPS are the lower energy consumption and the resulting lower emissions of pollutants. For electric and hybrid vehicles, the EPS also offers the advantage that a steering assistance can be provided in the ve-

1 Hydraulic Power Steering

2 axle parallel-Electric Power Steering

3 rack concentric-Electric Power Steering

hicle's electrical operation mode with an inactive internal combustion engine. The freedom in tuning and application of the servomotor makes it possible to implement comfort and safety enhancing driver assistance functions and also offers great potential for implementing a brand-specific steering feel. With respect to the feedback characteristics, the HPS has advantages over the EPS because the inertia within the steering system is significantly lower. The Dr. Ing. h.c. F. Porsche AG develops an own steering controller including the corresponding steering functions to achieve a Porsche typical steering feel (software in house development).

1.3 Target Values for Steering Response and Steering Feedback

Figure 3 a) shows the time ranges for the response behavior of a steering system. Reactions occurring in the range up to 50 ms after a steering input or an excitation are perceived as too direct or too "fierce". Reactions occurring in the range between 50 ms and 150 ms after the steering input or the excitation seem to occur at the same time. This range represents the optimum range of response. Reactions occurring later are perceived as delayed in the consequence as too late. In Figure 3 b), excitations of the steering system are classified into different frequency ranges. Excitations with frequencies below 3 Hz result from the steering inputs of the driver. In the range of 3 Hz to 20 Hz is the beneficial information relevant for the feedback. These frequencies contain information about the driving condition of the vehicle and the road contact or the side force reserve on the wheels and should be transmitted to the driver through the steering system. Excitation with frequencies above 20 Hz result from vibrations, imbalances or acoustic phenomena and are perceived as interfering. They should be suppressed by the steering system. [Brunn & Harrer, 2004] [Harrer, 2008] [von Groll, 2006] [Wolf, 2009]

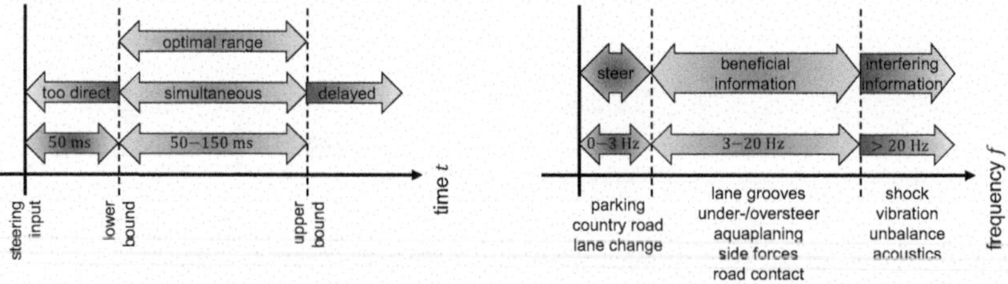

a) time domain for steering response b) frequency domain of steering feedback

Figure 3: Target values for steering response and steering feedback according to [Brunn & Harrer, 2004]

2 Feedback Behavior of the Steering System

In the following sections, the steering test bench used for the tests in this article is presented. Subsequently, the test bench tests for the objectification of the steering feedback are defined and the results are discussed. The chapter concludes with the creation of simulation models for the analysis of the feedback as well as their parameterization and validation.

2.1 Steering Test Bench

The use of test benches in the vehicle development is established for almost all components as well as for the entire vehicle. The aim of the test bench is the objective analysis of component and vehicle characteristics to support the development. The objectivity results from the high reproducibility of the test bench tests since these tests are not influenced by external weather conditions or human behavior and perception. Predefined and constant ambient conditions as well as the precise performance of the actuators ensure that the measurement results are comparable. Figure 7 shows the steering test bench used for the examinations of this article. By means of the two linear actuators the forces and displacements of the tie rods or the rack can be generated. The steering actuator is used to apply the steering torque or steering angles. The actuators used are capable of displaying all relevant test cases. [Düsterloh & Schrage, 2016]

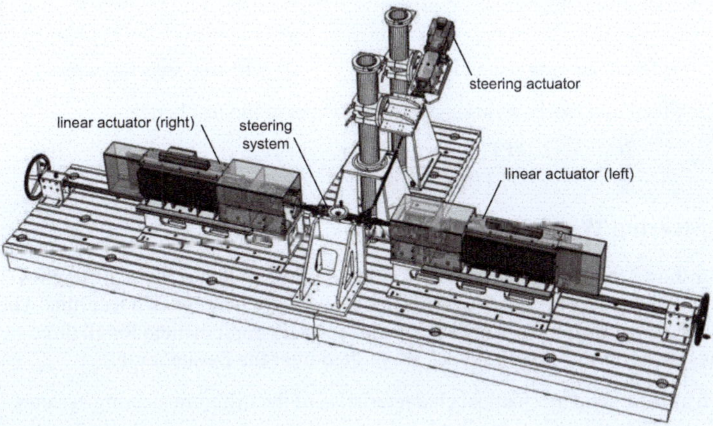

Figure 4: Steering test bench

2.2 Test Bench Tests for the Objective Measurement of Steering System Parameters

The feedback behavior of the steering system can be analyzed with the aid of the steering test bench in the frequency domain. For this purpose, a chirp signal for the force with an amplitude of 600 N at frequencies f of 1 Hz to 30 Hz at a frequency increase rate of 1 $\frac{\text{Hz}}{\text{s}}$ is impressed on the steering rack by means of one of the two linear actuators. According to [Schimpf, 2016] there are two different options. The steering wheel is either fixed in the zero position or it is connected to the steering rack freely swinging, see figure 5. For the evaluation, the frequency response diagram is calculated between the input signal rack force F_{rack} and the output signal torsion bar torque M_{tb} with the aid of the Fast-Fourier-Transformation (FFT), see equation (1). In the first instance, the option with fixed steering wheel is to be analyzed and afterward the option with freely swinging steering wheel. Finally, simulation models for the description of the steering system properties as well as the calculation of the associated parameters are presented.

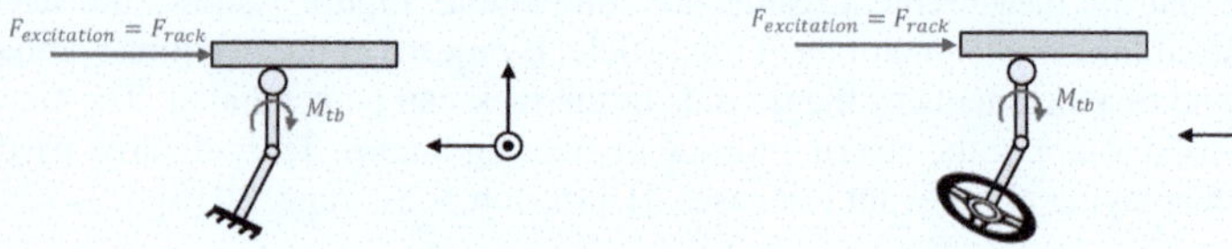

a) fixed steering wheel b) free steering wheel

Figure 5: Fixed and freely swinging steering wheel at the test bench

$$G = \frac{\text{FFT}(output\ signal)}{\text{FFT}(input\ signal)} = \frac{\text{FFT}(M_{tb})}{\text{FFT}(F_{rack})} \tag{1}$$

Fixed Steering Wheel

Figure 6 shows the measurement data for the steering system of three different vehicles with and without servo assistance with fixed steering wheel for two test runs. The Porsche in house steering software has been conceptually implemented for all three steering systems. The gain $|G|$ and phase $\angle G$ are plotted over the frequency f.

Figure 6 a) shows the mechanical characteristics of the different steering systems. It can be seen that the mechanical feedback behavior of the steering of vehicle A with respect to the gain $|G|$ is more distinct than the feedback behavior of the steering systems of vehicle B and C. This is due to the higher stiffness in the bearing of the servo unit and the lower inertia of the electric motor. In addition, the natural frequencies f_{nat} of the systems, which are located at the elevations, can be found in this diagram. Figure 6 b) shows the steering feedback with active servo assistance. It can be seen that the level of the gain $|G|$ is significantly lower than in figure 6 a). This circumstance is attributable to the fact that the servo assistance reduces the torsion bar torque M_{tb}. Moreover, it can be seen that the gain $|G|$ of the steering of vehicle A is higher than that of vehicle B and C. The steering of vehicle A thus has a more distinct feedback behavior and a higher basic hand torque than the other two systems. Especially in the feedback-relevant frequency range elevations of the gain $|G|$ can be recognized in figure 6 b).

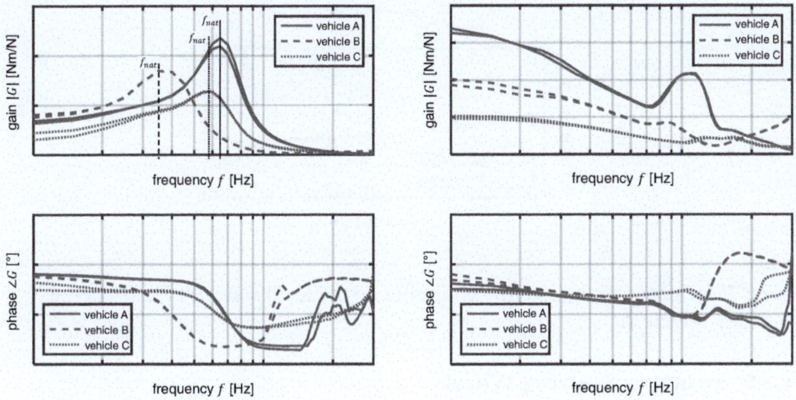

a) without servo assistance b) with servo assistance

Figure 6: Steering feedback with fixed steering wheel

A detailed analysis of the influence of the steering controller can be taken from figure 7 for the steering of vehicle A. It can be seen that the gain $|G|$ is characterized by a significant elevation for the activated controller.

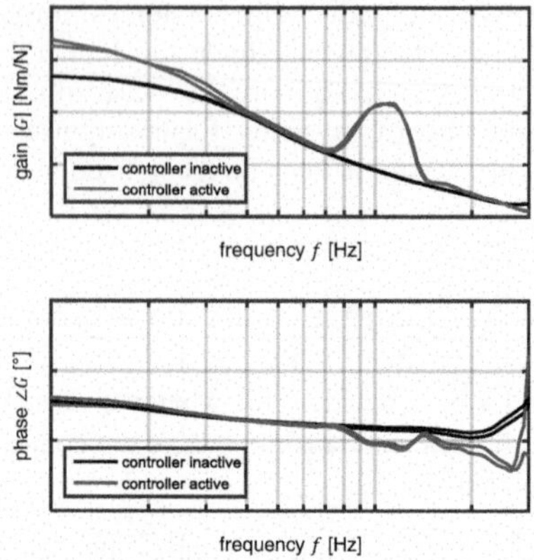

Figure 7: Influence of the steering controller on the feedback with fixed steering wheel for vehicle A

Freely Swinging Steering Wheel

Figure 8 shows the measurement data for the test option with freely swinging steering wheel, both with and without servo assistance for the three representative steering systems in analogy to figure 6. For all three systems, the same steering wheel is used for the tests. Consequently, it is measured with identical rotational inertia.

Compared to the measurement data of the fixed steering wheel in figure 6 a) the curves of the freely swinging steering wheel in figure 8 a) have a lower gain level $|G|$ as well as different position of the natural frequencies f_{nat}. The lower gain level $|G|$ results from the fact that the steering wheel can perform an “evasive” rotary movement when it is freely swinging. This rotation leads to a “relaxation” of the torsion bar and thus to a reduction in of the torsion bar torque M_{tb}. For the fixed option, this evasive movement and consequently the relaxation are not possible. The differences in the positions of the natural frequencies f_{nat} result from the influence of the steering wheel inertia, see also section 2.3. The reduction of the gain $|G|$ is recognizable in figure 8 b) with active servo assistance in particular in the region of low frequencies f. The elevations in the frequency-relevant range can be seen in analogy to the option with fixed steering.

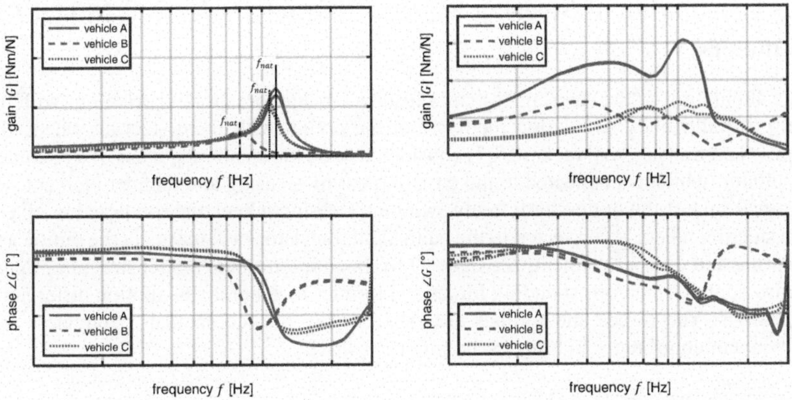

a) without servo assistance b) with servo assistance

Figure 8: Steering feedback with freely swinging steering wheel

The influence of the feedback controller is shown in figure 9 using the example of the steering of vehicle A. With active controller a clear increase of the gain $|G|$ in the frequency range from 7 Hz to 15 Hz is recognizable.

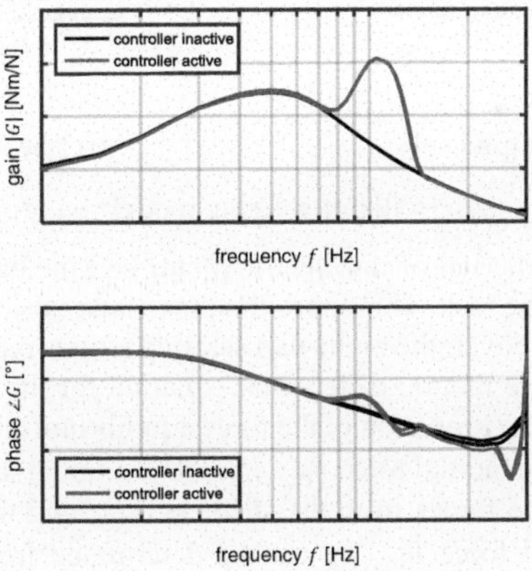

Figure 9: Influence of the steering controller on the feedback with freely swinging steering wheel for vehicle A

2.3 Calculation of Steering System Characteristics and Model Parameters

The simulation of the steering system's feedback behavior can be carried out with the help of multi body simulation models. For the option with fixed steering wheel, the equivalent translational rack mass is considered to be the oscillating mass of a one mass oscillator, which is connected to the environment by a spring and damper element, see figure 9 a). For the option with freely swinging steering wheel, the rotational inertia of the steering wheel is connected to the equivalent translator rack mass via the pinion and a spring and damper element, see figure 9 b). In addition to the model structure, figure 9 also contains the corresponding forces and torques to generate the motion differential equations. The forces and torques are visualized in red, component parameters and movement coordinates in blue. [Schramm et al, 2013]

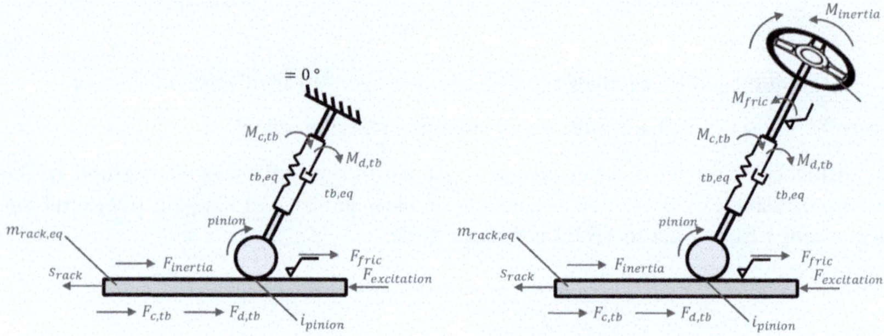

a) fixed steering wheel b) free steering wheel

Figure 10: One and two mass oscillator steering model

For the simulation, the inertial of the electric motor and the ball screw drive are combined with the mass of the rack. This approximation causes the number of motion equations to be reduced. Moreover, the multi-part steering strut is replaced by a single spring and damper element. The motion differential equation for the combined inertial mass $m_{ZS,eq}$ is calculated by the formation of the force equilibrium on the rack. The operating force components are the inertial force $F_{inertia}$, the excitation force $F_{excitation}$, the frictional force F_{fric} and the forces from the stiffness $F_{c,tb}$ and the damping $F_{d,tb}$ of the torsion bar. The inertial force $F_{inertia}$ can be determined from the product of mass $m_{rack,eq}$ and acceleration $\ddot{s}_{rack}$. Equation (2) represents this relation. Equation (3) shows the motion differential equation converted according to the rack acceleration $\ddot{s}_{rack}$.

$$F_{inertia} = m_{rack,eq} \cdot \ddot{s}_{rack} = F_{excitation} - F_{c,tb} - F_{d,tb} - F_{fric} \tag{2}$$

$$\rightarrow \ddot{s}_{rack} = \frac{1}{m_{rack,eq}} \cdot \left(F_{excitation} - F_{c,tb} - F_{d,tb} - F_{fric}\right) \tag{3}$$

The equivalent rack mass $m_{rack,eq}$ in the equations documented above represents the total inertial mass of the steering system. It is made up from the sum of the rack mass m_{rack} as well as the rotational inertia of ball screw drive J_{bsd} and electric motor J_{em}. The total system inertia $m_{rack,eq}$ can be calculated using the law of conservation of energy and the ratios of the ball screw drive i_{bsd} and the belt i_{belt}, see equations (4) to (6).

$$m_{rack,eq} = m_{rack} + m_{bsd,eq} + m_{em,eq} \tag{4}$$

$$\text{with } \frac{1}{2} \cdot {}_{bsd} \cdot \dot{\delta}_{bsd}^2 = \frac{1}{2} \cdot m_{bsd,eq} \cdot \dot{s}_{rack}^2 \quad \rightarrow \quad m_{bsd,eq} = J_{bsd} \cdot \frac{\dot{}^2_{bsd}}{\dot{s}_{rack}^2} = J_{bsd} \cdot \frac{1}{i_{bsd}^2} \tag{5}$$

$$\text{and } \frac{1}{2} \cdot {}_{m} \cdot \dot{\delta}_{em}^2 = \frac{1}{2} \cdot m_{em,eq} \cdot \dot{s}_{rack}^2 \quad \rightarrow \quad m_{em,eq} = J_{em} \cdot \frac{\left(\dot{\delta}_{b\ d} \cdot i_{belt}\right)^2}{\dot{s}_{rack}^2} = J_{em} \cdot \frac{i_{belt}^2}{i_{bcd}^2} \tag{6}$$

The force components $F_{c,tb}$ und $F_{d,tb}$ resulting from the torsion of the torsion bar are calculated with respect to the equivalent torsion bar stiffness $c_{tb,eq}$ and the equivalent torsion bar damping $d_{tb,eq}$ as well as the pinion ration $i_{pinion.}$ Additionally, the steering wheel angle δ_s and its speed $\dot{\delta}_s$ as well as the rack movement s_{rack} and its speed $\dot{s}_{rack}$ are required. The equivalent torsion bar stiffness $c_{tb,eq}$ and the equivalent torsion bar damping $d_{tb,eq}$ can be calculated assuming a linear connection of the individual parts of the torsion bar and steering column. Equations (7) and (8) represent the calculations. For the one mass oscillator model with fixed steering wheel, the steering wheel angle is $\delta_s = 0\,°$.

$$F_{c,tb} = \frac{{}_{tb,eq}}{i_{pinion}} \cdot \left(\frac{s_{rack}}{i_{pinion}} \quad \right) \qquad \text{with } c_{tb,eq} = \frac{{}_{tors\quad nbar} \cdot c_{column}}{{}_{tors\quad nbar} + c_{column}} \tag{7}$$

$$F_{d,tb} = \frac{{}_{tb,eq}}{i_{pinion}} \cdot \left(\frac{\dot{s}_{tb}}{i_{pinion}} \quad \cdot \right) \qquad \text{with } d_{tb,eq} = \frac{{}_{tors\quad nbar} \cdot d_{column}}{{}_{tors\quad nbar} + d_{column}} \tag{8}$$

In order to simulate the mostly nonlinear frictional effects on the rack F_{fric} or on the steering column M_{fric}, different approaches exist with different degrees of complexity for different use cases. For the studies in this article, a constant friction is assumed as a function of the direction of movement or rather speed, see equation (9).

$$F_{fric} = r_{rack} \cdot \text{sign}(\dot{s}_{rack}) \qquad M_{fric} = r_{column} \cdot \text{sign}(\dot{\delta}_s) \tag{9}$$

The torque equilibrium for the rotational inertia of the steering wheel J_s for the option with freely swinging steering wheel is shown in equation (10). The inertial torque $M_{inertia}$ can be determined from the product of rotational intertia J_s and angle acceleration $\ddot{\delta}_s$. Equation (11) shows the motion differential equation converted according to the steering angle acceleration $\ddot{\delta}_s$. The definition equations for the corresponding torque components can be found in the equations (9), (12) and (13).

$$M_{inertia} = J_s \cdot \ddot{\delta}_s = M_{c,tb} + M_{d,tb} - M_{fric} \tag{10}$$

$$\rightarrow \ddot{\ } = \frac{1}{\ } \cdot \left(M_{c,tb} + M_{d,tb} - M_{fric} \right) \tag{11}$$

$$\text{and } M_{c,tb} = c_{tb,eq} \cdot \left(\frac{s_{rack}}{i_{pinion}} \qquad \right) \tag{12}$$

$$\text{and } M_{d,tb} = d_{tb,eq} \cdot \left(\frac{\dot{s}_{rack}}{i_{pinion}} \quad \cdot \right) \tag{13}$$

The parameter used for the simulation are shown in table 1. These parameters (mass, inertia, ratio, stiffness, friction) can be determined with the help of further measurements. Only the damping constant $d_{tb,eq}$ cannot be determined experimentally with reasonable effort. By comparing measurement and simulation it can be adapted so that the simulation replicates the real behavior in an appropriate quality.

Table 1: Parameter of the simulation models

parameter / vehicle	vehicle A	vehicle B	vehicle C
rack mass m_{rack} [kg]	3	3,6	3,1
ball screw drive inertia J_{bsd} [kgm²]	0,000584	0,001185	0,000349
electric motor inertia J_{em} [kgm²]	0,000113	0,000298	0,000579
steering wheel inertia J_s [kgm²]	0,033	0,033	0,033
pinion ratio i_{pinion} [mm/rev]	57	54	52
ball screw drive ratio i_{bsd} [mm/rev]	9	8	6,5
belt ratio i_{belt} [ ]	2,49	2,85	1
torsion bar stiffness c_{tb} [Nm/°]	3,3	2,5	2,5
column stiffness c_{column} [Nm/°]	4,38	4,83	4,83
equivalent torsion bar stiffness $c_{tb,eq}$ [Nm/°]	1,96	1,65	1,65
rack friction r_{rack} [N]	200	250	300
column friction r_{column} [Nm]	0,2	0,2	0,2
equivalent mass $m_{rack,eq}$ [kg]	629,1	2227,7	870,2
equivalent torsion bar damping $d_{tb,eq}^{fix}$ [Nms/°]	0,0055	0,0135	0,01
equivalent torsion bar damping $d_{tb,eq}^{free}$ [Nms/°]	0,0018	0,0008	0,0008

Figure 11 shows the measured feedback behavior in analogy to the figure 6 and 8 as well as the simulated feedback behavior for the three representative steering systems. As the input signal for the simulation, the force excitation signal of the measurements is used. The simulated torsion bar torque M_{tb}, which can be evaluated according to equation (14), serves as the output signal. It can be seen that the courses of simulation and measurement have a high degree of correspondence. Remaining deviations can be attributed to simplifications in the modeling as well as inaccuracies in the parameterization.

$$M_{tb} = \left(F_{c,tb} + F_{d,tb}\right) \cdot i_{pin \quad n} = M_{c,tb} + M_{d,tb} \tag{14}$$

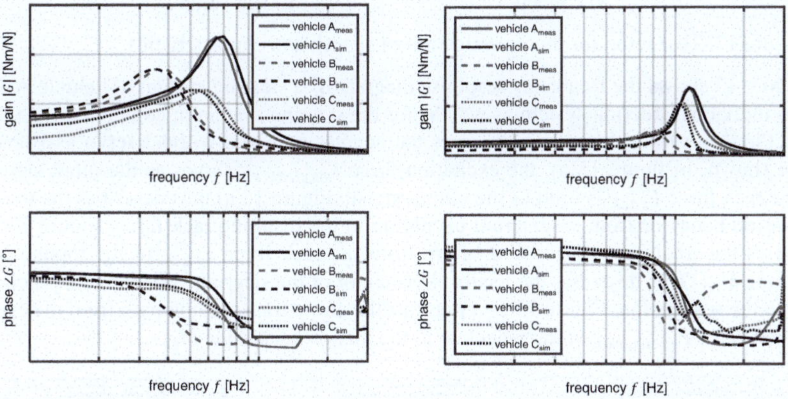

a) fixed steering wheel b) free steering wheel

Figure 11: Comparison of measurement and simulation

3 Feedback Behavior of the Suspension

The feedback behavior on the steering test bench in the previous figures relates exclusively to the steering system. At this point the characteristics of the suspension are not taken into account. In order to integrate the suspension into the transmission path, a special test setup was constructed within the scope of this article. The excitation of the steering system is generated indirectly via a torque on the wheel, see figure 12. The torque is realized by means of a linear actuator[4] via a lever arm, the steering wheel is integrated freely swinging into the test setup.

4 Due to space and cost reasons an EPS assumes the role of the linear actuator.

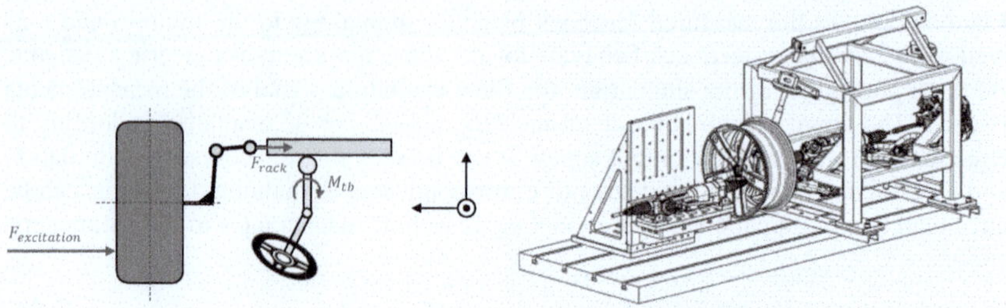

a) concept b) test setup

Figure 12: Test setup for the investigation of suspension characteristics

Figure 13 shows the frequency response diagram of a steering system of vehicle A for the individual steering system as well as for the steering system in compound with the McPherson suspension. For both cases the courses with and without servo assistance are shown. In figure 13 a), the excitation force $F_{excitation}$ serves as the input signal. The rack force F_{rack} is used as the output signal. Figure 13 a) thus describes the transmission behavior from an external excitation to the resulting rack force. Figure 13 b) shows the frequency response diagram from the rack force F_{rack} to the torsion bar torque M_{tb}. This describes the feedback behavior of the steering system in conjunction with the suspension. Figure 13 c) represents the entire transmission path from the excitation force $F_{excitation}$ at the wheel up to the torsion bar torque M_{tb}.

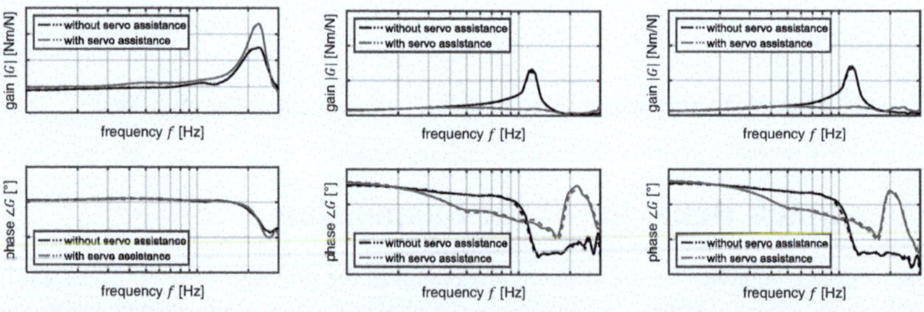

a) $F_{excitation} \rightarrow F_{rack}$ b) $F_{rack} \rightarrow M_{tb}$ c) $F_{excitation} \rightarrow M_{tb}$

Figure 13: Transmission behavior of the suspension

Figure 13 a) shows that the suspension ensures a constant force transmission at low frequencies f, independent of the servo assistance of the steering system. At higher frequencies, the transmission behavior has an elevation. However, this elevation has almost no influence on the feedback behavior of the steering, since the courses in the figures 13 b) and c) are nearly identical.

Figure 14 shows the feedback behavior of the steering system in conjunction with the suspension in relation to the feedback behavior of the individual steering system. In both cases the rack force F_{rack} is used as the input signal and the torsion bar torque M_{tb} is used as the output signal to generate the frequency response diagram. It can be seen that the behavior with respect to the gain $|G|$ as well as the phase $\angle G$ is similar in both cases. As a result, the analysis of the transmission behavior at the steering test bench allows correct conclusions to be drawn about the behavior including the suspension. [Düsterloh, 2015b] [Frey, 2016] [Scherhaufer, 2016] [Uselmann, 2017]

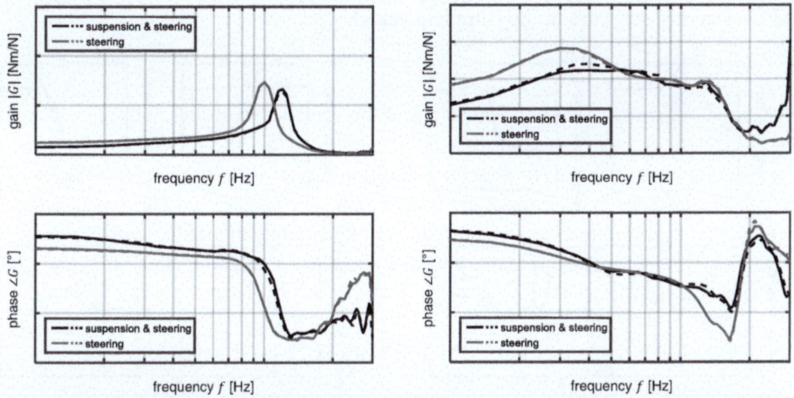

a) without servo assistance b) with servo assistance

Figure 14: Steering feedback in combination with the suspension

Driving Test to Evaluate the Steering Feedback

In the following, a driving maneuver is presented for examining the feedback behavior in the driving test. Therefore, at a constant velocity v, a steering wheel angle δ_s increase is required until the vehicle transitions into the understeering state. Figure 15 shows the measurement data of the driving maneuver with vehicle C for two different applications of the Porsche in house developed steering software. Figure 15 a) shows the measurement data with inactive steering controller. Figure 15 b) shows the data for with active controller. In the upper two diagrams the curves of the velocity v and the steering angle δ are plotted against the time t. The maneuver is performed at a constant velocity of approximately 80 $\frac{\text{km}}{\text{h}}$. The steering wheel angle δ is increased from 50 ° to 85 ° within about 1 s. In addition, the curves of the torsion bar torque M_{tb}, the yaw rate $\dot{\psi}$ and the lateral acceleration $\ddot{y}$ are shown.

By comparing the figures 15 a) and b), it can be seen that the torsion bar torque M_{tb} drop by 1 Nm during the transition to the understeering state, both when the steering controller is deactivated and when it is activated. However, when the controller is activated, the time of this drop is clearly ahead of the saturation of the rate $\dot{\psi}$ and the lateral acceleration $\ddot{y}$, whereas when the controller is inactive, the saturation of the vehicle dynamics occurs almost instantaneously. This time shift with activated controller allows the driver of the vehicle to sense the transition into the understeering state at an early stage by the fall of the torsion bar torque M_{tb}. The driving maneuver described can be used to objectively evaluate and qualify the application of the steering controller and the steering software in the complete vehicle.

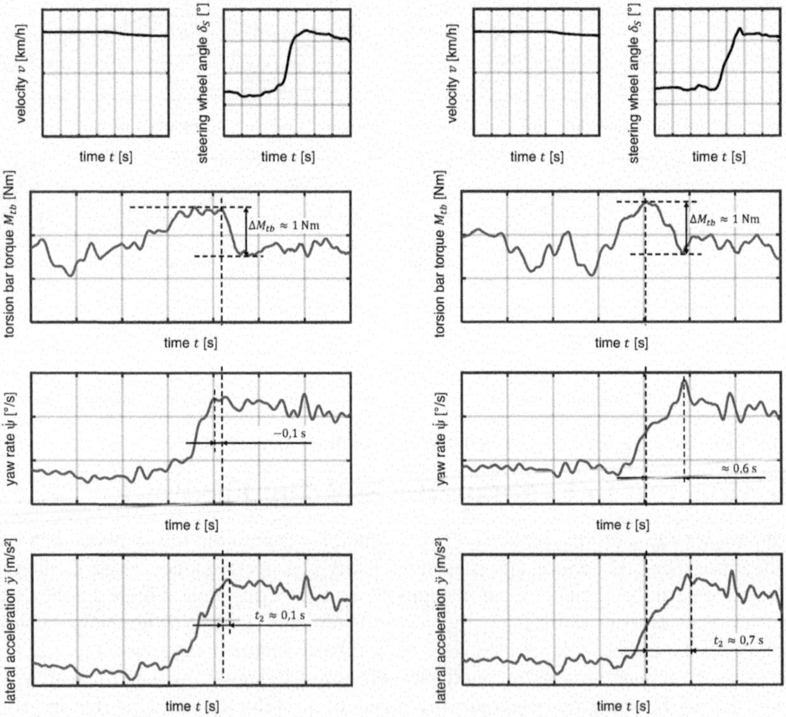

a) steering controller inactive b) steering controller active

Figure 15: Feedback in the driving test for vehicle C

5 Press Analysis

In the preceding chapters, objective analysis methods for the assessment of steering feedback are presented. Additionally, the influence of the Porsche in house feedback controller is presented. In the context of an extensive press analysis, the objective improvements by the steering controller can be determined on the basis of subjective evaluations of the steering systems of various vehicles of the Dr. Ing. h.c. F. Porsche AG. Evaluations from a total number of 684 press tests from 218 different sources were evaluated over a period from 02/2016 to 04/2018. The press tests contain 250 ratings regarding the steering feedback. Figure 16 shows the percentage of positive and negative ratings of five different Porsche vehicle types. Only the steering system of the types D and E use the Porsche in house steering software. It can be seen that the steering systems of these types with regard to the steering feedback are markedly better rated than the steering systems of the other vehicles. Thus, the objective optimization of the feedback behavior with the aid of the presented tests can also be confirmed subjectively by the test drivers and is consequently also perceptible to the customer.

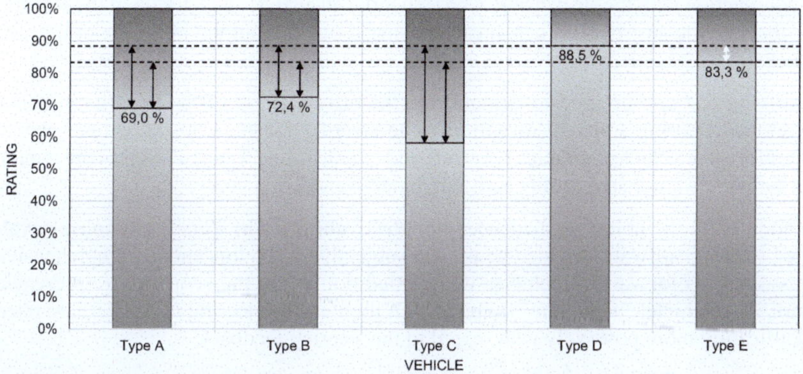

Figure 16: Press rating regarding steering feedback

6 Conclusion and Outlook

This article focusses on the objectification and analysis of steering feedback. Therefore, the state of the art is documented with regard to the guidance and feedback behavior of the steering system. Furthermore, the limitations of electric power steering systems regarding the steering feedback as well as the target values for guidance and feedback behavior are defined.

Subsequently, tests at the steering test bench in combination with the Fast-Fourier-Transformation are presented for the objective representation of the feedback characteristics of various steering systems. Both tests with a fixed steering wheel as well as a freely swinging steering wheel are carried out and the results are interpreted. Both methods allow the analysis of the feedback behavior. The mechanical characteristics of the different steering systems can be shown with the aid of the illustrations with deactivated servo assistance. These measurement results are of decisive importance for the parametrization of simulative steering models in order to be able to adapt the model behavior to reality by means of the damping. With the aid of the frequency response diagrams with active servo assistance, detailed analyzes of the influence of different variants of steering software and steering controller are possible.

When measuring and analyzing the feedback on the steering test bench, only the characteristics of the steering systems are considered. The suspension remains unconsidered. Therefore, a special test bench is presented to integrate the suspension into the feedback path. The measurements show that the transmission behavior of the separate steering system on the steering test bench is very similar to the transmission behavior in conjunction with the suspension. Consequently, an analysis and design of the steering feedback at the steering test bench can be carried out without the influence of the suspension.

A driving test can be used to demonstrate the effectiveness of the Porsche in house developed steering software to optimize the steering feedback in the entire vehicle. Therefore, the vehicle dynamics (yaw rate and lateral acceleration) as well as the torsion bar torque are compared during active and inactive steering controller.

In addition to the objective evidence, the effectiveness of the Porsche in house developed steering software can also be proven subjectively. To this end, a comprehensive press analysis for various vehicles was carried out within the scope of the presented work. The vehicles, which are equipped with the Porsche steering controller, are rated significantly more positively with regard to the steering feedback.

In the course of further work, the investigation of other suspension concepts such as, double wishbone suspension or a multi link suspension is of interest. In addition, an increase in the degree of detail of the simulation models is possible to increase the quality of the simulation. For example, this can be achieved by more complex friction models or the consideration of the steering bearing. A continuation of the press analysis for future vehicle projects is useful in order to be able to infer conclusions on the requirements for steering systems.

References

[Braess, 2004] **Braess, Hans-Hermann:** *Die schwierige Übung des richtigen Kurses*
Frankfurter Allgemeine Zeitung, 11.05.2004

[Brunn & Harrer, 2004] **Brunn, Philipp; Harrer, Manfred:** *Objektivierung der Lenkungsrückmeldung*
BMW AG, 2004

[Düsterloh, 2015b] **Düsterloh, Dario:** *Methode zur Darstellung einer Fahrzeugunabhängigen Fahrbahnrückmeldung durch ein elektromechanisches Lenksystem*
Technische Universität Braunschweig; Dr. Ing. h.c. F. Porsche AG, 2015

[Düsterloh & Schrage, 2016] **Düsterloh, Dario; Schrage, Benedikt:** *VDI-Berichte Nr. 2279: Lenkungsentwicklung am mHiL-Lenkungsprüfstand - Parameteridentifikation und Funktionsentwicklung am mHiL-Lenkungsprüfstand*
Verein Deutscher Ingenieure (VDI); Dr. Ing. h.c. F. Porsche AG; 18. SIMVEC Kongress: Simulation und Erprobung in der Fahrzeugentwicklung 2016 - Berechnung, Prüfstands- und Straßenversuch, 2016; ISBN: 978-3-18-092279-9

[Frey, 2016] **Frey, Katharina:** *Auswahl und Eignungsanalyse von Linearaktuatoren für den flexiblen Einsatz im Fahrwerkprüffeld unter besonderer Berücksichtigung der EPS-Lenkung*
Universität Stuttgart; Dr. Ing. h.c. F. Porsche AG, 2016

[Harrer, 2007] **Harrer, Manfred:** *Characterization of Steering Feel*
University of Bath, 2007

[Harrer, 2008] **Harrer, Manfred:** *Objektivierung des Lenkverhaltens*
Dokumentation Kraftfahrwesen e.V.; Aachener Kolloqium Fahrzeug- und Motorentechnik; Dr. Ing. h.c. F. Porsche AG, 2008

[Henning et al, 2015] **Henning, Josef; Schmidt, Steffen; Bodenstein, Jan:** *Effiziente Fahreigenschaftsoptimierung durch echtzeitfähige Mehrkörperachsen*
Automobiltechnische Zeitschrift (ATZ), 117. Jahrgang, Ausgabe 04/2015

[Pfeffer & Harrer, 2013] **Pfeffer, Peter; Harrer, Manfred:** *Lenkungshandbuch – Lenksysteme, Lenkgefühl, Fahrdynamik von Kraftfahrzeugen*
Springer Vieweg, 2. Auflage, 2013; ISBN: 978-3-658-00976-2

[Scherhaufer, 2016] **Scherhaufer, Julian:** *Analyse des Einflusses von Radaufhängung und Lenkung auf die Lenkungsrückmeldung von der Fahrbahn*
Hochschule Ravensburg-Weingarten; Dr. Ing. h.c. F. Porsche AG, 2016

[Schimpf, 2016] **Schimpf, Rico:** *Charakterisierung von Lenksystemen mit Hilfe eines Lenksystemprüfstands*
Technische Universität Wien, 2016

[Schindler, 2007] **Schindler, Erich:** *Fahrdynamik*
expert verlag, 2007; ISBN: 978-3-8169-2658-0

[Schramm et al, 2013] **Schramm, Dieter; Hiller, Manfred; Bardini, Roberto:** *Modellbildung und Simulation der Dynamik von Kraftfahrzeugen*
Springer, 2009; ISBN: 978-3-540-89313-4

[Uselmann, 2017] **Uselmann, Anton:** *Ein Beitrag zur funktionalen Entwicklung eines elektromechanischen Lenksystems für sportlich orientierte Fahrzeuge*
Universität Duisburg-Essen, 2017

[von Groll, 2006] **von Groll, Max:** *VDI-Fortschrittsberichte Reihe 12 Nr. 630: Modifizierung von Nutz- und Störinformationen am Lenkrad durch elektromechanische Lenksysteme*
Verein Deutscher Ingenieure (VDI); VDI-Verlag, 2006; ISBN: 3-18-363012-5

[Wolf, 2009] **Wolf, Hagen J.:** *Ergonomische Untersuchung des Lenkgefühls an Personenkraftwagen*
Technische Universität München, 2009

[Zschocke, 2009] **Zschocke, Alexander K.:** *Ein Beitrag zur objektiven und subjektiven Evaluierung des Lenkkomforts von Kraftfahrzeugen*
Universität Karlsruhe, 2009

Does the steering wheel have a future? A UX study

Dr. Alisa Lindner & Roland Greul

Robert Bosch Automotive Steering GmbH

P. Pfeffer (Ed.), *9th International Munich Chassis Symposium 2018*, Proceedings,
https://doi.org/10.1007/978-3-658-22050-1_34

1 Introduction

1.1 Motivation

The steering wheel has been nearly unmodified over the last decades. Nevertheless, there are many show cars, prototypes and small series models that use alternative devices to steer the car, e.g. by ZF Friedrichshafen AG (Buchmeier & Wilkerson, 2017), Continental AG (Fillenberg, 2016) and Jaguar (2017). The devices vary in size, position, shape and way of interaction with them. In the future, steer-by-wire systems might be an enabler for alternative steering devices also in series cars. Using these devices would lead to a massive change in the area around the driver seat, e.g. the dashboard and center console.

Depending on which of those devices is considered, different optimizations can be realized. A joystick next to the driver seat that is operated with one hand generates space in front of the driver. This space could be used for larger displays and to allow a more comfortable seating position. Therefore, alternative steering devices have great potential to improve the drivers' well-being. Nevertheless, it is unclear to what extent drivers are willing to use such a device instead of the well-known steering wheel. For that reason, the drivers' user experience must be considered.

1.2 What is user experience (UX)?

User experience (UX) includes all aspects of the user's perception of an offering, including the product itself and its functionalities. It includes the "user's emotions, beliefs, preferences, perceptions, physical and psychological responses, behaviors, and accomplishments that occur before, during and after use" (Turner, 2017, p. 13).

A positive or even outstanding UX can be achieved by adapting the product characteristics to satisfy all user needs. To achieve this goal, developers use a human-centered design proccss. This approach enables organizations to deliver desirable, viable and feasible solutions tailored on the specific necessities of the user. By offering additional functionalities or features that go beyond the expected offer, the perceived value of an offering increases and with this the customers' desire to buy and use this product. One key aspect that forms the UX of a product is its usability. It is defined as "[…] the effectiveness, efficiency and satisfaction with which specified users can achieve specified goals in particular environments" (Jordan, 1998, p. 5). Amongst that, the perceived aesthetics, joy-of-use and image-of-use regarding a certain product must be considered.

There are slightly different approaches to develop products that evoke a positive UX. Many of them are based on the human-centered design process as described in the ISO 9241 (ISO 2010). They are based on understanding user needs in the early phases of development and gathering user feedback in the last phases. They allow and even support multiple iteration within one step or even throughout the process.

One approach that has become more and more popular during the last years is design thinking. The design thinking process as described by the Hasso Plattner Institute is divided into five phases and starts and ends with direct input from users (Plattner et al., 2011). It includes analytical and creative thought processes and methods in order to solve difficult tasks and to find desirable solutions for clients (Ingle, 2013). By failing early and often, mistakes and weak spots of concepts can be spotted and eliminated quickly.

1.3 UX in the context of steering

Regarding UX in the vehicle, many different stakeholders have to be considered. From a supplier's perspective, the OEM as the customer in B2B is very important. All changes that are made within single components have to match the full vehicle. This is especially crucial for components such as the steering device that play an important role in interior and cockpit design.

Another obviously important user group are the drivers. When altering the design of the steering wheel, it must be considered that drivers are extremely adapted to it. Since maneuvering a vehicle is a highly automated, skill-based process (Rasmussen, 1983), drivers are mostly unaware of their inputs into the steering wheel. Nevertheless, it must allow them to steer comfortably, precisely and safely in order to generate a positive steering feel. It depends upon how drivers perceive the interaction with the steering wheel and the car's reaction (Wolf, 2009). Especially in dangerous situations, the driver's mental model that links the input into the steering wheel with the resulting vehicle reaction must be appropriate. Since the majority of drivers are only familiar with the steering wheel, any other steering device must be designed intuitively and self-explanatory. For example, drivers need to be able to understand how to operate it at first glance.

Thus, the success of such an alternative steering device depends on the design itself and the resulting steering feel. Based on this, we conducted a study to challenge the steering wheel and to evaluate the potential of an alternative steering device.

2 Methods and Results

2.1 UX process

The underlying process of this study is divided in five phases (see Fig. 1):

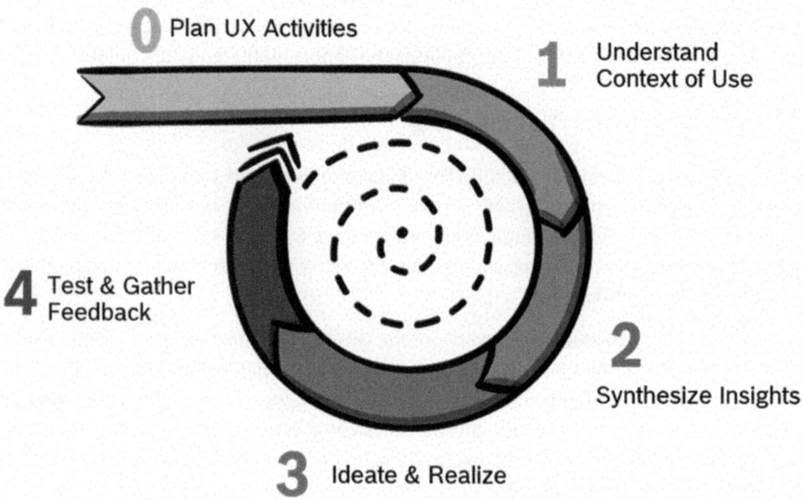

Fig. 1: UX process.

Phase 0: Plan UX activities
In the initial planning, the scope of the project is defined and stakeholders and relevant user groups are identified.

Phase 1: Understand context of use
The goal is to gain user insights and to understand their needs and situations. There is a variety of methods that can be applied, e.g. interviews or contextual inquiries.

Phase 2: Synthesize insights
The insights that were collected before are gathered, restructured and analyzed to create a bigger picture.

Phase 3: Ideate & realize
In this phase, new ideas are generated to meet the users' needs. Different kinds of idea workshops and creativity methods can be used. The most promising concepts are transferred into prototypes.

Phase 4: Test & gather feedback
Lastly, future users evaluate these prototypes. Common methods are AB comparisons or user tests. The collected results are used to identify the best concepts and to refine them further.

2.2 Phase 0: Plan UX activities

The project scope is defined by the so-called Question Zero. It is based on six questions that are answered within the project team:

- *What do we want to develop?*
- *What is it for?*
- *Where is it used?*
- *For whom is it?*
- *What is it in contrary with?*
- *How does it work?*

For this study, the Question Zero was: "A manual steering device that spotlights the innovative steering technology and is suitable for right and left handed drivers of passenger cars".

2.3 Phase 1: Understand context of use

In the next phase, the environment in which the product will be used was explored. By consulting special user groups, exceptional situations and outstanding opinions can be revealed. Based on them, general conclusions can be drawn for the entire field of interest. For this study, eight extreme users were recruited, e.g. a pilot and a ship captain.

The qualitative, semi-structured face-to-face interviews were conducted at the interviewees' work place. The duration of an interview was about 120 minutes. First, the interviewees were asked about steering or maneuvering in their specific context. Whenever possible, they were observed and interviewed while operating the vehicle or device. Afterwards, the interviewees were asked to transfer their specific expertise to the steering of cars and to comment on different alternative steering concepts.

2.4 Phase 2: Synthesize insights

A user board was prepared for every interviewee on which his pains and needs were collected. Two exemplary insights are presented: The pilot used a yoke to steer the plane. For him, it did not matter if he sits in the pilot's or co-pilot's seat. One key quote was that "a plane can be flown with the left or right hand". The ship captain, on the other hand, used joysticks for maneuvering. Although the ship he steered had very large dimension and reacted very slowly, he stated that "a ship can be maneuvered precisely".

After collecting all the individual input, the most important insights were restructured and analyzed qualitatively. Insights that are connected with each other were condensed to important, controversial and exciting themes. Eleven top findings were derived that represent only direct user feedback without interpretation. Examples are:

- The steering triangle: The steering device must have an ideal size, shape and position within in the car.
- Steering without thinking: It is important that the learning and adaption process is as short as possible so that drivers can easily use the steering device from the start.
- Less is more: Today's modern steering wheels are considered overloaded. Drivers want steering devices that only have the most important functions within reach.
- The one or two hand dilemma: Depending on the situation, drivers use one or two hands to steer the car. The steering device should adapt to these different settings.

2.5 Phase 3: Ideate & realize

These top findings were the basis for design activities in the third phase. In an ideation workshop, the team aimed to develop as many ideas as possible.

To begin with, a challenge between the classic steering wheel and other steering devices was initiated. Afterwards, scenarios of different steering situations were illustrated. In the following ideation process, the team created 110 new steering concepts. Those ideas were clustered into categories such as adaptive steering wheel dimensions, comfort steering (combined acceleration & brake) and adaptive rotation angle. By voting, top ideas were identified (see Fig. 2):

Fig. 2: Top ideas derived from ideation workshops.

Exemplary ideas for two-handed driving:

- Butterfly: The handles can grow and shrink based on the driving situation. An interactive display is included in the middle.
- Steering wheel within a steering wheel: The big wheel can be pushed back and a small formula 1 like steering wheel is available and can be used for dynamic steering.

Exemplary ideas for one-handed driving:

- Comfortable steering device: The driver is able to steer with minimal movement of one hand by wrist rotation. Accelerating and braking are also included.
- Push & pull steering device: The driver can steer by pushing one handle forward or pulling it back closer to him.
- Fast steer knob: A small knob can unfold out of the steering wheel and the driver is able to rotate the wheel faster and in a more precise way.

Some of the ideas were merged and translated into concepts. To operate the so-called "D-Steering" for example, the driver needs to rotate the D-shaped handle around the x-axis. The arm position that results from this concept might be particularly comfortable for long-term usage. The handles or wings of the "Adjustable Butterfly" can shrink into a sporty mode and completely disappear in the parking mode. An additional steering knob reduces the size of the movements the driver has to perform. Low-fidelity prototypes were built using materials like cardboard, paper, modelling material and styrofoam.

2.6 Phase 4: Test & gather feedback

The prototypes were then evaluated and improved in consecutive Rapid Iterative Testing and Evaluation (RITE) cycles. In this method, a product or system is tested with users and the identified weak spots are optimized directly after each test. This procedure leads to a quick and efficient continuous improvement of the product (Medlock et al., 2005). The users tested the prototypes' shape and size and commented on it. Within the tests, we used the "thinking aloud" method in order to get direct feedback about their impressions. The refined prototypes were transferred into CAD models and 3D printed. In the next step, they will be under evaluation in a driving simulator.

3 Conclusion

A human-centered study was conducted to develop concepts how alternative steering devices in the context of steer-by-wire systems could look like that lead to a positive UX. In user interviews, detailed information regarding the needs of drivers was collected. The insights gained were analyzed and summarized to top findings. The subsequent ideation workshop led to different concepts to meet the users' needs. Some of the concepts were elaborated further in iterative cycles and will be tested in a driving-simulator. Based on these results, further adjustments can be made. By keeping the user in mind, the concepts can be designed to finally meet the drivers' needs and to become a real hazard to the classic steering wheel.

Bibliography

Buchmeier, R., & Wilkerson, J. (2017): *ZF Develops Advanced Steering Wheel Concept for Automated and HMI Solutions.* https://press.zf.com/site/press/en_de/microsites/press/list/release/release_38080.html [14.05.18]

Fillenberg, S. (2016): *With One Single Swipe: Continental Integrates Gesture-Based Control into the Steering Wheel.* https://www.continental-corporation.com/en/press/press-releases/2016-05-10-hands-on-101874 [18.04.2018]

Ingle, B. R. (2013): *Design Thinking for Entrepreneurs and Small Businesses: Putting the Power of Design to Work.* New York: Springer.

ISO, 9241-210:2010 (2010). *Ergonomie der Mensch-System-Interaktion – Teil 210: Prozess zur Gestaltung gebrauchstauglicher interaktiver Systeme.*

Jaguar (2017): *"Hello, my name is Sayer. I am the steering wheel of the future".* https://media.jaguar.com/news/2017/09/hello-my-name-sayer-i-am-steering-wheel-future [14.05.18]

Jordan, P. W. (1998): *An Introduction to Usability*. Boca Raton: CRC Press.

Medlock, M. C., Wixon, D., McGee, M., & Welsh, D. (2005): The rapid iterative test and evaluation method: Better products in less time. In: *Cost-Justifying Usability*. Burlington: Morgan Kaufmann.

Plattner, H., Meinel, C., & Weinberg, U. (2009). *Design thinking*. Heidelberg: Springer.

Rasmussen, J. (1983): Skills, rules, and knowledge; signals, signs and symbols, and other distinctions in human performance models. *IEEE Transactions on Systems, Man, and Cybernetics*, 13(3): 257-66.

Turner, P. (2017): *A Psychology of xerience: Involvement, Affect and Aesthetics.* Cham: Springer.

Wolf, H. J. (2009). *Ergonomische Untersuchung des Lenkgefühls an Personenkraftwagen.* Dissertation, Technische Universität München. München.

NEW STEERING CONCEPTS

Steering on demand for dual-mode vehicles

Joe Klesing, Nexteer Automotive Corp.

Jeff Zuraski, Nexteer Automotive Corp.

Ayyoub Rezaeian, Nexteer Automotive Corp.

P. Pfeffer (Ed.), *9th International Munich Chassis Symposium 2018*, Proceedings,
https://doi.org/10.1007/978-3-658-22050-1_35

1 Introduction

Future Dual Mode Vehicles can be operated either in Manual or Autonomous Driving Modes (i.e. Conditional or High Automation). Drivers may want to re-assume control of the vehicle, after driving autonomously for a period of time. This may either be because they feel more comfortable to manage certain driving situations by themselves, e.g. on the highway at higher speeds, or because they enjoy the experience of driving.

Some of the key factors in the hand-over process are the avoidance of Mode Confusion (i.e. clarity who is in charge), establishment of Situation Awareness (i.e. the ability to perceive, comprehend and project a traffic situation) and full neuro-muscular control (i.e. the driver has developed a feel for the dynamic behaviour of the vehicle). The task therefore is to bring the driver back into the loop with full operational cognitive control as fast as possible.

The proposed concept ensures a safe and intuitive voluntary transition between autonomous and manual driving modes. The steering wheel will be retracted during autonomous operation—Quiet Wheel Steering—and returned to its original position for manual driving mode.

Steer-by-Wire helps to keep the vehicle stable independent from actual driver input, while giving the driver haptic feedback via the steering wheel. The operational excellence in managing driving situations can be improved during manual driving modes and during the transitions between driving modes. The Steering-on-Demand package allows automobiles with high automation levels to combine the comfort of autonomous driving with the excitement and experience of manual driving.

2 Driving Mode Transitions

2.1 A typical use-case

A typical use-case of autonomous driving might look as follows:

The driver would activate a Level 4 Highway Autopilot to watch a movie or work on a document with his laptop. Being immersed in this non-driving related activity for an extended period of time, the driver would be Out-of-the-Loop, i.e. he or she has lost awareness of the ongoing traffic situation and the feel for the dynamic behaviour of the vehicle. In particular, the driver would be un-aware of any changes in the traffic situation such as an additional vehicle appearing in the lane in front of him.

At some point, the driver would decide – for whatever reason – to assume control of the vehicle, again. And after a short transition phase the control is granted back to him via the vehicle control system.

Shortly afterwards, with the driver being in control just for a very few seconds, the driver needs to perform an emergency manoeuvre, e.g. because the vehicle in front brakes suddenly and un-expectedly.

As part of the transition from autonomous to manual driving, the driver needs to be prepared and supported to manage such a scenario. In this context "transition" is defined as the process and period of transferring responsibility of, and control over, some or all aspects of a driving task, between a human driver and an automated system /1/.

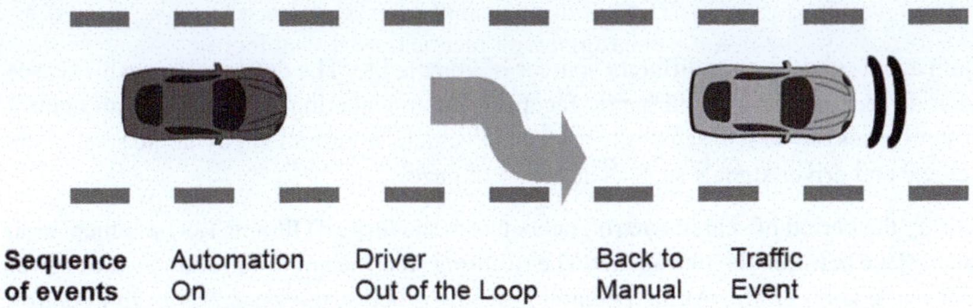

Fig. 1) Autonomous Use-Case

2.2 Driver/System-Communication and Shared Control

To achieve a safe and intuitive transfer of the control authority to the driver, we are proposing

- a communication between two agents, driver and path control system and
- a period of shared control between the two agents starting at the take-over request time TOR and finishing at the time T_{manual} and
- a driver assist function after T_{manual}.

To initiate the hand-over process, the driver needs to communicate his intent to the system. The system, on the other hand, needs to communicate the automation status to the driver at all times, evaluate driver awareness with regards to transition, evaluate driver inputs after the time TOR and resume driver assist functions after the time T_{manual}, when the driver is back in control of the vehicle.

1 Louw T. (2017), The human factors of transitions in highly automated driving, ITS

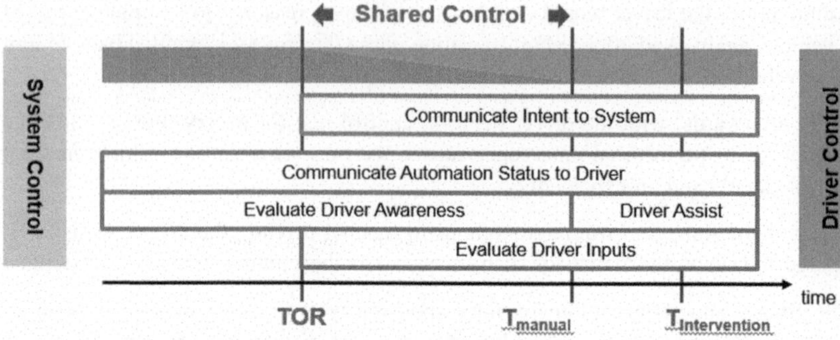

Fig. 2) Driver/System-Communication

Both agents make use of different sources of knowledge. The driver relies on his senses such as seeing, hearing and haptic feedback via the steering wheel. The path control system relies on vehicle sensors to assess the vehicle environment, dynamic state of the vehicle and driver signals such as direction of gaze.

During the period of shared control, i.e. between the times TOR and T_{manual}, a mediation takes place between the two agents. The resulting steering angle δ is a function of driver input and path control. At the beginning of this process, the system is in full control, i.e. $\delta = \delta_{path}$ and at the end of this process, the driver has full control, i.e. $\delta = \delta_{drv}$.

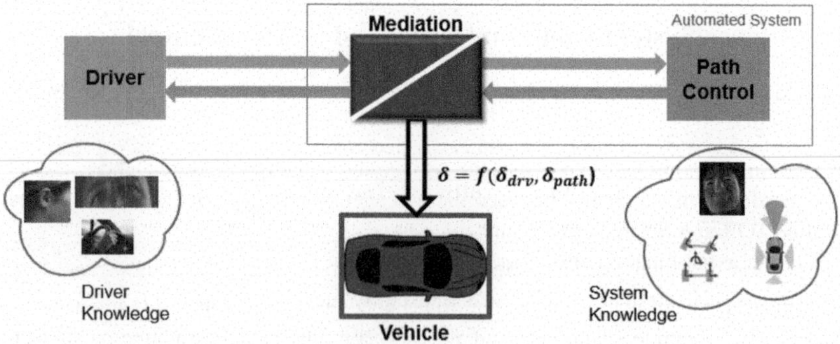

Fig. 3) Mediation between agents

2.3 Steering Performance Metrics

The human driver performs 2 roles with respect to steering: He acts as

- visual and tactile sensor to determine vehicle behaviour and as
- a controller of lateral vehicle direction.

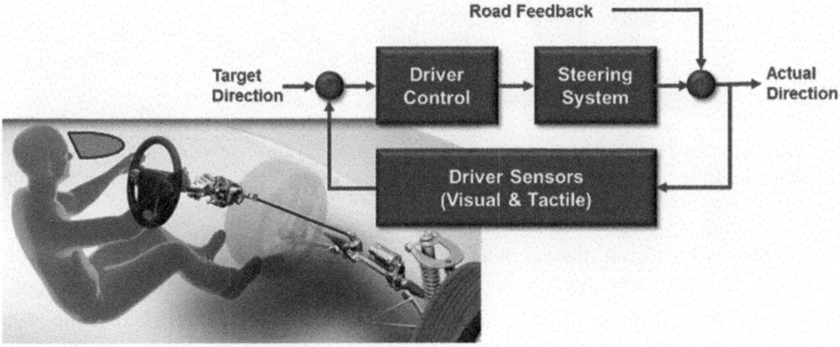

Fig. 4) Simplified steering control diagram

The handover from human control to control of the automated system can be achieved near instantaneously. However, the handover from the automated system to driver control requires the driver to regain situational awareness and cognitive physical control of the vehicle, i.e. a feel for how the vehicle responds to the driver's steering input.

Here are some performance metrics to determine the quality of transition to manual driving:

- Time to contact (e.g. to leading vehicle in the same driving lane)
- Lateral Position (of vehicle within driving lane)
- Time to Lane Crossing (i.e. time until lane markings are crossed)
- Steering Wheel Reversal Rate (i.e. rate at which the human driver changes direction within same driving manoeuvre)
- Steering Effort (i.e. Steering Torque)
- Situation Awareness (i.e. the perception, comprehension and projection of traffic events)
- Mode Awareness (i.e. the awareness whether the vehicle is being manually controlled by the driver or automatically by the system)

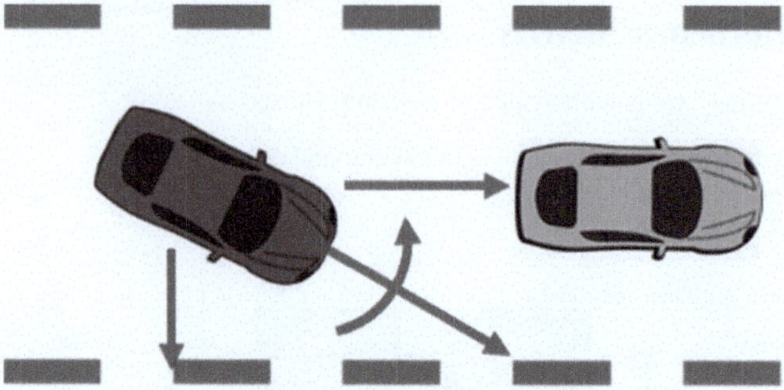

Fig. 5) Geometry for Time-to-Contact, Lateral Position, Time to Lane Crossing and Steering Wheel Reversal Rate

Recent work on transition management /1/, suggests that the loss of physical control and the loss of situation awareness can arise independently as result of vehicle automation and can lead to less effective return-to-manual performance. Importantly, the loss of physical control can also act on situation awareness, which can result in less effective performance. The further the driver is out-of-the-loop of the driving task, the worse his return to manual performance and his ability to deal with an ongoing traffic situation. The operational excellence with which the driver masters a driving situation after assuming manual control, can be measured for example via the lateral vehicle position in the driving lane and the steering wheel reversal rate /2/ such as during the manoeuvre outlined in 2.3.

The task therefore is to help the driver gain full operational cognitive control of the vehicle as fast as possible, so that the driver can master a traffic situation such as the one described in the use-case in the introduction above. This would minimize the period of time between the driver assuming control and having full operational cognitive control of the vehicle.

1 Louw T. (2017), The human factors of transitions in highly automated driving, ITS

2 Merat et. al. (2014), Transition to manual: Driver behaviour when resuming control from a highly automated vehicle. Transportation Research Part F: Traffic Psychology and Behaviour, 27

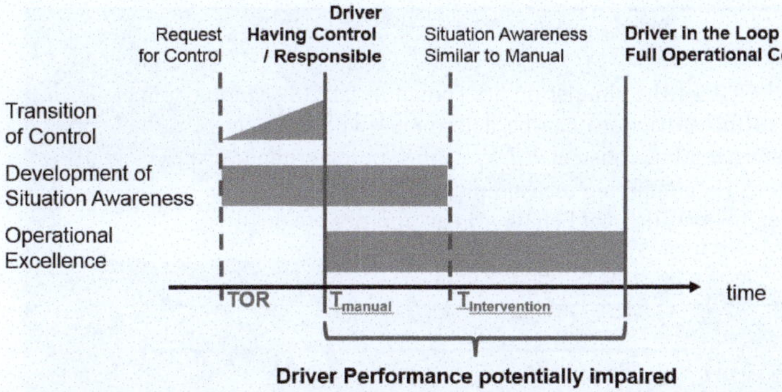

Fig. 6) Minimize time between driver having control and being in full operational cognitive control

3 Driver Interaction with Steering-Wheel & Column

3.1 Study set-up

The following study examines the interaction preferences of drivers during the transition from autonomous to manual driving as they relate to helping the driver get back to full operational cognitive control.

The study was conducted with over 30 participants with professional backgrounds other than engineering. The "drivers" were watching a pre-recorded video collected from the frontal camera of a vehicle while driving on a highway. A headset provided them with sounds from that highway drive and isolated them from potential background noises in the lab environment. The cockpit was furthermore simulated by an arrangement of steering wheel and column. To get the participants into an "out-of-the-loop" condition, they had the freedom of either reading a book on a provided tablet computer or to play with their own mobile phone.

Upon acoustic prompt, the drivers had to engage with the steering wheel and column to assume manual control. During manual control the drivers then had to respond to a traffic situation and correct a drift of the vehicle out of the desired driving lane.

During autonomous mode, the steering wheel was stowed to create more space for the driver. For manual driving, the steering wheel had to be brought back into to normal position, i.e. un-stowed.

This study examined multiple steering column and switch configurations including multiple speeds to deploy the steering wheel. The drivers' preferences and the gaze of their eyes were recorded during the transition.

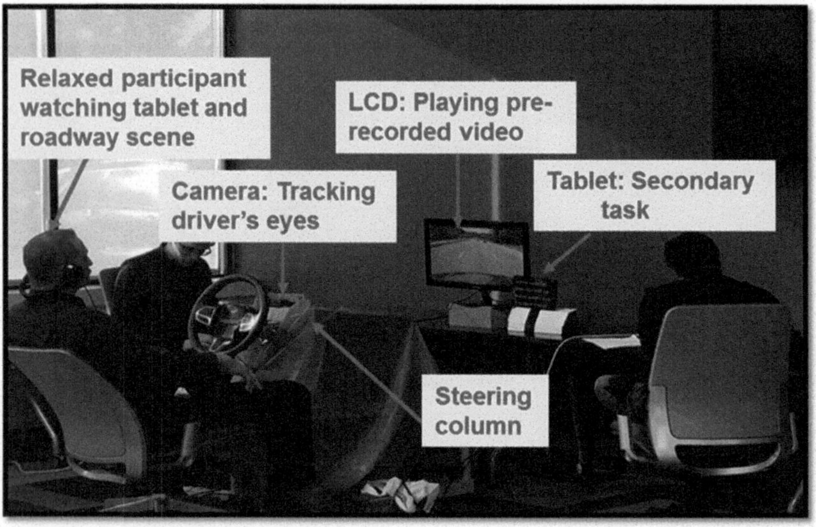

Fig. 7) Experimental setup

3.2 Study findings

A key finding regarding an interaction with the steering wheel and column was that it had to be intuitive to help the driver focus his attention on the traffic situation early. In particular, the usage of the stow / un-stow feature of the steering wheel helped to avoid mode-confusion, i.e. it was clear to the driver who was in control of the vehicle: Stow = autonomous mode / un-stow = manual mode.

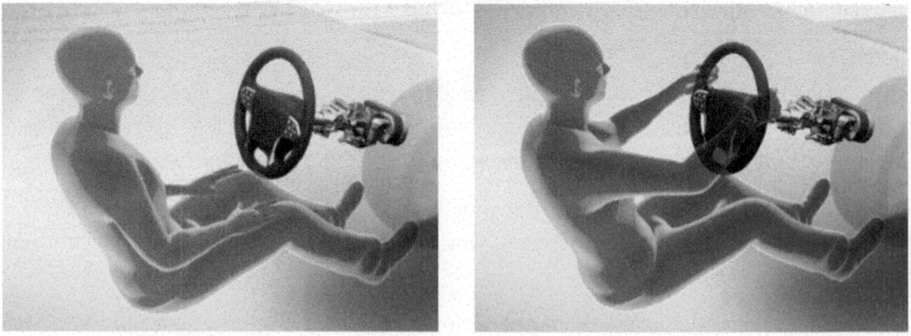

Fig. 8) Stow / un-stow feature to avoid mode-confusion

Furthermore, participants preferred and/or performed better, if the following conditions were met:

- Freedom to steer immediately as soon as hands were on the steering wheel, while interaction with the steering wheel would activate the stow/un-stow feature
- Elimination of the need to find and operate a switch
- Fast deployment of the steering wheel during transition to manual drive
- Mobile phone friendliness, i.e. the ability to operate the stow/un-stow feature without the need to drop the mobile phone
- Elimination of all potential to unintentionally deploy the steering wheel while encouraging driver head and eyes to focus on forward direction

4 Implementation of Shared Control

4.1 Evaluation of Driver Intent and Performance

During the phase of transition of vehicle control, the driver and system go through a phase of shared control as outlined in 2.2. In this phase the driver and the system communicate with each other e.g. via haptic input of driver and torque feedback of the system. Because of the steer-by-wire implementation, the system can send haptic cues to the driver without affecting the heading angle or stability of the vehicle.

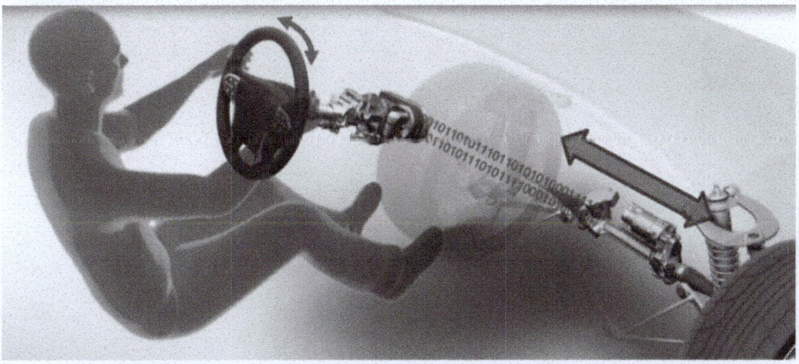

Fig. 9) Steer-by-wire enables independent haptic cues and steering intervention

This enables an evaluation of the driver readiness and ability to assume the control of the vehicle. Driver inputs are put into context of vehicle dynamic signals and environmental sensor data to calculate a confidence indicator α that represents the predicted ability of the driver to manage the control of the vehicle. This indicator determines the respective and potentially conflicting contributions of driver and system to the actual steering angle δ.

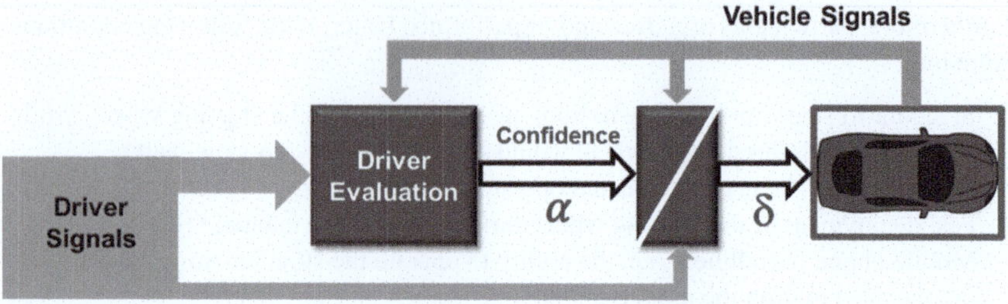

Fig. 10) Shared control based on evaluation of driver intent and performance

The driver inputs considered in this paper, consist of his steering action and direction of gaze. The vehicle signals consist of vehicle dynamics signals as well as signals that monitor the vehicle external environment of the vehicle for the purpose of advanced driver assist.

4.2 Model Predictive Framework

The driver evaluation is implemented with the help of a model predictive framework such as /3/. This framework predicts on optimal sequence of steering inputs as well as an optimal pattern of gaze to execute an intended driving manoeuvre such as lane change.

Fig. 11) Optimal trajectory for an intended lane change manoeuvre

3 Nguyen A.-T. et.al., "Sensor Reduction for Driver-Automation Shared Steering Control via an Adaptive Authority Allocation Strategy", IEEE/ASME Transactions on Mechatronics 2017

The driver's ability to adhere to a certain degree to an optimal trajectory is rewarded with a higher confidence factor and thus faster transition to full level of control. If the driver is violating requirements related to the optimal trajectory, the system gives feedback to the driver by haptic feedback via the steering wheel. In case of conflicting control targets of driver and system, the driver still retains the ultimate control authority. However, in order to assert that authority, he has to actively overcome a torque threshold and thus provide evidence and confirmation of true intent.

5 Summary

Shared control is a key concept in the transition of control authority from automated system to the driver. Between the Take-over-Request and the time when the human driver is in full control, a mediation between the two agents – driver and path control - determines the degree of human influence and helps the driver to stay safe during the transition.

The stow-able steering wheel creates interface transparency, i.e. it makes the state of vehicle automation obvious to the user and thus prevents mode confusion. As a result, the driver intuitively understands who is in charge of controlling the vehicle: If the steering wheel is stowed it is the system, if the steering wheel is deployed it is the driver.

The driver needs time to develop situational awareness and the feel for the behaviour of the vehicle when transitioning from autonomous to manual control. Until the driver has developed full operational cognitive control, haptic cues, steering-wheel torque and direction of driver gaze provide the means of vehicle to driver communication and vice versa. This improves the quality of the take-over.

Similar to fly-by-wire in aviation, with steer-by-wire the steering wheel merely becomes a Human Machine Interface element that interprets the intent of the driver. As steering wheel and steering position of the road wheels are decoupled, the driver input can be evaluated before translating to the actual steering motion. Hence, steer-by-wire increases the safety of the driver.

OEM perspective on decoupled steering systems

Riccardo Ficca, Jaguar Land Rover Limited, UK

This manuscript is not available according to publishing restriction. Thank you for your understanding.

P. Pfeffer (Ed.), *9th International Munich Chassis Symposium 2018*, Proceedings,
https://doi.org/10.1007/978-3-658-22050-1_36

Cockpit 2025 – an HMI concept for autonomous and manual driving

Dr. Thomas W. Heitz, Dr. A. Schacht, D. Kreutz, T. Bayer,
thyssenkrupp Presta AG, Liechtenstein

P. Pfeffer (Ed.), *9th International Munich Chassis Symposium 2018*, Proceedings,
https://doi.org/10.1007/978-3-658-22050-1_37

Introduction

One of the biggest trends in the automotive industry is vehicle automation. In recent years there have been many innovations in this area. Some examples include the introduction of cruise control, distance control radar, lane keeping assistance and traffic jam pilot. What this means is that drivers are handing over more and more driving tasks to the vehicle, the aim being to increase safety and comfort. This trend will continue in the coming years and will eventually lead to fully autonomous vehicles.

But there is still a long way to go with a number of hurdles to overcome. Many smaller development steps will therefore take place on the road to fully autonomous driving. The next of these development steps is SAE level 3 autonomous driving. The standard SAE J3016 classifies vehicles in relation to autonomous driving. At level 3, control is transferred completely to the vehicle on certain defined types of road, such as freeways, which corresponds to semi-autonomous driving. Compared with SAE level 2, the driver no longer has to monitor the vehicle completely and continuously. However, the restriction in level 3 is that the driver must always be ready to take over control of the vehicle at short notice. This prevents the driver from carrying out other activities during autonomous driving, such as surfing the internet.

Level 3 could involve certain changes with regard to the HMI steering system. For example, when the driver hands control over to the vehicle, rotation of the steering wheel could be reduced or even stopped altogether to increase comfort and safety and to achieve or indicate a clear separation between manual and autonomous driving. This would require a variable steering ratio from the steering wheel to the steering gear or mechanical decoupling of the steering wheel from the steering gear. A steer-by-wire system would be a solution for this.

The second potential change for the HMI steering system is the possibility of moving an electrically adjustable steering wheel away from the driver towards the engine compartment in order to create more space for the driver and at the same time highlight the separation of the two driving scenarios "manual" and "autonomous". The advantage here would be the clear separation of when the driver has to take control and when the system is operating autonomously.

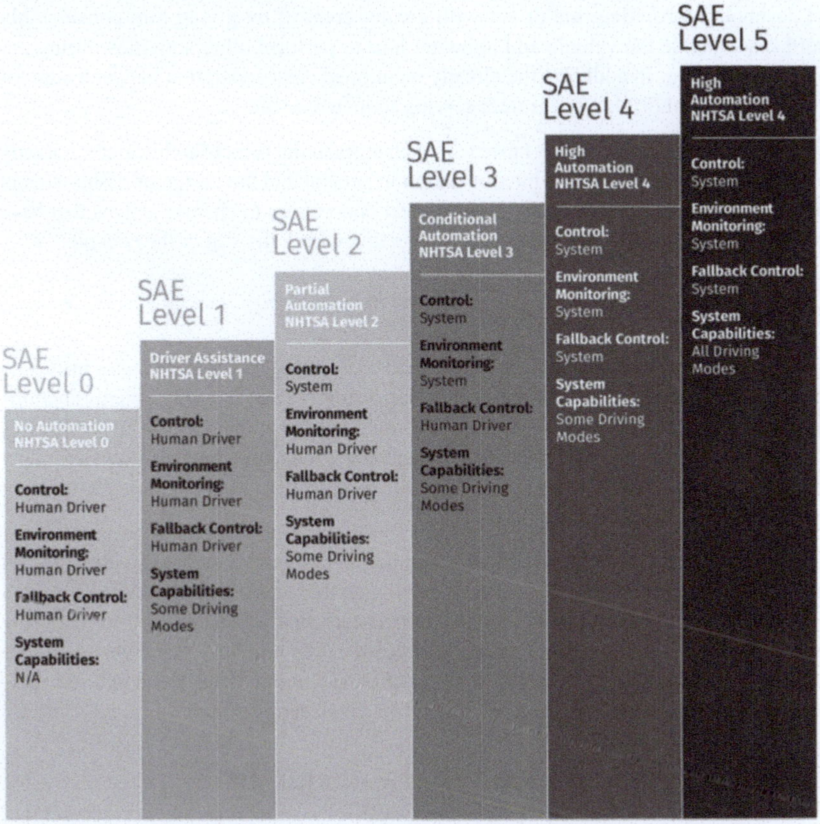

SAE Levels for autonomous driving (https://www.computerbase.de/bildstrecke/81913/3/)

The next automation step after that is SAE level 4. At this level the systems are so stable that on certain types of road control can be transferred to the vehicle and all situations can be controlled by the system without any intervention by the driver. The driver is notified minutes before the planned handover from autonomous to manual driving and thus has enough time to adjust. The driver therefore does not have to monitor the system at all times and does not have to be ready to take control of the vehicle at all times, as is the case in level 3.

Since the steering wheel is not needed for interventions at short notice during autonomous driving, it does not have to be within the driver's reach all the time. This allows the steering wheel to be moved further away from the driver and even stowed away in the dashboard, providing the driver with greater comfort by giving him considerably more space inside the vehicle and allowing him to perform other activities during autonomous driving. In addition, the already mentioned clear separation between autonomous and manual driving is ensured, serving to increase safety.

SAE4 in particular will result in far-reaching changes to the vehicle architecture, not only for the steering but also for the interior design in general and the design of dashboards in particular. However, this also represents a great opportunity to change designs that have been set for years, such as the design of the steering wheel, the seat or the cockpit.

Motivation for Presta Cockpit 2025

Such new vehicle concepts, especially at automation level 4, will have a major influence on the steering system and will require additional or modified steering system functions in the future. In order to identify these functions and their requirements at an early stage, it is necessary to build up a system understanding that goes beyond actual product boundaries. In the case of steering systems, the next higher system level is the steering wheel and the cockpit.

In order to understand the changed functions and requirements associated with increased vehicle automation (SAE level 3 and especially 4), thyssenkrupp Steering has set up a cockpit concept consisting of a steering column and steering wheel that can be completely stowed in the cockpit, a seat and a driving simulator. The aim is to show what a vehicle cockpit could look like in the future. It also makes it easier to identify and optimize new functions and requirements for the steering wheel and steering column.

Components and functions of Cockpit 2025

The cockpit concept makes it possible to examine the steering column in its overall context and in interaction with the driver. Since level 4 highly automated driving means that the driver no longer has to be ready to take control of the vehicle at all times, the steering wheel can be removed from the immediate vicinity of the driver. This can be achieved by a new steering column design to achieve larger adjustment travel. Such a new type of steering column is called a stowable steering column. It has significantly greater adjustment travel and thus allows the steering wheel to disappear completely into the dashboard.

A stowable steering column is the centerpiece of the cockpit concept setup. When changing from manual to autonomous driving, the steering column can be moved electrically away from the driver until the steering wheel disappears into the dashboard. This requires a rethink of the shape of today's steering wheels. In order to be able to stow the steering wheel and steering column in the dashboard and give the driver as much space as possible, it is necessary to move away from the current round shape of a steering wheel towards a compact and possibly no longer closed design. For this reason, thyssenkrupp Steering's cockpit concept uses a so-called horned steering wheel. When the steering column is stowed away in autonomous driving mode the horns can be folded in so that the steering wheel is as compact as possible. This is necessary in order to keep the recess in the dashboard as small as possible. In manual driving mode, the steering wheel horns fold out and the driver can operate the steering ergonomically.

Another function that becomes necessary when the steering wheel is stowed in the dashboard is to have a stationary steering wheel that doesn't rotate when in the dashboard. This function can be realized by a steer-by-wire system, for example. In a steer-by-wire system there is no permanent mechanical connection between the steering wheel and steering gear. The driver's steering commands are transmitted electronically (by wire) to the steering gear. The software determines when the driver's steering commands are transmitted. In addition, the additional use of a feedback actuator can generate a force at the steering wheel that is dependent on various vehicle parameters such as driving speed or wheel angle. This force at the steering wheel gives the driver feedback on driving conditions and makes safe maneuvering possible. In a conventional vehicle, this force is transmitted from the wheels to the steering wheel via the steering gear by means of a mechanical connection.

Another function of the cockpit concept is a driving simulator. This makes it possible to view and evaluate the individual components of the steering system with driver in the context of road traffic, also in cooperation and coordination with the customers. With the driving simulator, the driver drives the vehicle in manual mode over a virtual test course. The input signals for the virtual vehicle are provided via the steering wheel and the foot pedals. The corresponding force at the steering wheel is calculated via a vehicle dynamics model and adjusted via the feedback actuator. In certain areas of the test track, e.g. on overland routes, it is possible to switch to autonomous driving mode. The driver can activate this mode conveniently via the steering wheel. The steering wheel horns fold in and the steering column disappears into the dashboard. From the moment autonomous driving mode is activated, the vehicle is steered autonomously over the track. Both longitudinal and lateral control of the vehicle are fully autonomous.

In autonomous driving mode, the driver no longer has to pay attention to what is happening on the virtual road and can devote himself entirely to other activities. At SAE level 4 such non-driving activities will be completely normal, since the safety of the systems makes monitoring by the driver obsolete and he does not have to intervene at short notice.

For this purpose, the cockpit concept has another feature that enables the driver to go about other activities during autonomous driving. A table has been integrated into the steering column which can be moved out of the dashboard electrically when the steering wheel is stowed away. The driver can work comfortably on the table or he can place objects on it such as a tablet computer.

The cockpit concept in combination with the driving simulator allows initial evaluations of new functions and requirements as well as acceptance thereof. New functions in SAE level 4 vehicles include for example the use of a stowable steering wheel. An example of new requirements is the steering column adjustment speed during stowing. Compared with a conventional electrically adjustable steering column, the adjustment speed will most likely increase, as the driver should not have to wait too long before he can take control of the vehicle again. With the help of the cockpit, various parameters can be varied and examined with the help of test persons.

Cockpit in manual driving mode

Cockpit in autonomous driving mode

Insights

The cockpit has provided some important insights into the changes that will occur in the vehicle as a result of increasing automation.

For example, it was found that it is possible to move away from the current round shape of the steering wheel and to use other shapes such as horned steering wheels. With open steering wheels, the question of changing grip always arises when steering beyond a certain steering wheel angle. With a round and closed steering wheel this is possible without any difficulty, but not with an open steering wheel. When used in combination with a steer-by-wire system, however, the steering ratio can be varied so that the steering wheel can only be turned up to approx. ±160°, so changing grip is theoretically no longer necessary.

The trend in steering wheel design over the next few years is likely to continue towards reducing the size of the steering wheel to give the driver as much space as possible and make the steering wheel as compact as possible so that it can be stowed in or placed against the dashboard.

Furthermore, it could be shown that adjustment speed will increase significantly compared with today's electrically adjustable steering columns. Since the adjustment travel of a stowable steering column will be many times greater than that of today's steering columns, the adjustment speed must also change accordingly in order to achieve an acceptable steering wheel adjustment time. Otherwise it would take a long time for the steering wheel to move from the stowage area into the driver's operating area. With the help of the cockpit, various steering column adjustment speeds were examined to determine which were perceived as comfortable by the driver.

In addition, the number of adjustment cycles could also increase compared with today's steering columns since there will be more frequent switches from manual to autonomous driving mode. This means that the stowable steering column will have to be moved more often than today's electrically adjustable steering columns. This will influence both the design of the drive unit and the guidance system of the steering column.

With a stowable steering column that moves the steering wheel close to the dashboard or stows it inside the dashboard, detecting collisions with the driver will be important. There will be a risk of the driver becoming trapped between the moving parts during adjustment. It must be ensured that steering column or steering wheel adjustment is stopped as soon as an external force is applied, in order to prevent injury or jamming. Parameters for this collision detection were tested with the cockpit setup.

As a result of autonomous or semi-autonomous driving, the possibility of carrying out various other activities during drives could also change the interior of the vehicle. The built-in table, for example, has proven to be useful for holding a computer. In addition, touchscreens are getting bigger and bigger and the number of manual controls is decreasing.

Another area that will change greatly is the topic of safety. One goal of manufacturers of autonomous vehicles is to reduce the number of accidents. But there is still a long way to go before the systems for autonomous driving are that safe. Particularly in the transition phases from one automation level to the next, special attention must be paid to safety aspects. Since not all scenarios can be tested under test conditions, accidents will occur at least at the beginning of the introduction of a new automation level. That is why an airbag for passenger protection is indispensable for the time being. Due to the decreasing size or changing shape of the steering wheel, the conventional driver airbag in the steering wheel is likely to change.

Alongside the cockpit described here, thyssenkrupp Steering is also working on further concepts and solutions for the autonomous mobility of tomorrow.

DEVELOPMENT METHODS

Advanced data acquisition and real-time data analysis system for vehicle operational noise testing

Yuriy Kandinov
Senior Manager – Steering – Global NVH
ZF Group

P. Pfeffer (Ed.), *9th International Munich Chassis Symposium 2018*, Proceedings,
https://doi.org/10.1007/978-3-658-22050-1_38

Abstract

Electrically powered steering systems (EPS) are considered an auxiliary drive which is expected to deliver a function without annoying the user with comfort influencing noises.

Therefore, the noise emitted and transmitted by the EPS must be analyzed and solutions for remedial measures must be developed with the aim to reduce the noise to a minimum that is acceptable to the customer.

Especially in low vehicle speed maneuvers like parking or near standstill, steering noises become more detectable, e.g. at steering from one end lock to the other. These so called operating noises contain different noise components, relative to their noise sources inside the steering system. A careful understanding of these noise sources and their interactions is crucial for the development of an NVH improved steering system.

However, the main criterion for a well-designed noise comfort of a steering system is the subjective perception by the driver. To distinguish different noise components even well-trained evaluators can not only rely on the pure noise perception by human ear, but need to use tools such as measurement equipment and NVH analyses to assess the different noise components and track those back to their sources.

Current available methods can be used to do this, but generally lack of simplicity, speed of use and are difficult to standardize across test levels (vehicle, system and sub-system). The main time-consuming factors are application of measurement devices on the steering system and the vehicle as well as data handling and post processing until final report.

This paper describes a new approach to investigate operational noises at vehicle steering systems using customized hardware and software for data acquisition, analysis and reporting. Usability, speed of investigating, data integrity and ease of reporting has been significantly improved by the new method.

The performance of the new method is evaluated and proven to be significantly better within typical test cases. Furthermore, the new method even showed to be ready to be used in other scenarios outside vehicle measurements or even steering systems.

Introduction

ZF group's steering NVH team consists of a significant number of engineers that are occupied with design, product development, validation as well as process development. The team works on products that span across full spectrum of electrical power steering (EPS) systems. Testing steering system at different levels such as vehicle, system and sub-system is a significant portion of development process.

Test results are used to improve the design and validate compliance to customer requirements. For this reason, accurate and speedy conclusions based on test results are extremely important.

Over the years NVH team put priority on improving efficiency of testing, integrity of the data analysis and accuracy of validation reports. This activity resulted in many fine-tuned and improved processes. Even with improved processes the time it took from initiation of test request to submitted report was too slow. Accuracy of the analysis was still, at least to some extent, depended on human error.

More improvement was necessary. However, at some point the team found that any further effort to improve test methods would not offer any significant improvements in time and accuracy because the new test methods would still be limited by the realm of industry standard tools and software. At this point a new, more creative solution was necessary.

The team had identified several, most critical, areas of improvement, set targets and set out to find out-of-the-box solutions.

Issue definition

Vehicle operation noise testing consists of four major steps as shown below.

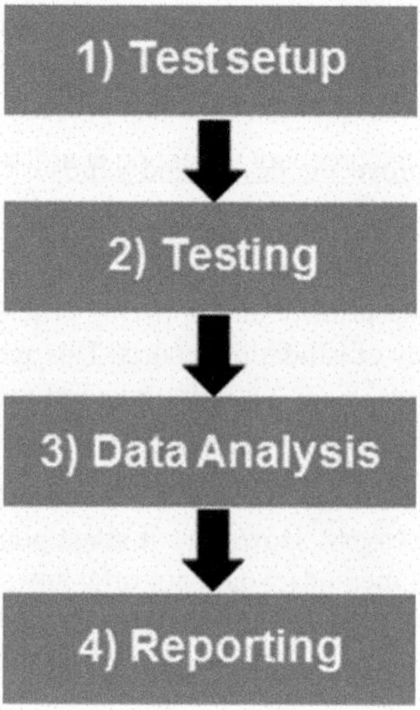

Each of the steps was time consuming and required manual input. Where there is manual input, there is a probability of human error. This is difficult to avoid. The only way to reliably avoid human errors is to avoid the requirement for human input all together.

Step 1 – Test setup. This step required the test/NVH engineer to setup the test vehicle with sensors. The set up included installation of sensors as well as capturing sensor sensitivities. These sensors typically include microphone, accelerometers and steering speed sensors. Other sensors are also used based on requirements. While microphone and accelerometer sensors are strait forward, steering speed sensors are not. There are several off-the-shelve solutions available, but all required some time to setup and typically required some time to post process the signal.

This step also required the user to enter test information and setup test folder structure. Since this is done manually, there was a possibility to enter incorrect information which may cause data misplacement in the future.

Step 2 – Testing. This step requires the user to conduct the test. The user must pay attention to make sure to follow the procedure. The procedure calls a specific test profile during which the test engineer must perform specific steering profile and must maintain a maximum speed variation.

Step 3 – Data analysis. Data analysis was performed using industry standard software packages/tools. These software packages include latest data processing techniques. These tools were off course designed to be applicable to a wide range of applications, but when applied to a very specific application such as EPS vehicle testing, the tools lacked flexibility.

The tools also lacked the ability to easily detect EPS specific noise issues and features critical to customer. The following analysis and issue detection were not possible to conduct on regular basis for due to being extremely time consuming or simply not possible due to the nature of vehicle testing:

- High resolution order tracking covering full speed range
- Order/resonance interaction covering full speed range
- Analysing EPS related noises without the interference of background noises
- Perform statistical analysis on all EPS related noise features

The team was not content with performing such analysis only on portion of samples because conducting such analysis on 100% of test samples would eventually improve customer satisfaction.

It is also worthwhile to mention that steering NVH team used only a small portion of the software functionality for standardized testing, but in most cases had no choice but to purchase full software package. This was not cost efficient.

Step 4 – Reporting. After the analysis, an NVH engineer must write reports and use the analysed data to support the conclusions. Conclusions are always highly dependent on the completeness and quality of data analysis. Both of which were vulnerable to human error.

Project objectives

The team defined and summarized objective based on the issue definition above. High level summary of the objective is table below.

Category	Objective
Time	• Reduce setup time • Get test results in real time instead of post processing
Detection Capability	• Improve noise detection • Improve repeatability
Waste	• Eliminate mistakes, rework and waste
Cost	• Reduce software and hardware cost
Global Standards	• Simplify testing process and adapt globally

Solution

All the objectives were met and exceeded with an innovative solution which consisted of four parts:

- Advanced data analysis methodology
- Real Time data analysis/reporting software
- Custom data acquisition hardware
- Wireless speed sensor

These four parts were developed specifically to meet a particular set of defined high and low-level objectives. The development work split into smaller parts and was assigned to different team members within the multinational team.

Advanced data analysis methodology. This is the most critical part of the solution because it enabled to easily and nearly instantly detect issues that previously required many hours of post data analysis.

The methodology consists of a collection of algorithms that analyse incoming data and report results in a standard easy to read format. The report format is not sensitive to the test profile and/or operator. The advanced data analysis methodology also produces data results that are more stable and repeatable regardless of operator experience level by detecting and ignoring "unusual" noises.

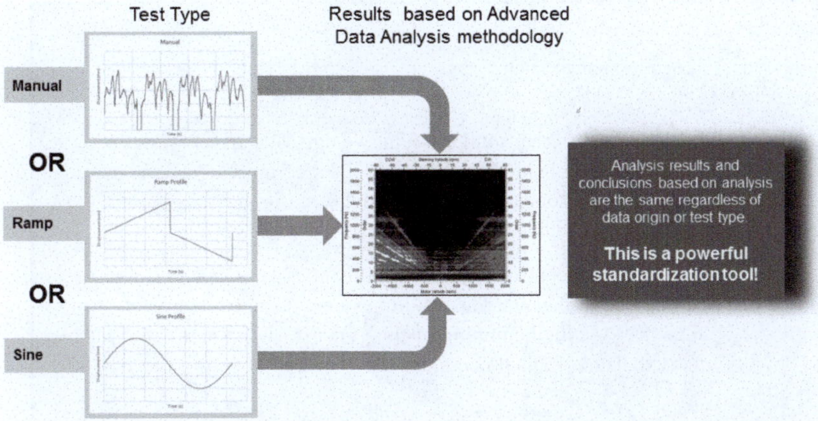

The Advanced data analysis methodology includes algorithms that detect different types of noises and report them in a format that can be used for reports and/or statistical analysis. Interaction of Order related noise (noises frequency of which is related to rotational speed) with resonances is also easily detected and reported.

Real time data analysis/reporting software. The software was developed internally and has powerful features such as:

- Real time data analysis. This feature provides analysis during the duration of the test allowing the user to hear the noise and instantly see the root cause.
- Report generation. The software generates reports and stores them in carefully organized standard folder structure.
- Automatic quality checks. The software continuously monitors progress and data quality. The software will sound a warning and provide suggestions for corrective action if it detects and issue or user's mistake. This feature effectively eliminates human errors.
- Cost effectiveness. The development cost of the software was a fraction of the cost of industry standard software packages.

Custom data acquisition hardware. The team had built custom data acquisition hardware that matches the requirements of the testing. The hardware is also easily configurable depending on test type. Functionally the hardware is not significantly different than what is available on the market. The major advantage of the hardware is the cost which is less than 50% of compatible of-the-shelve system and that it includes built-in wireless receiver for the speed sensor (see next paragraph).

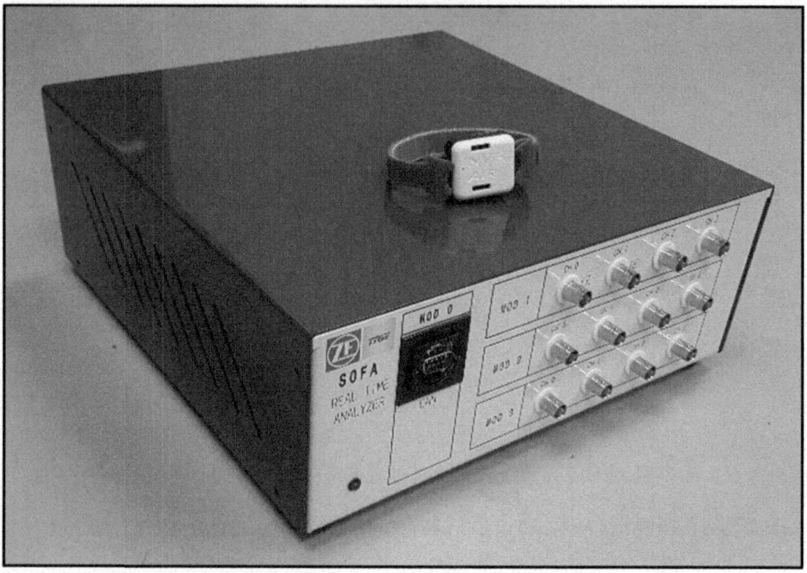

Wireless speed sensor. The sensor was developed internally. The sensor is much cheaper than what is available on the market, but it's best feature is the ease of operation. It takes just seconds to setup. The user is only required to turn the sensor on and attach it to the steering wheel via provided velcro tape. There is no specific orientation required. The sensor calibrates automatically and starts transmitting steering wheel velocity wirelessly to the data acquisition system. As shown in picture below.

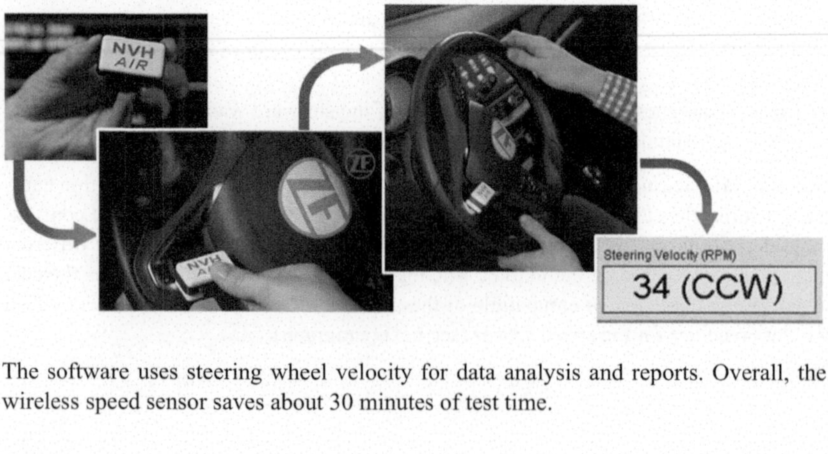

The software uses steering wheel velocity for data analysis and reports. Overall, the wireless speed sensor saves about 30 minutes of test time.

Summary

Vehicle operation noise testing process was significantly improved by the following:

- Significantly reducing test time by:
  - Introduction of wireless speed sensor
  - Allowing software to perform data analysis and provide reports
  - Eliminating rework due to mistakes
- Introduction of Advanced Data Analysis Methodology which improved:
  - Issue detection
  - Repeatability
  - Global standardization
- Significant reduction of human error by introducing software features that:
  - Analyse data and produce reports
  - Monitor data quality
  - Continuously monitor process
  - Flag mistakes and provide suggestion for corrective actions
- Reduced hardware and software cost

Image below illustrates how the process was mostly automated.

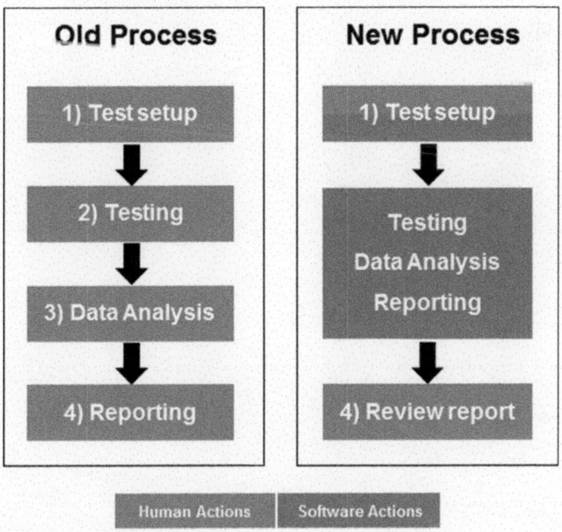

Development of a steering characteristics optimization process

Xabier Carrera Akutain, Kimiaki Ono, Alejandro Ocariz, Georgios Tsitouridis
TOYOTA MOTOR EUROPE, Brussels, Belgium

Shunsuke Iwamatsu
TOYOTA MOTORSPORT GmbH, Cologne, Germany

P. Pfeffer (Ed.), *9th International Munich Chassis Symposium 2018*, Proceedings,
https://doi.org/10.1007/978-3-658-22050-1_39

1 Introduction

The ability to estimate the vehicle states through the steering wheel or more commonly called 'steering feel' is one of the most controversial items during any vehicle program for several aspects: performance target and how it suits automaker's DNA, the suitability of the chosen application methodology as well as the necessary trade-offs provoked by communization in large scale programs with multiple vehicle variants.

Traditionally, steering feel development has relied on purely subjective assessment, which requires several iterations involving calibration and evaluation experts from the steering system and the vehicle manufacturers. The inherent nature of subjective processes often produces a time consuming exercise until the target is achieved. The increased complexity of Electric Power Steering (EPS) control algorithms and the ever growing amount of vehicle variants make the traditional approach no longer sustainable.

Simulation tools have been used for decades to provide a more agile and inexpensive design and calibration environment. More widespread deployment of Model Based Design (MBD) has enabled engineers to leverage vehicle and subsystem models, together with supporting software tools, in order to better recreate synthetically the calibration process and reduce the subjective part to the fine tuning on the last stage, when unachievable by MBD. More recently, XiL technologies have contributed further to facilitate the design and calibration processes. HiL ECU or full steering system benches, as well as Driver-in-the-Loop systems such as driving simulators are prominent examples.

The selection of appropriate control parameter values targeting specific metrics and following mathematical optimization has been typically difficult to achieve due to the complex objectification of the steering feel phenomena, which we challenge in this work. A high fidelity steering model and a new toolchain are set in place to convey a meaningful Design of Experiments (DoE), from which candidate settings are evaluated in a steering simulator and in a proving ground with positive results, confirming the validity of the process and the results. A Toyota vehicle currently on mass production is chosen for this proof of concept. We claim the innovative value of this rapid prototyping technology by an increased maturity of MBD results towards a streamlined scenario of more agile vehicle development programs, with subjective assessment restricted to the confirmation of intermediate and final results.

2 MBD process. General workflow and toolchain

Many recent developments claim to follow the V-shape approach, derived from the software engineering field. The traditional linear sequence is replaced by an interconnected combination of steps, aiming a more agile system development lifecycle. Model-Based Design or MBD has similar roots and supports the V-model cycle by setting a framework where the plant modelling and simulation become essential steps.

A calibration engineer with the expert knowledge of the simulation toolchain, control algorithm and performance sensitivity can leverage MBD tools to accelerate the outcome of his applied methodology. However, the multiple combinations of parameters values that influence the steering feel create an important performance dispersion in all driving areas that must be controlled. This problem holds true for the tuning of one single vehicle and it's significantly amplified if the number of vehicle variants is considered, therefore a more systematic approach is desirable. In [1, 2], DoE is applied for a more efficient tuning process.

On this work we present a highly automatized EPS calibration process, consisting of a sequence of distinct DoEs, weighed and combined for a holistic performance improvement, considering the sensitivity and trade-offs of several steering sub-performances resulting from the different control logic modules.

The general performance direction is translated into appropriate objective targets and specific parameterization of design variables. At this stage of the process setup, expert knowledge is essential for a sound selection of targets and mapping into system parameters. The DoE in question is set in place and after its completion, a parameter subset is fixed and carried over to the next DoE, as it'll be described in chapter 3. The resulting parameter sets are evaluated virtually on a driving simulator and on the physical vehicle (when available) to adjust possible deviations of the virtual development process.

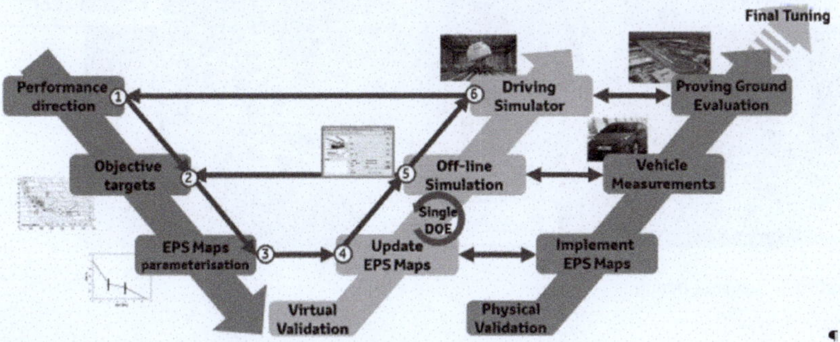

Figure 1. V-shape development cycle

2.1 Simulation model making

Recent development of steering feel characteristics using virtual tools becomes possible thanks to improvements on model fidelity and real time capability.

The vehicle's suspension elastokinematic properties are parameterized from kinematics and compliance (K&C) rig testing. In case of a pre-prototype stage, a virtual equivalent rig is employed for the same purpose [3]. Raw data is processed by an in-house software tool for automatic model making and translation into IPG CarMaker look-up tables.

Friction characteristics in a steering model have major influence on the steering feel performance [4] and these characteristics can be identified with precision from specific steering bench measurements [5, 6]

Figure 2 shows a general picture of the steering bench employed in this work. The usual scheme of one steering wheel actuator and two rack actuators is followed, and the dimensioning of the jig and actuators takes into account the torque, force and displacement values of the intended test sequence.

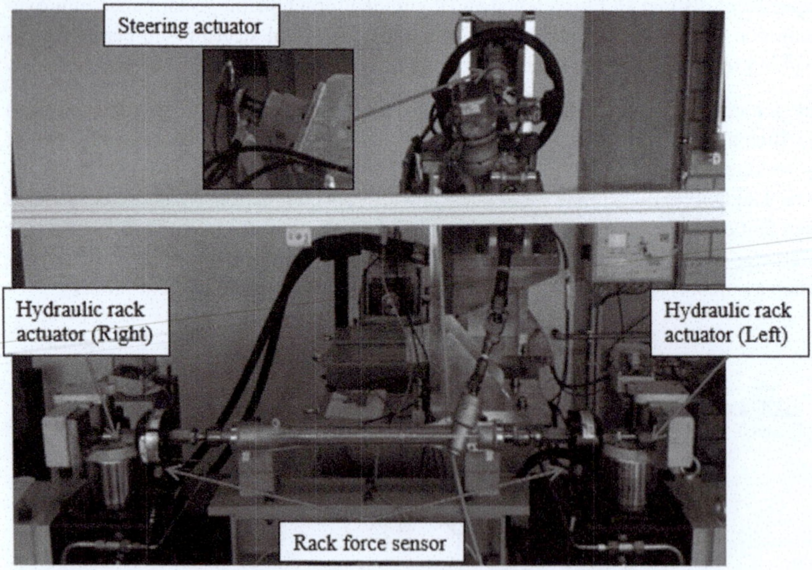

Figure 2. Steering bench

The vehicle subject to study features a steering system from the Toyota New Global Architecture (TNGA). In order to characterize the system components' friction, damping, inertia and stiffness in detail, the test sequence was exhaustive and comprised different steering assembly configurations: rack connected/disconnected; EPS assist on/off; EPS motor connected/disconnected/ fixed lower shaft, etc. Moreover, the system was subject to testing in brand new condition and repeated in two other occasions after going through durability cycles. On Figure 3, for the low velocity friction test, it can be observed how the column friction decreases from the brand new condition to the used condition, while exposing the system to further lifecycle emulation doesn't affect much the friction characteristics.

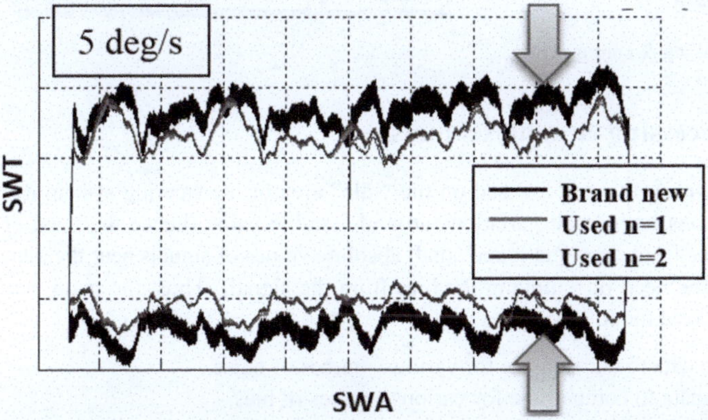

Figure 3. Column friction measurements

This thorough investigation allows a better understanding of modelling constraints and provides a valuable information for the cases in which the system characterization is not always possible at this level: only design data available, no prototypes existing, steering model simplification from K&C tests instead of steering bench, etc. The obtained data has been processed by a highly automatized script sequence, here generically called "Processing Tool" (Figure 4), for a trustable parameter values extraction. All the modelling elements have been combined in Simulink, with the steering mechanical model in FMU model exchange scheme, while the control logic model is accessible as S-function.

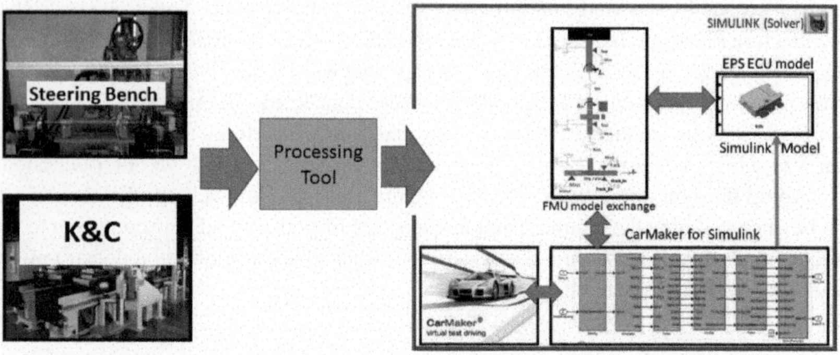

Figure 4. Model making & configuration

2.2 Data processing & analysis tools

Much of the post-processing is focused on the "idle" area of the steering system dynamic, which means both small variations around a stable value (hence the need of removing the noise from the signal), and small absolute values of signals near the neutral state (hence the need of removing the bias from the signal). Thus, the main pre-processing tasks were the following:

- Smoothing the data to compensate for various sources of noise.
- Offsetting the data to compensate for various sources of bias.

The smoothing was performed with an implementation of Chartrand's numeric differentiation algorithm [7]. The method's idea is to reconstruct the signal from a smoothed differential; thus, the signal itself will be smooth. The differentiation process is transformed into an optimization problem: minimize a linear combination of signal's total variation (as a measure of smoothness) and residual (as a measure of fidelity to the original signal); the linear coefficients' relative values determine the strength of the filtering. It provides good fidelity for both rapidly-varying and almost-constant signals, and allows robust numeric derivative calculations.

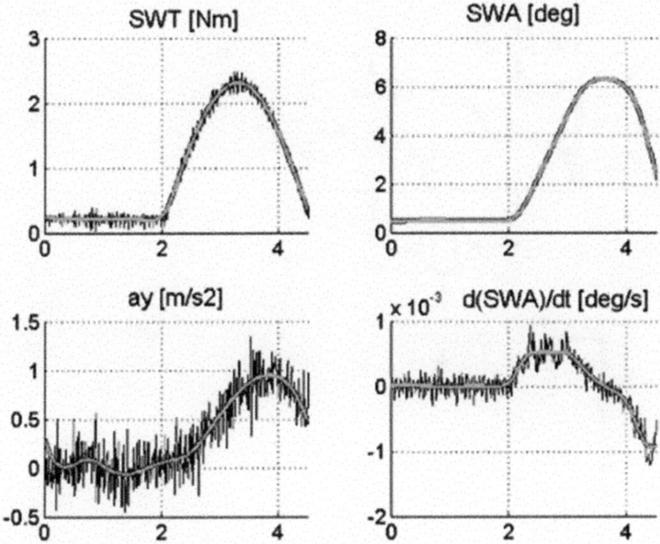

Figure 5. Total variation smoothing and differentiating

The essential assumption of the offsetting algorithm was that the bias is uniform throughout the entire signal history. Thus, the offsetting task is reduced to (a) detecting the part of the signal where the bias values are significantly larger than the values of the useful signal, (b) estimating the bias level in these areas, and (c) subtracting the bias level from the rest of the signal. The detection strategy combined finding the time slices where the signals are "small" (i.e. idle, near the 0 value) and quasi-constant for a minimal duration, and finding the signal slices where the time derivative is approximately 0 (which could be done robustly thanks to the smooth differentiation). The bias level is calculated as the signal's median calculated in the "idle" time slice; this increases the robustness of the estimation against the eventual idle-detection inaccuracies, and the bigger noise levels (Figure 6).

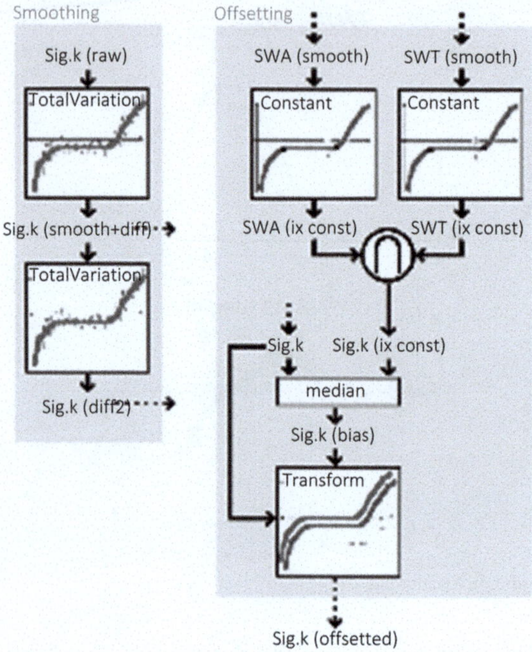

Figure 6. Algorithm schematic

The pre-processing tasks were done using an internally-developed toolchain, a collection of data container objects which are manipulated by processor objects. They may be organized into data processing workflows using chained step objects, which apply the same combination of processing objects to all their input container objects, and pass the results to the next step object(s).

The resulting workflow is displayed in Figure 7, where the smoothing, cutting and offsetting steps are parameterized for both simulation and experimental data cases to obtain comparable steering feel metrics. In the last step, the resulting curves are subject to break-down into different zones or stages of the steering manoeuvre, for a precise generation of metrics or key performance indexes (KPI).

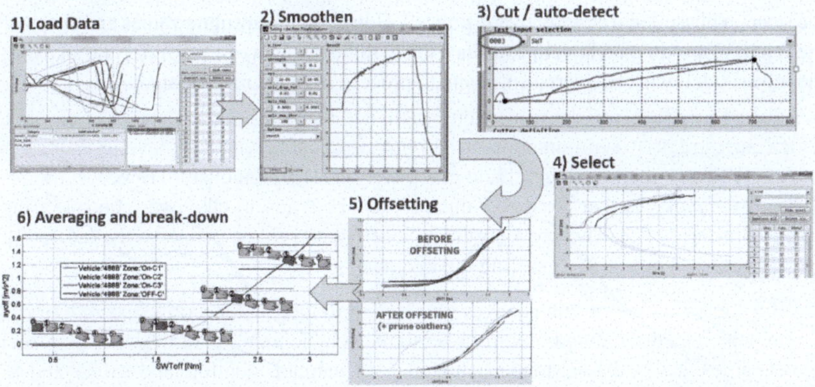

Figure 7. Data processing workflow

2.3 Validation results

The current mass production version of a Toyota vehicle is taken as reference. The simulated torque levels for the steering (angle) ramp input at 13.5 deg/s (Figure 8) show a good matching with the measured signals. The vehicle and road camber induced asymmetry are taken into account. The friction values of the "used" condition provide the better matching at the on-centre area.

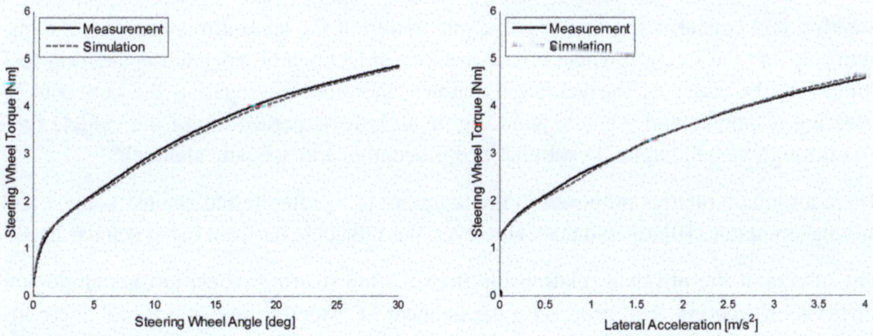

Figure 8. Ramp input results

Regarding the Lissajous curves of the sine input (Figure 9), the absolute values and the hysteresis level match well for the torque-angle curves, although some discrepancy can be observed on the torque-lateral acceleration curve, where the simulated torque values at the steer-in phase seem slightly overestimated. However, the absolute values and shape of the return phase do match well, and since this will be the area of interest for the analysis, the model fidelity is considered sufficient to proceed with the investigation.

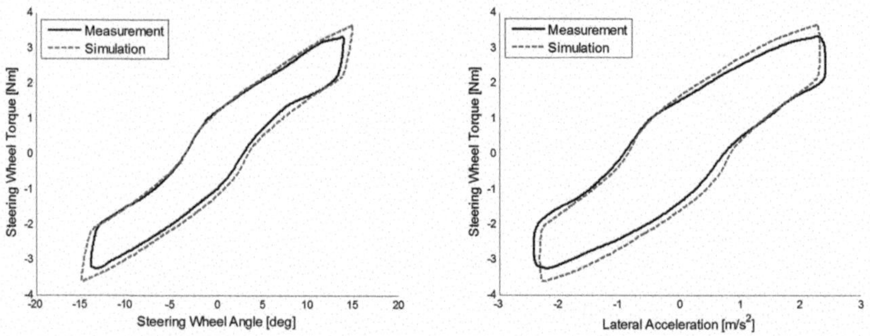

Figure 9. Sine input results

3 Defining steering feel metrics

As engineering standards progress, objective testing can better assess the performance metrics of vehicle development. However, steering feel has been characteristically resilient to meaningful performance metrics, due to the complex non-linear nature of the steering phenomena and the highly variable psychophysical aspects involved.

Steering feel concerns *sensu stricto* the interaction of the hand-arm system on the human side and the steering wheel on the machine side. On a broader sense, steering feel comprises the states of the vehicle dynamics, therefore aggregating the concepts of 'steering response' and 'vehicle handling' to an holistic perception of the vehicle motion through visual, haptic, vestibular, proprioceptive and acoustic channels.

There are many metrics proposed in the literature [8,9] often tested against vehicle and driving simulator [10] assessments. However, the subject is far from being standardized.

The effects of the physical relationship between the steering wheel torque, angle and stiffness on steering feel, proposing the concept of "steering reaction force" were investigated in [11,12]. Ono [13] identified the contribution of steering wheel torque and stiffness in the perceived steering reaction force for non-linear steering characteristics. For the "on-centre" or low steering wheel torque values, physical and perceived values have a logarithmic function relationship to satisfy Weber's law. In contrast, "off-centre" or the range of 2.5 Nm and above, shows concavity in the relationship between the physical and perceived values.

The work presented in this paper applies the steering reaction force as main target variable for the steer-in phase of the cornering manoeuvre. Furthermore, extensive research is been accomplished in extending the perceived reaction force concept to the other steering stages. In this section we will anticipate some of this work, aiming an integral performance tuning for the steer-in and out phases for a linear build-up, a smooth return and a good connection.

3.1 Steer-In

The steer-in phase is obviously fundamental in the steering operation. The new target of the perceived steering reaction force will be aimed with a two-step DoE approach and careful assessment of the results. Two candidate results, named 'Setting1' and Setting2 will be selected for further development.

3.1.1 Target KPIs

The perceived steering reaction force of the simulated model presents a good correlation in the on and off-centre areas. According to internal standards, this reference vehicle presents too much perceived reaction force on-centre, while the digressive rate on the off-centre area turns insufficient with respect to the target line (not displayed here).

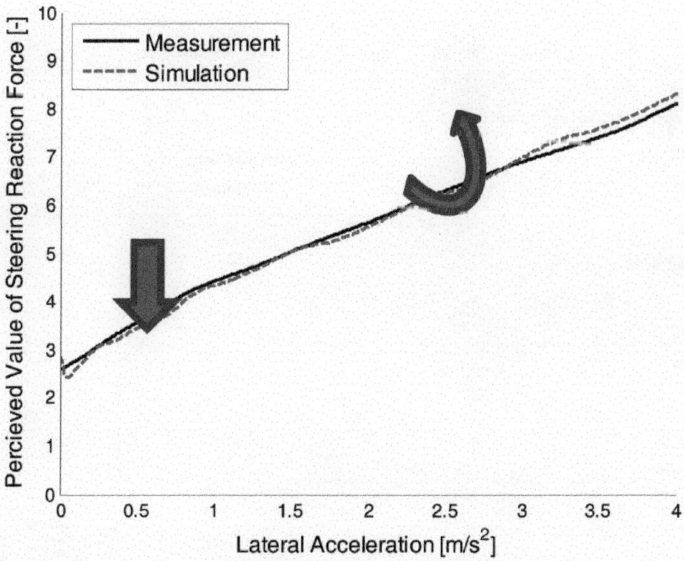

Figure 10. Perceived reaction force

3.1.2 Design variables and workflow

The flowchart on Figure 11 intends to explain the followed process. On the upper part, the first DoE is initiated exclusively for the Steer-in on-centre area. The parameterization of the basic EPS assist map in function of the measured torque is displayed on Figure 12, where two of the map points (P1, P2) have (limited) freedom on x,y axis. Any additional parameter or degree of freedom on the system creates a new dimension of the problem, exploding the number of simulations required for a suitable exploration of the design space. Therefore an economic and smart parameterization turns capital for the practical success of the task. This is the main reasoning behind the split into pseudo-independent DoEs. Assuming little cross-talk, chunks of the performance areas are split and tackled with the minimum possible set of parameters which possess most of the sensitivity.

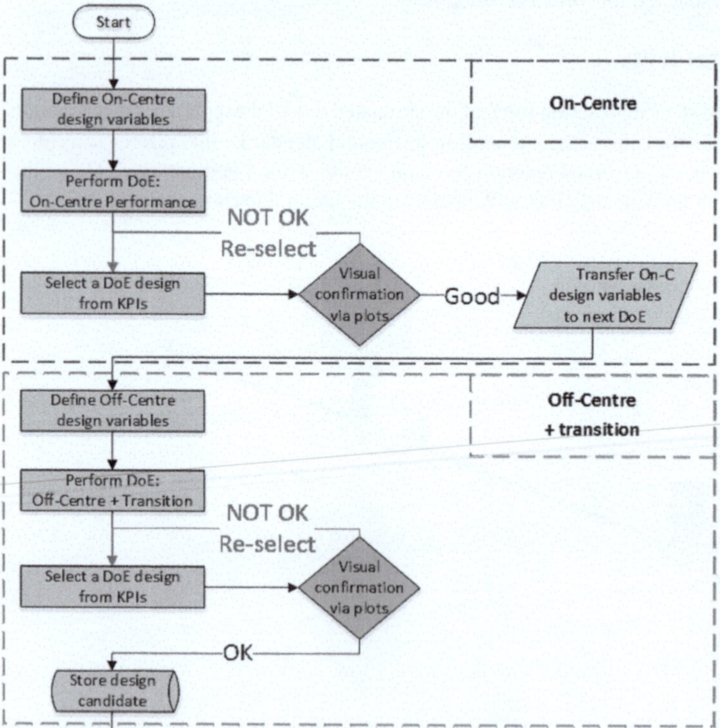

Figure 11. Steer-In workflow

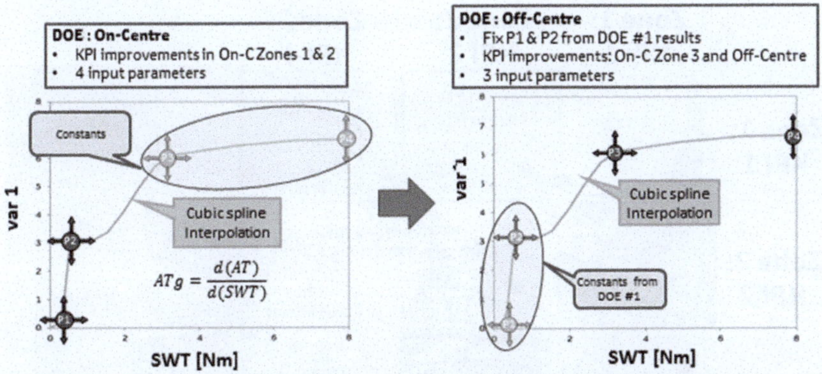

Figure 12. Steer-In parameterisation

The perceived reaction force and supplementary metrics are arbitrated, including a visual inspection of the overall plots of the selected candidates. The best two are picked and carried over for the later stage: off-centre and the transition in between (hereby called "on-centre Zone 3"). At this stage, an equivalent DoE is performed for the points P3 (x,y) and P4 (y direction), to further give shape to the settings under development.

3.1.3 Simulation results

A sample of the found correlation maps is displayed below for three KPIs. While it'd be desirable to increase KPI1, the intention is to minimize KPI2. As this presents quite a strong positive correlation (marker 'a') , inevitably we have to face a performance trade-off. On the other hand, KPI4 should be increased, which means a win-win between KPIs 2 and 4 (marker 'c'). KPI3 has a worsening tendency. Obviously these correlations are exclusive to particular mechanical topologies, hence this kind of exploration becomes useful for the steering feel assessment and calibration specialists, as the performance mapping gets significantly clarified.

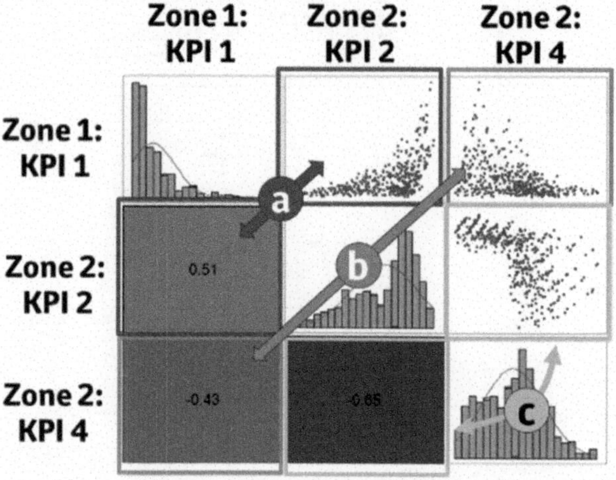

Figure 13. Steer-In KPIs correlation

Once the sub-performance trends and relationships get more clear, the traditional intersection plots are useful to proceed with the design parameter selection:

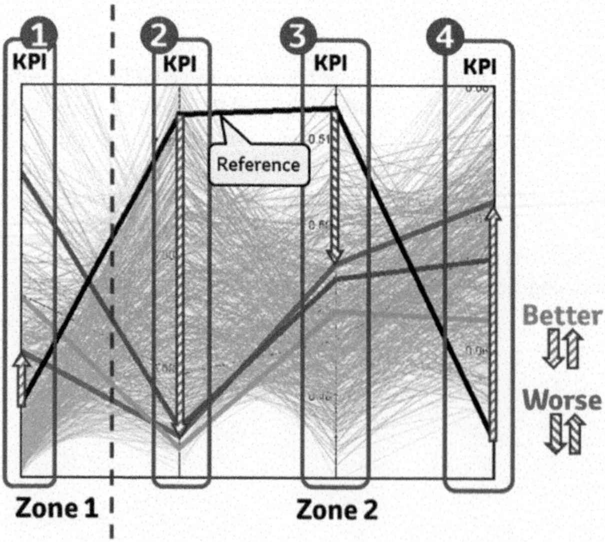

Figure 14. Design selection plots

The perceived steering reaction force is the main metric or KPI employed in this section, and Figure 15 shows the resulting plots, with Setting1 and Setting2 presenting little difference among them but a significant correction from the Reference values in the intended direction: the on-centre force is decreased and the off-centre force presents a linear-progressive shape instead of the original digressive one.

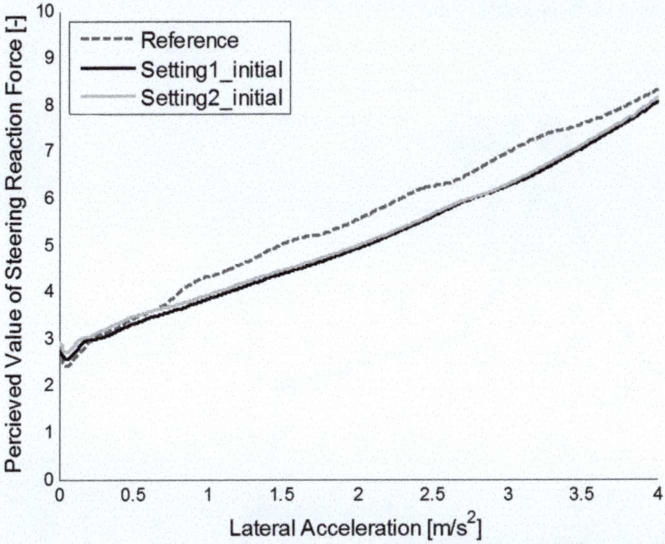

Figure 15. Proposed steering reaction force

3.2 Return and connection

Once the steer-in section is mostly fixed, the steer-out stages (mainly the return and connection phases) need to be designed, without affecting in excess the previously fixed steer-in performance. Setting1 and 2 need to be adjusted accordingly.

3.2.1 Target KPIs

The number of performance metrics for these stages increases even further. Therefore it is essential to focus on the most significant ones for an efficient usage of DoE approach. Here we focus on the "connection feel", which can be explained as the lateral acceleration level at which the stiffness is maximum. The sequence must be understood as follows: the driver starts the steer-out manoeuvre by passively absorbing the return of the steering wheel (moving on the positive sense of the abscise axis on Figure 16), while the vehicle is still moving on the previously set direction (negative lateral acceleration). The steering stiffness drops from the holding phase, where most of the hysteresis is overcome, to a significantly lower value as the steering wheel starts its return to neutral quickly; then it stiffens up before the vehicle motion turns positive. By that moment both the steering wheel torque and angle are already positive, therefore we assume that the closer to zero lateral acceleration this point, the more connected the vehicle will feel. The target will be to bring this peak as close as possible to the zero point.

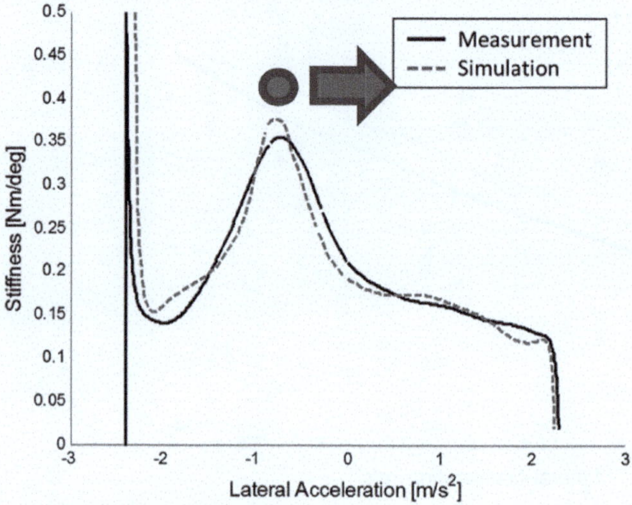

Figure 16. Stiffness at connection

A similar location and shape for both simulated and measured values can be observed. In addition, the Reference setting presents an overshooting behaviour at return because of lack of free damping, which will be tackled similarly.

3.2.2 Design variables and workflow

As indicated, the workflow for the steer-out phase starts where the steer-in ends (Figure 17). Furthermore, there is some cross influence on steer-in when tuning "specific" steer-out control modules, influence that needs to be minimized for an acceptable sub-optimum.

The parameterisation needs to take into account this, therefore 'P2' of Figure 12 is again subject to (moderate) modification together with damping and return control module parameters. The final settings selected present a practically identical perceived steering reaction force for steer-in as the initial selection of Figure 15.

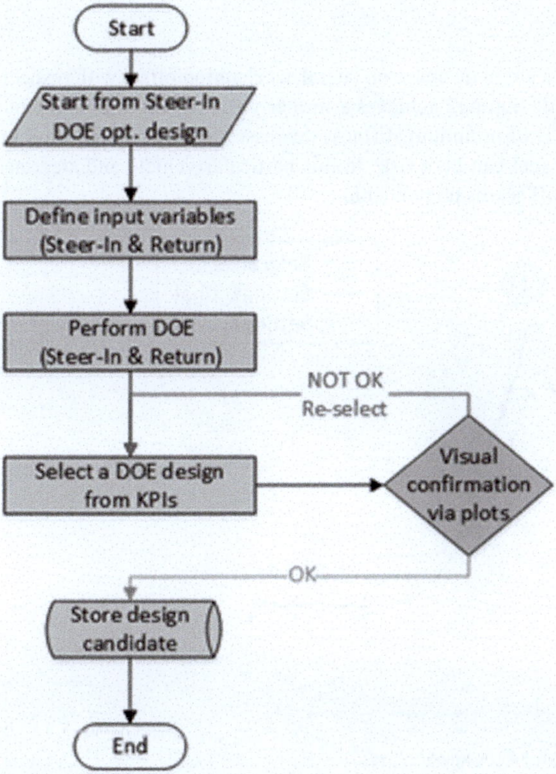

Figure 17. Steer-Out workflow

3.2.3 Simulation results

Point A on Setting1 presents a slight decrease on lateral acceleration offset with respect to the reference settings, while Setting2 achieves a further yet marginal improvement. On the other hand the point B of minimum stiffness decreases, moving closer into an area where "springy return" feel can be a risk. In this particular vehicle and steering construction a system trade-off seems unavoidable.

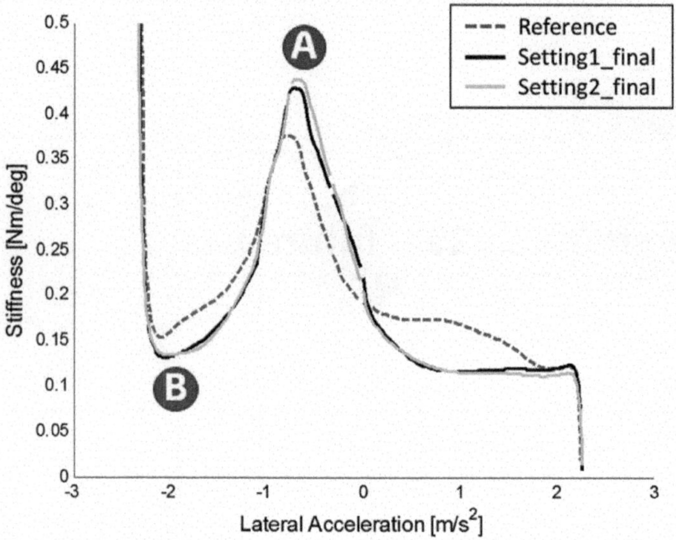

Figure 18. Simulated stiffness at return/connection

4 Objective validation of results

The chosen settings Setting1 and Setting2 are measured with the same vehicle unit as the Reference in the proving ground. Figure 19 and Figure 20 show comparable results to the simulated curves in Figure 15 and Figure 18.

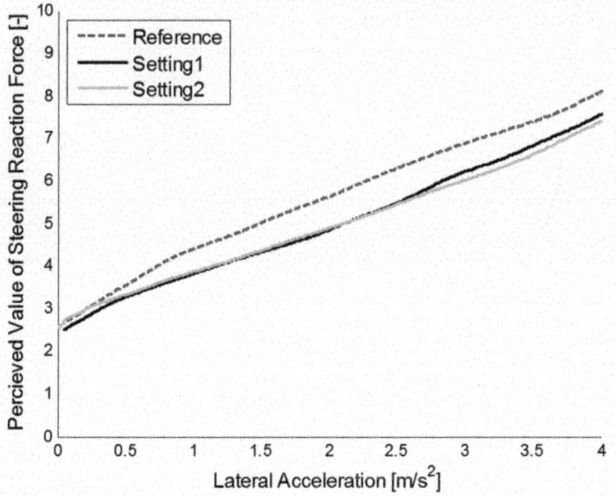

Figure 19. Confirmation of Steer-In

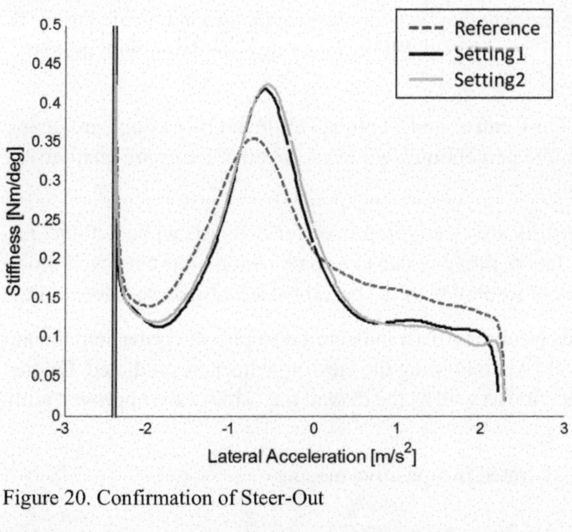

Figure 20. Confirmation of Steer-Out

The free damping is significantly improved by both new settings with respect to the Reference setting, as Figure 21 indicates, by removing the negative overshoot.

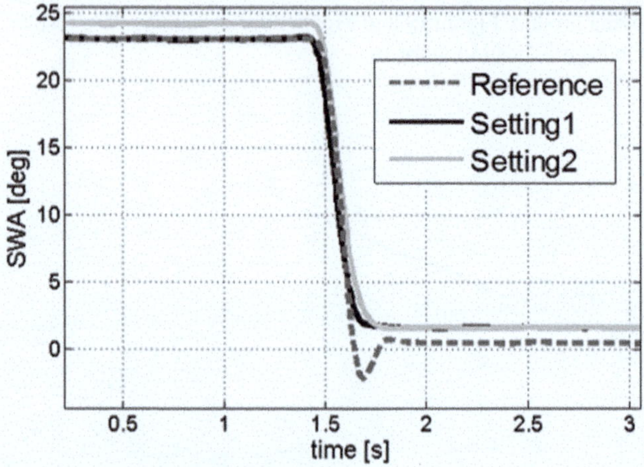

Figure 21. Free damping measurements

5 Subjective assessment

The proposed settings are given to two expert evaluators for assessment in two different scenarios: Driver A is driving the actual vehicle on the proving ground, while Driver B performs virtual assessment at a fixed-based driving (steering) simulator, with the same models used at the development stage.

The terms "Torque Gradient On-Centre" and "Torque Gradient Off-Centre" are quantitative, e.g. "higher score = more perception", while all the other items are qualitative, e.g. "higher score = better feel".

Dry friction and perceived linearity are clearly improved for Setting1, while Setting2 results are partially below expectation, possibly due to a combination of torque levels close to the heavy side limit and lack of resolution of the perceived steering force calculation.

Both drivers and environments (real & virtual) indicate a decrease of corner settlement, explained as "slight spring feel" when holding the steering wheel, as predicted. On the other hand , the return &connection as well as the overall traceability are improved with respect to the Reference setting.

The free damping item is clearly rated in a positive manner.

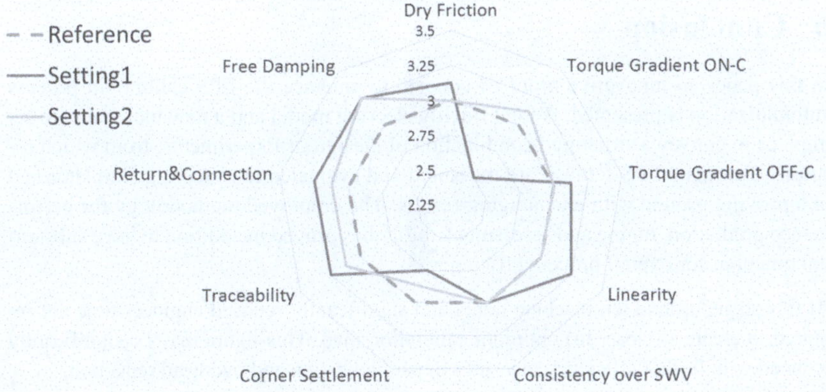

Figure 22. Assessment Driver A. Proving Ground

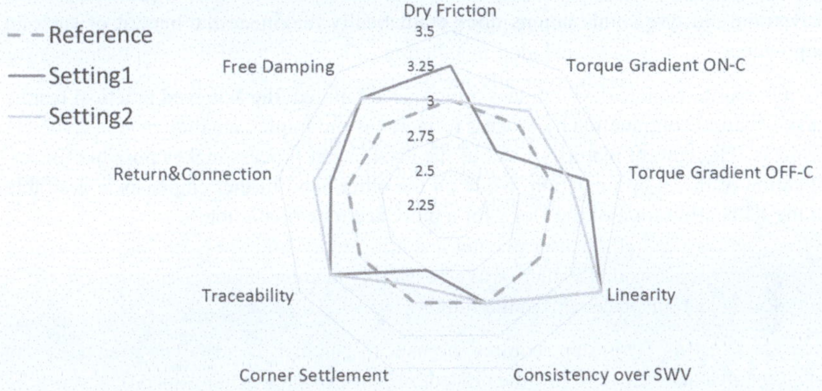

Figure 23. Assessment Driver B. Driving Simulator

6 Conclusion

In this paper an innovative proof of concept to enhance the EPS calibration process automatization is presented. A high fidelity steering model and a new toolchain are set in place to convey a structured combination of Design of Experiments, from which resulting candidate EPS settings are measured and evaluated in a steering simulator and in a proving ground with encouraging results. The employed methodology for performance prediction, tuning and assessment ability has proved in addition to have substantial potential for overall process efficiency.

At this stage, around ten machine hours and significantly reduced engineer time are required to create a (closer to) optimum calibration map. This figure may be significantly reduced with further efficiency measures in parallelization and overhead reduction.

While the handling and steering experts' contribution is essential in setting up such a process, their precious time could be spared to a great extent in the application programs. The case extension to multiple vehicle variants, with several weight, suspension, drivetrain and tyre combinations does dramatically maximize the benefit of such an approach.

As more agile vehicle development programs are aimed, the achieved practical results have a natural continuation by process refinement and implementation in actual vehicle projects. The current maturity level of the established process makes possible the application of direct optimization, while on the other hand further emphasis is desirable in the KPIs arbitration of the DoEs for a more accurate prediction.

References

[1] Wagner C., Flormann M, Meister T., Henze, R., Küçükay F. : Model-based online optimization of EPS controller using HiL test benches.
Proceedings of the 8th International Munich Chassis Symposium, 567-580, ATZ chassis.tech plus (2017)

[2] Contini A., Schoening T. : The new Hyundai i30's steering feel – integrated performance development by HiL and DoE..
Proceedings of the 8th International Munich Chassis Symposium, 555-566, ATZ chassis.tech plus (2017)

[3] Sundström P., Henningsson M, Carrera Akutain X., Hirano Y., Ocariz A., Iida H., Aikawa N., Andreasson J. : Virtual Vehicle Kinematics and Compliance Test Rig.
The First Japanese Modelica Conferences, Issue 124 Article 4, 29-34, Tokyo, Japan, (2016)

[4] Kushiro I., Koumura S., Kawai H. : Steering System for the On-center Handling.
Transactions of Society of Automotive Engineers of Japan, Volume 40, Issue 2, 233-238, (2009).

[5] Lin J., Pfeffer P.E. and Schick, W.: The Application of Advanced Steering System Models in Vehicle Dynamics Simulation.
Proceedings of Chassis.Tech plus 2011 Volume 2, p.441-451, Munich, Germany, Springer Automotive Media, (2011).

[6] Shin M.C, Kim S.H., Cho G.H, Chu C.N: Development of a steering-system model considering viscous friction and its verification.
Proceedings of IMechE Part D: Journal of Automobile Engineering, Vol. 228(2) 144-163, (2014).

[7] Chartrand, R., Numerical differentiation of noisy, nonsmooth data. ISRN Appl. Math., 164564 (2011)

[8] Harrer, M; Pfeffer, P. : Steering Handbook. 1st ed.
Berlin, Heidelberg: Springer (2016).

[9] Decker, M. : Zur Beurteilung der Querdynamik von Personenkraftwagen.
PhD Thesis, Technische Universität München (2009).

[10] Rothhämel M., IJkema J., Drugge L. : A method to find correlations between steering feel and vehicle handling properties using a moving base driving simulator.
Vehicle System Dynamics 49(12):1837-1854, (2011)

[11] Yamada D., Kuroyanagi H., Ono E., Kushiro I., Konomi K., Sato S. : Analysis of On-Center Steering Feel by a Steering Simulator.
Transactions of Society of Automotive Engineers of Japan, Vol. 40, No. 2, 239-244 (2009)

[12] Kushiro I., Yamamoto M.: Vehicle Behavior Under the Influence of Steering Dy-namics by Means of Low Frequency Torque Input SAE Technical Paper 2006-01-0557, (2006)

[13] Ono K., Kushiro I., Yamada D., Tomita T.: Steering feel design based on the driver's perceived value of steering reaction force.
Proceedings of the 7th International Munich Chassis Symposium, 595-616 (2016)

EPS SW integration test automation based on architectural design

Dipl.-Ing. Maciej Obszarny (presenter)
Team Manager EPS SW&System Testing, ZF Group

Dipl.-Ing. Peter Niggemeier
Chief Engineer EPS SW Platform Team, ZF Group

Dipl.-Ing. Wojciech Grygierek
Senior SW Test Engineer, ZF Group

Dipl.-Ing. Wojciech Makuch
Senior SW Test Engineer, ZF Group

P. Pfeffer (Ed.), *9th International Munich Chassis Symposium 2018*, Proceedings,
https://doi.org/10.1007/978-3-658-22050-1_40

Introduction

Electric Power Steering systems evolve to meet growing number of OEM requirements related to safety, functionality and driver comfort. To handle increasing design complexity ZF Steering Engineering selected model-based SW development approach where the source code is generated from a SW architecture model. UML based solution increased overall development efficiency, eased product maintenance and improved design portability among various projects. However, the new working scheme posed additional challenges for product verification, specifically in the area of interaction testing between modeled SW components. The paper discusses how dynamic SW architecture diagrams were used to automate SW integration testing activities to ensure back-to-back verification between model and its implementation. Key challenges in control and data flow monitoring are discussed together with selected solutions. Article concludes with project deployment report and its impact on relevant Automotive SPICE capability rating.

SW Architecture Model

Steering systems perform various functions on vehicle level ranging from basic reduction of driver steering effort to advanced steering assistance capabilities such as driving stability interventions and lane keep assist. These top-level features are defined on system level together with car manufacturers what is reflected in elicitation of product requirements and development of EPS system architecture. The role of SW organization is to derive own requirements out of the system view and design SW architecture that is consistent with previously performed steps[1].

Static Design

The SW architecture considers static and dynamic design aspects. In the initial step the static view is handled by decomposing EPS SW into major design blocks called SW features. A feature is defined as a high-level steering function that has an impact on the whole system; it groups all lower level design elements and requirements based on logical association. Each feature has corresponding behavior(s) that describe main objectives that have to be realized within, Figure 1. For example, Provide SW Rack End Stops can be identified as a SW feature with the following behaviors:

- Calculate SW End Stops Position, and
- Calculate SW End Stops Torque Demand.

1 Automotive SPICE Process Assessment/Reference Model ver. 3.0

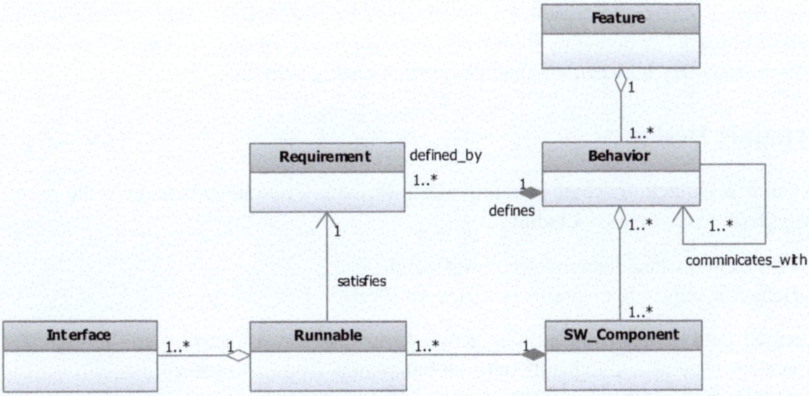

Figure 1. Simplified structure of architectural design elements.

Each behavior is implemented by one or more SW components (SW-C) which are composed of runnables and interfaces. A runnable groups sub-functions of a SW component and is called by a scheduler. Runnables can be relatively big in size and are broken-down further in the SW detailed design. Interfaces can have Client-Server or Sender-Receiver type and provide communication capabilities between various SW-Cs[2]. Static architecture includes a diagram that ties together these four design elements providing dependency information, Figure 2.

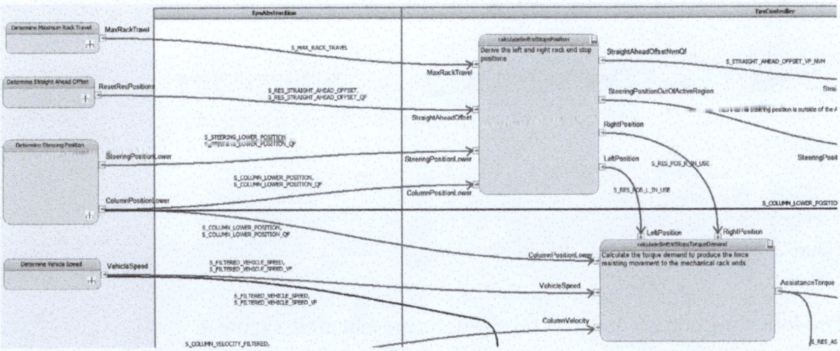

Figure 2. Provide SW Rack End Stops SW feature – behavioral dependencies (simplified).

2 See AUTOSAR interface, for example.

Definition of a SW component, its runnable(s) and interface(s) complete static design scope and mark the boundary between architectural and detailed design phases. It also forms a boundary between unit and integration testing activities.

Dynamic Design

Dynamic architecture design step provides information on interactions between previously defined elements, including:

- Feature sequence diagrams (top level), and
- Behavior sequence diagrams (component level).

Provided data differs only by abstraction layer under consideration and is logically equivalent. The article expands only on behavioral type of sequence diagram as it is simpler to explain taken approach using intuitive technical examples.

There may be multiple sequence diagrams assigned to a single behavior or feature as each runnable may be interacting differently with its environment depending on ECU state, Figure 3.

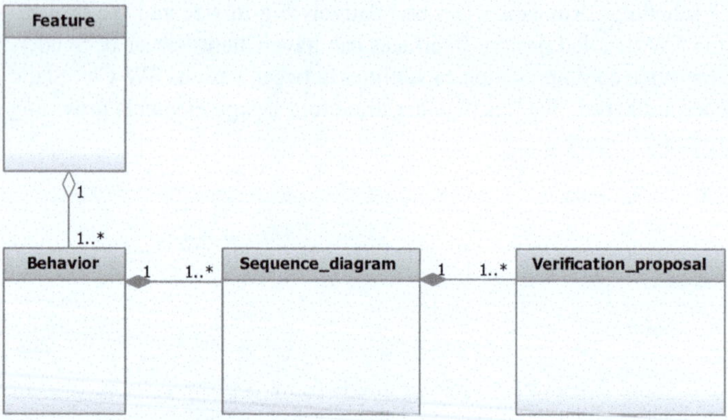

Figure 3. Structure of dynamic design elements.

The following control and data flow aspects are captured, Figure 4:

- Function call sequence starting from top level runnable,
- Data flow sequence using previously defined interfaces,
- Inputs and/or preconditions that have to be satisfied for the scenario to take place during SW execution, and
- Expected outputs as a result of combined control and data flow steps.

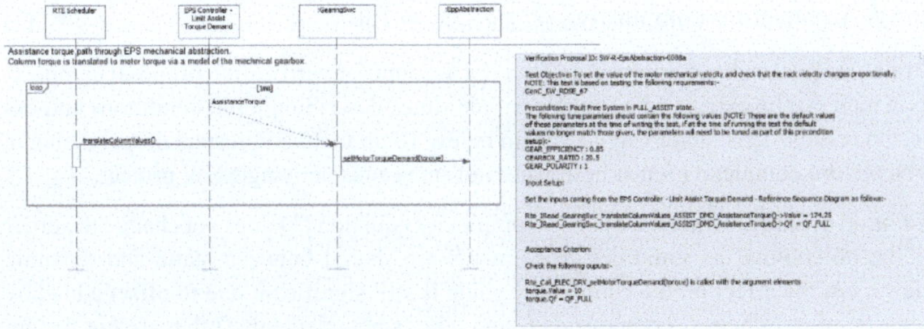

Figure 4. Sequence diagram example.

Both inputs and preconditions are defined by SW architect and collected under verification proposal design elements. List of all features/behaviors, their assigned sequence diagrams together with corresponding verification proposals form a test scope definition contract between SW architecture and SW testing departments.

State Space Explosion

Also known as combinatorial explosion, is a phenomenon defined by rapid growth of given problem complexity as a result of increasing numbers of problem elements that can be combined together[3]. State space explosion is applicable also in the context of SW architectural design as demonstrated below considering only the dynamic part (1):

$$\text{SW Architecture Complexity}_{(\text{dynamic view})} = \sum_{i=1}^{n} F(b, s, v)_i, \tag{1}$$

where F is a SW feature, b is a behavior, s is a sequence diagram and v is a verification proposal.

Based on Figure 3 we know that:

$$F(b, s, v) = b \times s \times v, \tag{2}$$

where b, s, v are numbers of feature design elements and $b, s, v \in N$.

Let's assume conservatively that a typical EPS SW architecture consists of $n = 25$ features, each feature has $b = 3$ behaviors, each behavior just $s = 1$ sequence diagram and $v = 2$ verification proposals. Substituting factors into (1) and (2) yields a complexity figure of 150 elements that SW architects need to design and maintain per each project (3).

3 Krippendorff, Klaus. "Combinatorial Explosion". Web Dictionary of Cybernetics and Systems

$$\text{SW Architecture Complexity}_{(\text{dynamic view})} = 150 \quad (3)$$

Furthermore, the above assumption does not account for verification proposal complexity which can be expressed as number of text lines describing input conditions and expected results. This factor can vary from below 10 up to 40 and results in real dynamic architecture complexity value in the order of thousands per single SW project.

Addition of static architecture part to complexity equation (1) does not change the order of the polynomial as some design elements are shared between static and dynamic views, but the effect on the complexity value is still significant due to offset added by addition of interfaces I, components C and other static elements O that are not feature exclusive and were not discussed here in detail (4).

$$\text{SW Architecture Complexity} \approx \sum_{i=1}^{n} F(b,s,v)_i + I + C + O \quad (4)$$

Such state space cannot be managed effectively in a multiple project business environment with short delivery dates and limited resources. ZF Steering Engineering mitigates this risk on SW architecture level by applying good design practices such as:

- Enforcement of modularity and low complexity,
- Strictly defined interfaces between components,
- Reusability of architectural elements between EPS platforms and projects that provide, in the best case, one-shot development effort and minor maintenance,
- Introduction of abstraction mechanisms in form of sequence diagram simplification. For example, a diagram may have valid control and data flow sequence for a range of input parameters and corresponding range of expected results (equivalence partitioning). The downside of this approach is a risk of incomplete modeling which can be mitigated only by formal architecture reviews.

Implementation of above stated approach allowed to manage design complexity but the challenge remained on SW integration testing side. Application of manual testing process would multiply development and maintenance costs in the following steps:

- Test specification creation,
- Test implementation,
- Test execution and result analysis,
- Reporting.

EPS SW testing organization decided to evaluate automation options using a combination of existing Software-in-the-Loop (SIL) test environment and tools developed in-house.

SW Integration Testing Scope

Industry standards such as ISO 26262:2011 and Automotive SPICE impose requirements on SW integration testing. The goal of this product development phase is to prove that the integrated SW components are consistent with corresponding architecture design and that the embedded SW performs its intended functionality without unwanted side effects or unreachable code.

Compliance with this target can be demonstrated by performing several types of tests that focus on the following aspects:

- Data flow correctness between SW components (interface test),
- Control and data flow verification on integrated SW against feature behaviors (interaction test),
- Data interpretation by SW components (functionality or requirement-based test),
- Back-to-back comparison between model and code.

EPS SW test organization utilizes all of the above criteria by performing integration testing on two levels:

1. Static interface test, where the data transport correctness by existing architectural interfaces is examined.
2. Dynamic interaction test, where all other aspects are checked.

Static Interface Test

Two step approach was introduced due to historical limitations both in the area of SW architectural design and testing. A few generations of products ago, SW architecture design was mainly serving documentation purposes and source code was manually implemented. On the testing side, all verification steps were also manual and both of those phases were highly prone to errors.

Improvements focused on generation of interfaces on the design side, while in testing effort was put into automated interface list parsing, test case generation, execution and reporting. Those activities resulted in release of first generation SW integration testing tools developed in-house by ZF Steering Engineering. The technology provides capability of end-to-end interface testing using equivalence partitioning and boundary value analysis techniques. It provides the following benefits:

1. Verification of interface implementation correctness (object names, data types),
2. Verification of data flow using equivalence partitioning and boundary value analysis depending on transferred data type (min/max values, enumerated lists etc.),
3. Easy to use, fully automated solution with short test cycle,
4. Partial satisfaction of industry standards.

Currently this test sub-level seems to be redundant in presence of AUTOSAR compatible SW development toolchain that generates interfaces completely and excludes human errors. Despite this, static interface testing is used today as data transport checking is still mandated by standards, it provides independent verification of AUTOSAR toolchain functionality and it is very economic to execute. Due to its relatively low complexity, the article does not expand on this test level and focuses on more sophisticated interaction testing solution.

Dynamic Interaction Test

Control and data flow verification against feature behaviors became possible once decision was made to include sequence diagrams (Figure 4) in SW architecture releases. The level of information enclosed in the design suggested that each diagram could be interpreted as a requirement and test specification at the same time. From a different point of view, numbers of documented behaviors combined with amount of control/data flow information made a perspective of manual test specification development (cloning) and further maintenance completely inefficient and unrealistic considering resource constrains and average SW delivery timings.

Development of static interface testing tools has proven that high degree of automation is possible, but further extension into dynamic testing field required solving of the following technical challenges:

1. UML design analysis and control/data flow sequence extraction,
2. Automated parsing of sequence diagram verification proposals for fetching of preconditions, inputs and expected results,
3. Test case generation by precondition and input setting in the SIL environment,
4. Source code execution tracing (profiling),
5. Evaluation of trace results against diagram reference and reporting.

Toolchain Architecture

During feasibility study technical goals were grouped into analysis and execution domains, and high-level interaction testing architecture was designed, Figure 5.

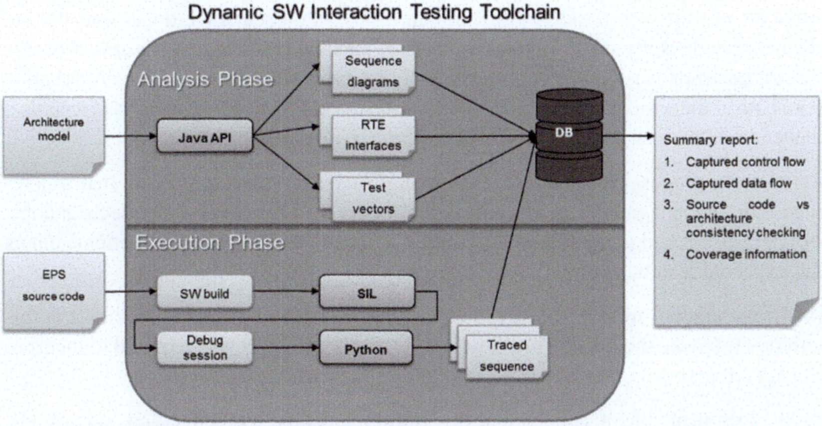

Figure 5. Model based interaction testing – simplified architecture.

Potential reward of having fully automated dynamic interaction testing technology eventually outweighed development costs and associated risks. Following chapters provide more detailed description on solutions to those challenges.

Architecture Design Analysis

Reaching UML diagram parsing capability was a key problem that could determine success or failure of R&D activities. There were three types of information required for an automated test case generation:

1. Control and data flow steps and sequence coming from sequence diagrams,
2. List of RTE interfaces used to perform data operations,
3. Test vectors containing preconditions, inputs and expected outputs from a modeled behavior.

Within those types there were two different technical challenges involved (Figure 4):

- Fetching any UML data from a native design file format by either exporting it to open XML files or attempting to use any form of modeling tool API.
- Lexical and syntax analyzing of the information provided within verification proposals, which in turn required definition of semi-formal description language to be followed by SW architects and development of parsing algorithms, similar to mechanisms used during compiler implementation.

Initial attempts to export native UML design format into open XML documents did not yield expected results and the alternative path of fetching required data via Java API was chosen. Based on feasibility study results, SQL database was selected to store information from both architecture analysis and SW execution trace, Figure 5.Solving of this technical challenge de facto enabled successful completion of all development activities.

Verification proposal formulation and parsing carried its own complexity that was related to both management and technical challenges. A group of SW architects and test tool developers had to agree on a description language for stating preconditions, inputs and expected results.

On the managerial front it was key to identify right staff members to participate in this activity and make them available. Afterwards, further support was required to incorporate agreed approach into SW architectural design process.

On the technical side, a semi-formal description language was agreed on and SW test organization developed lexical and syntax analysis algorithm that capture preconditions, inputs and expected results data in a deterministic manner, Figure 6.

```
Verification Proposal ID: SW-R-CSA-0006a

Test Objective: Verify StraightAheadOffset is set to the current Steering Wheel Postion when a calibrate
straight ahead request is made

Preconditions:
Fault Free Angle Sensor
Steering Wheel on the Mechanical Centre - Steering Wheel Position at 0deg
VehicleSpeed.Value <= VEHICLE_SPEED_THRESHOLD
ColumnTorque.Value < CALIBRATION_TRQ_THRESHOLD, ColumnTorque.Qf = FULL
ColumnVelocity < CALIBRATION_VEL_THRESHOLD, ColumnVelocity.Qf = FULL
ColumnPositionLower.QF = FULL

Step 1: Set GarageDiagnostics.CalibrateStraightAheadCommand to CAL_CALIBRATE
Step 2: Wait 1s

Acceptance Criterion:
Step 1:
CalibrateStatus = CAL_STATE_IN_PROGRESS
getCalibrateCommandStatus(status) returns CAL_STATE_IN_PROGRESS
CalibrateCmd = CAL_CALIBRATE
Rte_IRead_ColumnSwc_calculateColumnAngle_STR_AHD_OFFS_StraightAheadOffset()->Value = initial value
Rte_IRead_ColumnSwc_calculateColumnAngle_STR_AHD_OFFS_StraightAheadOffset()->Qf = UNAVAILABLE

Step 2:
CalibrateStatus = CAL_STATE_COMPLETE
getCalibrateCommandStatus(status) returns CAL_STATE_COMPLETE
CalibrateCmd = CAL_CALIBRATE
Rte_IRead_ColumnSwc_calculateColumnAngle_STR_AHD_OFFS_StraightAheadOffset()->Value = 0deg
Rte_IRead_ColumnSwc_calculateColumnAngle_STR_AHD_OFFS_StraightAheadOffset()->Qf = FULL
```

Figure 6. Verification proposal example following semi-formal description language.

Parser implementation is similar to one from C compilation process, where in the first step text stream is broken down into tokens belonging to a specific category (e.g. identifier, operator, value) which are later analyzed for dependencies during syntax analysis. For example, the following line (5) is recognized as a set of three tokens: identifier, operator (assignment symbol) and value.

Rte_IRead_ColumnSwc_StratightAheadOffset()->Value = 0 (5)

Results of verification proposal parsing are stored in SQL database for later comparison with SW execution data.

Test Execution

Once all required input information was made available, development attention focused on preparation of test environment and test cases for automated execution, consistency checking and reporting. The environment part did not require major effort due to reuse of existing Software-in-the-Loop technology from requirement based and static interface testing. Development of code profiling part was much more demanding as described in following chapters.

Software-in-the-Loop Environment

EPS SW Test organization uses Processor-in-the-Loop (PIL) and Software-in-the-Loop (SIL) test environments for SW verification and qualification activities. The PIL infrastructure uses target microcontroller (MCU) for code execution and simulates ECU circuitry and external signals, while SIL abstracts from MCU entirely by simulating operating system (OS) ticks with scheduling and some parts of HW peripherals (e.g. CAN interface, FLASH and RAM etc.). In generic terms, it is possible to state that the simulation part implements AUTOSAR Microcontroller Abstraction Layer (MCAL), Figure 7.

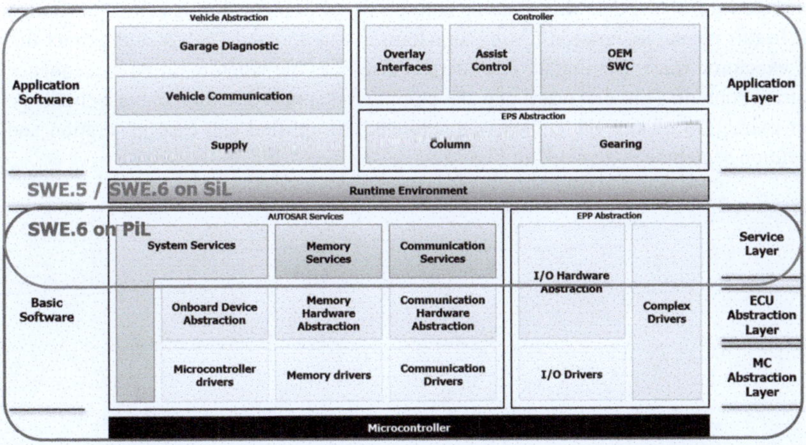

Figure 7. SIL environment architecture (simplified).

For the initial feasibility study and interaction testing technology deployment SIL environment was used due to reduction of complexity resulting from HW abstraction (microcontroller and ECU dependencies, timing etc.). In the first attempt, scope of dynamic interaction testing was narrowed down further to cover SW features realized only within AUTOSAR application layer.

Test Case Generation

There are two types of input information that must be applied to successfully prepare a test case for interaction testing (Figure 4):

1. Preconditions describing SW mode of operation and static SW configuration,
2. Control and data flow steps such as function calls, specific parameter values, global signals impacting a behavior, dynamic SW configuration etc.

Based on existing SIL infrastructure it was decided that precondition reaching will be achieved through external test framework tool that executes one of previously prepared test scenarios. A scenario is responsible for EPS SW static configuration and bringing it to a specific mode of operation. Toolchain automatically selects a valid file based on the sequence diagram data. Assumption was made that a complete set of start-up scenarios are prepared by SIL development experts and delivered with each environment release. From the execution perspective, this step is a direct reuse of requirement based testing approach that was internally defined to be compliant with Automotive SPICE SWE.6 process.

Dynamic control and data flow reading and setting takes place during source code tracing. In this phase, integration testing toolchain is able to run EPS SW and extract full SW execution trace. Manipulation of data is also possible when required by sequence diagram content. This step was a new development that posed second biggest challenge after extraction of key UML data was achieved. Simplified test case generation and execution sequence is depicted in Figure 8.

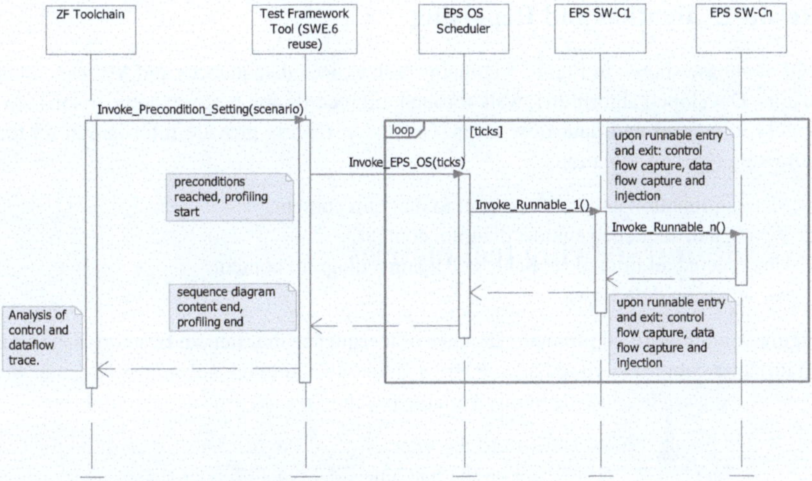

Figure 8. Simplified test execution sequence.

Source Code Tracing

The goal of this test activity was to gather control and data flow information during EPS embedded SW execution. It was also necessary to provide the ability to inject data at specific sequence step based on the information from SW architecture model. This could have been achieved by using one of the following strategies:

- Source code instrumentation that yields very accurate profiling data, but it modifies product under test and brings significant performance penalty that affects results. Timing and some parts of functional analysis can be impossible or not representative due to overhead this approach brings.
- Sampling which is a tracing strategy that periodically polls stack frames and extracts function call and data information. It is a less intrusive approach allowing to run programs nearly at full speed, but typically provides less accuracy.

Since the EPS SW runs on virtual SIL environment that executes the code in linear mode without preemptive scheduling, it was possible to utilize instrumentation approach without major functionality or performance penalties.

Automated instrumentation was developed that inserts trace markers at the beginning and before the end of each function execution. The running SW is halted at each marker allowing to 'record' the execution sequence and read or inject data when necessary. Tracing algorithm exports captured execution frame to SQL database.

Result Evaluation and Reporting

SQL database shown in Figure 5 contains both expected sequences and SW execution results. Developed algorithms 'walk through' all records and perform comparisons separately in control and data flow areas. Generated reports provide information on the following types of violations:

- Missing function calls with respect to diagram content,
- Wrong call sequence against diagram content,
- Unexpected/additional runnable call against diagram content,
- No or wrong data flow.

Figure 9 and Figure 10 provide examples of a sequence diagram and its corresponding control and data flow report.

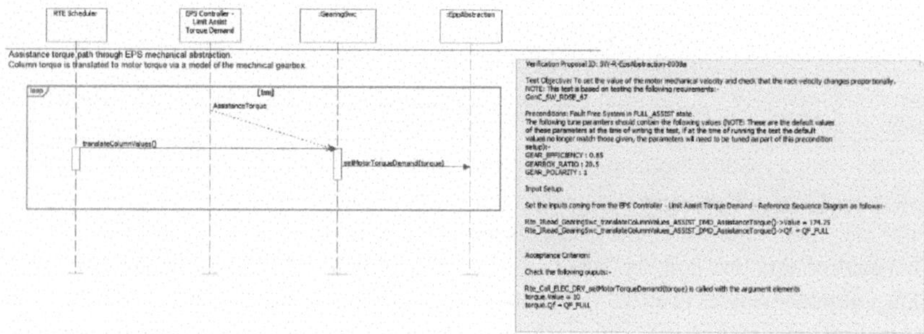

Figure 9. Determine motor torque demand behavior – sequence diagram.

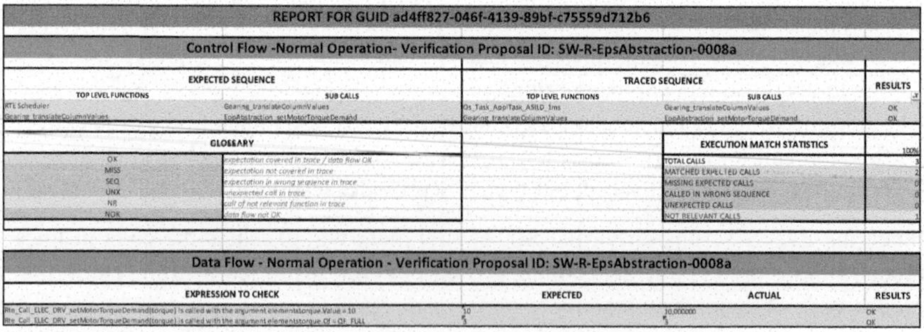

REPORT FOR GUID ad4ff827-046f-4139-89bf-c75559d712b6

Control Flow -Normal Operation- Verification Proposal ID: SW-R-EpsAbstraction-0008a

EXPECTED SEQUENCE		TRACED SEQUENCE		RESULTS
TOP LEVEL FUNCTIONS	SUB CALLS	TOP LEVEL FUNCTIONS	SUB CALLS	
RTE Scheduler	Gearing_translateColumnValues	Os_Task_AppiTask_ASILD_1ms	Gearing_translateColumnValues	OK
Gearing_translateColumnValues	EpsAbstraction_setMotorTorqueDemand	Gearing_translateColumnValues	EpsAbstraction_setMotorTorqueDemand	OK

GLOSSARY	
OK	expectation covered in trace / data flow OK
MISS	expectation not covered in trace
SEQ	expectation in wrong sequence in trace
UNX	unexpected call in trace
NR	call of not relevant function in trace
NOK	data flow not OK

EXECUTION MATCH STATISTICS	100%
TOTAL CALLS	3
MATCHED EXPECTED CALLS	2
MISSING EXPECTED CALLS	0
CALLED IN WRONG SEQUENCE	0
UNEXPECTED CALLS	0
NOT RELEVANT CALLS	1

Data Flow - Normal Operation - Verification Proposal ID: SW-R-EpsAbstraction-0008a

EXPRESSION TO CHECK	EXPECTED	ACTUAL	RESULTS
Rte_Call_ELEC_DRV_setMotorTorqueDemand(torque) is called with the argument elementstorque.Value = 10	10	10,000000	OK
Rte_Call_ELEC_DRV_setMotorTorqueDemand(torque) is called with the argument elementstorque.Qf = QF_FULL	3	3	OK

Figure 10. Determine motor torque demand behavior – interaction test result.

Quality Assurance

Standards and process assessment models like ISO 26262 and Automotive SPICE have put considerable emphasis on verification of work products. Completeness, consistency and traceability checking between items on opposite V-cycle sides was put on the spotlight[4]. Verification of integrated EPS SW against corresponding architectural design – as discussed in chapter on State Space Explosion – can carry significant effort and demand considerable expertise from test engineering departments, both in terms of product knowledge and span of design and test technologies involved. Development of model based verification technology addressed these concerns and provided additional efficiency gains compared with manual work.

From the quality assurance perspective introduction of standardized verification methods guarantees systematic approach that is easy to measure and demonstrate in front of customers. An automated solution provides further strengths due to efficiency, robustness and increased independence from human factors.

Technology developed by ZF Steering Engineering allows to create dynamic conditions depicted in SW architectural diagrams and verify consistency between model and implementation by observing the SW during its operation. Bidirectional traceability between architecture artifacts and test reports is guaranteed by design and can be graphically shown using commercially available tools. The solution is deterministic and does not require significant expert knowledge from engineering staff members. Requirements from ISO 26262 and Automotive SPICE standards are satisfied.

Deployment Summary

By the time when this article was written the interaction testing technology was put into use in two customer projects. Together with static interface testing it formed a comprehensive integration test process demonstrated in Automotive SPICE audits.

Currently there are still limitations to architecture coverage – the AUTOSAR application layer is fully in scope, but the HW related basic software (BSW) layer was partially omitted in the first toolchain release. Due to dependencies on the target HW and used code instrumentation technique for SW profiling it may happen that our technology will not be able to address all challenges. One potential solution would be to combine automated and manual approach to reach full coverage, but there's a chance that future toolchain releases could close this gap.

4 Consult ISO 26262-6:2011 Clause 10.4.3 and Automotive SPICE Process Assessment Model 3.0, SWE.5 Software Integration and Integration Test.

While technology development is predominantly a statement of R&D capabilities, management level had to be involved to support development effort and project rollout. Like any other algorithm based solution, our toolchain relies on standardized input. This calls for incorporating all toolchain requirements into SW architectural design and integration testing processes. Defined approach has to be consistently followed for all projects and all engineering locations – any discrepancies have to be rectified by work product reviews. All involved managers have to fully support this goal.

Technology rollout benefits are very lucrative – guaranteed consistency and traceability by design, high reuse of once developed toolchain by multiple projects, automated execution with Continuous Integration capability. On the other hand, these need to be cross checked with organizational structure and project needs. Small companies with simple product architectures may not see enough value to justify the development cost.

In case of ZF Steering Engineering it was clear from the beginning that exploding product complexity could not be followed up simply by expanding engineering teams and continuing manual approach. Technology development was selected as the most appropriate solution demonstrated here by the following example: just before a major customer audit we have identified that project pressure resulted in wrong architecture baseline being delivered for interaction testing. The problem was caught during test result review with two days before customer presentation. After some time a root cause was found and appropriate baseline was delivered for testing. Few hours later (including QA process work) our toolchain delivered new results – this would not be possible with manual approach. As an engineering organization we were ready to move forward and focus on custom and more complex issues, leaving 'standard' matters to automation mechanisms.

Dipl.-Ing. Maciej Obszarny (presenter)
Team Manager EPS SW&System Testing, ZF Group, Bielsko-Biala
maciej.obszarny@zf.com, +48 33-485-6422

Dipl.-Ing. Peter Niggemeier
Chief Engineer EPS SW Platform Team, ZF Group, Düsseldorf
peter.niggemeier@zf.com, +49 (0)211-584-931

Dipl.-Ing. Wojciech Grygierek
Senior SW Test Engineer, ZF Group, Bielsko-Biala
wojciech.grygierek@zf.com, +48 33-485-6313

Dipl.-Ing. Wojciech Makuch
Senior SW Test Engineer, ZF Group, Bielsko-Biala
wojciech.makuch@zf.com, +48 33-485-6299

BRAKE.TECH SECTION

FUTURE BRAKE SYSTEMS

Brake systems 2025 – future trends

Christian Vey

P. Pfeffer (Ed.), *9th International Munich Chassis Symposium 2018*, Proceedings,
https://doi.org/10.1007/978-3-658-22050-1_41

Introduction

The number of car sharing users in Germany exceeded the 2 million mark for the first time at the beginning of 2018. This represents an increase of 23% compared to the previous year.[1] Growing demand can be explained by the trend of "urbanization and digitalization". Access to mobility services is relevant. This contradicts the classic concept of private ownership of a vehicle. This example is an indication that changes are occurring in the area of individual mobility and the related automotive industry. With the topics of electro mobility and automated driving, there are other trends in the automotive industry that have an impact on the design of vehicles. New and changed requirements can be expected at the mobility level, vehicle level, system level and component level.

For the braking system as the vehicle's central safety system, the future requirements are not yet clear. But what should the braking system look like for electric vehicles, automated vehicles and connected vehicles? Which requirements change? Which developments are necessary in the area of the brake system? A uniform picture of when the described trends are relevant to the market is not existent. This results in a degree of uncertainty regarding development goals and development periods.

An analysis of the trends will be used to show which changes occur and what effects these have on future brake systems. Finally, possible solution concepts are presented.

Approach

In order to find out the relevant development goals for future brake systems, the approach shown in the following figure is chosen. This is a top-down approach. From the available trends and other influencing factors, scenarios for future mobility are created. These mobility scenarios illustrate the potential of each trend independently of their local and temporal occurrence. Based on the individual trends and scenarios, changes can be derived at the mobility level and the driving level. The knowledge about the change at the vehicle level allows a conclusion on the effects on the brake system. By evaluating those effects, it is possible to answer the question if the existing products are suitable for future requirements or new developments are needed. The identified need for development activities enables a targeted orientation.

[1] Bundesverband Carsharing e.V., 2018

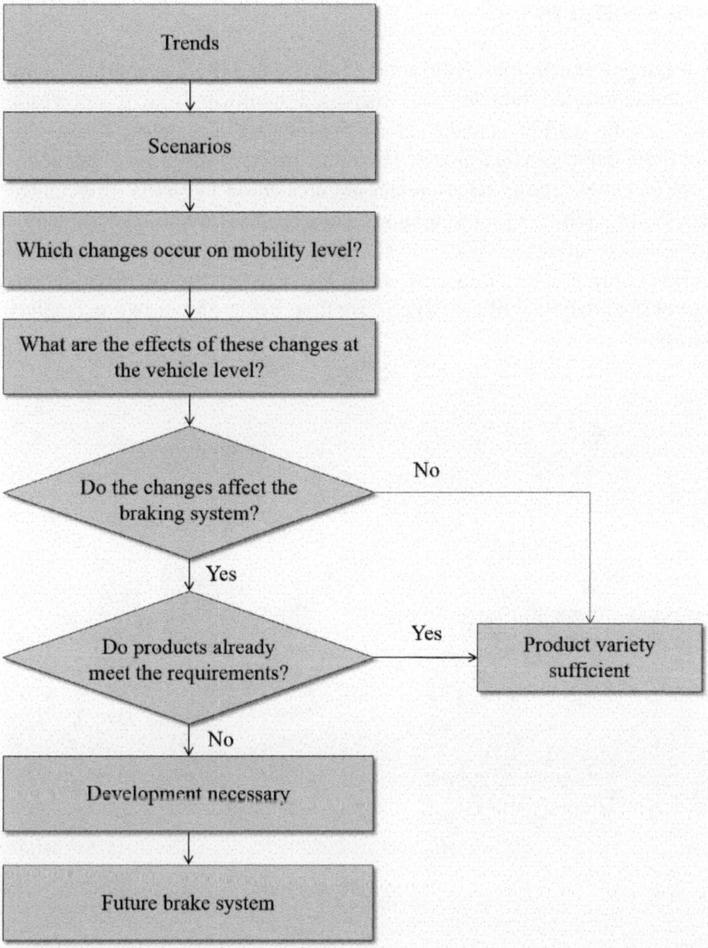
Trends
Scenarios
Which changes occur on mobility level?
What are the effects of these changes at the vehicle level?
Do the changes affect the braking system?
No
Yes
Do products already meet the requirements?
Yes
Product variety sufficient
No
Development necessary
Future brake system

Approach

Trends and scenarios

Research and development activities in the areas of electro mobility (e-mobility), automated driving and connected vehicles have increased significantly in recent years. These three areas are the key trends in the automotive industry. The potential and market relevance of these trends can be illustrated by future mobility scenarios. The mobility scenarios are potential future scenarios that are created on the basis of weighted influencing factors. In addition to those trends, factors such as environmental factors, economic factors or legislation are taken into account. The legislation in particular has a strong influence in the promotion of new technologies, as subsidies or regulations create a change of boundary conditions. The illustration below shows two exemplary mobility scenarios.

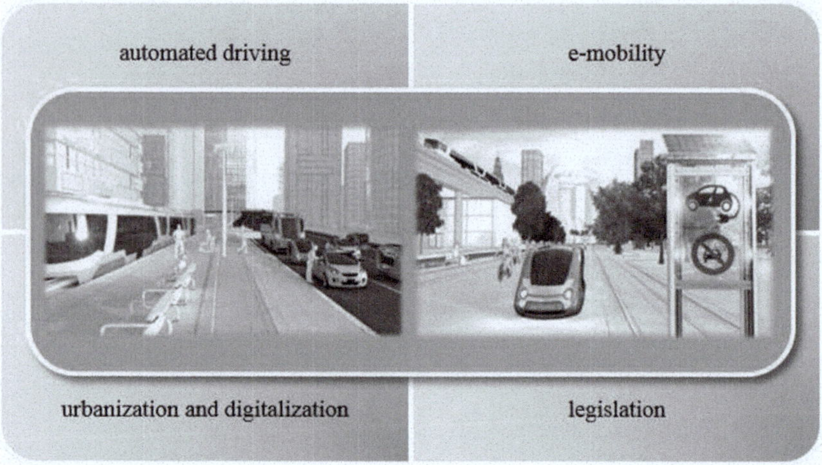

Trends and scenarios

The left picture shows a scenario called mobility transformation. It describes an urban environment with high economic performance. The population wants to be mobile. The choice of means of transport depends on the situation. The possession of a private car is no longer common, since a variety of mobility services are available. The car sharing possibilities are well developed and the automated driving is progressed.

The right picture shows a scenario titled ecology first. It shows an urban environment with strict regulation. This is to ensure the air quality. There are entry restrictions for combustion engine vehicles. Additional mobility offers are used by the population.

These exemplary scenarios highlight extreme changes at mobility level and vehicle level, and show the need to think about the effects on vehicle design.

For this purpose, the changes at the level of mobility as a function of the influencing factors are described below and the effects on the vehicle level and brake system levels are derived. Based on the brake system-relevant effects, corresponding requirements are made of the brake system.

Legislation

The increased demand for mobility in urban areas leads to an increase of traffic-related emissions. On the side of legislation, emissions will be regulated more extensively in the future in order to ensure a defined air quality. These two opposing developments of a higher output compared to a stronger regulation lead to a conflict of objectives. The solutions to this are diverse, although the effectiveness of the different measures cannot be clearly quantified. With regard to particulate matter, the Euro 6 standard show a limit values to a particulate mass (PM) of 4.5 mg/km[2]. A metastatic study on particulate matter emissions from vehicles shows that the particulate matter emissions caused by the brakes are between 2.0 and 8.8 mg / km[3]. This means that the fine dust emissions caused by braking systems are in the scale of the regulated exhaust emissions. A standardized method for the determination of fine dust emissions of the brake system is still state of the art. As soon as this process is completed, regulations can be expected. In order to meet this future legal requirement, measures are necessary in the braking system which lead to a reduction of the current emissions.

Effects of legislation

Mobility level	Vehicle level	Brake system level	New requirements
-	Reduced emissions	Reduction of fine dust	Measures for fine dust reduction

[2] Official Journal of the European Union: Commission Regulation (EC) No 692/2008, 2008

[3] Grigoratos, T.; Martini, G.: Non-exhaust traffic related emissions, 2014

e-mobility

The demand for an increased use of electrical vehicles is one of the potential solutions to reduce emissions and to ensure air quality. For full electric vehicles, no emissions are locally caused by the powertrain. To promote electric vehicles, incentives are provided by state subsidies in many countries. In addition, restrictions are introduced to restrict conventional technologies. In Germany for example there is a subsidy of up to 4000 € for the purchase of an electric vehicle. This increases the demand for electric vehicles and makes e-mobility a trend in the automotive industry.[4]

At the mobility level, the change is in a wider distribution of electrical vehicles. On the vehicle level, this results in some changes that have an impact on the braking system. Electric motors can also be used as generators. Thus, electric vehicles have the ability to convert kinetic energy into electrical energy by decelerating the vehicle. This process is called recuperation and it is used to optimize the range through the recovered energy. This reduces the deceleration that is provided by the hydraulic friction brake.[5] This leads on the one hand to a reduction of wear of lining and disc and on the other hand to an increase of the corrosion potential of the disc. The exact measurement of wear reduction is still state of the art. This means for the design of the brake that a change of brake pads during the lifetime (300,000 km) is no longer necessary, or by reducing the pad thickness, a cost and weight reduction is possible. The increased potential for corrosion requires measures for improved corrosion protection, since the corrosions free system is necessary to ensure the robustness and performance of the brakes.

For the interaction between recuperation and friction brake, the mechanical-hydraulic coupling must be interrupted in a defined way in conventional brake systems. Thus, the recuperation torque temporarily replace the friction brake torque or is combined with it. This function, called blending, is a requirements for the braking system.[6]

If limits are exceeded, the recuperative braking power is limited or even unavailable. In these cases, the deceleration must be done by the hydraulic friction brake. This means that the friction brake has to be designed in terms of performance regardless of the recuperation. A probability analysis of availability or even a guaranteed availability could significantly reduce the dimensioning of the friction brake.

[4] Bundesamt für Wirtschaft und Ausfuhrkontrolle: Förderung von elektrischbetriebenen Fahrzeugen, 2017

[5] Breuer, B.; Bill, K. H.: Bremsenhandbuch 2017, S. 49

[6] Breuer, B.; Bill, K. H.: Bremsenhandbuch 2017, S. 49

The range is one of the critical topics of electric vehicles. This depends on the battery capacity, which is the central cost driver of the technology. The dimension of the battery, is a compromise between cost, mass and range. Compared to conventional drive technologies the range of electric vehicles is lower. To increase the range, power losses have to be minimized. This means for the braking system, as much of the kinetic energy as possible should be recuperated and on the other hand, to minimize losses due to residual torque. Zero residual torque is therefore desirable.

Another measure to increase the range of electric vehicles is the reduction of the maximum speed, what reduces the maximum kinetic energy. This leads to a reduction of the maximum performance which has an effect on the dimensioning of the brake system. Due to the high mass of the vehicle battery an increased vehicle mass is possible, which contradicts the previously described reduction of the maximum performance. Thus, this effect has to be clarified on a vehicle-specific basis.

Effects of e-mobility

Mobility level	Vehicle level	Brake system level	New requirements
Increased use of electrical vehicles	recuperation	Reduction of wear	Thinner pads or longer lifetime
		Increase of potential of corrosion	Increased corrosion protection
		Ensure blending functionality	Decoupling brake pedal to brake
	Recuperation is not guaranteed permanently	Conventional design necessary	-
	Lower range	Reduction of losses	Reduction of drag torque
	Lower velocity vs. higher vehicle mass	Potential for lower performance brakes	Small sized disc brakes or drum brakes

Automated driving

Automated driving represents another trend in the automotive industry. The degree of automation is described over different levels. The grading criteria refer to the determination of the responsibility for the maneuver at normal operation and at fallback levels.

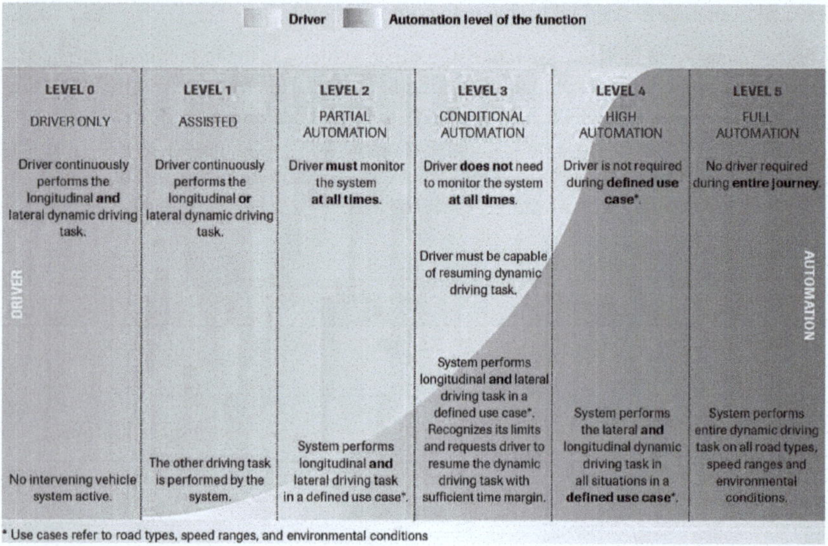

Levels of automation for automated driving[7]

From level 3, the responsibility of executing the maneuver is transferred from the driver to the vehicle. This is also valid to the fallback level in the event of a failure. Depending on the level of automation, the driver must be capable of resuming the dynamic driving task. To ensure safe vehicle guidance even in the event of a failure, redundancies are required in the system design. Thus, two separate systems for providing the braking energy are needed for the braking system. The activation for automated vehicles is independent of the driver, so that a coupling between the driver and the brake is no longer necessary. This type of brake systems are called brake-by-wire systems in which the brake pedal is decoupled from the wheel brakes.

[7] VDA: From Driver Assistance Systems to Automated Driving, VDA Magazine – Automation, 2015

For the actuation, the degree of automation is also important. Level 3 vehicles require a brake pedal because the driver must be capable of resuming the dynamic driving task. For level 4 vehicles, this is not mandatory and for level 5 vehicles, a brake pedal is no longer necessary because the system performs entire driving tasks.

Effects of automated driving

Mobility level	Vehicle level	Brake system level	New requirements
Automated vehicles, with and without driver	system performs driving tasks	Actuation independently of the driver	By-Wire-System necessary
		temporary availability at Fall-back level	Redundancies necessary
		Driver intervention possible	Brake pedal necessary
	system performs entire driving tasks	Availability at Fall-back level	Mechanical coupling of driver to wheel brake is not necessary

Urbanization and digitalization

A fundamental change in mobility behavior is imminent. The classic way of reaching a destination in a private car is no longer relevant. Rather, it's about having access to mobility. The choice here is made for comfort and time constraints. Multimobility is an interaction of different transport systems. The implementation of this flexible mobility is realized through the expansion of public transport by bus and train, and additional concepts like car sharing. The numbers cited in the introduction indicate an increased demand for fleet vehicles used in the car sharing sector. For urban use, small, agile city vehicles are needed without high demands for comfort. The maximum speed for those vehicles primarily used in the urban area will also be low. These factors affect the performance of the brake system and allow smaller sized wheel brakes.

Effects of Urbanization and digitalization

Mobility level	Vehicle level	Brake system level	New requirements
Increase of fleet vehicles	Small, light vehicles with low maximum speed	Low performance	Small sized disc brakes or drum brakes

Due to changes at mobility and vehicle levels the requirements of future brake system seems to be changing. In addition, the question must be asked whether the existing test and approval procedures are also relevant for future vehicles. Is a Groß Glockner test necessary for a fleet vehicle which is used only in the urban areas? Is an Auto-Motor-Sport-Test necessary for an electrical vehicle? Are those tests also valid for fully automated vehicles? This discussion should be made for future vehicles.

Design of future brake systems

In order to fulfill the requirements derived in the previous chapter, potential measures are highlighted below which focus on the development of future brake systems.

In principle, two approaches are conceivable for the reduction of fine dust emissions caused by the brake system. On the one hand, the formation of fine dust emissions can be reduced and, on the other hand, emissions to the environment can be contained.

A potential solution to reduce the formation of fine dust can be achieved by the material choice of the friction partners. The use of coated brake discs or aluminum discs can be an effective measure. Those discs show a reduction of wear and fine dust[8]. The exact potential is still to be determined in research projects. In addition, a cost-effective manufacturing process is necessary for coated discs and a consideration of the harmfulness of the emitted particles. Aluminum discs require research in the direction of manufacturability, temperature behavior and friction properties.

The reduction of emissions of particulate matter is conceivable through an enclosed system with a closed shape. Here, the drum brake in its encapsulated design has an advantage over the disc brake. With additional measures regarding the sealing of the

[8] Gulden, F; et al.: Fundamental investigations of aluminium matrix composite brake rotors for the use in passenger cars, 2017

drum, a low emission brake can be designed. Investigations are necessary for an encapsulated disc brake. The influence of encapsulation on the thermal properties of the disc brake has to be investigated. Other ideas, apart from encapsulation, are systems, which capture the generated particulate matter. Those concepts are as well conceivable.

The increased corrosion potential, through the reduced use of the friction brake in electric vehicles, can also be addressed by a choice of materials, as well as by the use of encapsulated systems. Systems that capture the generated fine dust don't increase the corrosion protection.

By using small fleet vehicles for the urban area, braking systems with lower performance are necessary. With regard to current brake types, disc brakes can be smaller sized or drum brakes are used, which has a lower performance compared to disc brakes.

The requirements arising from automated driving for decoupling the driver from the brake lead to brake-by-wire concepts. These can be designed hydraulically as well as electromechanically. In hydraulic brake-by-wire systems it is possible with little effort to represent a coupling between the driver and wheel brake via a switching valve. This is not possible with electromechanical brake-by-wire systems. This disadvantage does not apply to vehicles of the highest automation level. An operation of the dynamic driving task by the driver is not provided, whereby the hydraulic fallback level with a coupling between the driver and wheel brake is not needed. With the features of an active clearance adjustment, a good wheel-individual controllability and the absence of brake fluid, show electromechanical brake-by-wire system some advantage over hydraulically operated systems. With an active clearance adjustment the reduction of residual torque can be realized[9]. Thus, an electromechanical brake represents the consistent implementation of the brake-by-wire concept.

The dependencies of the brake system on the particular vehicle characteristics indicate the need for a system architecture. In such a system architecture the requirements regarding redundancy, functionality, fallback levels and interfaces to other vehicle components have to be considered.

[9] Breuer, B.; Bill, K. H.: Bremsenhandbuch, 2017, S. 408

Future brake systems

The electrification of wheel brakes has been a research topic in the field of brake technology for decades. In mass production, no vehicle is equipped with an electric brake until today. The braking system requirements for future vehicles, which show a higher complexity in terms of controllability and functionality, highlight the need for electrical brakes.

In some research and development projects, exemplary versions of electric disc brakes are shown. Based on a fist caliper, a housing that includes a motor gear unit, sensor and power electronics (wheel brake ECU) is directly mounted in the wheel.[10] The regulation of the brake is based on a tension force sensor. Thus, the EMB is suitable for the future requirements regarding the flexible control. A series design that is suitable in terms of packaging, weight, redundancy and cost is a key challenge for the electric disc brake. In terms of fine dust and corrosion problems, additional measures must be taken. The use of other materials in the friction pairing or encapsulation concepts has to be additionally integrated.

Example of an electric disc brake[11]

The drum brake with its closed design offers a shape of encapsulation. Thus, the requirements for the reduction of emissions and corrosion protection can be met. The lower performance of the drum brake over the disc brake is not an inconsistency for the

[10] Breuer, B.; Bill, K. H.: Bremsenhandbuch, 2017, S. 444
[11] Breuer, B.; Bill, K. H.: Bremsenhandbuch (2017), S. 444

lower performance requirement shown in the previous chapter. An electrification of the drum brake to enable the complex requirements in terms of controllability is analogous to the disc brake possible. The main future requirements can be fulfilled by such the concept of an electrical drum brake. The primary research effort is the development of an electromechanical actuation, a sensor concept and a brake controller.

Example of an electrical drum brake

An electric disc brake or drum brake is an adaptation of brake systems which are already suitable for the automotive industry. A view into other areas of use of brake systems reveals an interesting approach in the agricultural sector. Due to the very rough field of application a higher degree of robustness is required. With regard to the braking system, the company Knott designs a fully enclosed, multi-disc brake running in oil. The encapsulation ensures protection against the high level of environmental influences in the agricultural industry. The suitability for an automotive use is to investigate. The encapsulation offers the possibility to reduce the fine dust emissions to zero. The requirements of corrosion protection are also fulfilled by the encapsulation and the brake discs running in oil. For a suitably automotive use, topics such as performance, residual torque, dimensions and mass are to be investigated.

multi-disc brake running in oil [12]

Conclusion

The analysis of trends in the automotive industry shows that there is a change in mobility behavior. This influences the design on the vehicle, system and component level and leads to new and changed requirements. With respect to the braking system, the interaction and interfaces with other vehicle systems will increase. The resulting increase in complexity in the design of brake systems shows the need for system architectures to specifically address issues of redundancy or recuperation. At wheel brake level, the electrification of friction brakes and the reduction of emissions show two major deficits of the currently available brake systems with respect to future requirements. To be prepared for those requirements the development goals for future brake systems should address these deficits.

[12] Knott, Knott Bremsenübersicht

References

Bundesverband CarSharing e.V.: https://carsharing.de/zahl-carsharing-kunden-ueber-springt-2-millionen-marke, Retrieved 18.04.2018

Bundesamt für Wirtschaft und Ausfuhrkontrolle: Förderung von elektrischbetriebenen Fahrzeugen, Eschborn, 2017

Breuer, B.; Bill, K. H.: (Hrsg.) Bremsenhandbuch, Springer Fachmedien Wiesbaden, Wiesbaden, 2017

Gulden, F; et al.: Fundamental investigations of aluminium matrix composite brake rotors for the use in passenger cars, In: Eurobrake 2017, Dresden, 02.-05.05.2017

Grigoratos, T.; Martini, G.: Non-exhaust traffic related emissions, Luxembourg: Publications Office of the European Union, 2014

Knott: Knott Bremsenübersicht, https://www.knott.de/downloads/bremsentechnik/prospekte/KNOTT-Bremsenuebersicht-D_P07901_de_1115_web.pdf , Retrieved 18.04.2018

Verband der Automobilindustrie e. V.: From Driver Assistance Systems to Automated Driving, VDA Magazine – Automation, 2015

Official Journal of the European Union: Comission Regulation (EC) No 692/2008, 18.07.2008

Distributed brake-by-wire system for next-generation road vehicles

Beniamin Szewczyk

Alessandro Ciotti

Luca Cappelletti

Brembo S.p.A

P. Pfeffer (Ed.), *9th International Munich Chassis Symposium 2018*, Proceedings,
https://doi.org/10.1007/978-3-658-22050-1_42

1 Abstract

We are currently witnessing rapid transformations in mobility systems and vehicle architectures. Tight emissions requirements and the diesel crisis have accelerated the electrification of vehicles. Road safety targets and mobility system evolution is driving major vehicle automation. Braking systems, as one of the primary automotive safety systems, are called to fulfil new requirements and functions accordingly. The change has already begun with the gradual implementation of electrically boosted and so called one-box Brake by Wire systems. Brembo, as an expert in corner actuation brake systems and components, proposes its own innovative approach for the electrification of braking system components within distributed control systems. Brembo's proposal provides advantages for integration, new functions and enables new vehicle architectures.

2 Introduction, global trends and industry drivers

Automotive industry as a part of the global transport system are evolving in very rapid way. The evolution has to respond to new trends and challenges of globalization, metropolisation and earth population grow. The main global drivers derived from these changes are Environment improvements, Safety and Economical globalization scaling.

Industry players, car manufactures as first actors and OEMs following, creates technology and product development plans as consequence. Also industry consortiums and advisory councils such CLEPA and ERTRAC elaborates the industry roadmaps [1,2].

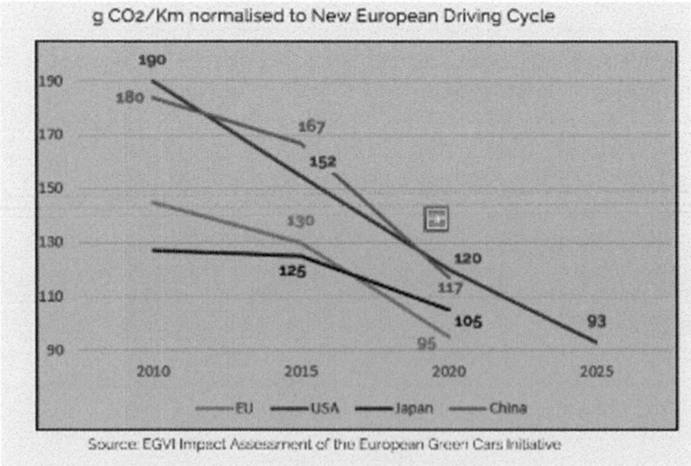

Figure 1 – CO_2 targets [1]

One of the main industry drivers is the global agreement on CO2 reductions setting the targets for 2020-2030 and beyond [3). The emission reduction requires car electrification for powertrain and other main systems in order to increase overall vehicle efficiency. Electric powertrain implementation also impacts the car nature emphasizing comfort and smoothness in use and changing specific targets in particular for the actuation systems in regards to the control accuracy and acoustics and vibration impact.

The second big industry driver is the Road Safety improvement. The latest accidents data show missing targets to set Vision Zero targets in Europe. One of the most promising way to improve the situation is an implementation of automated and connected mobility [4,5,6). Automated cars will require higher performance and faster actuation systems and open interface for increasing functionality.

The third industry driver is improvement of manufacturability and serviceability of the future cars. Increasing production automation requires modular easy to assembly systems with low environmental impact.

3 Next generation Brake by Wire

The automotive industry has been working on new braking system architectures since more than 20 years. A gradual evolution of the hydraulic system with implementation of new functionalities as ad-on blocks led to the present layout of the braking system with brake booster and ABS/ESC modulator. New generation of centralized systems that unifies booster-ESC functionality was recently proposed to car manufactures by suppliers. However such approach improves overall system performance and increases gradually number of functions but also takes all the disadvantages of the nature and architecture of hydraulic split braking system and hydraulic discreet pressure control. The solution proposed in this paper is based on distributed four independent braking corners system.

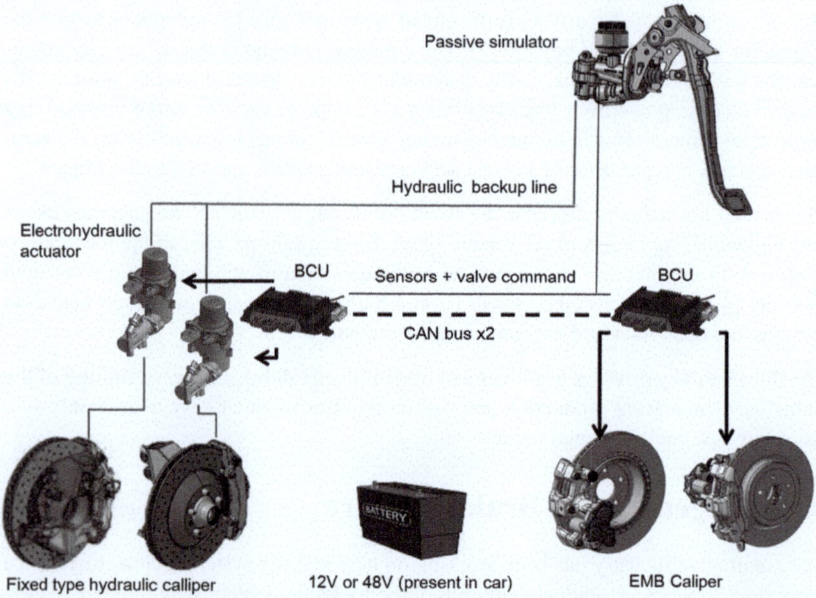

Figure 2 – Distributed Brake by Wire System

The system consists of Passive brake pedal simulator attached to the brake pedal and four actuated braking corners controlled by two independent ECUs. The brake pedal simulator is normally decoupled form the rest of the system and its main function is to give an appropriate brake feeling to the driver. The actuation system acts according to the sensors input form pedal simulator indicating driver demand for braking action. A front axle actuation is implemented by two independent electrohydraulic actuators acting on corresponding fixed type high performance brake callipers. The hydraulic actuators provides smooth pressure control in all the brake torque range using actuation and modulation form the electric motors itself without valves. A rear axle actuation is done by Electro-Mechanical calipers. Also rear calipers provide continuous control of the clamping force acting directly with electric motors. The rear calipers include also built-in parking brake functionality.

The ECUs are responsible to control the corners through electric motors and actuators' sensors according to the targets set by higher level vehicle control SW. The higher level vehicle control SW is also implemented in the ECUs providing as a base all standard braking system functionalities.

4 Regeneration and brake distribution related functions

Vehicle Kinetic energy recuperation during braking is a common way of executing braking action in other high electrified transport vehicles such as train or trams. Requirements for fine control and comfort in car industry are more strict due to relatively low vehicle weight. A common approach to simplify the regeneration function is to make it independent from the braking event with a regeneration brake torque portion added by the powertrain, staring when the driver lifts off the accelerator pedal. This kind of strategy could help in energy recuperation considering std driving cycles, but still a significant portion of the energy is left to waste in brake disks, especially for more realistic sporty driving styles. The simulation comparative results for both strategies implemented on a RWD FBEV of 2500kg are presented in the Table 1.

Table 1 – Energy recuperation comparison

Energy delta simulation [%]

	Regen Energy [%]			
CYCLE	Lift-off (0,25g)	Real blending		
US06	51,2%	52,2%	+	1,0%
NYCC	49,2%	52,4%	+	3,2%
FTP	51,9%	53,0%	+	1,1%
WLTP	51,2%	53,4%	+	2,2%
Fading2	10,3%	38,3%	+	28,0%
Nurburgring	9,8%	34,9%	+	25,1%
AMS	9,6%	36,0%	+	26,4%
Fading1	13,0%	42,0%	+	29,0%

Real optimization and maximization of regeneration challenges are related to the comfort impacts to the car and driver. The friction brake control must be smooth and reactive in all torque ranges in order blend perfectly with a regeneration electric torque provided by the electric motor. The proposed distributed BbW system gives full ability for such control. A real test example form 4WD BEV of 1600kg is presented in Figure 3.

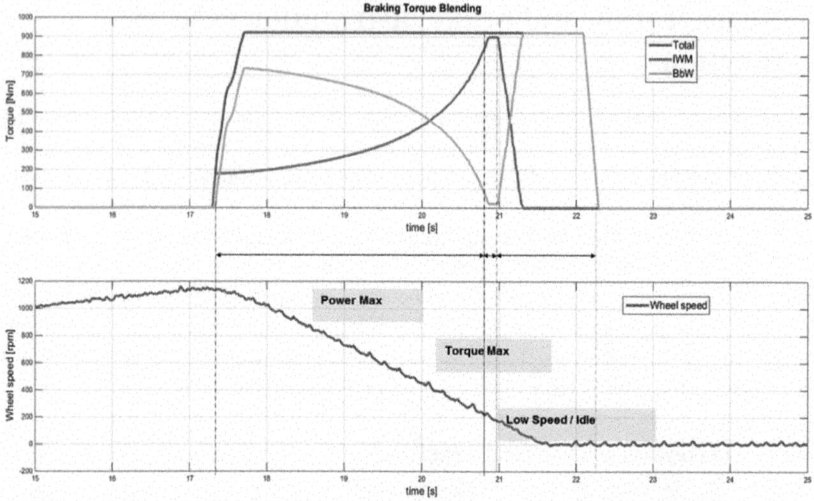

Figure 3 – Regen-friction torque blending

Without the constraints of the hydraulic actuation system, the distributed system enables to develop even higher functional integration for brake torque distribution including

- Ideal parabola distribution by SW only
- Mass adaptation
- Full blending
- Repartition between axles
- Rear to front torque shifting (ABS)
- Cornering EBD (left-right distribution)

The testing results with unladen and worst case laden 2800kg BEV Van are presented on Figures 4 and 5. The adaptation control algorithm dynamically calculates braking torque targets for each axle for different load situations. Low influence of vehicle weight and COG position changes to the braking deceleration response are demonstrated.

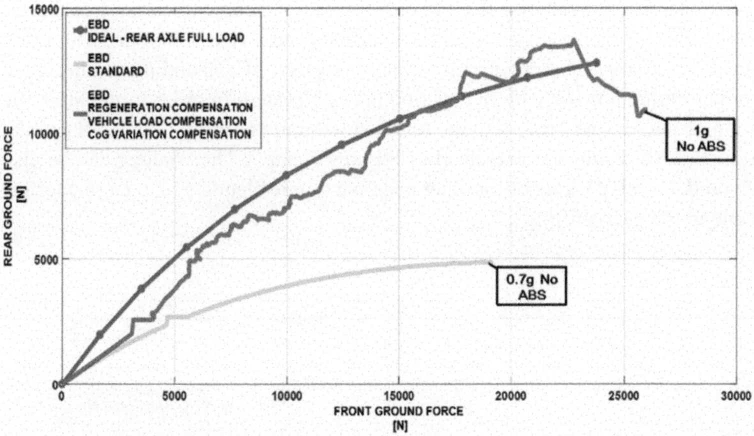

Figure 4 – Brake distribution control

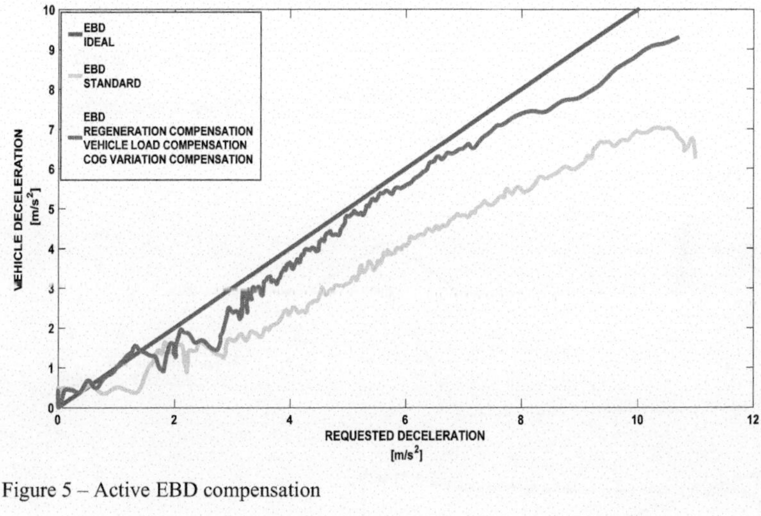

Figure 5 – Active EBD compensation

Another regeneration related function demonstrated during the system development was One Pedal Driving function. Serial production cars regenerate kinetic energy through electric motors often with very significant deceleration levels (up to 0,25g) when the accelerator pedal is released, but many of them has a limitation of brake regeneration capabilities at low speeds. A ramp down of the regeneration is necessary resulting with deceleration change from initial value to coasting with natural car deceleration. To complete the braking maneuver the driver has to use brake pedal to bring the car to the full stop. The developed strategy consists of a blending of the regen ramped down torque portion with the friction braking system in order to maintain the deceleration of the car and bring it to the full stop without necessity of using the brake pedal. Comparative results are presented of Figures 6 and 7. The strategy can be also used even on ICE on HEV cars or for other negative torque blending, e.g. engine torque changes due to the gear shifting.

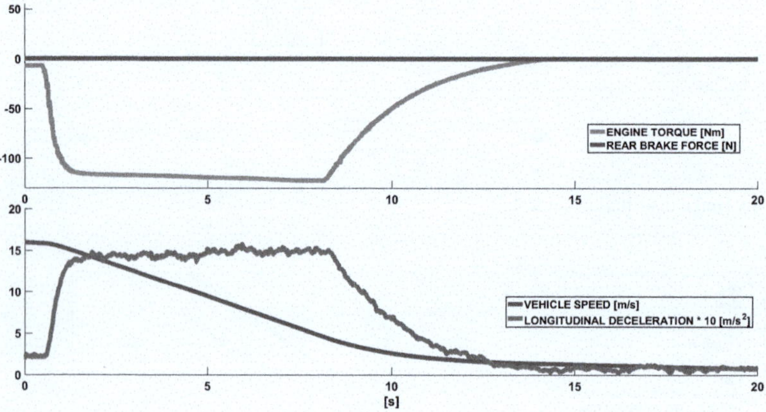

Figure 6 – One Pedal Driving function OFF

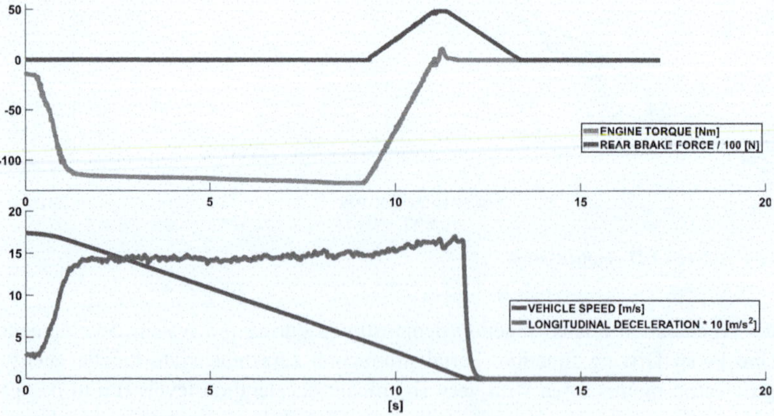

Figure 6 – One Pedal Driving function ON

5 Residual brake drag improvement

Brake residual drag or (braking rest torque) decreases the overall vehicle driving efficiency by wasting it into a heat during driving. It is a known common problem for all braking systems on all cars however electric vehicles are even more sensible to the problem due to energy to weight and cost higher ratio. The simulation comparative results for a RWD FBEV of 2500kg with 80kWh battery are presented in the Table 2.

For example the residual torque increase of 1Nm on each corner gives a delta energy consumption in FTP cycle of +2.4 % (which means same amount of the car range reduction) or increase of +15,6 kg of battery weight to obtain the same range (on BEVs, or same amount of increased fuel consumption or CO2 emissions on ICE or HEVs.

Table 2 – Brake drag influence to energy waste for different driving cycles

Residual Torque (corner) DELTA

	0Nm	1Nm	2Nm	3Nm	4Nm	5Nm
ArtMw130	0,0%	1,6%	3,3%	4,9%	6,5%	8,1%
ArtMw150	0,0%	1,6%	3,1%	4,7%	6,3%	7,8%
ArtRoad	0,0%	2,2%	4,4%	6,6%	8,8%	11,0%
ArtUrban	0,0%	2,1%	4,2%	6,4%	8,5%	10,6%
Constant	0,0%	1,6%	3,2%	4,8%	6,5%	8,1%
ECER15	0,0%	3,1%	6,1%	9,2%	12,2%	15,3%
EUDC	0,0%	2,4%	4,7%	7,1%	9,5%	11,8%
FTP	0,0%	2,4%	4,9%	7,3%	9,7%	12,2%
NEDC	0,0%	2,6%	5,1%	7,7%	10,3%	12,9%
US06_FTP	0,0%	1,6%	3,3%	4,9%	6,6%	8,2%
WLTP_class3	0,0%	2,1%	4,1%	6,2%	8,3%	10,3%

Reference Vehicle Setup

The only solution to the problem is keeping brake pads off from the disc with a retraction of the pistons (so called roll-back done with passive way by seals in std braking hydraulic caliper). For the standard braking systems with brake pedal coupled to the brake calipers by hydraulic connection the higher roll-back results in higher overall brake fluid consumption and has significant drawbacks for pedal feeling design, which is another important target of the overall braking system design. That requires fine tuning of the balance between optimum brake feeling and acceptable brake drag.

The proposed BbW system has a decoupled brake pedal from the rest of the system (in normal operation) and provides the capability for brake torque reduction up to near zero brake residual torque. For the front corners a passive retraction by the seals can be increased and managed with active control during different driving situations and on rear EMB calipers the piston position is directly controlled by the system.

6 Safety and automated driving

The proposed BbW distributed braking system with one power source and hydraulic back-up gives advantages in terms of built-in redundancies and incremental fall-back modes not possible with other braking system solutions. With the distributed architecture most of failures will lead to safe by-wire braking by four, three or two braking corners, depends on specific failure. Only few failures (eg. total power supply failure) results with a switch to the unboosted braking with only driver foot force through front brakes.

The test results for fall-back strategies implemented on a RWD FBEV of 2500kg are presented in the Table 3

Table 3 – BbW fall-back performance

	Mean dec [m/s^2]	Pedal force [N]
Hydraulic backup + Rear BBW	7,29	480
Three wheels braking	6,69	375*
Only BBW front	5,76	410*
Diagonal braking	5,26	350*
Only BBW rear	5,06	440*
Hydraulic backup	3,56	470

Increasing vehicle automation functionality puts new requirements to the braking system operability. From a Automation Level 3 (7) a driver cannot be considered ready to overtake vehicle control in case of failure or won't be present in the vehicle, or even the vehicle won't have any braking commands itself and the braking system has to provide minimum fall-back functionality to bring the car to safe stop. The failure operability ability for different automation levels are presented in Table 4.

Table 4 – Brake system solutions for SAE automation levels

2010 2020 2030

Automation level	0	1	2	3	4	5
Assistance level	NO	DAS Driver Assistance System	ADAS Advanced Driver Assistance System	Partial Automation	Conditional Automation	Full Automation
Function example		ABS, ESC, Cruise Control	AEBS, ACC	Highway autopilot	Valet parking	Robot taxi
Driver		In the loop		On board, as backup (10s)	Off the loop (or NO Driver)	
Actuation system	Fail silent	Fail silent	Fail silent	Fail operational (10s)	Fail operational	Fail operational
Brake system	STD, no ABS	ABS-ESC modulator	ESC modulator II gen, One-box, Brake by Wire	ESC modulator II gen + i-booster, One-box + backup, Brake by Wire	One-box + backup, Brake by Wire	One-box + backup, Brake by Wire
Actuation backup (Secondary braking)	by driver	by driver	by driver	by system (10s)	by system	by system

Adapted from SAE Standard J3016 (2014)

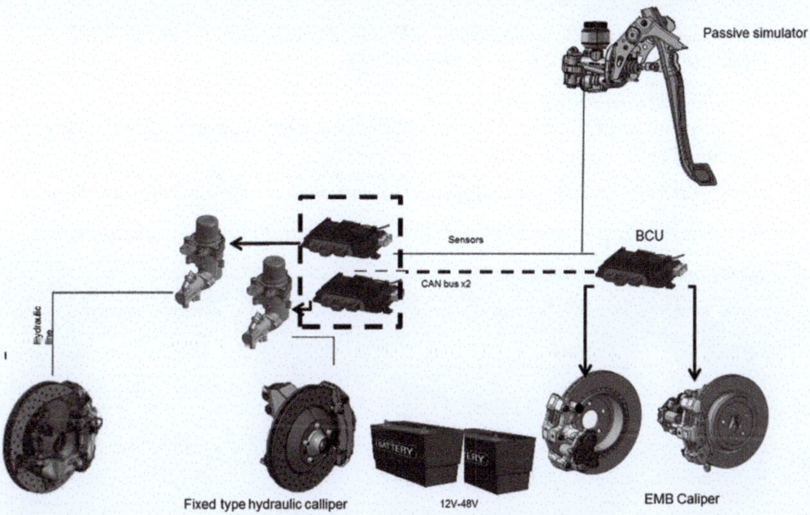

Figure 7 – Distributed BbW for HAD suitability

Providing power source redundancy to the system and split ECU design for single actuation control increases significantly failure operability of the system and applicability for Highly Automated Driving category 3 and 4. The system modularity enables further extension of redundancy also to the rear axle corners in order to obtain full redundancy against one point of failure and makes the system applicable for vehicle automation category 5.

7 Conclusions

The proposed Brake by Wire distributed braking system brings new functional and architectural alternatives for next generation vehicles. It is also suitable for modern ADAS functions and extendable by its modularity to the future requirements of braking systems for fully automated vehicles.

8 References

1 *CLEPA Position Paper on post-2020 CO2 emission reduction targets in Europe https://clepa.eu/mediaroom/clepa-position-paper-post-2020-co2-emission-reduction-targets-europe/*

2 *ERTRAC Research Agedba and Plans*
ERTRAC Roadmaps
http://www.ertrac.org/index.php?page=ertrac-research-agenda-and-plans
http://www.ertrac.org/index.php?page=ertrac-roadmap

3 *REGULATION (EU) No 333/2014 OF THE EUROPEAN PARLIAMENT AND OF THE COUNCILof 11 March 2014*

4 *Valletta Declaration on Road Safety*
https://www.eu2017.mt/en/Documents/Valletta_Declaration_on_Improving_Road_Safety.pdf

5 *https://etsc.eu/wp-content/uploads/2016_automated_driving_briefing_final.pdf*

6 *https://etsc.eu/wp-content/uploads/5th_rsap_2020-2030_etsc_position.pdf*

7 *SAE Standard J3016 (2014)*

Brake-control-based holistic truck motion control concept for automated driving

Dr. Falk Hecker, F. Stambrau, A. Mustapha,
Knorr-Bremse Systeme für Nutzfahrzeuge GmbH

This manuscript is not available according to publishing restriction. Thank you for your understanding.

P. Pfeffer (Ed.), *9th International Munich Chassis Symposium 2018*, Proceedings,
https://doi.org/10.1007/978-3-658-22050-1_43

BRAKE DUST

Brake particle emissions – a global challenge

Dr.-Ing. S. Gramstat, R. Waninger – AUDI AG, Ingolstadt

Dr.rer.nat. D. Lugovyy, M. Schröder – Horiba Europe GmbH

Dr. Theodoros Grigoratos – Joint Research Center

P. Pfeffer (Ed.), *9th International Munich Chassis Symposium 2018*, Proceedings,
https://doi.org/10.1007/978-3-658-22050-1_44

1 Introduction

Non-exhaust emissions of road vehicles are getting more and more in the public focus. Beside resuspension and tyre wear, a significant amount of those emissions is represented by brake wear [1], [2].

The friction brake has to fulfil several requirements, beginning with providing maximum stopping power, and ending up with various comfort issues. Since the foundation brake is recognized as a relevant particle emission source, numerous efforts were started to investigate the brake particle emission behavior [3].

With this paper a measurement test setup is proposed to investigate foundation brakes. This proposal includes the testing environment as well as the used measurement devices and test specimen. Furthermore the used test cycle is introduced.

Using such a measurement procedure it allows to characterize different friction couples. It was decided to compare the most popular brake lining formulations, namely NAO and low steel in combination with standard cast iron discs. Also an innovative hardmetal disc coating was investigated, combined with an adapted brake material composition.

This work intends to present testing experiences and recent results for established and also innovative friction couples.

2 A Test Setup Proposal

The proposed experimental setup is based on a brake dynamometer, which is used for performance testing of brake systems. Its customized design allows to operate below atmospheric pressure and specific ambient conditions can be considered as well. It must be underlined that the cooling air is pre-conditioned by the use of a HEPA filter (H13), also the possibility of temperature control between -30°C and +20°C is given. The intake air flow rates can be varied between 930 m^3/h and 3300 m^3/h to determine the influence of the cooling air speed. The Horiba GIANT 5200 owns an inertia simulation range from 10 kgm^2 to 200 kgm^2 and allows to consider a maximum speed of 2500 rpm at a maximum power output of 190 kW.

Furthermore, the sampling procedure and sampling location are crucial for brake particle emission testing. Since the brake system is more or less an open one, it was decided to sample in the exhaust duct, as it can be seen in Figure 1.

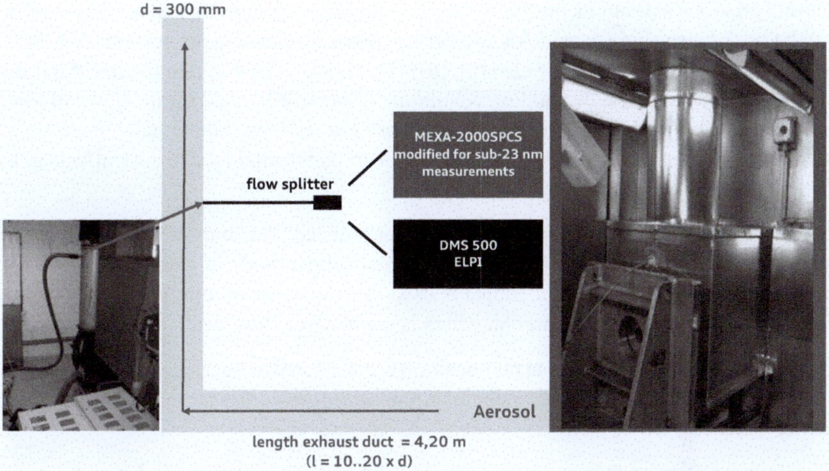

Figure 1: Scheme of the used experimental setup

Due to the fact, that the focus of the investigations is on particle number concentration (PNC) and particle size distribution (PSD), a homogenous mixture of the aerosol is strictly recommended. That means, that the distance L from the particle emission source (in other words the brake caliper) to the sampling location should be defined in dependency of the diameter D of the exhaust duct, which acts as a dilution tunnel. According to the GRPE-48-11 regulation (with L = 10 - 20 x D), the sampling location is defined as L = 14 x D. Moreover, it is worthwhile to be mentioned, that a sophisticated housing around the brake system (foundation brake) is installed. This allows a direct airflow around the brake and helps to avoid particle losses inside the test chamber (see also [1]).

The particle number concentration PNC is characterized for the complete measurement campaign by a MEXA 2000 SPCS system by HORIBA. It is a modified version to detect sub-23 nm particles in the range between 10 nm and 2,5 µm. Additionally, another device, namely a CAMBUSTION DMS500, was used to determine the particle size distribution PSD. Another device, an ELPI system from DEKATI, was used to determine particle mass distribution. The ELPI owns an impactor with 12 stages, which allows amongst others the preparation of filter samples according to a certain particle size range. Those filter samples were weighted under a certain conditioned state before and after the measurements to evaluate the difference of mass.

During previous measurements, the authors obtained very promising and reliable results by the use of a special adapted AK Master test procedure, where the sections 1-3, 5-7, 11 and 14-15 are excluded (see also [5], [6]). This cycle is able to characterize the friction under low and high demanding conditions respectively, what means in detail low temperature and pressure applications as well as fade sections. Additionally, it allows a synthetic pre-conditioning of the friction couple to understand load history effects in a better way.

Finally, the subjects of the investigations are introduced. The authors decided to use a 19" front axle brake system with a 6 piston fixed caliper made of aluminium. For reference reasons, a basic friction couple is used. This consists of a gray-cast iron brake rotor and a low steel lining material, which is copper-free (low steel).

Additionally, a second brake pad composition is considered, namely a copper-free NAO material (NAO). Those two formulations represent the worldwide majority of brake lining materials.

Beside those state-of-the-art friction couples, the authors investigated also an innovative rotor coating in combination with an adapted brake lining (HCD). This adequate brake lining material with a copper-content was selected that supports adhesive friction and ensures good braking performance.

3 Experimental Results

3.1 Comparison Low Steel – NAO

Since low steel and NAO brake lining materials represent the majority of pad formulations which can be found worldwide, it is relevant to compare them also in terms of their brake particle emission behavior.

In Figure **2** the particle number concentrations (determined by the SPCS devices) of both linings are presented for the entire modified AK Master cycle. It can be stated that the NAO shows lower emissions during section 8, where only low demanding brake applications are realized. For section 4 the particle number concentration is also lower for the NAO, but significantly increasing for the high performance brakings during sections 9 and 12.

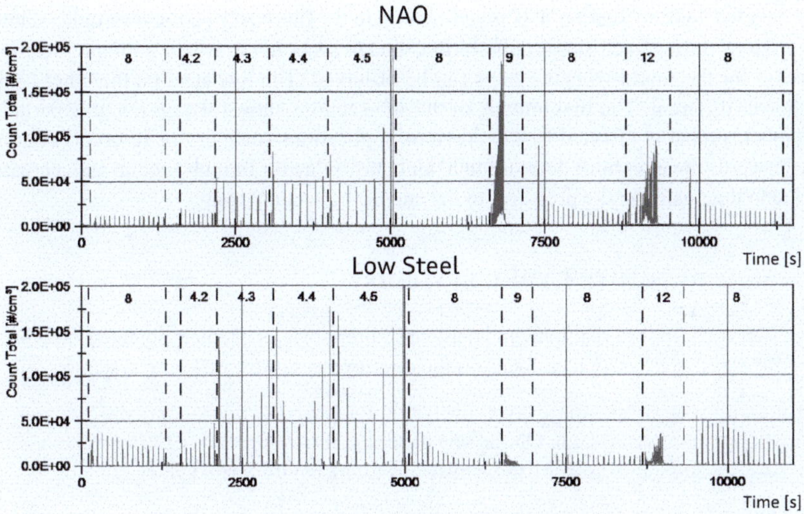

Figure 2: PNC for a low steel and NAO brake lining, modified AK Master

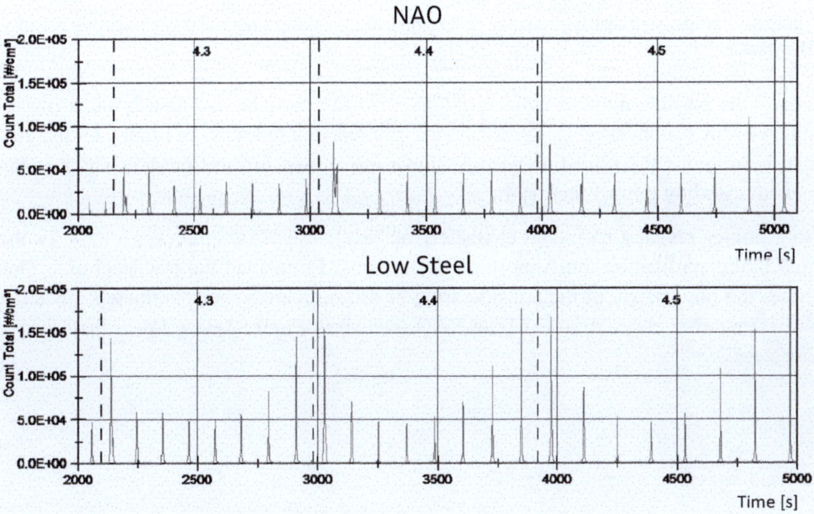

Figure 3: PNC for sections 4.3 to 4.5 for a low steel and NAO brake lining

A detailed look to section 4 is given in Figure 3. The NAO pad shows quite stable emissions during each single sub-section, except section 4.5, where an impressive increase for the final two brake steps can be observed. The low steel on the other hand behaves different. The first braking of the sub-sections shows always a higher particle number, probably affected by the previous applications (memory or reservoir effect). Afterwards the emissions decrease to a minimum value for the sub-section and increase (due to the raising brake pressure) to the end of the section again.

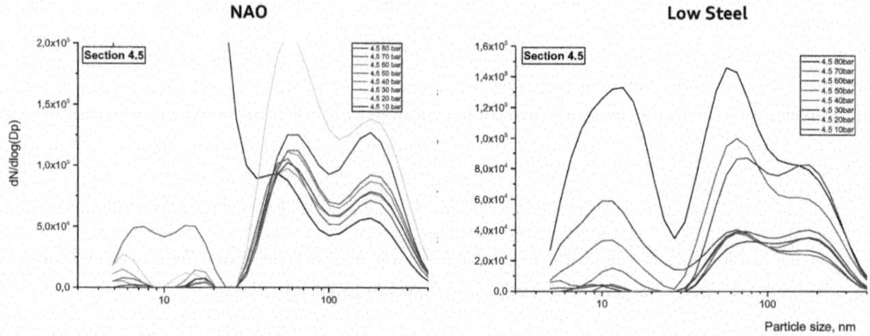

Figure 4: Particle size distributions for NAO (left) and low steel material (right), section 4.5 of AK Master

Beside the particle number concentrations, also the particle size distributions (determined by the Cambustion DMS 500 device) are relevant to know. In Figure 4 those size distributions for the ultra-fine particle range are shown for section 4.5, which is the most demanding part of section 4.

Both linings reveal a tri-modal characteristic, while the NAO emits especially for the final brake application much more particles (sub-23 nm) than the low steel pad. This shows the importance of the consideration of different analysing parameters since the particle number concentration of the NAO (shown in Figure 3) was lower than for the low steel.

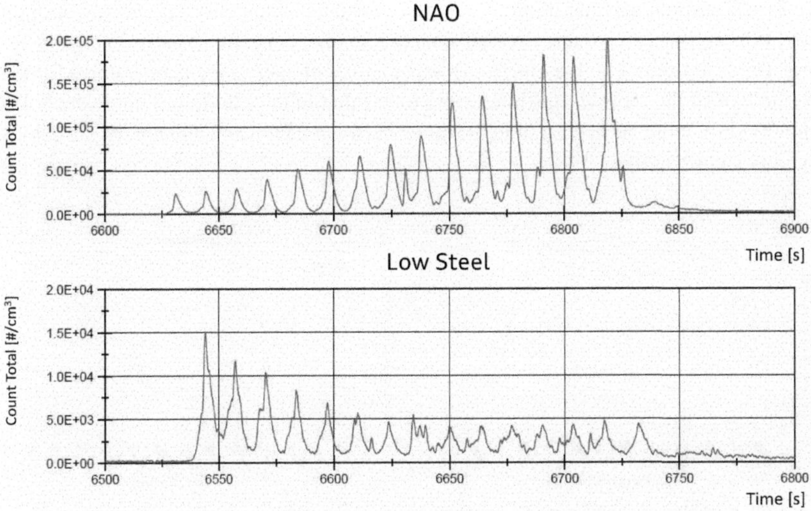

Figure 5: PNC of a low steel and NAO brake lining during a fade section

Having a closer look to the particle number concentration for the section 9 (fade section) in Figure 5 reveals for the NAO a constant increase during the entire section. Such a qualitative trend is not shown by the low steel, where the emissions are decreasing till the middle of the section, followed by a stabilization. In terms of the emitted numbers it can be seen that the NAO emits much more particle numbers than the low steel.

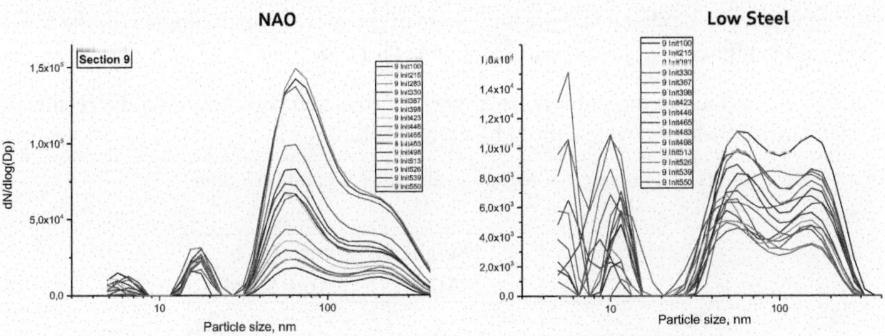

Figure 6: Particle size distributions for NAO (left) and low steel material (right), section 9 of AK Master

Again, the particle size distribution is analysed in detail (Figure 6). For the fade section the results confirm the picture derived from the number concentrations investigated before. The NAO shows higher emissions, especially with increasing temperatures for the second half of the section. It is interesting to see that in the sub-23 nm range the NAO emits slightly more particles, but in the range between 23 nm and 300 nm the emissions increase significantly.

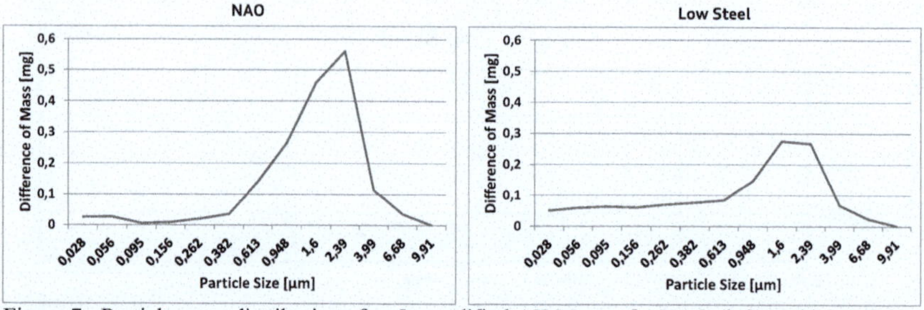

Figure 7: Particle mass distribution after 5x modified AK Master for NAO (left) and low steel (right) lining material

Another parameter is the particle mass distribution, shown in Figure 7. For this purpose, a test series of 5 modified AK Master cycles was used in combination with the cascade impactor of the Dekati ELPI system. It can be seen for both formulations that a monomodal characteristic exists with a maximum between 1 µm and 3 µm. These results confirm that the particle mass does not affect small particle sizes, especially ultra-fine particles.

Additionally, it can be stated that the total mass of the investigated samples for the NAO (1,695 mg) is higher than for the low steel pads (1,244 mg).

Finally, a closer look to the gravimetrical wear of disc and pads based on the results of one standard AK Master cycle should be given (Table 1).

Table 1: Wear results of complete AK Master cycle for NAO and Low Steel

	Weight Difference [g]	
	NAO	**Low Steel**
Disc	12	23
Inner Pad	23	12
Outer Pad	21	11

The NAO material reveals a very positive effect on the disc wear, where the mass loss is almost halved compared to the low steel. In terms of the pad wear on the other hand it can be stated that the wear of the NAO linings is almost doubled.

That means, that the NAO formulation helps to decrease disc wear, which could give an interesting hint for a convenient countermeasure to reduce brake particle emissions. However, this only works when the brake applications are not too stressful, meaning temperatures and brake pressures do not exceed certain values. In the case that high performance brake applications are considered, as it can be stated for the AK Master, an excessive pad wear is obtained as well as higher emissions of particle number and mass.

3.2 Effect of Hard-Metal Disc Coating

Hard-metal disc coatings became popular during the last two to three years. They show are very good performance in terms of gravimetrical wear for both friction partners. However, the question how this innovative friction couple performs in regards of particle emissions has still to be answered.

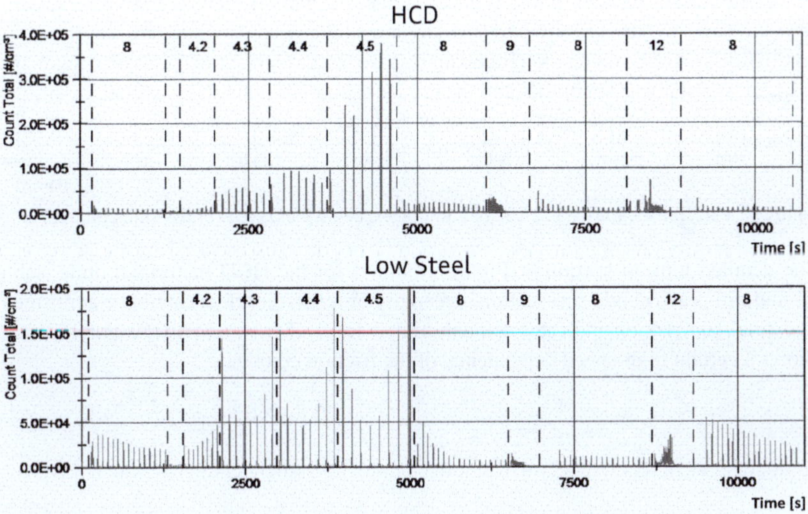

Figure 8: PNC of a low steel (below) and HCD brake lining (above), modified AK Master

In Figure 8 the time history of emitted particle number concentrations is given for the already introduced low steel material and the brake lining material (called HCD) specially adapted for the hard-metal coated disc.

It can be observed quite well that the HCD material shows very low emissions during section 8. For the remaining sections it seems that higher particle number concentrations are emitted, which should be investigated more in detail.

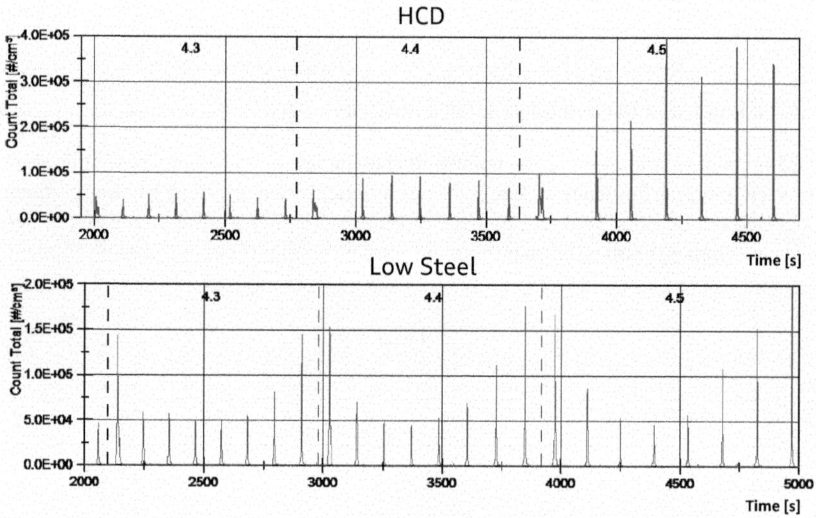

Figure 9: PNC for sections 4.3 to 4.5 for a low steel (below) and HCD brake lining (above)

The detailed look to section 4 (Figure 9) shows for the HCD material a quite stable trend during each single sub-section, a memory or reservoir effect cannot be confirmed. The final sub-section indicates a much higher level of emitted particle numbers and shows a certain high-speed dependency of the friction couple.

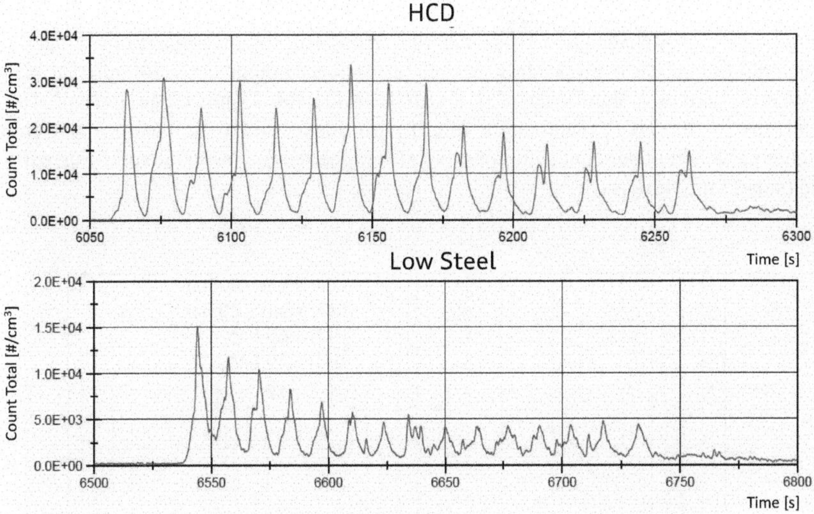

Figure 10: PNC of a low steel (below) and HCD brake lining (above) during a fade section

The particle number concentrations for the fade section (section 9) are depicted in Figure 10. The emission behaviour of the HCD is quite stable, even if for the first half of the section a higher level can be observed. In terms of the maximum values the HCD emits twice the particle numbers compared to the low steel material.

Finally, the particle size distribution for section 4.5 is investigated in Figure **11**, compared again for low steel and HCD material. While for low pressure applications (20 bar) both linings show low emissions in the particle size range up to 500 nm, for high pressure brakings a significant peak for HCD in the sub-23 nm range appears. This confirms for the HCD material for (high) performance brake applications not only higher particle number concentrations, but also higher emissions in the ultra-fine particle (sub-23 nm) range.

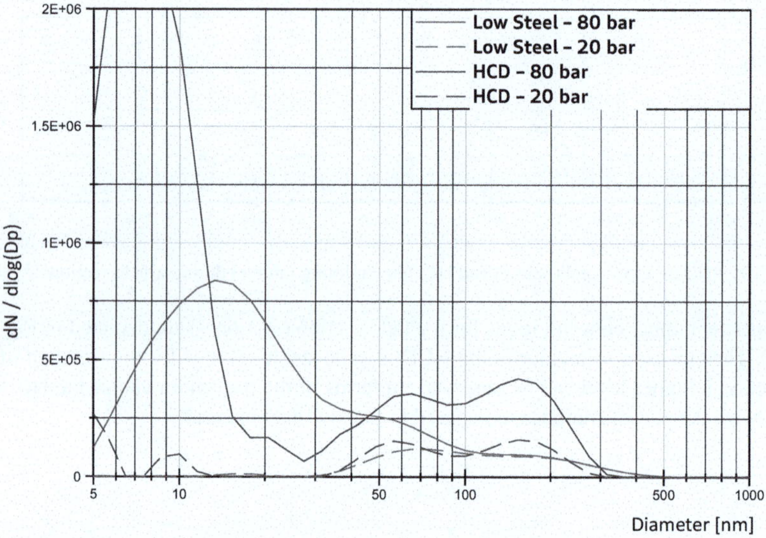

Figure 11: Particle size distribution of low steel and HCD for section 4.5

4 Summary and Conclusions

The authors introduce a proposal for a measurement setup. This considers the layout of the testing environment, which is represented by a full-scale brake dynamometer, an air conditioning / ventilation system and the sampling layout. Additionally, the used measurement devices to characterize particle number concentration and particle size distribution of particle number and mass are introduced. Another key to successful testing means the used test cycle, which is a modified AK Master. This procedure is a well-known industry standard, but modified to avoid excessive brake applications.

Using this measurement approach, a direct comparison between NAO and low steel brake lining materials is presented. The NAO formulation shows advantages during sections with lower braking power. Unlike to the low steel material, reservoir or memory effects are less impressive. High performance applications with high brake pressures and resulting temperatures lead to significant higher PNC than the low steel, also the amount of sub-23 nm particles increases impressively. Particle mass distributions shows for both linings a similar trend with a mono-modal characteristic between 1 µm and 3 µm. However, the NAO emits higher amounts of particle mass, what also can be seen from the gravimetrical wear behavior. On the other hand underline those results the positive effect of NAO compositions on the disc wear. It can be concluded, that NAO brake pads can help to reduce the emission of particle numbers and mass as well, when high temperatures (resulting of high brake pressures and velocities) could be avoided, for instance when considering urban scenarios.

The investigated innovative hard-metal disc coating in combination with a special dedicated lining material shows also very positive effects in terms of lower particle number concentrations for low performance brake applications. However, when brake pressure, speed and temperature are rising, also the emission of particles, also sub-23 nm particles, increases significantly. Since this kind of frictions couples is still quite new and not entirely understood yet, further investigations are strictly recommended

5 References

[1] Bukowiecki et al.: PM10 emission factors of abrasion particles from road traffic (APART). Swiss Association of Road and Transportation Experts (VSS), 2009

[2] Amato et al.: Traffic induced particle resuspension in Paris: Emission factors and source contributions, Elsevier Ltd., Atmospheric Environment 129 (2016) 114-124, 2016, doi: 10.1016/j.atmosenv.2016.01.002.

[3] Grigoratos, T., Martini, G: Non-exhaust traffic related emissions – Brake and tyre wear PM, Publications Office of the European Union, 2014, doi: 10.2790/21481

[4] Gramstat, S., Waninger, R., Lugovyy, D., Schröder, M. Investigations of Brake Particle Emissions – Testing Method, Vehicle Peculiarities and Friction Material Influence, FISITA, EB2017-VDT-016, proceedings Eurobrake 2017, Dresden

[5] Gramstat, S, Waninger, R., Lugovyy, D., Schröder, M.: Investigation possibilities of brake particle emissions on a brake dynamometer, 8th International Munich Chassis Symposium 2017: chassis.tech plus, pp. 629-653, 2017, doi: 10.1007/978-3-658-18459-9_44

[6] N.N., "Dynamometer Global Brake Effectiveness Superseding J2522 JUN2003", standard, SAE International, Warrendale, 2003

Real driving emissions measurement of brake dust particles

Prof. Klaus Augsburg, David Hesse, Toni Feißel, Felix Wenzel
Department of Automotive Engineering, TU Ilmenau

P. Pfeffer (Ed.), *9th International Munich Chassis Symposium 2018*, Proceedings,
https://doi.org/10.1007/978-3-658-22050-1_45

Real Driving Emissions Measurement of Brake Dust Particles

Department of Automotive Engineering
Prof. Klaus Augsburg,
David Hesse, Toni Feißel, Felix Wenzel

TECHNISCHE UNIVERSITÄT
ILMENAU
Department of Automotive Engineering

18.05.2018 Seite 1 www.tu-ilmenau.de TECHNISCHE UNIVERSITÄT ILMENAU

Outline

1. Motivation
2. RDE Requirements
3. Methodology of CVS Development
   3.1. CFD Modeling
   3.2. Validation Methods
4. RDE Measurement Results
5. Conclusions
6. Outlook

18.05.2018 Seite 2 www.tu-ilmenau.de TECHNISCHE UNIVERSITÄT ILMENAU

Motivation

Inertia Brake Dyno

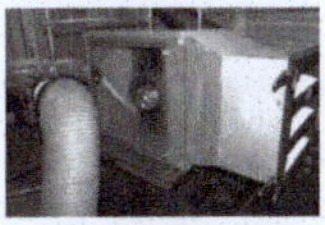

- *reproducible load parameters and environmental conditions*
- *stationary emission measurement systems*

Chassis Dynamometer

- *reproducible load parameters and environmental conditions*
- *maneuver based RDE-testing*
- *mobile (PEMS) and stationary emission measurement systems*

Road Tests / RDE

- *dynamic driving conditions*
- *investigation of driver influence*
- *mobile measurement systems (PEMS)*
- *influence of different road conditions*

- realistic emission behaviour
- complexity / poor reproducibility
- investigation costs

18.05.2018 Seite 3 www.tu-ilmenau.de TECHNISCHE UNIVERSITÄT ILMENAU

RDE Requirements

Aim: On-Road-Measurement of PN and PM
(#/cycle-km and mg/cycle-km)

High Measurement Efficiency

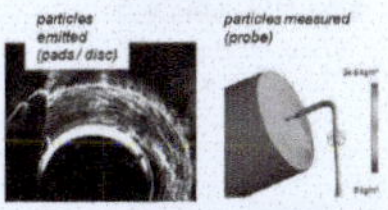

- *function of sampling- and transport efficiency*

$$\eta_M = \frac{p_{evacuated}}{p_{emitted}}$$

High Reproducibility

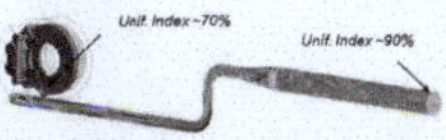

- *well premixed aerosol*
- *Steady air flow*
- *PN & PM concentration profile homogenously distributed across sampling plane*

Minimized Background Concentration

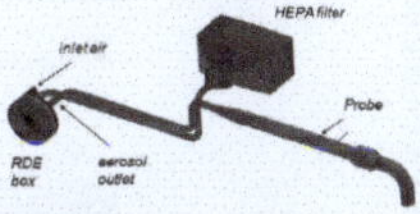

- *decoupling from the environment*
- *fully enclosed system*
- *inlet air filter required*

18.05.2018 Seite 4 www.tu-ilmenau.de TECHNISCHE UNIVERSITÄT ILMENAU

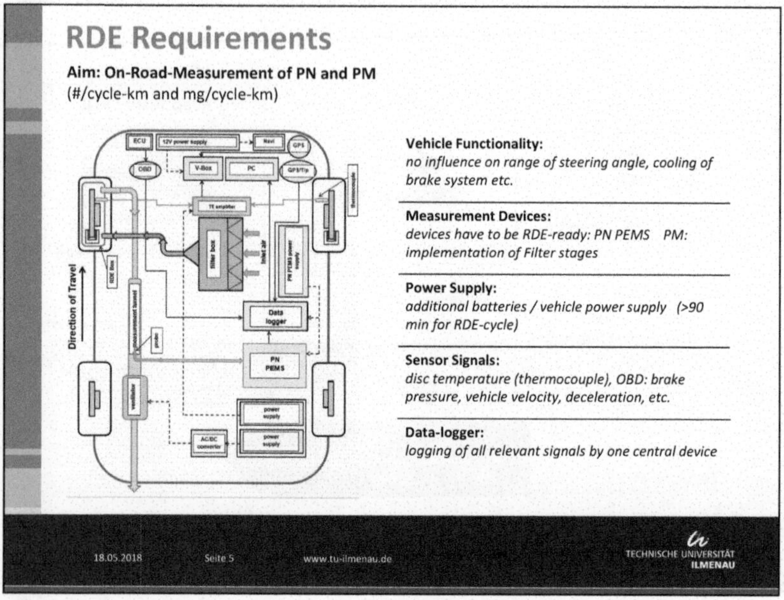
RDE Requirements
Aim: On-Road-Measurement of PN and PM
(#/cycle-km and mg/cycle-km)
ECU
12V power supply
Navi
GPS
OBD
V-Box
PC
GPS/Tip
TE amplifier
filter box
inlet air
PN PEMS power supply
RDE Box
Direction of Travel
measurement tunnel
probe
Data logger
PN PEMS
ventilator
power supply
AC/DC converter
power supply
Vehicle Functionality:
no influence on range of steering angle, cooling of brake system etc.
Measurement Devices:
devices have to be RDE-ready: PN PEMS PM: implementation of Filter stages
Power Supply:
additional batteries / vehicle power supply (>90 min for RDE-cycle)
Sensor Signals:
disc temperature (thermocouple), OBD: brake pressure, vehicle velocity, deceleration, etc.
Data-logger:
logging of all relevant signals by one central device
18.05.2018
Seite 5
www.tu-ilmenau.de
TECHNISCHE UNIVERSITÄT ILMENAU

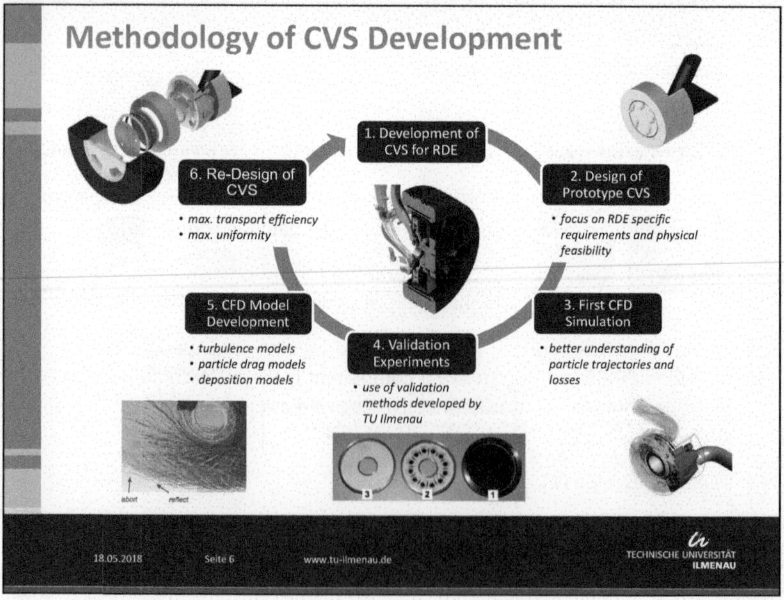
Methodology of CVS Development
1. Development of CVS for RDE
2. Design of Prototype CVS
• focus on RDE specific requirements and physical feasibility
3. First CFD Simulation
• better understanding of particle trajectories and losses
4. Validation Experiments
• use of validation methods developed by TU Ilmenau
5. CFD Model Development
• turbulence models
• particle drag models
• deposition models
6. Re-Design of CVS
• max. transport efficiency
• max. uniformity
abort
reflect
3
2
1
18.05.2018
Seite 6
www.tu-ilmenau.de
TECHNISCHE UNIVERSITÄT ILMENAU

CFD Modeling

Main influence factors for transport losses and particle uniformity:

- *volume flow*
- *direction of evacuation*
- *disc rotational speed*

Rotational speed is the only factor that varies during measurement:

- *determining initial particle velocity*
- *favors additional particle mixing*
- *effects particle deposition*

Challenge: *Indirect proportional correlation between particle mixing and transport efficiency.* → **compromise needed**!

Complex geometries require complex CFD modeling!

Variation of Air Volume Flow

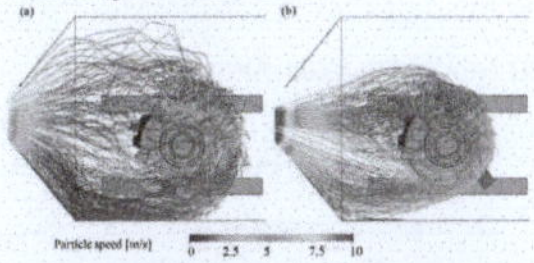

Variation of Evacuation Direction

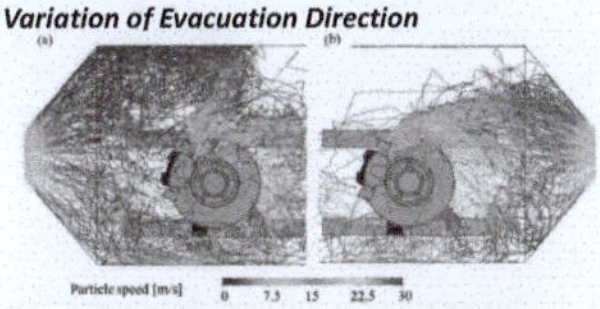

Variation of Disc Rotation Speed

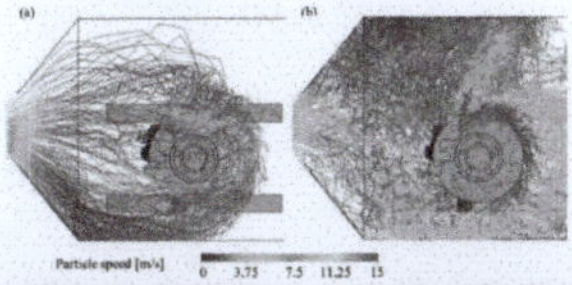

18.05.2018 Seite 7 www.tu-ilmenau.de TECHNISCHE UNIVERSITÄT ILMENAU

CFD Modeling

Design of Prototype CVS / First CFD Simulation

Design of Prototype

- *fully enclosed brake system (complete collection of brake dust)*
- *difficult geometrical conditions in proximity to rotating components (brake disk, caliper, rim)*

First CFD Simulation

- *determination of efficient air-inlet and aerosol-outlet positions and geometries*
- *air-inlet: backside of caliper (avoiding dead air spaces)*
- *aerosol-outlet: radial evacuation with offset of not more than 90° towards caliper in direction of rotation*
- *sampling plane position – highly uniform aerosol flow*

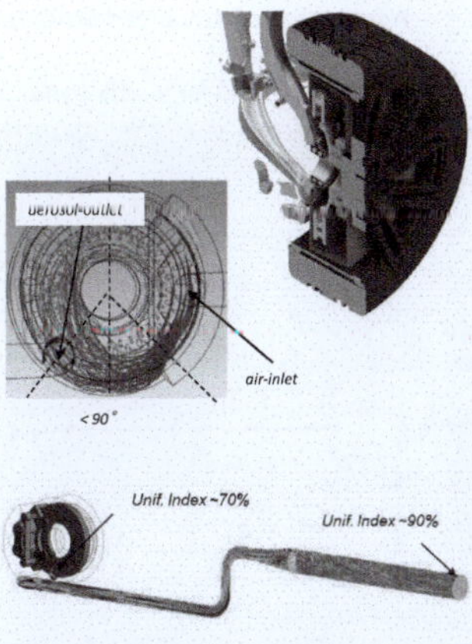

18.05.2018 Seite 8 www.tu-ilmenau.de TECHNISCHE UNIVERSITÄT ILMENAU

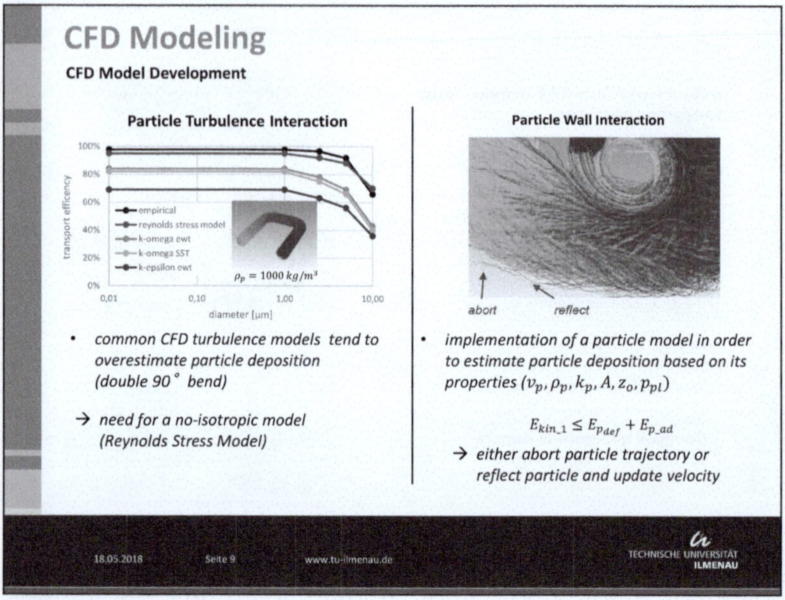
CFD Modeling
CFD Model Development
Particle Turbulence Interaction
100%
80%
60%
40%
20%
0%
transport efficiency
empirical
reynolds stress model
k-omega ewt
k-omega SST
k-epsilon ewt
$\rho_p = 1000\ kg/m^3$
0,01
0,10
1,00
10,00
diameter [µm]
Particle Wall Interaction
abort
reflect
• common CFD turbulence models tend to overestimate particle deposition (double 90° bend)
→ need for a no-isotropic model (Reynolds Stress Model)
• implementation of a particle model in order to estimate particle deposition based on its properties ($v_p, \rho_p, k_p, A, z_o, p_{pl}$)
$E_{kin\_1} \leq E_{p_{def}} + E_{p\_ad}$
→ either abort particle trajectory or reflect particle and update velocity
18.05.2018
Seite 9
www.tu-ilmenau.de
TECHNISCHE UNIVERSITÄT ILMENAU

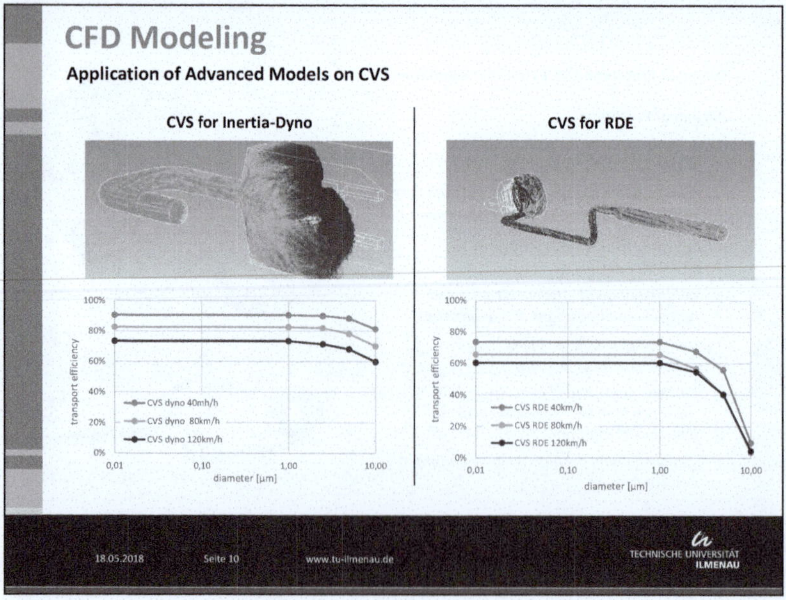
CFD Modeling
Application of Advanced Models on CVS
CVS for Inertia-Dyno
CVS for RDE
100%
80%
60%
40%
20%
0%
transport efficiency
CVS dyno 40mh/h
CVS dyno 80km/h
CVS dyno 120km/h
0,01
0,10
1,00
10,00
diameter [µm]
CVS RDE 40km/h
CVS RDE 80km/h
CVS RDE 120km/h
18.05.2018
Seite 10
www.tu-ilmenau.de
TECHNISCHE UNIVERSITÄT ILMENAU

Validation Methods

1. Mass Balance of Deposited Particles

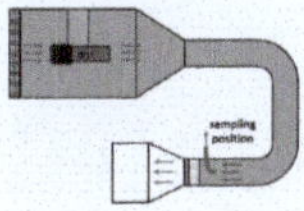

- *filter based methodology*
- *verification of the transport efficiency (PM) of a sampling system*
- *analysis of Influence factors on the transport efficiency*

2. Verification of the Concentration Profile

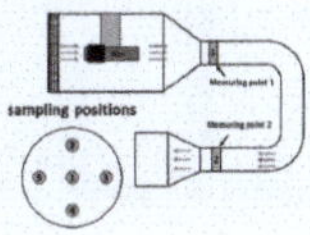

- *verification of the size-resolved number concentration and uniformity*
- *analysis of influence factors on the uniformity*

3. Verification of Particle Deposition Mechanisms

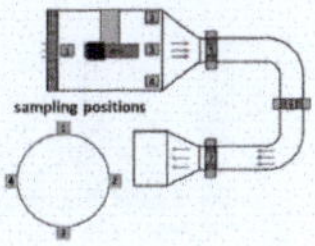

- *analysis of the uniformity of number concentration (PN)*
- *analysis of particle loss mechanisms*
- *analysis of the influence of material properties (electrostatics) or surface texture (roughness)*

Validation Methods

1. Mass Balance of Deposited Particles

Verification of the transport efficiency of a sampling system - weighing of deposited particle mass (without size resolution)

Methodology:

I. *integration of an outlet filter (fine dust - class M5) at the sampling position*
II. *execution of the test-cycle as specified (blower activated - evacuation of emitted particles)*
III. *determination of mass increase of the loaded outlet filter (Nr. 1)*
IV. *evacuation of deposited particles by compressed air (blower activated) and integrated filter (Nr. 2 – 5)*
V. *weighing of mass increase of the filter stage per section as well as loss of disk and pads*
VI. *calculation of the mass-based transport efficiency*
VII. *comparison of deposited particle mass vs. evacuated particle mass*

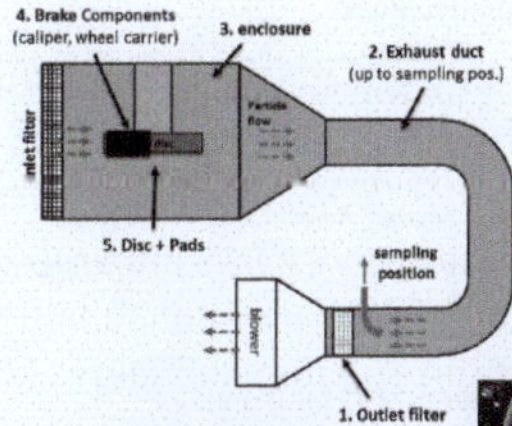

particle deposition inside CVS for RDE

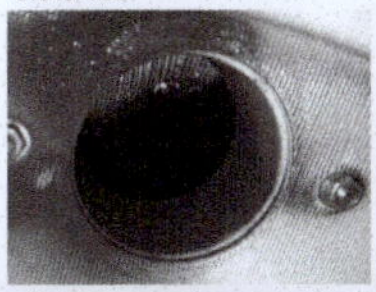

Validation Methods

1. Mass Balance of Deposited Particles

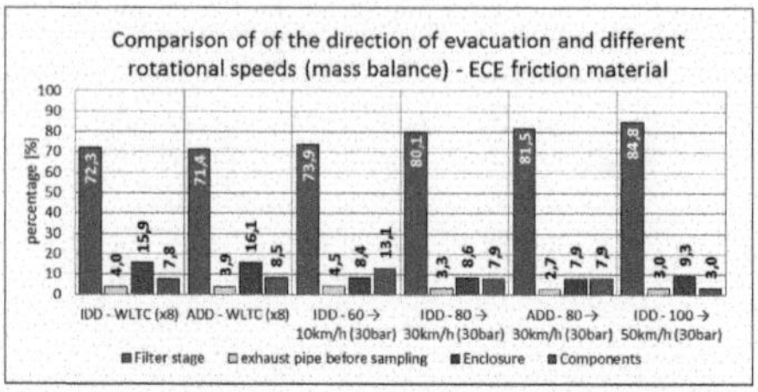

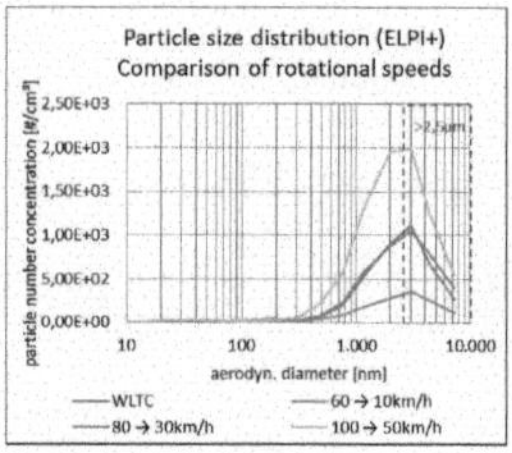

Findings (CVS for Dyno):

- *mass-based transport efficiency depends on the rotational speed and friction energy*
  - *change of the PSD with increase of the friction energy (red. of the CMD)*
  - *increase in efficiency of evacuation for smaller particle diameter*
- *deposition with variation of suction direction depends on the initial speed*
- *deposition at the components (disc caliper) depends on the rotational speed and the disc ventilation*

18.05.2018 Seite 13 www.tu-ilmenau.de TECHNISCHE UNIVERSITÄT ILMENAU

Validation Methods

2. Verification of the Concentration Profile

Methodology:

I. *sampling (isokinetic / isoaxial) at different points in the exhaust duct (position depends on the measuring target)*

II. *size resolved number concentration over the cross section: 4 positions offset by 90 °*

III. *detection at the central point (5) serves as benchmark*

Measurement device: Dekati ELPI+
Electrical low pressure impactor
Size range: 6 – 10.000nm

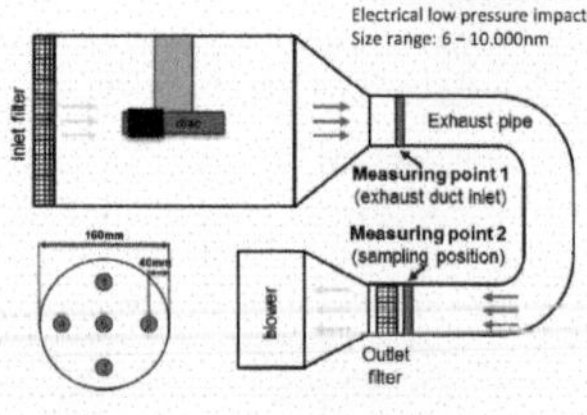

Findings (CVS for Dyno):

before double deflection
uniformity: 80,9%

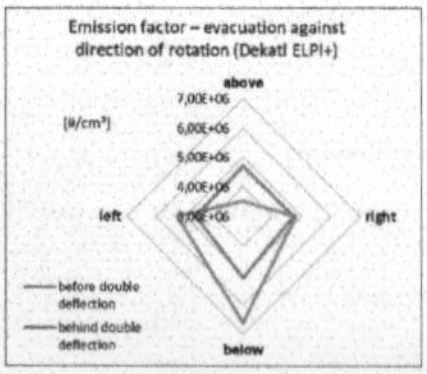

after double deflection
uniformity: 91,4%

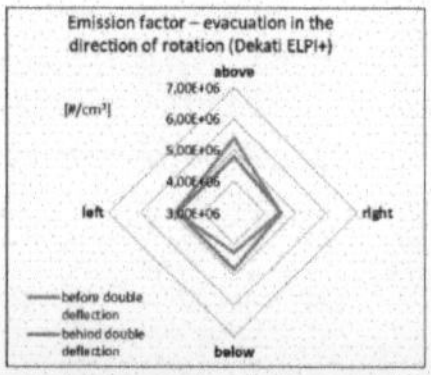

18.05.2018 Seite 14 www.tu-ilmenau.de TECHNISCHE UNIVERSITÄT ILMENAU

Validation Methods

3. Verification of Particle Deposition Mechanisms

Methodology:

I. *adaptation of sampling pads in the duct*

II. *analysis of particle size distribution at different positions: 2x4 pads offset by 90 °*

III. *execution of the test-cycle as specified*

IV. *removal of the sampling plates and performing a microscopic analysis (1000x)*

V. *analysis of microscopic images by software (e.g. scientific counter) - determination of the particle area and calculation of an equivalent diameter*

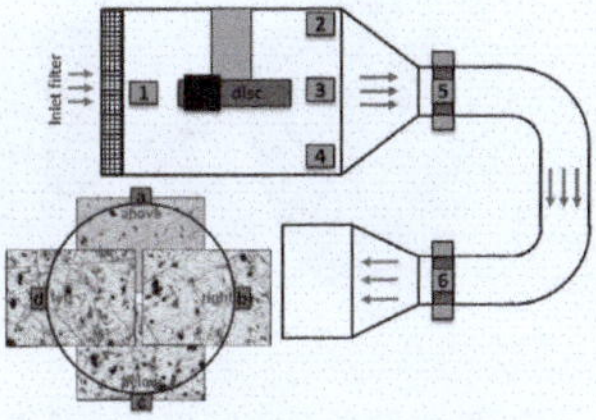

5.

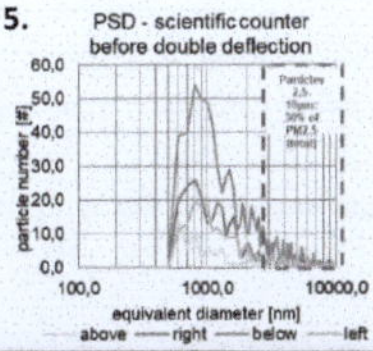

6.

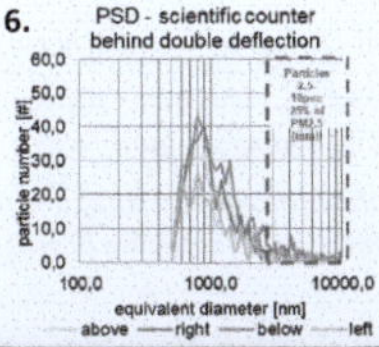

proportion of PM10 compared to PM2.5		
	before double deflection	behind double deflection
above	18,8	9,8
right	24,6	26,8
below	33,6	36,3
left	30,9	22,0
total	**30,0**	**25,4**

18.05.2018 Seite 15 www.tu-ilmenau.de TECHNISCHE UNIVERSITÄT ILMENAU

Validation Methods

1. Mass Balance of Deposited Particles

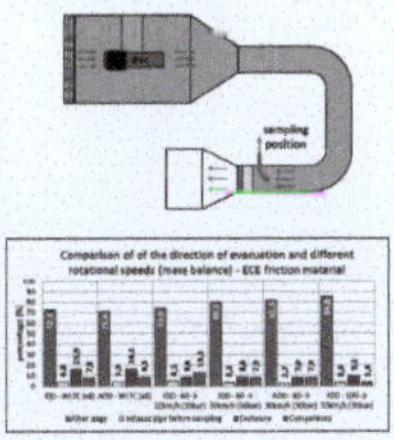

2. Verification of the Concentration Profile

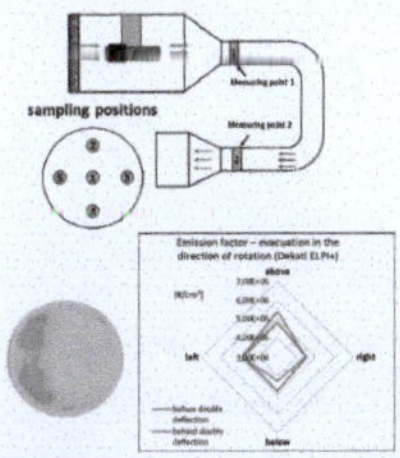

3. Verification of Particle Deposition Mechanisms

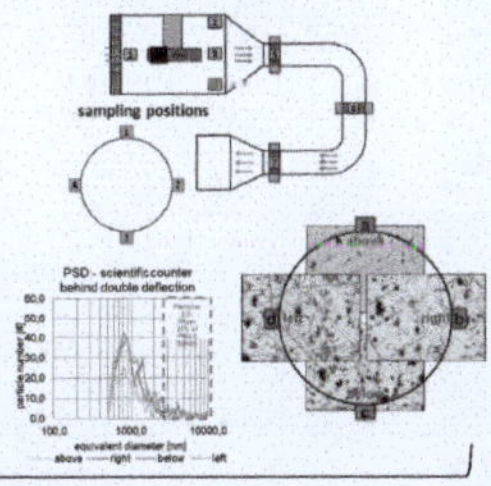

improvement and verification of CFD models

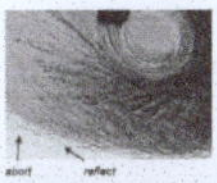

improvement of CVS design

18.05.2018 Seite 16 www.tu-ilmenau.de TECHNISCHE UNIVERSITÄT ILMENAU

RDE Measurement Results

Brake Dust Measurement under Real Drive Conditions

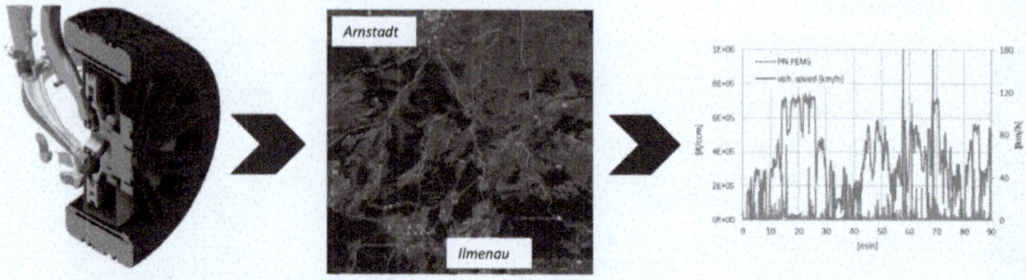

- *Implementation of CVS for RDE to test vehicle*
- *execution of test drives (test cycle which fulfills RDE norm requirements)*
- *first RDE measurement with PN PEMS - characteristic emission peaks depending from the intensity of the braking event (city, highway, motorway)*
- *need for more detailed information (brake pressure, deceleration rate, brake disk temperature) for a better understanding of the particle emission process*

18.05.2018 Seite 17 www.tu-ilmenau.de TECHNISCHE UNIVERSITÄT ILMENAU

RDE Measurement Results

Brake Dust Measurement under Real Drive Conditions

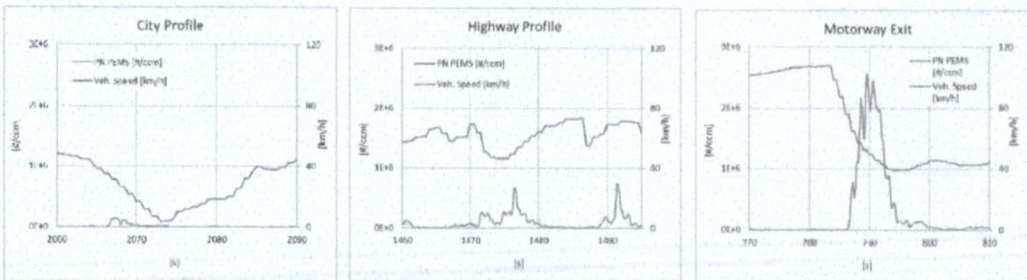

- *Implementation of CVS for RDE to test vehicle*
- *execution of test drives (test cycle which fulfills RDE norm requirements)*
- *first RDE measurement with PN PEMS - characteristic emission peaks depending from the intensity of the braking event (city, highway, motorway)*
- *need for more detailed information (brake pressure, deceleration rate, brake disk temperature) for a better understanding of the particle emission process*

18.05.2018 Seite 18 www.tu-ilmenau.de TECHNISCHE UNIVERSITÄT ILMENAU

Conclusions

Aim: On-Road-Measurement of PN and PM
(#/cycle-km and mg/cycle-km)

1. *definition of RDE-specific requirements for brake dust measurement is necessary*
2. *the transport efficiency of a sampling systems (CVS) depends from various influence factors*
3. *application, improvement and validation of CFD models is necessary for the development of a CVS*
4. *by use of the methods presented, the transport efficiency of different sampling systems (mobile or stationary) can be verified*
5. *first measurement results on the road give an indication of RDE brake dust emission characteristics*

18.05.2018 Seite 19 www.tu-ilmenau.de TECHNISCHE UNIVERSITÄT ILMENAU

Outlook

Aim: On-Road-Measurement of PN and PM
(#/cycle-km and mg/cycle-km)

CVS Development

1. *a validation method for size resolved transport efficiency has to be developed*
2. *findings regarding brake dust particle deposition mechanisms and particle properties can be used to improve CFD models*
3. *by using advanced models the CVS design can be enhanced*

RDE Measurement

1. *need for more detailed information (brake pressure, deceleration rate, brake disk temperature etc.)*
2. *Implementation of a corresponding RDE cycle to the dynamometer:*
   - *verification of on-road measurement results*
   - *application of stationary measurement devices*

18.05.2018 Seite 20 www.tu-ilmenau.de TECHNISCHE UNIVERSITÄT ILMENAU

Thank You for Your Attention!
Any open questions?
Please be so kind to ask or contact us!
Univ.-Prof. Dr.-Ing. Klaus Augsburg
klaus.augsburg@tu-ilmenau.de
Tel: +49 3677 69-3842
https://www.tu-ilmenau.de/kft/
Studieren mit
besten Aussichten
TECHNISCHE UNIVERSITÄT
ILMENAU
Department of Automotive Engineering
18.05.2018
Seite 21
www.tu-ilmenau.de
TECHNISCHE UNIVERSITÄT
ILMENAU

Low emission brakes – can the friction brake still be saved?

Donatus Neudeck
Head of Brake Development
Hydraulic & Actuation Systems
Dr. Ing. h.c. F. Porsche AG

P. Pfeffer (Ed.), *9th International Munich Chassis Symposium 2018*, Proceedings,
https://doi.org/10.1007/978-3-658-22050-1_46

1 Motivation

Urbanisation and the demand for mobility are increasing in society worldwide. Due to the ever increasing road traffic volumes and the associated environmental impact, technical solutions are required in order to make vehicles more environmentally compatible.

The limit values for particulate pollution are often exceeded, particularly in large cities. The focus of the associated public debate and, above all, options for reducing particulate pollution are now turning to areas such as tyres and brakes, which have received little attention to date.

Particulate matter are microscopically small solids, which are categorised according to various sizes in conjunction with air quality control. The most common definition originates from the US EPA (Environmental Protection Agency): the PM standard. Simplified, the particulates are differentiated according to particle size, for example PM_{10} (aerodynamic diameter 10 micrometers) or the significantly smaller $PM_{2.5}$.

The conventional method of braking a vehicle is by means of the friction brake. Here, the vehicle's kinetic energy is converted into heat. This also results in wear or abrasion of the brake pad and disc/drum friction partners. A portion of this abrasion becomes visible as unsightly soiling of the wheel rims. A further portion enters the environment as dust and particulate matter.

So what can be done in order to achieve low-emission and clean braking in the future? Does the friction brake still have a right to exist in future vehicles? Are alternative technologies available now or in the near future?

2 Overview

In this paper, options for the reduction of brake dust and its emission are presented and the implications of these for the conventional friction brake are outlined.

A distinction is made between the following approaches:

Friction-based brakes and particulate collection

Drum brakes

Brake dust collectors: e.g. Mann+Hummel, Tallano

Friction-based low-wear brakes (tribological approaches to the friction brake)

Low-wear brake lining compounds (so-called NAOs, common in the USA)

Carbon-fibre reinforced ceramic brake discs made from C/SiC, e.g. Porsche Ceramic Composite Brake (PCCB)

Coating of cast iron discs with low-wear substances, e.g. Porsche Surface Coated Brake (PSCB)

Non-friction-based brakes

Use of the braking torque generated by the E-drive through recuperation (on hybrid vehicles, so-called PHEVs and electric-motor-driven vehicles, so-called BEVs)

Avoidance of braking

Traffic recognition/guidance – not considered in this paper

3 Friction-based brakes and particulate collection

3.1 The drum brake

Those presumed dead live longer. Long before the triumph of the disc brake in motor vehicles, drum brakes were used as service and parking brakes. With this brake design, most of the generated dust remains within the drum and therefore does not constitute a direct environmental impact. When changing the brake linings, removal of the drum enables access to the brake dust, which can then be disposed of. In current vehicle development, the drum brake is again playing a more significant role, even in vehicle segments that have been the preserve of the disc brake over the last 20 years. In BEVs with reduced top speed, the brake performance requirements at the rear axle can also be fulfilled by drum brakes. At the same time, the drum protects the friction pairings against dirt and corrosive media. This is especially advantageous on BEVs as the friction brake is used significantly less often, which increases the risk of corrosion-induced functional impairment.

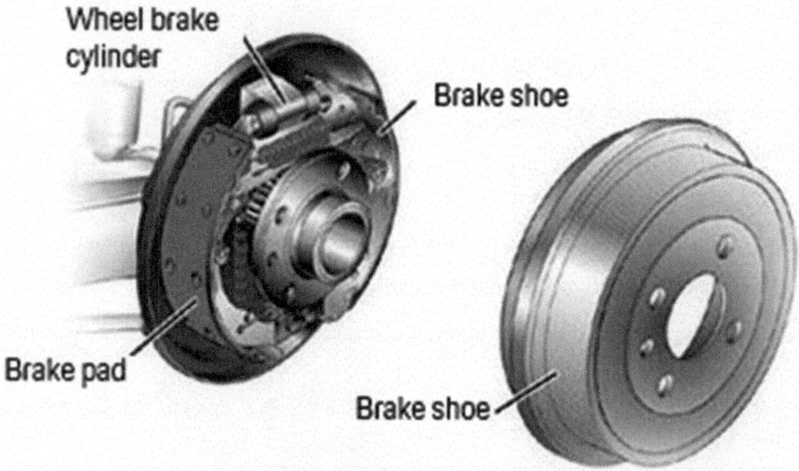

Image 1: Drum brake Trommelbremse – Räder & Fahrwerk – steinietrabi-skoda-fabia.de

3.2 Brake dust collectors

According to the current state of technology, no other production-ready solution is available for the collection of brake dust as an alternative to the drum brake. However, there are some companies and universities that are working on this topic. Numerous patents have already been registered on the subject of "brake dust collection".

In the already expired US5035304 patent, a device is described which is located on both sides of the brake disc and wipes the brake dust off the brake disc. This is collected in a container that retains the brake dust by means of a filter.

Patent DE202005006844U1 describes encapsulation of the brake disc. The housing is provided with an air inlet and outlet. The air guidance in the housing is arranged so that the abrasion is taken up by the air flow. The abrasion is subsequently collected through a filter.

Further patents, for example DE102006051972A1, describe a brake-dust-collection device that is directly connected to the brake calliper in the main rotational direction of the brake disc. The brake dust particles are released in this area and are best collected there.

Tallano Technologie is a French start-up which is active in the area of brake-dust collection. Using a suction device located behind the brake calliper at the brake disc, the brake dust is allegedly reduced by about 90%.

Image 2 shows examples of devices for collecting brake dust.

2016-01-1949
Published 09/18/2016
Copyright © 2016 SAE International
doi:10.4271/2016-01-1949
saepcmech.saejournals.org

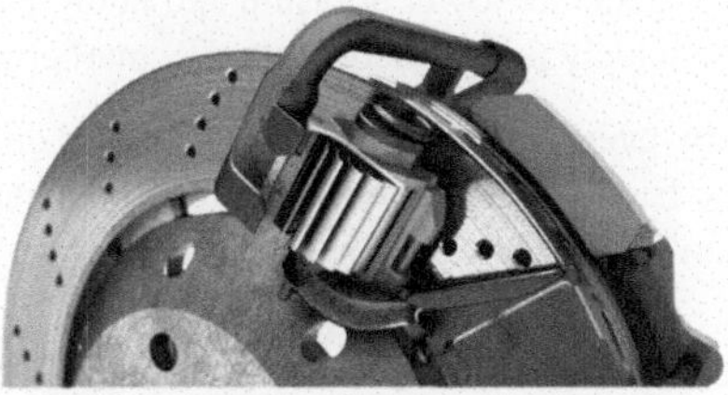

https://www.kfz.net/autonews/feinstaub-staubsauger

www.mann-hummel.com

Image 2: Examples for brake dust collectors

Conclusion: The presented solutions can collect up to around 90% of the brake dust volume, significantly reducing brake-induced particulate emissions. However, restrictions with regard to installation space, weight, costs and the negative influence on brake ventilation prevent widespread use in production according to current assessments.

4 Friction-based low-wear brakes

4.1 Use of low-wear brake lining compounds

A further option for reducing particulate emissions is the use of brake lining compounds that generate lower abrasion and additionally reduce disc wear. Linings in this quality are widely propagated on the US market as so-called NAOs (Non Asbestos Organics).

However, these linings are not yet used on the European market as the performance (i.e. the frictional behaviour at high component temperatures) falls significantly short of the current technological standard in Europe.

Conclusion:

The trade-off between brake performance and brake dust cannot be solved with the lining compounds known to date. The use of low-wear lining compounds is therefore not an adequate approach for reducing particulate emissions on the European market. This not only applies to the extremely high demands of Porsche on the performance of brake linings, but also to all European vehicle manufacturers.

4.2 Carbon-fibre reinforced ceramic brake discs made from C/SiC
e.g. Porsche Surface Coated Brake (PSCB)

Image 3: PCCB (Porsche)

The use of C/SiC brake discs, which are characterised by very high surface hardness and are therefore extremely low-wear, promises a significant reduction in brake dust emissions (approximately 90. In contrast to the use of NAOs, the trade-off between brake dust and brake performance has been solved here, as the ceramic discs are combined with high-performance brake linings.However, this technology also has its pitfalls. The market penetration is currently restricted to use in high-performance sports cars and premium and luxury-segment vehicles by the continuing high costs.

4.3 Coating of cast iron discs with low-wear substances
e.g. Porsche Surface Coated Brake (PSCB)

Similarly to the C/SiC brake discs, carbide-coated cast iron discs are characterised by a very high surface hardness and therefore wear resistance. The effectiveness with regard to reducing particulate emissions is consequently on a comparable level.

Image 4 shows the comparison of particle sizes and particle count for the iDisc (Bosch Buderus carbide-coated cast iron brake disc used at Porsche as component of the PSCB) and a conventional cast iron brake system. The comparison was made on the basis of bench tests at the Technical University of Ilmenau. In the case of the iDisc, the number of particles is lower by a factor of ten for all relevant particle sizes. The brake dust volume can therefore be considerably reduced through the use of carbide-coated cast iron brake discs. The significantly lower costs compared with the C/SiC discs offer potential for a higher market penetration.

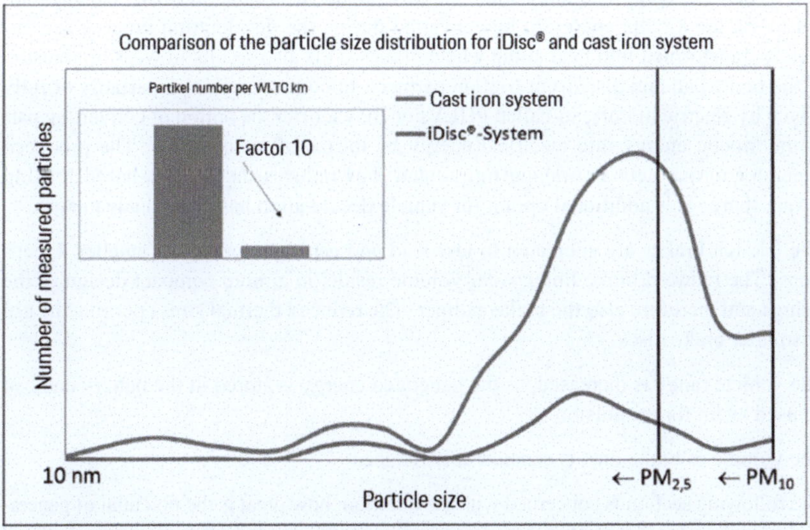

Image 4: Comparison particle size and number for iDisc and cast iron brake

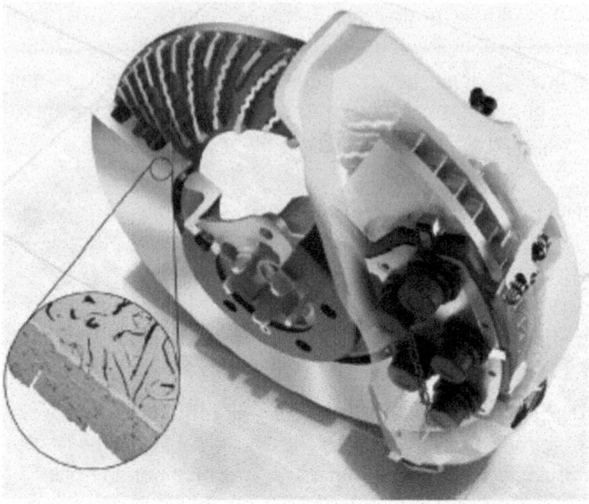

Image 5: PSCB (Porsche)

5 Non-friction-based brakes

On vehicles driven by combustion engines, apart from the intrinsic deceleration (<1 m/s²), the kinetic energy to be converted during the deceleration process is completely transformed into heat at the wheel brakes. This also results in wear or abrasion of the brake pad and disc/drum friction partners. In contrast, vehicles partially or fully driven by electric motors, so-called PHEVs or BEVs, offer the option of converting part of the kinetic energy into electrical energy by means of recuperation. The produced generator torque acts as braking torque and thus reduces the friction-based braking torque share. This additional option for vehicle deceleration has several advantages.

The friction brakes are subjected to less load and can therefore be of smaller dimensions. The reduced brake lining wear volume results in a more compact design of the linings and therefore also the brake calliper. The reduced thermal stress permits the use of smaller brake discs.

The vehicle range is increased, as the generated energy is stored in the battery and can be used again for propulsion.

The volume of brake dust is considerably reduced.

The following section is concerned with the question: how great is the potential of generator-based as opposed to friction-based deceleration with regard to the particulate reduction?

5.1 Brake torque curves during the deceleration process

If an electric motor is used for deceleration in a PHEV or BEV, the brake torque curve as a function of rotational speed is of decisive importance. In conjunction with the transmission ratio/gear selection, a potential is realised, which is mainly dependent on vehicle speed.

Image 6 shows the simplified generator and hydraulic braking torque curves, the driving speed and driver braking request, in this case a constant deceleration. It is evident that the driver's request can be expressed as the sum of both torques.

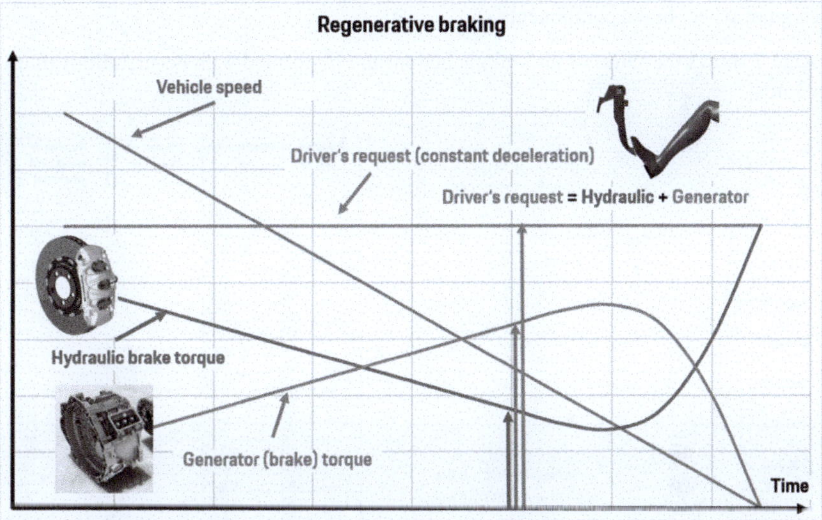

Image 6: Course of the braking torque and the driver's request torque

The non-constant hydraulic braking torque curve is made possible through the use of so-called blend-capable brake systems. Here, the required brake fluid volume is supplied and returned either directly to/from the wheel brakes or the pressure accumulator via the brake master cylinder depending on the generator torque and the driver's request. This occurs without the driver sensing it at the brake pedal.

A simpler option for using the generator torque for deceleration is so-called overrun recuperation. Here, the deceleration is initiated by releasing the accelerator pedal (similar to engine braking by shifting into a lower gear).

The power of the electric drive, the limitations of the blend-capable brake system, the battery as well as restrictions due to functional safety requirements restrict the maximum possible deceleration through recuperation to values of around 3 m/s². In specific cases, values of up to 5 m/s² are also possible.

The first good news for the friction brake: it cannot be completely dispensed with. However, for good braking distance results, decelerations of 12 m/s² are required.

5.2 Frequency distribution of braking decelerations

The specifications of the "AK-Bremse" (Brake Working Group) are often used in the design of wheel brake components.

Image 7 shows that 96% of all braking actions are performed with decelerations of up to 3 m/s².

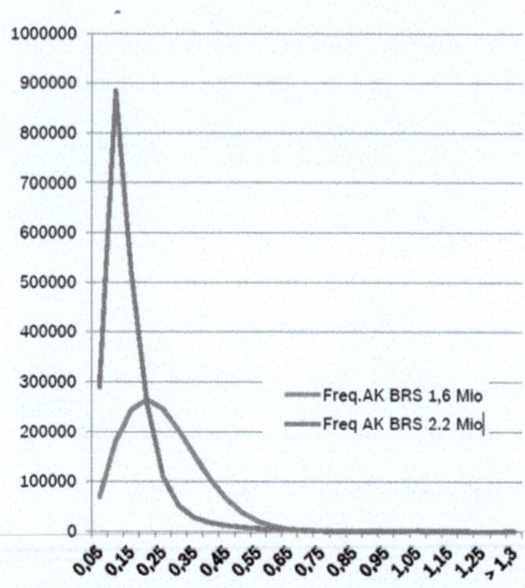

Pressure (bar)	Deceleration (g)	Number of brakings	Number of brakings at standstill
5	0,05	289602	84569
10	0,10	883633	243706
15	0,15	519370	115070
20	0,20	256359	56182
25	0,25	109616	25350
30	0,30	52059	16912
35	0,35	27234	12771
40	0,40	16978	9810
45	0,45	11686	7793
50	0,50	8036	5892
55	0,55	5837	4585
60	0,60	4219	3474
65	0,65	3184	2718
70	0,70	2550	2218
75	0,75	1811	1613
80	0,80	1158	1067
85	0,85	1072	995
90	0,90	980	917
95	0,95	883	831
100	1,00	780	739
105	1,05	672	640
110	1,10	559	534
115	1,15	440	421
120	1,20	315	301
125	1,25	185	175
130	1,30	114	105
> 130	> 1,3	668	612
Total		2200000	600000

Image 7: Cumulative frequencies of decelerations from AK specification 13.2

In the customer-relevant endurance runs performed at Porsche, the frequencies are even higher. Image 8 shows the evaluation of a noise endurance run in Spain. Here, over 99% of all braking actions are performed with decelerations of up to 3 m/s².

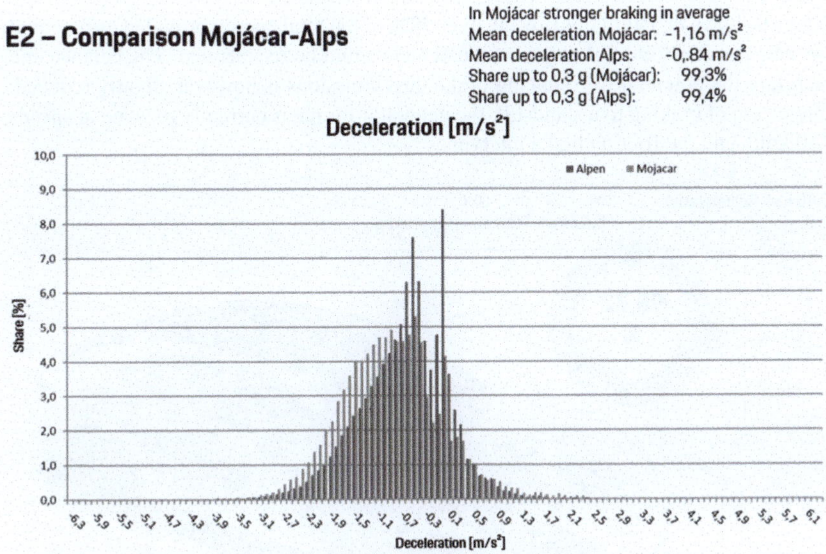

Image 8: Cumulative frequency in the Porsche noise endurance run

Based on the expected decelerations, virtually all braking actions could be performed via electric motors in customer-relevant operation. In PHEVs and BEVs, the friction brake is therefore mainly necessary for emergency braking, braking from high speeds and braking actions with high braking performance requirements.

5.3 Energy balance in vehicles driven by electric motors

For determination of the potential for particulate reduction through the use of electric vehicles, the simplified approach assumes a linear relationship between lining/disc wear and converted energy in customer-relevant braking manoeuvres. This is documented by numerous investigations, which are not mentioned individually here, as well as by brake lining, brake calliper and vehicle manufacturers. Braking manoeuvres with high thermal and mechanical loading are not considered.

In 2017, Bosch investigated a new blend-capable brake system on a Porsche Panamera E-Hybrid test vehicle. Data recordings included the drive and braking energy. This involved a customer-relevant trip on public roads over a distance of approximately 60 km and lasting around one hour. Speeds of up to 100 km/h and decelerations of up to 4 m/s² were achieved.

Image 9 shows that with this profile, over 98% of the total energy provided as potential by the generator for deceleration of the vehicle could be recuperated. The balance and therefore the potential for reducing brake dust emissions is however slightly reduced. In the case of braking to a standstill, the generator torque is currently reduced at speeds <10 km/h and the friction brake intervenes.

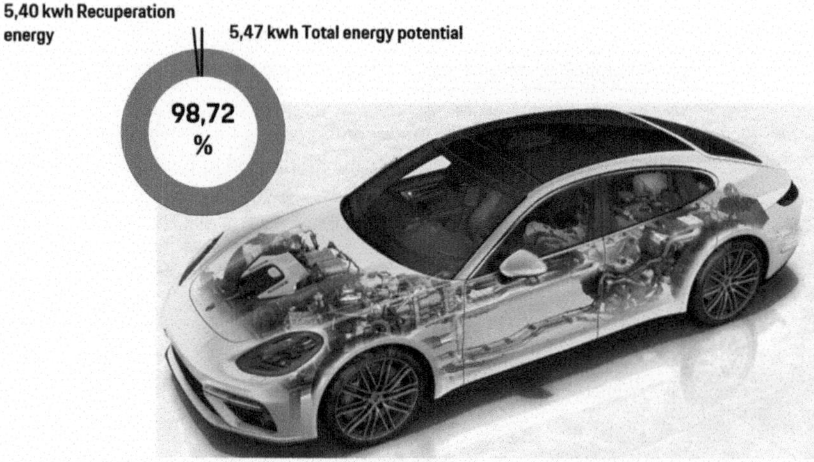

Image 9: Energy balance of a Panamera S e-hybrid in a customer relevant cycle

Conclusion: Electric-motor-driven vehicles offer great potential for reducing brake dust emissions.

6 Conclusion

Does the friction brake still have a future in motor vehicles if low-emission deceleration is desired? The answer is yes, but!

Technical options exist for friction-based and therefore brake-dust generating braking, but with collection of the abrasion. The good old drum brake is an example of this. Brake dust collectors for disc brakes are under investigation. A production-ready solution is however still not on the horizon.

Friction-based, low-wear braking is possible with special brake lining compounds, which however cannot be used on the European market due to inferior performance. In contrast, brake systems with C/SiC brake discs or carbide-coated cast iron discs could resolve this trade-off. The PCCB and PSCB from Porsche are technical solutions that have entered production.

Low-emission deceleration through friction braking is therefore absolutely possible. Up to 90% reductions in brake dust are possible. However, the consequence is additional technical effort and the associated higher costs.

Non-friction-based braking is the preserve of PHEVs and BEVs. Here, the electric drive is operated as a generator for deceleration. This method of braking is completely emission free and has the additional advantage that the kinetic energy of the vehicle is not converted into frictional heat, but intelligently into electric energy through recuperation.

However, the friction brake is also not obsolete in these vehicles. At decelerations >3 m/s^2 as well as in the case of emergency braking and braking actions with high performance requirements, the friction brake is the medium of choice.

Donatus Neudeck
Head of Brake Development
Hydraulic & Actuation Systems
Dr. Ing. h.c. F. Porsche AG
Porschestraße 911
71287 Weissach

BRAKE CONTROL SYSTEMS AND VIBRATION

Modal analysis of brake discs – effect of wear on the frequency spectrum

Dipl.- Ing. Lutz Pander
Univ.-Prof. Dr.-Ing. Ralph Mayer
Professorship Vehicle System Design,
Institute for Automotive Research,
Chemnitz University of Technology

P. Pfeffer (Ed.), *9th International Munich Chassis Symposium 2018*, Proceedings,
https://doi.org/10.1007/978-3-658-22050-1_47

1 Introduction

The eigenfrequencies of parts are an elementary design criterion and getting more important in later use in vehicles with respect to the NVH-behavior. In particular, the increasing use of new materials in vehicle construction, for example Fiber composites and metallic material mixes of different strengths in the car body construction (e.g. Audi A8, BMW 7 Series), leads to changes in the vibration characteristics.

The specification of corresponding values (1st eigenfrequency) on component drawings of brake discs, as a quality criterion for the series release, underlines this. Nevertheless, this information's only covers the new condition. However, various and impermissibly worn or corroded brake discs cannot be considered, although thereby in daily vehicle use corresponding comfort-reducing vibration and noise can occur.

Mechanical wear, heat input and corrosion occur together in reality and influence the eigenfrequency spectrum. However, the individual impacts of the mentioned influences are unknown, but relevant with regard to an optimal component design.

The present paper investigates the change of the natural frequencies as a function of different wear factors. Different brake discs are considered (massive brake disc, massive brake disc ventilated, light-weight design brake disc ventilated). The simulative and experimental modal analysis is an essential part of these researches and a means to estimate the modal properties of a system in advance. Additional consideration is given to the Q-factor, a test criterion that allows statements about the damping properties.

2 State of the art

2.1 NVH / vibrations

Brake disc vibrations are assigned to the topic of Noise Vibration Harshness (NVH). Thematically, this covers all occurring, subjectively perceived by vehicle occupants disturbances and targeted measures to prevent this occurrence, such as air suspension chassis or improved isolation of the vehicle interior [5]. Figure 2-1 compares the perception of vibrations to the corresponding frequency ranges. Low-frequency vibrations up to 15 Hz (Vibrations) are transmitted through the component and can only be felt. The category of harshness of 15-100 Hz includes noticeable and audible vibrations (e.g. droning). Higher located frequencies or high-frequency vibrations are assigned to the category of noise and are primarily audible [5].

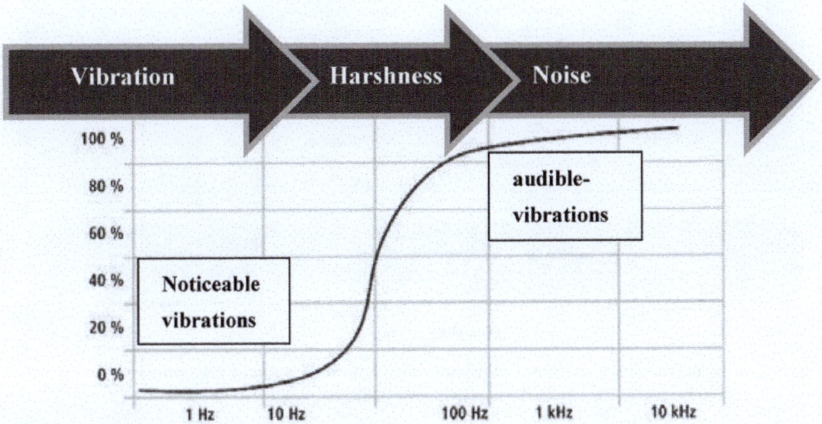

Figure 2-1: classification of NVH in relation to frequency and perception [5]

Vibrations induced by the brake system in the motor vehicle are forced (brake judder – category of vibration) or self-excited (brake squeal – category noise) [4].

Forced vibrations are caused by periodic braking torque fluctuations. Self-excited vibrations, such as brake squeal, are the result of dynamic instability of the brake system [1, 4].

The basis of the vibration properties of a component is the differential equation for the movement of the multi-mass oscillator consisting of mass matrix M, damping matrix D and the stiffness matrix C.

$$M\ddot{x}(t) + D\dot{x}(t) + Cx(t) = 0 \tag{2.1}$$

2.2 Modal analysis

Characteristic of the dynamic behavior of an oscillatory system are the modal parameters, eigen angular frequency, eigenvibration, modal mass and modal damping [6, 7].

The determination of these parameters of a solid is carried out by experimental modal analysis. In this case, a jerky excitation of the part takes place with subsequent evaluation of the impulse response. This time-dependent function is transferred in to an eigenfrequency spectrum using Fourier transformation [7], see Figure 2-2.

Referred to the brake disc, the modal analysis is used to identify acoustic problems and to verify the quality of the component [7]. For the experimental modal analysis of brake discs, the recommendations of the VDA 301 [2] and VDA 302 [3] apply.

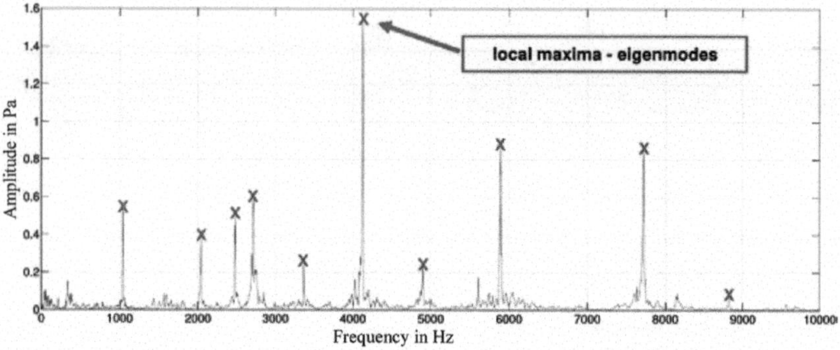

Figure 2-2: eigenfrequency spectrum of a brake disc with indication of eigenmodes

2.3 Eigenmodes of brake discs

Vibratory-capable systems form so-called eigenmodes when excited with their eigen angular frequencies ω_0 . In the process nodal lines (nodal circles and nodal diameters) occur at which the system remains at rest [9]. For identification, the eigenmodes occurring are named based on the number of nodal circles and nodal diameters (Figure 2-3).

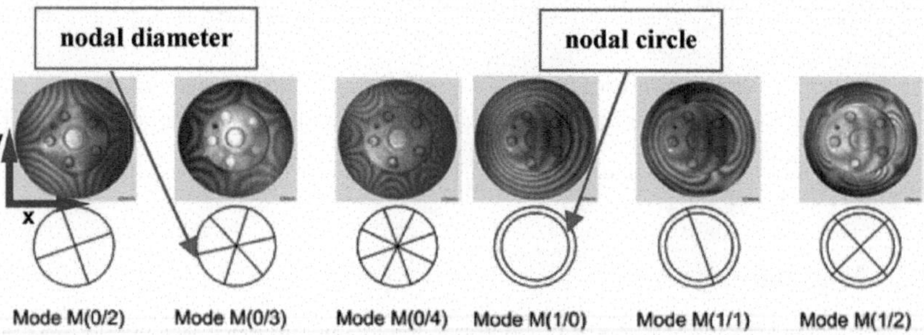

Figure 2-3: modes on brake discs to explain the mode designation [3]

For the present researches, modes without node circles, referred to in the further text as friction ring modes, are of interest. Eigenmodes can also be subdivided into in-plane modes (amplitudes within the x-y plane) and out-of-plane modes (amplitudes outside the x-y plane).

2.4 Q-factor

The quality factor Q (also resonance sharpness) of an oscillatory system represents the reciprocal of the damping d_0.

It is a measure of resonance overshoot in the case of resonance. In the case of resonance, a system of a high Q-factor answers to an excitation with a high amplitude [8]. For brake discs, the lowest possible quality factor in relation to each eigenmode is required. This means that the component should have a high attenuation. This ensures that emerging noise occurs less pronounced. The determination of the Q-factor takes place during the researches with the measuring device RTE SonicTC.Cube. The Q-factor of an eigenmode is determined by excitations of the frequencies and measurements of the response levels at different angular positions of the brake disc, depending on the mode segment size. The segment size depends on the sum of the node diameters per mode. The segment size for mode 0/4 is 45 degrees, since 8 knot diameters are created on the friction ring of the brake disk. The quality factor is then calculated using two methods implemented in the evaluation program.

2.5 Finite element simulation / meshing

The meshing of 3D CAD data is performed by tetrahedral or hexahedral solid elements, whose computation is following a linear or higher-order polynomial [11]. This can be happen either structured or unstructured. An unstructured meshing is usually generated automatically with tetrahedral elements. Otherwise, a structured meshing can reduce the computational effort by a smaller number of elements, but a manual meshing requires increased creation effort and is difficult to realize for complex geometries with fillets. It is principally possible to use both types of meshing in one model [12].

3 Simulative und experimental studies

3.1 General procedure and test samples

The test samples are three different brake discs see Table 3-1.

Table 3-1: overview of the examined brake discs

Light-weight-brake disc	**Massive brake disc ventilated**	**Massive brake disc**
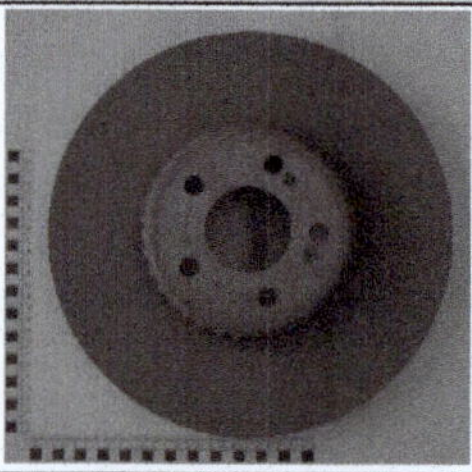	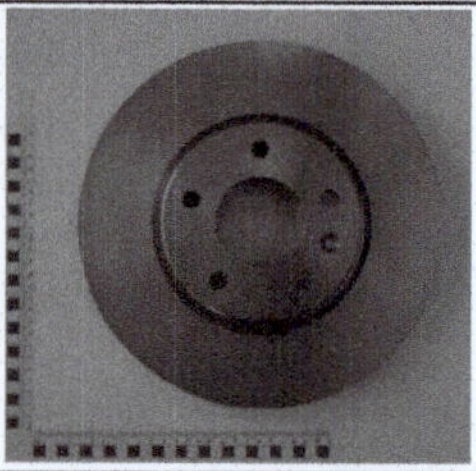	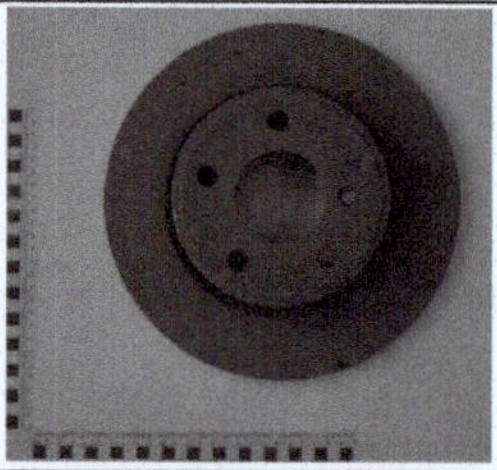
characteristics	**characteristics**	**characteristics**
Installation site: front axle Dimension: 342 x 32 Min. thickness: 29,4 mm	Installation site: front axle Dimension: 300 x 26 Min. thickness: 23 mm	Installation site: rear axle Dimension: 276 x 9 Min. thickness: 7,3 mm

These samples show a cross section in design and dimension of the currently in motor vehicles used variations. There are eight samples each.

The experiments are carried out according to the scheme in Figure 3-1. Then the test results are evaluated and compared with the simulation. Corresponding conclusions end the considerations.

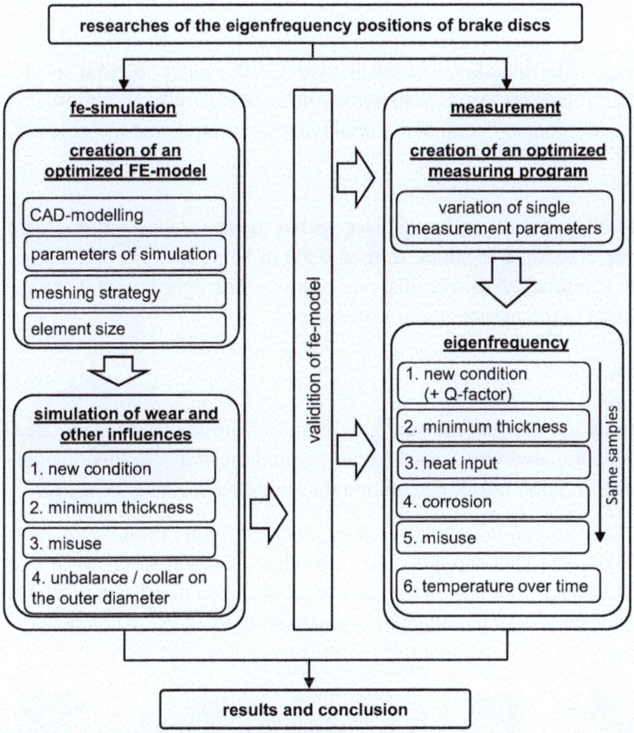

Figure 3-1: strategy for modal analysis researches of brake discs

3.2 Finite elemente simulation

The results obtained from the simulations later serve as a reference for the preparation of the measurement programs for the RTE SonicTC.Cube instrument.

Table 3-2: material parameters for modal analysis in ABAQUS

brake disc / material characteristics	**342x32**		**300x26 (resource Buderus)**	**276x9**
	brake bell	**friction ring**		
material	S355MC	GG15	GG15	GG15
specific gravity in $\frac{g}{cm^3}$	7,85	7,15	7,13	7,15
E-modulus in MPa	210.000	103.000	127.000	115.000
poisson`s ratio	0,29	0,3	0,27	0,3

Material parameters

For the eigenfrequency determination, the respective CAD model as well as the knowledge of the material characteristics density, modulus of elasticity and transverse contraction number are necessary. The final material parameters are listed in Table 3-2.

Boundary conditions

For the simulation, no clamping of the brake discs within the simulation environment is necessary, so that the eigenmodes can be formed without being influenced by other components [10]. The so-called "free-free analysis" corresponds approximately to the storage of the brake discs in the measurement instrument.

3.2.1 Meshing strategy

With regard to the simulation of different wear conditions, an automatically generated, unstructured mesh was compared with a partially structured mesh. From this, insights are gained regarding the creation effort, calculation duration and accuracy of results.

The creation of a partially structured variant enables identical model meshing in areas that are not affected by wear. This is applicable both when new and when worn out. However, due to the structured meshing in the wear area, the worn disc has a precisely defined difference in the number of elements in comparison to the mesh when new.

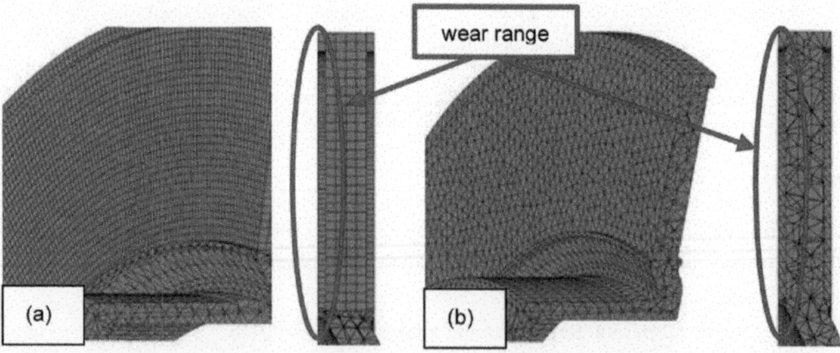

Figure 3-2: partially structured (a) and unstructured meshing (b) on the example 276x9 in the state of minimum thickness

A partially structured variant is achieved by a targeted decomposition of the geometry. This allows the assignment of structured element types in desired areas. Structured meshed areas are the friction ring and, in the case of a ventilated brake disk, the individual friction disks. Between the interfaces of structured and partially structured mesh, the FE software automatically generates a "TIE" relationship (gluing). Figure 3-2 shows the comparison of partially structured (a) and unstructured meshing (b).

3.2.2 Element size / sensitivity analysis

After determining the optimal meshing strategy, the element size optimization is done in four steps:

Step 1: automatic meshing through the FEM software (ABAQUS)
Step 2: specification of element sizes
Step 3: reduction of the element sizes
Step 4: specification of fine element sizes

After that, the results of each step regarding to convergence and calculation duration will be analyzed.

3.2.3 Simulation of wear conditions

New condition

3D CAD data are used in accordance with the drawing specifications, whereby maximum allowable unbalance machining are initially considered.

Minimum thickness

The basic data are adjusted according to the minimum thickness specifications of the respective brake discs. Further, on the minimization of the brake disc thickness is only done on the swept-over friction ride height. Depending on the brake disc, a characteristic, not worn collar forms on the outer diameter.

Influence of temperature and corrosion

The FE-models were not matched for temperature and corrosion. The state of minimum thickness is used as a comparison between measurement and simulation.

Misuse

The case of misuse depends on a continuous use of the brake disc, despite reaching the wear limit. For the brake discs 342x32 and 300x26, an additional reduction in thickness of 1 mm per friction ring and for the 276x9 of 0.5 mm per friction ring was assumed.

3.3 Experimental design

Figure 3-1 shows the individual, planned researches. In reality, wear, temperature input and corrosion occur in combination. Due to the planned separation of the influencing factors, their influence on the eigenfrequency position is researched separately. This allows a rating of the influencing factors of the eigenfrequency spectrum. First, the eigenfrequencies and the Q-factor in the new condition of the brake discs is determined. Subsequently, measurements are carried out in the state of minimum thickness (wear limit).

As a modification of the real heat input to brake discs, the samples are subsequently heated in five stages from 100°C to 500°C in an electrical chamber furnace. After each cooling phase, measurements are taken. This is followed by a 14-day outdoor weathering trial. This is intended to simulate an application in which a vehicle is outdoors for several weeks and is not moved. Finally, the samples are prepared beyond the permissible wear dimension (misuse).

In addition, researches are carried out which consider material changes over time after a temperature entry.

3.3.1 Sample conditions of the partial tests

New condition

The brake discs 342x32 and 276x9 are coated with zinc dust paint when new. The 300x26 is not coated. Further features are documented in Table 3-1and Table 3-2.

Minimum thickness

To realize the wear condition of minimum thickness, the test samples are over-revved. This is also used as a machining process within the brake disc production. It minimizes the heat input in the material removal, because the effects of locally occurring at the cutting edge of the turning tool temperatures cannot be compared with a large-scale temperature entry, as this occurs during real braking between friction lining and brake disc. In this sense, a separation of temperature and thickness decrease result.

Heat input

The samples are subjected to full through heating in five successive steps of temperature. Between each temperature step, there is a cooling phase under air ventilation until reach to room temperature and a following measurement of the eigenfrequencies. Additionally another series of measurements is made for all brake discs (new condition) with through heat temperatures of 500°C and 800°C. Measurements are taken directly after the cooling phase and up to 17 days after the temperature treatment to investigate the influence of possible relaxation processes.

Corrosion

The brake discs will exposed to a 14-day outdoor weathering in accordance with DIN EN ISO 11474 [13] – Corrosion of metals and alloys. To accelerate the corrosion, the samples are sprayed with a saline solution (NaCl, 30 g / l) every 3 to 4 days. Associated with that the recording of relevant weather data takes place.

Misuse

Analogous to the state of minimum thickness, all brake discs will be over-revved. For the brake discs 342x32 and 300x26 the thickness decrease is 1 mm per friction ring and for the brake disc 276x9 it is 0.5 mm per friction ring.

3.3.2 Specials

For the configuration of the measuring device software for eigenfrequency determination, the reconciliation was carried out with a Master brake disc of each brake disc group. This measure corrects excessive deviations from the simulation result and limits the search corridors of the desired modes.

The determination of the Q-factor is only carried out on samples of brake discs in new condition. Accidental inaccuracies of the results and the significant shortening of the measurement period due to the omission of the Q-factor determination are the reason for this strategy.

The samples are measured once for each condition. This result in a total of 216 individual measurements during the tests (nine different wear conditions for three different brake disks of 8 samples each).

4 Results

4.1 Simulation

4.1.1 Optimal meshing strategy

Meshing of brake discs for the determination of eigenfrequencies for different wear conditions, using an automatic, unstructured meshing with tetrahedral elements, is sufficiently accurate, see results in Table 4-1.

Table 4-1: summary of results for the meshing strategy for the modal analysis of brake discs for the state of minimum thickness

brake disc / meshing characteristics	**300x26 partially structured**	**300x26 un-structured**	**276x9 partially structured**	**276x9 un-structured**
element types				
brake bell / cool. channel	C3D10		C3D10	
friction ring	C3D20R		C3D20R	
total		C3D10		C3D10
global element size in mm	4	4	3	3
number of elements	183.237	233.279	99.012	164.729
number of nodes	348.014	366.595	193.108	263.953
calcalation time in sek	879,6	503,4	276,6	450,7
time effort	very high	low	very high	low
	mean value of the deviation from the simulation results when comparing partially structured to unstructured meshing in %			
		0,271		-0,006

The optimization of the meshing is done for the brake discs 300x26 (ventilated) and 276x9 (not ventilated) in the state of minimum thickness.

For the inspection of the deviation of the results of partially structured and unstructured variants, the calculation results for natural frequencies in a range of 0 to 10.000 Hz are used for the evaluation.

For the light-weight design brake disc 342x32 is a need of additional characteristics. Meshing of the brake disk bell with linear elements (C3D4) ensures the functionality of the simulation model. This solution reduces the accuracy of results for the brake disc bell. For the researches, however, only the modes of the friction ring are of importance.

To ensure a more precise coupling of the individual components (friction ring and brake disk bell) the toothing needs to be meshed finer.

The mentioned results for the calculation of natural frequencies leads to a lower effort in the creation of the models while high accuracy. The deviation of the accuracy of the calculation results between partially structured and unstructured meshing is less than 1%. In addition, the unstructured strategy leads to significant shortening of the calculation time, depending on the brake disc type.

4.1.2 Element size

Quantitatively the simulated eigenfrequencies differ only very slightly from each other, especially between step three and four, cf. Table 4-2.

Table 4-2: simulated values for mode 0/2 and 0/max for meshing steps 3 and 4

mode	brake disc	step 3	step 4
M 0/2	300x26	1.040,94	1.039,96
	342x32	954,90	951,62
	276x9	810,36	810,18
M 0/max	300x26	9.659,49	9.644,38
	342x32	9.573,39	9.515,86
	276x9	8.828,27	8.827,92

The results of the sensitivity study are summarized in Table 4-3 for each step. For all brake discs step 3 shall be regarded as sufficiently accurate.

A further reduction in element size and an increase in the number of elements lead to a significant increase in the calculation time, but only to a reduction in the deviation of the calculation results of less than 0.6%.

Table 4-3: summary of the optimal element size for the cross-linking of all brake discs

step 1		**step 2**	**step 3**	**step 4**
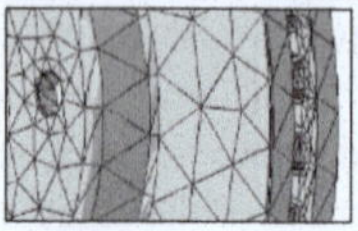			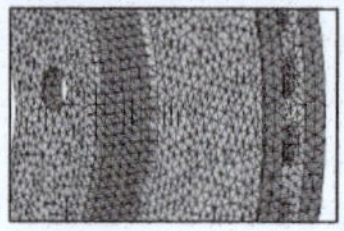	
300x26		300x26	300x26	300x26
global element size in mm (bb = brake bell; fr = friction ring)				
300x26	21	5	4	3
342x32	bb 9,8/9,8	bb 7/7	bb 6/5	bb 4/3
	fr 19/10	fr 7/7	fr 6/6	fr 4,25/4,25
276x9	17	5	2,5	2,25
factor of deviation for curvature 0,1				
number of elements				
300x26	26.923	145.829	236.969	429.994
342x32	113.258	227.364	273.359	464.314
276x9	5.092	44.674	286.664	386.585
calculation time in sec				
	41,7	263,1	583,7	1.617,2
		mean value of the deviation for the simulation results in comparison to the next larger element size (only friction ring modes) in %		
300x26		-3,31	-0,19	-0,13
342x32		-3,11	-0,13	-0,52
276x9		-2,86	-0,22	-0,01

4.1.3 Unbalance and collar on the outside diameter

The simulative researches lead to adaptations in the 3D-CAD modeling:

- elimination of the collar on the outer diameter
- elimination of the unbalance machining for 342x32 and 300x26

The influences of the simplifications used in relation to the real application case are very small for the brake discs 342x32 and 300x26 (deviation of the simulation values below 1%). However, the non-ventilated brake disc 276x9 shows a significant deviation greater ± 2 % for the simulation with and without collar, depending on the eigenmode.

The elimination of the unbalance is to be seen as a positive influence on the simulation mod-el towards the real application. It is not expedient to create 3D-CAD models after existing drawing documents with a maximum possible unbalance machining. The milled groove of the real samples is also geometrically smaller than the maximum specification on the drawings. Therefore, it is permissible to remove this for the simulation of the brake discs. In addition, a correct visual evaluation (clear visibility of the symmetry in the oscillation) of the simulation results for the further procedure is necessary, cf. Figure 4-1.

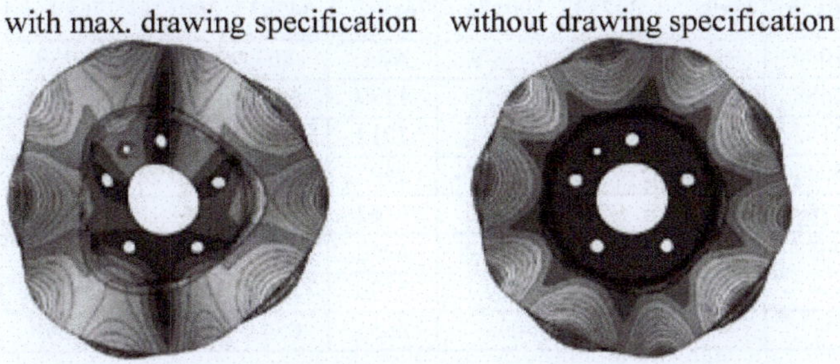

Figure 4-1: comparison of the friction ring mode 0/5 as a function of the unbalance machining

4.1.4 Simulated eigenfrequencies

Figure 4-2 exemplifies simulated eigenmodes in extracts.

M 0/2: 954,3 Hz	M 0/3: 2120,9 Hz	M 0/4: 3338,5 Hz	M 0/5: 4559,6 Hz
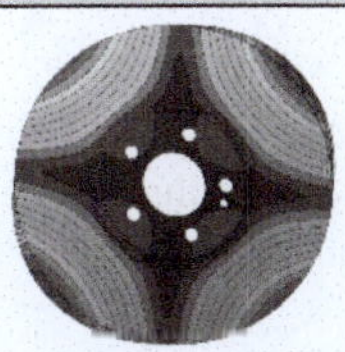	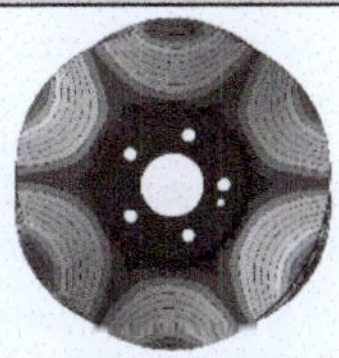	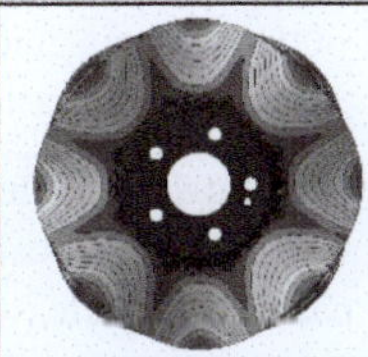	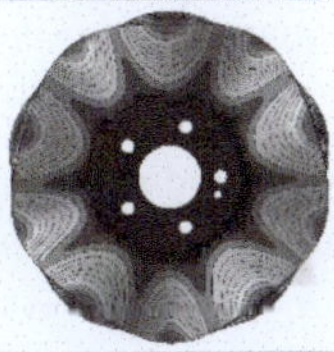

Figure 4-2: simulated friction ring modes 0/2 to 0/5 of brake disc 342x32 in state of new

Table 4-4 summarizes the simulated eigenfrequency values of all tested brake discs for the states of new, minimum thickness and misuse.

Table 4-4: simulated eigenfrequency values of all brake discs as a function of wear

mode	New condition brake disc f in Hz			minimum thickness brake disc f in Hz			misuse brake disc f in Hz		
	342x32	300x26	276x9	342x32	300x26	276x9	342x32	300x26	276x9
0/2	954	1040	811	916	979	784	863	953	789
0/3	2121	2488	1383	2018	2299	1189	1903	2202	1174
0/4	3339	4152	2136	3178	3820	1712	3018	3636	1664
0/5	4560	5942	3112	4337	5465	2407	4135	5192	2317
0/6	5787	7789	4289	5493	7166	3262	5244	6801	3120
0/7	7030	9658	5647	6651	8887	4262	6346	8425	4061
0/8	8289	-	7166	7818	-	5396	7445	-	5130
0/9	9573	-	8830	8995	-	6655	8543	-	6319

Depending on the brake disc and state of wear, there is a different high number of friction ring modes results, in the considered frequency band from zero to 10,000 Hz. The brake discs 342x32 and 276x9 shows up eight friction ring modes, while the 300x26 shows up six friction ring modes.

The prevailing friction ring modes can be clearly identified for all brake disc variants and the states mentioned. It can be seen a reduction in the frequency values as a function of the reduction of the brake disc thickness (by wear). This is understandable according to equation 2.1.

The height of the eigenfrequency deviations (Table 4-5) to the pre-state vary considerably depending on the brake disc. These fluctuations additionally vary within the respective eigenmodes.

Table 4-5: eigenfrequency value deviations in % depending on the state of wear

mode	deviation to the state new minimum thickness %			deviation to the pre-state misuse %		
	342x32	300x26	276x9	342x32	300x26	276x9
0/2	4,05	5,89	3,26	5,72	2,68	-0,63
0/3	4,86	7,59	14,04	5,71	4,26	1,23
0/4	4,82	8,01	19,87	5,01	4,80	2,78
0/5	4,88	8,03	22,65	4,65	4,99	3,76
0/6	5,08	8,00	23,96	4,54	5,10	4,35
0/7	5,38	7,98	24,53	4,60	5,20	4,71
0/8	5,69	-	24,70	4,77	-	4,92
0/9	6,04	-	24,63	5,02	-	5,04

Deviations between 4% and 6% can be detected for the 342x32, regardless of the state of wear and mode height. The 300x26 deviates from state of new to the state of minimum thickness by 5.89% for mode 0/2 to 7.98% for mode 0/7. The decrease of the eigenfrequency values for the state of misuse then amounts to 2.68% for mode 0/2 or 5.20% for mode 0/7.

The eigenfrequency shifts due to thickness reduction can be seen more clearly in the brake disk 276x9. Compared to the state of new, the values for mode 0/2 decrease by 3.26% and for mode 0/9 by 24.63%. This is a drastic compression of the eigenfrequency spectrum associated with the decrease in brake disc thickness. The reduction in the eigenfrequencies is much more moderate at 2.68% for Mode 0/2 and 5.20% for Mode 0/9 for the misuse case.

4.2 Experimental measurements

4.2.1 New condition

342x32

Similar to the simulation (see 4.1.4), the friction ring modes 0/2 to 0/9 can be identified. The mean values over all eight samples scatter with a standard deviation of up to 1.8%, depending on the mode. The identification of mode 0/3 in the frequency range around 2,000 Hz is not unique. Measurements show three together lying maxima in this area. According to the simulation, mode 1/0 and mode 1/1 also occur in the mentioned frequency range (both no friction ring modes). The evaluation logic of the measuring software cannot distinguish between these modes because of specifying a search frequency band. Mode 0/9 is also not clearly identifiable. The reason is a very low-level amplitude close to the ground noise of the measurement signal. Accordingly, the modes 0/3 and 0/9 were not evaluated for all tests. The measured values are below the simulated ones. For mode 0/2 the relative mean deviation is -2.39% and decreases to -0.52% until mode 0/9.

300x26

Similar to the simulation (see 4.1.4), the friction ring modes 0/2 to 0/7 could be identified. The mean values of all eight samples scatter with a standard deviation of up to 0.1% depending on the mode. The measured values are below the calculated simulation data. For mode 0/2, the relative mean deviation is -0.1% and increases to -1.0% until mode 0/7.

276x9

Similar to the simulation (see 4.1.4), the friction ring modes 0/2 to 0/9 could be identified. Depending on the mode, the mean values of all eight samples scatter with a standard deviation of up to 0.51%. The measured values are below the calculated simulation data. For mode 0/2, the relative mean deviation is -1.27% and increases to -4.31% until mode 0/7.

4.2.2 Minimum thickness

For all brake discs, the wear leads to a reduction of the eigenfrequency values of the friction ring modes. This coincides with the findings of the FE simulation. The mean values of the natural frequencies decrease by a maximum of 6.17% (342x32), 7.63% (300x26) and 16.77% (276x9). In comparison, Table 4-5 with 6.04% (342x32), 8.03% (300x26) and 24.70% (276x9).

4.2.3 Influence of temperature

The through heating of the brake discs in 100K increments up to 500°C has a small influence on the eigenfrequency spectrum. The mean values of the natural frequencies fluctuate with respect to the state of minimum thickness. For the 342x32-brake disc, the eigenfrequencies are reduced by up to 2.08%. For the two other brake discs, the values increase minimally (up to +0.25% at 300x26 and up to +0.61% at 276x9), cf. see Table 4-6. As the temperatures in the individual stages increase, color changes of the brake discs occurred. In addition, the brake discs are covered with layer of soot after the temperature of 500°C.

Table 4-6: relative mean deviation in % to the respective pre-state for three brake discs samples related to through-heating 100 °C to 500 °C

brake disc	100 °C	200 °C	300 °C	400 °C	500 °C	min. th. to 500°C
342x32						
min. deviation	-0,95%;	-1,25%;	-0,53%;	-0,32%;	-2,06%;	**-2,08%**
max. deviation	+0,13%	+0,96%	+0,56%	+1,53%	+0,03%	**+0,38%**
300x26						
min. deviation	-0,12%;	+0,09%	+0,03%	+0,04%	-0,21%;	**+0,11%**
max. deviation	+0,05%	+0,23%	+0,19%	+0,16%	-0,02%	**+0,25%**
279x6						
min. deviation	-0,08%;	+0,19%	-0,09%;	-0,03%;	+0,03%	**+0,34%**
max. deviation	+0,13%	+0,38%	+0,10%	+0,09%	+0,21%	**+0,61%**

In order to exclude time-delayed material alteration further brake discs of the dimensions mentioned are measured by way of example. After the determination of the eigenfrequency values in the state of new, through heating was carried out at 500°C and at 800°C for each test sample.

Subsequently, irregular measurements of the eigenfrequencies are made over a 17-day period. The change in the eigenfrequencies over time, based on the respective friction ring mode, are very small, cf. Table 4-7. A temperature impact of 500°C raises the eigenfrequencies minimal, but at 800°C a reduction of the eigenfrequency values can be seen with each brake disc variant.

4.2.4 Corrosion

The 14-day outdoor weathering with repeated treatment with a saline solution (NaCl, 30 g / l) lead to a significant surface corrosion of the friction surfaces. For the measurements, the surface of the friction rings is left as it is. For all brake discs, an increase in eigenfrequency aver-ages up to 1.92% (342x32), 0.79% (300x26) and 1.88% (276x9) can be seen. According to Table 4-7, an increase in the measured values, caused by time-delayed material alteration due to the temperature treatment cannot be completely excluded.

Table 4-7: relative mean deviation in % after 17 days (heat input) based on the state new

mode	0/2	0/3	0/4	0/5	0/6	0/7	0/8	0/9
temperature				**500 °C**				
342x32 (day 1)	-2,55	n.a.	n.a.	0,00	0,47	0,69	0,42	0,38
342x32 (day 17)	-2,13	n.a.	0,39	0,48	0,44	1,20	0,97	0,93
300x26 (day 1)	-0,08	-0,35	-0,15	-0,12	-0,14	-0,16	n.v.	n.v.
300x26 (day 17)	0,30	0,31	0,21	0,23	0,27	1,80	n.v.	n.v.
276x9 (day 1)	-0,19	-0,23	-0,38	-0,37	-0,45	-0,39	-0,78	-0,48
276x9 (day 17)	0,00	0,00	-0,11	-0,13	-0,19	-0,17	-0,32	-0,18
temperature				**800 °C**				
342x32 (day 1)	-7,97	n.a.	-3,73	-2,95	-2,27	-2,21	-2,09	-1,88
342x32 (day 17)	-7,80	n.a.	-3,22	-2,51	-2,04	-2,33	-1,72	-1,74
300x26 (day 1)	-0,98	-1,26	-1,15	-1,44	-1,44	-1,39	n.v.	n.v.
300x26 (day 17)	-0,83	-0,85	-0,98	-1,06	-1,14	-1,19	n.v.	n.v.
276x9 (day 1)	-2,53	-2,72	-2,83	-2,76	-2,86	-2,89	-2,93	-3,20
276x9 (day 17)	-2,33	-2,54	-2,60	-2,55	-2,59	-2,60	-2,81	-2,65

4.2.5 Misuse

For all brake discs, the mean values of eigenfrequencies are reducing analogously to the measurement of state of minimum thickness (cf. Table 4-8).

Table 4-8: eigenfrequency averages of all brake discs for friction ring modes for misuse

brake disc	342x32		300x26		276x9	
mode	mean value in Hz	σ in Hz	mean value in Hz	σ in Hz	mean value in Hz	σ in Hz
0/2	839,3	17,4	956,8	1,8	790,1	2,2
0/3	1875,6	33,3	2207,4	5,6	1170,7	3,0
0/4	3079,6	140,7	3633,9	11,7	1651,0	5,9
0/5	4106,3	64,3	5183,2	5,2	2293,3	7,3
0/6	5223,6	65,9	6785,0	13,4	3084,8	10,6
0/7	6361,1	58,7	8416,5	6,0	4009,9	14,3
0/8	7459,3	84,2			5071,3	21,2
0/9	8619,6	69,1			6247,6	21,8

There are relative mean decreases (cf. to corrosion) up to 5.22% (342x32), 5.64% (300x26), and 13.22% (276x9). The mean eigenfrequencies, see Table 4-8, are reduced analogously to the simulation results, see Table 4-4.

4.2.6 Q-factor / damping

The Q-factor determination is carried out on all brake discs for the modes 0/2 to 0/5 when new. Relevant is the respective Q-factor value of the given friction ring mode. This factor fluctuates partially considerable in test samples within the modes. The quality factor values of the brake disc 300x26 are higher than the values of the brake discs 342x32 and 276x9. Accordingly, these discs seems to be better damped.

5 Conclusion

5.1 Simulative researches

The present publication shows possibilities for the simulation of natural frequencies on brake discs. Accordingly, information about the material parameters, boundary conditions, the meshing strategy and element sizes are given. To be accurate as possible it is necessary to use simulation data based on very precise material characteristics, here the modulus of elasticity is of particular importance. The very small differences between simulation and measurement for the 300x26 type are due to this accuracy (see chapter 4.2.1).

Simulations are carried out without any boundary conditions (free-free). Meshing can be unstructured and with tetrahedral elements. Multi-part brake discs are more complex to model and it is important to mind contact definitions. Significant changes in the simulation results are not expected with a quadratic approach and a global element size less than 2.5 mm. The consideration of an unbalance machining does not seem necessary due to slightly changes of the result values. A collar at the outer diameter, based on real wear, should be simulated depending on the brake disc dimensions. The smaller the dimensions, the greater its influence.

A comparison of the simulation results with real measurement data (based on material characteristics and natural frequency measurements) is indispensable.

5.2 Experimental researches

In accordance to the topic, influences on the position of the eigenfrequencies of different brake discs are shown. For the friction ring modes values for the states of new, minimum thickness, heat input of 0-500°C and misuse are available. Figure 5-1 to Figure 5-3 show in summary for all brake discs the change of the eigenfrequency mean values of the modes 0/2 to 0/9 for the different states of wear. The reduction of the values by thickness decrease as well as the increase of the values by corrosion is clearly recognizable.

All influencing factors change the eigenfrequency spectrum of the various brake discs. Accordingly, relevant values shift positively or negatively:

1. The wear (minimum thickness and misuse) leads to a reduction of the eigenfrequencies. This reduction is very evident for the brake disc 276x9. This is due to the low geometric dimensions and in order to the influence of the several matrices (see equation 2.1)
2. The influence factor of heat input up to 500 ° C changes the eigenfrequencies very slightly, but a slight increase of the eigenfrequencies in the low single-digit percentage range is recognizable for a large part of the considered modes. A temperature of 800 ° C reduces the eigenfrequencies in the single-digit percentage range for all brake discs and friction ring modes. However, the time-delayed material alteration of the eigenfrequency values is very small.
3. The corrosion also leads to an increase in the eigenfrequencies in all modes, despite of the brevity of outdoor weathering.

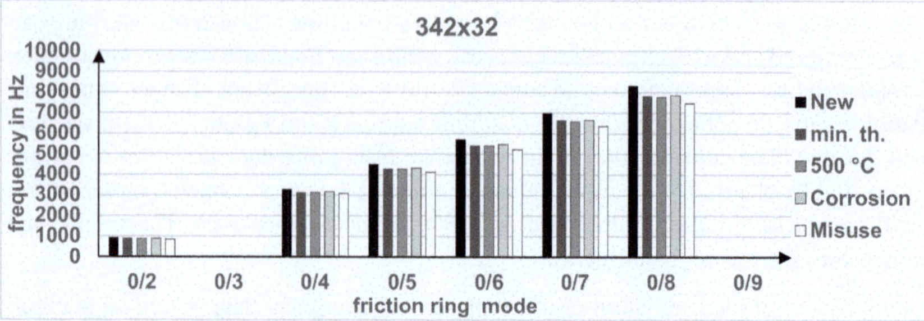

Figure 5-1: alteration of the eigen modes as a function of the wear state up to 10,000 Hz for the brake disc 342x32

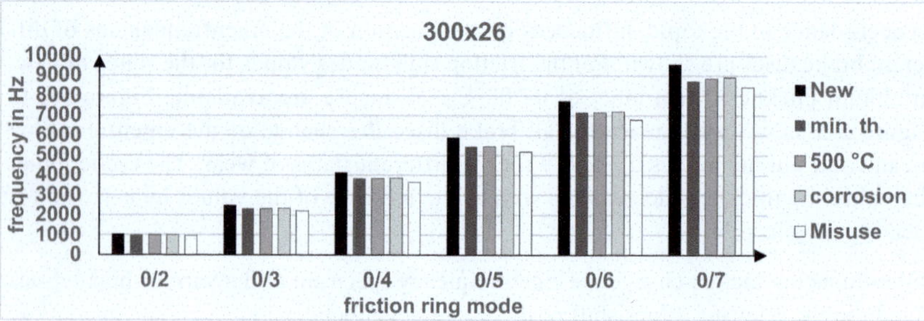

Figure 5-2: alteration of the eigen modes as a function of the wear state up to 10,000 Hz for the brake disc 300x26

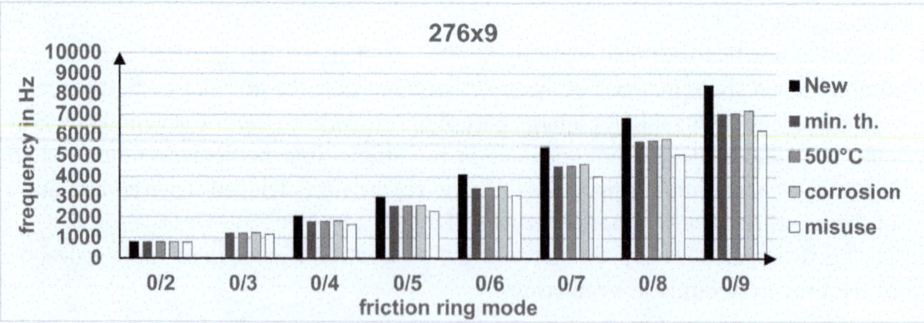

Figure 5-3: alteration of the eigen modes as a function of the wear state up to 10,000 Hz for the brake disc 276x9

The Q-factor determination for the state of new only permits limited statements about damping. In the event that results could be determined, this led to high quality factors, especially for the brake disc 300x26. This implicates a lower damping. Current evaluations do not allow any further statements on the quality factors of the samples in individual modes.

6 Open questions / outlook

In order to be able to better classify the present results of the research, further informations are necessary.

Conditioning of brake disc

1. The influence of material characteristics (modulus of elasticity over temperature and time after cooling) has to be considered (for simulation and experiment), so that simulated results become more accurate and conclusions can be drawn for temperature researches.
2. The performed through heating of the brake disc variants with a chamber furnace does not represent the real case (eg braking with subsequent, immediate parking of the vehicle). A significantly shortened heating by induction or tests with a brake dynamometer represent conditions that are more realistic. Corresponding to that, comparative values must be determined.
3. The influence of a corrosive layer on the eigenfrequency spectrum is shown. However, influences on the result values due to the changed surface structure must be ruled out. Long-term outdoor weathering and chamber tests are to be strived for comparatively. Furthermore, a further approach to reality is necessary (elimination of the corrosive layer after braking, corroded and fouled cooling channels).

Optimization of measurement

In order to increase the accuracy of the measurements and to obtain clearer results different settings can be varied.

1. Extensive adjustments of the measuring software with respect to time signal and Fourier transformation are to be aimed for.
2. Further measurements of the quality factor as well as evaluations are to be carried out for a better understanding.
3. Extension of the measuring device by a second microphone, acceleration sensors or distance measuring technology. This allows the measurement of real amplitude levels and the clear determination of modes (e.g. friction ring modes 0/3 and 1/0 or 1/1).

Bibliography

[1] ZELLER, P., *Handbuch Fahrzeugakustik.* Wiesbaden: Vieweg+Teubner Verlag, 2012. ISBN 978-3-8348-1443-2.

[2] Verband der Automobilindustrie e.V. 302, *VDA 302 Eigenfrequenzmessung und Modalanalyse von Bremsscheiben in der Entwicklung.* Frankfurt.

[3] Verband der Automobilindustrie e.V. 301, *VDA 301 Eigenfrequenzmessung und Modalanalyse von Bremsscheiben in der Entwicklung.* Frankfurt.

[4] MAYER, R. *Bremseninduzierte Schwingungen am Kraftfahrzeug. Analysemethodik und konstruktive Optimierung.* Düsseldorf: VDI Verlag, 2007. Fortschritt-Berichte VDI. Reihe 12, Verkehrstechnik/Fahrzeugtechnik. Nr. 641. ISBN 3183641127.

[5] TRELLEBORGVIBROACOUSTIC. *Schwingungstechnik im Automobil. Grundlagen, Werkstoffe, Konstruktion, Berechnung und Anwendungen.* Würzburg: Vogel Business Media, 2015. ISBN 9783834333575.

[6] MAGNUS, K.; K. POPP; W. SEXTRO. *Schwingungen. Grundlagen – Modelle – Beispiele.* 10., überarbeitete Auflage. Wiesbaden: Springer Vieweg, 2016. ISBN 978-3-658-13820-2.

[7] KOLLMANN, F.G., T.F. SCHÖSSER; R. ANGERT. *Praktische Maschinenakustik.* Berlin Heidelberg: Springer-Verlag Berlin Heidelberg, 2006. VDI-Buch. ISBN 9783540309758.

[8] VEIT, I. *Technische Akustik. Grundlagen der physikalischen, physiologischen und Elektroakustik.* 5., durchges. Aufl. Würzburg: Vogel, 1996. Vogel-Fachbuch. ISBN 3-8023-1707-6.

[9] HENN, H., G.R. SINAMBARI; M. FALLEN. *Ingenieurakustik. Grundlagen Anwendungen Verfahren.* 3., durchgesehene Auflage. Wiesbaden: Vieweg+Teubner Verlag, 2001. ISBN 3-528-28570-2.

[10] KHARATE, N.K.; S.S. Dr. CHAUDHARI. Researche of Natural Frequency and Modal Analysis of Brake Rotor Using Fea and Ema. In: *International Journal of Innovative Research in Science, Engineering and Technology.*

[11] KLEIN, B. *FEM. Grundlagen und Anwendungen der Finite-Element-Methode im Maschinen- und Fahrzeugbau.* 10., verb. Aufl. Wiesbaden: Springer Vieweg, 2015. Lehrbuch. ISBN 978-3-658-06054-1

[12] WAGNER, M. *Lineare und nichtlineare FEM. Eine Einführung mit Anwendungen in der Umformsimulation mit LS-DYNA®.* Wiesbaden: Springer Vieweg, 2017. Lehrbuch. ISBN 978-3-658-17866-6.

[13] DIN Deutsches Institut für Normung e.V. DIN EN ISO 11474:2014-06, *Korrosion von Metallen und Legierungen – Korrosionsprüfung in künstlicher Atmosphäre – Beschleunigte Außenbewitterung (Freibewitterung) mit intermittierendem Sprühen einer Salzlösung (SCAB-Test) (ISO 11474:1998).* Berlin: Beuth Verlag.

Virtualization of the brake noise development process to guarantee robust start quality

Dr. Martin Treimer, BMW Group,
Braking- and Actuating Systems, Simulation Longitudinal Dynamics

P. Pfeffer (Ed.), *9th International Munich Chassis Symposium 2018*, Proceedings,
https://doi.org/10.1007/978-3-658-22050-1_48

Abstract

This essay is covering the field of uncertainty quantification of brake systems concerning the occurrence of brake noise. With the aid of the presented procedure a better maturity level during the development process is given.

An algorithm is developed which characterizes the phenomenon of brake squeal for example with regard to its dependency on production-related scatter of the material parameter in simulation. A robustness assessment is done to ensure a noise free brake system over the entire term.

For the robustness assessment and definition of the maturity level different approaches are introduced. The parameter scatter for the simulation is identified out of the real distributions of the singular components. For the parameter scatter`s evaluation a design of experiment is created which allows a universally valid assessment of systems with a low computation effort. The analysis of the robustness is based on mathematical surrogate models for the characterization of the parameter`s influences. For the description of the results the surrogate model`s progressions, the parameter sensitivities and instability probabilities are used. The system`s characterization is done with a robustness index which draws conclusions about the maturity level.

During the virtual development of a brake system the algorithm provides a statement about its noise behavior under the given elements of uncertainty in the production of the singular components and their assembly. Therefor the development of a robust system is ensured so that there are no brake noises in the vehicle.

Introduction to Brake Squeal Simulation and Robustness evaluation[1]

Brake squeal is an important topic in NVH (Noise, Vibration, Harshness) investigations during the development process of a car [2,3] since it can cause high warranty costs and customer complaints [4,5]. Next to the safety and performance issues of a brake system there are important requirements on acoustic and comfort aspects [3,6]. In New York City brake squeal is one of the ten biggest noise problems and causes more than one billion of warranty costs in the United States [7]. Although the first developments in the field of brake squeal reduction started in the 1930s [7,8] there is no generalised theory for this complex phenomenon [9].

The excitation mechanisms for brake squeal are described in [10,11] and the corresponding minimal models in [12]. The noise is caused by a self-exciting oscillation [13] induced by friction [8,9,13]. This results in a mode coupling or lock-in [7,14] which causes high vibration related with noise radiation [5].

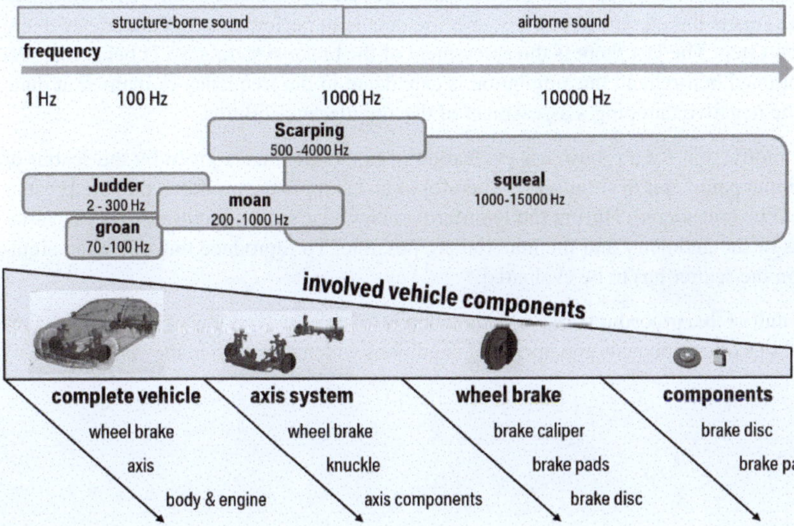

Overview of brake noises and the involved components.

1 See [1].

Brake squeal has to be distinguished from other noise emissions caused by brake systems [8]: Next to the brake squeal which occurs with frequencies ranging from 1 to 16 kHz [5] the most frequently cited noises are groaning for frequencies below 100 kHz and moaning in a range between 100 and 500 kHz [8].

There are different theoretical and experimental possibilities for the description of brake squeal [13]. In simulation brake squeal is evaluated typically in the frequency domain. Instead of a time-consuming transient analysis [15] the evaluation is carried out by reference to the system`s eigenfrequencies [16].

A method for the simulation of brake squeal for different frequency domains is proposed in [10]. Here the complex eigenvalues are calculated by means of a Finite Element model which is validated through a comparison with experimental results of the whole system and the analysis of the singular components` eigenfrequencies. Based on this work the quality of prediction is improved by further development [12,17]. On the basis of these methods effective countermeasures for the abatement of brake squeal are defined.

The simulation process is in general consists of three steps. In the first one the finite element model is set up with the necessary geometrical, material and environmental data. This is linked with the solver specific details to perform the CAE analysis in the second step. The last point is the assessment of the brake system`s NVH behaviour. As mentioned beforehand the simulation is calculated in the frequency-domain so it done by the negative damping`s assessment of the singular instabilities.

The motivation for a robustness evaluation of brake systems is given by the scatter of the input parameter in simulation. Therefor as an example the production related scatter has to be considered. During the long-term component`s manufacturing process variations of the geometry and the material are possible. To reproduce this effects in simulation the scatter has to be evaluated.

The aim of this procedure done by simulation is to have an overview of all possible combinations of components and operating conditions which are given in the vehicle fleet.

Stability Analysis[2]

The assessment of the brake system`s stability based on the Lyapunov theory [18,19] is carried out on the basis of a quasi-static analysis [4,8,12]. The system`s stability is assessed on the basis of an eigenvalue analysis.

The describing equation is given in the following notation.

$$\mathbf{M\ddot{x}+C\dot{x}+Kx=0}, \tag{1}$$

with **M** mass, **C** damping and **K** stiffness matrix. The asymmetric parts of the stiffness matrix K are linked to the friction mechanism and can cause instabilities. They are responsible for the mode coupling phenomenon.

Equation (1) is given in the state space by:

$$\mathbf{A\dot{y}=By}, \tag{2}$$

with $\mathbf{y}=\begin{pmatrix}\mathbf{x}\\ \dot{\mathbf{x}}\end{pmatrix}$, $\mathbf{A}=\begin{pmatrix}\mathbf{C} & \mathbf{M}\\ \mathbf{M} & \mathbf{0}\end{pmatrix}$ and $\mathbf{B}=\begin{pmatrix}\mathbf{-K} & \mathbf{0}\\ \mathbf{0} & \mathbf{M}\end{pmatrix}$.

Based on the approach

$$\mathbf{y}=\mathbf{\Psi}e^{st}, \tag{3}$$

in combination with equation (2) the complex eigenvalue problem is given,

$$(\mathbf{B}-\lambda_k\mathbf{A})\mathbf{\Psi}_k=\mathbf{0} \text{ with } 1 \le k \le 2n, \tag{4}$$

where λ represents the eigenvalue, Ψ the corresponding eigenvector and n the number of degrees of freedom of the Finite Element system. The eigenvalues are normalised by the following condition for the guarantee of the uniqueness:

$$\mathbf{\Psi}_k^T\mathbf{A}\mathbf{\Psi}_k=\mathbf{1}, \tag{5}$$

Prior to the calculation of the complex eigenvalue problem a subspace is created. For this the undamped and symmetric system is considered and all asymmetric terms are neglected. The eigenfrequencies form (4) set up the basis for the computation of the complex frequencies. For that reasons the matrices are protected onto the subspace.

$$\begin{aligned}\mathbf{M}^* &= \mathbf{\Psi}^T\mathbf{M}\mathbf{\Psi},\\ \mathbf{C}^* &= \mathbf{\Psi}^T\mathbf{C}\mathbf{\Psi},\\ \mathbf{K}^* &= \mathbf{\Psi}^T\mathbf{K}\mathbf{\Psi}.\end{aligned} \tag{6}$$

2 See [1].

With this projection the following equation is solved:

$$\mathbf{M}^*\ddot{\mathbf{x}}+\mathbf{C}^*\dot{\mathbf{x}}+\mathbf{K}^*\mathbf{x}=0, \quad (7)$$

Unstable frequencies are characterised by a positive real part of the eigenvalue and stable modes by a negative one. This analysis is used for the prediction of brake squeal.

Algorithm for robustness evaluation[3]

In the following an algorithm for the robustness assessment of the prediction of brake squeal on the basis of the real parts of the eigenvalues is presented. For the consideration of the input parameter scatter the Finite Element model of a brake system is described by a surrogate model. This is created by the usage of design of experiments. The surrogate model is used for the system's assessment.

General procedure

During the first step of the algorithm the relevant parameters are identified. The variables with a high influence to the propensity of brake squeal are chosen and additional parameters with a large scatter behaviour are taken into account. The parameters` boundaries of variation and distributions are described. These are based on experimental evaluations of several brake systems. Different distributions for each parameter are taken into account, a uniform or a normal distribution. The identified parameters are:

- Young`s modulus brake disc, E_d, with a normal distribution,
- Young`s modulus brake pad, E_p, with a normal distribution,
- Young`s modulus brake caliper, E_c, with a normal distribution,
- Young`s modulus knuckle, E_k, with a normal distribution,
- Friction value, μ, with a normal distribution,
- Brake pressure, p, with a normal distribution,
- wear, w, with a uniform distribution.

Based on the minimal and maximal values of the scatter and the distribution of the parameters an experimental design is created. Several designs are possible, for example full factorial or d-optimal. The choice depends on the parameter quantity, the sampling points and the prediction quality desired, with regard to the calculation effort.

In the next step the design of experiment is performed numerically on the basis of finite elements. For the later evaluation each experiment is assigned a value which identifies the occurrence of brake squeal. This is called objective function.

---

3 See [1].

Based on the assigned scalar value the design of experiment is evaluated. For an analytic discussion a regression analysis is performed. The resulting steady function is analysed regarding the behaviour on parameter changes, the dependencies and the influence factors. For the verification of the system´s reliability the correlation coefficient between numerical results and the results obtained on the basis of the analytical model is assessed.

The interpretation of the results is carried out with the overall scatter of the objective function, the parameter sensitivities regarding the objective function variation and an instability probability index.

Design of experiment

For the consideration of the identified parameter a d-optimal design of experiment is chosen. The d-optimal design [20,21,22] is a reduction of the full factorial design to the minimum number of experiments for calculating the surrogate model of a certain order. A consideration of all effects and interactions is possible. Unlike a Latin-Hypercube Sampling the values for the sampling points are not randomised and remain fixed. They are chosen before creating the design of experiment. For a second order model the minimum number of experiments is given by:

$$2p + \frac{p\,(p-1)}{2} + 1, \tag{8}$$

with p being the number of the main effects multiplied by 2 because of the quadratic terms. The number of the interactions between the parameters is given by p(p-1)/2. The last part is for the description of the constant term. The model is based on the quasi-linear polynomial ansatz

$$\mathbf{y} = \mathbf{X}\boldsymbol{\beta} + \boldsymbol{\varepsilon}, \tag{9}$$

with **y** being vector of the objective, $\mathbf{X} \in R^{n \times m}$ the design matrix constructed by the candidates for the experimental design, β the vector of unknown model parameters and ε the estimation error. The dimension n stands for the number of parameters and m for the number of experiments. To get the minimal and optimal design exchange algorithms are performed with the aim to minimize the expression $\det(X^T X)^{-1}$ based on a fixed number of experiments. The experiments in **X** are replaced iteratively until the best configuration is found. This is done by costly calculation algorithms [23,24] with the aim of the improvement of the design's quality of prediction. A full-factorial design is reduced to a minimum number of experiments but keeping the assessment`s possibility of all effects included in the model. The realised number of experiments is larger than the minimal one. This is needed to illustrate the variance in the experiments` results and is a compromise between the calculation effort and the quality of prediction.

Objective function

The analysis of the brake system is carried out by assessing the real Re and the imaginary parts Im of the eigenvalues λ_i by the following objective function $\mathbf{q}_{\mathrm{sim}}$:

$$\mathbf{q}_i = \frac{|\mathrm{Re}(\lambda_i)-\mathrm{Re}(\lambda_{i+1})|-|\mathrm{Im}(\lambda_i)-\mathrm{Im}((\lambda_{i+1})|}{0.5(\mathrm{Im}(\lambda_i)+\mathrm{Im}(\lambda_{i+1}))}. \tag{10}$$

For all neighbouring modes, i = 1, …, m-1, with m being the maximal mode number, the quotient $\mathbf{q}_i$ is calculated. The objective function for one experiment is described by:

$$q_{\mathrm{sim}} = \max_{i=1,\ldots,m-1} \mathbf{q}_i. \tag{11}$$

In practical applications with several instabilities in different frequency bands q_{sim} is calculated with dependency of the respective instabilities of interest. Therefor after the first the second maximum and so on is chosen.

For instabilities the objective function $\mathbf{q}_i$ is equal to the damping ratio of unstable modes d_i [25] which is a conventional squeal index:

$$d_i = -\frac{2\,\mathrm{Re}(\lambda_i)}{\mathrm{Im}(\lambda_i)}. \tag{12}$$

This belongs to the mode coupling phenomenon and its associated relations when two modes are coupled, $\mathrm{Re}(\lambda_i) = -\,\mathrm{Re}(\lambda_{i+1})$ and $\mathrm{Im}(\lambda_i) = \mathrm{Im}(\lambda_{i+1})$. That means:

$$q_i = \frac{|\mathrm{Re}(\lambda_i)-\mathrm{Re}(\lambda_i)|-|\mathrm{Im}(\lambda_i)-\mathrm{Im}((\lambda_i)|}{0.5(\mathrm{Im}(\lambda_i)+\mathrm{Im}(\lambda_i))} = \frac{2\,\mathrm{Re}(\lambda_i)}{\mathrm{Im}(\lambda_i)} = |d_i|\,. \tag{13}$$

An advantage of the objective function towards the damping ratio is the indication of stable modes with a negative value unequal to zero. The objective function allows a continuous evaluation of the mode coupling phenomenon even if the modes are not paired.

Regression model

The aim of a regression model is the generation of a surrogate model of first or second order out of experimental data. The experimental results are based on a design of experiment **X** and are given in the vector **Y**. The surrogate model is based on the experimental design **X** and expresses the linear approach:

$$\hat{\mathbf{Y}} = a + b\mathbf{X}, \tag{14}$$

with $\hat{\mathbf{Y}}$ being the vector containing the analytical results, a and b the coefficients of a linear function. The analytical results $\hat{\mathbf{Y}}$ are fitted to the experimental results **Y** by adjusting the coefficients a and b. The requirement for the estimation is the minimisation of the ordinary least squares

$$\text{ols}= \sum_{i=1}^{N} (\mathbf{Y} - \widehat{\mathbf{Y}})^2. \tag{14}$$

For comparison of the fitted function $\widehat{\mathbf{Y}}$ and the experimental data Y the coefficient of determination R^2 is calculated,

$$R^2= \frac{\sum_{i=1}^{N} (\overline{\mathbf{Y}}_i - \mathbf{Y}_i)^2}{\sum_{i=1}^{N} (\mathbf{Y}_i - \mathbf{Y}_i)^2}. \tag{14}$$

where $R^2 \in [0,1]$ and $\overline{\mathbf{Y}}$ representing the average of $\mathbf{Y}_i$. R^2 is a correlation coefficient between the experimental and analytical data. A R^2 near to 1 indicates a good fit.

Assessment of a Brake System[4]

The used brake system`s Finite Element model consists of 600.000 degrees of freedom and the elements are tetrahedron, pentahedron and hexahedron.

Considered brake system for further investigations.

For the evaluation a d-optimal design of experiment is used as described beforehand and 60 simulations are performed. As scatter variables the identified parameters are taken. The assessment is done by the given objective function. For this example an

4 See [26].

instability at 1.6 kHz is chosen. The calculated surrogate model is quadratic with interactions and has a coefficient of determination of $R^2 = 0.916$ which leads to a good prediction quality. For the assessment the parameter progression, the sensitivities and the instability probabilities are taken into account.

One has

$$q_{min} = -0.0713, \tag{15}$$

$$q_{max} = 0.0679. \tag{16}$$

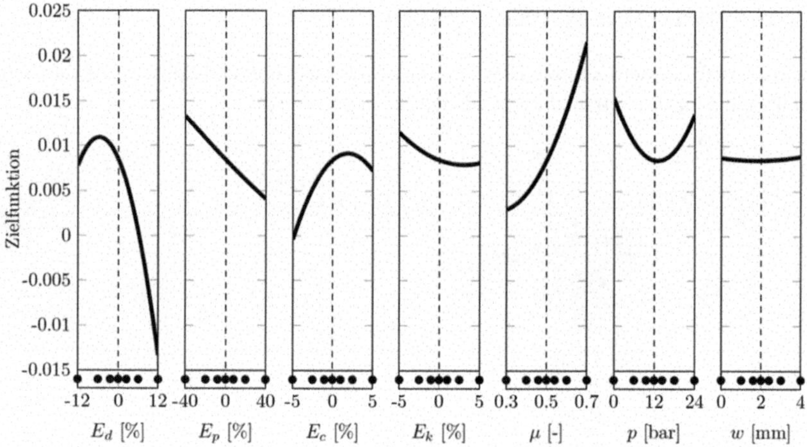

Visualisation of the parameter progression. The cut through the regression function is shown for nominal parameter settings. The dots at the bottom mark the sampling abscissa.

The parameter progressions show the biggest objective function`s influence by scatter of the brake disc`s Young`s modulus and the friction value. This conclusion is valid for the considered setting of the parameter which represent the average value.

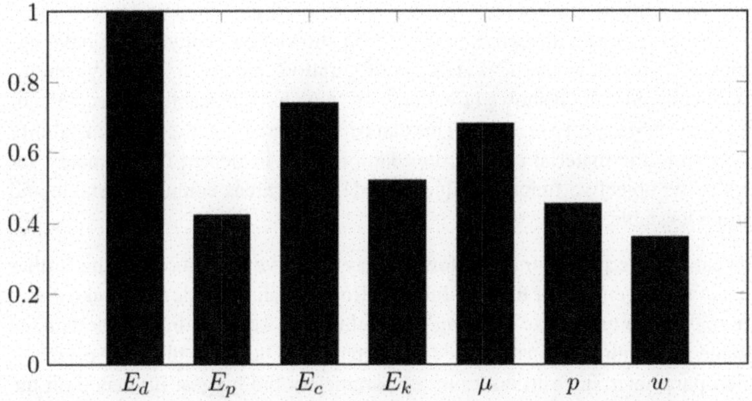

Visualisation of the parameter sensitivities causing changes in the objective function. A higher bar displays a greater influence to variations in the objective function. The values are normalised according to the parameter with the highest sensitivity.

For the sensitivity analysis an assessment of the parameter scatter in total is done. The biggest influence to the complete model is given by the brake disc`s Young`s modulus, the brake caliper component`s Young`s modulus and the friction value.

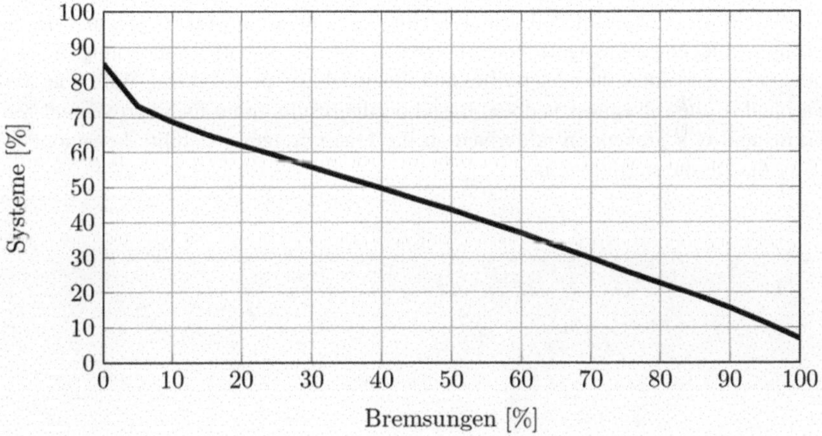

Entire instability probability with shows the ratio between systems with instable modes on the y-axis and braking events with instable modes on the x-axis. I.e. for 50% of the systems 40% of the braking events show instabilities.

The calculation of the instability probability depends on the defined parameter distributions. For an event instability probability, which shows how many events have unstable modes, 43% unstable and 57% stable configurations are given. If the system instability is considered, which shows how many systems have unstable modes, 85% of the brakes are unstable and 15% stable. This can be summarized in the entire instability probability. While the material and friction damping are neglected the boundary for instabilities is the objective function value 0.01. Higher values count as unstable and smaller values as stable.

The results can be interpreted in the following way and give a proposal for the further procedure. An improvement for the instability at 1.6 kHz can be made by a brake disc`s modification. A larger brake disc`s Young`s modulus shows a better behaviour than the nominal value concerning the instability`s characteristic. The brake disc is chosen as a modification parameter since it has from all parameters the biggest influence on the system. The surrogate model is calculated with interactions therefor by modification of one parameter all other parameter progressions are influenced.

The robustness of the brake system is improved by a constraint of the brake disc`s Young`s modulus. If this modification is not sufficient or possible an improvement by the modification of the brake caliper as the second largest influence parameter can be done. A robust solution can be obtained by the interaction`s utilization and a common modification. Changing the friction value is not possible since it is an environmental parameter.

For the classification of the procedure the following points are considered. By the calculation of an experimental design for the most important brake system`s influencing parameters a surrogate model is created. This is evaluated concerning the parameter progressions, the sensitivities and the instability probabilities. With it an assessment of the robustness is done. Based on the results room for improvement for the instability`s prevention is derived so the brake system`s stability behaviour is better than at the starting point.

Conclusion

This essay shows a useful tool to improve the noise behavior of a brake system and to reach faster higher maturity levels of the system. The aim is to ensure a noise free brake system over the entire term done by a reliability analysis.

The identification of the most important parameters through a sensitivity analysis and an analytical assessment combined with a reduction of the parameter set to them is carried out. For the robustness assessment and the definition of the maturity level the scatter and distributions of the parameters are identified on realistic examples and a design of experiment is created on basis of this scatter. To assess the design of experiment an objective function is derived which is based on the mode coupling phenomenon and allows to evaluate different frequency ranges.

Based on a calculated surrogate model the assessment of the brake system is done with low computation effort in a very short time and measures to improve the system can be derived. With the help of the regression model a reliability analysis and the calculation of instability probabilities are possible.

During the development process of a brake system this procedure is essential for a robust brake system and the maturity level management.

References

1. M. Treimer, B. Allert, K. Dylla, G. Müller, *Uncertainty quantification of the mode coupling phenomenon*, Journal of Sound and Vibration 388 (2017)
2. B.F. de Paula, G. Rejdych, T. Chancelier, G.V. des Roches, E. Balmes, *On the influence of geometry updating on modal correlation of brake components*, 18th symposium Vibrations, Shocks and Noise (2012)
3. R.F. Nunes, J. Will, V. Bayer, K. Chittepu, *Robustness evaluation of brake systems concerned to squeal noise problems*, Weimarer Optimierungs- und Stochastiktage, vol. 6 (2009)
4. N.M. Kinkaid, O.M. O`Reilly, P. Papadopulos, *Automotive disc brake squeal*, Journal of Sound and Vibration 267 (2003)
5. H. Lü, D. Yu, *Brake squeal reduction for vehicle disc brake system with interval parameters by uncertainty optimization*, Journal of Sound and Vibration 333 (2014)
6. N.P. Hoffmann, L. Gaul, *Friction induced vibration of brakes: Research fields and activities*, SAE Technical Paper Series 01-2579 (2008)
7. A. Akay, *Acoustic of friction*, Journal of the Acoustic Society of America 111 (2002)
8. T. Tison, A. Heussaff, F. Massa, I. Turpin, R.F. Nunes, *Improvement in the predictivity of squeal simulations: Uncertainty and robustness*, Journal of Sound and Vibration 333 (2014)
9. M. Nouby, D. Mathivanan, K. Sirinivasan, *A combined approach of complex eigenvalue analysis and design of experiment (DoE) to study disc brake squeal*, International Journal of Engineering, Science and Technology 1 (2009)
10. A. Buck, *Simulation von Bremsenquietschen (Brake Squeal)*, Shaker Verlag (2008)
11. F. Chen, C.A. Tan, R. Quagila, *Disc brake squeal – Mechanism, analysis, evaluation and reduction/prevention*, SAE International (2006)
12. B. Allert, G. Müller, *Brake-squeal simulation: An efficient approach to define countermeasures*, Proceedings of the 9th International Conference on Structural Dynamics, EURODYN (2014)
13. A.R. AbuBakar, H. Ouyang, *Complex eigenvalue analysis and dynamic transient analysis in predicting disc brake squeal*, International Journal of Vehicle Noise and Vibration 2 (2006)

14. A. Nobari, H. Ouyang, P. Bannister, *Statistics of complex eigenvalues in friction-induced vibration*, Journal of Sound and Vibration 338 (2015)

15. S. Oberst, J.C.S. Lai, *Nonlinear transient and chaotic interactions in disc brake squeal*, Journal of Sound and Vibration 342 (2015)

16. H. Ouyang, W. Nack, Y. Yuan, F. Chen, *Numerical analysis of automotive disc brake squeal: A contribution to the quality of prediction*. International Journal of Vehicle Noise and Vibration 1 (2005)

17. B. Allert, *Simulation von Bremsenquietschen*, Shaker Verlag (2014)

18. P.C. Müller, *Stabilität und Matrizen*, Springer (1977)

19. T.C. Sideris, *Ordinary differential equations and dynamical systems*, Atlantic Press (2013)

20. C. Ronniger, *Versuchsmethoden – Statistik* & DoE, BMW Group (2012)

21. A. Mitterer, *Optimierung vielparametriger Systeme in der Kfz-Antriebsentwicklung – Statistische Versuchsplanung und Künstliche Neuronale Netze in der Steuergeräteauslegung zur Motorabstimmung*, VDI Verlag (2000)

22. M.D. Morris, *Design of Experiments – An Introduction based on linear models*, Chapman & Hall (2011)

23. R.D. Cook, C.J. Nachtsheim, *A comparison of algorithms for constructing exact d-optimal designs*, Technometrics 22 (1980)

24. A. Heredia-Langner, M.W. Carlyle, D.C. Montgomery, C.M. Borror, G.C. Runger, *Genetic algorithms for the construction of d-optimal designs*, Journal of Quality and Technology 35 (2003)

25. A. Bajer, V. Belsky, S.-W. Kung, *Combining a nonlinear static analysis and complex eigenvalue extraction in brake squeal analysis*, SAE Technical Paper 01-3349 (2003)

26. M. Treimer, *Virtualisierung des Bremsgeräuschentwicklungsprozesses zur Sicherstellung einer robusten Anlaufqualität*, Shaker Verlag (2017)

Explicit non-linear model predictive control for vehicle stability control

Dipl.-Ing. Mathias Metzler, Dr. Davide Tavernini,
Prof. Dr. Aldo Sorniotti, Dr. Patrick Gruber

Centre for Automotive Engineering, University of Surrey, UK

P. Pfeffer (Ed.), *9th International Munich Chassis Symposium 2018*, Proceedings,
https://doi.org/10.1007/978-3-658-22050-1_49

Abstract

Nonlinear model predictive control is proposed in multiple academic studies as an advanced control system technology for vehicle operation at the limits of handling, allowing high tracking performance and formal consideration of system constraints. However, the implementation of implicit nonlinear model predictive control (NMPC), in which the control problem is solved on-line, poses significant challenges in terms of computational load. This issue can be overcome through explicit NMPC, in which the optimization problem is solved off-line, and the resulting explicit solution, with guaranteed level of sub-optimality, is evaluated on-line. Due to the simplicity of the explicit solution, the real-time execution of the controller is possible even on automotive control hardware platforms with low specifications. The explicit nature of the control law facilitates feasibility checks and functional safety validation. This study presents a yaw and lateral stability controller based on explicit NMPC, actuated through the electro-hydraulically controlled friction brakes of the vehicle. The controller performance is demonstrated during sine-with-dwell tests simulated with a high-fidelity model. The analysis includes a comparison of implicit and explicit implementations of the control system.

1 Introduction

The limit behavior of a vehicle is determined by tire nonlinearities. Therefore, model predictive control (MPC) implementations for vehicle stability control benefit from the inclusion of nonlinear system dynamics in the prediction model [1]. However, the resulting nonlinear optimization problem poses a major challenge for the real-time operation of the controllers, as demonstrated in [1] for an active steering system. Various alternatives to NMPC have been proposed in the literature, e.g., linear-time varying MPC [2], hybrid and switched MPC [3], or approximate NMPC [4]. [2] and [3] are based on the on-line solution of the MPC optimization problem.

In this study an explicit NMPC approach is used to solve the optimization problem off-linc, given the relatively limited number of parameters of the specific formulation. The obtained explicit solution is a sub-optimal multi-variable feedback law, which can be easily evaluated in real-time.

With respect to the published explicit NMPC work, such as [5], the contribution of this study is in the nonlinear vehicle model for control system design, which considers: i) the interaction of the longitudinal and lateral tire forces; and ii) the effect of the load transfers in cornering. i) and ii) are crucial to the exploitation of the benefits of NMPC for vehicle control at the limits of handling.

Moreover, the flexibility of the NMPC cost function formulation adopted in this study will allow ease of implementation of the controller on real vehicles, with different and usually rather complex performance requirements for the stability control function.

2 Prediction model

2.1 Lateral force and yaw moment balance equations

A double track vehicle model (Figure 1), with the yaw rate, $\dot{\psi}$, and vehicle sideslip angle, β, as state variables, is used as prediction model for the formulation of the optimal control problem. The longitudinal dynamics are neglected, as a constant speed, v, is assumed over the prediction horizon.

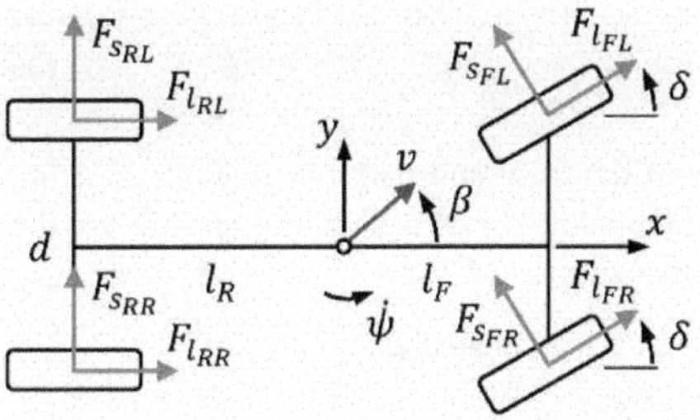

Figure 1 Notations and sign conventions of the proposed vehicle model formulation.

The Newton-Euler lateral force and yaw moment balance equations of the vehicle rigid body are:

$$\dot{\beta} = \frac{1}{mv}\left[\left(F_{l_{FL}} + F_{l_{FR}}\right)\sin(\delta - \beta) + \left(F_{S_{FL}} + F_{S_{FR}}\right)\cos(\delta - \beta) - \left(F_{l_{RL}} + F_{l_{RR}}\right)\sin\beta + \left(F_{S_{RL}} + F_{S_{RR}}\right)\cos\beta\right] - \dot{\psi} \tag{1}$$

$$\begin{aligned}\ddot{\psi} = \frac{1}{I_z}\Big\{&\left[\left(F_{l_{FL}} + F_{l_{FR}}\right)\sin\delta + \left(F_{S_{FL}} + F_{S_{FR}}\right)\cos\delta\right]l_F - \left(F_{S_{RL}} + F_{S_{RR}}\right)l_R \\ &+ \left(F_{l_{FR}}\cos\delta - F_{S_{FR}}\sin\delta + F_{l_{RR}}\right)\frac{d}{2} \\ &- \left(F_{l_{FL}}\cos\delta - F_{S_{FL}}\sin\delta + F_{l_{RL}}\right)\frac{d}{2}\Big\}\end{aligned} \tag{2}$$

where m is the vehicle mass; I_z is the yaw mass moment of inertia; l_F and l_R are the front and rear semi-wheelbases; d is the track width; δ is the steering angle, which is assumed to be equal on the left and right front wheels; $F_{l_{ij}}$ and $F_{S_{ij}}$ are the longitudinal and lateral tire forces, respectively, with the subscripts $i = F, R$ referring to the front and rear axles, and $j = L, R$ to the left and right sides.

As the vehicle stability controller ensures small sideslip angle values in any condition, and the relevant maneuvers imply rather limited values of steering angle, (1) and (2) are linearized with respect to the arguments β and $\delta - \beta$.

2.2 Vertical tire forces

The estimation of the vertical tire forces, $F_{z_{ij}}$, considers the load transfer associated with the lateral acceleration. The calculations are based on the constants defined in equations (3) and (4).

$$c_{Y_F} = \left(\frac{l_R}{l_F + l_R} \frac{h_F}{d} + \frac{K_F}{K_F + K_R} \frac{h'}{d} \right) m \tag{3}$$

$$c_{Y_R} = \left(\frac{l_F}{l_F + l_R} \frac{h_R}{d} + \frac{K_R}{K_F + K_R} \frac{h'}{d} \right) m \tag{4}$$

where h_F and h_R are the roll center heights of the front and rear suspensions; h' is the distance between the roll axis and the vehicle center of gravity; and K_F and K_R are the roll stiffness values of the front and rear suspensions.

From (3) and (4) the varying vertical tire loads are estimated as:

$$F_{z_{FL}} = F_{z_{FL},stat} - c_{Y_F} v\dot{\psi} \tag{5}$$

$$F_{z_{FR}} = F_{z_{FR},stat} + c_{Y_F} v\dot{\psi} \tag{6}$$

$$F_{z_{RL}} = F_{z_{RL},stat} - c_{Y_R} v\dot{\psi} \tag{7}$$

$$F_{z_{RR}} = F_{z_{RR},stat} + c_{Y_R} v\dot{\psi} \tag{8}$$

where $F_{z_{ij},stat}$ is the static vertical load on the individual tire. (5) – (8) neglect the effect of the sideslip angle rate contribution, $\dot{\beta}$. Future developments of this research will include consideration of the load transfers induced by the vehicle longitudinal dynamics.

2.3 Tire force model

The nonlinear behavior of the lateral tire forces is modeled with a simplified Pacejka Magic Formula [6]. The lateral forces in pure lateral slip conditions, $F_{s0_{ij}}$, are defined with nonlinear functions of the tire slip angles, α_{ij}.

$$F_{s0_{ij}} = D_{s_{ij}} \sin\left(C_{s_{ij}} \tan^{-1}\left(B_{s_{ij}} \alpha_{ij} \right) \right) \tag{9}$$

The coefficients in (9) are the lateral peak factor, $D_{s_{ij}}$; the lateral shape factor, $C_{s_{ij}}$; and the lateral stiffness factor, $B_{s_{ij}}$. $D_{s_{ij}}$ depends on $F_{z_{ij}}$, while $C_{s_{ij}}$ and $B_{s_{ij}}$ are constant.

The interaction between longitudinal and lateral tire forces in combined slip conditions is considered with a linear approximation of the tire friction envelope:

$$F_{s_{ij}} = F_{s0_{ij}}\left(1 + C_{ls_{ij}}\frac{F_{l_{ij}}}{D_{l_{ij}}}\right) \tag{10}$$

This means that each lateral tire force in combined slip, $F_{s_{ij}}$, is reduced by a factor, which depends on the current estimated longitudinal tire force, $F_{l_{ij}}$ (≤ 0 in braking), divided by its maximum value, $D_{l_{ij}}$, in pure longitudinal slip conditions. $D_{l_{ij}}$ is based on the estimated longitudinal tire-road friction coefficient, $\mu_{lp_{ij}}$, and the estimated vertical tire load:

$$D_{l_{ij}} = \mu_{lp_{ij}} F_{z_{ij}} \tag{11}$$

The constant $C_{ls_{ij}}$ in (10) is used as a tuning parameter.

The computation of the slip angle values, α_{ij}, neglects the track width of the vehicle and assumes small vehicle sideslip angles:

$$\alpha_{FL} = \alpha_{FR} = \delta - \beta - \frac{\dot{\psi} l_F}{v} \tag{12}$$

$$\alpha_{RL} = \alpha_{RR} = -\beta + \frac{\dot{\psi} l_R}{v} \tag{13}$$

2.4 Model parameters

Table 1 includes a selection of the prediction model parameters, representative of a Sport Utility Vehicle (SUV).

Table 1 Main parameters of the prediction model.

Symbol	Description	Quantity
m	Vehicle mass	1962 kg
I_z	Vehicle yaw mass moment of inertia	3382 kgm^2
l_F	Front semi-wheelbase	1.10 m
l_R	Rear semi-wheelbase	1.57 m
d	Front and rear track width	1.62 m
K_F	Front suspension roll stiffness	60000 Nm/rad
K_R	Rear suspension roll stiffness	50000 Nm/rad

3 Vehicle stability controller

The formulation of the NMPC vehicle stability controller is based on an optimal control problem defined by a cost function and constraints, in addition to the vehicle model equations in Section 2.

3.1 Optimal control problem

3.1.1 Cost function

The cost function, V, of the optimal control problem is defined as the integral of optimality criteria over the prediction horizon. More specifically, the longitudinal tire force distribution imposed by the stability controller minimizes the weighted deviation from reference targets for: i) the overall vehicle traction or braking force, which tracks $F_{x,ref}$; ii) the overall vehicle direct yaw moment, which tracks $M_{z,ref}$, i.e., the reference yaw moment to be generated by the longitudinal tire force difference between the left and right sides of the vehicle; iii) the braking ratio, which tracks b_{ref}, i.e., the desired ratio between the front and total longitudinal tire forces within the considered vehicle side; and iv) the yaw rate error, $e_{\dot{\psi}}$, and rear axle slip angle, α_R, bounded through the respective slack variables, $N_{e_{\dot{\psi}}}$ and N_{α_R}, discussed in detail in Section 3.1.3.

The prediction horizon is the interval between the current time, t_k, and the time at the end of the horizon, $t_f = t_k + t_s N_p$, defined by the number of prediction steps, N_p, and the discrete sampling time, t_s. The longitudinal tire forces are the control inputs to the prediction model. They can vary N_c times over the prediction horizon, where N_c is the number of control steps, and then they are kept constant from $t_k + t_s(N_c - 1)$ until t_f.

$$
\begin{aligned}
V(t_k) = \int_{t_k}^{t_f} \Bigg[& \frac{r_{u,F_x}}{{U_{sc,F_x}}^2}\left(F_{x,ref} - \left(F_{l_{FL}} + F_{l_{FR}} + F_{l_{RL}} + F_{l_{RR}}\right)\right)^2 \\
& + \frac{r_{u,M_z}}{{U_{sc,M_z}}^2}\left(M_{z,ref} - \frac{d}{2}\left(F_{l_{FR}} + F_{l_{RR}} - F_{l_{FL}} - F_{l_{RL}}\right)\right)^2 \\
& + \frac{r_{u,b}}{{U_{sc,b}}^2}\left(\left(1 - b_{ref}\right)F_{l_{FL}} - b_{ref}F_{l_{RL}}\right)^2 \\
& + \frac{r_{u,b}}{{U_{sc,b}}^2}\left(\left(1 - b_{ref}\right)F_{l_{FR}} - b_{ref}F_{l_{RR}}\right)^2 \\
& + \frac{q_{v,e_{\dot{\psi}}}}{{N_{sc,e_{\dot{\psi}}}}^2}{N_{e_{\dot{\psi}}}}^2 + \frac{q_{v,\alpha_R}}{{N_{sc,\alpha_R}}^2}{N_{\alpha_R}}^2 \Bigg] dt
\end{aligned}
\tag{14}
$$

The contribution of the different terms of $V(t_k)$ are weighted with the coefficients r_{u,F_x}, r_{u,M_z}, $r_{u,b}$, $q_{v,e_{\dot{\psi}}}$, and q_{v,α_R}. Appropriate scaling factors, i.e., U_{sc,F_x}, U_{sc,M_z}, $U_{sc,b}$, $N_{sc,e_{\dot{\psi}}}$, and N_{sc,α_R}, allow equivalent influences of the weights. The cost function formulation in (14) permits very different operating principles of the stability controller determined by the tuning choices, e.g., focused on both yaw rate and slip angle constraints, or the yaw rate tracking only, or the control allocation of the individual braking forces. The flexibility of this novel cost function formulation meets the diversified performance requirements of real vehicle implementations.

3.1.2 Reference generation

$F_{x,ref}$ is obtained from the demanded deceleration of the vehicle, $a_{x,ref}$. This is calculated from the driver input on the accelerator and brake pedals, and the condition of motion of the vehicle, e.g., in terms of yaw rate error and rear axle slip angle.

$$F_{x,ref} = m a_{x,ref} \tag{15}$$

For simplicity, in the preliminary implementation of this paper, focused on the sine-with-dwell test, $a_{x,ref}$ is set to 0, and will be considered as an additional parameter in the future.

The reference yaw moment, $M_{z,ref}$, is calculated based on the violation of the bounds on the yaw rate error and rear slip angle at t_k, with gains $K_{e_{\dot{\psi}}}$ and K_{α_R}, i.e., through proportional contributions. For appropriate signs, the terms relating the yaw rate error and rear slip angle to their bounds, $e_{\dot{\psi},max}$ and $\alpha_{R,max}$, are included.

$$M_{z,ref} = K_{e_{\dot{\psi}}} U_{sc,M_z} \frac{N_{e_{\dot{\psi}}}(t_k)}{N_{sc,e_{\dot{\psi}}}} \frac{\dot{\psi}_{ref}(t_k) - \dot{\psi}(t_k)}{e_{\dot{\psi},max}} - K_{\alpha_R} U_{sc,M_z} \frac{N_{\alpha_R}(t_k)}{N_{sc,\alpha_R}} \frac{\alpha_R(t_k)}{\alpha_{R,max}} \tag{16}$$

The ratio b_{ref} is considered constant in the current preliminary implementation of the controller.

For the desired handling of the controlled vehicle, a linear understeer characteristic with gradient K_{us} related to gravity g is considered, leading to the expression for the linear yaw rate reference, $\dot{\psi}_{lin}$, in (17). Both vehicle speed and steering angle are assumed constant over the prediction horizon.

$$\dot{\psi}_{lin} = \frac{v(t_k)}{l_F + l_R + \frac{K_{us}}{g} v(t_k)^2} \delta(t_k) \tag{17}$$

Taking into account the physical limitations of the vehicle, the maximum achievable yaw rate, $\dot{\psi}_{max}$, can be approximated in steady-state, based on the maximum lateral acceleration, $a_{y,max}$, in nominal friction conditions.

$$\dot{\psi}_{max} = \frac{a_{y,max}}{v(t_k)} \tag{18}$$

For smooth transition between the linear and maximum yaw rate, the use of a sigmoid function, with tuning parameters C_m and C_k, leads to a reference yaw rate, $\dot{\psi}_{ref}$, similar to the yaw rate associated with the understeer characteristic of the passive vehicle in nominal conditions.

$$\dot{\psi}_{ref} = C_m \dot{\psi}_{max} \frac{2}{\pi} \tan^{-1}\left(C_k \frac{\dot{\psi}_{lin}}{\dot{\psi}_{max}} \frac{\pi}{2} \right) \tag{19}$$

3.1.3 Constraint formulation

Soft constraints are used on the yaw rate error, $e_{\dot{\psi}} = \dot{\psi}_{ref} - \dot{\psi}$, with maximum and minimum bounds, $\pm e_{\dot{\psi},max}$:

$$-e_{\dot{\psi},max} - N_{e_{\dot{\psi}}}(t) \leq \dot{\psi}_{ref} - \dot{\psi} \leq e_{\dot{\psi},max} + N_{e_{\dot{\psi}}}(t) \tag{20}$$

As a consequence, the controller generates a direct yaw moment only when the yaw rate error exceeds the threshold indicated by positive values of the slack variable, $N_{e_{\dot{\psi}}}$.

Additional constraints are beneficial to vehicle stability for a wide range of driving conditions. For example, soft constraints are adopted on the rear slip angle, with maximum and minimum bounds, $\pm\alpha_{R,max}$:

$$-\alpha_{R,max} - N_{\alpha_R}(t) \leq \alpha_R \leq \alpha_{R,max} + N_{\alpha_R}(t) \tag{21}$$

The slack variables N_s, with $s = e_{\dot{\psi}}, \alpha_R$, represent the violation of the respective bounds, and are constrained to be positive:

$$N_{e_{\dot{\psi}}}(t) \geq 0 \tag{22}$$

$$N_{\alpha_R}(t) \geq 0 \tag{23}$$

Additionally, the longitudinal tire forces are constrained to be negative, since only braking torques can be applied by the friction brakes.

3.1.4 State transformation

The optimization problem, consisting of the cost function and the equality and inequality constraints, has the longitudinal tire forces, $F_{l_{ij}}(t)$, and slack variables, $N_{e_{\dot{\psi}}}(t)$ and $N_{\alpha_R}(t)$, as optimization variables in its continuous formulation. The sideslip angle, β, and yaw rate, $\dot{\psi}$, at the current time t_k, are required as initial conditions for the dynamic equality constraints. The vehicle speed, $v(t_k)$, and steering angle, $\delta(t_k)$, are considered as slowly varying parameters.

In the optimal control problem, the yaw rate error, $e_{\dot{\psi}} = \dot{\psi}_{ref} - \dot{\psi}$, and rear slip angle, α_R, are constrained. Therefore, a formulation of the optimization problem with the yaw rate error, $e_{\dot{\psi}}(t)$, and rear slip angle, $\alpha_R(t)$, instead of the sideslip angle, $\beta(t)$, and yaw rate, $\dot{\psi}(t)$, turned out to be beneficial. This is due to the fact that the state space exploration and splitting strategy of the explicit NMPC algorithm is based on orthogonal partitions. With the description using the new variables, the constrained variables are aligned with the coordinate axes of the orthogonal exploration space.

3.1.5 Settings

Table 2 reports the main settings for the optimal control problem of this study.

Table 2 Main settings of the optimal control problem.

Symbol	Description	Quantity
t_s	Sampling time	20 ms
N_p	Prediction steps	3
N_c	Control steps	1
$e_{\dot{\psi},max}$	Bound on yaw rate error	6 deg/s
$\alpha_{R,max}$	Bound on rear slip angle	4.5 deg

3.2 Multi-parametric nonlinear problem

The general optimal control formulation of Section 3.1 is reformulated into a form suitable for the numerical solution using the so-called direct methods.

According to the simultaneous approach, the ordinary differential equation constraints (1) – (2) are parameterized and discretized in time. The resulting algebraic equations are treated as additional nonlinear equality constraints. The forward Euler method, i.e., an explicit first order Runge-Kutta method with one stage, is applied as numerical integration scheme. The trajectories of control inputs and slack variables are parameterized by the vectors $U = \left[F_{l_{ij}}(t_k), \ldots, F_{l_{ij}}(t_{k+N_c-1})\right]$ and $N = \left[N_s(t_k), \ldots, N_s\left(t_{k+N_p}\right)\right]$, respectively, which are considered as unknown variables. In addition, the parameters of the state trajectory, described by $x(t) = [\beta(t), \dot{\psi}(t)]$, are summarized in the vector $X = \left[x(t_{k+1}), \ldots, x\left(t_{k+N_p}\right)\right]$, representing the intermediate states, and treated as additional unknown variables. The integral in the cost function (14) is approximated by a finite sum by applying a numerical integration method. The continuous inequality constraints are parameterized and relaxed to hold only at some discrete time instants, $\left\{t_k, \ldots, t_{k+N_p}\right\} \subset [t_k, t_f]$, leading to a finite number of discrete and parameterized constraint functions.

By implementing the finite parameterization and discretization on the continuous cost function and constraint functions, and by applying the integration scheme to the ordinary differential equations (ODEs), the following optimization problem, defined by the parameterized and discretized cost function V and constraint functions G, is obtained:

$$V^*\left(x_p(t_k)\right) = \min_{z} V\left(z, x_p(t_k)\right) \quad (24)$$

subject to:

$$G\left(z, x_p(t_k)\right) \leq 0 \quad (25)$$

The vector of parameters, $x_p(t_k)$, combines the initial states of the prediction model, $x(t_k)$, and slowly varying parameters, $p(t_k)$. The transformation, discussed in Section 3.1.4, leads to $x_p(t_k) = \left[e_{\dot{\psi}}(t_k), \alpha_R(t_k), v(t_k), \delta(t_k)\right]$ as vector of parameters. Similarly, the vector of optimization variables, z, combines the vector of input trajectory parameters, U, the slack variable trajectory parameters, N, and the state trajectory parameters, X, leading to $z = [U, N, X]$.

Assuming that it exists, the optimal solution of the receding horizon MPC strategy at t_k, denoted as z^*, represents the optimal open-loop trajectory over the prediction horizon. The longitudinal tire forces, $F_{l_{ij}}(t_k)$, are converted into reference hydraulic pressures for the electro-hydraulic braking system. At the next time instant, the optimization problem is solved with the updated parameter vector, making the MPC strategy a closed-loop approach.

The general formulation (24) – (25) is considered as a multi-parametric nonlinear problem with $x_p(t_k)$ as parameter, i.e., for which the optimal solution z^* has to be found over a range of values of x_p.

4 Explicit NMPC

Since the online solution of (24) – (25) poses significant challenges for the real-time implementation due to the involved nonlinearities, this study investigates an explicit solution of the multi-parametric problem. However, for multi-parametric nonlinear programming (mp-NLP) problems it is in general not possible to derive an explicit solution in an exact form. Therefore, an explicit NMPC algorithm, using multi-parametric quadratic programming (mp-QP) approximations of the mp-NLP, is developed to derive a sub-optimal solution with guaranteed levels of sub-optimality.

4.1 Algorithm using mp-QP approximations

The algorithm is a combination of the approximate mp-NLP algorithm incorporating global optimization tools, as proposed in [7] and [8], and the multi-parametric quadratic approximation (mp-QA) algorithm, as proposed in [9] and [8], and revisited in [10].

The multi-parametric nonlinear program is locally approximated with a multi-parametric quadratic program, leading to an approximate solution of the mp-NLP, consisting of the solution of local mp-QP sub-problems on orthogonal partitions. The accuracy of the approximation is controlled via iterative and recursive partitioning of the parameter space. The partitions are refined based on heuristic splitting rules in parts of the parameter space, to improve accuracy of the local mp-QP approximation, and meet the tolerances and accuracy specifications in terms of sub-optimality bounds on the cost, solution, and maximum constraint violation.

The local approximation of the mp-NLP with an mp-QP problem is derived from second order Taylor series expansion of the cost function, V, in (24), and first-order Taylor series expansion of the constraint function, G, in (25) about the linearization point $(z_0, x_{p,0})$, with z_0 corresponding to the optimal solution, $z_0 = z^*(x_{p,0})$.

The algorithm can be summarized as follows:

- Initialize the orthogonal partition, Π, to the hyper-rectangle in the parameter space, $\mathbb{X}$, to be explored, i.e., $\Pi := \mathbb{X}$, and mark the hyper-rectangle, $\mathbb{X}$, as unexplored.
- While there are unexplored partitions, select any unexplored hyper-rectangle $X_0 \subseteq \Pi$, compute the volume and center point, $x_{p,0}$, and search for an optimal solution, $z_0 = z^*(x_{p,0})$, to the problem (24) – (25) for x_p fixed to $x_{p,0}$ by solving an NLP.
- If a feasible solution for (24) – (25) was found at point $x_{p,0}$, search for an optimal solution for (24) – (25) for x_p fixed to some test points $x_{p,i}$ by solving an NLP; otherwise mark X_0 to be split.
- If the problem (24) – (25) has a feasible solution at all points $x_{p,i}$, compute the local mp-QP problem, derived by Taylor series expansion at $(z_0, x_{p,0})$; otherwise mark X_0 to be split.
- If a feasible explicit solution, $\hat{z}(x_p)$, is found for the local mp-QP problem, compute an estimation, $\hat{\varepsilon}$, of the cost error bound, ε, an estimation, $\hat{\rho}$, of the solution error bound, ρ, and an estimation, $\hat{\gamma}$, of the maximum constraint violation, γ, in X_0; otherwise mark X_0 to be split.
- If the estimates of the bounds, $\hat{\varepsilon}$, $\hat{\rho}$, and $\hat{\gamma}$, are bigger than user-defined tolerances, $\tilde{\varepsilon}$, $\tilde{\rho}$, and $\tilde{\gamma}$, mark X_0 to be split; otherwise mark X_0 as feasible and explored.

- While there are hyper-rectangles in Π that are marked to be split, select any hyper-rectangle $X_0 \subseteq \Pi$ marked to be split and split X_0 into new hyper-rectangles, $X_0, \dots, X_N$, by applying a heuristic splitting rule and mark the new hyper-rectangles, $X_0, \dots, X_N$, as unexplored.

In the implementation of the routine, the primal-dual interior-point algorithm with a filter line-search method IPOPT [11] is used to solve the nonlinear programming problems. For the solution of the mp-QP problems, the MPT 3.0 toolbox [12], an open source toolbox for parametric optimization, is embedded in the suggested algorithm.

The execution of the algorithm leads to an explicit approximate solution of the mp-NLP (24) – (25), including the solution of the mp-QP sub-problems over each orthogonal partition and consisting of piecewise affine solution functions and corresponding polyhedral critical regions. Therefore, the main computational effort is carried out off-line. The on-line computation reduces to the identification of the polyhedral region for a given parameter vector, x_p, and the evaluation of the associated multi-variable affine feedback law.

4.2 Post-processing and on-line evaluation

Appropriate post-processing generates and exports real-time executable code for the on-line evaluation of the explicit NMPC. The actual evaluation is a three-stage process:

- For a given parameter, $x_p(t_k)$, the first layer identifies the corresponding orthogonal partition and the associated local mp-QP sub-problem by traversing a memory-optimized binary search tree, generated based on [13] and implemented in the MPT 3.0 toolbox [12].
- Having determined the orthogonal partition, the second layer identifies the corresponding polyhedral critical region of the affine state-feedback law, by traversing a binary search tree or alternatively a sequential search data structure. It evaluates the feedback law, and returns the value.
- In the last layer, the feasibility of evaluation is reviewed and, in case of infeasibility, the control action applied at the previous time step is provided as reasonable alternative.

Since the dynamic system (1) – (2) is point symmetric with respect to the origin for a given speed, the exploration space can be reduced by bisection.

4.3 Implementation

The execution of the explicit NMPC algorithm, with suitable settings for the specific vehicle stability control application, leads to an explicit solution in the 4-dimensional parameter space consisting of 165 orthogonal partitions including 4886 polyhedral regions in total. The considered range of the parameters was chosen to keep the complexity of the explicit controller moderate, and needs to be reviewed for the implementation of the controller on a vehicle demonstrator.

5 Simulation results

5.1 Sine-with-dwell test

The behavior of the vehicle with the explicit NMPC based stability controller is simulated along the sine-with-dwell test, as shown in Figure 2, to demonstrate the fulfilment of the performance requirements of the regulation UN/ECE-R13H [14].

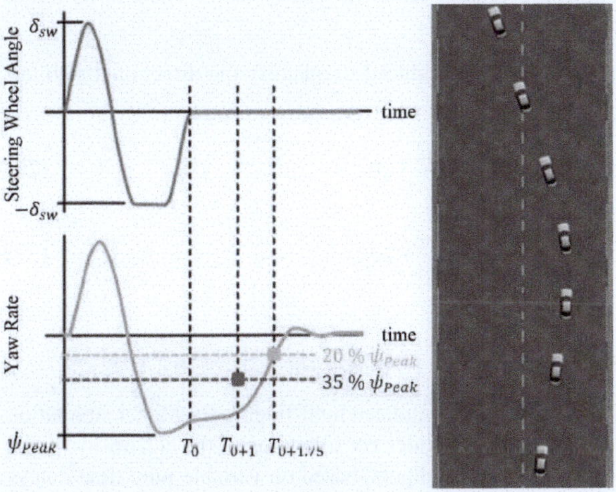

Figure 2 Sine-with-dwell test; figure adapted from [15].

The test conditions are explained in [14] and can be summarized as follows:

- Vehicle coasting in high gear at 80 km/h.
- $A_{\delta_{sw}}$, defining the steering wheel amplitude, is the steering wheel angle, δ_{sw}, at the lateral acceleration $a_y = 0.3$ g, determined during a slowly increasing steering maneuver at 80 km/h.
- In each sine-with-dwell test run of the test series, the steering wheel amplitude is increased from run to run by 0.5 $A_{\delta_{sw}}$, starting with 1.5 $A_{\delta_{sw}}$ for the initial run.

The directional or yaw stability criteria to be fulfilled are:

- The yaw rate at time T_{0+1}, i.e., 1 s after T_0, shall be < 35% of the yaw rate peak, $\dot{\psi}_{Peak}$.
- The yaw rate at time $T_{0+1.75}$, i.e., 1.75 s after T_0, shall be < 20% of the yaw rate peak, $\dot{\psi}_{Peak}$.

$\dot{\psi}_{Peak}$ is defined as the first peak value of yaw rate, recorded after the steering wheel angle changes sign, and T_0 is defined as the time at the completion of the steering wheel input.

Moreover, the following responsiveness criterion must be met by vehicles with a mass ≤ 3500 kg, such as the one of this study:

- The lateral displacement of the vehicle center of gravity with respect to its initial straight path must be > 1.83 m.

The performance indices J_1 and J_2 are introduced to quantify the directional stability criteria:

$$J_1 = \frac{\dot{\psi}(T_{0+1})}{\dot{\psi}_{Peak}} 100 \tag{26}$$

$$J_2 = \frac{\dot{\psi}(T_{0+1.75})}{\dot{\psi}_{Peak}} 100 \tag{27}$$

5.2 Results

The results are based on an experimentally validated high-fidelity CarMaker simulation model of a case study SUV. The model includes consideration of the electro-hydraulic braking system pressure dynamics at the calipers, based on variable pure time delays and first order dynamics, parameterized through experimental tests on a real plant. The steering wheel amplitude, $A_{\delta_{sw}} = 25.8$ deg, was determined according to [14].

Figure 3 reports the response of the passive vehicle, i.e., the vehicle without the stability controller, during a sine-with-dwell test with a steering wheel amplitude $\delta_{sw} = 103.2$ deg, corresponding to test run 6 of the test series. The large values of rear axle slip angle and vehicle sideslip angle show that vehicle controllability is compromised. Similar behavior can be observed for the following test runs of the passive vehicle, with a maximum steering wheel amplitude $\delta_{sw} = 270.9$ deg for the final test run 19.

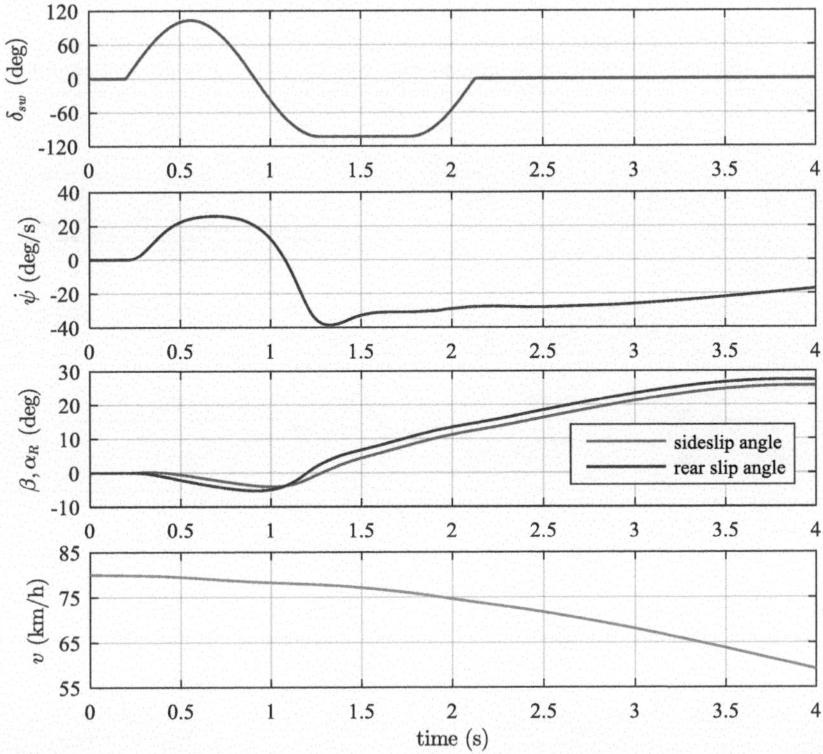

Figure 3 Passive vehicle.

Figure 4 shows the vehicle response of the controlled vehicle with the implicit NMPC based on the proposed formulation, and used as benchmark for the explicit NMPC. The plots are for test run 6, with the same steering wheel input as for the test with the passive vehicle reported in Figure 3. In contrast to the uncontrolled vehicle, the rear axle slip angle and vehicle sideslip angle have moderate values, which demonstrates the effectiveness of the proposed control system. The violation of the yaw rate error bound (indicated by the dash dotted lines in the relevant subplot of Figure 4) is significantly reduced by the control action, i.e., the yaw rate profile follows the steering wheel input profile. Moreover, the decrease of vehicle speed due to differential braking is at an acceptable level, and the final value of vehicle speed is higher than for the passive vehicle. The interdependence of the two main objectives, namely reducing the violation of the bound on the yaw rate error and rear axle slip angle, can be influenced by the appropriate selection of the weights in the cost function (14). In this setting, priority is given to the rear slip angle contribution, with moderate weights on the yaw rate error contribution.

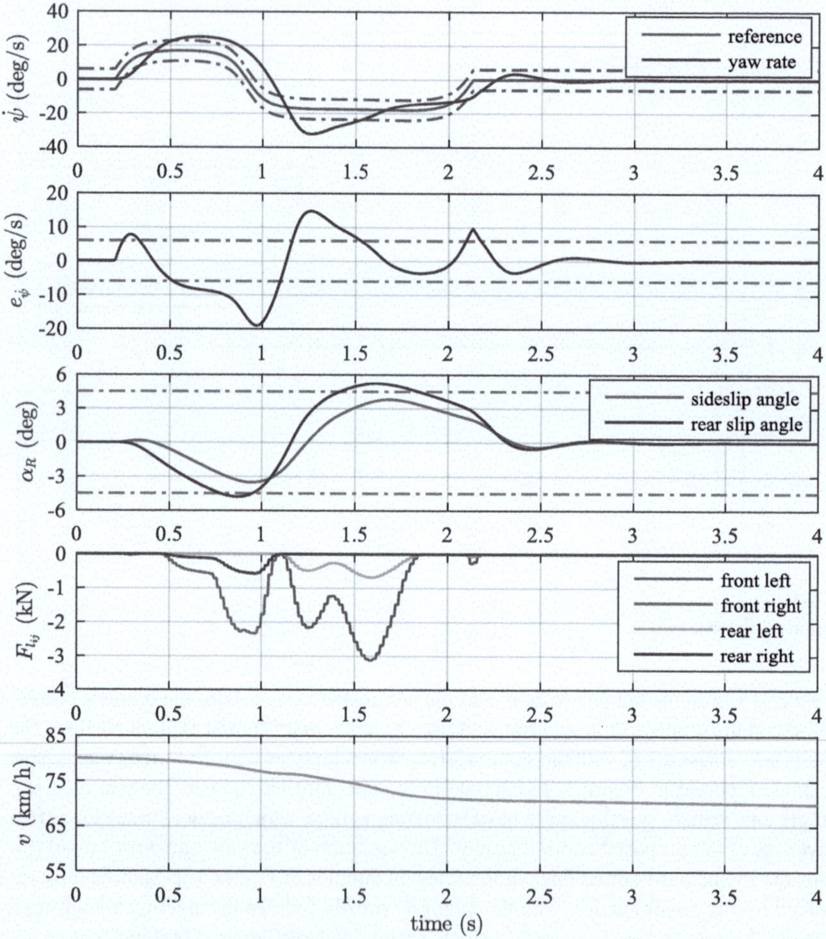

Figure 4 Vehicle with the implicit NMPC.

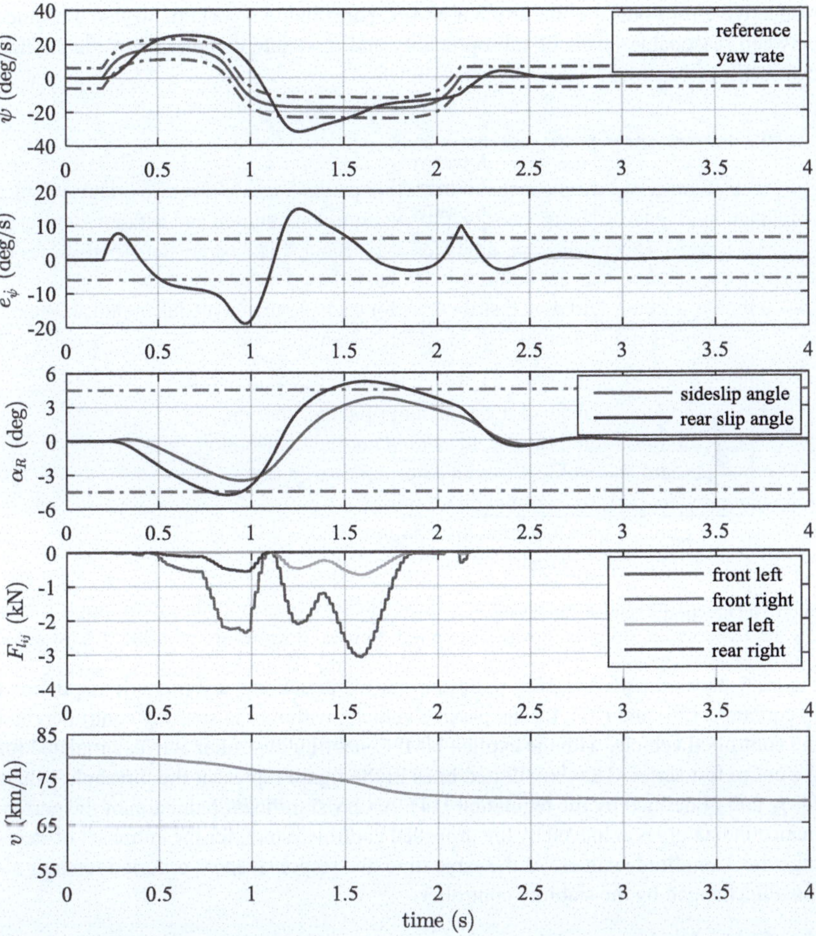

Figure 5 Vehicle with the explicit NMPC.

Figure 5 reports the vehicle response of the controlled vehicle with the explicit NMPC based stability controller. In comparison with the implicit NMPC in Figure 4, only a very marginal difference can be observed in the control action. More importantly, there is no observable difference in the vehicle response. The difference in results between implicit NMPC and explicit NMPC can be influenced by choosing the tolerances on the respective error bounds of the explicit NMPC partitioning algorithm, to find a balance between reasonable levels of sub-optimality and moderate complexity of the explicit controller.

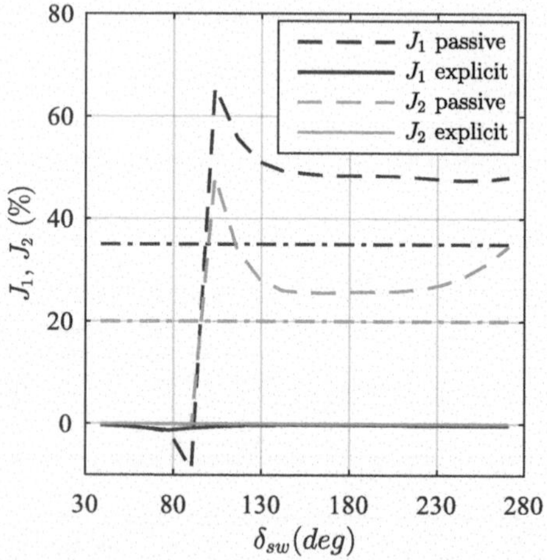

Figure 6 Directional stability criteria.

Figure 6 shows the performance indicators for the directional stability criteria, defined in equations (26) and (27), for the passive vehicle, without the stability controller, and the controlled vehicle, with the explicit NMPC stability controller active, simulated for the entire test series. The dash dotted lines in the figure represent the threshold values for J_1 and J_2 defined by the regulation [14]. A very significant reduction of the performance indicators is achieved by the proposed control system, for the entire set of steering wheel amplitude values. At the same time, the responsiveness criterion remains almost unchanged by the stability controller.

6 Conclusion

This study presented the design, implementation and performance assessment of an explicit NMPC for yaw and lateral stability control of a vehicle with an electro-hydraulic braking system. The preliminary simulations results demonstrate the effectiveness of the explicit NMPC approach, which does not imply any performance decay with respect to the corresponding implicit NMPC. Significant improvements of the vehicle response in sine-with-dwell tests are brought by the vehicle stability controller. The vehicle with the controller remains within the specified thresholds of the stability criteria of the sine-with-dwell test.

Future steps will focus on:

- The detailed analysis of the performance benefits of the individual nonlinearities included in the prediction model.
- The comparison with stability control systems based on classical control structures.
- The revision and extension of the current explicit NMPC based stability controller, to meet the practical requirements for its implementation on a vehicle demonstrator.

Acknowledgements

The Authors would like to thank MTS Ltd for the financial support to the project.

References

[1] P. Falcone, F. Borrelli, J. Asgari, H. E. Tseng, and D. Hrovat, "Predictive Active Steering Control for Autonomous Vehicle Systems," *IEEE Trans. Control Syst. Technol.*, vol. 15, no. 3, pp. 566–580, 2007.

[2] O. Barbarisi, G. Palmieri, S. Scala, and L. Glielmo, "LTV-MPC for Yaw Rate Control and Side Slip Control with Dynamically Constrained Differential Braking," *Eur. J. Control*, vol. 15, no. 3–4, pp. 468–479, 2009.

[3] S. Di Cairano, H. E. Tseng, D. Bernardini, and A. Bemporad, "Vehicle yaw stability control by coordinated active front steering and differential braking in the tire sideslip angles domain," *IEEE Trans. Control Syst. Technol.*, vol. 21, no. 4, pp. 1236–1248, 2013.

[4] M. Canale, L. Fagiano, and V. Razza, "Approximate NMPC for vehicle stability: Design, implementation and SIL testing," *Control Eng. Pract.*, vol. 18, no. 6, pp. 630–639, 2010.

[5] P. Tøndel and T. A. Johansen, "Lateral Vehicle Stabilization Using Constrained Nonlinear Control," in *European Control Conference (ECC)*, pp. 1887–1892, 2003.

[6] H. B. Pacejka, *Tire and Vehicle Dynamics*, 3rd ed. Oxford: Elsevier Ltd, 2012.

[7] A. Grancharova, T. A. Johansen, and P. Tøndel, "Computational Aspects of Approximate Explicit Nonlinear Model Predictive Control," in *Assessment and Future Directions of Nonlinear Model Predictive Control*, Berlin, Heidelberg:

Springer, pp. 181–192, 2007.

[8] A. Grancharova and T. A. Johansen, *Explicit Nonlinear Model Predictive Control, Theory and Applications*, vol. 429. Springer Science+Business Media, 2012.

[9] T. A. Johansen, “On multi-parametric nonlinear programming and explicit nonlinear model predictive control,” *Proc. 41st IEEE Conf. Decis. Control*, vol. 3, pp. 2768–2773, 2002.

[10] L. F. Domínguez and E. N. Pistikopoulos, “Recent Advances in Explicit Multiparametric Nonlinear Model Predictive Control,” *Ind. Eng. Chem. Res.*, vol. 50, no. 2, pp. 609–619, 2011.

[11] A. Wächter and L. T. Biegler, “On the implementation of an interior-point filter line-search algorithm for large-scale nonlinear programming,” *Math. Program.*, vol. 106, no. 1, pp. 25–57, 2006.

[12] M. Herceg, M. Kvasnica, C. N. Jones, and M. Morari, “Multi-Parametric Toolbox 3.0,” in *European Control Conference (ECC)*, pp. 502–510, 2013.

[13] P. Tøndel, T. A. Johansen, and A. Bemporad, “Evaluation of piecewise affine control via binary search tree,” *Automatica*, vol. 39, no. 5, pp. 945–950, 2003.

[14] UNECE, *Addendum 12-H: Regulation No. 13-H, Revision 2, Uniform provisions concerning the approval of passenger cars with regard to braking*. 2011.

[15] A. Lutz *et al.*, “Simulation methods supporting homologation of Electronic Stability Control in vehicle variants,” *Veh. Syst. Dyn.*, vol. 55, no. 10, pp. 1432–1497, 2017.

TIRE.WHEEL.TECH SECTION

WHEEL TECHNOLOGIES AND TRENDS

The Maxion flexible wheel with MICHELIN Acorus Technology

Ralf Duning, Maxion Wheels EAAP Holding GmbH;
Daniel Walser, Michelin Recherche et Technique SA, Switzerland;
P.-E. Sorel, MICHELIN Incubateurs Europe – ACORUS, France

P. Pfeffer (Ed.), *9th International Munich Chassis Symposium 2018*, Proceedings,
https://doi.org/10.1007/978-3-658-22050-1_50

What is pinch shock? What does it mean to tire and wheel?

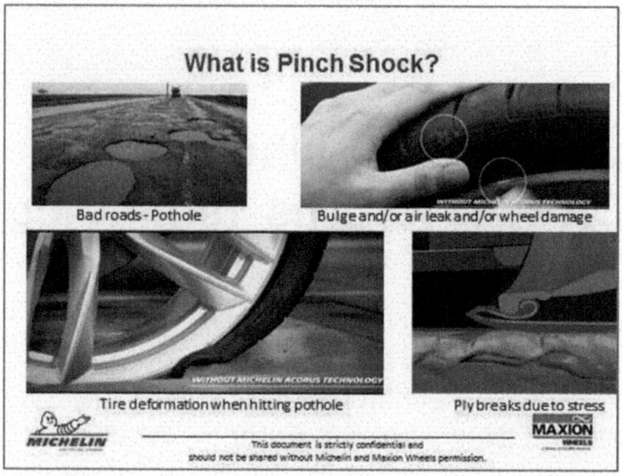

Pothole or road damage can happen everywhere, although today it mainly happens in emerging markets, due to a poor quality road network. It also happens in countries with poorly maintained roads or in countries with snow / ice, during spring. Roadworks, construction work or manholes can also damage wheels and tires.

When hitting a curb or a pothole the tire is hit very hard and its sidewall is deformed. It will pinch and stress the cables inside the tire up to rupture, creating a bulge on the sidewall or even a puncture. If the hit is severe, it can also damage the wheel, or create a small crack that after a few thousand miles will propagate and lead to an air leak.

The main factor for a pinch shock is the tire sidewall height. When the tire sidewall height is too small (<100mm for a car, 115 for an SUV, but the threshold can be lower depending on the country) then the tire is at risk, whatever the tire structure (one ply or double ply). For this reason, most tire manufacturers do not allow for the sale of those low sidewall tire in countries at risk.

But designing a car is all about proportions, and big wheels are part of premium car design. Customers ask for big wheels, in all countries. The just cannot understand why they cannot have them although the automotive press always shows those big wheels.

At the same time, pinch shock is a mobility issue: you do not want to get stuck at night because you hit a pothole.

Sidewall damage due to pinch shock is damaging customer satisfaction

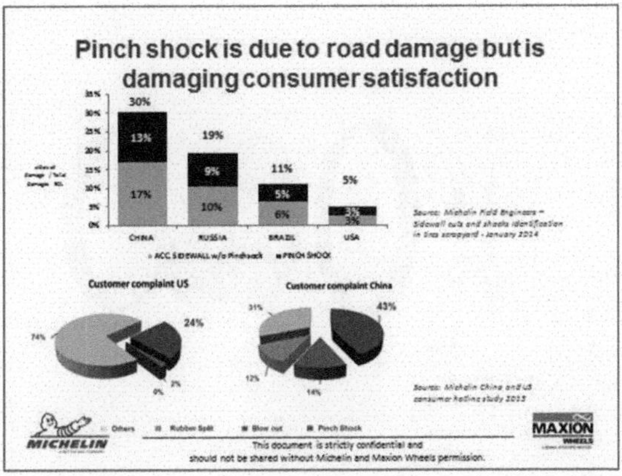

A few years ago, the Michelin Group started working to solve (on solving) that pinch shock issue, going beyond the tire. It appeared it was essential to car drivers.

At first Michelin wanted to know how bad the problem was. So we sent our field engineers into tire scrapyards to look at the damages on tires. And these are the tires that people use, not the big ones they wish to have used. It appeared that as much as 13% of tires in Chinese scrapyards have a pinch shock. And those tires have a high sidewall height, and also a reinforced ply to cope with the bad roads.

So pinch shock is a real issue, but it is also a source of customer dissatisfaction. Customer complaints to our hotline were focused on pinch shock (up to 43% in China). For them, it is not only a question of loss of mobility, but it can mean a lot of frustration or even endanger their personal security if they get stranded in an unknown or dangerous area.

Michelin and Maxion Wheels have decided to join forces for a new type of wheel

The design of the Acorus technology appeared to have some potential. Adding a rubber part to protect tire and rim in extreme cases proved to be a good idea.

Its functional elements are:

- A "wheel body" which is the aluminum part of the wheel. It is not technically a wheel because it does not touch the tire, but looks like it.
- Two rubber flanges that are identical. They are made of a tire structure (rubber, plies, beads) and are produced following a normal tire assembling technology. They are mounted on the wheel as a tire is mounted on a wheel.

Additional aesthetic elements can be added to make the wheel look nice. The reason why a car owner chooses a wheel over another is most of the time related to design. This wheel cannot look ugly!

This wheel design is the result of a joint work between Maxion and Michelin: both Michelin and Maxion designers and engineers joined forces to design a wheel that looks nice. It really is a design proof of concept.

How it works

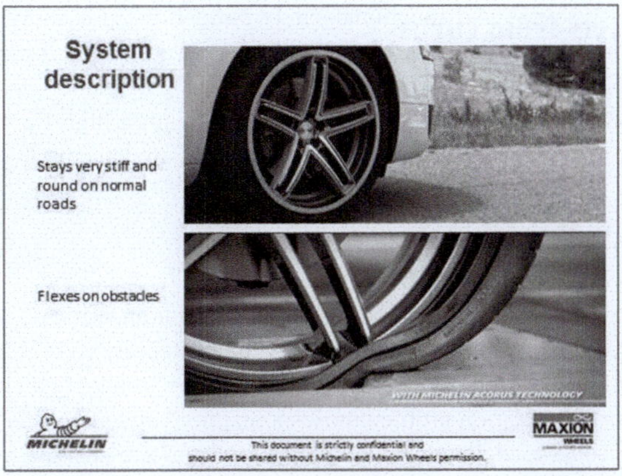

When the car hits a curb or a pothole very hard, the flange flexes. By doing so, it protects the tire that does not pinch. The cables and the ply inside the tire are not put at stress. The wheel is also protected, and the driver can continue its route.

In case of normal driving conditions, this wheel is actually not very "flexible"! It does not deform and stays firm. It behaves like a standard wheel.

Performance

Acorus protects from curbs

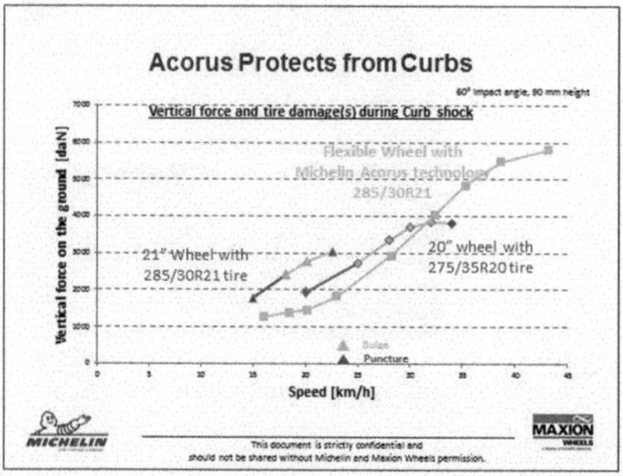

The following graph is a curb test – here a 90mm high curb, impacted with a 60° angle. The x axis is speed, while the y axis is the peak vertical force going up to the chassis, measured on the ground.

For a regular 275/35R20 tire and a 20 inch standard alloy wheel, the tire has a bulge when it hits at 24 km/h, and a puncture when it hits at 35 km/h.

For a 285/30R21 tire (same Overall Diameter than the previous 20 inch tire) and a 21" standard alloy wheel, the tire has a bulge at a lower speed (18 km/h) and a puncture at 22 km/h. For the same speed, the peak force going up to the chassis is significantly higher than with the 20 inch tire & wheel.

For a flexible wheel fitted with the same 21" tire, the test is stopped at 43 km/h, demonstrating the performance on curb test, largely beyond what is achievable with tire redesign. For the same speed, the peak force going up to the chassis is always lower than even the 20 inch wheel & tire.

Currently on a different curb test for a given car manufacturer, the threshold for bad roads countries was 40 km/h and 30 km/h for Europe. On a normal wheel the tire failed at 32 km/h. On a flexible wheel the test was stopped at 61 km/h to avoid the braking of suspension elements.

Acorus solves pothole issue

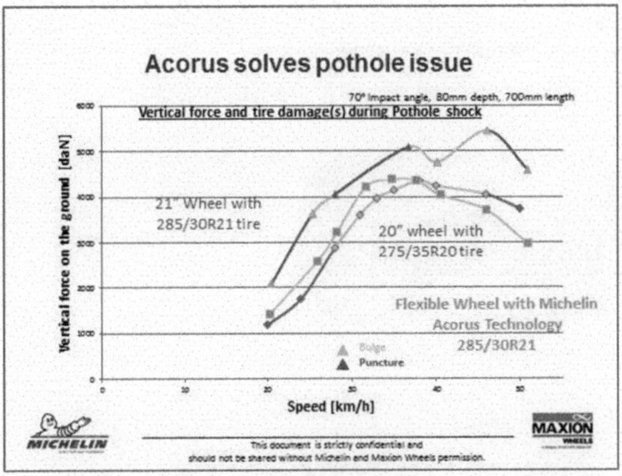

The graph here is the same as the previous one: speed and force going up to the chassis.

A pothole is very destructive because when driving over it, the suspension is pushing vertically down the wheel, tire and brakes. To pass it the wheel has to resist the pothole and come back to its original place in the wheel arch.

When driving over a pothole, if the speed is low, the tire and suspension work together and the pothole is passed with no damage. If the speed is very quick, the suspension does not have the time to push the wheel down. The wheel and tire fly over the pothole. But at a critical speed the tire is pinched very hard. This is what happens most of the time with potholes: drivers see it too late and tend to brake not enough.

As with the curb test, the flexible wheel passes the pothole with no damage. The tire and the wheel are protected.

Design Opportunity/Challenges

Bigger wheels are positively influencing the total car design by about 30%. Concept and first studio sketches are starting with super big wheels, while coming closer to reality the wheels are getting smaller again to compromise between design/cost as well as between comfort/performance.

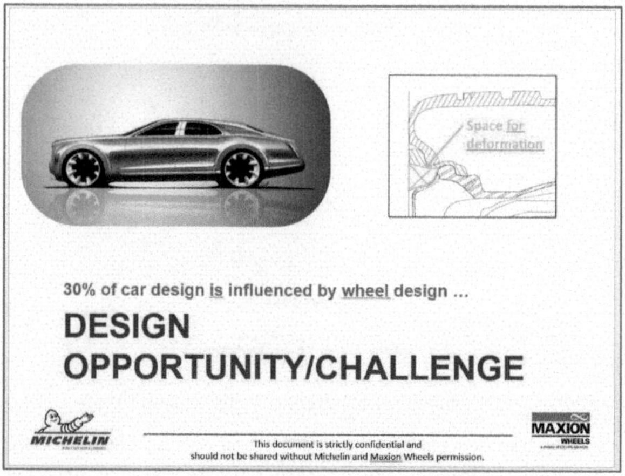

Big wheels are one of the important car design elements. First studio sketches are with big wheels for nicer car design. The Flexible wheel is a solution for bigger wheels in all countries.

The Flexible wheel needs space between the ends of the metal body spoke and the Acorus flange to allow for deformation during a pothole or curb stone shock event.

This opens a lot of new options to design the wheel, showing the technology or bridging the gap using styling elements for more conventional designs.

Design Opportunity/Challenges

For our first prototypes and show cars we used an actual and most common 5 spoke wheel design with styling elements to bridge the needed functional area for the Acorus flanges. As a result we see a nice big wheel with a circular trim, defining the bigger outer diameter of the wheel and a functional Star Shaped Trim as a spoke extender.

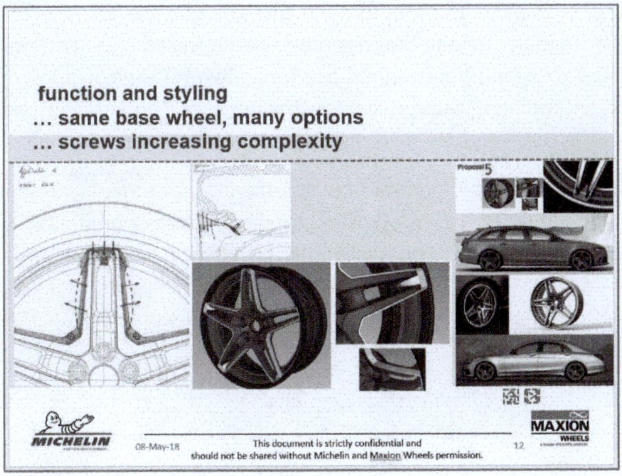

Patent pending for the function of the styling elements.
A circular trim is clipped to the Acorus flange to show the full and bigger outer dimension and a second star shaped trim is screwed to the metal body extending the length of the spokes.

As seen in our short movie, the plastic styling elements are flexible enough to deform during a pothole or curb shock event. They can bend and flex but don't break.

The screws are adding complexity to the wheel assembly and also a certain additional cost, which can be avoided. For our next generation of stylings and prototypes we used a more simple but effective solution to underline the bigger wheel diameter.

Design Opportunity/Challenges

The base for this solution is a simpler trim tooling, which allows for different spoke extension solutions. Changing only 10 to 15% of the trim tooling (using tool inserts), we are able to create different wheel designs. And the best thing is that it all works without any additional screws to fix the trim to the metal body. The trim will be "clipped on" the Acorus flange after the tire has been assembled.

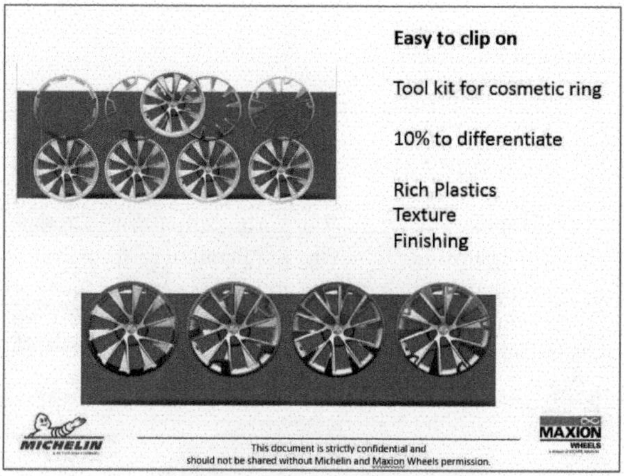

Easy "clip on" solution. Modular tool system to allow for differentiation using different shapes, material, textures and finishes. Using the same metal wheel body we can generate a family of different stylings.

This solution is creating value by using simpler trim tooling technology. Special design features are possible by using different rich plastics, colours, textures and special finishes. More styling options using the same high volume metal wheel body is, we believe, a clear advantage.

Value proposition

Customer complaints and consumer feedback generated the need to solve pinch shock and sidewall tire damage, especially in countries with bad or damaged road conditions. Acorus can avoid tire pinch shock damage. Acorus enables you to use "green", low rolling resistance tires in all countries.

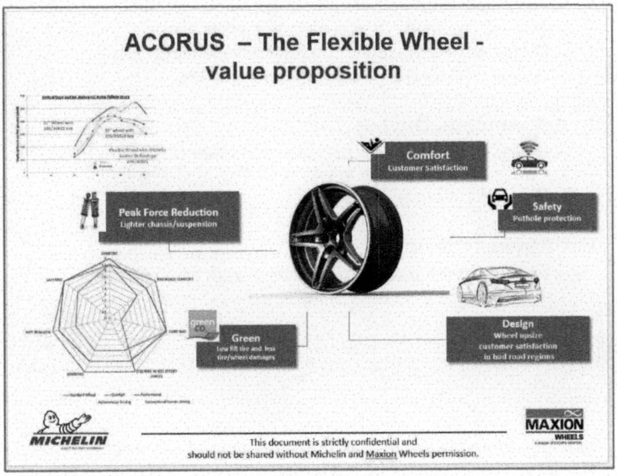

Acorus is adding technology to the wheel and new features are creating several value propositions to satisfy different customer needs. One System with multiple key value propositions:

Design – Green – Comfort – Safety

Acorus is more than a pothole protection system, allowing for bigger wheels and low rolling resistance tires in all markets. Acorus can also increase comfort and safety for autonomous cars: a mechanical solution for a digital challenge.

Summary and Outlook

Together with our OEM customers we see 3 steps to introduce Acorus into the mass market, using the same base function – **peak force reduction** during pothole and curb stone events.

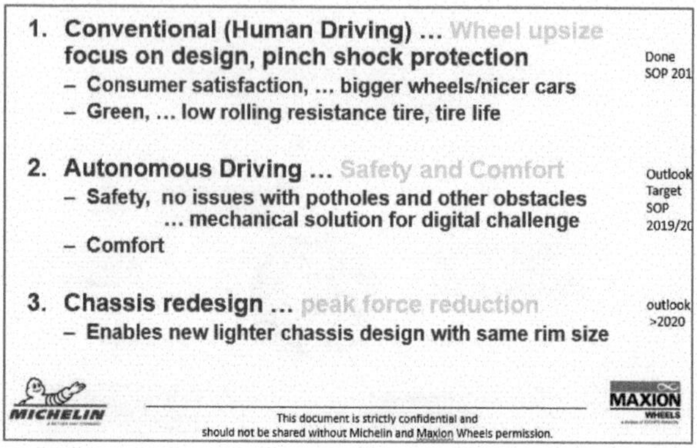

We developed two standardized Acorus versions. First for conventional cars focusing on driving performance. Second for autonomous cars focusing on comfort & safety.

Design – Green – Comfort – Safety

For conventional cars, the focus remains on Design, on wheel upsize, on driving performance and the ability to use "green" low rolling resistance tires in all countries. Maxion and Michelin developed the Acorus Flange and the wheel metal body structure to achieve high performance, best driving dynamics, including minimized body roll, a good steering feedback and centre feel for most applications.

For Autonomous cars the driving dynamics at high speed and the steering on-centre feel is less important. The Acorus characteristic will be used to maximize comfort and safety.

And as a step 3 the system may enable a chassis redesign with lighter suspensions due to reduced peak forces caused by shock events.

Guidelines for the testing and inspection of plastic wheels for passenger cars and motorcycles

Klaus Baltruschat, S. Dittmar, T. Tallafuß,
TÜV SÜD Product Service GmbH

P. Pfeffer (Ed.), *9th International Munich Chassis Symposium 2018*, Proceedings,
https://doi.org/10.1007/978-3-658-22050-1_51

Material-Specific Annex to:
Road Traffic No. 287, Section 30 StVZO; here; Directive for the testing and inspection of custom wheels for motor vehicles and their trailers

Contents:

1. Scope

This Annex applies in addition to the Directive. It is designed for standardised assessment of custom wheels with a load-bearing structure made of fibre-reinforced plastics and wheel designs made of at least 2 different materials (metallic material and fibre-reinforced plastic) within the scope of the preparation of expert reports for national type approvals for vehicle components in line with Section 22 StVZO and single vehicle approvals according to Section 21 StVZO.

The individual sections of this Annex cover the testing of custom wheels for vehicles in the following categories:

- M1, M1G and N1, N1G
- L3e

2. Terms and definitions

2.1 Fibre-reinforced plastics

Materials made of a plastic matrix reinforced by fibres.

2.2 Compound wheel constructions

A wheel whose wheel body is composed of at least two parts and consists of various materials in accordance with Section 1.

2.3 Structural failure

Loss of stiffness > 20 %. Loss of stiffness describes the difference of the wheel in condition as delivered and after the test (Note: In bi-axial wheel testing, the loss also refers to the delivery condition, not the condition after conditioning).

2.4 Crack

Describes the visible delamination of fibre layers or a visible incipient crack which does not penetrate a cross-section. The definition of "technical crack" applies to metal structural parts.

2.5 Fracture

Describes a visible separation of matrix and/or fibres penetrating through a cross section.

3. Overview of the test methods to be used

	Vehicles in categories M1, M1G and N1, N1G	L3e vehicles
4.1 Stiffness measurement	X	X
4.2 Electrical conductivity	X[4)]	X[4)]
4.3.1 P[1)]: Radial impact test	X	-
4.3.2 P: Impact Test ISO 8644	-	X
4.3.3 P: Tyre fitting 10 times	X	X
4.3.4 P: Stone chip and scribe marks	X	X
4.3.5 P: Corrosion testing	X	X
4.3.6 P: Env. cycle test acc. to VDA	X	X
4.3.7 P: Pressure test	X	X
4.3.8 P: Thermal	X	-
4.3.9 P: Steam jet test	X	X
4.4 Bending fatigue test	X[4)]	X[3)]
4.5 Bi-axial wheel testing	X[3)]	-
4.6 Rolling test	-	X[3)]
4.7 Impact Test 13°	X[4)]	-
4.8 Radial impact test	X[4)]	-
4.9 Impact test ISO 8644	-	X[3)]
4.10 Alternating torque test	X [2)- 4)]	X[3)]
4.11 Simplified brake test	X [5)_4)]	-
4.12 Mounting test	X[4)]	X[4)]
4.13 General requirements	X[4)]	X[4)]

1) P: Pre-damage; the listed sequence corresponds to the sequence of damage and shall be taken into account

2) Testing is necessary where there are technical concerns based on the cross-sections and number of spokes and on the design.

3) Pre-damaged 4) Not pre-damaged

5) Only for wheel-disc designs made of non-metallic materials

Overview of test methods for vehicles in categories M1, M1G and N1, N1G

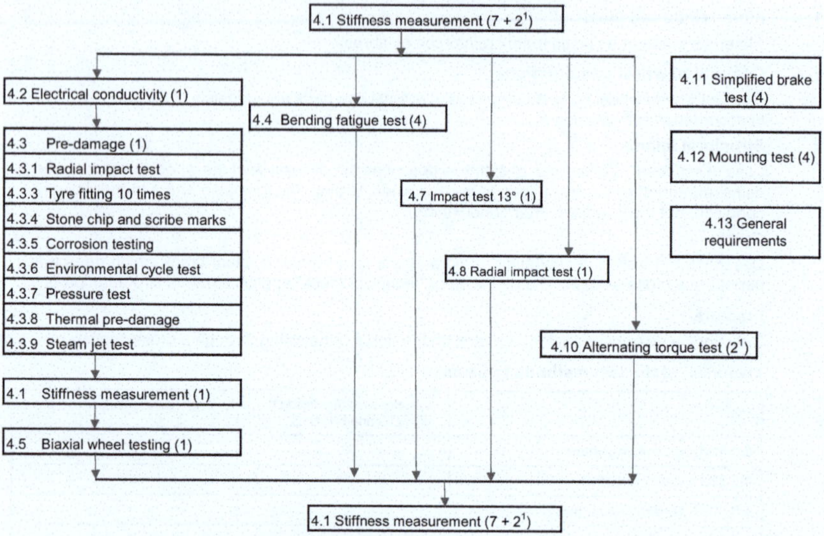

Overview of test method for vehicles in category L3e

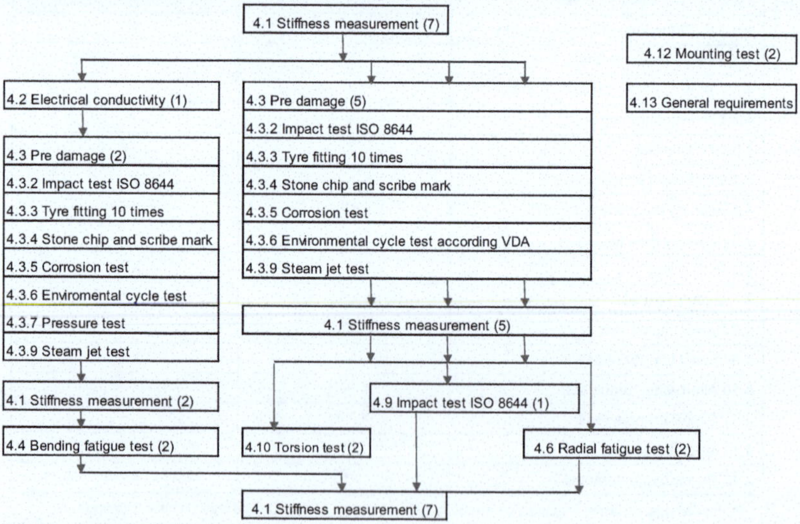

() Number of required wheels

2[1] Testing is necessary in case of technical concerns based on the cross-sections and number of spokes as well as the design.

4 Test methods

4.1 Stiffness measurement

4.1.1 Performing stiffness measurement

Stiffness measurement describes the attempt to verify the stiffness of a wheel design and its characteristics.

In axial direction: if possible on spokes at 90° to each other.

In radial direction: one measurement between two spokes, near the valve hole and additional one at one spoke

Assessment for structural failure requires testing to be repeated at least twice at each position (cf. overview of test methods) i.e. on wheel in condition as delivered, after pre-damage if applicable, and after strength testing.

The following measuring and test equipment is required:

- Adapters to take up the wheel
- Travel sensor, incl. evaluation unit (measuring accuracy ≤ 0.005 mm)

Testing is carried out without tyres, using the intended wheel fastening elements and the intended tightening torque at room temperature, RT = 20°C ± 5 °C.

The clamping device including the wheel connector shall be sufficiently stiff to prevent deformation during loading.

A full cycle shall be completed, start with pre-load (20 N) – impose test load (radially or axially) – unload (20 N)

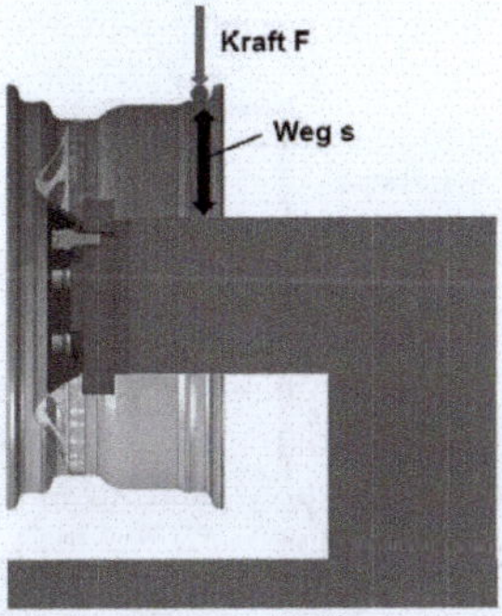

Drawing 1: Radial test setup

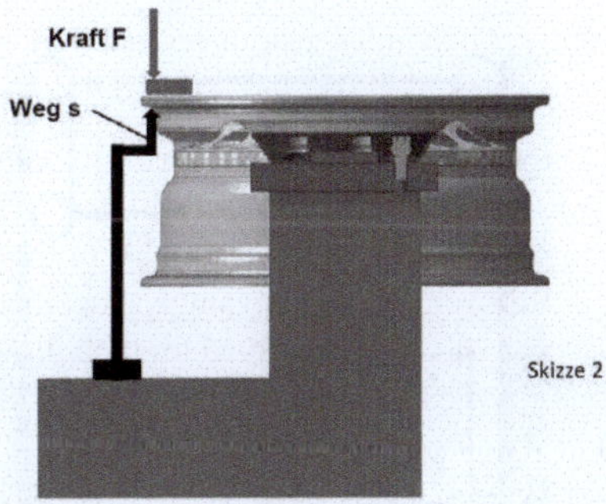

Drawing 2: Axial test setup

4.1.2 Test parameters

Vehicle categories	M1, M1G and N1, N1G		L3e
Direction of loading	Axial	Radial	Radial
Shape of test dolly	d = 24 mm Front side level	d = 20 mm cylinder	d = 12 mm cylinder
Load	0.8 * F_R	0.4 * F_R	0.5 * F_R
Speed of loading	3 mm/min ± 0.5 mm/min		

F_R permissible static wheel load [N]

4.1.3 Evaluation of stiffness measurement

Stiffness is given as the quotient of the maximum test load (kN) in relation to the maximum deformation path (mm) measured.
The arithmetic mean of the results for each position shall be calculated.

4.2 Electrical conductivity

Total conductivity of the wheel/tyre combination shall fulfil the requirements of WDK 110, as at August 2012 and shall be determined on a wheel/tyre combination.

4.3 Pre-damage

4.3.1 Radial impact test

This test is carried out between the spokes on the valve hole and simulates driving over a pothole or obstacle.
The tyre used is the widest tyre intended by the wheel manufacturer or in line with ETRTO (as amended).
Tyre inflation pressure 2.5 bar

$E = f * F_R$

f = 1.15
(F_R = wheel load in kg. The physical unit is not used for calculation, only the numerical value applies).

Schematic sketch of the test setup.

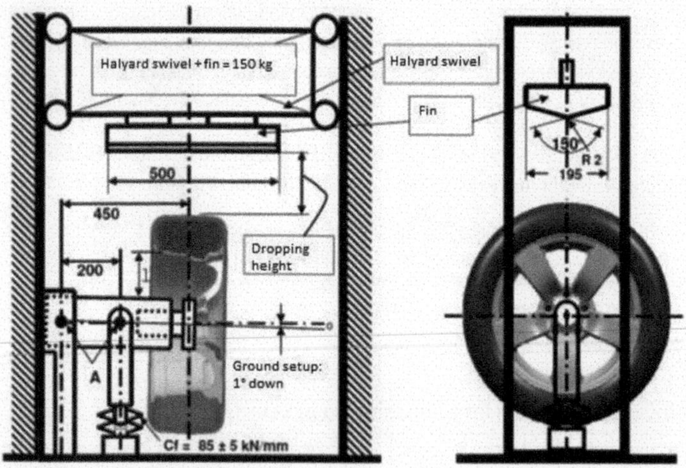

Test rig parameters and information:

The plastic deformation [unit l] that may occur at the location of impact shall be determined by measuring the difference (before/after) at the inner rim flange below the tyre seat and documented.

The halyard swivel shall be mechanically caught after impact; a second impact shall be prevented. Exceptions: wheel fracture and sudden loss of air

The weight of the halyard swivel including fin is limited to 150 ± 5 kg up to 2000 joules of impact energy. As of 2000 joules, the impact weight shall be increased in 5 kg increments to give an impact speed of 5 ± 0.2 m/s.

The calculated values of impact energy shall be set (with a tolerance of ± 2%). The test-rig parameters (drop height, weight) shall be adjusted accordingly.

Impact energy [J]	Drop height [mm]	Impact speed [s]	Drop weight [kg]
200	136	1.6	150
300	204	2.0	150
400	272	2.3	150
500	340	2.6	150
600	408	2.8	150
700	476	3.1	150
800	544	3.3	150
900	612	3.5	150
1,000	680	3.7	150
1,100	748	3.8	150
1,200	815	4.0	150
1,300	883	4.2	150
1,400	951	4.3	150
1,500	1,019	4.5	150
1,600	1,087	4.6	150
1,700	1,155	4.8	150
1,800	1,223	4.9	150
1,900	1,291	5.0	150
2,000	1,315	5.1	155
2,100	1,338	5.1	160
2,200	1,359	5.2	165
2,300	1,379	5.2	170
2,400	1,359	5.2	180
2,500	1,378	5.2	185
2,600	1,395	5.2	190
2,700	1,376	5.2	200
2,800	1,392	5.2	205
2,900	1,375	5.2	215
3,000	1,390	5.2	220
3,100	1,374	5.2	230
3,200	1,388	5.2	235
3,300	1,373	5.2	245
3,400	1,386	5.2	250
3,500	1,372	5.2	260
3,600	1,385	5.2	265
3,700	1,397	5.2	270
3,800	1,383	5.2	280
3,900	1,395	5.2	285
4,000	1,382	5.2	295
4,100	1,393	5.2	300
4,200	1,381	5.2	310
4,300	1,392	5.2	315

4.3.2 Impact test according to ISO 8644

This test simulates the wheel driving over a pothole or obstacle.
The tyre fitted for the test was the largest width intended for this wheel.
Tyre inflation pressure: 2.5 bar

Wheel-load increase factor f	1.5 (front wheel) 1.0 (rear wheel)

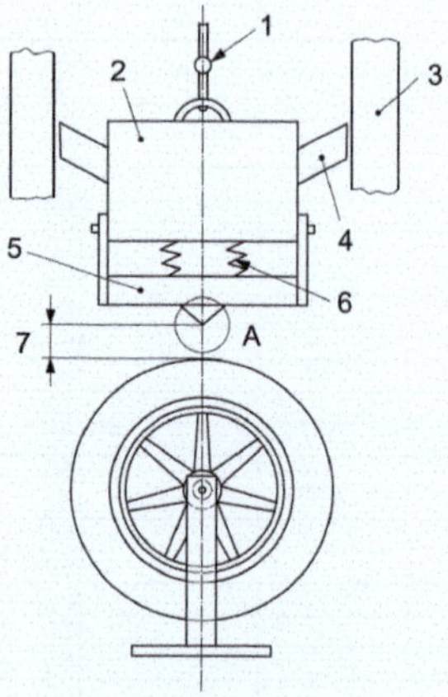

Rough outline of test setup

Key:

1 Trigger mechanism
2 Main weight
3 Slide rails
4 Guide rail
5 Additional weight
6 Coil springs (2 pieces)
7 Drop height

4.3.3 10 times tyre fitting in succession

Fit the tyre with the largest width and maximum radial stiffness, e.g. the tyre with minimum side-wall height at maximum width. Inflate with compressed air up to the "pop" pressure. Deflate, remove the tyre, repeat 9 times.

4.3.4 Stone chips and scribe marks

Stone chips: Multi-impact testing is to be performed according to EN ISO 20567-1, edition 2014-11, method B.
Scribe marks according to EN ISO 17872, ed. 2007.

The tests are carried out at the areas of the spokes exposed to maximum stress (position from FE calculation, load case: rotating bending test) and in the outer rim flange, deep scribe mark: ≥ 0.3mm

4.3.5 Corrosion testing

Salt spray testing as specified in NSS EN ISO 9227 shall be performed over 384 hours. The wheel, including all components such as metal screw valve, wheel bolts and wheel hub including brake disc, as well as any decorative elements that may cause contact corrosion, is subjected to testing in an "upright" position. Every 48 hours the wheel is turned through° 90°. Testing is carried out without tyres.

4.3.6 Environmental cycle test according to VDA

The environmental cycle test shall be carried out according to VDA 233-102 without part A (corrosive influence).
Test period vehicle categories M1, M1G and N1, N1G: 288 h
Test period vehicle category L3e: 96 h

4.3.7 Pressure testing

	M1, M1G and N1, N1G)	L3e
Pressure increase	0.5 bar / 3 s	0.5 bar / 3 s
Maximum pressure	8 bar	8 bar
Holding time, max. pressure	60 s	60 s

4.3.8 Thermal pre-damage (10 times)

Testing is carried out on a wheel with inflated tyre (minimum inflation pressure cold 2.0 bar) which will be tested standing (for the rim) and lying (for the wheel disc) at the initial condition at room temperature.
Heating up is effected via a circular heating element to map the actual temperature distribution. The circular heating element is positioned in the axial centre of the wheel; radial 30-50mm from the lowest point of the rim well.
For each test (lying and standing), circumferential closed circular heating elements of a defined size shall be used.
In the rim well, a temperature of at least 200°C shall be reached over 5 minutes, to be monitored with the help of temperature sensors attached to the rim well on the side of the brakes.

The temperature to be reached on the spoke, measured at a minimum of 1/3 of the spoke length seen from the centre of the wheel shall be at least 160°C over a period of 10 min.
As a criterion for the sensor, it shall have good contact with the wheel. The temperature sensor shall be firmly attached/bonded (2-component adhesive) to the wheel.
Let the wheel cool down at room temperature to at least 50°C without using additional cooling (blowers).

Both tests (standing and lying) shall be repeated 9 times each.

4.3.9 Steam jet test

Testing is carried out with conventional steam jets which shall comply with the following technical data:

Volume flow 8 l/min to 14 l/min
Pressure: 40 bar to 130 bar
Temperature: 20°C to 80°C

Testing:
Spray angle: 45°on scribe mark (4.3.4) both sides
Spray period 20 s to 30 s

Evaluation of the test is not necessary.

4.4. Rotating bending test:

4.4.1 Performing the rotating bending test

The bending fatigue test simulates lateral forces which act on the wheel when cornering. Every wheel is clamped rigidly to the test rig and exposed to a rotary bending moment Mb in a suitable manner via the wheel contact surface (e.g. with the help of a loading arm including a flange with the same connecting dimensions as the intended vehicle). The wheels are clamped on the inner rim flange with a clamping ring. If alternative types of clamping are used, their equivalence to the standard clamping method shall be proved.

The wheel fastening elements are tightened with the torque specified by the vehicle manufacturer and re-tightened after approx. 10,000 load cycles.4.4.2 Dimensioning equation:

Vehicle category	Bending moment
Category M1, M1G and N1, N1G	$M_{bmax} = f * F_R(\mu * r_{dyn} + e)$
Category L3e	$M_{bmax} = f * F_R(\mu * r_{dyn} + e)$

M_{bmax} Reference moment for load stages [Nm]

F_R Permissible static wheel load [N]

r_{dyn} Dynamic tyre radius of the largest intended tyre [m]

e Offset (according to ISO) [m]

l Half tread width [m] of the largest intended tyre

f Increase in wheel load factor [-]

μ Coefficient of friction between tyre and road 0.9

4.4.2 Testing is carried out according to the following criteria (a new wheel shall be used for each load stage):

Vehicle category	M1, M1G and N1, N1G		L3e	
			Front wheel	Rear wheel
Wheel-load increase factor f	2.0		0.8	0.6
Target load cycles per minute	As high as possible, but beyond self-resonance			
Load level	75% M_{bmax}	50% M_{bmax}	100 % M_{bmax}	
Number of test specimens	≥ 1	≥ 3	≥ 2	
Target load cycles	≥ 200,000	≥ 1,800,000	1,000,000	
Permissible drop of tightening torque	≤ 30 %			
Failure criterion	Structural damage, crack, fracture			

4.5 Biaxial wheel testing

The test is based on an existing program. The basic program consists of a synthetic circuit derived from European conditions of use. It comprises various load cases, such as driving in a straight line along rough road sections and motorways as well as various sharp right and left turns.

The tyre used is the tyre with the largest width intended by the wheel manufacturer or in line with ETRTO (as amended). If runflat-type tyres are intended, these tyres shall be used for testing.

Test program	See 4.5.1
Number of test specimens	≥ 1
Tyre inflation pressure	min. 4.0 bar
Target travel distance	7,500 km
Permissible drop of tightening torque	≤ 30 % (until first replacement of tyre and/or 500 km)
Failure criterion	Fracture, structural damage, loss of air

4.5.1 Test program

Basic population based on a wheel load of 425 kg, valid for external drum test bench

Sequ-ence	Cat. no.	Distance in m	F_{PR} [N]	Adjustment factor	F_{PA} [N]	Adjustment factor
1	5	1545	3750	Fv,c	150	Fl,c
2	22	129	11000	Fv,c	8000	Fl,c
3	5	1545	3750	Fv,c	150	Fl,c
4	21	291	9950	Fv,c	5350	Fl,c
5	9	670	9500	Fv,s	0	Fl,s
6	20	523	8300	Fv,c	5300	Fl,c
7	6	1379	4500	Fv,c	0	Fl,c
8	19	433	7500	Fv,c	5000	Fl,c
9	6	1379	4500	Fv,c	0	Fl,c
10	18	477	9000	Fv,c	4200	Fl,c
11	6	1379	4500	Fv,c	0	Fl,c
12	17	399	7850	Fv,c	4200	Fl,c
13	6	1379	4500	Fv,c	0	Fl,c
14	16	670	9500		3000	
15	1	670	9500	Fv,s	-3000	Fl,s
16	15	510	6350	Fv,c	4150	Fl,c
17	6	1379	4500	Fv,c	0	Fl,c
18	14	420	6000	Fv,c	3100	Fl,c
19	6	1379	4500	Fv,c	0	Fl,c
20	13	420	5200	Fv,c	3350	Fl,c
21	6	1379	4500	Fv,c	0	Fl,c
22	12	721	5300	Fv,c	2500	Fl,c
23	4	1271	2500	Fv,c	-1000	Fl,c
24	11	433	7300	Fv,c	2550	Fl,c
25	4	1271	2500	Fv,c	-1000	Fl,c
26	10	1214	4150	Fv,c	1650	Fl,c
27	3	1465	3400	Fv,c	-850	Fl,c
28	8	670	6000	Fv,c	1000	Fl,c
29	3	1465	3400	Fv,c	-850	Fl,c
30	7	1233	3850	Fv,c	1000	Fl,c
31	2	670	6000	Fv,c	-1000	Fl,c
32	7	1233	3850	Fv,c	1000	Fl,c

F_{PR} = Radial test force (wheel contact force) F_{PA} = Axial test force (lateral wheel force)

The test forces for radial and axial load shall be extrapolated using the calculation method outlined below and corresponding to the real wheel load including factors for cornering (Fv,c / Fl, c):

Fields highlighted in grey shall be extrapolated using the factors for straight-line travel (Fv, s / Fl, s). The test program shall be converted for internal drums, e.g. using the Schwendemann procedure.

The test loads apply to wheel loads of up to 1000 kg. For higher wheel loads, the load values shall be determined with the help of specific measurements on the target vehicle.

The wheel-load adjustment factors are calculated with the help of the following formula:

Factors:		Calculation formula
Fv,c	(Cornering)	= (1.8112 x F_R +3450 / 11000
Fl,c	(Cornering)	= (1.4281 x F_R +2050 / 8000
Fv,s	(Straight line)	= (2.4995 x F_R -920 / 9500
Fl,s	(Straight line)	= (0.6121 x F_R +450 / 3000

F_R = Permissible static wheel load [N]

4.6 Rolling test

The rolling test performed on an external drum test rig with a minimum diameter of 1.7 m simulates the stresses acting on a wheel while driving in a straight line. The test shall be performed with a tyre of the largest nominal width intended for the wheel.

4.6.1 Dimensioning equation

L3e	$F_P = f * F_R$

F_P = Test load [N]

F_R= Permissible static wheel load [N]

f = Wheel-load increase factor [-]

4.6.2 Testing is carried out in line with the following criteria:

Direction of travel	Straight-line
Wheel-load increase factor f	2.25
Tyres	Standard; in case of several optional tyres, the tyre with the largest nominal width
Test speed [km/h]:	60 – 100
Tyre inflation pressure [bar]	4.5
Running distance (wheel revolutions)	3,000,000
Number of test specimens	≥ 2
Failure criterion	Structural damage, crack, fracture, air loss

4.7 Impact test 13°

Wheel fracturing behaviour when driving over an obstacle on the outer rim flange shall be verified at the critical positions. To verify adequate fracturing behaviour, an impact test according to ISO 7141 shall be performed. The most critical position shall be selected as impact position.

4.7.1 Dimensioning equation:

$D = 0.6{*}F_R / g + 180$ kg

D = drop weight [kg]

F_R = Permissible static wheel load [N]

g = Drop acceleration 9.81 m/s²

Number of test specimens	≥ 1
Test setup, tyres, impact position	See Directive for the testing and inspection of custom wheels for motor vehicles and their trailers
Failure criterion	Complete pressure loss within one minute Fractures outside the direct influence zone of the impact plate Permissible damage: Structural damage, cracking

4.8 Radial impact test (factor 4.3)

$E = f * F_R$

$f = 4{,}3$

Description of test setup see 4.3.1.

Failure criterion	Fracture not permissible Cracks in the influence zone of the wedge are permissible up to 50 % of the total rim width.

4.9 Impact test according to ISO 8644

Testing is carried out in line with the following criteria:

Number of test specimens	≥ 1
Tyres	The widest permissible tyre according to E.T.R.T.O.
Position of impact	Between 2 spokes on the valve hole
Wheel-load increase factor f	1.5 (front wheel) 1.5 (rear wheel)
Failure criterion	Fracture, loss of air

4.10 Alternating torque test

The tangential forces which act as stresses on the wheel during braking and acceleration are simulated in a vibration fatigue test. Two wheels shall be tested for each load stage. Each wheel is rigidly clamped to the test rig on one rim flange and suitably loaded, e.g. via the intended brake disc or other fittings, with an alternating test torque M_T. The wheel mounting elements are tightened using the torque specified by the wheel manufacturer.

Dimensioning equation:

$M_T = f{*}F_R * r_{dyn}$

M_T = Alternating test torque[Nm]

f = Increase in wheel load factor

F_R = Permissible static wheel load [N]

r_{dyn} = Dynamic tyre radius of the largest intended tyre [m]

Testing is carried out in line with the following criteria:

Vehicle category	L3e	M1, M1G and N1, N1G
Wheel-load increase factor f	1.0 (front wheel) 1.1 (rear wheel)	1.0
Min. number of load cycles at 100 % M_T	$1*10^6$	
Min. number of load cycles at 90 % M_T		$2.0 * 10^5$
Min. number of load cycles at 45 % M_T		$2.0 * 10^6$
Number of test specimens	≥ 2	> 2[1)]
Permissible drop in tightening torque	≤ 30 %	
Failure criterion	Crack, fracture, structural damage	

1) Testing is necessary in case of technical concerns based on the cross-sections and number of spokes as well as the design.

4. 11 Simplified brake test

Carrying out a simplified brake test (only for wheel disc designs made of non-metallic materials).

Selection of the critical brake in similar vehicles and models of one vehicle manufacturer, according to approval.	Selection of the test vehicle that has the least favourable ratio of braking performance (type, mass, diameter, ventilation, etc.) to drive power and permissible gross vehicle weight rating (GVWR). This is generally the heaviest vehicle with the lowest brake-disc mass (in accordance with ECE R-90, Section 5.3.6) Test vehicles are selected in a traceable manner by the Technical Service in agreement with the German Federal Motor Transport Authority (Kraftfahrtbundesamt, KBA).
Vehicle preparation	Load the test vehicle in line with the maximum GVWR (ECE R-13 H 1.4.1.2.1). In terms of minimum documentation, the temperature on the braking path of the disc shall be recorded during testing by using a thermometer with ground joint at the external exit point of the brake caliper. In addition to the above, temperatures shall be recorded on the wheel near the wheel attachment face (temperature of wheel bolts / head of the wheel nut), after the road test and after the post-heating phase (around 10 minutes of standing phase).
Testing of the braking function under thermal load.	According to ECE R-13H, type 1
Assessment	No failure of the brake linings The requirement of ECE R13 H shall be fulfilled in case of hot braking (min. 75% of the deceleration of a cold brake).

4.12 Mounting test

The mounting test has to be done according to the "Guidelines for the testing and inspection of custom wheels for motor vehicles and their trailers", page 18.

4.13 General requirements

- Metallic inserts at the wheel mounting area are always required (not in the case of wheel mounting areas made of metals). The structure of the inserts shall be described in the information document.
- Adhesion of the adhesive balancing weights shall be permanently ensured. Where appropriate this shall be proved by comparative testing. Clamp weights are not permissible.
- The materials and pairs of materials selected, e.g. wheel, hub, valve and bolts and inner reinforcement elements, shall be non-corrosive. Galvanic series are also taken into account.
- The functioning of directly measuring tyre pressure monitoring systems (TPMS) shall be ensured. The Technical Service shall assess the type of the wheel in this respect.
- The wheel manufacturer shall define the criteria for wear in the area of the tyre seat / rim flange in the information document.
- The information document on the wheel shall provide information on both the materials used and the structure (e.g. position and design of the fibre layers).
- In addition to Clause 6 Conformity of production (CoP) of the Directive for the Testing of Custom Wheels of Motor Vehicles and their Trailers, the following test records shall be submitted at least once annually or following the production of 10,000 wheels.
  M1, M1G and N1, N1G –vehicles:
  Rotating bending test (once 50% and once 75% M_{bmax})
  Radial impact test, one wheel

  L3e vehicles
  Alternating torque test, one wheel
  Rolling test, one wheel
  Impact test ISO 8644, one wheel

Porsche 911 Turbo carbon wheel

Gerd Burk / Manager Chassis Engineering Wheels

Dr. Ing. h.c. F. Porsche AG

P. Pfeffer (Ed.), *9th International Munich Chassis Symposium 2018*, Proceedings,
https://doi.org/10.1007/978-3-658-22050-1_52

1 Porsche Turbo carbon wheel

An exclusive innovation for enhanced driving dynamics with spectacular looks. Porsche is the world's first vehicle manufacturer to offer lightweight wheels with braided carbon fibres. These are optionally available for the Porsche 911 Turbo S Exclusive Series. Under a protective clear coating, the characteristic black high-tech material carbon provides a striking appearance. The innovative wheels weigh around 8.5 kilos or 20 percent less in total than the standard alloy wheels, while also being 20 percent stronger. Through the reduction in unsprung masses, the tyres follow the road surface better and optimally transfer the longitudinal and lateral forces. The lower rotating masses also mean more spontaneous acceleration and braking, resulting in enhanced driving dynamics and driving pleasure.

The wheel, which is made exclusively of carbon-fibre reinforced plastic (CFRP), basically comprises two parts. The wheel centre is made from carbon-fibre mats. Here, more than 200 individual components are cut and joined. The second part, the rim base, is made from braided carbon fibres produced on the world's largest carbon-fibre braiding machine with a diameter of around nine metres. The wheel centre is then braided into the rim base. The completed wheel is impregnated with resin and pre-hardened under high pressure and at high temperatures. Curing of the finished wheel takes place at high temperatures, followed by slow cooling. The central lock is then inserted into the finished wheel, before the wheel is protected with clearcoat.

This extremely complex technology is being used for the first time worldwide in automotive manufacturing by Porsche. Compared to the usual manufacturing method using pre-impregnated carbon-fibre mats, the braid technology offers decisive benefits: The material structure of the carbon is significantly denser and tighter using this manufacturing technique. This results in higher strength. Through the efficient use of materials, less scrap is also produced. In total, the new carbon wheel consists of carbon fibres with a length of 18 kilometres or eight square metres of carbon-fibre mats.

The new carbon wheels will be introduced onto the market in the sizes 9 J x 20 for the front axle and 11.5 J x 20 for the rear axle as an option for the 911 Turbo S Exclusive Series at the beginning of 2018.

2 Milestones in Porsche wheel development Motivation: Weight reduction

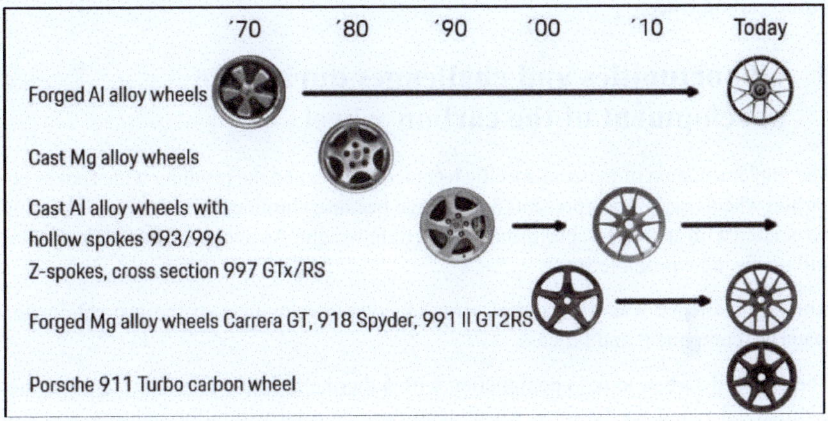

Figure 1: Some milestones in Porsche wheel development

Porsche has been supplying weight-optimised wheels using a variety of manufacturing processes and materials for more than 50 years. For a summary of some of the highlights, see Figure 1. The weight-reduction efforts began with the forged aluminium alloy wheels, also known as Fuchs-style wheels. This technology has become established throughout all the model series and derivatives, and is in use today. The number of forged wheels is increasing continuously and will sooner or later exceed the number of flow-formed cast wheels.

The 959 and 964RS featured cast magnesium alloy wheels. The combination of magnesium alloy and casting is no longer used at Porsche today, however, magnesium alloy is still used as a material in wheels.

Further weight reductions have been achieved with cast wheels using a variety of patented geometries. Particularly noteworthy here are hollow spoke wheels, as well as the Z-spoke geometry, which have been used on various derivatives. Although the styling of the Porsche wheels is of course already tailored towards lightweight construction, we nonetheless continue our work on innovative geometries to achieve further weight reductions.

Recently, the use of magnesium alloy for wheels has been revived, but manufacture now takes place using the forging process. The Carrera GT featured design-forged wheels, the 918 Spyder and the current 991 II GT2RS in production are equipped with forged/milled wheels to achieve even greater weight savings.

The most recent highlight in terms of Porsche wheel development today is the 911 Turbo carbon wheel.

3 Opportunities and challenges during the development of the carbon wheel

The implementation of material-specific technical styling and the resulting wide spokes on carbon wheels certainly represents a challenge because slimline, delicate spokes remain associated with sportiness, performance and light weight. An acceptance of wide spokes still needs to be promoted.

Another challenge was the parallel, simultaneous development of the test catalogue, material, component and process.

The exposed carbon look in conjunction with a thermally stable matrix system and the geometrical requirements for a wheel centre are unusual within the industry and are highly complex.

During implementation of a wheel with a new or different material within a given vehicle structure, the geometrical requirements can be of significance, particularly if target rigidities have to be achieved and a larger installation space is required for this purpose with the new material system.

Moreover, depending on the process chain, the standards for die-cast geometries are increased to meet the geometric requirements.

The two challenges for achieving the target weight and temperature stability are described in detail below.

Despite all the challenges, the wheels present one major technical opportunity – weight reduction. For the customer, the visual appearance of course also represents a new opportunity for individualisation.

4 Weight

4.1 Assumptions for a weight comparison

Assumption 1 is the implementation of a material-specific technical styling. In the case of metal wheels, this means slimline, high spokes, in some cases with machine finishing.

Assumption 2 is an unrestricted installation space. Depending on the material, more or less installation space is required to achieve sufficient strength and rigidity.

Assumption 3 relates to identical requirements. Wheel load, rigidity in the spoke area and the rim base must not undershoot the zero line (forged wheel). Moreover, vehicle-specific requirements such as race track usability and drag coefficient must equally be taken into account.

The objective at Porsche was to achieve a 20% weight reduction with the carbon wheel compared to a forged wheel, while maintaining the same or achieving greater rigidity.

4.2 Weight/rigidity comparison of different wheel materials and technologies in numbers

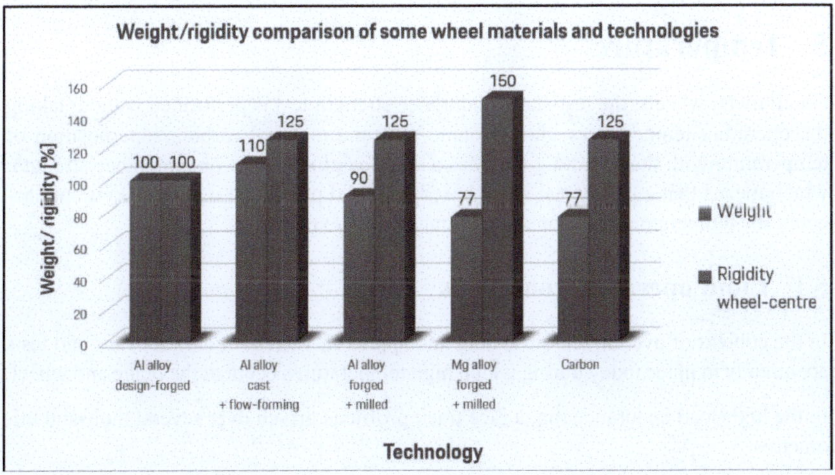

Figure 2: Weight/rigidity comparison of some wheel materials and technologies

As shown in Figure 2, design-forged aluminium alloy wheels are used for the zero line during the weight/rigidity comparison. Both the weight and the flexural rigidity of the wheel centre are set to 100%. The rigidity of the rim and inner flange are set to be identical for the various materials and technologies.

When using cast aluminium alloy wheels with flow-forming, the weight increases by 10%. Owing to the lower material performance and the resulting larger spoke cross-section, rigidity is increased by 25%.

With milled, forged aluminium alloy wheels, the weight can be reduced by 10% compared to design-forged wheels (zero line) as no process-related restrictions such as height-to-width ratio have to be taken into account. Because there are no restrictions with regard to the process in terms of spoke geometry, the rigidity can be increased by 25%. When using magnesium rather than aluminium alloy with the same technology, a further weight reduction of 13% can be achieved, i.e. compared to the 23% zero line. Owing to the higher requirements for achieving endurance strength, which are reflected in the geometry and spoke height, the rigidity is increased by 50% compared to an aluminium alloy wheel.

With the carbon wheel, in the case of implementation of identical rigidities to a forged aluminium alloy wheel, weight savings of 23% can be achieved in the rim base and flange as with a forged magnesium alloy wheel.

5 Temperature

On all metal wheels, the test-stand endurance strength test is performed without taking the operating temperatures into account. With the material carbon, superposition of temperature with the operating load has a significant influence on endurance strength, which means that a purely test-stand based approval procedure without taking into account the temperature loads on the test stand is not expedient.

5.1 Limit operating conditions

In the context of overall vehicle testing and approval, various endurance runs and tests are already in place today during which high temperatures occur at the brake and wheel.

In the high-load endurance run, a race track profile is driven over several thousand kilometres.

Downhill driving is less relevant for a sports car, but may be significant for the relevant vehicles during trailer operation.

During “race track usability” testing, laps are driven against the clock, i.e. as fast as possible.

Beyond this, there are race tracks with a high braking requirement/km. Here, the maximum temperatures occur compared to the previous tests. These are so-called stop-and-go tracks with straights on which the maximum possible acceleration and deceleration take place, i.e. tracks with a low proportion of flowing bends. As with the race track usability testing, driving is as fast as possible against the clock.

To determine the max. temperatures, the above-mentioned tests were performed to saturation of the temperatures at the wheel and brake. Subsequently, a stationary heat build-up took place without cooling-down laps, which must be evaluated as an improper procedure.

5.2 Temperature characteristic during race track operation

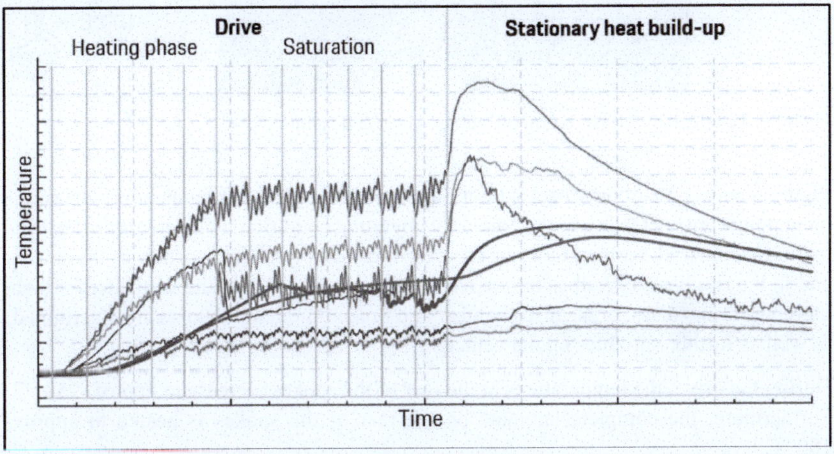

Figure 3: Temperature characteristic during race track operation

Figure 3 shows an example of the temperature characteristic on a race track at various measurement points on the wheel, here over 12 laps. During driving, the phases “heating up” and “saturation” can be clearly distinguished. Through the stationary heat build-up, the temperatures additionally increase in nearly all areas.

5.3 Temperature measurement on carbon wheel and result

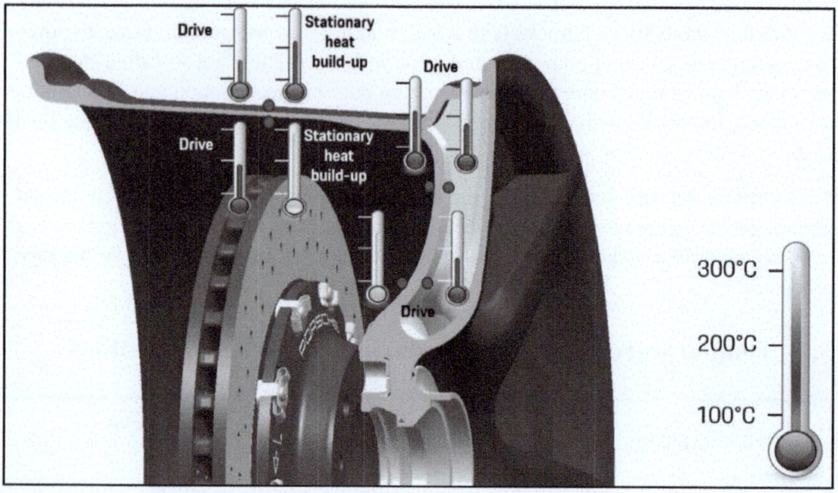

Figure 4: Temperature measurement on carbon wheel and result

Figure 4 shows a carbon wheel with its spoke structure and the temperature measurement points during testing of the boundary operating conditions. To determine the heating of the material, temperature sensors were attached on the rear surface of the spokes and in the rim base as well as on the respective opposite side of the laminate. Furthermore, the brake discs, brake disc chamber and brake fluid temperatures are determined, as well as further variables on the vehicle.

During the drive, the temperatures at the rear of the spokes increase to approx. 250 °C. Furthermore, the complete laminate on the rear of the spokes is heated to approx. 180 °C.

The temperatures at the rim base are approx. 200 °C and increase to just under 300 °C through the stand heating.

All the temperatures can vary strongly due to boundary conditions such as the distance to the brake or brake ventilation and must be determined again according to the specific vehicle and derivative.

It is certain that with the available matrix system, the temperature loads indicated, superimposed with the mechanical loads, cannot be realised, meaning that protection of the material system is required.

5.4 Measures for temperature resistance

Figure 5 shows the rear spoke side of the 911 Turbo carbon wheel as well as the brake-side rim base with the measures for achieving temperature resistance implemented. For thermal protection, a heat shield made from aluminium is used on the rear of the spokes. In the rim base, heat-conducting fibres from the aerospace industry are used, which prevent hotspots and distribute the locally generated high temperatures in the rim base.

Figure 5: 911 Turbo carbon wheel with heat shield and fibres from the aerospace industry

6 Rigidity characteristic with superimposed temperature and operating load

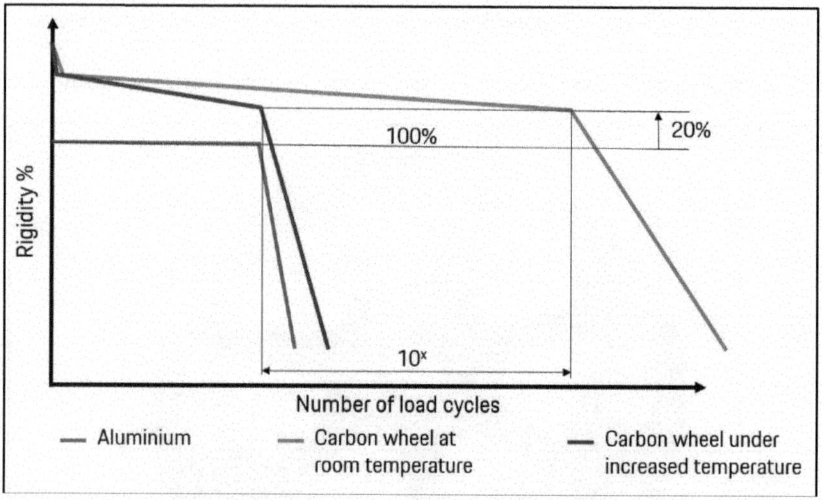

Figure 6: Rigidity characteristic with superimposed temperature and operating load

Figure 6 shows the rigidity over the load cycle numbers. In the case of aluminium alloy wheels, the rigidity of the wheel remains constant until a crack occurs. With the progress/growth of the crack, the rigidity is reduced. The dip represents the technical crack. Here as well, the forged aluminium alloy wheels are used as the zero line for rigidity, so that a constant rigidity characteristic of the aluminium alloy wheel is marked as 100%.

In the case of carbon wheels, initial settling takes place owing to the material-specific micro-fibre cracks. Next, a linearly stable area in the rigidity occurs in which the rigidity gradually reduces. After a certain number of load cycles, depending on the temperature, de-laminations and fibre brakes occur, which results in a fall in rigidity at a now higher gradient. This dip can be referred to as a technical insipient crack. If the load cycles occur under the effect of temperature, the technical crack occurs faster than at room temperature. The 911 Turbo carbon wheel has a 20% higher rigidity at the time of the technical crack than the forged aluminium alloy wheel.

A comparison of load cycle numbers achieved before technical cracking of a wheel at room temperature with those for one at increased temperature, the resulting quotient can be between one and several powers of ten depending on the boundary and operating conditions.

7 Conclusion/outlook

From the Porsche point-of-view, follow-up operating load testing or vehicle endurance test and partly further strength tests with superimposed temperature and operating loads must be performed. Furthermore, superimposition both in dynamic, i.e. during driving, as well as in the static range, i.e. during stationary heat build-up must be performed.

Depending on the boundary conditions, measures for temperature reduction to protect the material system are required. Moreover, a material-appropriate design must be ensured.

On the Porsche side, an approval process has been developed and approved, which will be used for future carbon wheel developments.

During future projects, further weight reductions can be achieved through further development of the material system. Furthermore, the conflict of objectives between "reduction of flow resistance" and "weight saving" can partially be resolved.

A further objective is a test-stand based approval procedure for carbon wheels.

8 Results

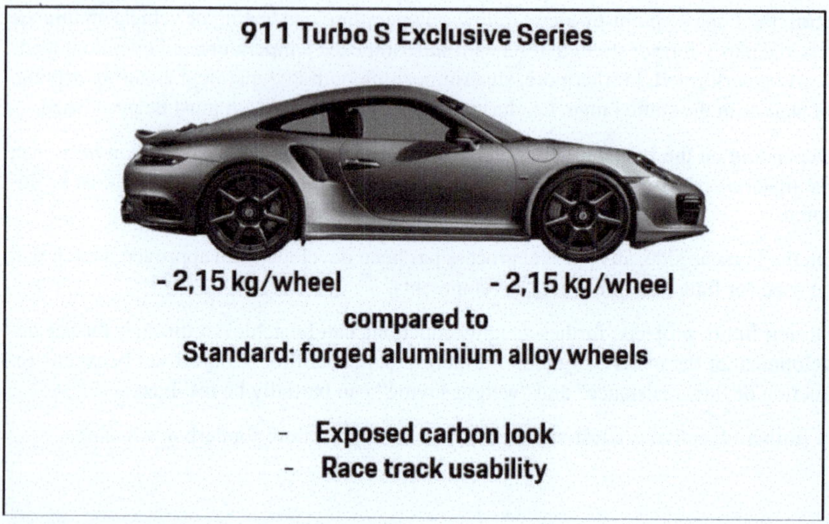

Figure 7: Porsche 911 Turbo carbon wheel outcome

The carbon wheel available for the 911 Turbo S Exclusive Series saves 2.15 kg/wheel, i.e. 8.6 kg for the vehicle compared to the standard production forged aluminium alloy wheels. Furthermore, an incomparable exposed carbon look has been achieved in this sector. This is completed by the day-to-day and race track usability of any 911.

TIRE TESTING AND SIMULATION

European and international harmonized tire regulations – impact on vehicle development

Lars Netsch, TÜV SÜD Auto Service GmbH; J. Burghardt, AUDI AG

P. Pfeffer (Ed.), *9th International Munich Chassis Symposium 2018*, Proceedings,
https://doi.org/10.1007/978-3-658-22050-1_53

TÜV SÜD
Auto Service
Choose certainty.
Add value.
ATZ live
chassis.tech plus 2018
12-06-2018 – 13-06-2018 – Munich, Germany
Section tire.wheel.tech
European and international harmonized tire regulations/ Impact on the vehicle development
Lars Netsch, TÜV SÜD Auto Service GmbH
Jörg Burghardt, AUDI AG
Audi
TÜV SÜD Auto Service GmbH
2018-06-13
Chassis Tech 2018
Slide 1

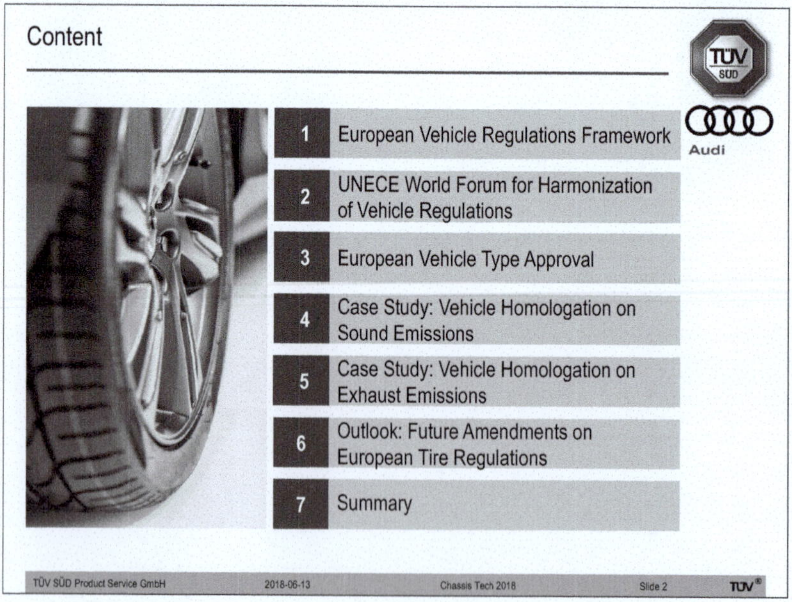
Content
TÜV SÜD
Audi
1 European Vehicle Regulations Framework
2 UNECE World Forum for Harmonization of Vehicle Regulations
3 European Vehicle Type Approval
4 Case Study: Vehicle Homologation on Sound Emissions
5 Case Study: Vehicle Homologation on Exhaust Emissions
6 Outlook: Future Amendments on European Tire Regulations
7 Summary
TÜV SÜD Product Service GmbH
2018-06-13
Chassis Tech 2018
Slide 2

Introduction – European Vehicle Regulations Framework

Audi

Introduction:

The HLG (High Level Group on the Competitiveness and Sustainable Growth of the Automotive Industry in the European Union) stresses the importance of **global technical harmonization** under the United Nations Economic Commission for Europe (UNECE) framework as a key factor in strengthening global competitiveness, reducing redundant development and testing costs and avoiding duplication of administrative procedures.

In the case of tires and vehicle fitting, EU through the adoption of UNECE Regulations still has the lead and has the strictest regulatory requirements.

from "GEAR 2030 High Level Group on the Competitiveness and Sustainable Growth of the Automotive Industry in the European Union", Final Report – 2017, DG GROW – Internal Market, Industry, Entrepreneurship and SMEs

TÜV SÜD Auto Service GmbH 2018-06-13 Chassis Tech 2018 Slide 3

UNECE World Forum for Harmonization of Vehicle Regulations

UNECE

Audi

- The World Forum is a subsidiary of the United Nations Economic Commission for Europe (UNECE)
- "1958 Agreement" concerning the adoption of uniform technical prescriptions for wheeled vehicles, equipment and parts which can be fitted and/or be used on wheeled vehicles and the conditions for reciprocal recognition of approvals granted on the basis of these prescriptions (52 Contracting Parties "CPs", currently 143 UNECE Regulations issued)
- Agreement open to all nations of the UN, GOs and NGOs
- Decisions on adoption taken by Governments of CPs
- CPs are free to be bound by all, some or no Regulation
- Mutual recognition of the type-approvals granted: Type approval by country "CP-A" acc. UNECE provisions is accepted by all countries (CPs).

TÜV SÜD Auto Service GmbH 2018-06-13 Chassis Tech 2018 Slide 4

UNECE World Forum for Harmonization of Vehicle Regulations

UNECE

Audi

- "1998 Agreement" concerning the establishing of global technical regulations (GTRs) for wheeled vehicles, equipment and parts which can be fitted and/or be used on wheeled vehicle (36 Contracting Parties, 19 GTRs in force)

Source: UNECE.org

- CPs can decide not to apply the UN GTRs or transpose it with amendments
- Global technical regulation GTR No. 16 - Global technical regulation on tires.

TÜV SÜD Auto Service GmbH 2018-06-13 Chassis Tech 2018 Slide 5 TÜV®

European Vehicle Type Approval

Audi

Adopting UNECE Regulations by the European Union and other states

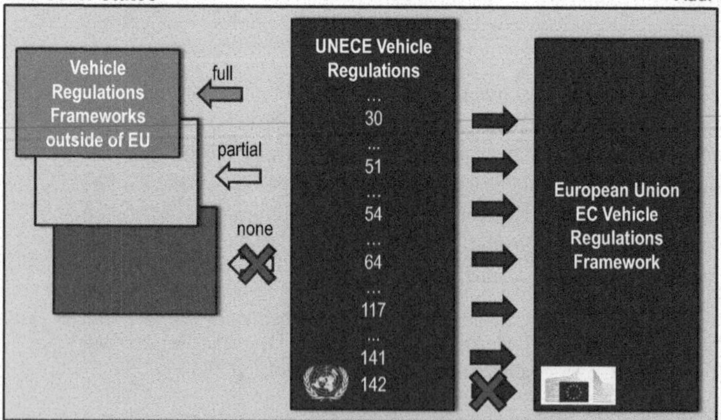

TÜV SÜD Auto Service GmbH 2018-06-13 Chassis Tech 2018 Slide 6 TÜV®

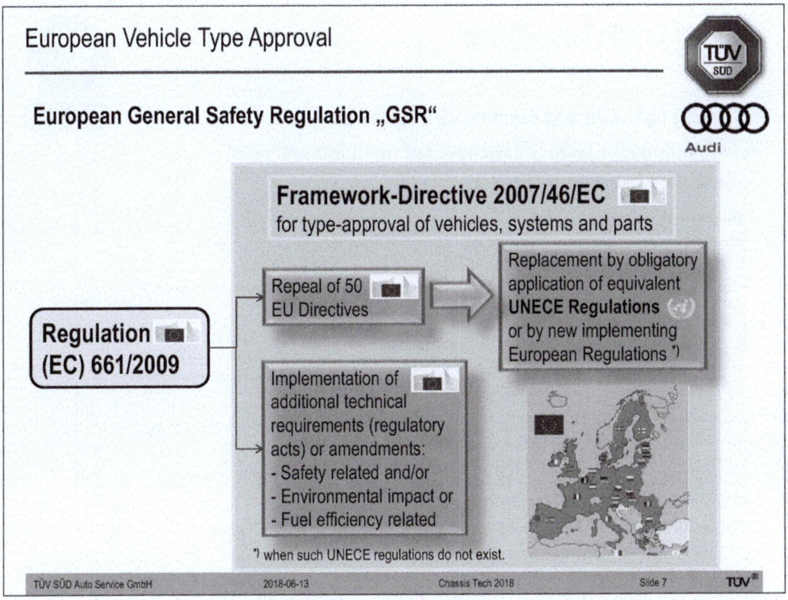
European Vehicle Type Approval
TÜV SÜD
Audi
European General Safety Regulation „GSR“
Framework-Directive 2007/46/EC
for type-approval of vehicles, systems and parts
Regulation
(EC) 661/2009
Repeal of 50
EU Directives
Replacement by obligatory application of equivalent UNECE Regulations or by new implementing European Regulations *)
Implementation of additional technical requirements (regulatory acts) or amendments:
- Safety related and/or
- Environmental impact or
- Fuel efficiency related
*) when such UNECE regulations do not exist.
TÜV SÜD Auto Service GmbH
2018-06-13
Chassis Tech 2018
Slide 7
TÜV®

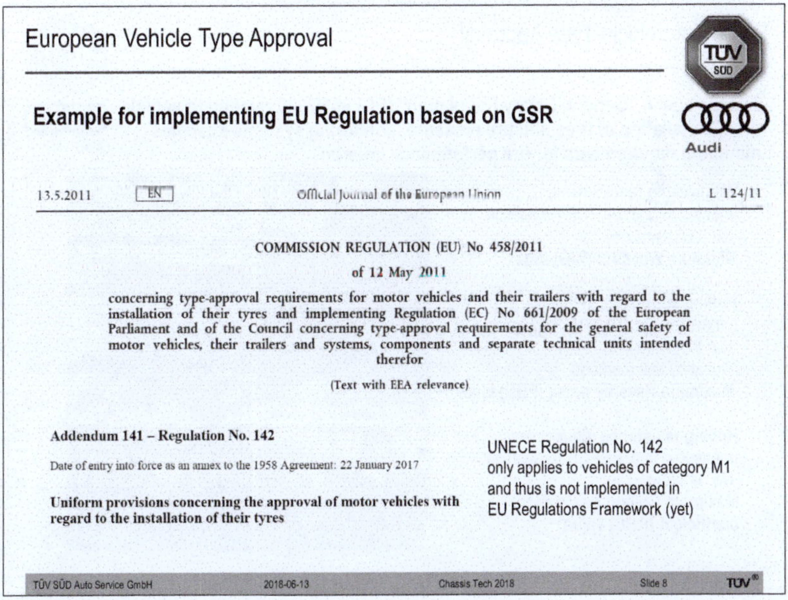
European Vehicle Type Approval
TÜV SÜD
Audi
Example for implementing EU Regulation based on GSR
13.5.2011
EN
Official Journal of the European Union
L 124/11
COMMISSION REGULATION (EU) No 458/2011
of 12 May 2011
concerning type-approval requirements for motor vehicles and their trailers with regard to the installation of their tyres and implementing Regulation (EC) No 661/2009 of the European Parliament and of the Council concerning type-approval requirements for the general safety of motor vehicles, their trailers and systems, components and separate technical units intended therefor
(Text with EEA relevance)
Addendum 141 – Regulation No. 142
Date of entry into force as an annex to the 1958 Agreement: 22 January 2017
Uniform provisions concerning the approval of motor vehicles with regard to the installation of their tyres
UNECE Regulation No. 142
only applies to vehicles of category M1
and thus is not implemented in
EU Regulations Framework (yet)
TÜV SÜD Auto Service GmbH
2018-06-13
Chassis Tech 2018
Slide 8
TÜV®

European Vehicle Type Approval

GSR (EC) 661/2009 and Framework 2007/46 EC

- **Obligations for OEM (Passenger car manufacturer) and Supplier (Tire manufacturer)**

Obligations - Vehicle Manufactuerer (OEM)	Obligations - Tire Manufacturer
Installation of tires with certification acc. UNECE-R 30	Approval of component tire acc. UNECE-R 30
Installation of tires with certification acc. UNECE-R 117	Approval of component tire with regard to rolling sound emissions and to adhesion on wet surfaces and to rolling resistance acc. UNECE-R 117
Installation of tires according REGULATION (EU) No 458/2011	not involved
Approval of vehicle with regard to equipment, which may include a temporary-use spare unit, run-flat tires and/or a run-flat system acc. UNECE-R 64	Approval of component tire when intended for use as part of a temporary-use spare unit in accordance with UNECE-R 30
Approval of vehicle with regard to sound emissions: permissible sound level acc. REGULATION (EU) No 540/2014 or acc. UNECE-R 51	not involved
Approval of vehicle with regard to Tire Pressure Monitoring System (TPMS) acc. UNECE-R 141 (former UNECE-R 64)	not involved
Approval of vehicle with respect to emissions acc. REGULATION (EU) 2017/1151 (WLTP / UN ECE GTR No. 15 and RDE)	not involved

Table: Tire related vehicle type approval (homologation) requirements

TÜV SÜD Auto Service GmbH 2018-06-13 Chassis Tech 2018 Slide 9

European Vehicle Type Approval

The European "General Safety Regulation" (EC) 661/2009 relating to vehicle tires, installation of tires and tire pressure monitoring systems specifies minimum requirements for tire performance criteria:

- **Wet braking** requirements (measured acc. UNECE Regulation No. 117, Annex 5) **Minimum Wet Grip (WG)-Index**
- **Rolling sound** requirements (measured acc. UNECE Regulation No. 117, Annex 3, "Coast-by test method") **Maximum Pass by noise (PBN) level**
- **Rolling resistance** requirements (measured acc. UNECE Regulation No. 117, Annex 6) **Maximum Rolling Resistance coefficient (RRC) value**

TÜV SÜD Auto Service GmbH 2018-06-13 Chassis Tech 2018 Slide 10

European Vehicle Type Approval

European General Safety Regulation (EC) 661/2009 has adopted UNECE tire regulations (applying on a compulsory basis)

UNECE Regulation No. 30 - Tire structure and load-speed performance
UNECE Regulation No. 54 - Tire structure and load-speed endurance (trucks)
UNECE Regulation No. 64 - TPMS etc. (from 2017: additionally Regulation No. 141)
UNECE Regulation No. 117 - Wet grip, rolling sound emission and rolling resistance.

Further requirements related to tire characteristics mandatory for the purpose of vehicle type approval acc. Framework-Directive 2007/46/EC

Sound emission homologation: Permissible sound level acc. REGULATION (EU) No. 540/2014 or UNECE Regulation No. 51 (recognised as an alternative).

Type approval of motor vehicles with respect to emissions (Euro 5 and Euro 6) acc. REGULATION No. (EU) 2017/1151 (emission tests WLTP/RDE).

TÜV SÜD Auto Service GmbH 2018-06-13 Chassis Tech 2018 Slide 11 TÜV

Case Study: Vehicle Homologation on Sound Emissions

UNECE-R117 tire homologation		
valid	Since 1.11.2016	
procedure	Rolling @ 80km/h	
limits	< 185 mm	70 dB(A)
	>185mm and < 245mm	71 dB(A)
	>245mm and < 275mm	72 dB(A)
	>275mm	74 dB(A)

Most of the tires

No real correlation

UNECE-R51.03 vehicle homologation		
valid	Phase 3 coming 1.7.2024	
procedure	Calculation based on rolling @ 50km/h and acceleration @50km/h	
limits	PMR < 120	68 dB(A)
	120 < PMR < 160	69 dB(A)
	PMR > 160	71 dB(A)

Most of the vehicles

@Audi:
All tires are tested regarding UNECE-R51.03 <u>on pure E-vehicles</u> on an ISO certified track.
Only by fullfilling Audi-defined limits they get a release and vehicle homologation will pass

TÜV SÜD Auto Service GmbH 2018-06-13 Chassis Tech 2018 Slide 12 TÜV

Case Study: Vehicle Homologation on Exhaust Emissions

COMMISSION REGULATION (EU) 2017/1151 ("WLTP")
supplementing Regulation (EC) 715/2007 of the European Parliament and of the Council on type approval of motor vehicles with respect to emissions from light passenger and commercial vehicles (Euro 5 and Euro 6)

The emissions test procedures require for road load determination and dynamometer setting a particular tire selection. For using the interpolation method for individual vehicles two test vehicles shall be selected from the interpolation family: Test vehicle H producing the higher, and preferably highest, cycle energy demand of that selection, test vehicle L one producing the lower, and preferably lowest, cycle energy demand of that selection. Road load relevant characteristics are e.g. mass, aerodynamic drag and **tire rolling resistance**.

Tire rolling resistance is measured according to Annex 6 of UNECE-Regulation No. 117.02 series of amendments and shall be aligned and categorised according to the rolling resistance classes in Regulation (EC) No 1222/2009 = Fuel Efficiency Classes in "EU-Tire Label"

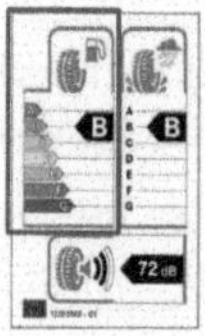

European Commission

TÜV SÜD Auto Service GmbH 2018-06-13 Chassis Tech 2018 Slide 13

Case Study: Vehicle Homologation on Exhaust Emissions

- Coast down on road is basis for adjusted force on drum for WLTP cycle
- Variation in RRC values can lead to different interpolation curves
- All calculated cars based on this curves => influence on fleet consumtion

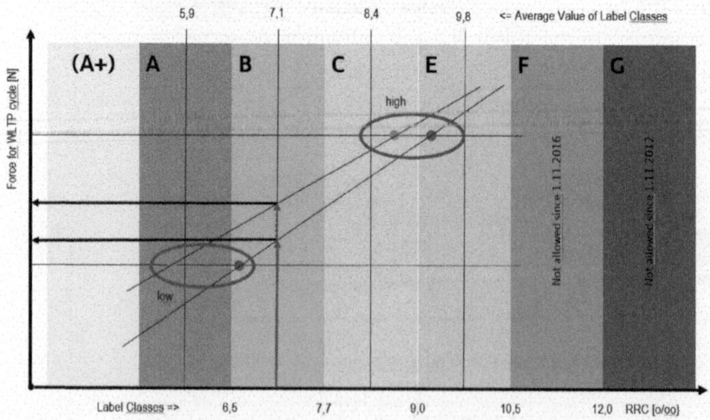

TÜV SÜD Auto Service GmbH 2018-06-13 Chassis Tech 2018 Slide 14

Case Study: Vehicle Homologation on Exhaust Emissions

Physical identical tires meassured on different RRC machines regarding ISO28580 and calculated to official alligned values

	Alligned Values		RRC Label	
	tire 1	tire 2	tire 1	tire 2
LAB1	5,8	8,8	A	C
LAB2	6,2	8,9	A	C
LAB3	6,1	8,9	A	C
LAB4	6,4	9,2	A	E
LAB5	6,2	9,0	A	C
LAB6	6,6	9,2	B	E
LAB7	6,6	9,2	B	E
LAB8	6,2	8,8	A	C
LAB9	6,1	8,7	A	C
LAB10	6,3	8,9	A	C
LAB11	6,3	8,9	A	C
LAB12	6,1	9,2	A	E

Example for bilateral correlation between Audi and suppliers

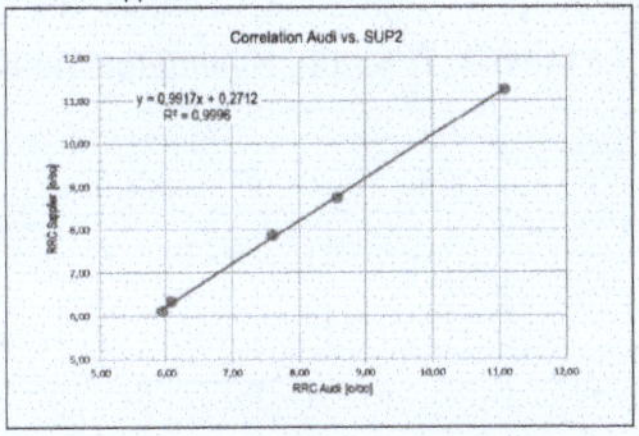

For WLTP it is neccassary to ensure identical RRC values independend from Testing machine !

1. Procedure for alligned values must be reliable and deliver same values
2. Using bilateral comparison between Audi machine and supplier´s

Bilateral correlation formulas between Audi and suppliers

RoWi-Measurement @ ISO 28580			measurement	calculated	factors	
Supplier	formula	coefficient of determination	@ Audi	@ SUP	a	b
SUP1	= 1,0033 x Audi + 0,3005	R² = 0,9986	7,00	7,32	1,0033	0,3005
SUP2	= 0,9917 x Audi + 0,2712	R² = 0,9996	7,00	7,21	0,9917	0,2712
SUP3	=1,0336 x Audi + 0,2422	R² = 0,9991	7,00	7,48	1,0336	0,2422
SUP4	=0,9922 x Audi + 0,1107	R² = 0,9922	7,00	7,06	0,9922	0,1107
SUP5	= 1,036 x Audi + 0,1803	R² = 0,9996	7,00	7,43	1,036	0,1803
SUP61	= 0,8703 x Audi + 0,1887	R² = 0,9983	7,00	8,28	0,8703	0,1887
SUP62	= 1,0008 x Audi + 0,6733	R² = 0,9998	7,00	7,68	1,0008	0,6733
SUP7	= 1,0018 x Audi + 0,2616	R² = 0,9945	7,00	7,27	1,0018	0,2616
SUP81	= 0,9673 x Audi + 0,5659	R² = 0,9966	7,00	7,34	0,9673	0,5659
SUP82	= 1,0897 x Audi - 0,3820	R² = 0,9952	7,00	7,25	1,0897	-0,382
SUP9	= 1,0188 x Audi + 0,6099	R² = 0,9817	7,00	7,74	1,0188	0,6099

TÜV SÜD Auto Service GmbH 2018-06-13 Chassis Tech 2018 Slide 15

Outlook: Future Amendments on European Tire Regulations

Proposal for amendments regarding tires with run-flat capabilities
Amendments of UNECE Regulations Nos. 30&64 intended to provide a legal framework for Extended Mobility Tires (EMT), as an emergency mobility equipment.

- New test procedure shall be introduced, based on a revised technical standard ISO 16992 Passenger car tires - Spare unit substitutive equipment ("SUSE")
- Intention: Standardization of "Soft Run-flat tire" for replacement of spare wheel, initiated by vehicle industry
- Technology already available in market – shall now be standardized in UN framework
- Proposal prepared by the tire standardization organisation ETRTO , the "European Tyre and Rim Technical Organisation".

Positive impact on vehicle development:
- Mobility
- Weight saving (no spare wheel or other emergency equipment necessary)
- Chassis development (current Run-flat tires have disadvantages regarding performance)
- Comfort (current Run-flat tires have disadvantages on NVH behaviour).

TÜV SÜD Auto Service GmbH 2018-06-13 Chassis Tech 2018 Slide 16

Outlook: Future Amendments on European Tire Regulations

Proposal for amendments to UNECE Regulation 117:

Stage 3 and Stage 4 - tightening of the EU tire limit values for all Performance criteria, Wet grip, Rolling resistance and **Rolling noise** at the same time.

table I Rolling resistance. Suggestion for future Stage 3 and 4 limits for standard tyres[a,b]

Tyre type	Current limit			Stage 3 Short term (e.g. 2020)				Stage 4 Longer term (e.g. 2030)			
	Data analysis			Data analysis			Suggested limit	Data analysis			Suggested limit
	Limit (kg/ton)	compliant label values	% tyres compliant	label values analysed	connected limit (kg/ton)	% tyres compliant	(kg/ton)	label values analysed	connected limit (kg/ton)	% tyres compliant	(kg/ton)
C1	≤10.5	A,B,C,E	89%	A,B,C	≤ 9.0	59%	≤ 9.0	A,B	≤ 7.7	19%	≤ 8.0

Target: reduce RRC limit from **10,5 to 8,0**

table II Wet grip. Suggestion for future Stage 3 and 4 limits for standard tyres[a,b].

Tyre type	Current limit			Stage 3 Short term (e.g. 2020)				Stage 4 Longer term (e.g. 2030)			
	Data analysis			Data analysis			Suggested limit	Data analysis			Suggested limit
	Limit (G)	compliant label values	% tyres compliant	label values analysed	connected limit (G)	% tyres compliant	(G)	label values analysed	(G)	% tyres compliant	(G)
C1	≥1.1	A,B,C,E	100%	A,B	≥ 1.4	79%	≥ 1.45	A	≥ 1.55	26%	≥ 1.6

Target: rise min. WG index from **1,1 to 1,6**

table III External Noise. Suggestion for future Stage 3 and 4 limits for standard tyre[a,b].

Tyre type	Current limit			Stage 3 Short term (e.g. 2020)				Stage 4 Longer term (e.g. 2030)			
	Data analysis			Data analysis			Suggested limit	Data analysis			Suggested limit
	Limit (dB(A))	compliant label values	% tyres compliant	Change vs current limit	noise values analysed	% tyres compliant	(dB(A))	Change vs current limit	noise values analysed	% tyres compliant	(dB(A))
C1 (A-E)	≤70-74	A,B	93%	-1dB	≤69-73	58%	≤69-73	-3dB	≤67-71	16%	≤67-71

Target: reduce PBN limit from **70-74 to 67-71**

[Source: Informal document UNECE, GRB-66-01, September 2017, by the Netherlands]

TÜV SÜD Auto Service GmbH 2018-06-13 Chassis Tech 2018 Slide 17 TÜV®

Outlook: Future Amendments on European Tire Regulations

Proposal for amendments to UNECE Regulation 117:

Stage 3 and Stage 4

Potential impact on vehicle manufacturer:

Conflict of interest on balance of performance, in particular

- Overall safety performance of the car affected if single tire performance parameters are overrated
- Proposed noise limits are increasingly corresponding to "Slick"-tire characteristics
- Wet behavior (Aquaplaning) in future to be achieved with low tread depth pattern (resulting from lower noise and lower weight / rolling resistance technologies)
- New challenges for extended mobility tire solutions (Run-flat technology).

TÜV SÜD Auto Service GmbH 2018-06-13 Chassis Tech 2018 Slide 18 TÜV®

Summary

Summary

- The vehicle manufacturer is facing several challenges to be solved concerning tire related legal requirements (in addition to obligations for the tire manufacturer)
- Harmonization of tire regulations should also consider alignment of requirements in different fields, as full vehicle noise and exhaust emissions; tire regulations and full vehicle regulations should be based on same test conditions and comparable
- Tire parameter values must particularly be reliable when based on self declaration
- Evolution in tire regulations can also support development of vehicles
- When implementing new thresholds for tire performance the approach of balancing overall performances needs to be considered.

TÜV SÜD Auto Service GmbH 2018-06-13 Chassis Tech 2018 Slide 19

Efficient parameterization of a user-friendly tire model

Dipl.-Ing. Ronnie Dessort
Dr.-Ing. Cornelius Chucholowski

TESIS GmbH

P. Pfeffer (Ed.), *9th International Munich Chassis Symposium 2018*, Proceedings,
https://doi.org/10.1007/978-3-658-22050-1_54

1 Introduction

Early simulation is indispensable to save costs and shorten development times in the automotive industry. The tire as a highly complex component with many variants and designs not only determines the driving dynamics, but also influences vibration behavior and comfort and is ultimately responsible for safe driving behavior. Tire behavior is influenced by its states and properties, which may change over time. Therefore, it is a major challenge, even impossible, to provide a generic simulation model for all applications. In this contribution we show how the unique nature of the FTire can be used to fit the less sophisticated but faster and easy to use tire model TMeasy along with comparison of component and full vehicle simulation results. In this context a model extension to take varying inflation pressure into account is presented.

2 Tire models

Vehicle simulation requires precise tire models. Results are always as accurate as the data fed to the models. In theory, mathematical black-box models can be parameterized perfectly with unlimited measurements. However, the effort involved in the parameterization must be reasonable and the models should be able to extrapolate even non-measured operating points or plausibly reproduce parameter dependencies. As a physically-based tire model, FTire is able to generate additional insights and information [9], which less complex tire models like TMeasy can attempt to represent.

2.1 FTire – the supervisor

FTire (Flexible Structure Tire Model) is a full 3D nonlinear in-plane and out-of-plane tire simulation model with sophisticated 2D and 3D rigid and flexible road surface description models. It explains most of the complex tire phenomena on a mechanical, thermodynamical, and tribological basis and serves as a sophisticated tire force element. A detailed thermal model is also provided optionally. FTire is used for vehicle ride and durability investigations, or high frequency vehicle dynamics simulations on even or uneven surfaces. The tire belt is represented by a slim ring, that can be displaced and bent in arbitrary directions relative to the rim: vertically, longitudinally, and laterally.

Figure 1: FTire used for ride comfort analysis in DYNA4 [5]

2.2 TMeasy – the learner

TMeasy represents a handling tire model "easy-to-use" based on a semi-physical model. It includes a massless force element acting between the road and the wheel. The unevenness of roads is approximated by small local planes in the contact region of the tire. Based on a generalized force computation as depicted in figure 2, TMeasy generates all components of the contact force vector and contact torque vector including first order tire dynamics. The wheel modeled by a rigid body must incorporate mass and inertia properties of the rim and the tire. TMeasy is available as part of the vehicle simulation framework DYNA4 by TESIS GmbH [1] and is very successful used in handling applications – offline and in realtime.

TMeasy consists of a manageable number of physically based parameters. The big advantage, especially in situations of limited data availability, comes with its easy to estimate parameters: even with a crude knowledge of size, payload as well as friction property of the tire-road combination a first guess gives feasible results – good enough for simulation of extraordinary tires, [2]. Of course, the parameters can be adjusted by curve fits to meet given tire measurements or vehicle dynamic results more precisely, [3] and [4]. Another advantage lies in the ease with which tire properties can be scaled to represent different road and tire conditions. Nevertheless, this model is currently not capable of depicting any influence of time-variant tire properties like temperature or inflation pressure.

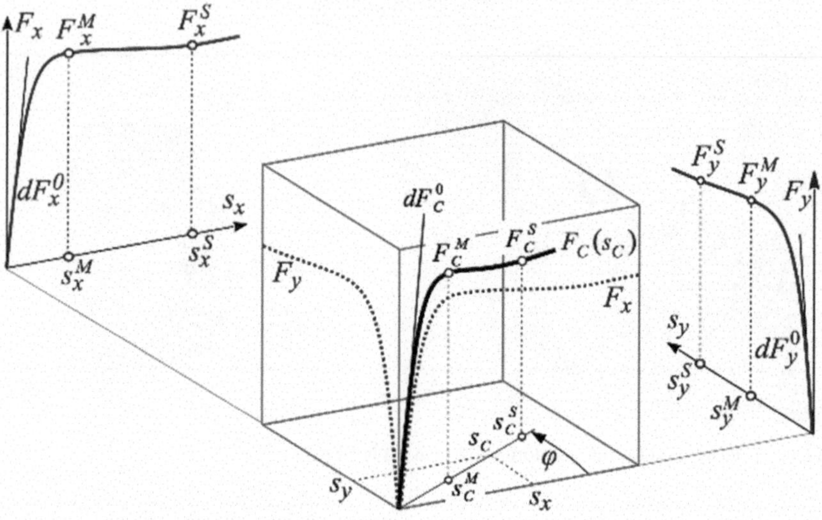

Figure 2: Combined tire forces in TMeasy

3 Influence of a varying inflation pressure

Typically, the first step of research is the observation of the system behaviour. In this contribution the same tire as used in [5] is examined, where the impact of inflation pressure on static and dynamic vertical tire properties was considered to improve the performance of a suspension controller model. In general plausibility of simulation results can be verified by deploying the brush model, a well-known approach to model tire forces, [6], [7] and [8]. As far as possible, all signals are named and defined according to the TYDEX standard [11]. All test cases are conducted on a virtual tire test rig being part of the simulation framework DYNA4 and are further used for parameter fitting purposes in chapter 4.

3.1 Static tire stiffness

To determine the static tire stiffness in longitudinal and lateral direction as well as the torsional stiffness in pure bore motion, the tire is deflected in a variation of constant wheel load by translational or rotational displacement of the ground in the corresponding spatial directions. Figure 3 shows in the first row the resulting force and torque curves as dependency of the corresponding degree of freedom (column-wise) for a certain wheel load and varying tire inflation pressure. With increasing quasi-static deflection (LONGDISP,

LATDISP, STEEANGL), the longitudinal and lateral force (FX, FYW) and the bore torque (MZW) increases quasi-linearly with adhering profile particles until a continuous transition into a sliding movement and thus constant force or torque occurs. The second row of figure 3 shows the change in the determined stiffnesses (gradient of dotted lines in first row) against the inflation pressure variation. A degressive behaviour of the longitudinal and lateral stiffness can be observed, whereas the torsional stiffness tends to decrease with increasing pressure.

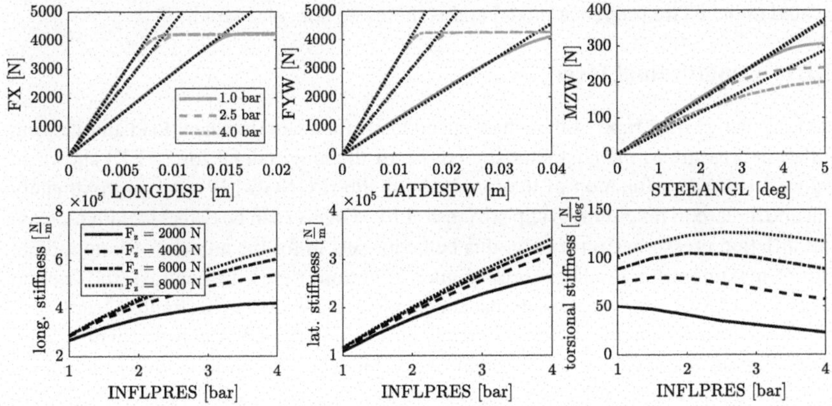

Figure 3: Longitudinal, lateral and torsional tire stiffness (column-wise)

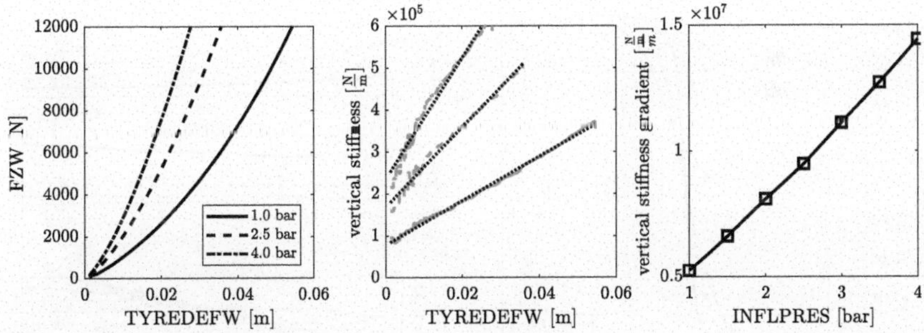

Figure 4: Vertical tire stiffness for different inflation pressures

By a quasi-static vertical compression of the tire, the corresponding stiffness can be determined. The tire behaves analogously to a gas pressure spring. The use of a quadratic polynomial as a regression function of the wheel load (FZW) over the spring deflection (TYREDEFW) is very well suited, as can be seen from the linearization of the gradient (figure 4 center). The rate of change of the vertical stiffness over the deflection shows a linear behavior with respect to the inflation pressure (figure 4, right).

3.2 Quasi-static tire characteristics

In the following, the quasi-static tire characteristics and their behavior at different inflation pressures will be examined. A potential influence due to heating has been deactivated in the FTire reference model and is therefore not considered.

3.2.1 Longitudinal force

A slow change over time of the wheel speed leads for a certain wheel load and varying inflation pressure to the characteristic course of the longitudinal force (FX) shown in figure 5 (left) as a function of the longitudinal slip (LONGS-LIP). An approximately linear decrease in longitudinal slip stiffness (LFLSGRFX) can be detected with increasing inflation pressure. Conversely, this behavior sets in for the amount of longitudinal slip at maximum longitudinal force (figure 5, bottom right). In contrast, the friction potential (LGFCCOEF) moves almost constantly in a narrow range (figure 5, right center). The level of maximum force remains almost unchanged at constant wheel load. Overall, thus, the acceleration potential of a vehicle is reduced in the longitudinal direction with higher inflation pressure or such a tire tends to slower behavior while reducing rolling resistance, respectively.

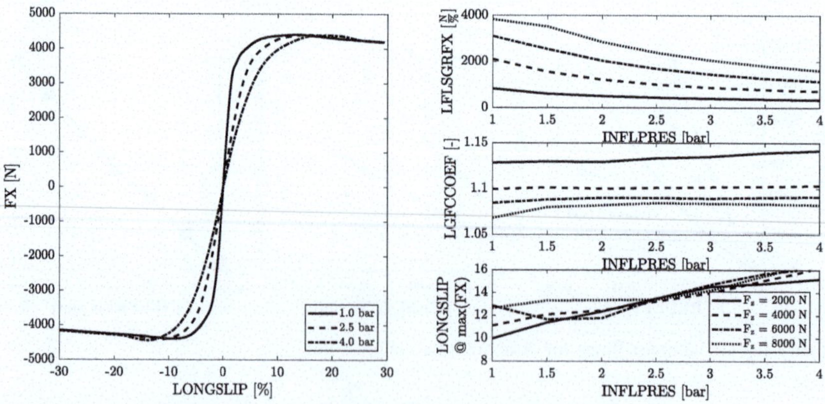

Figure 5: Static longitudinal force for different inflation pressures and wheel loads

3.2.2 Lateral force and self-aligning torque

A free-rolling wheel is slowly steered from left to right under a certain wheel load and thus leads to the characteristic curve of the lateral force (FYW) as a function of the slip angle (SLIPANGL) shown in figure 6. In contrast to the longitudinal force behaviour, cornering stiffness shows an increasingly degressive behaviour via the inflation pressure (figure 6 top right). The friction potential (LTFCCOEF) is, analogous to the specific lateral force maximum, slightly and approximately linearly increasing (figure 6 right center), whereas the required slip angle across all wheel loads and inflation pressures strives towards a similar value (figure 6 bottom right).

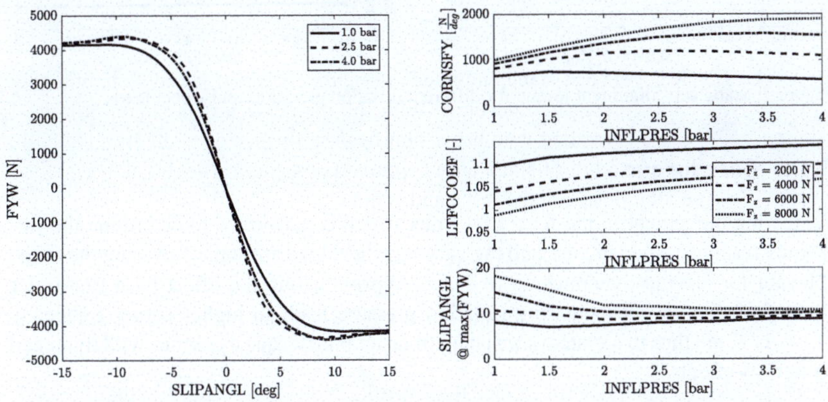

Figure 6: Static lateral force for different inflation pressures and wheel loads

In addition, the behavior of the self-aligning torque is obtained from the same test case as shown in figure 7 (left). Due to the decreasing gradient around the zero position (ALIGNSMZ), the build-up of the self-aligning torque is slower and with decreasing amplitude despite increasing cornering stiffness (figure 7 right center). The maximum is reached almost unchanged at a similar side slip angle (figure 7 bottom right) and before the lateral force maximum.

3.3 Dynamic tire characteristics

In the last test sequence, the dynamic behavior under different pressure conditions is examined in more detail, with the focus being placed at this point on the bore torque (frequency range) and the lateral force build-up under step excitation (time range).

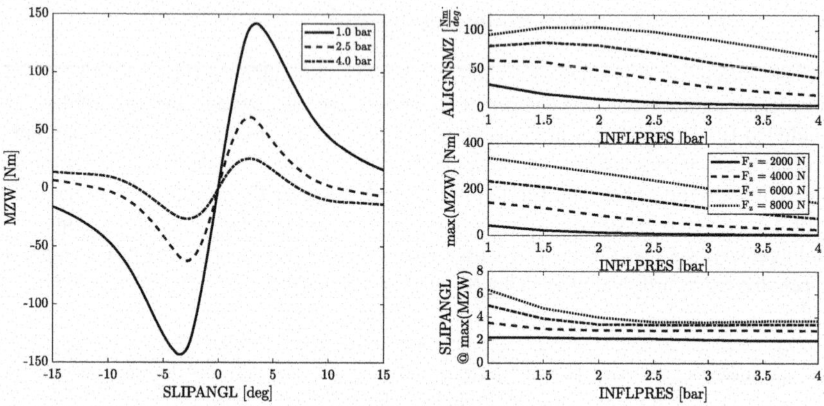

Figure 7: Static self-alignment torque for different inflation pressures and wheel loads

3.3.1 Bore torque

A steering movement in the form of a sinus sweep is performed to determine the frequency response when the tire and the varying wheel load are static, covering the range relevant for driving dynamics up to 3 Hz. Approximating the effect by a first-order dynamic behavior, an increasing inflation pressure leads to higher corner frequency $\omega_0 = 1/T$ and thus to a system consisting of faster bore torque dynamics with tightened hysteresis effects.

3.3.2 Step response behavior

A freely rolling tire is subjected to an abrupt steering excitation and the resulting lateral force is assessed in terms of its time behavior. The overshoot ratio of the lateralforce, which increases approximately linearly above the inflation pressure, is shown in figure 9 (top right) as an objective quantity for characterizing the step response. Typically, less complex models such as TMeasy react with first-order dynamics and thus are not capable of depicting such oscillations. In this context, the response time (figure 9 bottom right; here, for example, the time to reach the maximum) is of greater importance. Here are, amongst others, the cornering stiffness as well as the lateral stiffness decisive factors, which lead to a reduced reaction time due to their mutual increase with increasing inflation pressure. Furthermore, a tire with higher inflation pressure provides a more direct steering feel with its faster system dynamics in terms of lateral dynamic handling.

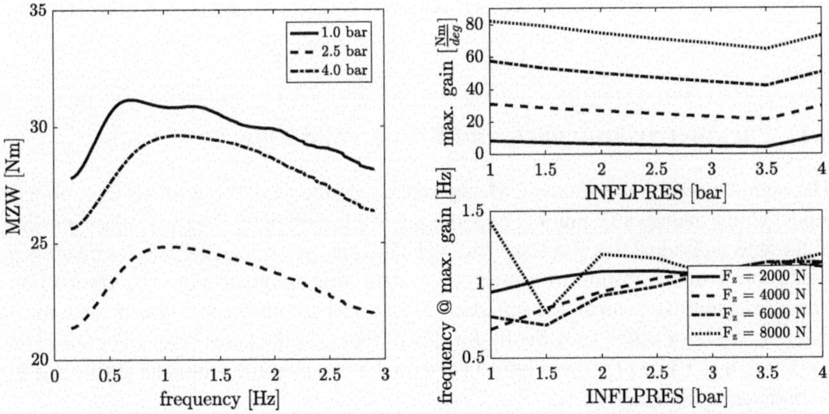

Figure 8: Frequency behaviour of bore torque for different inflation pressures and wheel loads

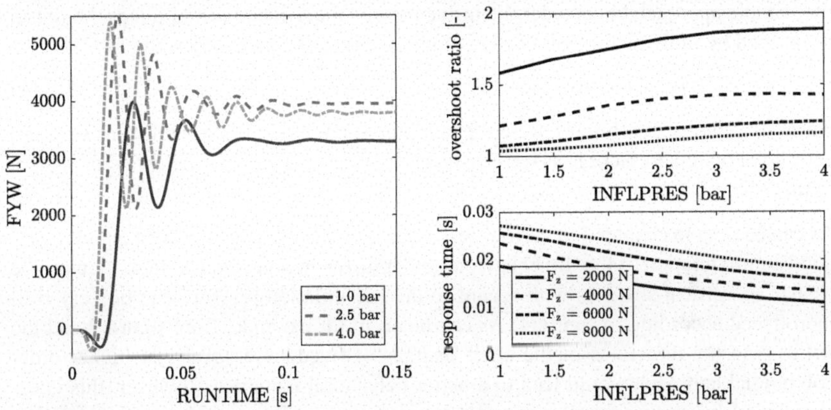

Figure 9: Step response of lateral force for different inflation pressures and wheel loads

4 TMeasy model extension

Based on the tire behavior under variable inflation pressure, which was analyzed in chapter 3, an empirical approach to the extension of the tire model TMeasy will be presented in the following. The model parameter fitting process and its measurement depicting quality are discussed in more detail. A subsequent test as part of an overall vehicle simulation in DYNA4 [1] discusses the effects of a changed inflation pressure in steady-state circular driving compared to the reference simulation with the physically-based tire model FTire [9]. For the sake of simplicity, the focus here is on the lateral behavior.

4.1 Parametric approach and fitting validation

The semi-empirical tire model TMeasy represents the philosophy of an easy-to-use model with a manageable number of comprehensible parameters and a clear assignment of these to individual tire characteristics [3]. Classically, tire models are parameterized on the basis of the results from a specific measuring program with a reference tire. These reference data can be generated either on a real test bench or in a virtual environment, as DYNA4 offers specifically for this purpose. In the latter case, a complex tire model such as FTire [9] allows realistic data to be obtained for parameter fitting due to its high depicting quality.

An important feature of TMeasy is the possibility to create a plausible data set even without the availability of specific measurement data [2]. Following this basic idea, a first simple approach to consider the influence of a modified tire inflation pressure can be defined by individual linear change of all system parameters according to

$$\mathcal{P}(p_t) = \mathcal{P}(p_{nom}) + g_{\mathcal{P}} \cdot (p_t - p_{nom}) \tag{1}$$

where $\mathcal{P}$ denotes an arbitrary system parameter at nominal inflation pressure p_{nom} and currently applied pressure p_t as well as the scalar value $g_{\mathcal{P}}$ represents the rate of change of this specific value.

As can be seen in chapter 3 in various test cases, this approach is initially a qualitatively good approximation of the tire behaviour under different wheel loads. In particular, the change in vertical stiffness with regard to the tire inflation pressure can be very well reproduced according to figure 4. As can be seen, for example, from figure 3 and the stiffness in the directions of the other degrees of freedom, a second or higher order polynomial approach would lead to a better behavioral mapping. However, this is accompanied by more complex parameterization and is diametrically opposed to the model idea of easy access for users. The objective of this model extension is less the high-precision mapping of tire behavior over a wide range of tire inflation pressures, since specialized data sets can also be generated for this case. Rather, the intention lies in a plausible representation of this effect in order to enable the user to easily integrate corresponding deviations from e.g. ECU functions (TPMS) in his test scope with his precise data set at nominal pressure.

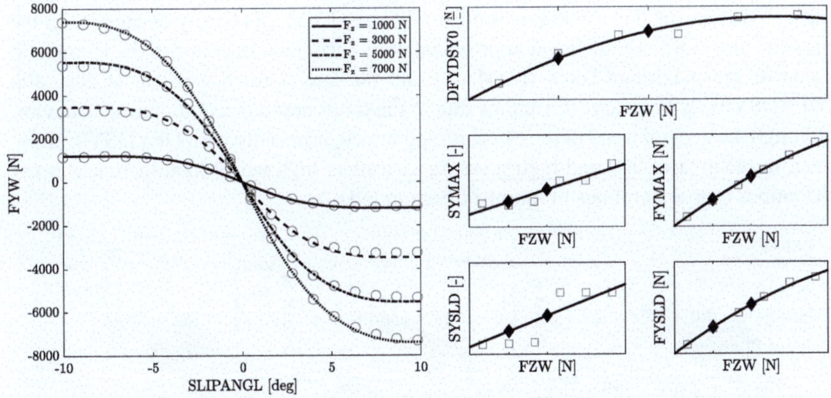

Figure 10: TMeasy fitting result of static lateral force at nominal inflation pressure

The gradients are determined in a multi-stage fitting process. For this purpose, virtual reference measurements are first carried out using the tire model FTire at different constant inflation pressures. These sequences include static (stiffnesses), quasi-static (longitudinal, lateral and torsional tire characteristics) and dynamic (running-in behavior) test cases. With this reference data, individual TMeasy data sets for discrete inflation pressures (here in the range of 1 to 4 bar) are generated. A detailed parameter fitting procedure is described in [3]. Figure 10 (left) shows an example of the TMeasy fitting result of the lateral quasi-static tire behavior at nominal pressure (2.5 bar) and different wheel loads. Here a very good correlation to the FTire reference data (circle markers) can be seen. The right-hand side shows the parameter identification for describing the static characteristic curve for individual wheel loads (square markers) and a corresponding polynomial approximation with the parameter values ultimately stored in the data set as grid points for two reference wheel loads (diamond markers). Also here, a very good correlation is evident with regard to the linear and quadratic functions for parameter approximation above the wheel load.

In a second step, as shown in figure 11 (first row), a data point cloud is generated from these individual fitting results (square markers of figure 10 at discrete inflation pressure and individual wheel load). The starting point for a pressure-dependent change are the already determined parameter support points at nominal pressure (see diamond markers in figure 10), i.e. the precise depicting quality of TMeasy remains unchanged in this standard tire condition. Based on this, the (identical) rate of change of these two points is determined according to the method of least squares in order to approximate the given data points optimally with this simple regression approach. The second row of figure 11 shows the projection onto the parameter-inflation pressure-plane for selected wheel loads. You can see that the maximum lateral force (FYMAX) hardly changes with the pressure and therefore the linear approximation is sufficient. In contrast, the lateral slip value of the maximum force (SYMAX) and the lateral force gradient at zero slip (DFYDSY0) show clearer deviations due to the more pronounced nonlinear behavior, although the general trend of the effect is maintained, especially with DFYDSY0. However, in the area of low and high pressure as well as high wheel load there are larger deviations (upper two lines in figure 11 bottom left).

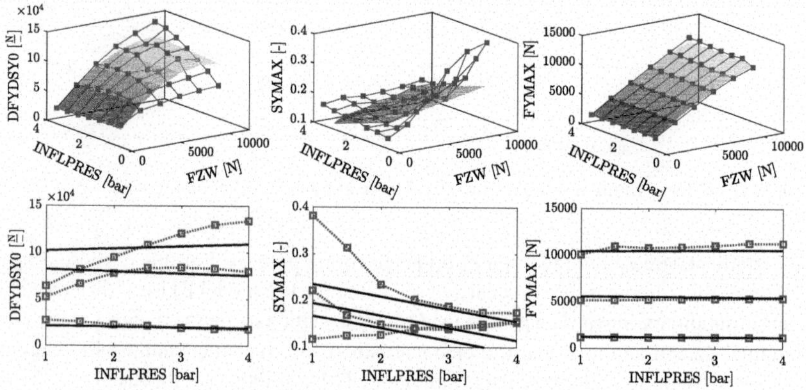

Figure 11: Lateral parameter fitting of TMeasy considering inflation pressure influence

4.2 Full vehicle simulation

The behavior of the extended TMeasy tire model to take a variable inflation pressure into account is examined in more detail by deploying the data set determined in chapter 4.1 as part of a full vehicle simulation. As a test case, the maneuver of a steady-sate circular driving according to ISO-4138 was used, with data sets for the FTire and TMeasy reference model configured with pressures of 1.5, 2.5 (nominal pressure) and 3.5 bar. Figure 12 shows the results regarding the self-steering behavior (left column) and the applied steering torque (right column).

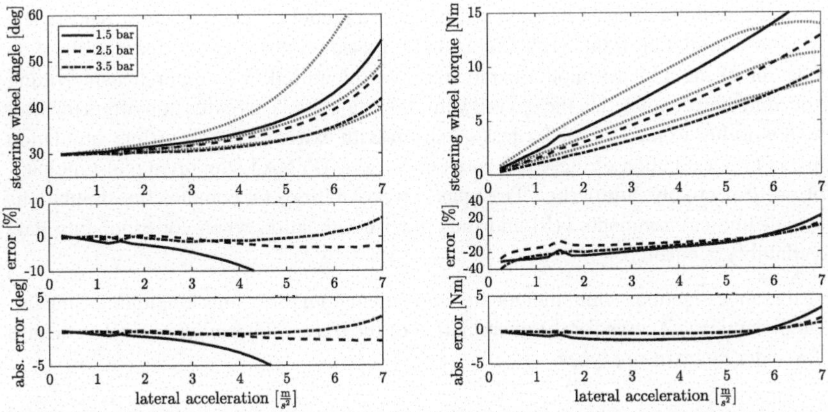

Figure 12: Steady-state cornering behavior of FTire (dotted) and enhanced TMeasy for different inflation pressures

The self-steering behavior at nominal pressure with an absolute maximum deviation of 3.2% or 1.4 deg has a very good correlation to the reference. The same quality is also achieved at increased inflation pressure (3.5 bar), where the deviations at the limits are 5.7% and 2.3 deg, respectively. At reduced inflation pressure, the behavior in the linear range with a deviation of less than 10% or 5 deg is still acceptable, but then deteriorates significantly towards the limits. All in all, the extended TMeasy model shows the global effect in self-steering behavior, which is caused by a change in inflation pressure, plausibly and especially at medium and high pressures very well. The applied steering torque in the virtual test vehicle shows a higher deviation across all pressures, but at the limits it is of similar quality to the curve at nominal pressure. This allows conclusions to be drawn about a lower nonlinear characteristic of the pressure-dependent behavior for parameters that characterise the self-aligning torque of the tire.

5 Conclusion

„Make everything as simple as possible, but not simpler“ (Albert Einstein). A philosophy that is reflected in the real-time tire model TMeasy. Characteristic of this is a model with a manageable number of physically interpretable parameters, which provides a highly accurate correlation to virtual or real measurements of a reference tire. Alternatively, it also allows the user to manually generate a plausible data set if no access to measurements is available. Following this credo, this paper presented the potential of a simple model extension to take into account a variable tire inflation pressure. To this end, the tire behaviour in different static, steady-state and dynamic situations was first examined with the help of the physical-based tire model FTire [9] on a virtual tire test rig in the simulation framework DYNA4 [1]. Based on these results, the modification behavior of the TMeasy model parameters was identified in a comprehensive fitting procedure and evaluated in the overall vehicle simulation. This extension offers the user the possibility to map a broader test spectrum with extremely short offline simulation times or the execution of these test cases on various DYNA4-supported real-time platforms, respectively. Similarly, FTire opens up possibilities for complex tire simulations in real-time environments [10] and thus precise validation when corresponding data availability is ensured.

Further investigations could include the implementation of various longitudinal and lateral dynamics test cases for the examination of the transient behavior as well as the use of a wider range of tire types.

References

[1] https://www.tesis.de

[2] G. Rill. An engineer's guess on tyre parameter made possible with TMeasy. in: Proceedings of the 4th International Tyre Colloquium (P. Gruber and R. S. Sharp, eds.), University of Surrey, GB, 2015.

[3] R. Dessort, C. Chucholowski, G. Rill. Parametrical approach for modeling of tire forces and torques in TMeasy 5. In: Bargende M., Reuss HC., Wiedemann J. (eds) 16. Internationales Stuttgarter Symposium. Proceedings. Springer, Wiesbaden, 2016

[4] W. Hirschberg, F. Palacek, G. Rill and J. Sotnik. Reliable Vehicle Dynamics Simu-lation in Spite of Uncertain Input Data. Proceedings of 12th EAEC European Auto-motive Congress, Bratislava, 2009.

[5] R. Dessort, C. Chucholowski. Explicit model predictive control of semi-active suspension systems using Artificial Neural Networks (ANN). In: Pfeffer P. (eds) 8th International Munich Chassis Symposium 2017. Proceedings. Springer Vieweg, Wiesbaden, 2017

[6] H. Dugoff, P. S. Fancher, and L. Segel. Tire performance characteristics affecting vehicle response to steering and braking control inputs. final report. Technical Report. Highway Safety Research Institute, Ann Arbor, Michigan, 1969.

[7] H. B. Pacejka. Modelling of the pneumatic tyre and its impact on vehicle dynamic behavior." Technical Report i72B. Technische Universitätt, Delft, 1988

[8] J.Y. Wong. Theory of Ground Vehicles. 3rd edition, John Wiley & Sons, New York, 2001

[9] M. Gipser. FTire and Puzzling Tire Physics: Teacher, not Student. 4th Tyre Colloquium, University of Surrey, 2015

[10] https://www.cosin.eu/wp-content/uploads/FTire_Artikel_VehicleDynamic_1601213.pdf, accessed May 9, 2018

[11] https://www.fast.kit.edu/download/DownloadsFahrzeugtechnik/TY100531_TYDEX_V1_3.pdf, accessed May 9, 2018

Managing the variety of potential tire / wheel sizes in the early vehicle development process

Francesco Calabrese*, Luca Dusini+, Dr. Manfred Bäcker*,
Axel Gallrein*

* Fraunhofer ITWM
+ Maserati S.p.A, Italy

P. Pfeffer (Ed.), *9th International Munich Chassis Symposium 2018*, Proceedings,
https://doi.org/10.1007/978-3-658-22050-1_55

Introduction

Reducing the development time is always an important goal for the car manufacturers. Maserati's product and development team targets to shorten the time needed to produce a car starting from the kick-off by one year. In order to reach this target, many different actions have been defined.

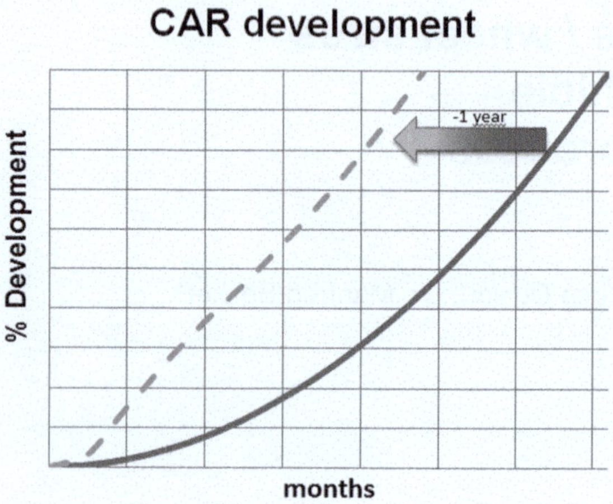

Figure 1. Car Development: time reduction challenge.

The best way to reduce the development time span is speeding up the concept phase before the experimental phase. The quality of the final product is extremely important for a luxury brand such as Maserati, and it can be guaranteed only by testing the car on the track, analysing all the different performances in depth. Using simulation tools, it should be possible to modify the gradient of increasing maturity of the project – without affecting the quality. For this reason Maserati is putting considerable effort into improving its simulation tools to increase the accuracy of the prediction done at the early stage of the development.

At the beginning of the project (but after the target setting phase), Maserati vehicle dynamics team starts to work on the different areas which can improve the performance and reach the high level target for such kind of cars:

- Kinematic and compliance of the suspension
- Tuning of springs, dampers and stabilizer bars
- Active systems
- Tires

For sure, the tire is one important aspect to be considered when evaluating the performance of the car. Using the tire model CDTire/3D, a new approach has been developed in Maserati in order to predict the performance of the tire before the first physical prototypes are actually available.

The key questions in the early design phase of a car related to a tire are [5]:

1. Which tire and rim size to choose?
2. What is the optimal tire pressure setup?
3. How close can we come to the key performance indicators by improving the tire and how much have been done on the vehicle side?

Maserati uses the CDTire/3D model to answer these important questions. The main idea is to start from a reference tire for which tire measurements are available. For example, this can be a tire from a predecessor. Because of the fact that CDTire/3D is a structural tire model which has a strict separation of geometry and material properties, one can create tire models for different tire and rim sizes out of this reference model by using a morphing technique to adapt the geometry. This methodology is the enabler to investigate the first question before actual physical tires are available for all tire and rim sizes under consideration.

In a previous article [1] the authors already showed that CDTire/3D is able to predict the tire behaviour physically at any pressure, starting from 0 bar up to very high pressure without changing the tire model parameters. This capability makes it possible to optimize the tire pressure setup in a very efficient and economical way.

The last question, we want to address in a future paper. The most important ingredient to investigate this topic is the suitability of CDTire/3D for a fast virtual prototyping of a tire in combination with the capability to be able to run in a full vehicle simulation scenario with a good performance. Based on this, changes in the tread properties – but also structural changes – can be realized very fast and assessed by full vehicle simulations afterwards.

In the first chapter of the paper we introduce CDTire/3D from a structural point of view an explain why this model is suitable to address the questions above.

In the second chapter, we are showing a tire standalone validation using tire test rig measurements for the pressure variation capability as well as for the tire and rim resizing methodology.

In the third chapter Maserati presents an evaluation and validation of the vehicle performance by modifying the tire dimensions, using the CDTire morphing methodology.

CDTire/3D

The model

Fraunhofer ITWM has developed the structural tire model CDTire/3D [3,4,6] used in the vehicle industry for comfort, durability and advanced handling scenarios. This model is based on a spatial finite difference (FD) formulation of the tire modeled as a shell. The functional layers of a tire (like cap ply, belt plies and carcass/body plies) are accumulated into the shell properties during tire initialization, but are accessible through their respective material parameters in the parameter file. The modeling of each cord-reinforced layer includes a non-linear part in the elastic component of the material description due to different behaviour under compression and tension regimes. The geometric formulation of the material behaviour allows for very large deformations. The dissipative parts of the material description combine viscous-elastic and inner friction behaviour. The tread is modeled by a brush-type contact formulation, allowing for local stick-slip effects. Figure 2 sketches the principle functional components of this structural MBD tire model:

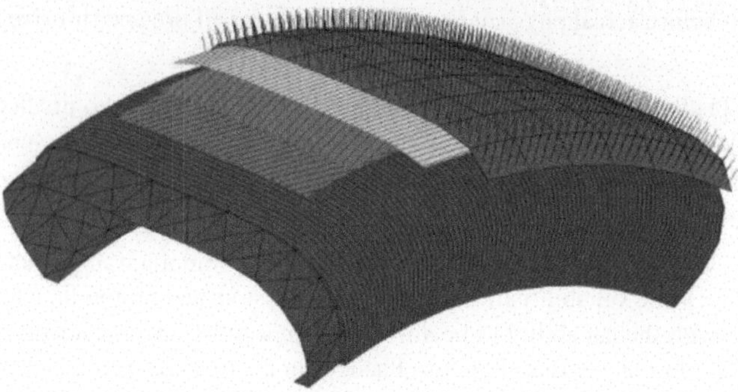

Figure 2. Functional layers of tire in CDTire

The model features a strict separation between material and geometric properties with the inflation pressure applied correctly onto the inner liner surface of the tire.

The model is set up with the help of a Construction Assistant.

The Construction Assistant

The Construction Assistant utilizes the inflated cross section geometry to parameterize the tire based on its constructional properties, such as the properties of the functional layers of tread, cap ply, belt and carcass layers – as well as geometric properties, such as position of the functional layers. See Figure 3 for a layout of the Construction assistant.

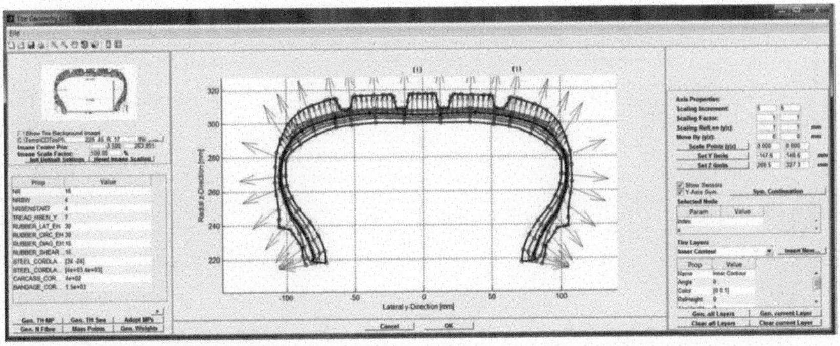

Figure 3. Construction assistant

From the material properties of the functional layers and its geometric dimensions, a discrete shell model is derived.

The Morphing

As the model features a strict separation between material end geometric properties, Fraunhofer ITWM developed a method to morph an existing tire based on nominal size specifications

- Tire nominal width (e.g. 225 mm)
- Tire aspect ratio (e.g. 45 %)
- Rim nominal diameter (e.g. 17'')
- Rim nominal width (e.g. 7.5')

The morphing algorithm adapts (morphs) the geometric description of the reference configuration of the cross section of the tire according to changes to any of these 4 nominal descriptive parameters without modifying the material description. It also adjusts the mass distribution according to the change in dimension. In short, the tire can be morphed e.g. from a 225/45 R 17 (x7.5) to a 235/40 R 18 (x8). Figure 4 shows this morphing in principle:

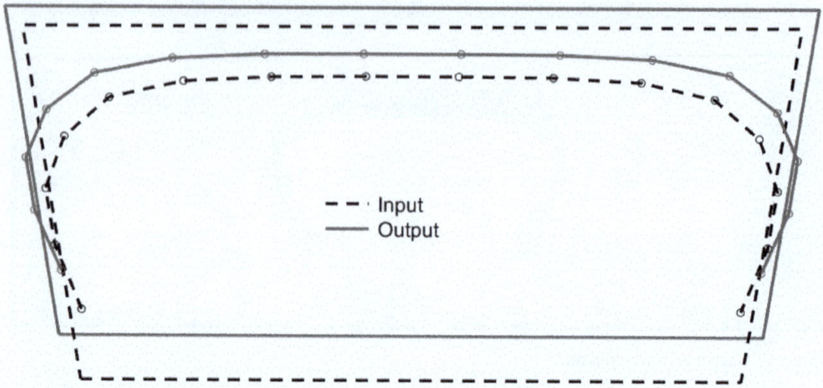

Figure 4. Tire morphing

In figure 4, the trapezes are made up of the nominal size parameters and shows the effect on the shell mid-plane geometry of the base and morphed model.

Tire stand-alone test rig validation

As it has been explained in the introduction, the utilization of different tire/rim sizes and inflation pressures in the vehicle performance optimization is a challenging task to solve. In the following paragraphs we analyse some of the modelling ingredients that can help to improve the process. These focus on the predictive capability of the tire model for situations that have not been measured directly. Specifically, we will focus on the 'effect of the inflation pressure' and the 'morphing capability' (adaption of base model to different sizes).

Effect of the inflation pressure

Many authors already in the past described the importance of the effect of inflation pressure on the tire characteristics [2]. The inflation pressure is an important element in the tire and needs to be described in a physical way. The pressurized air is acting on the inner liner and loads the tire structure. This force is absorbed mostly from the cord layers (steel wires, carcass and cap plies) that represent the load bearing structure of the tire. As logical consequence in order to predict the pressure effect a tire model needs:

- An exact description of the inflation pressure p
- An exact description of the volume V in any operational condition of the tire

By reading the first chapter of the article is clear that CDTire/3D has predictive capabilities with respect to inflation pressure variations because:

- The tire is modelled from bead to bead by using a shell approach that takes into account different rubber zone, carcass, steel cords, cap plies.
- The structural part of the tire is well described as CDTire models all the functional layers present in a tire.
- The inflation pressure is physically described as pressure that act normally on the locally deformed surface.
- As consequence the interaction between volume, pressure and structure is physically interpreted from the model.

In this first part of this paragraph we have explained the motivation of the key modelling ingredient that make CDTire/3D a high-level physical tire model respect to pressure.

In a previous article [1] the authors already showed how only measurements at one inflation pressure are needed for the tire parameter identification and how the model is able to predict the tire behaviour physically at any pressure (starting from 0 bar up to very high pressure). In the following the validation of some characteristics pressure dependent are shown:

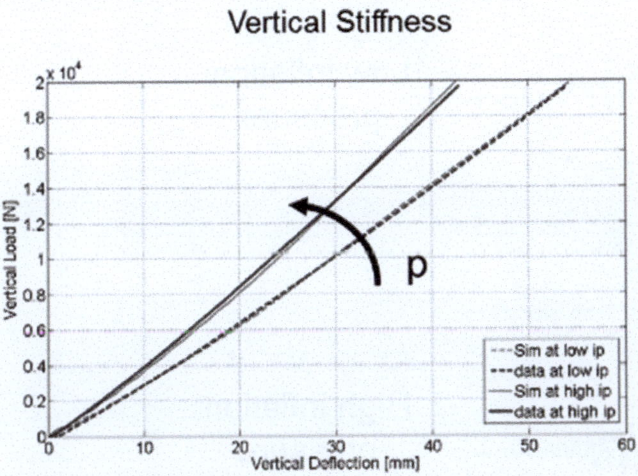

Figure 5. Vertical Stiffness: Pressure effect on a truck tire.

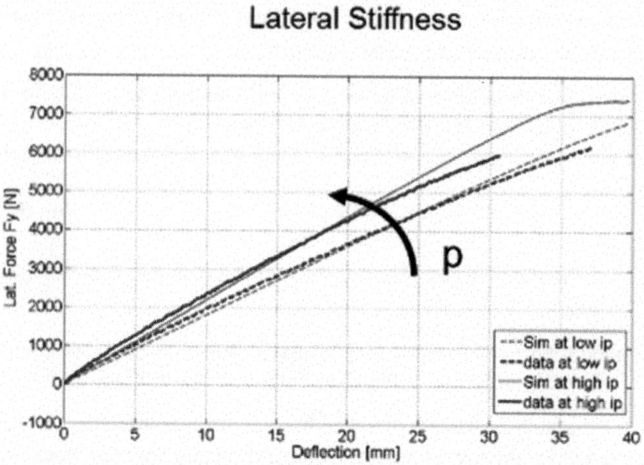

Figure 6. Lateral Stiffness, pressure effect on a truck tire.

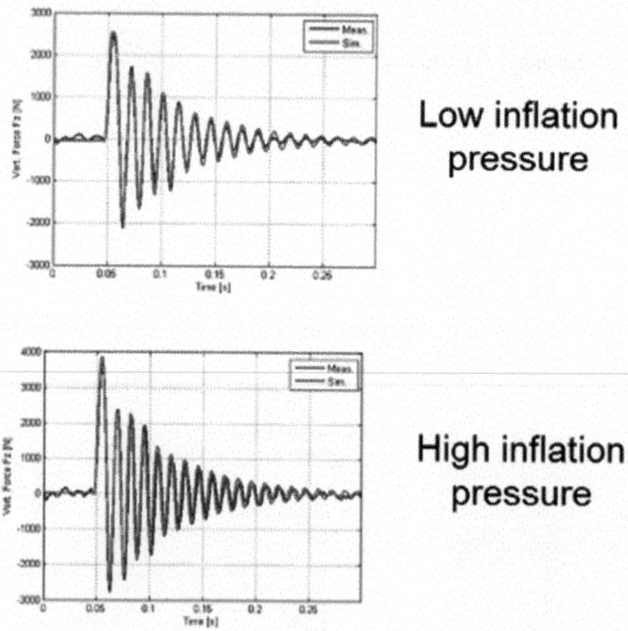

Figure 7. Vertical spindle force during cleat run, pressure effect on a truck tire.

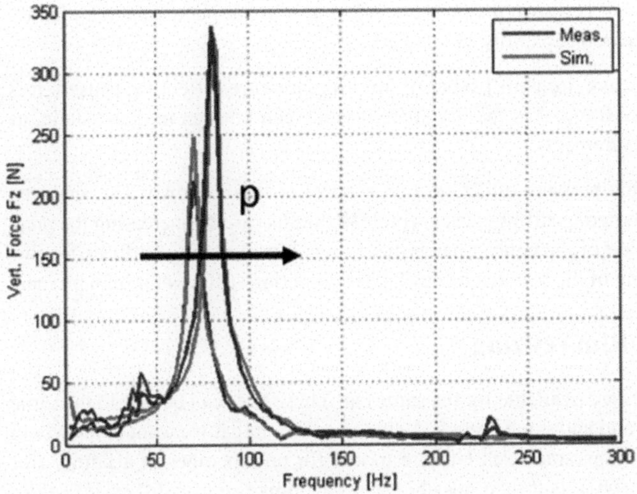

Figure 8. FFT of vertical spindle forces (Figure 7)

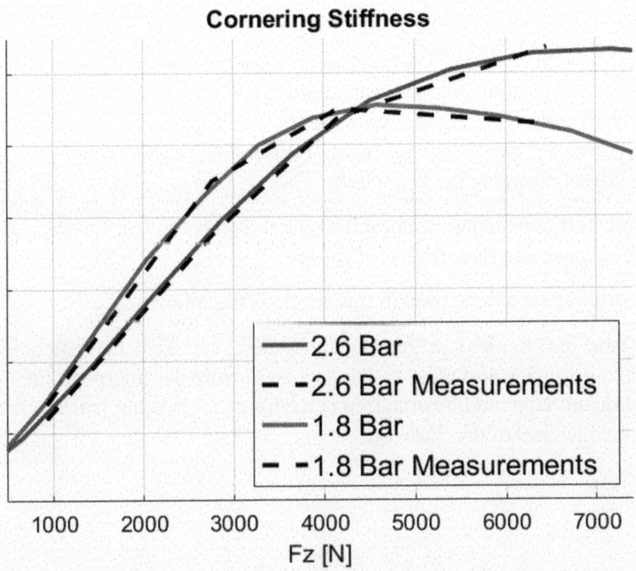

Figure 9. Cornering stiffness versus preload: Pressure effect on a passenger tire.

In the previous figures it is shown the pressure effect on some fundamental tire characteristics and the prediction capability of CDTire.

From figure 5 to 8 a truck tire with 1.6 bar of inflation pressure difference is used. It is clear that inflating the tire makes the tire stiffer, which can be seen in quasi-static, as well as dynamic measurements.

In figure 9 the inflation pressure effect on a passenger car tire is shown. The analyzed characteristic is the cornering stiffness versus preload. In this case by increasing the pressure there is a reduction of the contact patch that brings less cornering stiffness for low preload and a stiffening of the belt that brings higher cornering stiffness for high preload.

Morphing: Tire/Rim resizing

Tires of different sizes are available for the same car. These tires can belong to the same tire 'family', same brand and same branch. Now the question that we want to answer in this chapter is: given a parametrized CDTire model for one member of a family (for example a 235/55 R19 8J), is it then possible to re-size this tire to a different tire/rim size (for example 255/50 R19 9J)? And how accurate is the model prediction respect for the new size?

The general meaning of the morphing capability is: starting from a parametrized tire model called 'predecessor tire' or 'base tire':

a) To have a tire model that allows to predict the qualitative behavior of a tire that does not exist yet in the market by re-sizing the 'base tire' to the 'target size'
b) To predict the qualitative difference between tires of different sizes
c) To have the possibility of changing the rim width

CDTire/3D is predictive in 'tire resizing' because the model features a clear separation of mass, geometrical and material property.

The meaning of the morphing technique used in this article is the following:

- Starting from the 'base tire' to obtain a 'morphed tire' that keep the same material properties but update geometry and mass to the new size. Now the 'morphed tire' will have the same quantitative and quantitative behavior of an existing real tire if this has also the same material of the 'base tire'.

From the application point of view this approach can be used in the early stage of the tire design with the following strategies or final aim:

a) Starting from the 'predecessor tire' to create a model of a 'virtual tire' with a different tire/rim size that does not exist yet in the market or it has never been equipped on this vehicle before. This is important because the vehicle maker has to have an idea of the qualitative behaviors on the new 'target tire'.
b) Understand the qualitative spread of performance between different tires of different sizes: for obvious cost reasons it is not possible in the early stage of development to make testing on vehicle of all the tire size combinations. For the same reason it is not possible to parametrize all the possible tire and make this tests virtually. The idea is to predict the spread of performance qualitatively by using the morphing techniques and evaluate his effect on the vehicle performances via simulation.

In the next figures some validation examples of the morphing are shown.

235/55 R 19 8J (Base tire) **255 50 R19 9J(Morphed tire)**

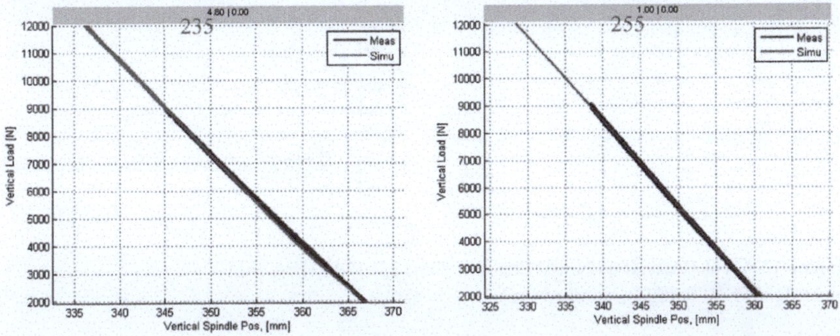

Figure 10. Dynamic vertical stiffness at 120 Km/h. Left fig.: 235/55 R 19 8J (base tire) right fig. 255/50 R 19 9J (morphed tire)

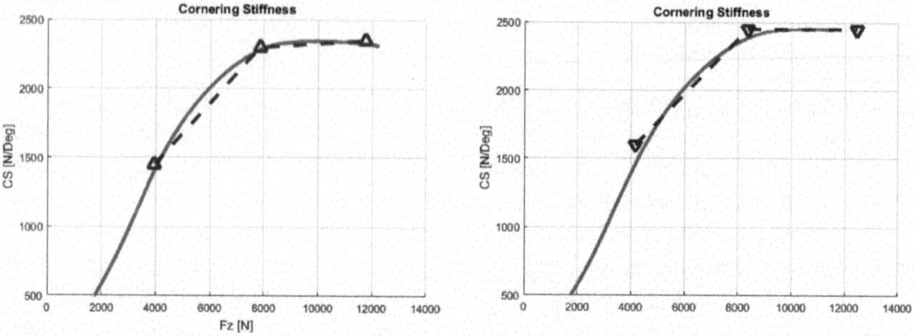

Figure 11. Cornering Stiffness vs Fz. Left fig.: 235/55 R 19 8J (base tire) right fig. 255/50 R 19 9J (morphed tire)

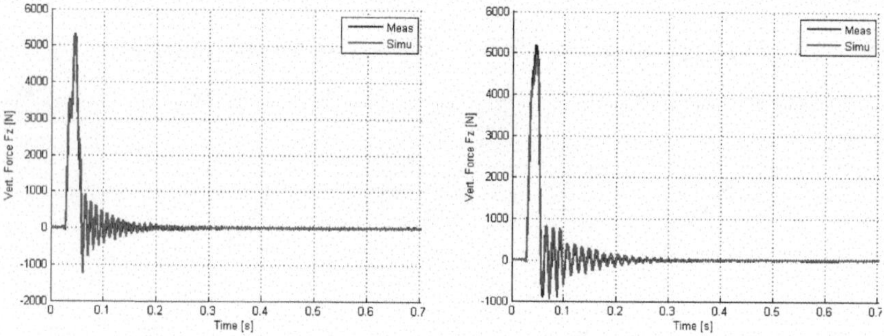

Figure 12. Cleat 10x20 at 60 Km/h. Left fig.: 235/55 R 19 8J (base tire) right fig. 255/50 R 19 9J (morphed tire)

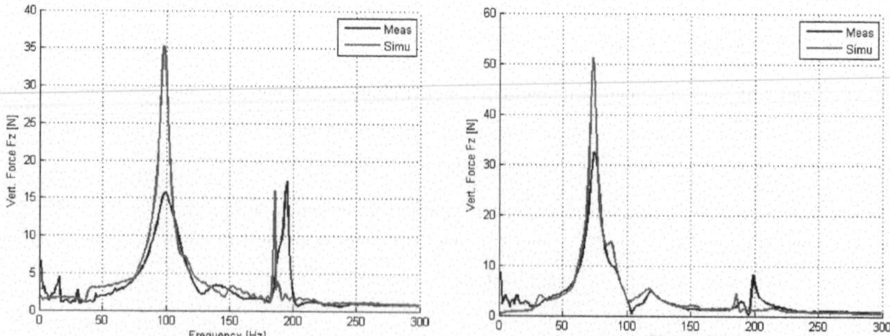

Figure 13. FFT of the cleat 10x20 at 60 km/h. Left fig.: 235/55 R 19 8J (base tire) right fig. 255/50 R 19 9J (morphed tire)

From figure 9 to 13 some comparative results of the morphing are shown. The validation examples show a 235/55 R 19 8J as base tire and a 255/50 R 19 9J as the morphed tire. The same tires are used in the last chapter on the vehicle. It is clear how the model can reproduce the geometry effect, the mass effect and the rim effect, as result for this case:

- The vertical stiffness slightly increase
- The cornering stiffness increase for high preload because of the stiffer belt
- The typical frequency of the bigger tire decrease (2 kg more mass)

The model give an accurate prediction of these tendencies.

Evaluation of the performance modifying the tire dimensions

At the beginning of the development a reference tire has been measured (235/55 R 19 in a quadro setup, same tire front and rear axle). Maserati uses the CDTire/3D model to predict (starting from this reference tire) the performance of the car if the tire is changing. In this way it is simple to understand how much the performance can be improved by making modifications on the tire – and what must be done working in different areas. In this way, it is clear from the beginning if some targets cannot be reached by improving only the tire performance and there is still time to work on the other aspects of the car. The parameters evaluated in this phase are:

- Rim width
- Tread stiffness
- Inflation pressure

An additional advantage of the CDTire/3D is its morphing capability, which can help the car manufacturer to predict the spread of performance of the car considering the whole tire grid. This is what will be analyzed in this paper by comparing the performance of the car starting from a 235/60 R 18 tire (front & rear) to a 235/55 R 19 on front and 255/50 R 19 on the rear axle.

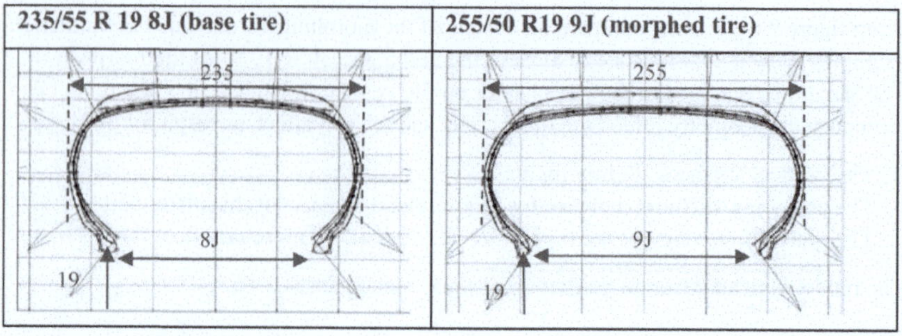

Figure 14. Construction assistant

Starting from the reference 235/55 R 19 tire and using the morphing tool available in the CDTire/3D suite, it is possible to modify the tire model without making new measurements (of a possibly non-existing tire) and predict the performance of the modified tire when changing its dimensions.

First of all, a comparison between the simulation analysis and the measurements of the car must be done. In figure 15 is possible to evaluate the good correlation between the model and the measurement in the steady state manoeuvre.

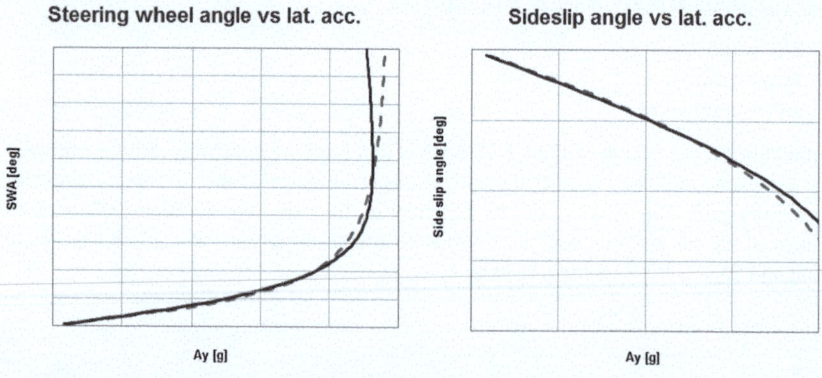

Figure 15. - - - - simulation, ___ measurement

After this first step, the morphing of the tire was been performed, and the prediction of the performance has been evaluated. In figure 16 and 17 is possible to evaluate the variation of the performance in the steady state and in the sweep analysis, as predicted by the simulation tools in the case of 235/55 R 19 8J at the front axle and 255/50 R 19 9J at the rear axle.

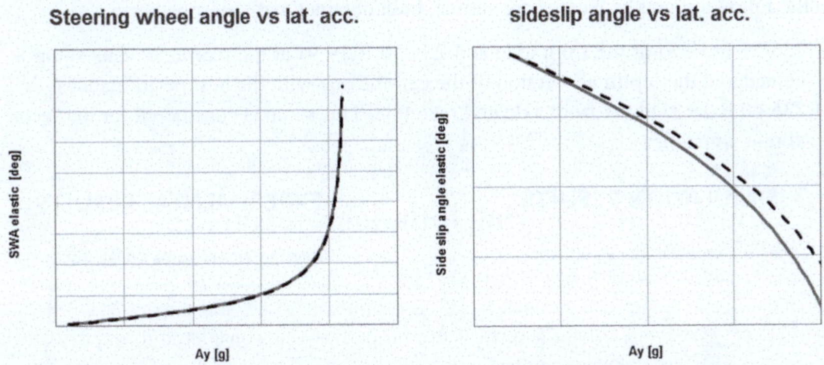

Figure 16. - - - - 235/60 R 18 front & rear, ___ 235/55 R 19 front & 255/50 R 19 rear

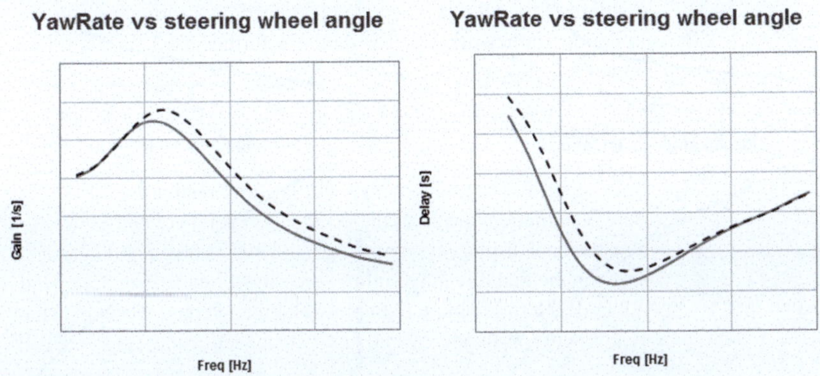

Figure 17. - - - - 235/55 R 19 front & 255/50 R 19 rear, ___ 235/60 R 18 front & rear

Comparison between the simulation and the experimental results

In order to evaluate if the tool is working properly, a comparison between the simulation prediction and the experimental results has been done, once the new tires were available. As can be seen in figures 18 and19, the difference in performance due to tire dimension change between measured and predicted performance indicators is very similar and this confirms that the approach can be used also at the early development phase without many measurements available.

In the following graphs 2 cases are shown, both obtained with morphed tires:

a) 235/55 R 19 8J at the front axle and 255/50 R 19 9J at the rear axle. This setup is considered the ‘optimal’ version of the car, the one with the best performances.
b) 235/60 R 18 8J at the front axle and rear axle. This setup is considered for the ‘economic’ version.

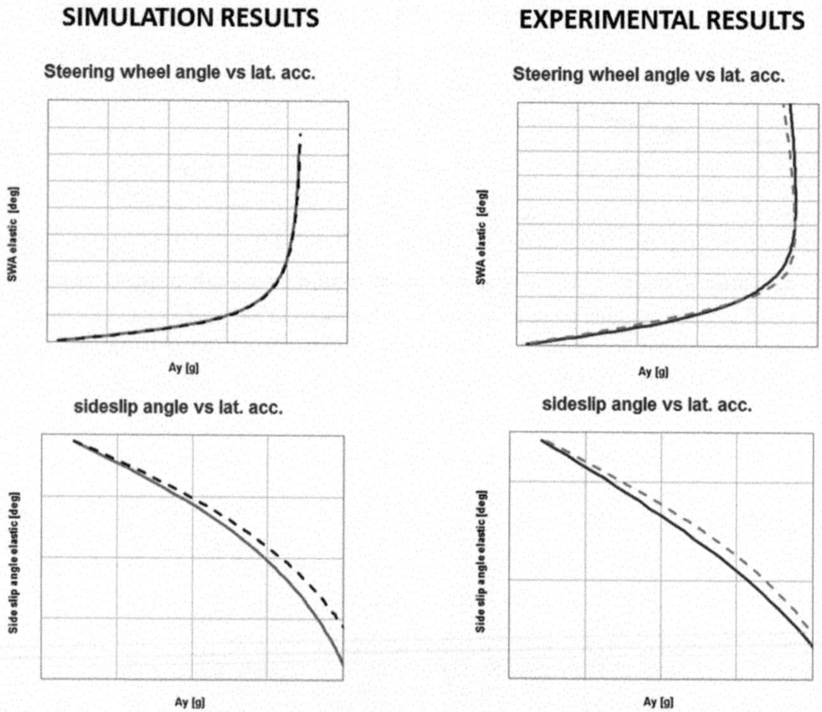

Figure 18. - - - - 235/55 R 19 front & 255/50 R 19 rear, ___ 235/60 R 18 front & rear

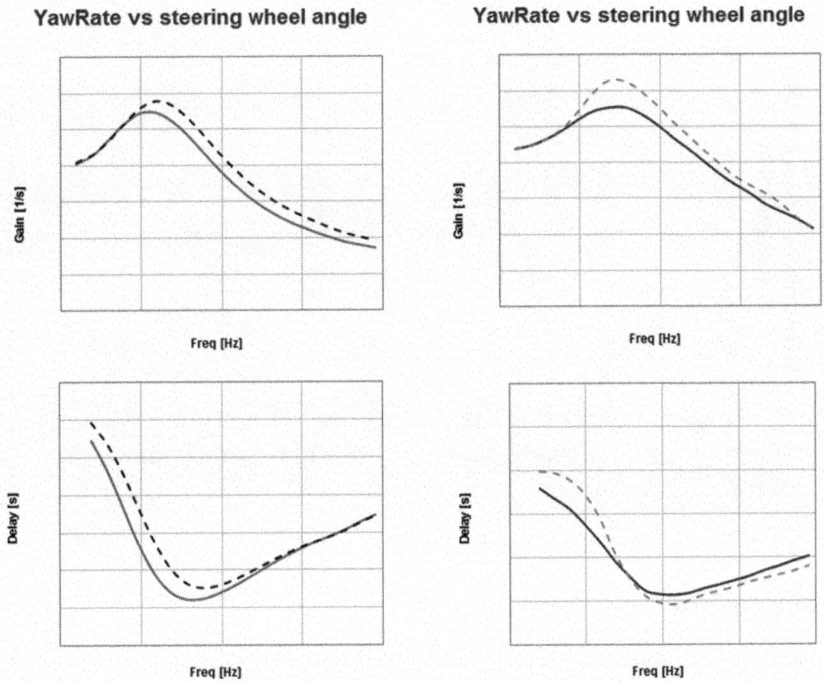

Figure 19. - - - - 235/55 R 19 front & 255/50 R 19 rear, ___ 235/60 R 18 front & rear

In figure 20, a comparison between the % of improvement measured and the percentage of improvement predicted by the simulation has been done. The comparison has been done using all the parameters used to evaluate the target achievement of the handling performance. It can be seen that the prediction gives always the same tendency measured, with a residual sum of squares over 0.75, and the accuracy of the prediction is under 10 % of error.

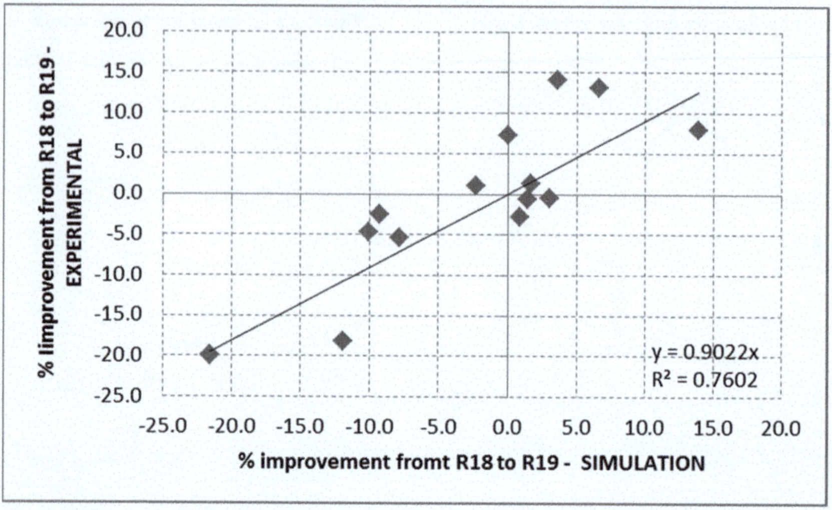

Figure 20.

Conclusion

In this paper, the authors show how CDTire/3D has been used to shorten the development time of a new car by strengthening and improving the tire and rim related pre-development process. The paper is focussing on the virtual prototype capability of CDTire/3D with respect to changes in tire and rim dimensions, as well as changes in inflation pressure.

In the first chapter, the model structure of CDTire/3D is described, paying particular attention to the question why this model can be used to accurately predict tire behaviour when changing inflation pressure and why and how the model can used to utilize a resizing of tire and rim in a predictive way.

In the second chapter, some selected validation results are shown for the pressure variation and tire resizing capability. This validation is done by comparing typical tire standalone measurements against simulations. In both categories the validation show good predictions results and the therefor the prediction capability of CDTire/3D can be seen as proofed.

Based on these positive results in tire standalone validation, Maserati performed full vehicle measurements to validate the process they have in mind using CDTire/3D in the pre-development phase of a new car. These full vehicle based validations investigate the question whether CDTire/3D can be used to pre-assess tire and rim dimension variations in a development phase where the physical tires for the different sizes under consideration are not yet available. In this validation, the CDTire/3D tire models for the different tire and rim sizes have been produced by using a morphing technique based on a predecessor tire that has been parametrized using standard tire measurements. Full vehicle measurements using the different tire sizes have been done and comparisons with full vehicle simulations using a simple vehicle model are shown in chapter 3. The investigation was not focusing on absolute quantitative result because the vehicle model used has been intentionally chosen as a simplified model as in the early design phase the vehicle parameters are not complete. The investigation was more related to the question whether or not the tire resizing technique by morphing can predict the correct relative trends which are needed to decide for the optimal tire and rim sizes.

This comparison has been done using all key indicators of the handling performance. It is shown that the prediction results always have the same tendency as the measurements – with a residual sum of squares of 0.76, and the accuracy of all predictions is under 10% error.

Based on these results, the validation under productive conditions was successful and looks very promising. Maserati implements the tire model resizing capability with CDTire/3D in their development process.

As a future prospective, we also think to include (in addition to the pressure variation and resizing capability) the possibility of slight changes on the tire material side (for example to enable the change the storage modulus of the tread or the bead material). This would be an anticipation in the direction of tire pre-development.

References

1. Bäcker M., Gallrein A., Calabrese F., Leister G., "Simulation of a tire blow-out in a full vehicle scenario. " In: Pfeffer P. (eds) 7th International Munich Chassis Symposium 2016. Proceedings. Springer Vieweg, Wiesbaden
2. Koutny, F., "A Method for Computing the Radial Deformation Characteristics of Belted Tires," Tire Science and Technology, TSTCA, Vol. 4, No. 3, Aug. 1976, pp. 190-212
3. Gallrein, A., Baecker, M., Gizatullin, A.: "Structural MBD Tire Models: Closing the Gap to Structural Analysis – History and Future of Parameter Identification", SAE Technical Paper 2013-01-0630, 2013, doi:10.4271/2013-01-0630.
4. Baecker, M., Gallrein, A. and Haga, H. "A Tire Model for Very Large Tire Deformations and its Application in Very Severe Events" SAE Int. J. Mater. Manuf. 3(1): 142-151, 2010
5. Leister G., Fahrzeugreifen und Fahrwerkentwicklung: Strategie, Methoden, Tools (ATZ/MTZ-Fachbuch) Gebundene Ausgabe – 11. Dezember 2008, ISBN-10: 3834806714
6. A. Gallrein, J. DeCuyper, W. Dehandschutter, M. Baecker: "Parameter Identification for LMS CDTire", 3rd Int. Tyre Colloquium, Tyre Models for Vehicle Dynamics Analysis Proceedings, Vehicle System Dynamics Vol.43, Supplement, 2005, pp.444-456.

Contact Information

Dr. Manfred Bäcker, Axel Gallrein, Francesco Calabrese
Fraunhofer ITWM
Fraunhofer-Platz 1
67663 Kaiserslautern
Germany

http://www.itwm.fraunhofer.de

manfred.baecker@itwm.fraunhofer.de
axel.gallrein@itwm.fraunhofer.de
francesco.calabrese@itwm.fraunhofer.de

Francesco Calabrese*, Luca Dusini+, Dr. Manfred Bäcker*, Axel Gallrein*

* Fraunhofer ITWM, Fraunhofer-Platz 1, 67663 Kaiserslautern, Germany
+ Maserati S.p.A, 41100 Modena, Italy

TIRE PERFORMANCE AND TIRE SLIP ANALYSES

Real-time high-resolution road condition map for the EU

Per Magnusson, NIRA Dynamics AB

Håkan Frank, NIRA Dynamics AB

Torbjörn Gustavsson, Klimator AB

Esben Almkvist, Klimator AB

P. Pfeffer (Ed.), *9th International Munich Chassis Symposium 2018*, Proceedings,
https://doi.org/10.1007/978-3-658-22050-1_56

Abstract

This paper outlines the main challenges with creating a high-resolution road condition map for the EU with focus on road friction and presents results from proof-of-concept evaluations done during the last two winters (2016/2017 and 2017/2018) where vehicle data from a fleet of roughly 500 vehicles has been used to measure road friction and combine these with local road weather models to create a high-resolution friction map.

The results show that longitudinal slip-based friction potential estimations from vehicles can be used to build up a friction map that describes the friction potential distribution on road segments also in challenging weather such as temperatures varying around zero degrees. It is also shown that it is possible to improve friction potential accuracy by taking into consideration each vehicle's tire properties. Furthermore, it is shown that local weather models can be used to provide friction map coverage for areas where there are few or no vehicles providing friction measurements.

Background

Knowledge about road conditions such as the friction and roughness of the road, speedbumps and potholes are important properties that can enable both a safer and a more convenient and efficient travel. For self-driving vehicles, this knowledge will be a necessity. Imagine a self-driving vehicle cruising on autobahn at high speed without awareness of the road being slippery. Knowledge about road condition can also be used to improve road maintenance such as salting, ploughing, mending potholes etc.

NIRA Dynamics together with Klimator are developing a service called Road Surface Information (RSI), that will provide up-to-date information of road conditions with the aim of saving lives, improving the environment and enabling automated driving. RSI will be provided as a cloud based road condition map with a set of map layers describing the road surface properties such as road friction, road roughness, speedbumps and potholes enabling vehicles to adjust for upcoming road conditions in advance. RSI can also be used locally within each vehicle.

Providing safety critical information such as road friction to autonomous vehicles also puts very high requirements on security and data authenticity so that data cannot be tampered with as well as high availability. The RSI infrastructure has therefore been designed to be secure, horizontally scalable, highly redundant and portable. The RSI infrastructure is also compliant with the HERE Open Location Platform which provides a vendor neutral and open platform for location data such as RSI.

Road friction use cases

Road friction can be used by vehicles for a wide range of use cases:

- Local Hazard Warning is a way to inform the driver of for example slippery conditions in advance.
- Emergency Braking can be improved by taking into account the slipperiness of the road when estimating the braking distance.
- Adaptive Cruise Control can account for the slipperiness of the road as input to the automatic distance control.
- Autonomous driving needs to know that the road is not slippery in order to drive safely at high speeds.
- The braking distance can be reduced especially during slippery conditions if the initial brake pressure at the start of the ABS control activity can be limited and not set too high.
- In electric vehicles friction information allows safer and more effective electric motor recuperation.

Road friction can also be used by road maintenance organizations to optimize salting and ploughing of the roads by knowing in advance if the road will become slippery. By being able to better predict which roads will become slippery it is possible to both reduce the risk for slippery roads by salting and ploughing as well as only put out salt where it is needed, which will reduce the negative environmental impact.

RSI is aimed to support all of these areas but in this paper the focus will be on how to support autonomous vehicles with knowledge about the road slipperiness.

Terminology and concepts

Friction

Estimating the friction between a tire and a road surface during normal driving is a technically complex task. To avoid confusion, it is essential to be aware of the underlying physics and agree on some definitions.

Friction is a phenomenon which always involves at least two bodies in contact with each other. As a consequence, on the same road vehicles with different tire characteristics will experience different friction potential resulting in different minimum braking distance, given all other circumstances equal.

There are two causes to the tire grip (friction): molecular adhesion and road roughness effects, [8].

- The molecular adhesion is caused by Van deer Waals bonding between the molecules of the tire rubber and the road surface. It requires that the distance between the tire rubber and road surface is less than 1 nm. That means that road must be clean and dry.
- The road roughness effects refers to that the road irregularities induce an excitation of the tire rubber between 10^2 and 10^6 Hz. The rubber is deformed around road spots and due to the hysteresis properties of the rubber a tangential force is generated, see [8] for more details.

Friction potential

The friction potential is described as the average in-plane deceleration with a maximum ABS breaking in a straight line, normalized by standard gravity with a vehicle on a specific road surface using specific tires. It is a measurement of the maximum friction coefficient μ between a certain tire and a given road surface. See [1] section 4.2 for details and test procedure.

Experienced Friction

Describes the friction potential distribution on a road segment that different tires will experience based on aggregating measured friction received from one or more vehicles.

Nowcasted friction

Combining experienced friction with weather data to estimate friction potential distribution for road segments on a very short time horizon (<10 minutes).

Forecasted friction

Combining experienced friction with weather data to estimate friction potential distribution for road segments on a time horizon up to 24 hours.

Normalization/De-normalization

Normalization transforms the individual friction potential values from vehicles with different tire characteristics to a normalized friction potential value that corresponds to an artificial reference tire on a given road segment.

De-normalization transform the individual friction potential values from vehicles with different tire characteristics to a tire specific friction potential on a given road segment.

Reference Friction

A physically correct measurement of the maximum friction coefficient between a road surface and a tire would require the exact measurement of forces while driving.

As this in practice impossible for larger numbers of series production vehicles in normal use, an alternative method is being used.

If the traffic situation allows, the driver will carry out an emergency braking maneuver until standstill with all four wheels under ABS-control. The average deceleration during the maneuver divided by $g = 9{,}81\ m/s^2$ is then regarded as proportional to the maximum average coefficient of friction for all four tire positions on the road segment.

The resulting measurement of the maximum friction coefficient is then referred to as Reference Friction which is valid for the current road surface, circumstances and the used tires only.

Road segment

Describing continuous road conditions such as friction, roughness, water amount for a road in a map layer is typically done by dividing the road into segments. The length of a road segment varies depending on use case but is typically between 25-500 meters. The length is chosen based on use case requirements. Use cases with high safety and performance requirements such as autonomous driving typically require shorter segment lengths due to higher accuracy requirements.

If there are separate lanes for each driving direction, these can be described as different road segments. Also if there are multiple lanes in each driving direction these can be described as different road segments.

RSI currently describes road segments for each direction as separate segments but treat multiple lanes in the same direction as the same segment. As soon as vehicles robustly can determine and report the travelled lane, each lane will be treated as separate road segments.

Describing the friction on road segments

Vehicles may experience very different friction potential driving over the same road segment because of a number of reasons where the most prominent ones are tire characteristics and spatial variations.

The experienced friction potential is caused by interaction between the tires and the road surface. Therefore both the tire properties and the road surface will affect the experienced friction potential. Soft winter tires will experience significantly higher friction potential than hard summer tires during winter conditions and the opposite during summer conditions.

Spatial variations may be caused by many different reasons. Vehicle tracks forming after a snowfall will typically cause vehicles to experience lower friction outside the tracks. Local situations such as oil or mud spills will typically cause vehicles to experience significantly lower friction on only parts of the segment.

A friction distribution therefore seems to be a suitable way to describe the probability that a vehicle will experience a certain friction potential on a specific road segment.

Based on use case, different friction potential values can be extracted from the friction distribution.

Safety related functions will use a lower bound with a specific confidence level and performance related functions will use a probable value or an upper bound.

Friction based on vehicle data

Vehicle functions as ABS, TCS, etc. provide accurate friction estimates. However, these functions are only active when longitudinal acceleration and/or lateral acceleration are close to their maximum limits. Therefore, in the majority of driving situations they are not active. This is illustrated in Figure 1 below.

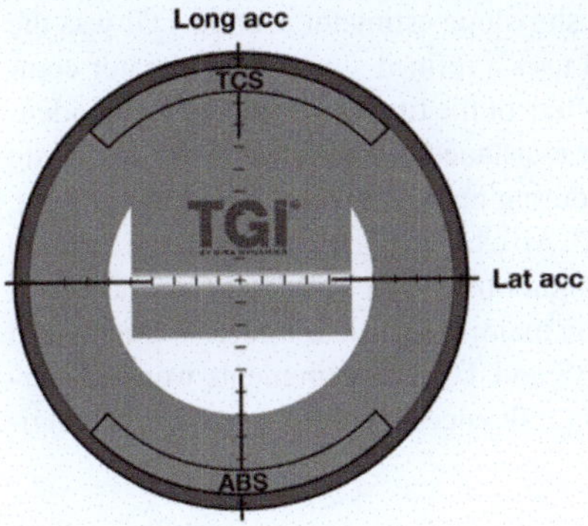

Figure 1: Kamm's Cirle

Many methods have been proposed in order to fill the inner parts of the Kamm's circle, two good surveys are [2] and [6]. NIRA's answer to the problem is named TGI (Tire Grip Indicator). TGI contains different modules to estimate the friction when ABS, TCS or ESP is active. For the area in the middle, a longitudinal slip-slope estimation module is used.

The principle of longitudinal slip-slope friction estimation is described below.

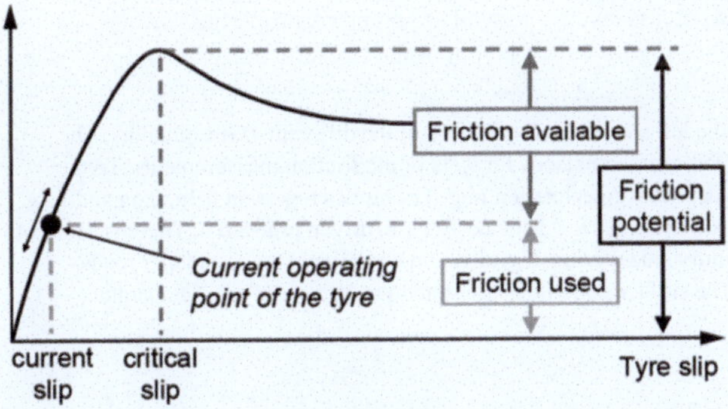

Figure 2: Longitudinal slip-slope estimation

The slip for each wheel is defined as

$$s_i = \frac{\omega_i r_i}{v_i} - 1 \qquad (1)$$

where ω_i is the angular rate, r_i is roll radius and v_i is the speed of wheel, the index i denotes the wheels. By observing the current operating point, the friction potential is derived under the assumption that the friction potential and slope is correlated.

The starting point for the development of the slip-slope estimator used in TGI was the work by F. Gustafson [3] and [4]. Today TGI uses a refined slip-slope estimator combined with a tire properties model. The parameters of the tire properties model are identified automatically by TGI without any driver assistance. This enables TGI to estimate the friction potential with high accuracy also during normal customer relevant driving, i.e., when very small traction forces are applied. An example is shown in Figure 3 below which shows 8.6 h of test driving in winter conditions with a VW Golf GTE. The longitudinal slip-slope estimator contributes to the majority of friction measurements and most of them are below 0.3g. Most of the ABS and TCS measurements were deliberately provoked by the test driver in order to get references measurements for the longitudinal slip-slope estimator.

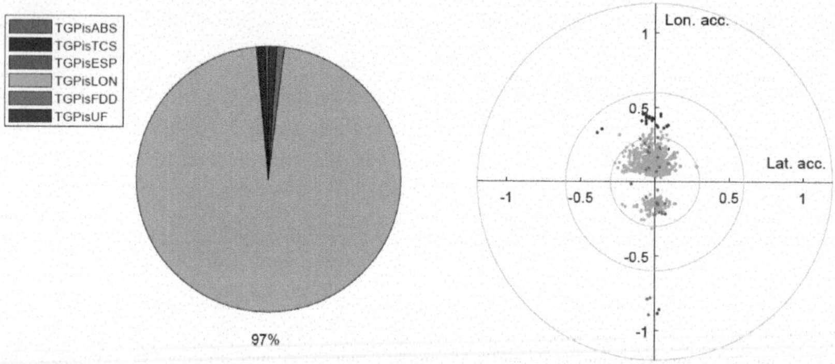

Figure 3: The pie chart to the left shows the distribution of the different TGI estimators. The longitudinal slip-slope (LON) estimator stands for 97% of the friction measurements. To the right the Kamm's circle is depicted. Acceleration in g. The circles represent 0.3g, 0.6g and 1.2g. The test vehicle is a VW Golf GTE, driven distance 709 km, driven time 8.6 h. There is a mix of driving scenarios: city, country side driving, highways, main/national roads, smaller roads, asphalt and gravel roads. The surface conditions are salted/wet/dry/snow/ice. Temperature -4°C to +4°C

During the period of 6th to 9th of February 2017, the Norwegian Public Road Administration (NPRA) and Luleå University of Technology conducted a test campaign [7]. Friction was measured both by TGI installed in a Volvo V70 and NPRA's friction measurement systems. One of these is the ROAR Mk III system from Norsemeter which uses a traditional wheel to measure the friction, see Figure 4. The testing was conducted in winter condition with snow and ice on road surfaces.

Figure 4: Picture of the ROAR system used for the test campaign in [7].

The TGI and ROAR measurements shows good agreement, see table from [7]. The average RMSE is just 0.045 μ from all runs.

Table 1: Comparing TGI and ROAR, table from [7]

	TGI (μ)			ROAR (μ)				
Run	Max	Average	Min	Max	Average	Min	RMSE	Section
1	0.31	0.29	0.26	0.37	0.26	0.11	0.027	Airstrip
2	0.38	0.30	0.26	0.43	0.33	0.1	0.030	
3	0.30	0.24	0.22	0.37	0.29	0.16	0.051	
4	0.38	0.29	0.23	0.37	0.29	0.17	0.001	
5	0.34	0.28	0.26	0.58	0.27	0.1	0.012	
6	0.43	0.41	0.39	0.65	0.47	0.34	0.062	
7	0.32	0.19	0.12	0.32	0.23	0.19	0.047	Moto-cross
8	0.31	0.27	0.23	0.56	0.29	0.18	0.012	
9	0.36	0.33	0.31	0.36	0.30	0.24	0.028	
10	0.36	0.29	0.23	0.36	0.24	0.1	0.049	
11	0.32	0.30	0.26	0.35	0.27	0.1	0.023	
12	0.37	0.27	0.17	0.49	0.26	0.1	0.005	Road to Bjorli
13	0.68	0.67	0.60	0.71	0.48	0.26	0.180	
14	0.32	0.25	0.22	0.35	0.22	0.12	0.036	
15	0.31	0.28	0.25	0.37	0.21	0.17	0.069	
16	0.28	0.24	0.20	0.34	0.21	0.17	0.039	Road to Rånå
17	0.38	0.36	0.31	0.35	0.27	0.17	0.083	
18	0.42	0.26	0.200	0.53	0.24	0.15	0.027	
19	0.36	0.24	0.19	0.5	0.23	0.17	0.014	
20	0.40	0.32	0.26	0.69	0.44	0.29	0.122	
21	0.54	0.52	0.52	0.66	0.49	0.27	0.038	Brøste

The main purpose for TGI is to be a friction source for the RSI system. However, the estimated friction potential can also be used locally in the vehicle, e.g. for gearbox and engine adaption to improve performance.

The drawback with only using TGI locally in the vehicle is that the available friction information is for the road behind the vehicle, but it is the friction ahead the vehicle that is most interesting. In a cloud based solution, as RSI, the friction ahead is available from other vehicles.

Road friction map based on vehicle and weather data

Creating a high quality friction map with good coverage is a challenge:

- Vehicles travelling over the same road segment may experience different friction potential due to tires with different characteristics.
- Vehicles travelling over the same road segment may experience different friction potential due to spatial variations on the segment (ex: snow with tracks).
- Friction measurements from vehicles alone cannot provide coverage on all roads at all time due to that the coverage is related to vehicle density. This means for instance that coverage will be sparse during the night hours when the vehicle density is low on most roads.

NIRA together with Klimator has therefore evaluated how these challenges can be met by combining high resolution weather data on segment level, created from advanced road climate models, with vehicle data to create a friction map with both high quality and good coverage. The evaluation was performed using data from a taxi fleet consisting of roughly 500 vehicles during the winters 2016/2017 and 2017/2018.

Three different map layers were created and evaluated:

- Experienced friction based on vehicle data
- Nowcasted friction based on a combination of weather and vehicle data
- Forecasted friction based on a combination of weather and vehicle data

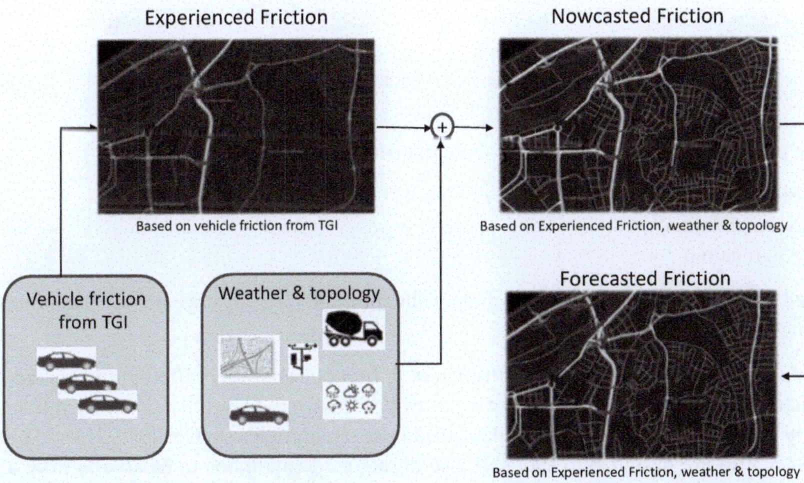

Figure 5: Evaluated friction map layers

Map layer for experienced friction

The friction on roads where vehicles have driven is described by a map layer called "Experienced Friction", which receives friction measurements from vehicles and combines them into a friction distribution for each road segment. The friction distribution describes the probability that a vehicle will experience a certain friction potential on a specific road segment. The friction distribution is updated continuously to always provide an up-to-date and accurate description of the friction potential on the road segment. The Experienced friction layer is based on friction vehicle data, initially NIRA's own friction estimation algorithm and in the future possibly also other vehicle friction sources.

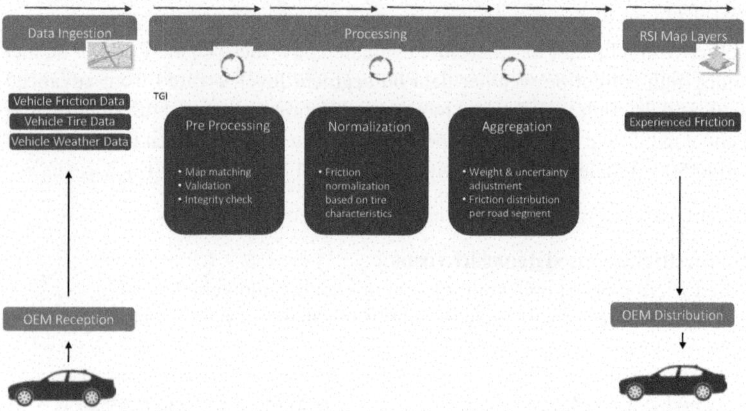

Figure 6: Processing pipeline for experienced friction

The processing pipeline consists of three main steps:

- Pre processing
- Normalization
- Aggregation

Pre Processing performs map matching, validation and integrity check of the incoming vehicle data.

Normalization compensates for differences in friction caused by different tire characteristics. The reason is that different types of tires may experience friction very differently on the same road surface as described in several articles such as [3], [5] and [8]. Winter tires are designed to stay soft also during cold conditions to be able to have a good grip of the road. Summer tires and all-season tires becomes hard at cold temperatures and will have problems to get a good grip of the road.

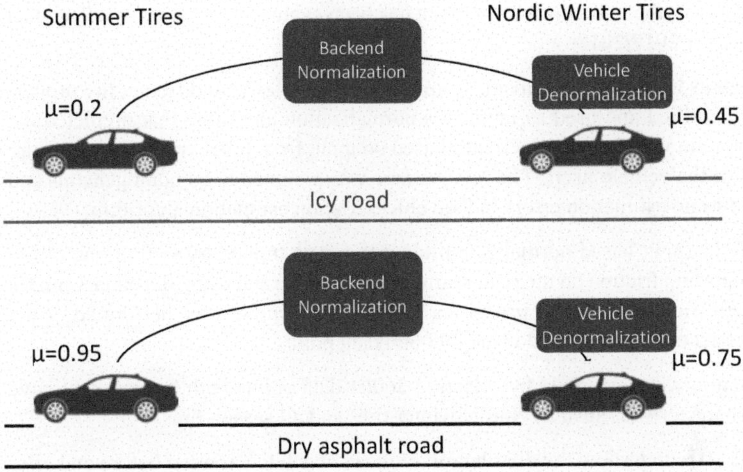

Figure 7: Normalization and denormalization for different tires and road weather

The principle is that the backend will normalize the measured friction potential based on tire characteristics and current road weather conditions to an artificial reference tire (normalized friction) which will be used as input to aggregation. Receiving vehicles equipped with TGI can then de-normalize the friction to the vehicles own tire characteristics. Vehicles without TGI can adjust to a suitable friction depending on use case. One example can be to de-normalize to representative tires for the season.

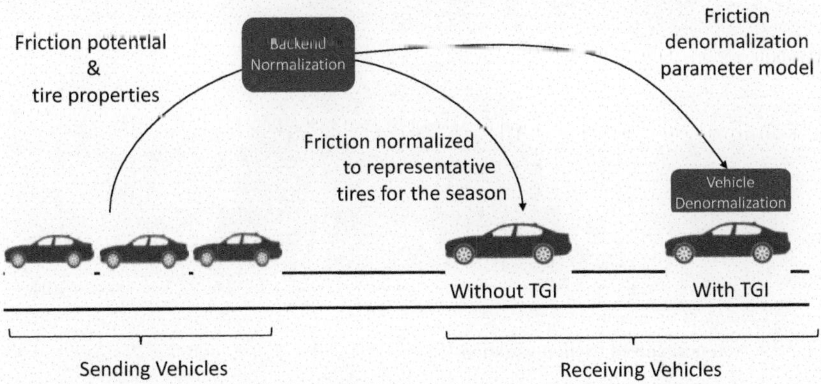

Figure 8: Normalization and denormalization model

The feasibility of the normalization concept has been evaluated based on taxi fleet data from the winter 2017/2018.

Road segments having measurements from multiple vehicles with different tire properties were identified and used as input to a normalization algorithm that predicted the friction a certain vehicle will experience based on input from other vehicles and the tire properties of the vehicle itself. The road surface was assumed to be homogeneous and that there was no estimation errors in the vehicle friction estimation algorithm.

A machine learning based normalization algorithm was trained by setting one of the vehicles reporting friction on the road segment as reference and using the other vehicles experienced friction and tire properties as input. Training was performed using 1 795 000 data points and tested using 395 000 data points.

The evaluation was performed by selecting 36064 data points with tire sets that have different characteristics such as Nordic winter tires and all season tires.

Figure 9 & 10 below show the distribution of mean absolute variation before and after normalization for the selected tire sets. The absolute mean variation was reduced from 0.20 to 0.12 μ by performing normalization. Ideally there would be no variation left but due to model error as well as non-homogenous surface and measurement errors there will always be a remaining variation.

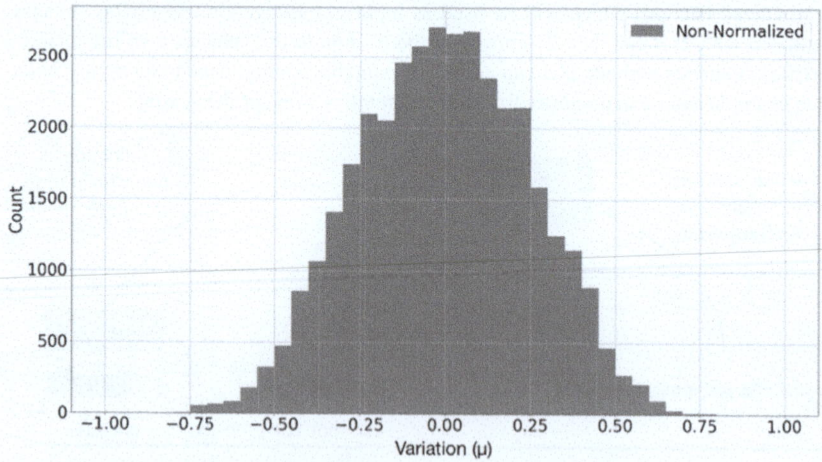

Figure 9: Non-normalized friction variation

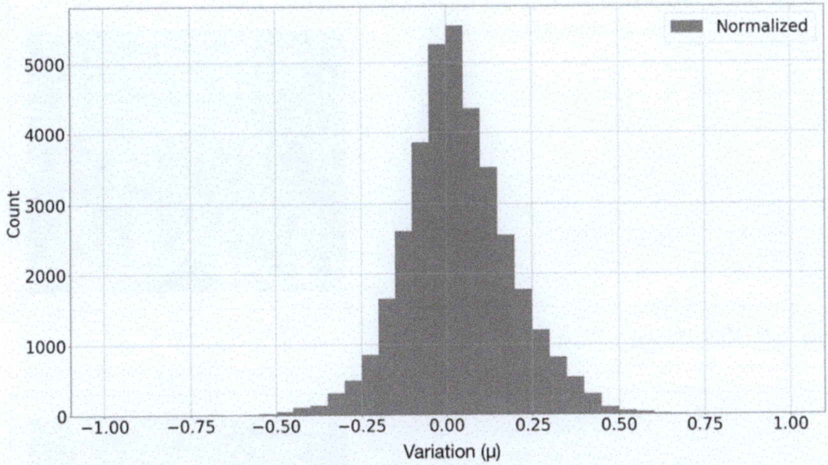

Figure 10: Normalized friction variation

Aggregation combines friction from each vehicle to a friction distribution per road segment. The reason for a distribution is to capture uncertainties in the friction measurements from the vehicles, spatial variations on a road segment and possible changes since last vehicle measurement due to changes in weather or local situations.

Furthermore, representing friction as a distribution enables support for different use cases based on the same aggregation. Depending on requirements for quality and coverage, different metrics can be extracted such as a lower friction value that can be used for safety related purposes and an upper value that can be used for performance related purposes.

Figure 11, 12 and 13 show three distribution examples taken from taxi fleet data during the winter 2017/2018. The distribution examples are from the same street during different weather conditions and aggregated for a road segment with 25 meters length.

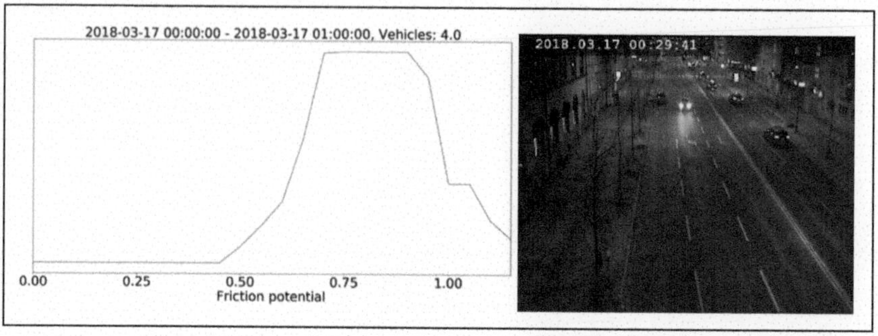

Figure 11: High friction

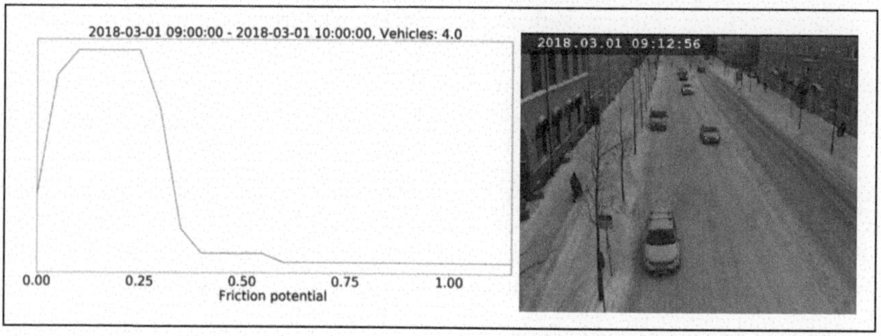

Figure 12: Low friction

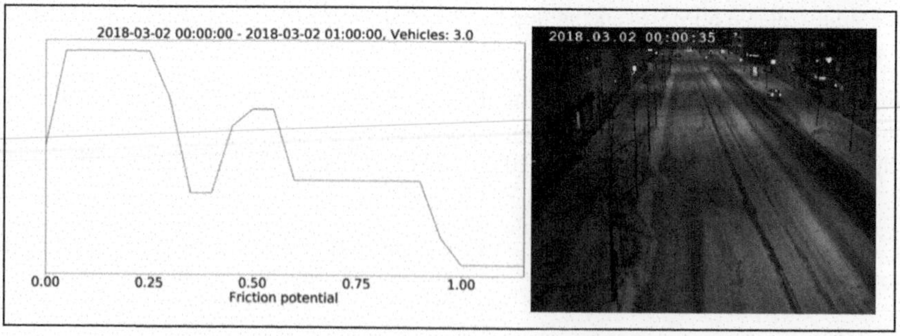

Figure 13: Multi-modal friction

Figure 13 shows winter conditions where the vehicles have created tracks. It clearly shows that a road segment can exhibit virtually all kinds of friction.

The feasibility of aggregation has been evaluated by using fleet data from the winter 2017/2018 and comparing these with reference friction observations. The data was collected during the period 1st of March 2018 to 16th of March 2018, which was a period with very challenging weather, i.e., temperatures around zero degrees Celsius with precipitation in the form of both water and snow.

The reference friction observations were performed by ABS brakings and TCS interventions with dedicated test vehicles and measuring these with the TGI ABS and TCS module in TGI. The road segment distribution was aggregated using fleet data measurements from the TGI longitudinal module using a sliding window aggregation period of one hour. The reason for such a long aggregation period was to achieve sufficient coverage.

The comparison was performed on 547 road segments where both fleet measurements and reference measurements were available. The experienced friction is assumed to be valid if (2) is true which is illustrated in figure 14. The reference friction uncertainty μ_{Delta} is 0.15.

$$\begin{aligned} \mu_{ExperiencedUpper} &> \mu_{Ref} - \mu_{Delta} \\ \mu_{ExperiencedLower} &< \mu_{Ref} + \mu_{Delta} \end{aligned} \tag{2}$$

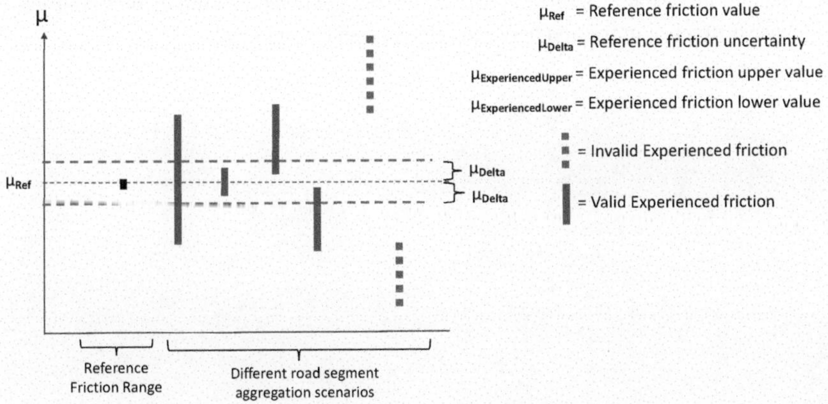

Figure 14: Comparing experienced friction with reference friction observations

Table 2: Reference measurements vs experienced friction

	Valid (%)		
	At least 1 fleet vehicle	At least 2 fleet vehicles	At least 3 fleet vehicles
Reference ABS	91.3	94.3	99.7
Reference TCS	90.4	95.9	100.0

Table 2 shows that for road segments with at least 1 fleet vehicle, the experienced friction was valid in approximately 91% of the cases and with at least 3 fleet vehicles, the experienced friction was valid in almost 100 % compared to reference vehicle measurements.

The main reasons for the increased correctness when using input from more fleet vehicles are:

- Normalization between differences in tire characteristics between vehicles has not been included.
- Winter conditions such as snow with tracks may require measurements from multiple vehicles to capture the complete friction distribution.
- Changes in weather captured by the fairly long aggregation window of 1 hour may cause difference in friction between reference and fleet vehicles.

Map layers for nowcasted and forecasted friction

For roads where there has been very few or no vehicle friction measurements, weather data is used to model the road weather conditions on a road segment level. This requires detailed knowledge about topographical conditions for each road segment such as height over sea, which hours the sun can shine on the road, exposure to wind from different directions etc. Weather forecasts and weather observations (radar, satellite etc.) are also needed but also local weather observations from RWIS (Road Weather Information System) stations and vehicle weather data. In addition, data from winter maintenance actions like salting and ploughing are beneficial to decrease the uncertainty of the friction estimates.

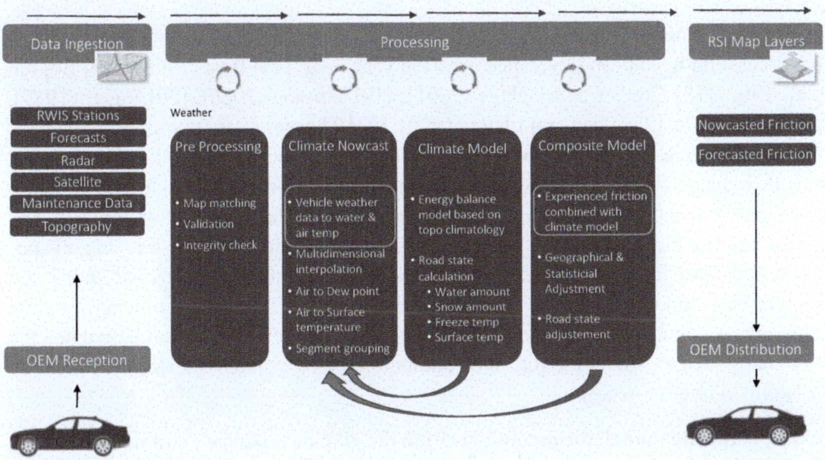

Figure 15: Processing pipeline for nowcasted and forecasted friction

The weather processing pipeline consists of the following main steps:

- Pre processing
- Climate Nowcast
- Climate Model
- Composite Model

In countries with winter weather related road safety problems the roads are monitored by a relatively dense network of RWIS stations; in Sweden, the average distance between stations is 15 km. RWIS stations are designed to provide the most relevant and accurate weather information for road conditions. To determine the road condition between the stations plenty of research has been conducted to estimate the temperature variation by thermal mapping (e.g. [9] Bogren and Gustavsson, 1991; [10] Gustavsson et al., 1998; [11] Gustavsson, 1999) or from geographical parameters (eg. [12] Bogren et al., 2000; [13] Chapman et al., 2001a; [14] Chapman et al., 2001b; [15] Hu et al., 2015).

The **climate nowcast** is derived from the results from this research and will use the RWIS station data along with the vehicles, which act as mobile weather stations, to estimate the weather at each road segment. The climate nowcast also bases the calculation on geographical data such as topography and urban density.

The conditions for each segment are identified (temperature, road status, topography, etc.) and are used to group segments together with similar topo-climatological conditions which is then used as input to the climate model

Several zero-dimensional energy balance and one-dimensional heat conduction models for forecasting have been developed over the years, e.g. [16] Best (1998), [17] Bogren et al. 1992, [18] Crevier and Delage (2001), [19] Jansson (2006), [20] Rayer (1987), [21] Sass (1997), [22] Shao and Lister (1996), [23] Thornes (1984). Similar to many of the models above, the **climate model** has an energy balance model to handle radiation and other energy fluxes to and from the road surface, a one-dimensional heat conduction model that calculates a ground profile of temperature and heat flux within the road, and a mass balance model to handle water processes such as precipitation, freezing, evaporation and frost formation. The climate model will calculate a new road status and weather based on input from the climate nowcasting. The freezing point determination takes into account any available maintenance data. The climate model calculates the road status such as water amount, snow amount, freezing point and surface temperature for up to 24 hours forecasts.

In the **composite model**, the predictions from the climate model are co-interpreted with the TGI friction measurements to compensate for the uncertainty of factors that are inherently hard to predict, such as the exact effect of traffic or precise snow amount. In this context, even sparse measurements have a large potential. The generated data is referred to as **nowcasted friction**.

For longer forecasts, the predictions are interpreted in light of past situations, resulting in empirically based probable distributions. This is the **forecasted friction**.

Accuracy for nowcasted and forecasted friction

One critical factor that determinates the friction is the surface temperature. In Figure 16, the result from evaluating the prediction of surface temperature from the winter 2016/2017 is presented. The surface temperature was cross validated by predicting for road segments with RWIS stations, but completely removing the predicted station from the nowcast, thus making the forecast completely independent from the observed data at the station and thereby representing an arbitrary road segment. The RWIS station data was instead used as reference to compare the predicted and observed surface temperature. This was done for all stations in the test area; a total of 294 stations. The investigated temperature interval was between -5 °C and 5°C, in which errors would be relevant for slipperiness. During the winter 2016/2017 no vehicle data was included.

As can be seen in Figure 16, the Mean Absolute Error is very low, typically below 1 degree Celsius for up to 10 hour forecasts. The error increases in March when the solar elevation is higher and solar influx larger, which leads to larger and more rapid surface temperature variation compared to other winter months.

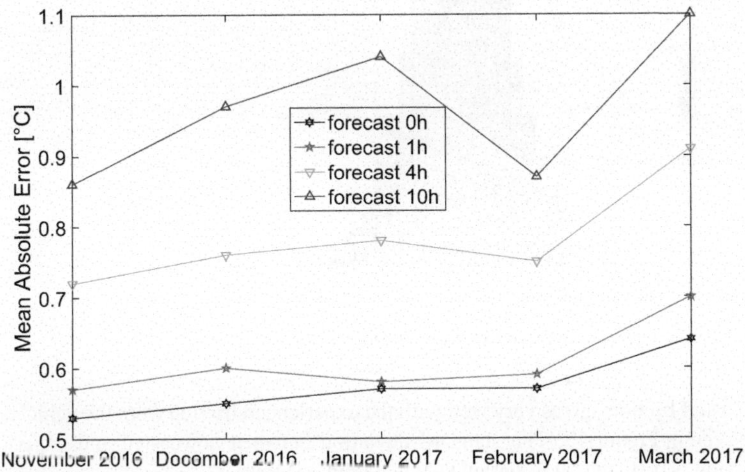

Figure 16: Mean Absolute Error for the surface temperature for now-casting (forecast 0h) and three forecasting examples, from the winter 2016/2017.

Climate nowcast compared to experienced friction

The nowcasting of friction was evaluated in a case study from 2018-03-06 to 2018-03-08 before and after two snow storms passing southern Sweden during this 3-day period. The nowcasted friction was cross validated by removing all vehicle measurements from a single map tile with a size of 11 x 11 km and nowcasting the friction within that tile based on experienced friction from the other tiles. Then the nowcasted friction was compared to the experienced friction within the map tile. This was done similarly for all map tiles in the test area. Thereby the quality of interpolating/extrapolating friction measurements to surrounding roads without vehicle measurements could be tested.

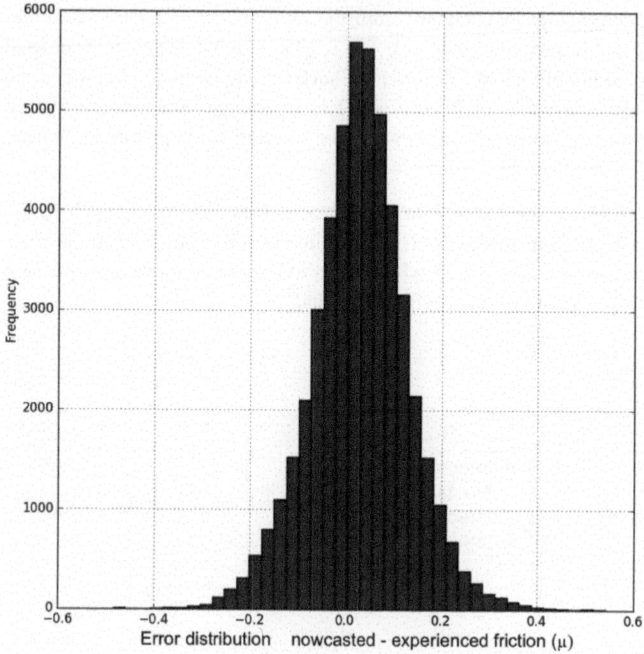

Figure 17: Comparing nowcasted and experienced friction

The nowcasted friction agrees very well with the experienced friction from the vehicles as can be seen in Figure 17. The standard deviation between nowcasted and experienced friction is 0.10 and 99.2% of all data is within 3 sigma.

Conclusions

Key aspects in providing a real-time high resolution friction map for EU are quality and coverage. The results from the evaluation performed during the winters 2016/2017 and 2017/2018 show that combining friction measurements from vehicles with local weather models indicates that both high quality and high coverage can be achieved.

The quality of friction measurements from a vehicle equipped with TGI has been compared with reference measurements from ROAR showing an average error of 0.045 μ which clearly indicates that vehicles can provide friction measurements with high quality.

A vehicle fleet of roughly 500 vehicles has been equipped with TGI and used as input to a cloud based service aggregating friction on road segments with high resolution (down to 25 meter length). Results show that the accuracy of aggregated friction is high compared to reference measurements also during challenging weather such as temperatures around zero degrees combined with precipitation. Accuracy close to 100% can be achieved by aggregating data from multiple vehicles on a road segment.

In addition, by taking into consideration each vehicle's tire characteristics it is possible to reduce the friction uncertainty, up to 0.08 μ for vehicles with significant differences in tire characteristics.

To provide coverage on roads with low or no traffic, weather data was used to model road weather conditions. The results from winter 2016/2017 show that it is possible to estimate the surface temperature with a precision below 1 degree Celsius also during winter conditions. Furthermore, results from winter 2017/2018 show that nowcasted friction without vehicle data is able to estimate the average road friction with good accuracy: the standard deviation between nowcasted and experienced friction is 0.10 and 99.2% of all data is within 3 sigma

References

[1] VDA FAT 299 Unfallvermeidung durch Reibwertprognosen, 16 Mai 2017

[2] Acosta M., Kanarachos S. and Bundell M. (2017). Road Friction Virtual Sensing: A Review of Estimation Techniques with Emphasis on Low Excitation Approaches. Applied science (www.mpdi.com) 2017

[3] Gustafsson F. (1997) Slip-based Tire-road friction coefficient estimation, Automatica 1997, 33, 1087-1099

[4] Gustafsson F., Drevö M., Forsell U., Löfgren M., Persson N. and Quicklund H. (2001) Virtual sensors of tire pressure and road friction. SAE Paper number 2001-01-0796

[5] Carelson C.R., Gerdes J.C. (2005) Consistent Nonlinear Estimation of Longitudinal Tire Stiffness and Effective Radius, IEEE Transactions on Control Systems Technology Volume 13, Issue 6, Nov. 2005.

[6] Khaleghian S, Emami A, Taheri S A technical survey on tire-road friction estimation

[7] Andersson E, Validation of Friction Estimating System, B.Sc Thesis Luleå University of Technology, June 2017

[8] The tyre grip, Société de Technologie Michelin, 2001

[9] Bogren, J. and Gustavsson (1991). Nocturnal air and road surface-temperature variations in complex terrain. International Journal of Climatology, Vol. 11(4), p443-455.

[10] Gustavsson, T., Karlsson, I. M., Bogren, J., and Lindqvist, S. (1998). Development of temperature pattern during clear nights. Journal of Applied Meteorology, Vol. 37, p559-571.

[11] Gustavsson, T. (1999). Thermal mapping – a technique for road climatological studies. Meteorological Applications, Vol. 6(4), p385-394.

[12] Bogren, J., Gustavsson, T., Karlsson, M., and Postgåard, U. (2000). The impact of screening on road surface temperature. Meteorological Applications, Vol. 7, p97-104.

[13] Chapman, L., Thornes, J.E., Bradley, A.V., 2001a. Modelling of road surface temperature from a geographical parameter database. Part 1: Statistical. Meteorol Appl 8, 409-419.

[14] Chapman, L., Thornes, J.E., Bradley, A.V., 2001b. Modelling of road surface temperature from a geographical parameter database. Part 2: Numerical. Meteorological Applications 8, 421-436.

[15] Hu Y, Almkvist E, Lindberg F, Bogren J, Gustavsson T (2015) The use of screening effects in modelling route-based daytime road surface temperature. Theoretical and Applied Climatology:1-17. doi:10.1007/s00704-015-1508-9

[16] Best, M. J. (1998) A model to predict surface temperatures. Boundary Layer Meteorol. 88: 279–306.

[17] Bogren, J., Gustavsson, T., and Lindqvist, S. (1992). A description of a local climatological model used to predict temperature-variations along stretches of road. Meteorological Magazine, Vol. 121(1440), p157-164.

[18] Crevier, L.-P., and Y. Delage.(2001) METRo: A New Model for Road-Condition Forecasting in Canada. Journal of applied meteorology Vol. 40, p2026-2037.

[19] Jansson, C., Almkvist, E., and Jansson, P.-E. (2006). Heat balance of an asphalt surface: Observations and physical based simulations. Meteorological applications, Vol. 13(2):203-212.

[20] Rayer, P. J. (1987) The Meteorological Office road surface temperature model. Meteorol. Mag. 116: 180–191.

[21] Sass, B. H. (1997). A numerical forecasting system for the prediction of slippery roads. Journal of Applied Meteorology, 36:801-817.

[22] Shao, J. & Lister P. J. (1996) An automated nowcasting model of road surface temperature and state for winter road maintenance. J. Appl. Meteorol. 35: 1352–1361.

[23] Thornes, J. E. (1984) The prediction of ice formation on motorways in Britain. Unpublished PhD thesis, University of London.

Longitudinal tire slip curve identification from vehicle road tests

Bissoli Alberto Giovanni,
Vehicle Dynamics Performance Competence Center, FCA Italy S.p.A.

Ceccarini Lorenzo, C.S.I. S.p.A., Automotive Division

Gaspari Yuri, Università di Pisa

Murgia Stefano,
Vehicle Dynamics Performance Competence Center, FCA Italy S.p.A.

P. Pfeffer (Ed.), *9th International Munich Chassis Symposium 2018*, Proceedings,
https://doi.org/10.1007/978-3-658-22050-1_57

Introduction

In the tire development process the carmaker shares with the tire supplier some information about the performances that a tire must reach. One of these regards the longitudinal behaviour of the tire during braking, given as stopping distance performance. But this information, which is very important for carmakers on development phase and for customers and magazines once the vehicle is already on the market, depends not only on the tire but also on the vehicle intended as the sum of all its subsystems (in particular braking system and ABS).

In addition stopping distance measurement is affected by many sources of variability that are not easy to be controlled in a restricted time and with limited means, and thus have to be considered as noise factors. Two of them in particular are tire wear resulting from braking and ambient/tarmac temperature during tests.

For these reasons, having a simple method to measure tire longitudinal slip curve on a vehicle without the need of a bench is very important under different points of view:

- It allows to compare different types of tires;
- It allows to study the behaviour of a tire in different conditions of temperature and wear and to understand how these two factors influence stopping distance measurement results;
- It allows a better communication among the carmaker and the tire supplier, sharing with him an information focused on the tire;
- It could help to have more accurate virtual analysis models;
- It could help to tune the ABS system in a more efficient and effective way, bringing benefits in terms of time saved and quality.

For these reasons a method to measure the longitudinal tire slip characteristic with an on vehicle test will be presented in the following work. Some results will be then discussed.

Mathematical Model

The considered mathematical model is the braking vehicle one (Figure 1). Some hypotheses were made:

- Flat surface;
- Homogeneous adherence surface;
- Negligible pitch oscillation and thus negligible rate of change of sprung mass angular moment;
- Aerodynamics forces applied on the centre of gravity, with the simplified expression (4);
- Negligible downforce, since the vehicle considered is not characterized by strong aerodynamic qualities;
- Negligible wheel inertia.

With these hypotheses the model is obtained with the following equilibrium equations:

$$-X_1 - X_2 - D = m\dot{u} \tag{1}$$

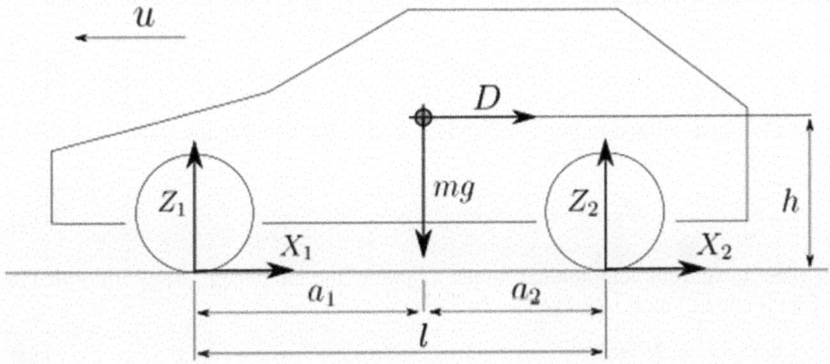

Figure 1

$$Z_1 + Z_2 - mg = 0 \tag{2}$$

$$(X_1 + X_2)h + Z_2 a_2 - Z_1 a_1 = 0 \tag{3}$$

In which D represent the aerodynamic drag defined as:

$$D = \frac{1}{2} C_x S \rho u^2 \tag{4}$$

The system has four unknown parameters (X_1, X_2, Z_1, Z_2) and three equations.

Combining them it is possible to find the expressions of vertical loads:

$$Z_1 = mg\frac{a_2}{l} - \frac{h}{l}m\dot{u} - \frac{h}{2l}\rho SC_x u^2 \tag{5}$$

$$Z_2 = mg\frac{a_1}{l} + \frac{h}{l}m\dot{u} + \frac{h}{2l}\rho SC_x u^2 \tag{6}$$

Which can be re-written as:

$$Z_1 = Z_1^0 + \Delta Z - \xi u^2 \tag{7}$$

$$Z_2 = Z_2^0 - \Delta Z + \xi u^2 \tag{8}$$

where Z^0_i is the static load acting on the axle, ΔZ is the load displacement during braking and

$$\xi = -\frac{h}{2l}\rho SC_x \tag{9}$$

Considering the longitudinal forces it is likewise obvious that the braking force must not exceed the grip limit:

$$|X_1| \leq \mu_1 Z_1 \tag{10}$$

$$|X_2| \leq \mu_2 Z_2 \tag{11}$$

where the longitudinal friction coefficient μ_i is the ratio between the peak longitudinal force and its corresponding vertical load:

$$\mu_i = \frac{X_i}{Z_i} \tag{12}$$

In order to find the longitudinal forces X_1 and X_2 the momentum equilibrium of the wheel (Figure 2) must be solved:

$$X_i R_i - Z_i s - M_{f_i} = J_r \dot{w}_i \tag{13}$$

that means to know the braking force acting on the wheel or the braking bias of the vehicle.

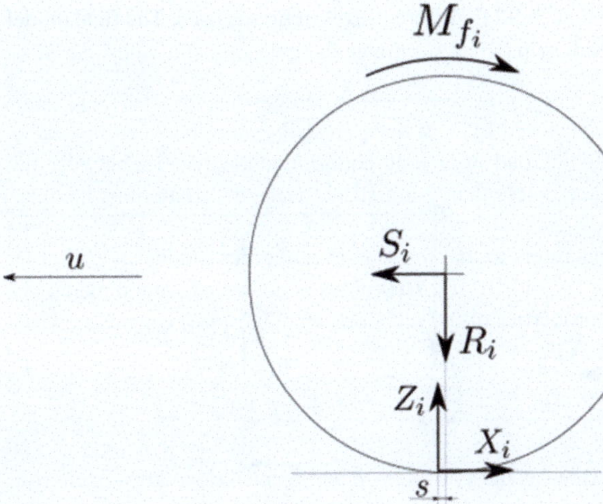

Figure 2

But, as already mentioned, the purpose of this work is to determine the longitudinal tire slip curve with a lean test setup. For this reason all the measured quantities are taken from the CAN network of the vehicle without additional instrumentation.

With this setup is not possible to determine neither the braking force acting on the single wheel nor the brake bias and thus the rear brakes of the vehicle were removed from the test vehicle, in order to simplify the equations that describe the system as follows:

$$X_1 = -m\dot{u} - \frac{l}{h}\xi u^2 \tag{14}$$

$$Z_1 = Z_1^0 + \Delta Z - \xi u^2 \tag{15}$$

$$Z_2 = Z_2^0 - \Delta Z + \xi u^2 \tag{16}$$

and the adhesion coefficient becomes:

$$\mu = \frac{X_1}{Z_1} \tag{17}$$

that is the ratio between the longitudinal force and the vertical load. This parameter is not the friction coefficient of the tire, but it is useful and easy to be used to compare the results of the analysis because it is non dimensional.

Since the expressions of the forces depend on the geometry of the vehicle, all the tests must be performed on the same vehicle.

For this reason a test vehicle that could fit different sizes of tires is used. The final model used is represented in Figure 3.

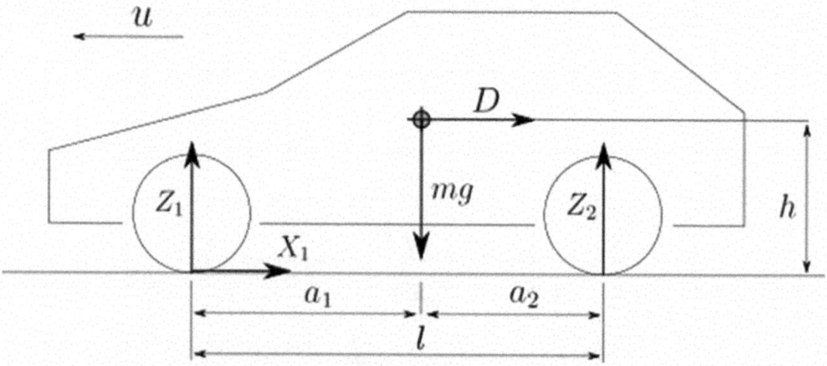

Figure 3

Vehicle Setup and Tests

As already mentioned signals are all acquired from vehicle CAN network. They are listed in Table 1.

Table 1 – Signals acquired from vehicle CAN Network

Vehicle Speed	[kph]
Wheel Speeds	[kph]
Longitudinal Acceleration	[m/s^2]
Lateral Acceleration	[m/s^2]
Master Cylinder Pressure	[bar]
Yaw Rate	[deg/s]
Steering Wheel Angle	[deg]

In addition, to calculate all the parameters needed for the analysis, some measurements are done on the vehicle before tests:

- Track width;
- Wheelbase;
- Vehicle mass and mass distribution (front and rear);
- Centre of gravity height.

As already mentioned, the vehicle is prepared removing rear brakes in order to calculate all quantities of interest without knowing brake repartition levels between front and rear.

In addition this allows to measure the longitudinal speed of the vehicle from the on board rear wheel speed sensors, without the need to equip the vehicle with a GPS system. Vehicle speed becomes:

$$V_x = R_r \omega_r \tag{18}$$

where ω_r is the mean value of rear wheels angular speed and R_r is the rolling radius.

With the signals coming from on board wheel speed sensors is also possible to calculate the longitudinal slip of the tires, which is defined as:

$$k_x = \frac{V_x - \omega_c R_r}{V_x} \tag{19}$$

where ω_c is the angular velocity of the rim.

Equipping the vehicle with the same size tires both in front and rear it is possible to calculate the longitudinal slip (neglecting the rolling radius dependence from the vertical load) as:

$$k_x = \frac{\omega_{rear} - \omega_{front}}{\omega_{rear}} \tag{20}$$

In order to cover all levels of longitudinal slip from 0 to 100% (i.e. front wheels locked) the ABS system is deactivated.

Test Maneuver

The maneuver performed to identify the longitudinal tire slip characteristic is a slow progressive braking (master cylinder pressure rate ≈ 20 bar/s) in order to have negligible vehicle body pitch inertia effect. Starting speed is ≈ 130 kph and the front wheels reach deep slipping (i.e. k_x = 100%) at a speed ≈ 50 kph. It is executed with the vehicle in neutral gear so that also the inertia of the engine is negligible. As soon as the front wheels lock the driver must release the brake pedal in order to avoid to wear excessively testing tires. An example of master cylinder pressure input (and consequent longitudinal acceleration of the vehicle) and wheel speeds behaviour can be seen in Figure 4 and Figure 5.

This maneuver is repeated at least 3 times for each run, in order to have a sufficient amount of data to be fitted, in particular at slips values greater than the one corresponding to the maximum of the parameter μ (as defined in equation 17).

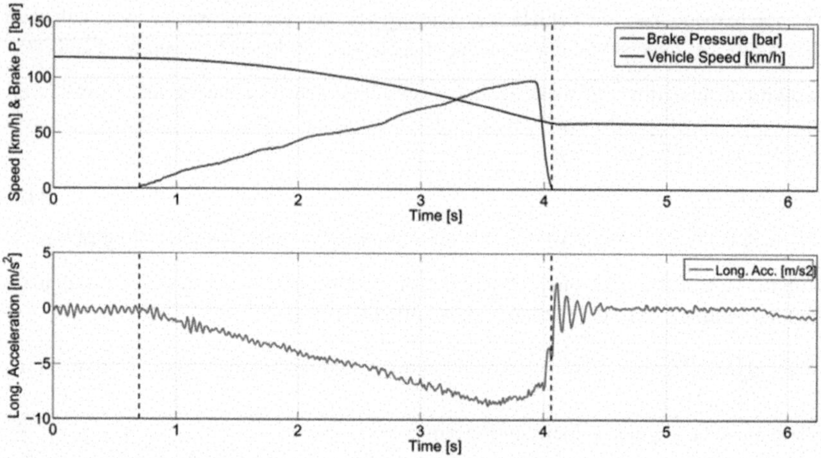

Figure 4

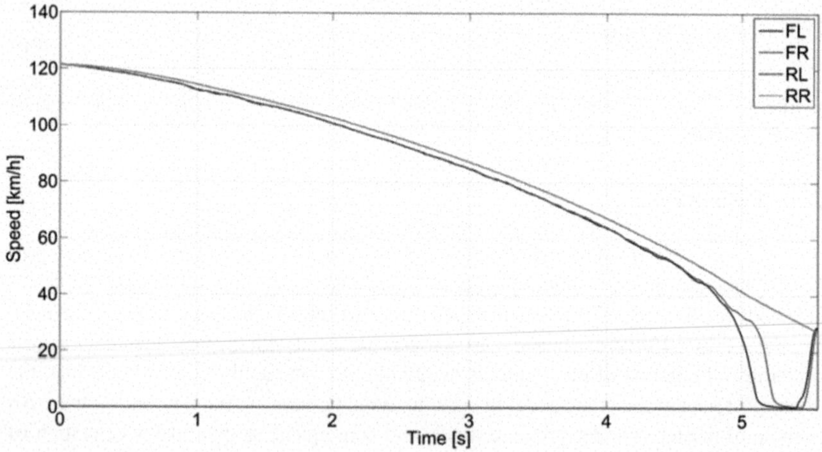

Figure 5

Data Analysis

Longitudinal Slip Curve Characteristic Identification

Acquired data are plotted in a graph in which longitudinal slip is reported in the x axis and longitudinal acceleration on the y axis (Figure 6). Longitudinal acceleration is directly obtained from CAN network, while longitudinal slip is calculated using wheel speed signals, that are also acquired from CAN network, with equation 19.

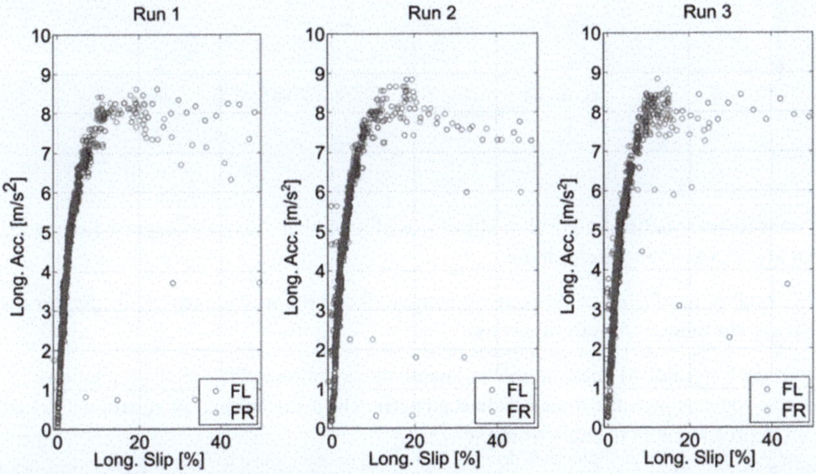

Figure 6

The points are then grouped in a single graph in order to have a sufficient amount of data to be analyzed (Figure 7).

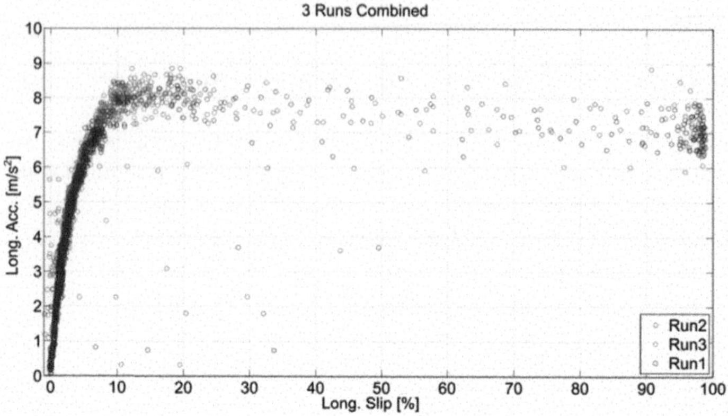

Figure 7

A conveniently defined algorithm eliminates all data that are out of expected range, or that are far from the general trend.

The result of pre-fitting could be seen in Figure 8. All points that are outside the thresholds are eliminated from the analysis.

From the longitudinal deceleration and using the equations of the model (equations 13, 14 and 15) it is possible to calculate the coefficient μ (as defined in equation 17) and then to proceed with the data fitting.

The chosen expression for data fitting is the Pacejka formula at four coefficients:

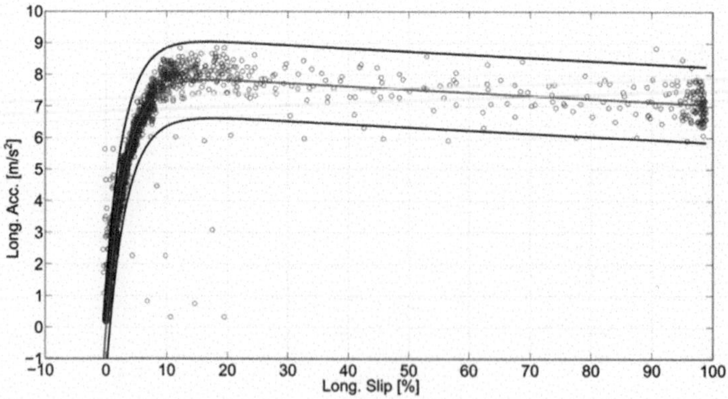

Figure 8

$$y(x) = D\,sin\left(C\,tan^{-1}\left(Bx - E\left(Bx - tan^{-1}(Bx)\right)\right)\right) \tag{23}$$

where:

- B is the stiffness factor;
- C is the shape factor;
- D is the peak value;
- E is the curvature factor.

The algorithm used for data fitting is the least square method that minimize the following function:

$$F(X,k,r) = \sum_{i=1}^{n} (y(X,k_i) - r_i)^2 \tag{24}$$

where:

- y is the function chosen for the fit (i.e. the Pacejka formula at 4 coefficients);
- k_i is the independent variable (in this case the longitudinal slip);
- n is the sample number;
- r_i is the acquired data to be fitted;
- X is the coefficient array to be determined.

Fitting the data with this formula it is possible to obtain good results but, having a large amount of points in the first part (i.e. low k_x) and less at higher slip values, the correlation is not optimal in the zone of maximum μ, which is one of the most useful parameters to be obtained with this procedure. For this reason the fitting was improved using another function to be minimized:

$$W(X,k,r,w) = \sum_{i=1}^{n} w_i(y(X,k_i) - r_i)^2 \tag{25}$$

This new function presents the array of weights w, whose length is equal to the length of the array of data r. It multiplies the single square distances, modifying in this way the function W. The idea is to assign an higher weight to data near to maximum μ and to data at high k_x so that the minor concentration of data in these regions of the graph could be compensated. In this case the weight coefficients are proportional to the corresponding value r_i raised to the power of p, where p has to be defined. In conclusion, having $w_i = r_i{}^p$ the function becomes:

$$W(X,k,r,w) = \sum_{i=1}^{n} r_i^p\,(y(X,k_i) - r_i)^2 \tag{26}$$

This type of fitting was used to analyse data of this work.

Statistical analysis shows the repeatability of tests done on different tarmac surfaces and on the same surface at different temperatures. This allows to evaluate which is the parameter that is influencing the results. In particular results are considered reliable with the following tolerances:

- 1% for the value of μ_{max}
- 5% for the value of coefficients, k_x and BCD (i.e. the slope at the origin)

These values were obtained from experimental tests, an example of which can be seen in Table 2.

Table 2

Run	μ_{max}	K_x @μ_{max}	BCD
Run 1	1.119	11.51%	0.349
Run 2	1.132	12.11%	0.323
Run 3	1.125	12.81%	0.319
Run 4	1.117	12.21%	0.308
Run 5	1.112	12.21%	0.313
Run 6	1.098	12.21%	0.317
Run 7	1.117	13.31%	0.314
Run 8	1.117	12.71%	0.307

	μ_{max}	K_x @μ_{max}	BCD
Std. Dev.	0.01	0.54	0.013
Var. Coeff. [%]	0.887	3.43	4.2
Mean	1.117	12.4	0.319

Results

With the method presented in previous paragraphs it is possible to identify the longitudinal tire slip characteristic of a tire fitted in a mule vehicle, using a lean instrumentation setup (only CAN network signals are used for the analysis). Once the mule vehicle is prepared, the test itself takes less than 0.5 hours to be completed. At this effort the time needed to equip the vehicle with test tires must be added. For data analysis a specific tool in Matlab® environment was developed, making data analysis very simple and fast.

Results can then be used to make different types of evaluations (and in general to improve tire testing from different points of view):

- Comparison on a single tire at different wear and/or tarmac temperature conditions;
- Comparisons between different tires at the same boundary conditions;
- Comparison between on vehicle tests and bench tests on the same tire;

- Analysis of the influence of noise factors and their influence on stopping distance results;
- Improved communication with the tire supplier, giving them an information more focused on the tire (longitudinal tire slip characteristic) instead of an information at vehicle level (stopping distances) for target setting;
- More effective ABS tuning;
- Useful data for virtual analysis models.

Comparison between different tires

A comparison between three different tires is presented. For each tire three progressive braking were done and the longitudinal tire slip characteristic was obtained. Results can be seen in Figure 9 and Figure 10. Figure 9 shows the entire characteristic from 0% to 100% slip. The differences are significant on all the parameters of the Pacejka fit (i.e. longitudinal stiffness, maximum adhesion coefficient μ_{max} and its corresponding longitudinal slip value). Winter tires (identified as Tire 1) are less stiff than the other two and μ_{max} is smaller. In addition the longitudinal slip at which μ_{max} is reached is higher (as expected for a winter tire), also considering the fact that test were performed with tarmac temperature equal to 50°C.

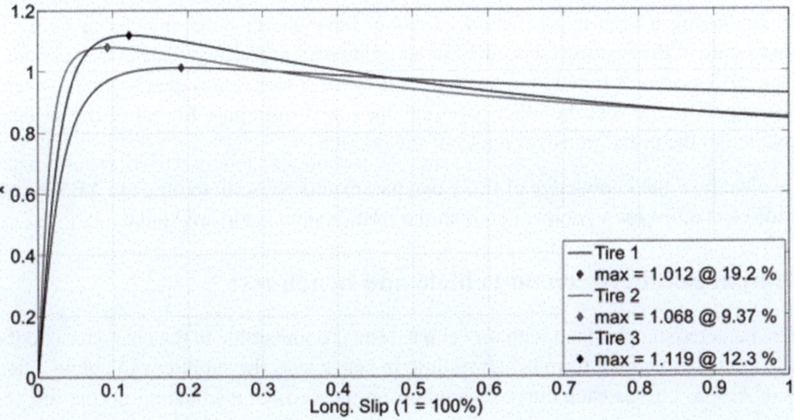

Figure 9

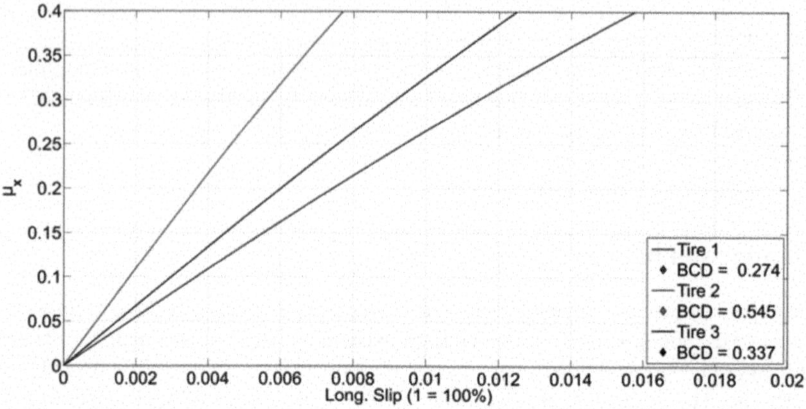

Figure 10

Important differences can also be appreciated between Tire 2 and Tire 3, both of which are summer tires. In particular Tire 3 has an higher μ_{max} value but is less stiff than Tire 2.

Both of these factors (μ_{max} and stiffness) are very interesting considering stopping distances: having a high μ coefficient allows to have shorter stopping distances in the steady state of the braking (short MFDD stopping distance value). This is true of course if the ABS system is tuned to allow the tire to work at slip values near to the μ_{max} value of the specific tire. On the other hand having a stiff tire allow to reduce the distance travelled in the initial transient phase of the braking.

It is clear how the knowledge of these two parameters is useful to tune the ABS system or to select a tire for a vehicle in which the ABS system is already tuned.

Comparison between on vehicle and bench test

The characteristic obtained with on vehicle tests is comparable to the characteristic that the supplier can obtain from bench testing. In bench tests the supplier can obtain a family of $F_x - k_x$ curves each curve depending on the vertical load acting on the tire (F_z) that it is possible to control during the test on the bench (Figure 11).

With on vehicle tests it is not possible to control the vertical load acting on the tire because it depends on vehicle's geometric characteristics and load displacement during braking (and for this reason on vehicle tests must be done on the same mule vehicle to obtain equivalent results).

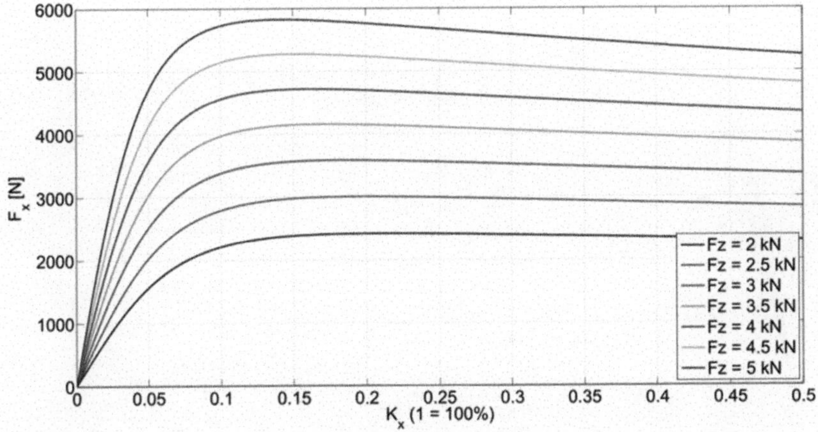

Figure 11

To make a comparison is therefore necessary to manage bench tests results as follows: having a certain value of the longitudinal force F_x the model of braking described in previous paragraphs (equations 14, 15 and 16) gives a single value of longitudinal acceleration, at which corresponds a single value of load transfer for the mule vehicle and thus a single value of vertical force F_z (corresponding to one of the curves obtained from a bench test).

That means that a single point for each one of the curves coming from bench test can be selected and an equivalent curve can be determined (Figure 12). Passing then from the longitudinal force F_x to the adhesion coefficient μ it is possible to compare bench tests with on vehicle tests.

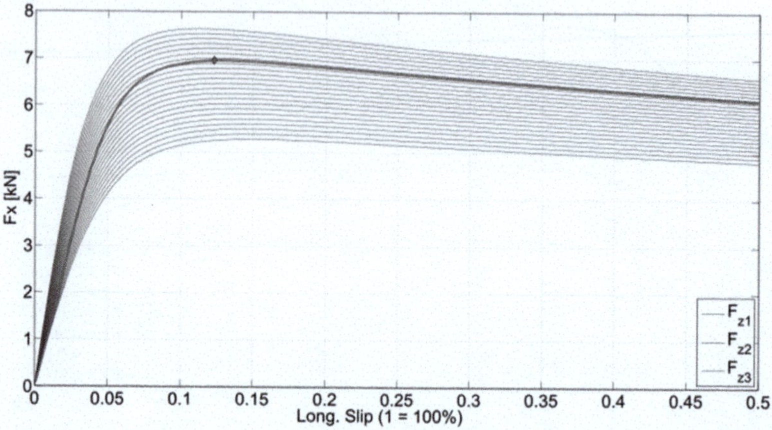

Figure 12

An example of comparison done on the same tire can be seen in Figure 13, where the on vehicle test was repeated two time in two different tracks of the proving ground. It is possible to notice slight differences on two of the three most important parameters, the longitudinal stiffness and the μ_{max} value, while there is almost no difference in the value of k_x at which the maximum is reached. It is possible to notice that anyway such differences are minimal and that there is also a certain repeatability of on vehicle tests performed on different tracks.

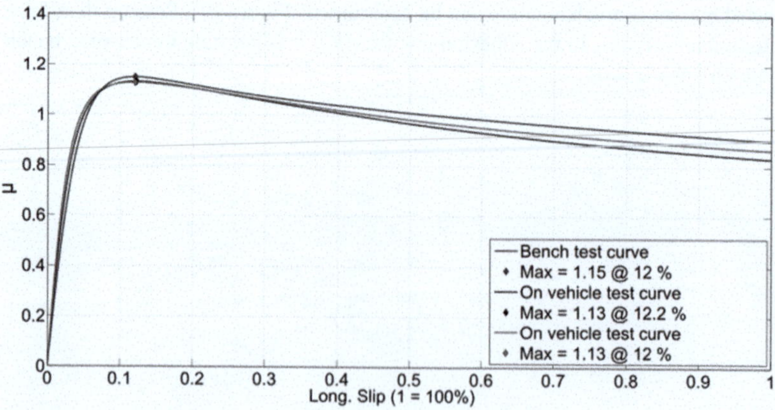

Figure 13

Major differences can instead be seen in the right portion of the characteristic (i.e. at k_x values higher than that corresponding to the μ_{max} value). This is a limit of the conversion of bench tests to be compared with on vehicle tests: with the model of the braking vehicle it is possible to reach a maximum F_z (the one corresponding to the maximum one reachable on the vehicle, i.e. front static load plus a weight displacement corresponding to rear static weight) and select the corresponding bench test curve obtaining one point. An additional point for this part of the characteristic is the one corresponding to k_x = 100%. This means that after the point corresponding to the μ_{max} is possible to select only an additional point (the one at k_x = 100%) leading to an approximation of that part of the characteristic.

This is of course a limit of this procedure, but it does not invalidate the results, since the interesting part of the characteristic for the purposes of this study is the part that goes from 0 to μ_{max}.

Behaviour of the tire with tarmac temperature

In Figure 14 a comparison of an all season tire in different tarmac temperature conditions on the same track can be seen. Red dots data are obtained from a single run (one progressive braking), while the blue markers are obtained from the three runs. Dotted line is the fitting of the blue points. Black lines represent the standard deviation of the single test at the same temperature.

From graphs of this kind is possible to estimate the variation of stopping distances with tarmac temperatures for a certain category of tire (i.e. summer, winter or all seasons tires).

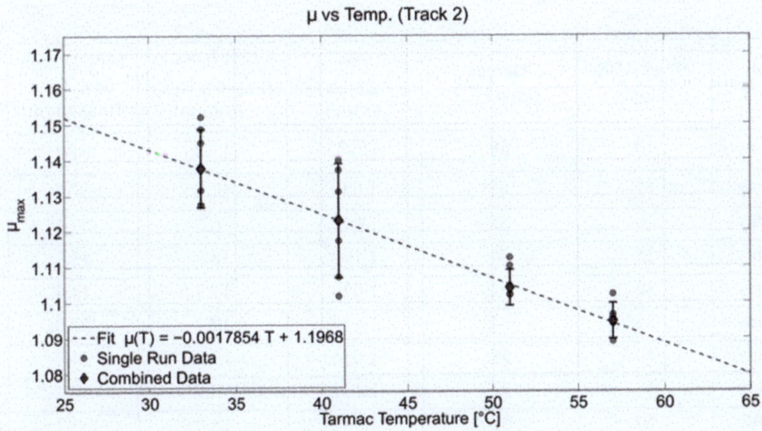

Figure 14

Behavior of the tire with tarmac temperature and tire wear

In order to analyze the behavior of the tire at different conditions of wear (intended as number of braking performed) and tarmac temperature, tests were conducted as follows: the vehicle was equipped with a brand new tire on which nine consecutive test sessions were performed alternating between morning and afternoon in order to have a wide range of tarmac temperatures (the study for an all seasons tire is reported in **Table 3**).

On each test session both progressive braking (three braking) and ABS braking (five braking) were performed in order to have tire behaviour data (adhesion coefficient trend) and stopping distance values. This procedure was repeated for different tire types (summer, all seasons and winter tires were tested).

In Figure 15 the slip curves for tests conducted at tarmac temperature of about 30°C (morning sessions) are reported. It can be seen how increasing the number of braking on the same tire (wear) the μ_{max} value and the longitudinal stiffness of the tire increase. The same happens for tests performed at 55-60°C (afternoon sessions, Figure 16), but with absolute values of stiffness and μ_{max} lower than the corresponding ones of the morning sessions. This confirms what seen in tests conducted at different tarmac temperatures in which μ_{max} decreases with temperature increase.

In **Table 3** it is also possible to see the trend of stopping distances (calculated with MFDD method) which follows, as expected, that of the coefficient μ_{max}. Notice that the high absolute values of stopping distance measurements are a consequence of the braking, performed with activated ABS system, but using only front brakes.

Table 3

Date (2017)	N° of ABS Stops	Tarmac Temp.[°C]	μ_{max} (Pacejka D factor)	Stiffness (Pacejka BCD factors)	MFDD stop. Dist. from 100 kph [m]
13th July	10	58	0.9733	17.2	61.4
14th July	15	35	1.038	17.8	57.4
14th July	20	55	1.047	16.3	57.3
17th July	25	30	1.100	18.5	53.5
17th July	30	54	1.079	17.2	54.4
18th July	35	34	1.114	19.1	53.0
18th July	40	58	1.084	18.0	54.9
19th July	45	32	1.131	18.9	50.3
19th July	50	58	1.081	18.0	53.2

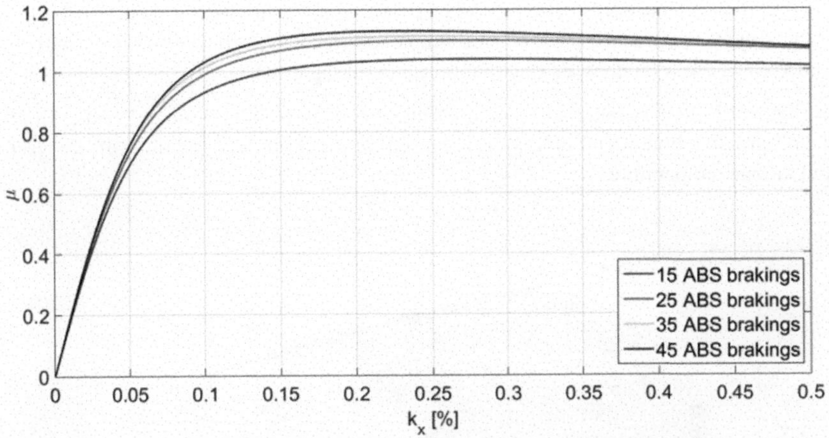

Figure 16

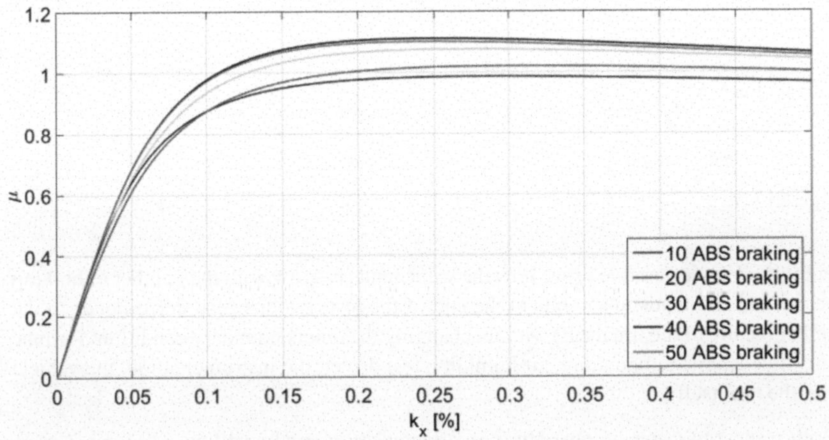

Figure 15

Grip coefficient analysis with multiple polynomial regression model

In order to analyze separately the effect of tire wear and tarmac temperature a multiple polynomial regression model, that takes into account also the interactions between the two predictors (i.e. tire wear and tarmac temperature) was used.

In fact, while is quite easy to evaluate the effect of the solely tire wear, conducting the test at the same tarmac temperature, it is not simple to evaluate the effect of tarmac temperature keeping the wear constant (the tire wears out during tests).

In Figure 17 the maximum adhesion coefficient (μ_{max}) obtained through Pacejka fit of the progressive braking (**Table 3**), is analysed for the different conditions of tire wear and tarmac temperature.

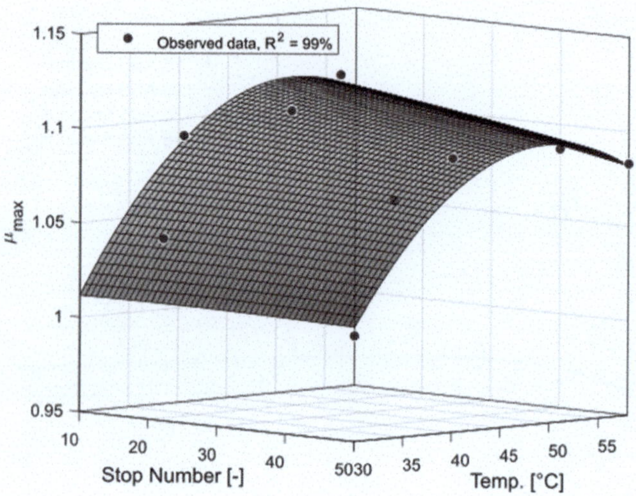

Figure 17

In Figure 19 it is possible to see how the value tends to grow with the number of braking performed and tends to saturate at the end of the test sessions (that corresponds to the end of the life of the tire itself) without changing the others boundary conditions (tarmac and atmospheric temperature, air humidity, test driver, the instrumentation, the vehicle and the tire itself).

With respect to tarmac temperature μ_{max} decreases in an almost linear way when temperature grows (Figure 18).

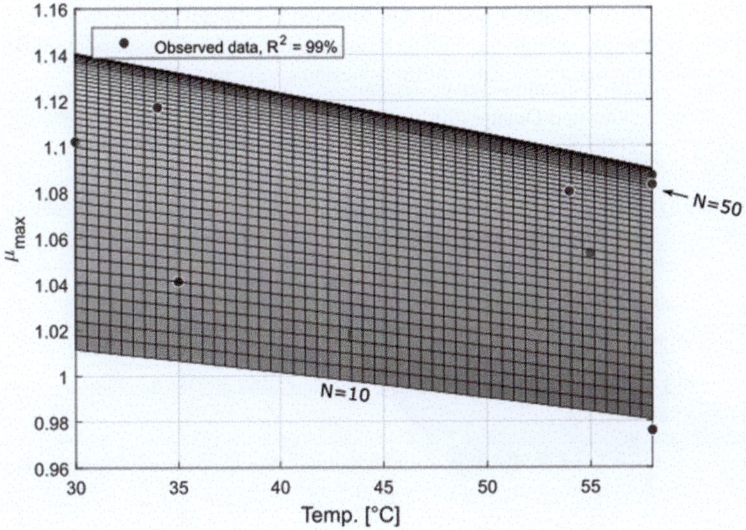

Figure 18

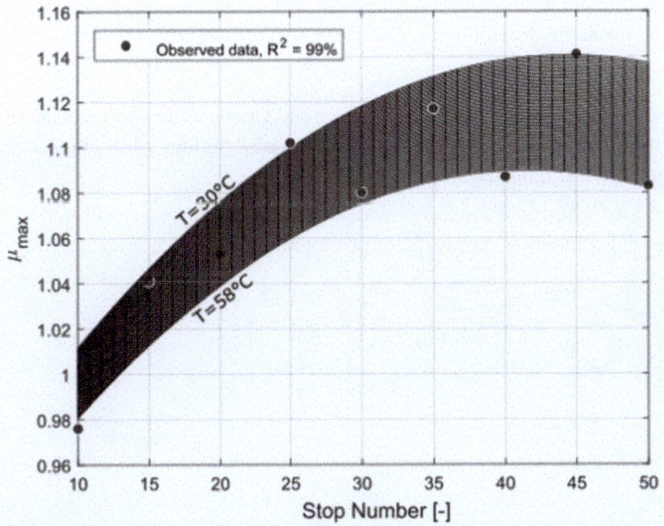

Figure 19

The same analysis can be done considering the adhesion coefficient corresponding to a longitudinal acceleration equal to the MFDD (μ_{MFDD}) value of the corresponding ABS braking, instead of the maximum value of the adhesion coefficient. (Figure 20).

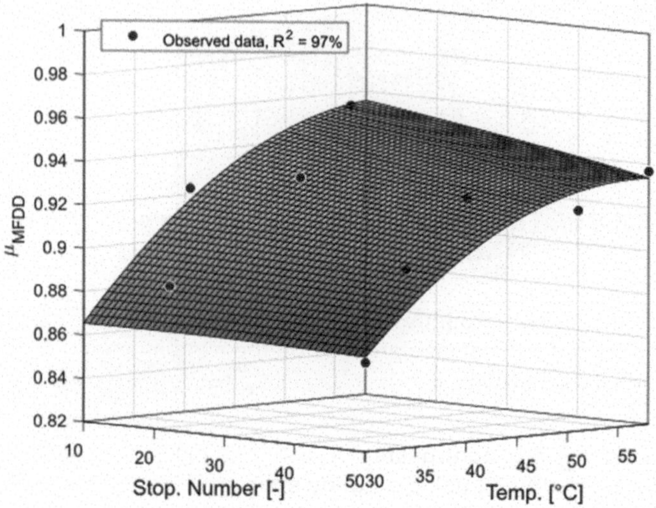

Figure 20

For this trend the same considerations done for μ_{max} trend can be done. An increase of the coefficient μ_{MFDD} can be seen with the number of ABS braking, while there is a decrease with temperature (Figure 22 and Figure 21).

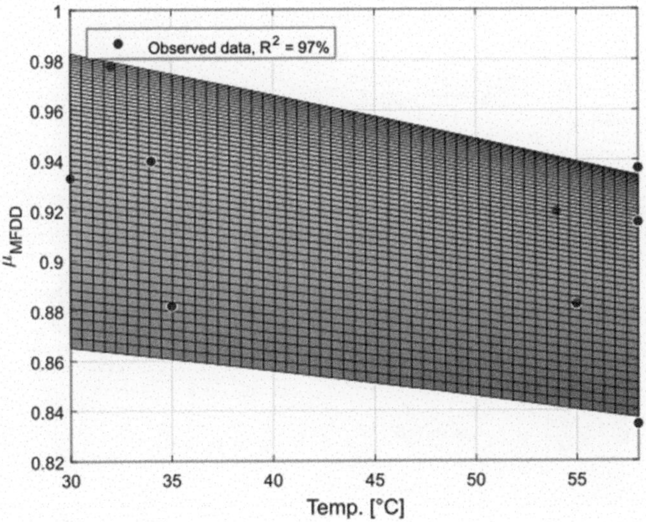

Figure 21

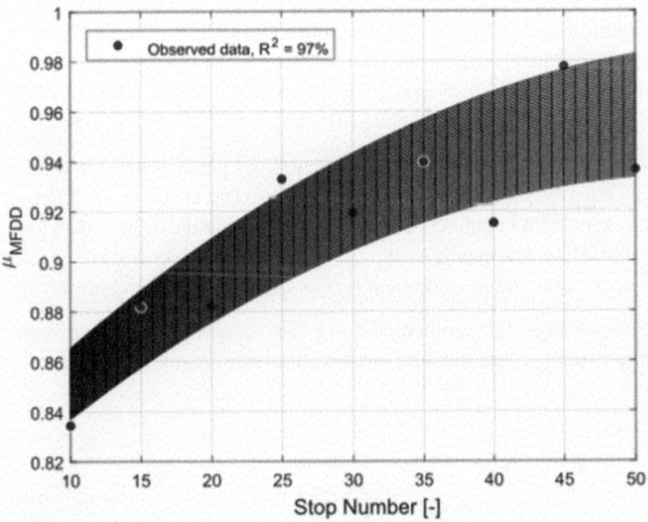

Figure 22

In Figure 23 the residual of the model for μ_{MFDD} are shown.

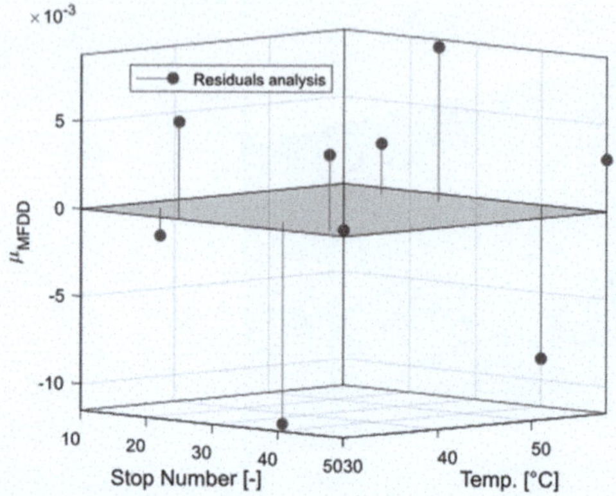

	Observed data	Model output	Residuals
1	0.83	0.84	-2.6660e-03
2	0.88	0.88	-1.5822e-03
3	0.88	0.88	2.8425e-03
4	0.93	0.93	5.5590e-03
5	0.92	0.91	8.8321e-03
6	0.94	0.95	-1.1518e-02
7	0.92	0.92	-8.7890e-03
8	0.98	0.97	4.3586e-03
9	0.94	0.93	2.9634e-03

Figure 23

Conclusions

With the method presented in previous paragraphs it is possible to determine in a simple and fast way the longitudinal slip characteristic of a tire equipped on the test vehicle. Data analysis is done with a specific software tool and a standard procedure was introduced in the company. Test repeatability and statistical reliability were evaluated with a positive outcome.

The new procedure can be used to evaluate the longitudinal performance of the tire fitted in the vehicle. Data obtained are useful to make comparisons between different tires, to make comparisons with bench tests, to analyze the influence of boundary conditions on the most important parameters of the tire itself, with an increase of the efficiency and effectiveness of the tire development process and ABS system tuning.

The method will be used also to define one model to adjust results of stopping distance measurements done in different conditions of tarmac temperature and tire wear.

Something from (almost) nothing: an overview of tire modeling in F1 using minimal data sets

Dr Vasilis Tsinias

Tyre Performance Section Leader
Renault Sport Racing
United Kingdom

P. Pfeffer (Ed.), *9th International Munich Chassis Symposium 2018*, Proceedings,
https://doi.org/10.1007/978-3-658-22050-1_58

Abstract

We live in an era where our ability to collect and analyse data is progressing exponentially. Terms such as 'machine learning' and 'data mining' are on the engineering cutting edge and dictate the way forward in numerous fields, from fundamental research to industry and everyday life. However, the ability to extract information from minimal data sets is – still – a valuable asset in a range of engineering applications, and Formula 1 is not an exception. The purpose of this work is to shed some light on the challenges faced by Formula 1 tyre engineers in developing an understanding of race tyres using limited testing resources.

Introduction

Tyre performance is, arguably, one of the most critical factors in all forms of motorsport. Since the introduction of the single tyre manufacturer, Formula 1 teams are in a constant loop of improving their understanding of tyre behaviour which would allow for unlocking extra performance on the track.

In the last several years, cost cutting regulations have been introduced which resulted in severely limiting the tyre testing that Formula 1 teams are permitted to perform. A direct consequence of this limitation is that, effectively, each track becomes a tyre testing laboratory and each practice session is considered – as far as the Tyre Engineer is concerned – as tyre testing time.

However, with track time being so valuable, it is inevitable that non-tyre related tests must be performed as well, sometimes simultaneously with a tyre test. This means that the acquired data from a tyre track test is usually contaminated in the sense that any observed effects are the outcome of the tested tyre change, any other test that is running in parallel, and secondary changes (for example, track evolution, change in car balance, etc).

Track testing limitations

On top of these effects, there is also the fundamental difference between a test performed in a laboratory and one performed at the track. In the former case, quantities such as the vertical load applied on the wheel and its camber angle may be adjusted separately to yield the full matrix representing the tyre. On the contrary, in the latter case, there is strong coupling between most of these quantities (for example, high speed data will always correspond to high vertical load due to downforce) which means that only part of the tyre operating envelope may be explored. This is normally not an issue, as the data set used to identify any tyre properties will correspond to the real conditions of the particular car of interest (that is, we do not actually need to know the variation of any tyre property outside these conditions).

However, this limitation is introducing a certain uncertainty when new concepts are being investigated, either for current car development or for future car concepts. This means that the car manufacturer and/or the racing team is exploring various new ideas which will change the car operating envelope and, ultimately, skew the coupling between the various quantities that affect tyre performance.

For example, the relation between car speed, axle vertical loads, and camber angles is fixed for a given car equipped with given springs and a given downforce level. In a simplified fashion, the coupling between lateral acceleration and suspension kinematics will dictate the camber profile along a corner. Therefore, the tyre characteristics may be captured accurately enough for that combination of, speed, vertical load, camber angle, etc, which allows for updating an existing tyre model.

The above-mentioned limitation applies when the coupling between two or more of the model inputs changes towards a previously unexplored area of the car performance envelope. A typical example is to investigate different suspension layouts which would result in different suspension kinematic characteristics and, hence, different camber angle profile for everything else being equal. Assuming that these new suspension layouts exist only in the virtual world, which normally is the case, the tyre model will have to operate in "extrapolation" mode and it is vital to ensure that the model output is sensible as it might have a detrimental effect on car development if that is not the case.

The purpose of this work is to demonstrate how a seemingly contaminated tyre test could yield valuable information on tyre behaviour even if it has been conducted on the track instead of a purposed-build laboratory. Two, quite typical, test cases will be presented:

- Identification of tyre forces and moments from track data
- Identification of tyre loaded radius from track data

Identification of tyre forces and moments from track testing

In modern Formula 1, a tyre model that correctly represents the actual race tyres is the cornerstone of any simulation work, both offline and with a driver in the loop. As mentioned before, laboratory tyre testing has been severely limited which means that racing teams resort to track testing for measuring tyre forces and moments. There are various pieces of equipment available but they all fall under two categories:

- force/moment measurement
- slip measurement

In the first category, the objective is to measure the forces and moments acting on the contact patch of all four tyres. This is usually done either via measuring the force applied to each of the suspension members (using strain gauges) or via measuring the forces and moments applied to the wheel hub using what is typically called a wheel force transducer (WFT, illustrated in the following picture[1]):

Although these methods allow for a direct measurement of tyre forces and moments, they are not problem-free. In the first case, having to instrument all suspension members means that losing one strain gauge will make the data collected from that particular corner useless; contact patch forces cannot be calculated accurately if the force of one suspension member is unknown. On the other hand, WFT are generally very intrusive in terms of Aerodynamics which means that running WFT on a car will not allow for any other parallel test.

Measuring car body slip velocity with respect to ground is generally easier than measuring tyre forces and normally involves a sensor which is mounted on the car and measures the speed at which the road is "travelling" (relative to the car of course).

Having acquired the tyre forces/moments of one corner, and the associated slip, leads back to the limitations discussed in the previous section. Even if the acquired data is grouped in vertical load bins (which is not an unrealistic case provided that the push/pullrods are equipped with strain gauges), the measured forces (or moments) are contaminated by the coupling between various quantities such as:

- tyre temperature
- inflation pressure
- camber angle
- pure vs combined slip

These quantities are easily controlled at a flat track machine, which means that the associated sensitivities of the generated tyre shear force can be feasibly identified, but that is certainly not the case during track testing.

However, there is still knowledge to be gained in the latter case by applying the principles of fundamental tyre mechanics in practice. The following figure shows a typical (albeit completely made up) cloud of lateral force, and hence lateral friction coefficient, vs slip angle for a particular vertical load gating.

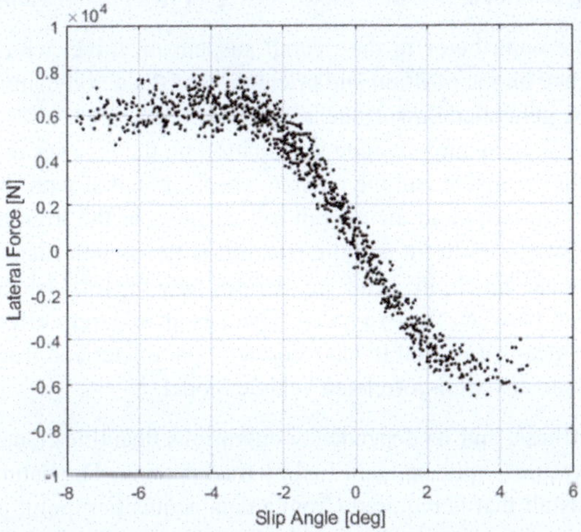

As mentioned, the above cloud is contaminated by a number of factors that do not allow for a direct model fit. However, one could apply different data gatings (assuming the resulting bins contain enough data points) or correct the position of the original data points based on known sensitivities (for example, the friction coefficient sensitivity on inflation pressure and tyre temperature).

A typical example of the first case is the distinction between pure and combined slip data. By gating for low slip ratio, the resulting cloud will correspond to pure lateral slip and will, essentially, be consisted of data located in the upper part of the original cloud as the introduction of longitudinal slip has a degrading effect on the lateral coefficient of friction.

Having split the original cloud into two sub-clouds (based on the slip ratio corresponding to each data point), the next step is to try to minimise the spread of points by exploiting the (either known or assumed) tyre force sensitivity on some fundamental factors. For instance, it is well known that cornering stiffness depends on the tyre "bulk" temperature. If that sensitivity is known, all the data points in the linear region will collapse into a significantly narrower cloud, which will correspond to the cornering stiffness at a reference temperature. The same process could be followed for any other factor affecting the generated tyre force until the width of the data cloud is narrow enough to allow for a reasonable model fit.

Identification of tyre loaded radius from track data

Identifying the parameters of a tyre loaded radius model, so that it correctly predicts the tyre behavior in a broad operating envelope, is vital in motor racing for two main reasons:

- Tyre squash is a considerable contributor to the overall suspension travel which means that it has a big influence on the position and orientation of the aerodynamic platform and consequently the generated downforce in all corner phases.
- A tyre loaded radius model will, by nature, include information on the tyre vertical stiffness. The stiffness of both the inside and the outside wheels, coupled with the torsional stiffness of the anti-roll bar, yield the overall roll stiffness of the front or the rear axle. The balance between the front and the rear roll stiffness will dictate the mechanical balance of the car which, in cornering, defines how the total lateral weight transfer is proportioned between the two axles. This has an apparent effect on overall car balance which means it is vital to have correctly parameterized front and rear loaded radius models in any well-correlated vehicle model.

Once again, the tyre vertical stiffness, and its dependency on various quantities (such as vertical load and inflation pressure), must be identified by track testing. The nature of such testing is fairly simple as all that is required (from a data acquisition point of view) is to log the tyre deformation in various points within the performance envelope of the car. To isolate the said tyre deformation from the overall chassis vertical motion, Formula 1 teams employ lasers that are mounted on each upright and, once corrected for camber, measure the vertical distance between the wheel centerline and the ground, as shown in the following diagram:

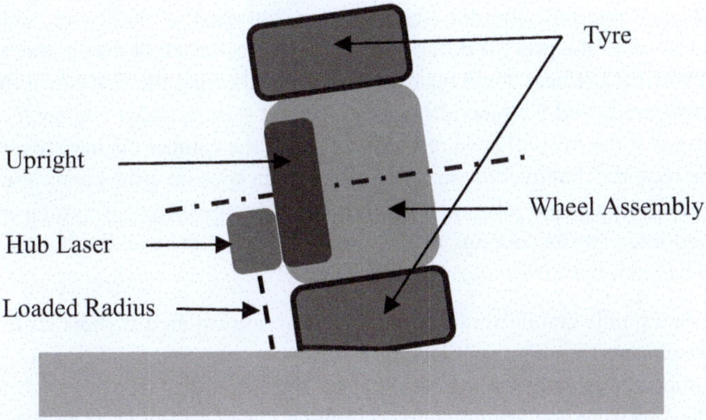

The challenge in parameterizing a tyre loaded radius model is that most of the quantities typically affecting tyre vertical stiffness are coupled to each other:

- For a high downforce racing car, speed (dictating the centrifugal stiffening of each tyre) is coupled with the vertical load applied to each wheel which, in turn, is coupled with the camber angle via the suspension kinematics and compliance.
- Another strong coupling that influences tyre vertical stiffness is the one between tyre bulk temperature and inflation pressure. Rather obviously, higher inflation pressure means higher vertical stiffness while higher bulk temperature will reduce vertical stiffness. However, both quantities migrate in the same direction on track; for instance, if bulk temperature is increased (because the driver is pushing for example) pressure will increase as well. In the above scenario, the two opposing trends (higher bulk temperature → lower vertical stiffness, higher inflation pressure → higher vertical stiffness) will not allow for a clean sensitivity identification between these two quantities and the overall vertical stiffness.

Based on the above, regular track running will not provide clean enough data to isolate the various sensitivities of the tyre vertical stiffness. To overcome these limitations, Formula 1 teams employ some relatively simple techniques:

- To by-pass the coupling between inflation pressure and tyre bulk temperature, one option is to perform a baseline run at baseline inflation pressure and temperature, followed by another run at the same temperature but different pressure (or vice versa). There are various practical concerns in employing the above practice (for example, the effect of the inflation pressure change on overall car balance will not be negligible and might contaminate the acquired data) but it should provide correct sensitivities once applied properly.

- By-passing the coupling between speed, vertical load, and camber angle is a bit more elaborate as it requires more time-consuming setup changes. However, and following a variation of the process described above, the influence of the camber angle on the tyre vertical stiffness could be identified by performing the first run at the standard camber angle and the second run at a different static camber angle. Once again, it is likely that the car will need re-balancing after the camber change, so a third run might be required, but overall the coupling between speed, vertical load, and camber angle is broken and the relevant sensitivity may be identified. A similar approach may be adopted for the coupling between speed and vertical load by changing downforce level from the first to the second run.

The above two points demonstrated ways of identifying isolated vertical stiffness sensitivities despite performing track testing when most relevant quantities are coupled. However, another challenge is that the logged signals (either from the hub lasers or from standard instrumentation) contain a combination of steady state responses, high frequency content, and noise, as the car is normally under acceleration (lateral, longitudinal, or both) running on a non-smooth track. Trying to identify a steady state response (which will reveal the underlying sensitivity of the tyre vertical stiffness on some other quantity) from such a signal is not always straightforward.

One possible way to acquire lower frequency data is to perform a series of constant speed runs. Essentially, the driver is asked to keep the car at a straight line while maintaining a constant speed. This process could be repeated several times at different speeds. Depending on the speed and the available straight length at a particular track, different speeds may be tested in the same lap to eliminate any data contamination from tyre pressure and temperature drifting.

Summary

Overall, this paper is not presenting anything groundbreaking or innovative. However, its purpose has been to demonstrate how people in motorsport work with minimal data sets, focusing on the particular case of tyre model parameterisation.

Track testing, which in Formula 1 is heavily relied upon for data acquisition, is posing a number of different challenges in terms of identifying tyre sensitivities. The first class of challenges derives from the very nature of this type of testing as there is very little control over the test conditions. A typical example of this class is test track evolution; any tyre performance difference spotted between a baseline and a test run will always be contaminated by track evolution. Therefore, the identification of any tyre sensitivity should always account for changes other than the intended one. The second class of track testing limitations includes the various couplings found in a modern Formula 1 car. In particular, and for a given setup, not all areas of the performance envelope may be explored. These limitations should be considered when parameterising a tyre model and, more importantly, when using the said model outside the performance envelope used to identify its parameters.

In terms of modelling tyre forces and moments, Formula 1 teams employ specifically developed test equipment such as strain-gauged suspension members and wheel force transducers. Running these sensors will result in acquiring the forces and moments applied from the ground to the vehicle. Careful gating and application of fundamental tyre mechanics theory will allow for clean and specific test data that can be used to parameterise such a model.

Another fundamental class of tyre models in Formula 1 is that of models that predict tyre loaded radius changes. Having a valid loaded radius model is critical as it allows for accurately predicting what the aerodynamic platform of the car is doing but also for calculating the mechanical balance of the car. Similar to the case of tyre forces and moments, Formula 1 teams use specifically designed sensors ("hub lasers") to isolate the tyre vertical motion from the overall chassis vertical motion. However, the ever-present coupling between the various quantities of interest is posing an obstacle in obtaining clean and robust dependences between the coupled quantities and the tyre vertical stiffness. A relatively simple, although expensive in terms of test time, way of bypassing that coupling is to perform setup changes after a baseline run. Another technique for ensuring high quality steady state data, which will help in robustly identifying how tyre vertical stiffness is changing from one loadcase to the next one, is to perform constant speed runs. Running at a specific speed for several seconds allows for obtaining a reasonably accurate loaded radius reading and performing more than one constant speed runs in the same straight section will provide a direct measurement of tyre vertical stiffness, unaffected by tyre evolution.

References

[1] Photograph by LAT Images©, as published on motorsport.com on 30 November 2017.

The manufacturer's authorised representative in the EU is Springer Nature Customer Service Centre GmbH, Europaplatz 3, 69115 Heidelberg, Germany. If you have any concerns regarding our products, please contact ProductSafety@springernature.com

Printed and bound by CPI Group (UK) Ltd, Croydon, CR0 4YY
30/09/2026
02225313-0001